Corporate Liability: Work Related Deaths and Criminal Prosecutions

Corporate Liability: Work Related Deaths and Criminal Prosecutions

Third Edition

General Editor

Gerard Forlin QC

Lincoln's Inn, LLB (Hons) (LSE), LLM (LSE), MPhil (Trinity Hall, University of Cambridge), Diploma in Air and Space Law (London Institute of World Affairs)

Editor

Louise Smail

BSc (Hons), MA (Health and Safety Law and Environmental Law) CMIOSH, PhD, MBCS, CITP

With specialist contributors

Bloomsbury Professional

Bloomsbury Professional Ltd,
Maxwelton House,
41–43 Boltro Road,
Haywards Heath,
West Sussex RH16 1BJ

A CIP Catalogue record for this book is available from the British Library.

ISBN: 978 1 78043 115 4

Typeset by Columns Design XML Ltd, Reading, Berkshire
Printed and bound in Great Britain by Hobbs the Printers, Totton, Hampshire

Foreword by Mr Justice Burnett

The third edition of this book comes at a timely moment. General Editor Gerard Forlin QC, Louise Smail and the specialist contributors are to be congratulated on producing an up-to-date and accessible work which covers all aspects of the subject. There has been much going on in the world of health and safety, both on the legal front and also on the political front.

The book introduces the changes made to reflect the Government's desire to reduce burdensome and costly regulation whilst not compromising the protection that health and safety regulation provides. Work related deaths in the United Kingdom continue on a downward trajectory although reported injuries appear to be increasing. The Health and Safety Executive has undertaken a general regulatory review and Parliament introduced legislation (the Enterprise and Regulatory Reform Act 2013) which it hopes will achieve those policy objectives. Both are discussed in the new edition.

The Corporate Manslaughter and Corporate Homicide Act 2007 has been used in a small number of prosecutions (which are taking far too long to investigate and bring to trial). The developing case law is discussed. So too is the impact of the Definitive Guideline from the Sentencing Guidelines Council of February 2010 relating to health and safety offences. There has been a perceptible upward trend in the levels of fines imposed upon corporations for serious health and safety breaches.

As before, in addition to valuable chapters on general matters of health and safety, this edition contains detailed discussion of the law relating to individual industries.

This work will be a valuable addition to the libraries not only of lawyers who practise in the field, but also health and safety professionals whose task it is to ensure compliance with regulation and, more importantly, continue to squeeze unavoidable injury out of the workplace.

The Hon Mr Justice Burnett
President of the Health and Safety Lawyers' Association
January 2014

Dedication

To my mum, Moira, lots and lots of love and thanks for everything!

Also in memory of my father, Professor Forlin, who although a scientist and mathematician, which I am most certainly not, always provided support, enthusiasm and inspiration in whatever I did. I miss him loads!

Preface

There have been many changes since the last edition of this book some four years ago.

I have attempted to indicate as many of these as possible in the space allowed. There has also been an expansion of the areas covered by the book, including new chapters on the Media, Travel, Sport and the Emergency Services and an expanded International section.

Whilst there is no separate Oil and Gas chapter in this edition, certain elements of this vast topic have been covered in various interlinking chapters.

As so often happens with works of the present kind, in the immediate run-up to publication we have seen the release of both the triennial Temple Report in January 2014 and the judgment of the Court of Appeal Criminal Division in *R v Sellafield Ltd; R v Network Rail Infrastructure Limited* [2014] EWCA Crim 49.

This case is clearly a very important development in the law relating to the sentencing of large corporations for breaches of health and safety. Unfortunately, it was too late to incorporate it all within the text, but where possible it has been covered in the final draft stages. Only time will tell if this decision, coupled with the inability of corporates and other non-human organisations to recover their costs, will lead to either more guilty pleas or more trials.

Since the last edition, the Coalition Government has introduced myriad changes to the law and policy in this area, some of which are covered in this book. The future of Fees for Intervention is also a very hot topic, and it will be interesting to see if the procedure survives.

The number of fatalities continues to fall, and the UK has an enviable record when compared to the rest of the world.

It is, however, dangerous to be complacent and there is clearly a lot of work still to be done. My real hope is that this downward trend of fatalities should continue. One arguable risk is that the more changes to the system that occur, the higher the chance that the framework might open up gaps and the fatality rate start to rise again.

In a speech in late 2013 that I gave for the Westminster legal policy forum, chaired by Lord Browne of Ladyton, I reminded the audience that, rather like the Kerplunk game, we must be very careful that the stick we pull out does not cause the marbles to drop to the bottom, thereby losing the game. We must all stay vigilant to this risk. In fact, just before this book went to print, David Cameron said in a conference speech to the Federation of Small Businesses that 'we need to be a country that celebrates enterprise and backs risk takers'. Striking the right balance will not be easy.

We have also seen a number of corporate manslaughter prosecutions and there are more in the pipeline. We are still waiting for a large organisation to be prosecuted for the offence, and many of the provisions of the Corporate Manslaughter and Corporate Homicide Act 2007 still have to be interpreted by the higher courts. More individuals are also being sent to prison for manslaughter and health and safety offences. On 16 January 2014 the DWP published the latest figures relating to sentencing in the lower courts since the introduction of the Health and Safety (Offences) Act 2008. This makes interesting reading, including the fact that fines for breaches of both health and safety regulations and the HSWA 1974 have increased some 25% since the law changed.

There have been a number of changes to the courts' interpretation of the Health and Safety at Work etc Act 1974 and, where possible, these developments have been included.

In the last few years, I have increasingly travelled widely abroad for work, representing global organisations. This has shown me, time and time again, that procedures need to be as uniform as possible across the whole entity; differences between countries need to be justified. Further, advisers need to consider not only the instant case but also how the conduct of that case can affect the organisation as a whole. To put it another way: increasingly, legal (and other) advisers need to factor in not just the battle but how that knocks on to the war! The Court of Appeal Criminal Division in *R v Bodycote HIP* [2010] EWCA Crim 802 reiterates that position. Further, regulators globally are co-operating and joining the dots, to a greater extent, to pluck out organisations and industrial sectors whose health and safety falls below par in an international setting.

We have also seen a complete overhaul of the system of coroners, and 'prevention of future death' reports will become increasingly important in the future. This area of the law is in a time of change, and it will be fascinating to plot the course of where it will be in another four years or so.

On a personal note, there are so many people to thank, including my mother, family, Sonia, Mary, friends, clients and my Chambers (both here and in Australia and Singapore), Elliot, Michelle and all the clerks. I also thank Dr Louise Smail and all the contributors who have done a sterling job to fit this project into their busy lives and careers. I thank the Hon Mr Justice Burnett for finding the time to provide the eloquently drafted foreword again and for acting as President of the Health and Safety Lawyers' Association (of which I am the Vice-Chair). I also thank all at Bloomsbury, especially Kiran Goss, Peter Smith, Jenny Lank, Victoria Daniels, Jane Bradford, Martin Casimir and the rest of the team. This has been a mammoth task and I am very grateful. I thank Liz for all her work typing up the manuscript; and Anne Davies, Andrew Clinton and others for reading extracts of the manuscript. I also thank His Honour Judge Hodge QC for reading the preface and coming up with so many helpful comments.

This book was written with both lawyers and non-lawyers in mind, and it attempts to reflect matters (but by no means all) as at the end of 2013.

Gerard Forlin QC

Cornerstone Barristers, London
Denman Chambers, Sydney
Maxwell Chambers, Singapore

gerard@gerardforlin.com
www.gerardforlin.com

Gray's Inn
January 2014

List of Contributors

General Editor

Gerard Forlin QC was educated at Trinity Hall, Cambridge and the LSE; he was called to the Bar by Lincoln's Inn and appointed Queen's Counsel 2010 in both Civil and Criminal.

He was previously Senior Crown Counsel, Hong Kong.

Gerard is called to the Bar in Australia and is a member of Denman Chambers in Sydney. He is also a member of Maxwell Chambers in Singapore where he has his own room.

He is admitted as a Barrister of the Eastern Caribbean Supreme Court, British Virgin Island Circuit.

Gerard has just undertaken a series of lectures for the Foreign and Commonwealth office on the UK Bribery Act in Australia, Singapore and Korea (2013). He will be conducting more in 2014, in many other countries in Asia.

He has recently been appointed to an International Mediation and Arbitration panel of 14 worldwide Members for the Royal Aeronautical Society. There are only two QCs on this global panel.

He is Vice-Chair of the Health and Safety Lawyers' Association.

Gerard's areas of expertise, in which he also acts globally, include the following: regulatory, specialising in health and safety and corporate and gross negligence manslaughter, disaster litigation, aviation, railways, shipping safety, diving, oil and gas, environmental crime, construction, professional discipline, public inquiries, fraud, regulatory offences, product liability, healthcare, commercial fishing (EU), consumer crime, licensing, corruption and bribery, waste offences, human rights and inquests.

The *Legal 500* continues recognition of Gerard's practice with the most recent edition (2013) ranking him as a leading Silk under 'Health and Safety' and 'Product Liability'. Having taken Silk, Gerard was ranked in the newly amalgamated grouping 'Consumer and product liability' and 'Health, safety and environment'. In consumer law, Gerard was singled out as showing 'impressive judgment and strategic skill'. In the 2011 edition, Gerard was described as having an 'encyclopedic knowledge of Health and Safety law', and in relation to consumer and product liability it said, 'Technical and industrial product claims specialist Gerard Forlin QC is blessed with an encyclopedic knowledge of the law and the inner workings of the prosecution'. In the 2009 edition, he is described as a 'formidable advocate with strength in technical and industrial product claims', 'managing the evidence well and is very good with juries'. In the 2014 edition, Gerard was ranked in both Product Liability and Health and Safety.

In the September 2008 edition he is described as a 'manslaughter specialist', clients praising his 'great industry knowledge' and 'incredible work ethic'; recent cases include acting for Balfour Beatty on the Hatfield crash. 'Forlin's robust style of advocacy, easy client manner and market awareness all combine with the above to merit a first-tier ranking.'

Chambers and Partners 2014 ranks Gerard as a leading Silk in Health and Safety, describing him as 'an able, experienced practitioner, and a good advocate'. 'He is one to have on your side during any investigation into a workplace death.'

In *Chambers and Partners 2013* it is described that he 'brings bags of charm, has an extraordinary mastery of his brief and produces a lovely turn of phrase'.

In *Chambers and Partners 2012* he is ranked as a leading Silk in Health and Safety. He is admired by peers for 'the flamboyance with which he presents his cases'. He is a specialist in numerous fields and has notable expertise of the rail and aviation sectors. His health and safety practice has led him to undertake the defence of many corporate manslaughter prosecutions. Solicitors say: 'He is exceptional as he is both very good with clients and very good at keeping a team together'.

In *Chambers and Partners 2011* he is ranked as a leading new Silk in Health and Safety. Interviewees describe him as 'an old-fashioned jury advocate with excellent knowledge of his subject and a great feel for his cases'. 'A seasoned defender of international blue-chip companies in fatality cases, he is a tireless performer.'

In *Legal Experts 2011* he is ranked as a leading new Silk in Consumer and Health and Safety. In May 2011, Gerard was awarded a *Finance Monthly* Law Award. These prestigious awards span many different fields and jurisdictions throughout the legal world. Gerard was voted as a winner in the Health and Safety category. Andrew Palmer of *Finance Monthly* observed that 'The 2011 *Finance Monthly* Law Awards recognise all those firms [and individuals] that have dedicated their resources to innovation, built on their depth of expertise and performed outstandingly over the year'.

In the leading independent directories, he has also been described as 'Mr Manslaughter', 'The Manslaughter Guy', 'redoubtable', 'powerful orator', 'one of the most compelling advocates you could ever see in a courtroom', 'great at planning', 'ahead of everyone else', 'excellent advocate', 'unrivalled', 'well-deserved reputation as leading counsel in this field', 'unique combination of rhetorical skill and superb technical knowledge' and 'a master of thinking on his feet and exudes a compelling charisma that makes people want to listen'.

In the last few years, he has been ranked in no fewer than five areas in both the leading independent directories, including:

- Health and Safety;
- Crime;
- Public Inquiries and Inquests;
- Consumer Law; and
- Environmental Law.

Gerard is also:

- Special Adviser to the Bar Council on the Corporate Manslaughter Bill (2005);
- Consultant to a recent Global Aviation Concordat on Aviation Safety;
- Standing Counsel to numerous Plc's, Unions and Government Departments in the UK and abroad, and Acts for Crown Departments on Crown Censure (Occasional AG nomination);
- Consultant to the Army and various police forces; and
- Licensed to be instructed directly by members of the public

Gerard has been involved in over 250 fatality cases and has appeared in many of the pivotal cases in this field in the last 10 years, including:

- Watford train crash;
- Southall train crash;
- Paddington train crash;
- Hillgrove School/James Porter: leading case on 'risk';
- Teebay train disaster;
- Barrow Legionnaires case;
- Hatfield train crash;
- Purley train crash;
- Faversham coach crash;
- Britannia air crash;
- Lift fatality in the City of London;
- Catamaran incident in Greece; and
- Inquest in the Falkland Islands (death in South Georgia).

He has also acted in over 150 jury inquests both in the UK and overseas.

His client list includes:

- many global companies and organisations incorporated both in the UK and elsewhere;
- acting for the Environment Agency on policy issues;
- acting for numerous local authorities, including Kensington & Chelsea, Hammersmith & Fulham, Enfield, Islington, Test Valley, Dorset, Harrow, Camden, Welwyn Hatfield, Forest Heath, South Northamptonshire, Suffolk Coastlines and Chelmsford;
- acting for sporting governing bodies, eg International Motor Sports (Formula 1);
- acting as a disaster legal risk consultant all over the world;
- acting as an emergency planning consultant in many countries;
- acting in fraud and DTI cases;
- consultant to Confederation of British Industry on regulatory matters;
- consultant to the English and Scottish Police Training Schools; and
- consultant to the English and Scottish Prison Services.

Gerard also lectures and writes:

- member of a General Bar Council delegation to China and Seoul (2013);
- member of a General Bar Council delegation to South Korea and Singapore (March 2012);
- member of a General Bar Council delegation to the Middle East (September 2011);
- visiting lecturer to numerous universities including King's College

London, Murdoch, University of California, Loyola, Yonsei, Singapore, UMIST, Loughborough, Cranfield, Salford and Aberdeen; and

- numerous TV, radio and educational video appearances.

After taking Silk in March 2010, Gerard has been involved in over 14 fatality cases. In May 2010 and in January 2012 (Inquest) he acted for Thyssen Krupp in a lift fatality case in the city of London. Inter alia he has acted for four large local authorities. Another recent case was where he acted in a catamaran inquest, the case dating back to 2003. In 2012/2013 he acted in a lengthy inquest in the Falkland Islands and South Georgia. All of these were high-profile matters.

He has also been retained in other global matters involving catastrophic incidents. These include cases involving fatalities in construction (two), aviation, hospitals, gas, water pollution, agriculture, funfairs, driving, mining, shipping and swimming pools.

He has also lectured and/or consulted in the last two years in over 20 countries, including the USA, Australia, Canada, Korea, Brunei, The Falkland Islands, The British Virgin Islands, China, the Netherlands, the Czech Republic, Singapore and the Middle East. He has lectured, consulted or practised in over 40 countries.

Gerard was invited, together with the Chief Executive of Zurich, to participate in a debate with Lord Young at the inaugural seminar on the Coalition Government's proposed changes to the regulatory regime.

Languages spoken: English, French, Italian and basic Cantonese.

Editor and Contributor

Louise Smail has extensive experience in the field of risk management with a wide range of organisations, including local government, police, manufacturing and many train operating companies. Between 1994 and 1996, she acted as Advanced Works Safety Manager for Union Railways Limited (a government-owned company charged with the task of deciding the route for the Channel Tunnel Rail Link). Following the takeover of the successful consortium for the Rail Link project, Louise went to work for Bechtel, based in Warrington, looking at issues related to construction risk for clients in the utilities sector and compliance with legislation concerning construction and environmental issues. Louise has worked as Principal Consultant and Director of two international consulting groups, and she now runs her own consultancy, Ortalan. In Ortalan, Louise works on high-profile projects, looking at business risk evaluation and performance at board level and the impact of changes in legislation in both the United Kingdom and Europe and has lectured in Risk Management at the Emergency Planning College; clients include the emergency services and local authorities and a number of public and private sector companies. Louise also works for United Utilities, a FTSE 100 company.

Louise is a member of the British Computer Society – Chartered Information Technology Practitioner, and is a Chartered Member of the Institute of Occupational Safety and Health (CMIOSH).

Louise writes and lectures extensively and is the Consultant Editor of *Health and Safety at Work Newsletter* (published by Bloomsbury Professional). She also regularly contributes to other publications on legislative issues related to construction.

Contributors

Michael Appleby is a solicitor and ranked as a star individual in the legal directory *Chambers and Partners 2014* for health and safety, which says he is noted for inspiring clients. One client is quoted as saying: 'He has impressed us with his loyalty, tenacity and approachability. The quality of his work and care is exemplary'.

In relation to the Hatfield train derailment, he acted for Balfour Beatty when it was acquitted of corporate manslaughter in 2005. The company then pleaded guilty to a breach of section 3 of the HSWA 1974. He successfully appealed the £10 million fine, reducing it to £7.5 million.

He acted for the managing director in one of the leading appellate cases on directors' health and safety duties. More recently, he acted for Austin and McLean Ltd in its acquittal of corporate manslaughter (2013). The company then pleaded guilty to a breach of section 3 of the HSWA 1974 and was fined £60,000. Its co-defendants, Esso, were fined £100,000 for a health and safety breach.

Ruth Barber was called to the Bar in 1996, qualified as a solicitor in 1998, and became a higher courts (criminal) advocate in 2005. She was accepted onto the Attorney General's C list of Prosecution Advocates in 2006 and the B list of Health and Safety Advocates in 2012. She has specialised in crime since qualification, representing clients charged with offences ranging from murder, rape and robbery, to serious fraud and arms dealing.

She taught on the Bar Council Human Rights Programme from 1999 to 2005 and regularly writes and lectures on regulatory crime, criminal procedure and confiscation law.

In 2004, she became a solicitor agent for the Health and Safety Executive and has since practised primarily in the area of regulatory crime as an agent for the Health and Safety Executive.

She has a particular interest in offshore safety, hazardous installations and explosives, and has completed the offshore safety course.

She currently practises as a self-employed consultant.

Philip Bonner is an associate within Squire Sanders' Litigation team and has experience in a wide range of commercial litigation disputes, as well as health and safety and environmental prosecutions.

Aisling Butler is a solicitor with William Fry, one of the leading law firms in Ireland. She practises in the area of regulatory law, including health and safety law, and has wide-ranging experience advising companies in both the public and private sectors on health and safety compliance issues. Her work also includes liaising with the Health and Safety Authority and defending health and safety prosecutions. Aisling is the ex-Chair and a founding member of the Health and Safety Lawyers Association of Ireland. She lectures extensively on the subject, including at the Law Society of Ireland and on the HDip and MSc in Health & Safety Law at University College Dublin.

Andrew Clinton is the managing partner of asb law LLP, which is recognised as a leading law firm in the travel and aviation sectors. He is a highly experienced litigator with more than 20 years' experience in large-scale complex commercial disputes with particular expertise in travel-related disputes, and he represents clients at inquests following fatal accidents.

His experience includes aviation disputes, injunctive work and claims involving travel companies, much of which involves cross-jurisdictional issues. He has frequently worked with lawyers in other jurisdictions to protect his clients' interests following incidents abroad.

Andrew is an Independent Expert appointed by Nominet to determine domain name disputes and is a trained mediator. He is a trustee and company secretary of Mind in Brighton & Hove.

Anne Davies is a health and safety specialist based in Withers' London office. She advises both company and individual directors on their criminal and civil liabilities arising out of an investigation by the HSE, Food Standards Agency, DEFRA, local authority, Environment Agency, Trading Standards or the police. Her highlight cases include defending against a group litigation representing McDonald's in the 'Hot Drinks' litigation; this case remains one of the leading decisions under the Consumer Protection Act 1987.

Her practical and proactive approach is held in high regard by her clients and peers. In *Chambers & Partners 2014*, Anne is described as 'able, switched-on' and also recognised for being 'tenacious in her approach, but retains a pragmatic and business-aware overview in dealing with health and safety enforcement authorities', whilst in the 2011 edition she won praise from market sources who say she has '… good judgement, a sound grasp of the relevant legal principles and a very good relationship with the HSE'. *Legal 500* (2013) describes Anne as 'very creative' and 'practical', whilst the 2012 edition describes her as 'very sensible, straightforward, knowledgeable and easy to work with' and also 'an expert in her field'.

Anne provides bespoke training to companies and boards of directors on all aspects of health and safety law, with a focus on how to respond if an incident occurs. She is a member of the Health and Safety Lawyers' Association (HSLA) and the Institution of Occupational Safety and Health (IOSH).

Mike Elliker is a Legal Director at Addleshaw Goddard LLP. His experience has included all types of High Court and County Court litigation. He sat as a Deputy District Judge from 1987 to 1997. He currently handles all types of criminal cases, including environmental and trading standards cases. Mike prosecutes on behalf of the Health and Safety Executive and is accustomed to appearing in the Crown Court and magistrates' courts.

Rob Elvin is a partner and the European Head of the Environmental, Safety and Health Group at Squire Sanders (UK) LLP and Managing Partner of the firm's Manchester office. His particular expertise includes: defending health and safety and environmental prosecutions; noise abatement, air, water and ground pollution cases; appeals of improvement and prohibition notices; product safety; and inquest law. He deals with a wide range of other business crime and is vastly experienced in the chemicals, power generation, construction and nuclear energy industries. Rob is a solicitor-advocate and lectures on handling regulatory prosecutions, the Bribery Act 2010 and corporate manslaughter.

David Firth is a partner at Capsticks Solicitors LLP and head of the firm's Leeds office. His practice covers the regulation of health and social care providers by, among other agencies, the Care Quality Commission, and responding to criminal investigations and prosecutions against corporations and individuals.

David provides a wide range of health and safety advice to NHS, charity and private sector clients, from reviewing risk management arrangements and compliance with directors' duties to responding to serious incidents requiring close liaison between client, the police and the HSE. Over the last 15 years, he has represented clients nationally in response to prosecutions arising from deaths in the workplace (clinical and non-clinical). He is a regular speaker at conferences on healthcare issues, and took part in the Health Service Journal's first web-based seminar 'Board Talk' on the Corporate Manslaughter and Corporate Homicide Act 2007. David was assisted in the preparation of the Healthcare chapter by **Janice Smith**.

Mark Gay is a partner of Burges Salmon LLP. He was educated at Lady Margaret Hall, Oxford where he read Jurisprudence, and trained at Linklaters & Paines. He was made a partner of Herbert Smith in 1995 and joined Burges Salmon LLP in 2011. His specialisation is regulation and dispute resolution in the sports industry. In that regard, he has advised many of the most prominent bodies in the sports industry, including the Premier League, the Football League, the Racecourse Association, Sahara Force India F1 Team, the Motor Sports Association, the Federation Internationale de l'Automobile and numerous other sports governing bodies, clubs and teams. He is ranked in Category 1 as an expert in Sports Law in the current *Chambers and Partners* Directory. He is also a qualified Solicitor Advocate and regularly appears before disciplinary and arbitral tribunals.

David Gordon is a partner in Squire Sanders' Environmental, Safety and Health Group and is dually qualified with a scientific background and PhD in Environmental Toxicology. As a recognised national expert in dealing with environmental compliance, sustainability and negotiation with statutory authorities, David has particular experience in dealing with the negotiation and apportionment of environmental liability on large multi-jurisdictional mergers and acquisitions and real estate transactions, and advises on both technical and legal issues arising from EU environmental regulation and enforcement under the UK Contaminated Land Regime. He is recognised by the Law Society for his chemical sector expertise, having gained a national profile advising on chemical regulatory issues such as REACH and CLP, and is described by both the *Legal 500* and *Chambers and Partners* as a leading individual in environmental law.

Gilles Graham is a Partner and Solicitor Advocate at Simpson & Marwick Solicitors. He is ranked in *Chambers and Partners 2014* for health and safety, and is Safety Group Coordinator of the Scottish Committee of the Health and Safety Lawyers Association. He has a long experience of litigating in the courts, and lectures the profession regularly on health and safety and corporate homicide.

Bruce Hodgkinson is the Head of Denman Chambers in Sydney. Bruce is one of Australia's leading health and safety Senior Counsel and has been practising as a barrister in that area for 25 years. Bruce has appeared in major health and safety cases that have been run in Australia, including the Gretley mine disaster and the Kogarah gas explosion. Bruce has appeared for the Australian Federal and State governments, statutory corporations, large and small public and private corporations as well as individuals.

Bruce has been an active contributor to the debate on health and safety reform in Australia at both the Federal and State levels. In addition, Bruce has presented at numerous local and international conferences on all aspects of health and safety regulation and compliance. He regularly advises boards on risk management and health and safety issues. Bruce also practises in other areas involving government regulation, contracts, sports and appellate law.

Bruce has extensive experience in the charitable and not-for-profit sector. He is Chairman of Cancer Council NSW, Director of Cancer Council Australia and the Chairman of the Rugby Union Players' Association in Australia.

Piero Ionta is a Senior Solicitor for the London Borough of Hammersmith and Fulham and the Royal Borough of Kensington & Chelsea. His practice covers both civil and criminal litigation resulting from actions brought by or against each local authority in connection with its various statutory functions. He has a Diploma in Local Government Law and Practice and over 11 years' experience acting for and advising various local and public authorities.

Nick Kettle is the Strategic Manager for the Metropolitan Police Service Safety and Health Risk Management Team. Nick is a Chartered Environmental Health Practitioner and a Chartered Safety and Health Practitioner. In his current role, he is responsible for the delivery of corporate safety and health risk management support, including the delivery of an operational risk-based safety capability, provision of health and safety assurance, continuous safety improvement and operational interface with enforcing authorities. Since joining the service, he has also deployed to support safety delivery on international and London-based operations.

Matt Kyle is an associate within the Regulatory Team at Burges Salmon LLP. Prior to qualifying as a solicitor, Matt was a police officer, serving for five years with the Avon and Somerset Constabulary. Matt advises both corporate clients and individuals who are subject to criminal and regulatory investigations on health and safety grounds. Matt conducts his own advocacy at hearings, particularly at inquests. His experience includes providing representation for individuals investigated following the explosion at the Buncefield oil storage depot and a multiple fatality at an oil refinery in Pembroke. Matt also assisted with safety advice in relation to the claim against West Yorkshire Police concerning the legality of levying charges for Special Police Services brought by Leeds United, led by Burges Salmon partner, Mark Gay.

Caroline May is a partner at Norton Rose Fulbright LLP and heads the Environment, Safety and Planning practice. She is ranked as a Tier 1 leading practitioner for environment in *Chambers and Partners* and is a recognised leading expert in health and safety matters: 'Norton Rose Fulbright has a splendid practice under Caroline May' (Chambers UK 2014). Caroline has over 20 years' experience in advising corporate clients in defence of both criminal and civil proceedings in relation to major environmental and health and safety incidents, and in representing clients at coroner's inquests, public inquiries and permitting/licensing appeals. Caroline has been advising on a number of major incidents, both on and offshore, and the team at Norton Rose Fulbright provides a 24/7 incident response service. The Norton Rose Fulbright global Environment, Safety and Planning team consists of specialist teams of lawyers operating across 25 jurisdictions. The team is often called upon to respond to global regulatory issues, including product safety and recall.

Barney Monahan currently works in-house as Editorial Legal Counsel at News Group Newspapers Limited, publisher of The Sun newspaper. He was admitted as a solicitor in 1998 and has, during that time, practised commercial and media law both in-house and in private practice. Barney has advised and worked for a number of large media companies on pre- and post-publication matters.

Niav O'Higgins is a partner and head of the construction and engineering group in Arthur Cox, one of the top law firms in Ireland, where she also heads up the firm's health and safety group. She advises regularly on health and safety, particularly as it affects those working in construction. She advises public and private sector employers, contractors, consultants and employees on compliance with all health and safety requirements, as well as on enforcement action taken by the Health and Safety Authority, and has acted on the successful defence of a number of prosecutions. She has an MA in Health and Safety and Environmental Law and is one of the founding members, and current chair, of the Health and Safety Lawyers' Association of Ireland. She is a lecturer at Trinity College Dublin on health and safety in construction, and regularly presents on health and safety topics.

Mark Oliver is a serving police officer, a career detective with 30 years' service in the Metropolitan Police Service and Humberside Police. He is an accredited Senior Investigating Officer on Major Incident Teams, and has led homicide, kidnap and Serious and Organised Crime investigations. He formerly worked on the MPS Murder Review Group, including Criminal Cases Review Commission cases, and the ACPO Homicide Working Group. Mark has developed UK Counter Terrorism joint exercising. He is currently head of the Humberside Police Anti-Corruption Unit.

Mark is the UK's most experienced Senior Identification Manager (SIM) and a founder member and trainer of the UK Disaster Victim Identification (DVI) Team. He has been deployed as a SIM to: Kosovo 2000 (deputised as the Head of the British Forensic Team); Sri Lanka 2005 (led the identification of all foreign nationals); Air France 447 Brazil 2008 (agreed international collaboration identification protocols); Afriqiyah Airlines Tripoli 2010; and Philippines Typhoon Yolanda 2013 (advised British Ambassador on UK involvement).

Mark currently trains members of the UKDVI team, including Senior Identification Managers and Reconciliation Managers.

Paul Rice heads up the Pinsent Masons LLP Energy and Environment practice. He specialises in all aspects of UK, EU and international energy and environmental law and his work includes advising and assisting waste management clients on the various aspects of waste projects, waste planning and permitting, as well as enforcement, landfill tax and environmental bodies, producer responsibility and carbon/emissions issues.

Paul has a Masters Degree in environmental laws, is a Member of the Chartered Institution of Wastes Management and the UK Environmental Law Association Waste Working Party.

He is recognised as a leading UK environmental lawyer by Chambers Legal Directory, Legal 500, Who's Who International and Legal Experts.

Maria Saraceni is a barrister practising at Francis Burt Chambers in Perth, Western Australia. She specialises in regulatory law, particularly health and safety and industrial relations; public inquiries and coronial inquests. Previously, Maria headed the health and safety practice at Norton Rose in Perth.

Maria was also the President of the Law Society of Western Australia and a Director of the Law Council of Australia.

In addition, Maria is an Adjunct Professor at Murdoch University Law School, where she lectures in health and safety and employment law. She is also a director of the Industrial Foundation for Accident Prevention (IFAP).

Robin Stacey is assistant news editor at The Times and, as such, has observed the way in which numerous companies have grappled with an adverse media environment with varying degrees of success.

Graham Stoker is one of the highest placed persons from the UK in the leadership of world sports federations. He holds the position of Deputy President for Sport in the Federation Internationale de l'Automobile (FIA), the international world federation governing motorsport – including F1 – based in Paris and Geneva, where he is involved with top-level decision making and the work of the World Motorsport Council as deputy to the President, Jean Todt. With an LLB and LLM from the London School of Economics in international law and politics, he is a member of the English Bar, the Middle Temple and Lincoln's Inn, and was a Winston Churchill pupillage prize holder. A member of chambers at Cornerstone Barristers, he is also a member of the Parliamentary Bar and the Royal Institute for International Affairs (Chatham House).

With 29 years working with national and international sports federations, including a period as a member of the FIA Court of Appeal, Graham has extensive experience advising at the highest level on sports law and sports administration; and, as a long-standing member of Sport Dispute Resolution UK, he has resolved numerous disputes by ADR in athletics, skiing, parachuting, motorcycling, swimming, football, and other sports, including team selection for the Commonwealth Games. He is also a trustee of a major sport charity, the FIA Foundation, based in London.

Michael Tooma heads up Norton Rose Fulbright's global occupational health and safety practice. He is the author of 17 books on occupational health and safety law. Michael is a Senior Visiting Fellow of the University of New South Wales and an Honorary Fellow of the Australian Catholic University.

Debbie Venn is a commercial lawyer at travel specialist law firm, asb law LLP. Her vast knowledge of the travel industry and of the onerous regulations impacting travel companies mark her out as an expert in her field. Her status as such is confirmed by regular plaudits in both *Chambers and Partners* and *Legal 500*. She frequently advises on travel terms and conditions, regulations, agency and distribution arrangements, data protection and privacy issues.

Debbie has presented on the legal aspects of the travel industry at annual ABTA-sponsored training events for the past four years. She has more than 10 years' experience and particular expertise in the protection, licensing management and exploitation of intellectual property rights complementing expertise in information technology, e-commerce, data protection and information security issues.

Contents

Table of Cases

References are to paragraph numbers.

C

R

S

Table of Statutes

References are to paragraph numbers.

Table of Statutory Instruments

References are to paragraph numbers.

Table of European Legislation

References are to paragraph numbers.

Table of Foreign Legislation

References are to paragraph numbers.

SECONDARY LEGISLATION

Chapter 1

Criminal responsibility for work related accidents

Gerard Forlin QC, Barrister, Cornerstone Barristers, London; Denman Chambers, Sydney; Maxwell Chambers, Singapore

Dr Louise Smail, Risk Consultant, Ortalan, Manchester

INTRODUCTION

1.1 This introductory chapter is a basic overview and is not intended to cover all the aspects of the vast subject matter. Many of the recent developments are dealt with within the parameters of the various chapters. Environmental issues are also crucial; often, these are interlinked to health and safety offences including legionella, asbestos and water pollution.

1.1 *Introduction*

Since the last edition, there have been myriad changes brought about, particularly by the Coalition Government. Set out below are a few examples of the enormous amount of change that has taken place since the last edition.

1.2 In February 2013, Judith Hackett reiterated that health and safety must be a core value. At an IOSH Conference she said:

> 'I hear so many companies that say health and safety is their number one priority … Health and safety needs to become a core value in all businesses. Whether it has to become a bigger priority in its own right – I would really question.'

This basically means that health and safety cannot be a bolt-on, but must live at the very heart of the entity.

1.3 The number of fatalities continues to fall, from 171 in the previous year to 148 in the current year. The number of people who die from exposure to harmful substances still remains about 12,000 a year. There is a clear downward trend in fatal accidents.

1.4 In 2010, fresh from victory, Lord Young was appointed to 'investigate on and report back on the rise of the compensation culture over the last decade coupled with the current low standing that health and safety legislation now enjoys and to suggest solutions'. This resulted in 'Common Sense, Common Safety' in 2010, and health and safety reforms were passed to the DWP.

In 2011, 'Good Health and Safety, Good for everyone' was published.

1.5 Professor Löfstedt of Kings College, London was appointed to review the current health and safety landscape. His report[1] was accepted by the Government, and has reinforced the Government stance by the 'red tape challenge' to scrap, review or improve 84 per cent of health and safety legislation.

1 *Reclaiming Health and Safety for All: An Independent Review of Health and Safety Legislation.*

1.6 His programme to undertake sector-specific consolidations into petroleum, mining explosives and genetically modified organisms is underway and due to be completed in 2014. Professor Löfstedt stated at the outset his basic approach as to what he had found. He said, 'There's nothing wrong with the legislation, it's how it's interpreted'. He also said:

> 'In addition, there is also a need to stimulate a debate about risk in society to ensure that everyone has a much better understanding of risks and its management.'

See the very helpful progress chart set out in *Safety Management* magazine.[1]

1 January 2014 issue, at pp 21–27.

1.7 At the end of November 2013, the DWP announced that 17 of the 22 areas raised by Professor Löfstedt, and 30 of Lord Young's 35 recommendations, have been addressed by the Coalition Government.

1.8 One such example is that the Health and Safety (First-Aid) Regulations 1981[1] have been amended, removing the need for HSE approval of first aid training from 1 October 2013. Employers are still required to have adequate first aid. There is new HSE guidance on the website regarding these duties.

1 SI 1981/917.

1.9 There is also a new independent evaluation of the Construction (Design and Management) Regulations 2007. Further, in an attempt to assist businesses on what is 'reasonably practicable', the HSE has started a new Health and Safety Toolbox which has had about half a million website visits already. There are intentions to introduce new CDM Regulations in 2015. These amendments will 'deliver significant savings to businesses by removing some of the bureaucracy and some of the unnecessary processes like the specific CDM co-ordinator role which may not be necessary in all cases'.[1]

The HSE has also reviewed and, on occasion, withdrawn a number of their Approved Codes of Practice (ACOPs). This process is ongoing. It has also removed over 20 various Regulations, and many more are due to be revoked. In 2013, the HSE released some 'updated and clarified' ACOPs including legionella, COSHH and the workplace regulations. The HSE said '... it will help employers understand the regulatory requirements on key issues such as temperature, cleanliness, workstations and seating, toilets and washing facilities'.

1 Judith Hackett, 5 December 2013.

1.10 The HSE, DWP and FCO are also currently working on harmonising legislation in health and safety within the European Union, including in the ergonomics and electromagnetic sections. There has also been recently, inter alia, a triennial review of the current role of the HSE included in the new triennial review, by Martin Temple of the EEF. The report (published in January 2014) looked at a number of areas of concern. Since Fee for Intervention (FFI) was introduced, the report has found that stakeholders are greatly concerned about the relationship between HSE inspectors and businesses: this was not so much about the scale of the fees as about how this has had a negative impact on the relationship with the HSE inspectors. The report states:

> 'I recommend that, unless the link between fines and funding can be removed or the benefits can be shown to outweigh the detrimental effects, and it is not possible to minimise those effects, FFI should be phased out.'

The report also considered the time that is taken to investigate incidents. The report makes the following recommendations:

> 'I recommend that HSE continues to improve its performance on the length of time taken to complete its investigations. It should aim for 95% of non-fatal accident investigations to be completed within 12 months of the accident. In addition, there should be a suitable target for the completion of fatal investigations once HSE has assumed primacy.'

The report also looked at the inconsistency in the way in which regulators carried out enforcement. The report therefore states:

> 'I recommend that HSE should actively review LAs annual returns on their inspection and advisory activities. Where there is evidence of significant deviations from the norm they should explore the reasons with the outliers. HSE should draw the attention of the appropriate political leader of those LAs where its performance is significantly out of step of the potential risk this may pose.
>
> I recommend that the National Local Authority Enforcement Code is reviewed in 2014 in the light of experience to identify areas for change and amendment.
>
> ...

> I recommend that HSE's LA Liaison Groups should be strengthened and maintained and that HSE's role in those Groups should be to provide expert professional guidance, constructive challenge and leadership.
>
> I recommend and value LAs working together and in partnership with HSE to ensure value for money. Ideally, there should be a senior champion and a single point of contact and single regulatory organisation in each LA or grouping of LAs. But what works well locally and local political accountability is just as important.'

Overall, the report has listened to the comments of stakeholders – it is an opportunity to improve a relationship which has becomes difficult, so that the HSE can be seen as a positive force for making the workplace safer.

1.11 Very importantly, the HSE published the National Local Authority Enforcement Code, which sets out priorities and targets for local authority enforcement. A series of guidance documents has also been published. There have been a number of other changes, such as the Myth Busters Challenge Panel (which has received about 200 cases) and the Independent Regulatory Challenge Panel (which has, to date, received one case). See generally **Chapter 11**.

1.12 The exempting of various health and safety laws for self-employed people who pose no potential harm to others, such as lawyers and accountants who work from home, are also cited as examples of these ground-breaking changes.

1.13 These modifications are all part of the Deregulation Bill which is being debated in Parliament. We are not clear about the entirety of the changes. We have also recently been informed by the DWP that '... the HSE will develop clear guidance for businesses on how the exemption will apply, to be published, subject to successful passage of the Bill'. This relates to self-employed people.

1.14 Additionally, the law relating to the Reporting of Injuries, Diseases and Dangerous Occurrences Regulations (RIDDOR, 2005) has been amended, and came into force on 15 October 2013. There are now eight categories entitled 'Specified Injuries'. The reporting trigger, based on time away from work, has also been extended from three to seven days.

Interestingly, despite this amendment, the number of dangerous occurrences has increased, despite the death rate falling to approximately 148 a year. The latest figures from the Labour Force Survey indicate that the number of employees reporting being injured at work has hit a three-year high: between April 2012 and March 2013, 646,000 were injured at work. This represents an increase of 55,000 on the 2011 to 2012 figures.

Further, the HSE introduced on 1 August 2013 a new version of HSG 65 entitled 'Managing for Health and Safety'. A printed version will be available in January 2014. According to the HSE, 'this newest and most significant development is the "Plan, Do, Check, Act" approach ...'. This differs from the old HSG 65 approach. The IOD/HSE Guidance to Management has also been amended in 2013.[1]

The 2013 IOD/HSE guidance states that it sets out:

- 'a four-point agenda for embedding the essential health and safety principles;
- a summary of legal liabilities;
- a checklist of key questions for leaders; and

- a list of resources and references for implementing this guidance in detail.'

It continues that:

> 'the agenda consists of:
>
> - core actions for boards and individual board members that relate directly to the legal duties of an organisation. These actions are intended to set a standard;
> - guidelines that set out ways to give the core actions practical effect. These guidelines provide ideas on how you might achieve the four core actions; and
> - case studies selected to be relevant to most sectors.'

1 INDG417 (Rev 1); see below.

1.15 According to the HSE, this new approach is about identifying the key actions needed in each part of the cycle and relating them back, when appropriate, to leadership, management, worker involvement and competence. It:

> '... achieves a better balance between the systems and behavioural aspects of management. It also treats health and safety management as an integral part of good management generally, rather than as a standalone system.'

1.16 At the same time, the HSE withdrew the Approved Code of Practice to the 1999 Management of Health and Safety at Work Regulations. In effect, the revised HSG 65 became the new blueprint to management's approach to safety.

1.17 This has been a mere snapshot of the many changes, some of which are explored further in the chapters that follow.

CORPORATE MANSLAUGHTER AND CORPORATE HOMICIDE ACT 2007

1.18 It has been about five years since the introduction of the Corporate Manslaughter and Corporate Homicide Act 2007 (CMCHA 2007). As had been predicted, there has been a relatively slow start to the prosecution process.

In the early days, there has been an interim time for investigations to be completed. These complicated cases can naturally take a period of time to come to court. The number of corporate manslaughter investigations by the Crown Prosecution Service rose 40%, from 45 in 2011 to 63 in 2012. Since records began in 2009, 141 corporate manslaughter files have been opened. The first corporate manslaughter conviction in 2011 related to a 2008 fatality. The second conviction in 2012 related to a 2010 fatality, while the third conviction in 2012 related to a fatality four years prior, in 2008. There are up to 500 cases in special casework. Some of these are corporate manslaughter cases. The evidence appears also to show that prosecutions are speeding up, particularly since the Crown was somewhat chastised, by the trial judge in the *Lion Steel* case, for the delay in prosecution. See below for some recent examples. Further, in January 2014, District Judge Barr (sitting at Blackpool Magistrates' Court) granted a stay of proceedings against Blackpool Victoria Hospital for delay, where a summons had been issued 21 months after an accident. The judge said:

> 'I feel there has been an abuse of process. There has been a significant delay in this matter putting it lightly. There is also evidence Mr Earnshaw may have been the architect of his own misfortune.'

1.19 The first case was *R v Cotswold Geotechnical (Holdings) Limited*.[1]

In essence, a young employee died when an excavation trench collapsed on him. He was tragically not found for a few days.

The Managing Director, Mr Eaton, had also been originally charged, but the prosecution discontinued after he became very seriously ill with cancer.

The conduct of Mr Eaton was taken into account by the jury. This had been described as 'extremely irresponsible and dangerous'. The company was found guilty.

1 [2011] EWCA Crim 1337.

1.20 The trial judge, Mr Justice Field, stated that the company had committed a 'grave offence'. Despite being in a stricken financial condition, they were fined £385,000 to be paid over a ten-year period. This represented about 114% of the company's annual turnover.

The learned judge stated:

> 'It may well be that the fine in the terms of its payments will put this company into liquidation. If that is the case it is unfortunate but unavoidable … but it is a consequence of the serious breach.'

1.21 The company appealed the level of fine to the Court of Appeal.[1] The Lord Chief Justice (sitting with Beatson and Bean JJ) in an ex tempore judgment said:

> '… it was unfortunate but "unavoidable and inevitable" that the company would probably have to go into liquidation to pay the fine.
>
> In the circumstances, it was plainly foreseeable that the way the company conducted operations could cause serious injury or death. The reality was that the Judge had taken the view, and had been right to take that view, that in the circumstances the fact that the company could be put in to liquidation was unfortunate but inevitable. There could be no justifiable criticism of the sentence imposed. The company was faced with manslaughter causing death as a result of a gross breach of duty following a system of work that was unsafe with the potential for causing death.'

1 [2011] All ER (D) 100.

1.22 In May 2012 in Northern Ireland, JMW Farms Ltd were fined £187,500 after pleading guilty to corporate manslaughter.

In this case, a farm worker was killed when a metal bin that he was cleaning toppled over, killing him.

The company turnover was approximately £1.38 million with a declared dividend of £200,000. The Recorder of Belfast imposed the fine after allowing a 25% discount to reflect the guilty plea. He also allowed the company six months to pay. Costs of £13,000 were also imposed.

1.23 It is also interesting to note that there has been a further very recent case in Northern Ireland. In *R v J Murray & Sons Ltd*,[1] a company pleaded guilty to corporate manslaughter. The case concerned a farm meat mixer that had its safety guards removed. It had been described as 'safe as houses' by management.

The learned judge said:

> 'The danger was so obvious and so serious that the man who principally operated the machines himself could not conceivably have been unaware of its dangers.'

The guilty plea received a 33% discount. The total fine was £100,000 to be paid over five years, and £10,450 costs were also allowed.

1 [2013] (October 2013), NICC.

1.24 The next major prosecution was that of *R v Lion Steel Equipment Limited.*[1] Steven Berry, a maintenance man employed by Lion Steel, was killed on 29 May 2008 after falling through a roof light at his employer's premises.

The company was charged with corporate manslaughter and under s 2(1) of the Health and Safety at Work etc Act 1974 (HSWA 1974). Three company directors were also charged with gross negligence manslaughter, and under s 37 of the HSWA 1974 and contravening the Work at Height Regulations 2005.

The thrust of the Crown's case was that Mr Berry had not been trained as a roofer, and there had been previous warnings and no adequate system had been introduced. There were a series of preparatory hearings before Judge Gilbart QC, the Recorder of Manchester (as he then was).

1 Unreported (20 July 2012), Manchester Crown Court.

1.25 Importantly, the learned judge ruled that the corporate manslaughter charges be heard separately from the gross negligence counts. This was because, although previous conduct was relevant to the health and safety charges, it was not admissible in the corporate manslaughter charge if it occurred before the commencement date of the Act. The regulatory charges were also stayed, as it was felt they added nothing to the trial.

At half time, the gross negligence charges were dismissed against two of the directors and a s 37 count against one of them. Below are extracts from the ruling given by the learned judge which, although not binding, give a very helpful indication of the current judicial stance on certain of the provisions:

> 'The Crown cannot look to evidence of "activities" or whether they involved a "breach" or "gross breach" where such activities, gross breach occurred before the date of commencement, save in so far as they are relevant to the existence of a duty on or after that date, or whether a breach after that date was a "gross breach".' [para 27]

> 'It is right to say that the Crown seemed to accept that LSEL could have been charged with an offence at common law. I do not accept the common view of the Crown and of LSEL that, in the circumstances of this case, LSEL could be prosecuted at common law for manslaughter where the death only occurred after the common law offence had been abolished. (And it should be noted that section 20 does not just prevent prosecution; it abolishes the offence in its application to corporations). In my judgment nothing in section 27 enables a prosecution to be brought against a company in the circumstances of this case for the common law offence of manslaughter by gross negligence, where the death occurred after the commencement date.' [para 39]

'If that approach be correct, does section 27(3) to (5) prevent the adduction of evidence of what happened before the commencement date? I do not consider that subsection (3) is concerned with the admissibility of evidence, but with the existence of liability, or by its reference to investigation, penalties etcetera, to clarify that an offence committed before then can still be investigated, charged and indicted under the common law.' [para 40]

'Its purpose and effect is to define the date after which the matters described in section 1 constitute an offence – i.e. whether 'the way in which its activities are managed or organise causes a person's death' and whether that amounts to a gross breach of a relevant duty of care owed by the organisation to the deceased. It does not seek to prescribe or proscribe the evidence which may be called to prove that a duty of care existed, nor whether any breach of gross.' [para 41]

'I accept (and agree) that section 1 does not bite on activities or gross breaches occurring before the commencement date. But that is a different question from the admissibility of evidence about the nature or qualitative assessment of those breaches (i.e. whether gross or not). Nothing in section 20 or 27 prevents such evidence being adduced, if relevant and otherwise admissible.' [para 42]

1.26 On the relevance of 'bad character' evidence, the judge said:

'If those Defendants were not being tried together with LSEL, that evidence would not be admissible in the trial of LSEL, unless an application were made under sections 98ff of the Criminal Justice Act 2003. Given that the Crown's case is that the conduct of the other three Defendants was reprehensible (and is therefore evidence of bad character within section 98), it must follow that it could only be admitted against LSEL if it passed through a gateway in section 100 of the Act. Section 101 is not relevant as none of them would be a Defendant if LSEL were tried alone. Similarly, as the evidence of past conduct is by definition not conduct falling within "activities" in section 1, it is not excluded from the scope of "bad character" by section 98(b) of the 2003 Act.' [para 50]

'I am prepared to accept that it could amount to important explanatory evidence under section 100(1)(a) or (b), which read

"100(1)

(a) It is important explanatory evidence,
(b) It has substantial probative value in relation to a matter which –
  (i) Is a matter in issue in the proceedings, and
  (ii) Is of substantial importance in the context of the case as a whole."

But its relevance is not unrestricted. It only passes muster if it goes to an issue germane to Count 1, i.e. whether the directors had the relevant knowledge on or after 6 April 2008, and whether any breach on or after 6 April 2008 was gross.' [para 52]

'But now one asks what happens when Count 1 [Corporate Manslaughter] is tried with Count 2 [Gross Negligence Manslaughter v Directors]. It is important to note that the issue in a general sense of whether there was or were a gross breach/breaches of duty by those directors is common to all four Defendants. The critical difference between LSEL on the one hand, and the Second to fourth Defendants on the other, is that the court is not concerned in the case of Count 1 whether the breaches before April 2008 are to be regarded as gross; indeed they are irrelevant, save for the purposes I have referred to already. It follows that the jury will have to be directed that while it can and must consider whether the conduct of any or all of the Second to Fourth Defendants on Count 2 in the years before April 2008 amounted to a gross breach insofar as the case against him or them is concerned, it must not treat that as determinative of whether there was as gross breach after 6 April 2008, and because of the terms of section 1(3) about senior management, must consider separately what the state of mind was of the same Defendants as at 6 April 2008 and subsequently.' [para 53]

1.27 The court found that there was a case to answer against the Managing Director of Lion Steel Ltd. At that juncture, the company pleaded guilty to corporate manslaughter, and all other charges were dropped. The company was fined £480,000, and had to pay £84,000 towards the prosecution costs. There was a 20% reduction for the guilty plea, and the costs were reduced by 40%, given the delay by the prosecution in bringing the charges.

1.28 On 22 November 2013, Princes Sporting Club Limited pleaded guilty to corporate manslaughter after an 11-year-old girl died from serious leg injuries when she tragically fell off a banana boat into a lake in Bedfont, West London.

The charges against the company director were discontinued by the prosecution.

The boat driver had no UK-recognised qualifications, there had been no 'spotter', and the boat and the colour of the equipment made it difficult to see the girl in the water.

The CPS press announcement stated:

> 'Today's guilty plea to corporate manslaughter is an acknowledgement that there were significant failings in the way water sports were organised at the Club. This was a gross breach of the duty of care owed to Mari-Simon which could have been avoided by having a competent adult in the towing boat acting as an observer and we are pleased this company has been held criminally accountable for this significant failing.'

The company was fined £134,579 including costs, which represented the entire assets of the company. His Honour Judge McCreath said, 'I propose to fine this company every penny it has'. He also imposed the first ever publicity order on the company.

1.29 The Marine Accident Investigation Board said that the water park procedures were 'flawed at every level'. They also recommended that Hounslow Council should introduce a licensing regime for commercial operations, which was implemented straight away by the local authority.

1.30 Rather surprisingly, we are still a long way from knowing what many of the CMCHA 2007 main provisions actually mean. Until the Court of Criminal Appeal (and Supreme Court) have an opportunity to rule on the various provisions of the Act, we are somewhat in the dark. The 'old' law is still crucial, as it helps us to interpret what will happen when the cases go on appeal in the future.

We also appear to be some way off from the prosecution of a large corporation. One factor, to be considered later on, is whether the amendment to the provisions on defence costs – if organisations successfully defend cases and do not recover their costs – will cause defendant companies to plead guilty more often in the future. This aspect arguably may be of greater significance to smaller rather than larger entities. Further, the very recent appeal regarding Sellafield and Network Rail heralding far higher fines in the future may either lead to more trials, or it may encourage earlier guilty pleas. Time will inform us. See **1.153** below.

1.31 There are a number of other prosecutions in the pipeline for corporate manslaughter, including: PS & JE Ward, due to start in March 2014; MNS Mining, also due to commence in March 2014; and Mobile Sweepers (Reading) in early 2014. Cavendish Masonry has also been charged with corporate manslaughter; and another company, Stericycle Ltd, has recently been charged – other charges under s 7 of the HSWA 1974 have been preferred against other employees of the company.

HEALTH AND SAFETY (OFFENCES) ACT 2008

1.32 We have also seen the first conviction under the Health and Safety (Offences) Act 2008, which allows for the imprisonment of individuals convicted of certain health and safety offences. A Mr Dutton, the Health and Safety Manager of a metal distribution firm called South Essex Stockholders Ltd, was given a four-month term of imprisonment, suspended for two years. He pleaded guilty to one count under s 7 of the HSWA 1974. Mr Dutton had asked a junior employee to bring him a can of chemical to use as an accelerant to burn debris in a skip. On pouring it in, it (somewhat predictably!) exploded and seriously burned him. He spent over three months in hospital and required major skin grafts. His Honour Judge Hayward Smith QC commented that he had been 'extremely foolish … you were severely burned. You were in flames. Those who saw what happened suffered trauma'. Given the risk to himself and others who may also have been injured, a suspended sentence was imposed. In mitigation, it was pointed out that he had, perhaps fortuitously, been the only injured person. He also had to pay £5,000 towards the prosecution costs. His employer was convicted after a trial under s 2 of the HSWA 1974 and fined.[1]

1 See further **Chapter 2**.

1.33 There have been a series of other convictions, particularly for gas safety breaches, in 2012 and 2013 that have led to imprisonment.

A builder, Patrick Regan, was jailed for 12 months for illegally carrying out substandard gas work in a pensioner's house, after he pleaded guilty. The court was informed that he had left a connected and unstable gas fire flue laying across the floor and routed into a floor void. There were also other gas leaks in the house.

An unregistered gas fitter, Mr Carter, pleaded guilty to three gas safety offences, including one charge relating to illegal work on over 85 properties, which included work on gas boilers, fires and cookers. He had let his Gas Safe registration lapse, and was sentenced to four months' imprisonment.

Another gas fitter, Christopher Johnson, was sentenced to nine months' imprisonment for Gas Safety regulation breaches and under s 3 of the HSWA 1974.

REGULATORY REFORM (FIRE SAFETY) ORDER 2005

1.34 Prosecutors are also making more use of the Regulatory Reform (Fire Safety) Order 2005 which is triggered by the possible risk of fire. The Order came into effect on 1 October 2006 and imposes duties on defined 'responsible persons' to protect anyone who may be on or in the vicinity of the premises. Substantial fines include £300,000 against Shell plc and £400,000 imposed on New Look Ltd.[1]

1 *New Look Retailers Ltd v London Fire & Emergency Planning Authority* [2010] EWCA Crim 1268.

1.35 In July 2011 a hotel manager and an external fire risk assessor were fined and jailed for eight months under this Order. The learned judge said:

> '… [the time has come] … to send out a message to those who conduct fire risk assessments and to hoteliers who are prepared to put profit before safety.'

1.36 The Regulatory Reform (Fire Safety) Order 2005 replaced regulations made under the Fire Precautions Act 1971. This order was made pursuant to the Regulatory Reform Act 2001 to be in compliance with the EU Directive on fire safety in the workplace and business premises. There have been a series of recent cases where the fines have been gradually ramping up. Arguably, the first relevant case fell under the previous legislation. In *ESB Hotels Ltd*, owners of a hotel pleaded guilty to two counts of contravening the requirements of a fire certificate, contrary to s 4 of the Fire Precautions Act 1971. Bed mattresses had been stored in various corridors. Investigators found that the hotel staff did not appear to have a complete understanding of the potential fire hazard. The seat of the fire was the ignition of mattresses which had been caused by a deliberate act of an employee. The fine was reduced from £400,000 to one of £250,000.

1.37 In *R v Shell International Ltd*, Shell was fined £300,000 in 2009 for the breaches of the 2005 legislation. The Shell building employed about 2,000 people. There had been two small fires in three weeks in the Waterloo Shell Centre. There was no appeal against sentence.

1.38 In *New Look Retailers Ltd*[1] the Court of Appeal looked again at the relevant law and previous decisions. The case concerned an appeal against a £400,000 sentence imposed by HHJ Rivlin QC for two offences under the Regulatory Reform (Fire Safety) Order 2005. The national clothing retailer had a serious fire in its Oxford Street premises. The fire probably started in a store room on the second floor. Thirty fire appliances attended and Oxford Street was closed for two days. Four hundred people had required emergency evacuation. Article 8(1)(a) of the Order requires that the responsible person under the Act must take such fire safety precautions as will ensure, so far as is reasonably predictable, the safety of his employees. Article 8(1)(b) also requires the responsible person to take such general fire precautions as may reasonably be required to ensure that the premises are safe.

1 *New Look Retailers Ltd v London Fire & Emergency Planning Authority* [2010] EWCA Crim 1268.

1.39 The original indictment had 35 alleged breaches of duty, but the actual indictment set out two counts, namely (a) a failure to carry out an adequate risk assessment, and (b) a failure to ensure that the employees were provided with adequate training. The Court of Appeal, in determining the appropriate sentence, looked at a number of factors and made some interesting observations.

One such observation was:

> 'Contrary to the submissions made to us, *ESB* [the case last cited], in our view, provides support for the judge's observation that assessing fines in these cases is, first and foremost, a fact sensitive exercise.'

The court also quoted and agreed with what HHJ Rivlin QC had said:

> '.., when it comes to fire, one does not have to think very deeply in order to appreciate the potential for disaster.'

1.40 The court went on:

> 'What the sentencing judge was entitled to recognise was the fact that the nature of the risk against which employees and others were to be protected was the risk from death or serious injury in a fire. Fire can be indiscriminate in its effect and, in the case of an organisation which in the centre of a large city undertakes responsibility for a large number of visitors to its premises, breach will usually be a very serious matter.'

1.41 The court continued:

> 'What the fire served to illustrate was the magnitude of the risk which the appellant ran with public safety. Exactly the same considerations would have been relevant if, in the case of a near miss, investigation had revealed wholesale disregard by Balfour Beatty and Railtrack [see *R v Balfour Beatty Rail Infrastructure Services Ltd* [2006] EWCA Crim 1586] for their responsibilities towards rail passengers. Fines would in that eventuality have been imposed for the magnitude of the risk knowingly taken and not for the causation of any tragic consequences.'

It is perhaps of some interest that the court used a health and safety case as an analogy.

1.42 In conclusion, the court said:

> 'However, we share the judge's scepticism, expressed during argument, that the appointment of a single fire safety advisor for a group of 600 and more shops was a sufficient response to the magnitude of the obligation.'

The court felt that:

> 'The breaches of duty acknowledged by the appellant fell into two distinct categories, first, deficiencies in the appellant's provision and maintenance of fire safety precautions and, secondly failure to provide any adequate training and retraining schemes not just for essential health and safety staff but employees generally. We share the judge's view that the appellants' performance of its fire safety duty in a large departmental store in the centre of London was lamentable. The fines were, we recognise severe, but they were not in our judgment manifestly excessive and the appeal is dismissed.'

This judgment heralded a more stringent approach by the courts, and penalties are increasing in severity.

1.43 In April 2010, Tesco was fined £95,000 at Wood Green Crown Court and ordered to pay £24,000 in costs after pleading guilty to five breaches of the Regulatory Reform (Fire Safety) Order 2005 arising from various breaches at their store in Barnet after a small fire in the staff kitchen. London Fire Commissioner, Ron Dobson, said:

'Fire safety is a key part of good business management and the general public should feel safe from fire when they are out shopping. London Fire Brigade will continue to take action when businesses, large or small, do not take their fire responsibilities seriously. Failure to comply with the law can, as this case has shown, result in a prosecution.'

1.44 Also in April 2010, the Co-operative Group was fined £210,000 and ordered to pay £28,000 costs. The company pleaded guilty to six breaches of the Regulatory Reform (Fire Safety) Order 2005 relating to its store in Southampton. These related, inter alia, to failure to keep the rear emergency exit doors unlocked for easy egress in an emergency, and the fact that they had fitted a lock between the retail and storage area which required a security code to unlock it.

The judge said that the case demonstrated a lamentable approach to fire safety and that the Group had been responsible for a potential death trap, given the severity of the fire safety failings.

1.45 In 2010, a hotel in Cheshire was fined £75,000 and an order for £52,000 costs imposed after pleading guilty to three breaches of the Regulatory Reform (Fire Safety) Order 2005. During renovation, a prohibition notice was served on the hotel. There was no operable automatic detection system and there were faulty smoke detectors. The hotel was allowed to reopen four days after the remedial action was taken.

1.46 In July 2011, a hotel manager of two hotels and an external fire risk assessor, Mr O'Rourke, were both jailed for eight months for fire safety offences. Nottingham Fire and Rescue Services found, in a routine inspection, that fire precautions in the sleeping areas were inadequate, including at one hotel where officers found both staircases terminated in the same ground floor area with no alternative escapes or separation. There were also blocked exit routes and a locked fire door. Mr O'Rourke also pleaded guilty to two counts for failing to provide a suitable fire risk assessment.

1.47 Further, at Blackfriars Crown Court, a hotel owner has been fined £210,000 following a fire. It is believed to be the first time that a jury, rather than a judge or a magistrate, has convicted an organisation for breaches under the Regulatory Reform (Fire Safety) Order 2005.

1.48 In 2011, the HSE prosecuted two waste recycling companies after a major fire on an industrial estate near Preston. More than 100 fire fighters and 25 engines were needed to put the fire out. It spread to more than 10,000 square metres, and surrounding roads were closed for a day. The companies were prosecuted under ss 2(1) and 6(1)(a) of the HSWA 1974, and were fined £87,000 with an order for costs of £136,000. The Chief Fire Officer of Cheshire stated:

'This was a major incident which could have been a lot worse if crews had not prevented flames from reaching cylinders containing 25 tonnes of liquid petroleum gas close to the original site of the fire. It is extremely fortunate that there were no injuries to members of the public or fire fighters.'

1.49 In 2011, a London building owner was sentenced to a six-month suspended prison sentence after being convicted of seven offences under the Regulatory Reform (Fire Safety) Order 2005. He was also sentenced to 150 hours of community service and £13,000 in costs. The property was used as a takeaway restaurant whilst the upstairs was used as sleeping accommodation.

The officers found a range of fire safety breaches, including no fire doors or emergency lighting. There was also no alternative means of escape from the sleeping accommodation.

1.50 In 2011 the owners of a Cornish hotel were fined £80,000 and £62,000 costs. The fire in Newquay, described as the worst British hotel fire in 40 years, killed three people. There were failings relating to checking that the alarms and detectors were working and a failure to make an adequate risk assessment. Further, recently, a north Devon landlord in Ilfracombe was fined £11,500 and ordered to pay £3,000 costs after a major fire occurred in flats. No one was injured.

1.51 In June 2012, ASDA Stores Ltd were fined £40,000 and £15,647 costs after pleading guilty to two charges under the Regulatory Reform (Fire Safety) Order 2005. An inspection by Royal Berkshire Fire Authority identified serious breaches, including two fire exit doors that were chained and locked shut and fire exit doors wedged open. The Fire Authority said:

> 'Staff and customers are entitled to feel safe when working at, or visiting, a supermarket or any other premises. We will continue with our efforts to ensure that any business owner, or manager who refuses to take obligations seriously will be brought before the courts.'

1.52 More recently, after a fatal fire in the Midlands, Northamptonshire Fire and Rescue Services prosecuted a number of owners, and an agent for breaches of the Order in relation to Houses in Multiple Occupation (HMOs).

The owner was sentenced to nine months' imprisonment and the managing agent to suspended prison sentences. This sends out a clear message to landlords and their agents.

The Chief Fire Officer said:

> 'We proactively look for suspected HMO's because we know these are the highest risk premises housing vulnerable people who are being exploited by landlords who won't fit fire doors or fire alarm systems. These simple precautions are affordable and would save the lives of their tenants should a fire occur ... where we find systemic non-compliance or a severe breach, putting lives at risk we will take enforcement action and in the serious cases we will prosecute those responsible.'

1.53 There have been a series of other similar cases. In October 2013 a landlord, Mr Sanghera, was fined £7,150 after a fire broke out in a house accommodating tenants. The judge said:

> 'If the fire had occurred a few hours later, when the residents were asleep, the outcome could have been much worse. Luckily enough one of the residents heard the fire developing and was able to raise the alarm.'

The Area Manager said:

> 'Derbyshire Fire and Rescue Service would like to remind all landlords and managing agents of rental accommodation of their legal responsibility to protect their tenants against the risk of fire.'

1.54 An important case was heard by the Court of Appeal in October 2013. In *Michael Wilson v R*,[1] the Court of Appeal (Gross LJ, MacDuff J and the Recorder of Chester) reviewed all the authorities relating to whether the indictment was a nullity, and other discrete areas.

They further looked at the construction of an indictment concerned with the Regulatory Reform (Fire Safety) Order 2005. The court did not have to look at the facts and merely said:

> 'On 6 December 2011 in the Crown Court at Blackfriars, the Appellant Michael Wilson was convicted of six offences charged under Article 32(8) of the Regulatory Reform (Fire Safety) Order 2005 before HH Judge Richardson and a jury. He was later sentenced by way of substantial fines with terms of imprisonment in default. Other consequential orders were made which we need not consider.'

1 [2013] EWCA Crim 1780.

1.55 The court added:

> '… There was a co-defendant Chumleigh Lodge Hotel Limited which was also convicted of six offences under different articles of the Order.'

The court goes on:

> '… we do not need to consider the background facts in any detail. Chumleigh Lodge Hotel, situated in Finchley, London N3, was owned by the Appellant. He was the sole director of the company. On 18 May 2008, a fire broke out in the premises. This had been started by a guest in one of the bedrooms carelessly disposing of a lighted cigarette. The Fire Brigade attended and extinguished the fire. The circumstances of the fire were investigated and the company and the appellant were charged with the offences.'

1.56 Interestingly, the London Fire Brigade press release set out some of the facts, stating that the hotel was fined £30,000 plus costs and Mr Wilson £180,000 plus costs. It said:

> 'The blaze had spread quickly from a first floor guest bedroom, up a staircase to the floor above and along a corridor. Three people escaped from the fire, two by using the stairs and a third by climbing out of a second floor window … Following the fire, London Fire Brigade Fire Safety Inspectors visited the hotel and raised a number of serious fire safety concerns. These included defective fire doors, blocked escape routes and no smoke alarms in some of the hotel's bedrooms. Mr Wilson was also unable to produce a suitable and sufficient Risk Assessment and was found not to have provided staff with adequate fire safety training.'

1.57 The Chairman of the London Fire and Emergency Planning Authority stated:

> 'Business owners have a clear responsibility under fire safety law to ensure that both the public and their employees are as safe as possible from the risk of fire. This verdict sends out a clear message that if these responsibilities are ignored we will not hesitate in prosecuting and people will face serious penalties.'

1.58 The Court of Appeal, when dismissing the appeal on the technical nullity argument (set out in paras 72–80) stated:

> 'However, not every error in an Indictment renders it a nullity.'

The court added:

> 'The Indictment; it was defective, but the defect occasioned neither unfairness nor prejudice. This appeal must be dismissed.'

1.59 In conclusion, these cases herald a severe approach to issues relating to fire risk. Organisations and individuals (including their professional advisers) who do not have adequate systems and assessments in place run the risk of large fines and terms of imprisonment. Prevention is now, more than ever, the touchstone of prudence.

ENTERPRISE AND REGULATORY REFORM ACT 2013

1.60 On 25 April 2013 the Enterprise and Regulatory Reform Act 2013 came into effect. The aims of the Act are:

> 'to cut the costs of doing business in Britain, boosting consumer and business confidence and helping the private sector to create jobs.'

The guide to the Act, set out in June 2013, further states:

> 'small, medium and micro businesses will benefit from this Act in particular from the lower costs involved in settling disputes and from red tape challenge measures.'

1.61 In relation to health and safety, the Act limits the right to claim for civil action compensation to where it can be proved that an employer has actually acted negligently. As the guide puts it:

> '… this means in future, if a claim is made, the employer will have the opportunity to defend themselves on the basis of having taken reasonable steps to reduce the risk of an accident.'

1.62 The DWP explained its stance by stating:

> 'an approach to targeting each strict liability duty would be much more complex to achieve, requiring a large number of changes to many sets of regulations. A single change to the HSW Act achieves the same overall policy objective and will be significantly easier for employers and other stakeholders to understand. It also provides a consistent approach to civil litigation across health and safety legislation. It is therefore like to have more impact in changing perceptions of the "compensation culture".'

In essence, this means that a non-public sector employee must actually show fault and consequence rather than just relying on a breach of law or practice. This will make it much harder for employees to bring civil cases in the future. The Jackson reforms, the legal aid cuts and the reduction in 'uplifts' for claimant lawyers will also have a major effect going forward.

'RISK' AND 'FORESEEABILITY' UNDER THE HSWA 1974

1.63 Further, since the previous edition of this book, there has been a series of cases decided relating to 'risk' and 'foreseeability'. These follow on from *R v Porter*[1] and *R v Chargot Ltd.*[2]

It is not possible to go through every possible case in this overview chapter. We must however remind ourselves that, in essence, Lord Justice Moses in *Porter* said that the test for 'risk' is that it had to be real, not hypothetical or fanciful, risk. The court also put it another way, namely that it was one of the risks attendant on everyday life.

In essence, the Court of Appeal was creating clear blue water from the previous test of 'possibility of danger', as set out in the *R v Board of Trustees of the Science Museum*[3] case.

1 *R v Porter (James Godfrey)* [2008] EWCA Crim 1271, [2008] ICR 1259.
2 [2008] UKHL 73, [2009] 1 WLR 1.
3 [1993] 1 WLR 1171, [1993] 3 All ER 853.

1.64 The House of Lords in *Chargot* approved the *Porter* case. Lord Hope in the leading speech stated (at para 27):

'… when the legislation refers to risk it is not contemplating risks that are trivial or fanciful. It is not its purpose to impose burdens on employers that are wholly unreasonable. Its aim is to spell out the basic duty of the employer to create a safe working environment …

The law does not aim to create an environment that is entirely risk free. It concerns itself with risks that are material. That, in effect, is what the word "risk" in the statute means. It is directed at situations where there is a material risk to health and safety, which any reasonable person would appreciate and take steps to guard against.'

1.65 The Court of Appeal in *Tangerine Confectionery Ltd and Veolia ES (UK) Ltd v R*[1] reviewed a series of previous authorities. The court said (at para 26):

'… Their Lordships approved the decision of this Court in *R v Porter* [2008] ICR 1259, describing it as an exceptional case which made an important point. There a child had fallen in the school playground when jumping down some steps. The decision was that although there may always be some possibility (thus, we observe, **some** risk) that a child may fall, or jump unwisely, there was no risk such as section 3 identifies arising from the conduct of the school undertaking. The question of reasonable practicability was not reached. That decision is thus one focussing on the first (risk) stage of the enquiry and not on the second (reasonable practicability) stage. The judgment of Moses LJ does not employ the expression "foreseeability". It concentrates upon whether there was or was not a real risk, as distinct from a fanciful or hypothetical one. It holds that this is a question of fact in each case, and that a number of indicia may be relevant to answering it. In that case, they included the fact that there had never previously been any accident on the steps, or even elsewhere in the playground, from jumping, that the steps were not in any way faulty, and that a properly prepared risk assessment made before the incident had not identified any risk from jumping or from the differing levels of the playground. Underlying the decision was, plainly, the fact that any risk of injury as no different in the playground from what it inevitably and unavoidably is in any other place that a child might be.'

1 [2011] EWCA Crim 2015.

1.66 Further on, the court, citing the Supreme Court case of *Baker v Quantum Clothing Group Limited and Others*,[1] said:

'the conclusion which we draw is that *Baker* does apply to sections 2 and 3 of the HSWA. Foreseeability of risk (strictly foreseeability of danger) is indeed relevant to the question whether a risk to safety exists. That accords with the ordinary meaning of risk, as is demonstrated by the concept of a risk assessment, which is itself an exercise in foresight. Whether a material risk exists or does not is, in these cases, a jury question and the foreseeability (or

> lack of it) of some danger or injury is a part of the enquiry. None of this, however, means that in a prosecution under either section it is incumbent on the Crown to prove that the accident which occurred was foreseeable. That would convert the sections into ones creating offences of failing to take reasonable care to avoid a specific incident. It means no more than that the sections are concerned with exposure to risk of injury, and that the extent to which injury is foreseeable is part of the enquiry into the level of risk. The sections do not command an enquiry into the likelihood (or foreseeability) of the events which have in fact occurred. They command an enquiry into the possibility of injury. They are not limited, in the risks to which they apply, to risks which are obvious. They impose, in effect, a duty on employers to think deliberately about things which are not obvious. In most cases, absent the sort of time factor which obtained in *Baker v Quantum*, it is likely that consideration of foreseeability will add little to the question whether there was a risk. In most cases, we think, the principal relevance of foreseeability will be to go to the defence of all reasonable practicable precautions having been taken. We note that this defence does not impose on an employer the duty to take every feasible precaution, or even every practicable one; it imposes a duty to take every **reasonably practicable** one. What is reasonably practicable no doubt depends on all the circumstances of the case, including principally the degree of foreseeable risk of injury, the gravity of injury if it occurs, and the implications of suggested methods of avoiding it.'

1 [2011] UKSC 17.

1.67 Dismissing the appeals, the Court of Appeal said (at para 37):

> 'In both these cases before us we have not the slightest doubt that injury was a foreseeable incident of the activity of the employer which was in question. The risk of injury plainly existed in both cases.'

1.68 In *R v Merlin Attractions Operations Limited*,[1] the Court of Appeal, when considering an appeal against sentence, revisited the issue of 'foreseeability'. A visitor to Warwick Castle died after falling some 14 feet from a parapet bridge. The drop was obscured by trees and bushes. The wall to the bridge was only 15 inches high. Twenty million visitors had been to the site.

The jury convicted the company on two counts. The judge sentenced them to £350,015 and costs of £145,000. The Court of Appeal looked at the sentencing guidelines and a number of other authorities. When dismissing the appeal, the court said the following:

> 'the Appellant accepts that by its verdict, the Jury necessarily concluded that use of the bridge did expose people to a material and foreseeable risk of harm, and that it was reasonably practicable for the appellant to have done more than it did to reduce that exposure – whether by barriers or warning signs …'

1 [2012] EWCA Crim 2670.

1.69 The court also added (at para 60):

> 'the Judge was clearly entitled to conclude, as he did, that there had been a serious breach of the Appellant's health and safety systems in relation to the Bear and Clarence Bridge and that had resulted in an obvious danger of at least serious injury (not so much in relation to adults, but in relation to children) to which a very large number of people had been exposed over

many years and which had been the product, in part, of a failure to react appropriately to advice in 2003.'

1.70 In *R v Willmott Dixon Contractors & Others*,[1] which concerned asbestos exposure during a refurbishment of two stores, the Court of Appeal made clear that there was an inherent risk from asbestos and the known danger of mesothelioma. Dismissing the appeal, Mr Justice Hedley said:

> '… the purpose of the legislation is to protect those who otherwise cannot protect themselves. It does not impose absolute liability but it does impose strict obligations in terms of justifying the steps that are taken. Accordingly, the social policy imperative may give rise to criminal liability in circumstances which, absent that context, might appear surprising.'

1 [2012] EWCA Crim 1226.

1.71 Marks and Spencer, which operated the two stores, were fined £1 million and costs of £600,000. Willmott Dixon, the contractors, were fined £50,000 and £75,000 costs.

1.72 In December 2013, Mr Justice Simon dismissed a charge of health and safety against a director of a firework display company, at the half-time stage. The learned judge said:

> '… the prosecution case was heavily weighted on hindsight and there was not sufficient evidence to show that [the Defendant] ought to have foreseen that smoke from the display could have drifted and mixed with fog to create thick smog.'

1.73 This state of flux between the various cases set out above (and others) increasingly makes it harder for lawyers and advisers to advise on whether to contest these types of case and for the potential defendants to decide on the best course of action.

Further, even if successful, their defence costs will generally not be returned to them. This is another factor to be taken into consideration by defendants, their advisers and insurers.

FEE FOR INTERVENTION (FFI) REGIME

1.74 This scheme, which came into effect on 1 October 2012, was triggered by regs 23–25 of the Health and Safety (Fees) Regulations 2012.[1]

1 SI 2012/1652.

1.75 In essence, the HSE charges for carrying out its regulatory functions from those found to be in 'material breach' of health and safety law.

A material breach occurs when there has been a contravention of health and safety law. This is decided by the HSE Inspector. This has to be applied using the Enforcement Management Model and Policy Statement.[1]

The work carried out by the HSE for gathering evidence including expert evidence and recording conclusions etc can all be invoiced by the HSE.

1 See the various guidance documents on the HSE website.

1.76 The findings and costings can be disputed. There are two levels. Firstly, by an independent HSE Senior Manager; and then by a panel of HSE personnel, together with an independent representative to the panel.

These appeal processes, if unsuccessful, can also be charged for by the HSE. The scheme will be reviewed after 12 and 36 months respectively. It awaits to be seen how many investigations will occur.

1.77 With government cuts to various departments, the FFI scheme provides a further source of funding. The scheme does not apply to local authority enforcement.

In June 2013 the HSE, which charges £124 per hour, reported that the scheme had charged about £2.7 million for 5,766 invoices rendered.

In a speech in December 2013, Judith Hackett said:

> '... We've raised close to 15,000 invoices this year raising in the region of £8 million, so the average cost of an invoice is around £470. So far, in that number of 15,000, the number that have been queried is about 400, 150 of them because people think it's taken an unreasonable amount of time, 87 because we think we've aimed at the wrong duty holder, ... so there have only been 13 that have gone to Level 1 disputes and we've only had one case out of 15,000 that's gone all the way to Level 2 dispute.'

The HSE estimates that the fee ranges from approximately £750, for an inspection that results in a letter, to about £1,500 where an enforcement notice is secured, and thousands of pounds for a full investigation. See the recent recommendations as to the future of FFI at **1.10** above.

1.78 It is also of some interest whether organisations, on a short-term basis, may agree to pay for the service, forgetting that these breaches may form evidence in current and future enforcement action. They may well form the basis of a 'bad character' application in the future. There will also be complicated issues relating to internal documentation and legal professional privilege.[1]

Organisations will have to consider not just winning a 'battle' but also the consequences for the future to their national and international operations.

1 See further **Chapter 3**.

1.79 Further, organisations need to factor in the importance of the decision in *R v Bodycote HIP*.[1]

The Court of Appeal, in an appeal against sentence for a double fatality, took into account what the defendant's sister company had failed to do in the United States. The court stated:

> 'the serious aggravating feature in this case was that a similar incident had led to the deaths of two employees at a HIP plant operated by another company in the same group. That occurred in May 2001 in Tarzana, California. These two employees had also died from asphyxiation as a result of inhaling argon and nitrogen. That may have prompted the appellant to either adopt or to expedite the introduction of its permit system, but, as the prosecution was able to show, not with the rigour that the dangers demanded and the appellant allowed the other defaults in its safety procedures to continue or to get worse.'

1 [2010] EWCA Crim 802.

1.80 Further on, the court said:

'the other aggravating features of particular note were that there had been two deaths, and the appellant had not adequately heeded the warnings from the failures in the Californian plant.'

1.81 Increasingly, courts and regulators are factoring in the conduct of organisations (and industries) in other jurisdictions. This is an important recent development. Decisions about tactics need to be decided, arguably, at a global level.

CORONERS AND JUSTICE ACT 2009

1.82 On 25 July 2013, various provisions of the Coroners and Justice Act 2009, which received Royal Assent in November 2009, came into force. There are still a number of provisions – such as those relating to treasure trove and powers of entry – to be implemented.

This overview chapter cannot highlight all of the important changes, but will restrict itself to the most important.

1.83 Firstly, a Chief Coroner (HH Judge Thornton QC) has been appointed. His role is to provide leadership, guidance and training, and to author an annual report. He also approves all new coroner appointments.

All reports made with a view to preventing future deaths[1] are also to be made to him.

1 See Coroners and Justice Act 2009, Sch 5, para 7.

1.84 Further, the Chief Coroner must look into any investigations lasting over 12 months.[1] Another major change is that a senior coroner, who is made aware that a body of a deceased person lies within the area, must, as soon as practicable, conduct an investigation into the person's death. This varies from the previous 1988 Act, which stated that an inquest needed to occur.

Not all investigations will lead to a full-blown inquest. We will have to see what this means in reality. Some coroners believe there will be fewer inquests.

1 Coroners and Justice Act 2009, s 1(1).

1.85 Documents and other evidence need to be provided to the coroner on their request, *even* at an investigation stage.[1]

1 Coroners and Justice Act 2009, Sch 5.

1.86 Further and very importantly, rule 43 of the Coroners Rules 1988 has now been replaced by paragraph 7 of Schedule 5 to the Coroners and Justice Act 2009. This requires that the coroner complete an 'action to prevent other deaths' report. This is now a duty to report rather than a discretionary exercise. These detailed reports are found on the Ministry of Justice website under Regulation 28 reports dating back to July 2013.

1.87 Under the new provisions, investigations need to be completed within six months.[1] This will mean that interested parties to an inquest will have to immediately react and prepare in the event of a death in their organisation. Investigations taking longer have to be reported to the Chief Coroner for him to investigate. There are no 'teeth' built into the new law – save that individual coroners will be asked about delays by the new Chief Coroner.

In theory, however, no longer will the parties have the luxury of many years between the workplace death and the inquest, which increasingly will be with a jury. This means that organisations need to have a system in place and ready to spring into action. These inquests are becoming more complex and lengthier, particularly with the engagement of Article 2 of the European Convention on Human Rights and the increasing use of narrative verdicts.

1 Coroners (Inquests) Rules 2013, SI 2013/1616, r 8.

1.88 We have also seen some major inquests being heard, including those involving Mark Duggan, the Staffordshire Trust (which has just announced that a further Inquiry is to take place), the Lakanal House inquest (where six people died in a fire in a tower block in South London) and the Hillsborough inquest (due to commence in March 2014).[1]

1 See further **Chapter 3**.

GLOBAL ACCIDENTS

1.89 On the evening of 20 April 2010, a gas release and subsequent explosion occurred on the Deepwater Horizon oil rig working on the Macondo exploration well for BP in the Gulf of Mexico. Eleven people died as a result of the accident and others were injured. The fire burned for 36 hours before the rig sank, and hydrocarbons leaked into the Gulf of Mexico before the well was closed and sealed. The accident involved a well integrity failure, followed by a loss of hydrostatic control of the well. This was followed by a failure to control the flow from the well with the blowout preventer (BOP) equipment, which allowed the release and subsequent ignition of hydrocarbons. Ultimately, the BOP emergency functions failed to seal the well after the initial explosions. BP put aside $41bn to pay for the spill, two and a half times more than BP's entire profit in 2009. However, the costs for the explosions are yet to be finalised. Further, very recently the 5th Circuit Court has ruled that BP must pay compensation to some claimants who had in fact suffered no loss from the oil spill. The civil trial is ongoing. The criminal fines for BP are over £2.6 billion after they pleaded guilty to 14 counts.

1.90 The Fukushima Daiichi nuclear disaster was initiated by the tsunami of 11 March 2011. The damage caused by the tsunami produced equipment failures, and without this equipment a loss-of-coolant accident followed, with nuclear meltdowns and releases of radioactive materials beginning on 12 March. A Japanese parliamentary panel delivered a verdict after investigating the accident at the Fukushima nuclear plant. Although such an event should have been predicted and planned for, the panel said, it found gaping holes in safety standards and emergency procedures: 'Our report catalogues a multitude of errors and wilful negligence that left the Fukushima plant unprepared for the events of March 11'. It found serious deficiencies in the response to the accident by Tepco, regulators and the government. The inquiries are still ongoing.

1.91 In December 1999 a family of four died in a gas explosion in Lanarkshire, and Transco were found guilty of breaches of health and safety. During sentencing the judge said '... the corporate mind of Transco has little or no remorse for this tragedy which, they ought at least now to accept, was exclusively their own creation'. A series of companies (and individuals) have been summonsed. The public inquiry into the explosion at the Scottish Plastics factory reported in September 2009, and was chaired by Lord Gill, a senior judge.

1.92 In the early hours of Sunday 11 December 2005, a number of explosions occurred at Buncefield Oil Storage Depot, Hemel Hempstead, Hertfordshire. At least one of the initial explosions was of massive proportions and there was a large fire, which engulfed a high proportion of the site. Over 40 people were injured; fortunately, there were no fatalities. Significant damage occurred to both commercial and residential properties in the vicinity, and a large area around the site was evacuated on emergency service advice. The fire burned for several days, destroying most of the site and emitting large clouds of black smoke into the atmosphere. A four-month trial at St Albans Crown Court concluded in July 2010, with five companies being found guilty and ordered to pay a total of £9.5 million in fines and costs. One of the companies involved is currently appealing against conviction and sentencing. A report published in February 2011 concluded that fundamental safety management failings were the root cause of the disaster.

The trial judge, Sir David Calvert-Smith, said:

> 'Had the explosion happened during a working day, the loss of life may have been measured in tens or even hundreds. The failures which led in particular to the explosion were failures which could have combined to produce these consequences at almost any hour of any day. The fact that they did so at 6.01 on a Sunday morning was little short of miraculous. So too was the fact that not one of the few people on the site or in the surrounding area on that Sunday morning lost their lives.'

See also *West London Pipeline and Storage Ltd v Total UK Ltd.*[1]

1 [2008] EWHC 1729 (Comm).

1.93 After the Piper Alpha disaster, Lord Cullen found that Occidental Petroleum had 'significant flaws in the quality of [its] management of safety'.[2] In the Inquiry into the Clapham Junction train crash, Mr Justice Hidden QC criticised British Rail for letting its working practices 'slip to unacceptable and dangerous standards'.[3] Mr Justice Fennell QC in his Inquiry into the Kings Cross fire criticised London Underground for 'collective failure from the most senior management level downward over many years to minimise the outbreak of fire'.[4] Lord Cullen in his report concerning the Ladbroke Grove train crash wrote of the 'incompetent management and inadequate procedures' of Railtrack, the then infrastructure controller.[5]

1 US Chemical Safety and Hazards Investigation Board, *Investigation Report*, Refinery Explosion and Fire Report No. 2005–04-I-TX March 2007.
2 The Hon Lord WD Cullen *The Public Inquiry into the Piper Alpha Disaster* (Cm 1310, 1990).
3 A Hidden QC *Investigation into the Clapham Junction Railway accident* (Cm 820, 1989).
4 D Fennell QC *Investigation into the King's Cross Underground Fire* (Cm 499, 1988).
5 The Rt Hon Lord WD Cullen PC *The Ladbroke Grove Rail Inquiry, Part 1 Report* (2001), available at www.rail-reg.gov.uk/server/show/nav.1204.

CORPORATE MANSLAUGHTER

1.94
The recent developments of the CMCHA 2007 have been set out above.[1]

1 See **1.18–1.31**.

1.95 Corporate manslaughter is a criminal offence. A corporation can be convicted of corporate manslaughter when someone is killed as a result of the way the organisation is managed or organised and the failings by senior managers add up to a gross breach of the 'relevant duty of care' owed to the deceased person. The offence of corporate manslaughter builds on key aspects of the common law offence of gross negligence manslaughter in England and Wales and Northern Ireland. However, rather than relying on the guilt of one or more individuals, liability for the offence depends on a finding of gross negligence in the way in which the activities of the organisation are run. The offence is now committed where, in particular circumstances, an organisation owes a duty to take reasonable care for a person's safety, and the way in which activities of the organisation have been managed or organised amounts to a gross breach of that duty and causes the person's death. How the activities were managed or organised by senior management must be a substantial element of the gross breach.

1.96 For corporate manslaughter or corporate homicide, a person to whom a duty of care is owed has to be killed, and that death has to be due to senior management failures which amount to a gross breach of the 'relevant duty of care'.

The offence of corporate manslaughter

1.97 The offence is set out in s 1(1):

> 'An organisation [corporate or other relevant body] ... is guilty of an offence [of corporate manslaughter] if the way in which its *activities are managed or organised–*
>
> (a) causes a person's death, and
> (b) amounts to a gross breach of a relevant duty of care owed by the organisation to the deceased.'

Section 1(3) states:

> 'An organisation is guilty of an offence under this section *only if* the way in which its activities are managed or organised by its *senior management* is a substantial element of the breach referred to in subsection (1).'

Causation

1.98 The relevant 'management failure' need not be the sole cause of death; it need only be *a* cause. Compare this with the 'common law' manslaughter offence, where the law required the failure to be a *substantial* cause of death. The courts might imply a 'substantiality' requirement into the new offence. We will have to wait and see.

Relevant duty of care

1.99 The organisation must owe to the deceased a duty of care connected with the organisation's activities. The relevant duties are set out in s 2. It is for the trial judge (and not the jury) to determine whether such a duty exists. Thus, the same position exists under the Act as under common law – see *R v Khan and Khan*.[1]

1 [1998] Crim LR 830 (CA).

Senior manager

1.100 The failure must be at *'senior manager'* level. A 'senior manager' is defined in s 1(4)(c). The definition identifies those whose management *responsibilities* relate to *the whole* of an organisation's activities or to a *substantial part* of them. It is intended that management conduct be considered collectively as well as individually (thus, failures of different managers can be aggregated, which under the current common law test they cannot be). The definition is intended to identify two strands of management responsibility: the taking of decisions about how activities are to be managed or organised; and actually managing those activities. This is a very wide definition and very far from the previous directing mind test pursuant to common law. It also implies that aggregation is possible. It is also important to factor in the recent line of cases reviewing s 37 of the HSWA 1974.[1]

1 See *R v P* [2007] EWCA Crim 1937.

Gross breach

1.101 This is set out in s 1(4)(b):

> 'a breach of duty of care by an organisation is a *"gross" breach* if the *conduct* alleged to amount to a breach of that duty *falls far below what can reasonably be expected* of the organisation in the circumstances.'

1.102 Factors for the jury to consider are set out in s 8. The jury *must* consider whether the organisation failed to comply with the health and safety legislation and, if so, (a) how *serious* that failure was and (b) how much *risk* is posed. The jury *will be able to* consider health and safety guidance and the organisation's safety culture (this includes the 2013 IOD/HSE guidance – see above). The jury is not precluded from considering any evidence it thinks relevant to the issues. Defence submissions to exclude prejudicial evidence will become more difficult. This is arguably the most important aspect of the new legislation and certainly very dangerous for those being prosecuted.

1.103 Even if someone is very seriously injured and there are gross management failings, a charge of corporate manslaughter cannot be brought because no one has been killed.

1.104 If a person is killed and the organisation has failed to comply fully with its duties under the HSWA 1974, it does not automatically mean that there will be a charge of corporate manslaughter. This offence is reserved for the most serious breaches that lead to a death.

The elements of the corporate manslaughter offence are as follows:

- The organisation must owe a 'relevant duty of care' to the victim.
- The organisation must be in breach of that duty of care as a result of the way in which the activities of the organisation were managed or organised. An organisation cannot be convicted of the offence unless a substantial element of the breach lies in the way the senior management of the organisation managed or organised its activities.
- The way in which the organisation's activities were managed or organised must have caused the victim's death. The usual principles of causation in the criminal law will apply to determine this question. This means that the management failure need not have been the sole cause of death; it need only be a cause.
- The management failure must amount to a gross breach of the duty of care. Section 1(4)(b) sets out the test for whether a particular breach is 'gross'. The test asks whether the conduct that constitutes the breach falls far below what could reasonably have been expected. This reflects the threshold for the common law offence of gross negligence

manslaughter. There is no question of liability where the management of an activity includes reasonable safeguards and a death nonetheless occurs.

1.105 Further, how far does a company's management failure need to fall below what can reasonably be expected? The answer is uncertain, as it will take considerable time and a series of appellant decisions to build up significant binding case law but perhaps as an indication we can revisit the old law. In 1999 brother and sister Stephen and Julie Bowles (directors of a family haulage company) mentioned at **2.225** were convicted of manslaughter and received custodial sentences of 15 months and 12 months respectively, suspended for two years. Evidence was given to the court that the hours worked by the company's drivers were in line with the haulage industry as a whole. Despite this, the directors were still found to be grossly negligent. Thus, pointing to what is common practice in the type of business the company is in will not necessarily amount to a defence. A driver employed by their company fell asleep at the wheel of his lorry, having regularly spent more than 60 hours per week on the road. He was involved in a seven-vehicle pile-up on the M25, killing two people. The prosecution argued that the defendants should have known about his excessive hours and the effects this would have on his driving. This is also the test pursuant to s 37 of the HSWA 1974. In 2004, Melvyn Spree, director of Keymark Services, another road haulage company, was convicted of manslaughter and conspiracy to falsifying driving records. He received a custodial sentence of seven years. Two deaths occurred after a lorry careered across the M1 during a driver's 18-hour shift. There was found to be widespread tachograph fraud in the company. The company was also fined £50,000 for manslaughter. See also the later section dealing with workplace road deaths.

Further recent manslaughter cases

1.106 In June 2013, a father and son were convicted of gross negligence manslaughter and fraud. They were sentenced to seven and four years' imprisonment respectively. The McMurrays ran AJ Haulage in Daventry. An employee fell asleep at the wheel after driving 13 hours in a 19-hour period. The trial judge, His Honour Bright QC, said '... the excessive hours for which he was driving caused him to be so tired he was a danger to himself and other road users'.

1.107 Also in 2013, a director, Mr Turnbull, was sentenced to three years' imprisonment after a jury convicted him of manslaughter. The deceased was knocked off a cherry picker (on which he was working) by a falling girder. The company and another director were convicted on health and safety charges by the Newcastle jury and were fined and costs imposed.

1.108 Another builder, Mr Collier, was sentenced to two years' imprisonment after a wall that he was constructing fell over and killed a three-year-old girl. He had pleaded not guilty to gross negligence manslaughter. He was charged following the death of the girl in July 2008 who had been walking home with her mother when a wall constructed by the builder's company collapsed on top of her. The company pleaded guilty to breaches of health and safety legislation. The builder was found guilty by a jury following a three-week trial and, in addition to receiving the custodial sentence, was disqualified from being a company director for seven years.

1.109 The CPS said:

'On the day of the incident George Collier and Parcol Developments Ltd backfilled the wall despite knowing that retaining walls should be designed by a specialist structural engineer. Neither had engaged a structural engineer to design a safe retaining wall.

This case is an important reminder to those working in construction to make sure that design work is done by competent people and building is done to the appropriate standard. It also highlights the importance of ensuring that members of the public are kept away from construction work.'

1.110 In 2013 a haulier, Mr Napier, failed to lower the stabilising legs of his lorry, which fell over killing a co-worker and seriously injuring another worker. The jury convicted him and he received 12 months' imprisonment and was fined £50,000.

1.111 In 2012, a consultant urologist, Dr Garg, pleaded guilty to one count of manslaughter. On the patient being admitted to hospital, he failed to order a standard ultrasound. He later altered his medical notes. He was sentenced to two years' imprisonment. Globe J said his conduct had been a 'disturbing picture of a failure to take action'. The alteration to the medical records was described by the judge as 'blatant attempt to disguise and conceal evidence of [his] failure'. See also *R v Shaw* and *R v Barrass*.[1]

1 [2012] Cr App R (S) 80.

1.112 In 2013, a doctor, David Sellu, pleaded guilty to manslaughter for failing to notice symptoms of severe diabetes; instead, he had diagnosed depression. He was jailed for two and a half years.

1.113 Further, a surgeon, David Sells, was convicted in November 2013 of gross negligence manslaughter after James Hughes died from a perforated bowel when in hospital for knee surgery. The jury heard that the doctor had simply ignored the urgency of the situation and carried on with his scheduled lists. The surgeon was sentenced to two and a half years' imprisonment.

1.114 A train guard, Mr McGhee, was sentenced to five years' imprisonment after a 16-year-old girl who was intoxicated fell under a moving train after he signalled it was authorised to move away. The Court of Appeal dismissed the appeal against sentence.

1.115 Another example occurred in 2009 where a builder was jailed for three years after a 15-year-old boy working under his supervision was killed by a falling wall in April 2007. Colin Holtom was convicted of gross negligence manslaughter at the Old Bailey, for failures which included not supervising the 15-year-old and not providing him with safety equipment and guidance. The site contractor, Darren Fowler, was also jailed for nine months at the same hearing, for running a company whilst disqualified from being a director.

1.116 It was widely predicted that some of the first prosecutions under the Act would involve company drivers. The fact that about 900 people die driving whilst at work, when compared to 200 other workplace deaths, indicates this is very likely. Superintendent Bird of the Metropolitan Police said:

'Just as employers would make sure that employees are safe in the workplace so they should whilst they are on the roads. Businesses must face up to their duty of care responsibilities and realise that they are responsible for employees' welfare when on the road for business purposes whether they are driving a company car or not.'[1]

1 (2007) The Times, 25 October.

1.117 The Road Death Investigation Manual is in its third edition and, given the large cost and complexity of corporate manslaughter investigations and prosecutions for driving at work deaths are relatively simple, quick and cheaper to prosecute. These cases have not yet materialised under the Act, but it is merely a question of time.

1.118 For instance, a haulage firm Translact was fined £40,000 after a driver, who was pressured into working long hours, killed two people when he crashed his lorry into their car. The company, instead of employing more drivers, expected their existing staff to drive longer hours. The company directors were also fined individually £7,200 and £4,500 respectively.

1.119 Further, in July 2008 the Sentencing Guidelines Council recommended that those drivers who cause a death whilst using a mobile phone can be sentenced to up to seven years' imprisonment.

1.120 In May 2009, Susan Love was sentenced to two years' imprisonment when she fell asleep at the wheel, killing a man working on his car which had broken down. Evidence indicates that she had been awake for over 16 hours when the accident happened. There was no evidence of alcohol and drugs.

1.121 Recent research has shown that the CMCHA 2007 has really brought home the need to have adequate systems relating to driving at work, including mobiles and Blackberry use. Companies should also review the types of vehicles used, what type of driving record the driver has (including whether they have a licence!), his health and the type of driving undertaken. Often, there will be an organisational systems failure.[1]

1 See *R v Yaqoob* [2005] EWCA Crim 1269. See also 'Driving at work: Managing work-related road safety' which can be found on the HSE website at www.hse.gov.uk.

Who is covered by the Act

1.122 The Act applies to companies incorporated under companies legislation or overseas and other corporations including: public bodies incorporated by statute such as local authorities, NHS bodies and a wide range of non-departmental public bodies, organisations incorporated by Royal Charter, limited liability partnerships, all other partnerships, trade unions and employers' associations (if the organisation concerned is an employer), and Crown bodies such as government departments and police forces.

1.123 The CMCHA 2007 does not change the law regarding the prosecution of individuals, who are in any event increasingly being imprisoned following conviction for manslaughter. For example, in *R v Shaw*,[1] following a reference by the Attorney General, a director who had been given a two-year suspended sentence for pleading guilty to manslaughter after a hung jury at his trial was immediately sentenced by the Court of Appeal to 15 months' imprisonment. This trend is also shown by other recent cases, including *R v Connolly*,[2] arising from the Teebay case where four men working on the railway were killed by a runaway trailer located with Railtrack (now Network Rail). The brakes of the trailer had been removed. The guilty party was sentenced to nine years' imprisonment, reduced to seven years on appeal. There are other examples above and in **Chapter 2**.

1 [2006] EWCA Crim 2570.
2 [2007] EWCA Crim 790.

1.124 There is a long-established legal doctrine that means that Crown bodies (such as government departments) cannot be prosecuted. Section 11(1) makes it clear that this principle does not apply to prosecutions under the CMCHA 2007. The vast majority of government departments have now lost their immunity shield. This heralds a big change as so many have hitherto enjoyed a climate safe from prosecution. In law, government departments operate in the name of the Crown; the Act has a number of technical provisions to ensure that the offence operates in respect of Crown bodies in the same way that it operates for corporations. This is dealt with in ss 11 and 12. Although they are not Crown bodies, similar issues arise for police forces, and this is addressed by s 13. Prisons lost their immunity in 2011. See **1.148** below.

Jurisdiction

1.125 The Act applies across the UK. The offence can be prosecuted if the harm (actus reus) resulting in death occurs in the UK, in the UK's territorial waters (for example, in an incident involving commercial shipping or leisure craft), on a British ship, aircraft or hovercraft, on an oil rig or other offshore installation already covered by UK criminal law.

1.126 It is important to recall that, if the death of a British national occurs outside the UK there will still need to be an inquest (or fatal accident inquiry, in Scotland) which may bring a verdict of unlawful killing. Such a verdict will necessitate the police formally opening a manslaughter investigation.

1.127 On 20 December 2005 the Home Affairs and Works and Pensions Committees of the House of Commons also saw no reason why this offence should not be extraterritorial, especially within the Eurozone where Eurowarrants are already possible. (Likewise the Scottish expert advisory group.) There are also many other precedents where the UK criminal law had extraterritorial reach, including terrorism and certain sexual offences. Of course, individuals can currently be extradited for manslaughter within time. It is very likely that the Act will eventually become extraterritorial, in that linkage to the UK will become the prosecution threshold rather than a simple jurisdiction issue. There have still been no developments about this.

The 'directing mind'

1.128 Although the CMCHA 2007 has dispensed with the 'directing mind' principle, the 'old' law will still have its place when courts are grappling with the meaning of the CMCHA. It is also still relevant for individuals.[1]

1 See Appleby and Forlin, *A Guide to Health and Safety Prosecutions* (Thomas Telford, 2007).

1.129 For instance, in 2005, Gillian Beckingham, a council architect who was head of the Design Services group at Barrow Borough Council, was prosecuted as the directing mind of the council. The case was brought as a result of Britain's biggest outbreak of Legionnaires' disease, which occurred in the summer of 2002. It was traced back to the air conditioning unit in Forum 28. The prosecution alleged that she was instrumental in cancelling the maintenance contract and renegotiating a new contract that did not provide for any water treatment regime.

1.130 The case failed against the council because the trial judge ruled that Ms Beckingham was not a 'directing mind' of the local authority.

1.131 The council subsequently pleaded guilty to a breach of s 3 of the HSWA 1974. The jury was unable to reach a verdict in relation to the manslaughter charges against Ms Beckingham in her own right. At the retrial in 2006, she was acquitted of manslaughter but convicted of a breach of s 7 of the HSWA 1974. This trial took three months.

1.132 In *Tesco Supermarkets Ltd v Nattrass*,[1] which concerned a prosecution pursuant to the Trade Descriptions Act 1968, the House of Lords held that a company can only be held criminally liable for the acts of:

> '... the Board of Directors, the Managing Director and perhaps other superior officers of the company ... [who] ... carry out the functions of management and speak and act as the company.'

1 [1972] AC 153.

1.133 See also *HL Bolton (Engineering) Ltd v TJ Graham & Sons*[1] and *El-Ajou v Dollar Land Holdings PLC*[2] which found that the directing mind could be someone other than the nominal person in control.

See also the case of *Ferguson v British Gas*,[3] where the Court of Appeal has again dramatically limited the doctrine of incorporation and the corporate veil. In *Prest v Prest*,[4] the Supreme Court recently looked at 'piercing the corporate veil' (despite certain judges not approving of the term). In essence, the court found that the husband had provided funds to offshore companies for his own behalf, and not the companies themselves. The companies therefore held monies on trust for the husband, and his divorced wife was entitled to a share of those funds. The courts are increasingly willing to pierce the corporate veil once proceedings have commenced.

1 [1957] 1 QB 159.
2 [1994] 2 All ER 685.
3 [2009] EWCA Civ 46.
4 [2013] UKSC 34.

1.134 Further, the important case of *Chandler v Cape plc*[1] has also greatly increased legal corporate responsibility for their subsidiaries.

The claimant was exposed to asbestos dust in 1952 and 1962. He was diagnosed with asbestos-related diseases in 2007. During his employment, his company was a wholly owned subsidiary of Cape plc. The Court of Appeal dismissed the company's appeal, citing *Smith v Littlewoods Ltd*.[2] The court rejected the notion that the case was attempting to pierce the corporate veil – but turned simply on whether the corporation could be said to have taken on a direct duty for the health and safety of the employees of the subsidiary.

1 [2012] EWCA Civ 525.
2 [1987] 1 AC 241.

1.135 Arden LJ set out some general parameters as to when, in civil law, health and safety responsibility was passed onto a parent company:

(a) that the businesses are basically the same;
(b) the parent has, or ought to have, knowledge of health and safety in the particular sector;
(c) the parent ought to have known that the subsidiary system was unsafe; and
(d) the parent knew or ought to have known or foreseen that the subsidiary

or its employees would rely on its sector or industry information or knowledge to protect employees.

1.136 The Court of Appeal said there was a close enough connection and the appeal was dismissed. Permission to appeal to the Supreme Court has been sought.

This decision has far-reaching implications for organisations, especially arguably ones with an international dimension.

1.137 In *R v Her Majesty's Coroner for East Kent, ex p Spooner*,[1] which was a judicial review of a decision of the coroner at the inquest into the deaths following the sinking of the *Herald of Free Enterprise*, Bingham LJ said:

> 'It is important to bear in mind an important distinction. A company may be vicariously liable for the negligent acts and omissions of its servants and agents but for a company to be criminally liable for manslaughter … it is required that the mens rea [the mental element of the crime] and the actus reus [the act/omission of the crime] should be established not against those who acted for or in the name of the company but against those who were to be identified as the embodiment of the company itself.'

1 (1989) 88 Cr App R 10.

1.138 The prosecution of P&O followed after the inquest finished.[1] The trial collapsed in 1990 when Turner J directed the jury that, as a matter of law, there was no evidence upon which they could properly convict the company of manslaughter. The main reason for his decision was that, in order to convict of manslaughter, one of the individual defendants who could be 'identified' within the company would have to himself be guilty of manslaughter; and, since there was insufficient evidence on which to convict any of the individual defendants, the case against the company had to fail. During the course of his judgment, Turner J said:[2]

> 'Since the 19th century there has been a huge increase in numbers and activities of corporations whether nationalised, municipal or commercial which enter the private lives of all or most of 'men and subjects' in a diversity of ways. A clear case can be made for imputing to such corporations social duties including duty not to offend all relevant parts of the criminal law. By tracing the history of the cases decided by the English courts over the period of the last 150 years, it can be seen how first tentatively and finally confidently the courts have been able to ascribe to corporations a "mind" which is generally one of the essential ingredients of common law and statutory offences. Indeed, it can be seen that in many acts of Parliament the same concept has been embraced … once a state of mind could be effectively attributed to a corporation, all that remained was to determine the means by which that state of mind could be ascribed and imputed to a non natural person. That done, the obstacle to the acceptance of general criminal liability of a corporation was overcome … there is nothing essentially incongruous in the notion that a corporation, through the controlling mind of one of its agents, does an act which fulfils the pre-requisites of the crime of manslaughter, it is properly indictable for the crime of manslaughter.'

1 *R v P&O European Ferries (Dover) Ltd* (1991) 93 Cr App R 72.
2 (1991) 93 Cr App R 72.

1.139 Turner J emphasised the need to prove that the person prosecuted as the 'directing mind' (or 'controlling mind' as it is also referred to) had sufficient responsibility for safety to establish guilt. It was not permissible to 'aggregate' the faults of a number of different individuals, none of whose faults on their own would merit individual prosecution.

1.140 The charge of manslaughter in the *P&O* case[1] was brought on the basis of recklessness. However, as said earlier, the appropriate form in work related deaths is gross negligence manslaughter.

1 *R v P&O European Ferries (Dover) Ltd* (1991) 93 Cr App R 72.

1.141 Great Western Trains were prosecuted for corporate manslaughter, based on gross negligence, in relation to the Southall train crash of 1997, where a safety device on the train was inoperative which it is likely would have prevented it. Counsel for the prosecution argued that, as the test in *R v Adomako*[1] is objective, and so does not require a mental element to prove the crime (ie mens rea does not have to be proved), it was not necessary to prosecute a director of the company as a 'directing mind'. The trial judge, Mr Justice Scott-Baker, rejected this argument, saying:[2]

> 'In my judgement it is still necessary to look for … a directing mind and identify where gross negligence is that fixes the company with criminal responsibility … Accordingly I conclude that the doctrine of identification which is both clear, certain and established is the relevant doctrine by which a corporation may be fixed for manslaughter by gross negligence …'

1 [1995] 1 AC 171.
2 Transcript of the hearing at the Central Criminal Court on 30 June 1999, pp 26 and 27.

1.142 The Court of Appeal agreed that this was the correct position – see the Attorney-General's appeal to the Court of Appeal on 15 February 2000.[1]

1 *Attorney-General's Reference (No 2 of 1999).*

SERIOUS MANAGEMENT FAILINGS

1.143 The offence of corporate manslaughter addresses the issue that organisations could only be convicted of manslaughter (or culpable homicide in Scotland) if a 'directing mind' at the top of the company (such as a director) was also personally guilty of manslaughter. Decision-making in large organisations often does not reflect this pattern. The offence allows an organisation's liability to be assessed on a wider basis, providing a more effective means of accountability for very serious management failings across the organisation. In effect, the CMCHA 2007 allows the prosecution to aggregate management failure. The 'old' law did not allow this to happen.[1]

1 See *R v P&O European Ferries (Dover) Ltd* (1991) 93 Cr App R 72.

DUTY OF CARE

1.144 A duty of care is an obligation that an organisation has to take reasonable steps to protect a person's safety. These duties exist for employees, and the work systems and equipment that they use, the condition of premises and worksites occupied by organisations, and of products and services supplied to customers. The CMCHA 2007 does not create new duties – they are already owed in the civil law of negligence and the offence is based on these. Under the offence, it will be a question of law for the judge to determine. See *R v Wacker*,[1] where the Court of Appeal looked at certain civil law principles in the context of gross negligence manslaughter.

1 [2002] EWCA Crim 1944, [2003] QB 1207, [2003] 2 WLR 374.

Duty of care of directors

1.145 The fact that an individual is a director of a company does not, of itself, give rise to a duty of care on that person's part to someone who is injured or killed by the company's activity. This point was highlighted by Mr Justice Scott Baker in his ruling on the corporate manslaughter charges against Great Western Trains.[1] He said:

> 'The requirements of foreseeability, proximity, fairness, justice and reasonableness must be satisfied in order to satisfy such a duty. The law is careful as to the circumstances in which a director acting in his capacity as such is personally liable. It would ordinarily require that he procured, directed authorised a commission of the tort in question.'

See also *R v P*[2] and *Wilson v R*.[3]

See also *R v Gidney*[4] regarding the Solway harvester accident.

1 Unreported (30 June 1999).
2 [2007] EWCA Crim 1937.
3 [2013] EWCA Crim 1780.
4 Unreported (18 May 2013), Isle of Man Deemster Mr Moran QC.

Directors

1.146 The government originally proposed that individuals (directors or officers) who could be shown to have had some influence on, or responsibility for, the circumstances in which a management failure falling far below what could be reasonably expected was a cause of someone's death should be automatically disqualified from acting in a management role in Great Britain. This proposal was dropped. However, such an individual might still be disqualified under the Company Directors' Disqualification Act 1986, where appropriate. A report[1] looking at the use and effectiveness of the Company Directors' Disqualification Act 1986 as a legal sanction against directors convicted of health and safety offences found at least ten cases in the period of the study of a director who was disqualified after conviction for health and safety breaches, although the policy of document shredding meant that a true historical picture was not possible. There are some recent examples of this in **Chapter 2**.

1 *A survey of the use and effectiveness of the Company Directors Disqualification Act 1986 as a legal sanction against directors convicted of health and safety offences* (HSE RR597, 2007).

EXEMPTIONS

1.147 Sections 3–7 of the Act set out specific exemptions. This means that the offence will not apply to deaths connected with the management of particular activities. There are two types of exemption: comprehensive, and partial. Where a comprehensive exemption applies, the offence will not apply in respect of any duty of care that the organisation may owe. Examples of these are public policy decisions, including strategic funding decisions and other decisions which involve competing public interests, military combat operations including violent peacekeeping operations, dealing with terrorism and violent disorder. It also exempts support and preparatory operations, such as hazardous training (including Special Forces). This exemption extends to police operations which deal with terrorism and violent disorder and their support and preparatory activities and training. Deaths in custody are now no longer exempt: the new Coroners' Rules also include deaths in detention, which includes mental hospitals.

1.148 Partial exemptions from the offence do not apply unless the death relates to the organisation's responsibility as employer (or to others working for the organisation) or as an occupier of premises. These include policing and law enforcement activities (s 5(2)), the emergency response of fire authorities and other emergency response organisations, NHS trusts (including ambulance trusts) – this does not exempt duties of care relating to medical treatment in an emergency, other than triage decisions – the Coastguard, Royal National Lifeboat Institution and other rescue bodies and the armed forces. Further, carrying out statutory inspection work (s 3(3)), child protection functions or probation activities (s 7) are also covered. The exercise of 'exclusively public functions' (s 3(2)) – for example, the government using prerogative powers such as acting in a civil emergency, or functions that require a statutory (or prerogative) power – is also exempted. See also **1.124** above.

1.149 The Metropolitan Police were found guilty over the shooting of Jean Charles de Menezes on 22 July 2005 under s 3 of the HSWA 1974, which places general duties on employers and the self-employed towards persons other than their employees. Interestingly, this case probably could not have been brought under the CMCHA 2007, as the police operation was an anti-terrorist one and would have come under a comprehensive exemption from the Act.

INVESTIGATION AND PROSECUTION

1.150 The police will lead investigations relating to suspected cases of corporate manslaughter or corporate homicide. Prosecution decisions will be made by the Crown Prosecution Service (England and Wales), the Crown Office and Procurator Fiscal Service (Scotland) and the Director of Public Prosecutions (Northern Ireland). Protocols currently exist to ensure that the knowledge and expertise of the regulatory enforcing authorities (such as the Health and Safety Executive, the Office of Rail Regulation, Food Standards Agency and local authorities) are properly utilised. The Rail, Air and Marine Accident Investigation Branches will continue to be responsible for separate investigations to determine the cause of an incident and to issue reports.

1.151 Proceedings for the offence will be the responsibility of the general prosecuting authorities: the Crown Prosecution Service in England and Wales, the Public Prosecution Service in Northern Ireland and the Procurator Fiscal in Scotland. Proceedings cannot be brought by regulatory bodies. No proceedings in England and Wales can begin under the CMCHA 2007 without the specific consent of the DPP. In Scotland, it is the Crown Office and Procurator Fiscal Service (Scotland) and the Director of Public Prosecutions (Northern Ireland). See examples of recent cases above.

PENALTIES

1.152 Sections 9 and 10 of the Act provide for remedial orders requiring an organisation to remedy the breaches of which it has been convicted within a specified period of time. A remedial order can only be made where the prosecution applies for one. These powers already exist in s 42 of the HSWA 1974. This application must be accompanied with the proposed terms of the order. Before the prosecution makes the application, it must consult with the appropriate regulatory authority such as the HSE, Office of Rail Regulation, Food Standards Agency or local authority. As the relevant enforcement expert, it is expected that the regulatory body will be closely involved in drafting the proposed terms of the order, and suggesting a period in which the necessary measures need to be taken. It is expected that the relevant regulator will have been involved in the case from an early stage and will have used their existing enforcement powers to address issues before the case comes to court; however, a judge will still have the power to impose the order. If a corporation fails to comply with a remedial order, it can be prosecuted and this would be the responsibility of the prosecuting authorities; and, on conviction, an unlimited fine could be imposed. These orders were not previously available for organisations convicted of manslaughter/culpable homicide, although they can be imposed under health and safety legislation.

Convicted organisations will also be able to be given a publicity order, which is an order requiring them to publicise in a specified manner their conviction, particulars of the offence, the amount of the fine and the terms of the remedial orders imposed. Princes Sporting Club had one imposed in 2013. This is believed to be the first ever such order.

1.153 Although it is a health and safety case, the Court of Appeal Criminal Division has very recently handed down a major judgment (in *R v Sellafield; R v Network Rail Infrastructure Ltd* [2014] EWCA Crim 49) which raises issues of principle in relation to the level of fines to be imposed for breaches of safety and environmental protection legislation on very large companies – Sellafield Limited (Sellafield Ltd) with a turnover of £1.6 billion, and Network Rail Infrastructure Ltd (Network Rail) with a turnover of £6.2 billion:

i) Sellafield Ltd was fined £700,000 in the Crown Court at Carlisle on 7 February 2013 for offences arising out of the disposal of radioactive waste; and

ii) Network Rail was fined £500,000 in the Crown Court at Ipswich on 27 June 2013 for an offence arising out of a collision at an unmanned level crossing, causing very serious injuries to a child.

Both companies had appealed the fines as they believed them to be excessive. The appeals were dismissed. Sellafield Ltd submitted that the level of fine equated to a major public disaster or loss of life, a significant nuclear event or an unmitigated environmental pollution incident. Network Rail submitted that a starting point of £750,000 would only be appropriate where there was more than one fatality, a public disaster, or where the defendant was convicted of corporate manslaughter.

The court concluded that a fine of this size against Sellafield Ltd, which equates to around 2% of its weekly income, would:

> 'achieve the statutory purposes of sentencing by bringing home to the directors of Sellafield and its professional shareholders the seriousness of the offences committed and provide a real incentive to the directors and shareholders to remedy the failures which the judge found existed at the site.'

With regard to the Network Rail appeal, the court agreed that a significant fine would, unlike Sellafield Ltd, inflict no direct punishment on anyone and might harm the public, because the company's profits are reinvested into the rail infrastructure. However, it still felt that the fine served other purposes, including the need for all Network Rail directors to 'pay much greater attention to their duties' in relation to accidents at level crossings.

The court also concluded that, had the sentencing judge imposed a bigger fine, 'there would have been no basis for criticism'. See further consideration of this case at **2.17**.

OLDER SUCCESSFUL PROSECUTIONS[1]

1.154 In July 2007, the Concrete Company were fined £75,000 and costs of £89,000 after pleading guilty to the manslaughter of Christopher Meachen, who was killed after he became entangled in an unguarded slew conveyor. The owner and company director was jailed for 12 months, and his area manager was jailed for nine months. The company was also convicted of breaches of health and safety regulations.

1 See above for recent examples.

1.155 In February 2007 a boat owner was found guilty of manslaughter and breaches of the HSWA 1974 and jailed for the death of a worker. This followed an explosion in the boatyard.

1.156 In December1989 the CPS prosecuted two directors of David Holt Plastic Ltd.[1] A workman at the company had been dragged into a machine devised to 'crumble' plastic. He was sliced by a 20–50 inch blade revolving at 1,200 revolutions per second. The machine should have been only able to operate with its lid down. However, the police investigation revealed that someone had removed the lid bolt, thereby allowing the machine to operate with its lid up.

1 (1989) unreported.

1.157 One of the directors was present at the time of the incident. He pleaded guilty to manslaughter. No evidence was offered against the other director on the basis that, although he knew as much as his fellow director about the machine, he was not physically present at the time of the incident.

1.158 In November 1994, OLL Ltd became the first company in English legal history to be convicted of manslaughter.[1] This was under the old law. Peter Kite also became the first director to be given an immediate custodial sentence arising out of the operation of a business. This case concerned a canoeing accident in 1993 in which four students died.

1 (1994) unreported.

1.159 A teacher, a group of eight students and two instructors from OLL Ltd tried to canoe across the sea from Lyme Regis to Charmouth, a distance of about one-and-a-half miles. The teacher got into difficulties, and one of the instructors went over to assist. The other instructor proceeded with the students. The group drifted out to sea, where the students drowned.

1.160 Mr Kite was one of two directors of the company. He had the primary responsibility for devising, instituting, enforcing and maintaining an appropriate safety policy. The jury found that he failed in his responsibility, and this was a substantial cause of the deaths.

1.161 In September 1996, Jackson Transport (Ossett) Ltd and its managing director, Alan Jackson, were convicted for the manslaughter of James Hodgson (aged 21) who died in May 1994 after cleaning chemical residues at the rear of a road tanker.[1] He died when he used steam pressure to clean a valve in a tanker blocked with highly toxic chemicals. When the incident happened, he was wearing ordinary overalls and a baseball cap. He was not provided with the appropriate personal protective equipment or supervision or adequate training.

1 (1996) unreported.

1.162 In August 2001, English Brothers Ltd, a Wisbech-based construction company, pleaded guilty to the manslaughter of Bill Larkman, a gang foreman, who died in June 1999 when he fell from a height of over eight metres through a fragile roof to his death.[1] The prosecution had earlier accepted a plea of 'not guilty' from Melvyn Hubbard, a director of the company. The court heard that, in 1997, inspectors from the HSE had seen Bill Larkman working at another English Brothers site without using correct safety equipment, and had spoken to the company about its safety failings.

1 (2001) unreported.

1.163 In February 2003, Teglgaard Hardwood (UK) and John Horner, Managing Director, pleaded guilty to gross negligence manslaughter in relation to the death of an 18-year-old labourer.[1] The company imported timber. The deceased was crushed to death when a pile of poorly stacked hardwood packs toppled over on top of him. The underlying cause of the incident was the failure to carry out a suitable risk assessment covering the storage of the hardwood packs. The timber packs were frequently off-set and protruding from the pile, making them unstable. Further, the steel binding on some of the packs, used to keep the packs together, was damaged, resulting in reduced stability.

1 (2003) unreported.

RISK[1]

1.164 With the HSE's guidance entitled *Directors' responsibilities for health and safety*[2] and the *Turnbull Report*,[3] major health and safety risks are matters that are expected to be dealt with at board level. Thus, the issue of how a company and its board deal with risk is likely to become more relevant to corporate manslaughter cases. See the IOD Guidelines.[4] An illustration of this perhaps is the prosecution of Euromin and its director, Mr Martell, following the death of Simon Jones in April 1998.[5]

1 See the new HSE Guidance on 'alarp' at **1.273** below.
2 IND(G)343 (2001), available at www.hse.gov.uk/pubns/indg343.pdf – see below for the 2013 amended version.
3 N Turnbull (chair), Institute of Chartered Accountants for England and Wales, *Internal control: guidance for directors on the combined code* (the *Turnbull Report*) (1999).
4 *Leading Health and Safety at Work – Leadership Actions for Directors and Board Members*, available at www.iod.com/hsguide.
5 (2001) unreported.

1.165 In June 2013, the IOD and HSE published a new version of *Leading Health and Safety at Work*.[1] This has basically updated the publication to tie in with the new HSG 65 mentioned earlier.

1 See INDG17 (Rev 1) June 2013.

1.166 Simon Jones, aged 24, died on his first day at work at Shoreham Docks while unloading cargo from a ship when the jaws of a grab closed around his neck. It transpired that the crane operator could not see inside the ship's hold and the person responsible for communication between the crane driver and the hold was a Polish seaman and could not speak English.

1.167 Ultimately the case was unsuccessful, with the company and its director being acquitted of manslaughter at the Old Bailey in November 2001.[1]

It should be noted that the manslaughter investigation did not start until six weeks after the incident, which meant there were evidential problems. Further, the trial judge seemed to have concerns that the error of the crane driver broke the chain of causation.

1 (2001) unreported.

1.168 However, the case is significant for the judicial review by the deceased's family of the original decision by the CPS not to prosecute.[1] At the judicial review the Director of Public Prosecutions argued that there was no evidence that the managing director had been warned of the dangers of the system of work being operated. The judge said:

> 'The point can be put in this way. A clear and potential … major criticism [of the managing director and the company] is that they set up the unsafe system … it was unsafe because, *inter alia*, it arguably presented a danger of death – the danger that in fact had eventuated. Such a system requires detailed precautions to be taken to ensure the incipient dangers did not in fact eventuate.'

The CMCHA 2007 also allows the jury to look at the corporate culture and does not require individuals to be convicted for manslaughter.[2]

1 *R v DPP, ex p Jones* [2000] IRLR 373.
2 See above, and see further **Chapter 3**.

1.169 Another way of looking at this is to say that the company should have carried out a risk assessment and that failure to do so was a cause of the death. It would be for the jury to consider whether that was a substantial cause and if the failure amounted to gross negligence.

1.170 There are exceptions to this general strict approach. For instance, see *Hampstead Heath Winter Swimming Club v The Corporation of London and the HSE*.[1]

Another example occurred when a child was seriously injured at a birthday party on a bouncy castle. The couple who had hired the castle were accused of having inadequate supervision, and their insurers faced a £1 million payout as a result. The injured boy was 11 years old when he asked for permission to play on the bouncy castle in September 2005. He had spied the castle in the grounds of the local school after football practice. When the one of the couple that had hired the castle's back was turned, a 15-year-old accidentally kicked the boy in the head while turning a somersault, causing 'very serious and traumatic brain injury'. The judge ruled that the couple should compensate the seriously brain damaged boy, who suffers from Asperger's Syndrome, because they were liable. The judge said, although one of the couple was present throughout, her attention was diverted as she supervised a bungee run for the party. The case went to the Court of Appeal, where the couple were successful.

1 [2005] EWHC 713 (Admin).

1.171 At the conclusion of the judgment, Lord Phillips stated:

> 'We have given our reasons at some length, but to a large extent a case of this nature properly turns on first impressions … The issue is whether a reasonably careful parent would have acted in the same way as the defendant … Each of us had the same reaction to the facts … The manner in which she was supervising activities on the bouncy castle and the bungee run accorded with the demands of reasonable care for the children using them.'

CAUSATION

1.172 Although the new law has modified much of the old law, issues concerning causation will still arise. It is also important that, when construing the CMCHA 2007, the courts will continue to look at old case law for guidance. This issue has yet to be ruled upon by the senior courts (see above).

1.173 Most work related deaths occur when there is a human error by a front-line worker. The traditional approach of the police has been to view the error as breaking the 'chain of causation' with regard to any failure on the part of a director of a company. This approach to causation does not sit comfortably with the HSE's view of the root causes of such incidents: accidents, ill health and incidents are seldom random events and generally arise from management and organisational failures.[1]

1 *Successful health and safety management* HS(G)65 (2013) – see **1.14** above.

1.174 In the case following the Southall train crash, the judge, when sentencing Great Western Trains on 27 July 1999 for the health and safety offence to which it had pleaded guilty, did not view the fact that the driver in its employment had passed a red signal as breaking the chain of causation. He stated:[1]

> '… But a substantial contributory cause was the fact that the defendant company permitted the train to run from Swansea to Paddington at speeds of up to 125 m.p.h. with the Automatic Warning System (AWS) isolated …
>
> The company should have applied its mind to the risk created by allowing a high speed train to travel a journey of this length and this speed without the AWS operating. It maybe that the likelihood of a driver passing a red signal was relatively small. But if the event occurred the consequences were going to be, as in the event they were, appalling.'

1 (1999) unreported.

1.175 Difficulties with causation occur in other areas of law. For instance, in 2002 the House of Lords heard appeals in cases concerning personal injury claims relating to the contraction of mesothelioma – *Fairchild v Glenhaven Funeral Services Ltd*.[1] The main issue before the court was causation.

1 [2002] 3 All ER 305.

1.176 Lord Bingham said that a traditional approach 'denies recovery where our instinctive sense of justice … tells us the victim should obtain compensation'. In his judgment[1] he referred to the environmental prosecution case of *Empress Car Co (Abertillery) Ltd v National Rivers Authority*,[2] in which there was an argument as to whether the act of an unknown person broke the chain of causation.

1 *Fairchild v Glenhaven Funeral Services Ltd* [2002] 3 All ER 305.
2 [1998] 1 All ER 481.

1.177 The company was prosecuted for polluting a river contrary to s 65(1) of the Water Resources Act 1991. The House of Lords said what had to be determined was whether the event, the act of the unknown person, was ordinary (if so, the company was guilty and the chain of causation was not broken) or extraordinary (if so, the company was not guilty).

1.178 Lord Hoffmann in this case said:[1]

> 'Questions of causation often arise for the purpose of attributing responsibility to someone, for example, so as to blame him for something which has happened or to make him guilty of an offence or liable in damages. On such cases, the answer will depend on the rule by which responsibility is being attributed.'

See also *R v Pitts*.[2]

1 *Empress Car Co (Abertillery) Ltd v National Rivers Authority* [1998] 1 All ER 481.
2 (1842) cases on the Western Circuit S Vict. 284.

1.179 In 2000, Lord Cullen chaired the Public Inquiry into the causes of the Ladbroke Grove train crash of October 1999 that killed 29 passengers and the drivers of both trains involved in the collision. The driver of the Thames Turbo train, Michael Hodder, passed signal SN109 at red. This signal had been passed on eight previous occasions in just over five years, and each time the incident was recorded as 'driver error'. Three days after the crash, HM Railway Inspectorate (part of the HSE) served a prohibition notice on Railtrack, the then operators of the railway infrastructure.

1.180 Prior to the start of the Inquiry, the police announced there would be no manslaughter prosecutions. Lord Cullen in his Report[1] declined to blame Mr Hodder for the crash. He concluded that Mr Hodder passed the signal because of poor 'sighting' of the signal, both in itself and with other signals on the gantry on which it was situated.

1 The Rt Hon Lord WD Cullen PC *The Ladbroke Grove Rail Inquiry, Part 1 Report* (2001), available at www.rail-reg.gov.uk/server/show/nav.1204.

1.181 Lord Cullen found that there was a 'lamentable failure on the part of Railtrack' to respond to recommendations of inquiries into two serious incidents in the Paddington area involving signals passed at danger (SPADs). However, the recognition by Railtrack of a SPAD problem in the area led to the formation of a number of groups to deal with the problem. But Lord Cullen said little was achieved because they were so 'disjointed and ineffective'. He accused Railtrack of not dealing with the problem in a 'prompt, proactive and effective manner'. Lord Cullen quoted the comments of the Zone Director of Railtrack who said: 'The culture of the place had gone seriously adrift over many years'.

1.182 After publication of the report, lawyers for the bereaved and injured campaigned for a reconsideration of corporate manslaughter charges. Following the Crown Prosecution Service obtaining advice concerning, in part, causation from a leading academic in this field of law, in May 2002 it was announced that the police investigation would be re-opened to reconsider the case against Railtrack. One of the main reasons given by the Crown Prosecution Service for the original decision not to proceed against Railtrack and any of its senior people was that any fault on the part of Railtrack could not be linked to the collision.[1]

1 It should be noted that, in October 2003, the HSE prosecuted Thames Trains, the employer of Mr Hodder, for health and safety breaches. Executives of Railtrack were not prosecuted, but Network Rail (which succeeded Railtrack as the network operator) was fined £4 million for health and safety failures.

1.183 On 17 October 2000, there was a train derailment at Hatfield which killed four people. It was established that a broken rail was the cause of the accident. In July 2003, Balfour Beatty, the company that maintained the track at the time of the accident, and Network Rail, who took over from Railtrack as owners of the infrastructure, were summonsed for corporate manslaughter.

Senior people from both companies were charged with manslaughter in their own right and were identified as directing minds.

1.184 After a nine-month trial, the judge dismissed the manslaughter charges. Balfour Beatty pleaded guilty to health and safety charges, and Network Rail was convicted of health and safety breaches. Balfour Beatty had a £10 million fine reduced to £7.5 million on appeal.[1] See also *R v Morgan*.[2]

1 See further **Chapter 2**.
2 [2007] EWCA Crim 3313.

GROSS NEGLIGENCE MANSLAUGHTER

1.185 Negligence can form the basis of criminal liability in involuntary manslaughter; however, the negligence has to be *gross*. In work related death cases, recent history suggests that gross negligence manslaughter is the most appropriate offence for which to prosecute an individual whose conduct it is alleged has materially contributed to the death.

1.186 The law relating to gross negligence manslaughter was radically altered by the House of Lords in *R v Adomako*.[1] The case concerned an anaesthetist in charge during an eye operation. He failed to notice that an endotracheal tube (to allow the patient to breathe normally) had become disconnected for a period of about six minutes. As a consequence, the patient died. The defence conceded at trial that the doctor had been negligent but denied that this negligence was so bad that it should be deemed a criminal offence. This case has recently been invoked in two recent prosecutions of doctors for gross negligence.[2]

1 [1995] 1 AC 171.
2 See **1.112** and **1.113** above.

1.187 The only speech was given by the then Lord Chancellor, Lord Mackay of Clashfern, who said:[1]

> '… in my opinion the ordinary principles of the law of negligence apply to ascertain whether or not the defendant has been in breach of a duty of care towards the victim who has died. If such breach of duty is established the next question is whether that breach of duty caused the death of a victim. If so the jury must go onto consider whether that breach of duty should be characterised as gross negligence and therefore as a crime. This will depend on the seriousness of the breach of duty committed by the defendant in all the circumstances in which the defendant was placed when it occurred. The jury will have to consider whether the extent to which the defendant's conduct parted from the proper standard of care incumbent upon him, involving as it must have done, a risk of death [to the deceased], was such that it should be judged criminal.'

1 *R v Adomako* [1995] 1 AC 171 at 187.

1.188 *R v Adomako*[1] establishes that, in order to convict someone of gross negligence manslaughter, the jury must be satisfied that:

1 the defendant owed a duty of care to the deceased;
2 he or she was in breach of that duty;
3 the breach of duty was a *substantial* cause of death; and
4 the breach was so grossly negligent that the defendant can be deemed to

have had such disregard for life of the deceased that it should be seen as criminal and deserving of punishment by the state.

1 [1995] 1 AC 171.

1.189 Note that evidence as to the defendant's state of mind is not a prerequisite of a conviction (see *Attorney-General's Reference* (*No 2 of 1999*),[1] which related to the Southall train crash of 1997).

1 [2000] 2 Cr App R 207.

1.190 In many ways, this offence is a civil case in a criminal arena: it has to be proved that the defendant's negligence was one of the causes of the death, as in a civil case. The differences are:

- The standard of proof (ie beyond reasonable doubt in a criminal case, as opposed to on the balance of probabilities in a civil case).
- The negligence must be a *substantial* cause in a criminal case.
- The negligence must be *gross*.

Duty of care

1.191 The ordinary principles of the law of negligence apply to determine whether the defendant owed a duty of care and was in breach of that duty. In *R v Khan and Khan*[1] it was said that it is for the trial judge to decide whether the facts of the case are capable of giving rise to a duty of care and to direct the jury accordingly. If such a direction is given, the jury goes on to decide whether there was a duty. This is still the case, as the trial judge needs to satisfy himself that there is a duty of care (see above).

1 [1998] Crim LR 830, CA.

1.192 There has been conflicting authority as to whether the jury should decide if there is a duty of care, or whether it is a matter of law for the judge to direct to the existence of a duty. In the case of *R v Willoughby*,[1] the Court of Appeal held that whether the duty of care exists is a matter for the jury once the judge has decided that there is evidence capable of establishing a duty. However, the court did add that there may be exceptional circumstances, such as between a doctor and a patient, where a judge can direct a jury that there is a duty of care.

1 [2004] EWCA Crim 3365.

Substantial cause

1.193 In *R v Adomako*,[1] causation was not an issue. However, in work related deaths there are often a number of causes. For a defendant to be guilty of gross negligence, a prosecution does not have to prove that the defendant's breach of duty was the only cause, or even the main cause, but that it was a substantial cause. Therefore, the issue will be whether the individual's conduct was a substantial cause of death.

1 [1995] 1 AC 171.

1.194 The prosecution for gross negligence manslaughter in the case of *R v O'Connor*[1] concerned the sinking of a fishing vessel, the *Pescado*, where all six crew members died. The prosecution's case against Mr O'Connor, who was a managing agent, was that he was grossly negligent in allowing the vessel to go to sea in an unseaworthy condition. This was compounded by the absence of specific life-saving equipment which, it was argued, led to the loss of life. There was no direct evidence on the cause of the sinking or when exactly the boat sank.

1 [1997] Crim LR 516, CA.

1.195 The case centred on whether the vessel had capsized as a result of inherent instability or, as was suggested by the defence, as a result of a collision with another ship.

1.196 There had been a number of breaches of the Fishing Vessels (Safety Provisions) Rules 1975.[1] This included a breach that only one life raft had been provided instead of the prescribed two. The prosecution was able to prove that Mr O'Connor's gross negligence had been a substantial cause of the deaths. The Court of Appeal approved the trial judge's ruling that proving a substantial cause was sufficient.

1 SI 1975/330.

1.197 There is old authority for saying that liability for manslaughter does not arise where the defendant's conduct is not a direct and immediate cause of death. However, given the terms of *R v Adomako*[1] and the modern-day understanding of the causes of workplace accidents (see, for example, the HSE's *Successful health and safety management*),[2] this is likely not to be accepted as a limitation on determining whether a defendant's conduct was a substantial cause.

1 [1995] 1 AC 171.
2 HS(G)65 (1997).

1.198 If the deceased's own negligence contributed to his or her death, this will not necessarily defeat a prosecution. The issue is whether the defendant's negligence was a substantial cause.

Gross negligence

1.199 As to what constitutes gross negligence, Lord Mackay said in *R v Adomako*:[1]

> 'It is true that to a certain extent this involves an element of circularity, but in this branch of the law I do not believe that is fatal to its being correct as a test of how far conduct must depart from accepted standards to be characterised as criminal. This is necessarily a question of degree and an attempt to specify that degree more closely is I think likely to achieve only spurious precision. The essence of the matter, which is supremely a jury question, is whether, having regard to the risk of death involved, the conduct of the defendant was so bad in all the circumstances as to amount in their judgement to a criminal act or omission.'

1 [1995] 1 AC 171 at 187.

1.200 To set this test in context, it is worth looking at the direction given by the trial judge in *R v Litchfield*,[1] which was approved by the Court of Appeal. Mr Litchfield was both master and owner of a square-rigged schooner that foundered off the north Cornish coast. Consequently, the schooner was blown onto rocks at Rump Point, just outside Padstow Bay, causing it to break up. Three of the 14 crew died.

1 [1998] Crim LR 507, CA.

1.201 *Gross negligence manslaughter*

1.201 The prosecution argued that Mr Litchfield had steered an unsafe course, too close to a notoriously dangerous shoreline, and had plotted the course relying on engines that he knew, or should have known, were liable to fail through fuel contamination.

1.202 The direction given was as follows:[1]

> 'Before you could convict this defendant of manslaughter, the negligence established must go way beyond the mere matter of compensation between parties. It must be more than just some degree of fault, or mistake, or error of judgement, or carelessness even though that led to death. It must be such as to demonstrate a reckless disregard for the lives of others of such a nature, and to such an extent, that in your judgement the negligence is so bad that it can properly amount to a criminal act.'

1 *R v Litchfield* [1998] Crim LR 507, CA.

1.203 The test is very high. It is entirely a matter for the jury to say whether the conduct is so negligent (using the ordinary meaning of the word) that penal sanction should follow. The Court of Appeal in *R v Misra and Srivastava*[1] said that the risk must be one of death. To put it another way, the risk of injury, even if serious, is not sufficient.

1 [2005] Crim LR 234.

Omission manslaughter

1.204 Most work related deaths occur due to negligent omissions rather than negligent acts. As Lord Mackay says in relation to gross negligence, the offence includes omissions as well as acts. The case of *R v Khan and Khan*[1] confirms that manslaughter by omission is not a separate entity, but is part of gross negligence manslaughter. See also *R v Singh*.[2]

1 [1998] Crim LR 830, CA.
2 [1999] Crim LR 582.

DIRECTORS' RESPONSIBILITIES FOR HEALTH AND SAFETY

Guidance upon directors' responsibilities for health and safety[1]

1.205 The Institute of Directors and Health and Safety Commission published a document about leading health and safety at work aimed at company directors, entitled 'Leading Health and Safety at Work'.[2] This was updated in 2013.

1 See also above.
2 *Leading Health and Safety at Work: Leadership actions for directors and board members* (HSE, June 2013) – see **1.14** above.

1.206 Supporting the original guidance, Health and Safety Minister Lord McKenzie of Luton said:

> 'The health and safety of employees is a moral and ethical obligation for each and every employer and this must be driven home from Board level. Only this way will we ensure that health and safety is taken seriously. This guidance clearly sets out the agenda for effective leadership of health and safety.'

1.207 The Chair of the HSC, Judith Hackitt, agreed:

> 'It is visible leadership from the top of an organisation which truly makes for an effective health and safety culture which in turn delivers good health and safety performance and much more. I am still confounded by the number of people who see "health and safety" as a barrier to doing things, as experience and evidence shows that the reverse is true. The challenge before us is changing behaviour. This guidance makes it clear what directors need to do but it is their action and delivery which will really count.'

1.208 Director General of the IoD, Miles Templeman, added:

> 'The Institute of Directors (IoD) believes that it's vital that board members lead the approach of their organisation to health and safety, whatever the environment they operate in. Too often health and safety are words used as excuses by organisations that have not developed their thinking in this area. The IoD hopes that the new guidance can help organisations integrate health and safety into business decisions in an appropriate way, not one that stifles appropriate activity.'

1.209 The Guidance sets out three essential Principles (in ten bullet points):

Strong and active leadership from the top – visible, active commitment from the board, establishing effective 'downward' communication systems and management structures and integration of good health and safety management with business decisions.

Worker involvement – engaging the workforce in the promotion and achievement of safe and healthy conditions, effective 'upward' communication and providing high-quality training.

Assessment and review – identifying and managing health and safety risks, accessing (and following) competent advice, monitoring, reporting and reviewing. This document is crucial as the guidance actually states that it can be used by the court, particularly when reviewing what the corporate culture was pursuant to the CMCHA 2007.

1.210 In 2001 the Health and Safety Commission (HSC) issued guidance for board members of all types of organisation, in both the public and private sectors, entitled *Directors' responsibilities for health and safety*.[1] This was published following a consultation exercise earlier in that year. Originally it was intended that there should be a Code, but the resulting document has guidance status only.

1 IND(G)343, available at www.hse.gov.uk/pubns/indg343.

1.211 The guidance sets out five action points:

(1) The board needs to accept formally and publicly its collective role in providing health and safety leadership in its organisation.
(2) Each member of the board needs to accept his or her individual role in providing health and safety leadership for their organisation.
(3) The board needs to ensure that all board decisions reflect its health and safety intentions, as articulated in the organisation's health and safety policy statement.
(4) The board needs to recognise its role in engaging the active participation of workers in improving health and safety.
(5) The board needs to ensure that it is kept informed of, and alert to, relevant health and safety risk management issues. The guide

recommends that one of the board members be appointed as the 'Health & Safety Director'.

1.212 The guide states that, by appointing a health and safety director, the company will have a board member who can ensure health and safety risk management issues are properly addressed. During the consultation, employers' organisations and trade associations failed to persuade the HSC not to have reference to a named health and safety director. The Confederation of British Industry, Construction Confederation, Engineering Employers' Federation and Institute of Directors all opposed the appointment of a health and safety director, arguing that it would promote a 'blame culture and detract from collective responsibility'. The fear is that, if an incident occurs, the named director would become a 'scapegoat'.

1.213 On publication of the guidance, the HSC chairman, Bill Callaghan, said:

> 'Health and safety is a boardroom issue. Good health and safety reflects strong leadership from the top.'

HEALTH AND SAFETY REGULATION[1]

1.214 The Robens Report of 1972[2] was the first comprehensive review of health and safety law in the UK. The report recommended that, in place of what it described as the 'haphazard mass of ill assorted and intricate detail' of the existing legislation, there should be a 'comprehensive and orderly' set of revised provisions under a new enabling Act. This Act would contain a clear statement of the basic principles of safety responsibilities and be supported by regulations and by non-statutory codes of practice.

1 See above for the recent changes brought in by the Coalition Government.
2 Lord Roben's report on *Safety and Health at Work* was published in June 1972 (Cmnd 5034, 1972).

1.215 As a consequence of this recommendation, the HSWA 1974 was born. This sets out a number of general duties upon employers. It also set up the Health and Safety Commission (HSC) and its operating arm, the Health and Safety Executive (HSE). However, despite the Act, much of the previous law remained in place which it had been intended it should replace.

1.216 The original aim of a single set of regulations did not come about until some two decades later, when a number of European Health and Safety Directives were enacted into UK law as regulations through HSWA 1974. The bulk of these regulations appeared in 1992. They are often referred to as the 'six-pack'. The regulations are:

- the Management of Health and Safety at Work Regulations, now the Management of Health and Safety at Work Regulations 1999 (MHSWR),[1]
- the Workplace (Health, Safety and Welfare) Regulations 1992,[2]
- the Personal Protective Equipment at Work Regulations 1992,[3]
- the Provision and Use of Work Equipment Regulations, now the Provision and Use of Work Equipment Regulations 1998,[4]
- the Manual Handling Operations Regulations 1992,[5] and
- the Health and Safety (Display Screen Equipment) Regulations 1992.[6]

Most of these regulations are supplemented by Approved Codes of Practice that place meat on the bones of the law in the regulations.

1 SI 1999/3242.
2 SI 1992/3004.
3 SI 1992/2966.
4 SI 1998/2306.
5 SI 1992/2793.
6 SI 1992/2792.

1.217 The 'framework' regulations of the six-pack are the MHSWR,[1] which enacted the Framework Directive.[2] The MHSWR has as its bedrock the principle of *risk assessment*. In the last three decades, there has been a change in the way safety legislation is framed in a number of industrialised countries. There has been a move away from laws that specify the way in which safe working should be achieved, to laws that set out the attainment of safety goals.

1 SI 1999/3242.
2 Council Directive 89/391/EEC.

1.218 The focus of regulatory supervision has shifted towards an examination of the way health and safety is managed and standards are set, implemented and monitored. The traditional approach was strong on where the ultimate responsibility lay, but did not seek to establish what kind of continuing external monitoring was necessary.

1.219 It is said that a scheme of monitoring and auditing by a safety regulator, which examines an organisation's management, control and design arrangements (and looks at the criteria on which decisions are made), can exert a positive influence to reduce risk without supplanting or diminishing the prime responsibility on the company.

1.220 With risk assessment-based legislation, rather than prescriptive legislation which sets out what action needs to be taken, companies are required to carry out risk assessments to set out their own goals in reducing risk. The emphasis is upon reducing the probability of accident occurrence and, if an accident does occur, reducing the harmful consequences of the accident.

See also, for example, the CDM Regulations 2007, the Work at Height Regulations 2005 and the various fire reform orders.

The Health and Safety at Work etc Act 1974[1]

1.221 The essential purpose of HSWA 1974 is to secure the health, safety and welfare of employees at work and to protect the public from the activities of a company's business.[2] The HSWA 1974 sets out a number of general duties upon employers and their employees. If one of these duties is breached, this may amount to a criminal offence.

1 See above for some recent examples.
2 HSWA 1974, s 1.

1.222 The main duties upon employers are set out in ss 2 and 3 of the HSWA 1974. A breach of one or both of these duties normally forms the basis of most prosecutions of employers pursuant to HSWA 1974. See also s 37 of the HSWA 1974.

1.223 HSWA 1974, s 2(1) states:

> 'It shall be the duty of every employer to ensure, so far as is reasonably practicable, the health, safety and welfare at work of all his employees.'

1.224 HSWA 1974, s 3(1) states:

'It shall be the duty of every employer to conduct his undertaking in such a way as to ensure, so far as is reasonably practicable, that persons not in his employment who may be affected thereby are not thereby exposed to risks to their health or safety.'

1.225 The other general duties under HSWA 1974 are ss 4 (concerning duties of employers to people not in their employment in relation to premises in their control) and 6 (in relation to manufacturers as regards articles and substances for use at work).

1.226 The main duty upon an employee is contained in s 7(a) of the HSWA 1974, which states:

'It shall be the duty of every employee while at work … to take reasonable care for the health and safety of himself and of other persons who may be affected by his acts or omissions at work …'

1.227 The important words in ss 2(1) and 3(1) of the HSWA 1974 are 'so far as is reasonably practicable' (often referred to as SFAIRP). The test as to what is reasonably practicable was set out in the case of *Edwards v National Coal Board*.[1] This case established that the risk must be balanced against the 'sacrifice' (whether in money, time or trouble) needed to avert or mitigate the risk. By carrying out this exercise, the employer can determine what measures are reasonable to take. This is effectively an implied requirement for a risk assessment.[2]

1 [1949] 1 KB 704. Lord Atkin stated: ' "reasonably practicable" is a narrower term than "physically possible" and seems to me to imply that a computation must be made by the owner, in which the quantum of risk is placed on one scale and the sacrifice in the measures necessary for averting the risk (whether in money, time or trouble) is placed on the other; and if it be shown that there is a gross disproportion between them the risk being insignificant in relation to the sacrifice the [person on whom the duty is placed] discharges the onus on them [of proving that compliance was not reasonably practicable].'

2 *R v Chargot* [2008] UKHL 73, [2009] 1 WLR 1.

Health and Safety (Offences) Act 2008

1.228 This Act came into force on 16 January 2009, and makes three changes to existing safety law: the current upper limits of penalties for most safety offences tried in the magistrates' court will increase from £5,000 to £20,000; there will be the option for custody in instances that are especially serious; and it is likely that more cases will be tried at the Crown Court.

1.229 The Robens Report, which led to the HSWA 1974, recommended that criminal sanctions for health and safety breaches be reserved for offences of a 'flagrant, wilful or reckless nature'. This did not make its way into the 1974 Act.

1.230 An individual, whether a worker or manager, can be convicted of breaching s 7 of the HSWA 1974, for failing 'to take reasonable care' for their and others' health and safety, ie being criminally negligent.

1.231 A director can be convicted under s 37 of the HSWA 1974 if his/her company's breach was 'attributable to any neglect' on their part. In the case of *R v P Ltd and G*,[1] the Court of Appeal held that a director could sometimes be guilty of neglect, even if he/she did not know of the unsafe practices of their company. Under the Act, both would carry a maximum term of up to two years' imprisonment. The test is what they ought to have known, rather than what they said they knew (ie the test is objective and not subjective). See further the section on the Fire Reform Orders above.

1 [2007] All ER (D) 173 (Jul).

1.232 The DWP Minister at the time, Lord McKenzie, said:

> 'It is generally accepted that the level of fines for some health and safety offences is too low. These changes will ensure that sentences can now be more easily set at a level to deter businesses that do not take their health and safety management responsibilities seriously and further encourage employers and others to comply with the law. Furthermore, by extending the £20,000 maximum fine to the lower courts and making imprisonment an option, more cases will be resolved in the lower courts and justice will be faster, less costly and more efficient. Jail sentences for particularly blameworthy health and safety offences committed by individuals, can now be imposed reflecting the severity of such crimes, whereas there were more limited options in the past.'[1]

1 DWP Newsroom, 16 October 2008, available at www.dwp.gov.uk. See further **Chapter 2**.

1.233 Under the Act, if an individual as an employer is convicted of breaching s 2 or 3 of the HSWA 1974 (ie failing to ensure the health and safety of employees or non-employees), he/she can receive a custodial sentence of up to two years. In such cases, the prosecution only has to prove that there has been an exposure to risk, and then it is for the employer, under s 40 of the HSWA 1974, to prove that he/she took all reasonably practicable steps to avoid the exposure. This reverse burden of proof can be difficult.

1.234 Any provision that imposes a reverse burden of proof is an inroad into the presumption of innocence enshrined in Article 6(2) of the European Convention on Human Rights (ECHR). Despite this, if the reverse burden is necessary, justified and proportionate, it will not be incompatible with that presumption of innocence. The courts have already held that the burden imposed by s 40 of the 1974 Act is compatible with the ECHR. The case of *R v EGS*[1] specifically states that s 40 defences need to be highlighted in any defence statements; failure to do so can prevent a defence team from using s 40 at trial. The question that arises in the context of this Act is whether the addition of imprisonment, as a possible penalty when a person is convicted of an offence to which the reverse burden applies, means that this is no longer the case. This issue was reviewed in the House of Lords case of *R v Chargot*.[2]

1 [2009] EWCA Crim 1942.
2 [2008] UKHL 73, [2009] 1 WLR 1.

1.235 The change to present arrangements set out in Schedule 1 to the HSWA 1974 has made imprisonment an option for most health and safety offences in both the lower and the higher courts. Previously, imprisonment was an option only in certain cases. In the lower courts, it was an option only for failure to comply with an improvement or prohibition notice, or with a court remedy order, and for offshore offences. In the higher courts, it was an option only for failure to comply with licensing requirements or explosives provisions, or disclosures in breach of the Act. These are all serious offences. However, the 2008 Act extends the option of a custodial sentence to a greater range of offences, and that responds in part to the fact that judges have remarked in several cases over the years on the lack of imprisonment as an option and that they would have jailed the offender had they been able to do so. It is likely that imprisonment will continue to occur only in the most serious of cases, and that there will be only a very small increase in the number of offenders going to prison under the revised legislation. These will involve very serious breaches resulting in fatalities, where profit has been put before safety.

1.236 The Act extends to England and Wales and Scotland. Health and safety is outside the legislative competence of the devolved administrations. This Act does not have any special effect on Wales and does not affect the National Assembly for Wales.

Risk assessment[1]

The requirement for risk assessment

1.237 In *R v Willmott Dixon Construction Limited*,[2] the Court of Appeal looked at risk in terms of an asbestos situation. The Court of Appeal, dismissing the appeal, restated that the risk of inhaling a single fibre could lead to mesothelioma (see above). Marks and Spencer plc were fined £1 million and ordered to pay £600,000 costs. Other contractors were also fined, and Willmott Dixon Ltd were fined £50,000 and ordered to pay £75,000 costs.

1 See above, and see further **Chapter 2**.
2 [2012] EWCA Crim 1226.

1.238 2013 saw an important judgment in the case of *Kennedy v Cordia (Services) LLP*.[1] In 2010, Mrs Kennedy incurred a serious wrist injury when she slipped over in icy conditions on a pavement. She was a home carer. She was on her way to see a terminally ill patient. The court said she had to go out that night as she was on 'a mercy mission'. The court found against her employers.

1 2013 CSOH 130 (Outer House, Court of Session).

1.239 The court found that her employers should have identified and risk assessed the risk and provide suitable footwear to minimise the risk. This case has far-reaching implications for employers under various regulations including the Personal Protective Equipment at Work Regulations.

1.240 The main requirement for risk assessment is contained in the MHSWR.[1] The Regulations require the company to assess the risks and, in effect, to act to neutralise them. Failure to carry out an adequate risk assessment may amount to a criminal offence.

1 SI 1999/3242.

1.241 Regulation 3 of the MHSWR states:

> '(1) Every employer shall make a suitable and sufficient assessment of–
>
> (a) the risks to the health and safety of his employees to which they are exposed whilst they are at work; and
>
> (b) the risks to the health and safety of persons not in his employment arising out of or in connection with the conduct by him of his undertaking,
>
> for the purpose of identifying the measures he needs to take to comply with the requirement and prohibitions imposed upon him, by or under the relevant statutory provision and by Part II of the Fire Precautions (Workplace) Regulations 1997.'[1]

1 SI 1997/1840, Pt II.

1.242 Regulation 4 of the MHSWR sets out the way in which the risk should be approached:

> 'Where an employer implements any preventive and protective measures he shall do so on the basis of the principles specified in Schedule 1 to these Regulations.'

1.243 Schedule 1 to the MHSWR details the following:

'(a) avoiding risks;
(b) evaluating the risks which cannot be avoided;
(c) combating the risks at source;
(d) adapting the work to the individual, especially as regards the design of workplaces, the choice of work equipment and the choice of working and production methods, with a view, in particular, to alleviating monotonous work and work at a predetermined work rate and to reducing their effect on health;
(e) adapting to technical progress;
(f) replacing the dangerous by the non-dangerous or the less dangerous;
(g) developing a coherent overall prevention policy which covers technology, organisation of work, working conditions, social relationships and the influence of factors relating to the working environment;
(h) giving collective protective measures priority over individual protective measures; and
(i) giving appropriate instructions to employees.'

1.244 Note the hierarchy of requirements. Thus, for example, it would not be appropriate for an employer to give instructions to an employee on how to deal with the risk, as a control measure, if it was reasonably practicable to avoid the risk altogether.

1.245 Risk assessment appears in three ways in the health and safety legislation. The first way is as a general risk assessment (that is, to assess the risks from hazards in a workplace). A general risk assessment is required by MHSWR, reg 3.

1.246 Another way in which a risk assessment can be used is by supporting another piece of legislation which sets out a safety goal. For example, reg 11 of the Provision and Use of Work Equipment Regulations 1998[1] (PUWER) requires employers to ensure that measures are taken which are effective to prevent access to any dangerous part of machinery, for example, providing a guard on the machine. In order to assess the possible control measure, a risk assessment under MHSWR, reg 3 will need to be carried out.

1 SI 1998/2306.

1.247 A third way in which a risk assessment can be carried out is as a specific risk assessment. For example, a general risk assessment under MHSWR, reg 3 may highlight the fact that in the workplace there are a number of manual operations (i.e. lifting) hazards. These are covered by the Manual Handling Operations Regulations 1992[1] (MHOR). Regulation 4 of MHOR requires manual handling operations be avoided so far as is reasonably practicable. If they cannot be avoided, reg 4 requires a specific risk assessment to be carried out of the manual handling operation in order to ensure that the risk is to reduce to the lowest level as far as is reasonably practicable.

1 SI 1992/2793.

What is a suitable and sufficient risk assessment?[1]

1.248 Risk assessment essentially considers the question: 'What if?'. It is not sufficient to base a risk assessment solely on an employer's own accident experience. In the case of *Walker v Bletchley Fletons Ltd*[2] the court found the employer liable in negligence. It was said in the judgment the fact that an accident had never happened did not necessarily diminish the chances of one happening. Thus a company is not entitled to wait for a near miss or accident before it considers a potential risk.

1 See further **1.14** above in relation to the new HSG 65 that came into operation in 2013.
2 [1937] 1 All ER 170.

1.249 As to the meaning of risk, in *R v Board of Trustees of the Science Museum*[1] the Court of Appeal said that risk means the *possibility* of danger and not *actual* danger. The test has now been altered by the case of *R v Porter*,[2] where the Court of Appeal has redefined the test of the meaning of 'risk', which is the trigger point of the offences under ss 2, 3, 7 and 37 of the HSWA 1974.[3]

1 [1993] 1 WLR 1171, [1993] 3 All ER 853.
2 *R v Porter (James Godfrey)* [2008] EWCA Crim 1271, [2008] ICR 1259.
3 See above in relation to other cases.

1.250 In the court's judgment, Moses LJ said:

> 'In our view it is not necessary to provide any paraphrase of the statutory concept of risk, even though judges have in the past felt it necessary to do so: see, for example, *R v Board of Trustees at the Science Museum* [1993] 1 WLR 1171, 1171D, in which the court referred to the concept of risk as containing the idea of "a possibility of danger". What is important is that the risk which the prosecution must prove should be real as opposed to fanciful or hypothetical: see *R v Chargot Ltd (trading as Contract Services) and Others* [2007] EWCA Crim 3032 at para 26. There is no obligation under the statute to alleviate those risks which are merely fanciful.'

See also *Tangerine Confectionery Ltd and Veolia ES (UK) Ltd v R*[1] on risk assessments being required to deal with risks beyond the obvious.

1 [2011] EWCA Crim 2015 – see **1.65** above.

1.251 The *Porter* decision has arguably extended the law to adopt the realities of day-to-day life, and appears to have gone a substantial way to redeeming the current status quo. A crucial point is that it states that the lack of a similar accident is a vital factor to be considered. It is also an important case, inter alia, for prosecution authorities in deciding whether to prosecute for health and safety (and corporate manslaughter), for defence teams advising their clients and for judges arbitrating on trials and *Newton* hearings. In the future, it will be real and not fanciful risks that will be considered.[1]

1 See Patrick Harrington QC and Gerard Forlin, 'Child's Play' (2008) *New Law Journal*, 1 August. See also *R v EGS Ltd* [2009] EWCA Crim 1942.

1.252 The *Porter* test (see above) was approved by the House of Lords in *R v Chargot*,[1] where the court said:

> 'In *R v Porter (James Godfrey)* the Court of Appeal set aside the conviction of the headmaster of the school where one of the pupils lost his footing on a step which gave access from one playground to another whilst he was being supervised, with tragic consequences … That was an exceptional case, but makes an important point. The law does not aim to create an environment that is entirely risk free. It concerns itself with risks that are material. This is

in effect is what the word risk that the statute uses means. It is directed at situations where there is a material risk to health and safety, which any reasonable person could appreciate and take steps to guard against.'

1 [2008] UKHL 73, [2009] 1 WLR 1, [2009] 2 All ER 645, [2009] ICR 263.

1.253 The case of *R v EGS Ltd*[1] concerned the tragic death of a nine-year-old boy crushed by a gate after leaning through a small gap and pressing an egress button. The case was a prosecution appeal pursuant to s 61(4)(a) of the Criminal Justice Act 2003, after the trial judge allowed a half time submission to succeed by the defence. The Court of Appeal (Dyson LJ, Davis and Lloyd Jones JJ) allowed the appeal and remitted the case back to the Crown Court.

1 [2009] EWCA Crim 1942.

1.254 In doing this, the court has reviewed the latest authorities and clarified certain aspects arising out of *R v Chargot* and issued some practical guidance on case management. Dyson LJ said in relation to risk, citing *R v Porter* and *R v Chargot*:

> 'In other words, it is helpful to ask whether a reasonable person would appreciate and guard against the risk in deciding whether the risk is more than trivial or fanciful.'
>
> '… it is a question of fact and degree whether a risk is trivial or fanciful. It is pre-eminently a jury question.'

He also went on to say:

> 'The prosecution did not have to prove that the risk was appreciable or foreseeable. They had to prove that the risk was not fanciful and was more than trivial.'

Finally he said:

> '… as Lord Hope said at [21] in *Chargot*, once the result that section 3(1) describes is not achieved, a prima facia case of breach is established and the onus passes to the defendant to make good the section 40 defence, if that defence is raised. He made the same point at [22], where he said that where an injury is caused, the facts will speak for themselves. If the judge had focused on this simple way of putting the case, he would surely have rejected the submission of no case to answer. It may well be that he was distracted by the complicated way in which the prosecution was presented.'

The case appears to further narrow the defence being able to do little more than call their evidence in rebuttal as soon as the prosecution has proved that the accident happened in a work context.

1.255 It should also be noted that a risk assessment should consider not only the probability that the hazard will occur, but also the severity of harm if the hazard materialises.[1] Essentially, if the risk of the hazard occurring is small but the severity of the consequence is great, there must be control measures put in place (if the hazard cannot be avoided).

1 See para 11 of the original ACOP.

1.256 This was highlighted in the sentencing remarks of Mr Justice Clarke following health and safety prosecutions in relation to the collapse of a walkway onto a ferry at Ramsgate Port in 1994 which killed six passengers (*R v Fartygsentreprenader AB, Fartygskonstructioner AB, Port Ramsgate Ltd and Lloyd's Register of Shipping*).[1] He said:

'... if thought had been given to its responsibilities especially having regard to the provision of the 1974 Act [HSWA 1974], Port Ramsgate could have appreciated that there were potential risks, albeit, perhaps very small risks ... Further, once it was appreciated that there were potential risks, it would have been appreciated that such risks should have been guarded against because of the catastrophic consequences if anything went wrong.'

1 (28 February 1997, unreported).

1.257 There are two types of risk assessment: qualitative and quantitative. The former is an assessment that is judgement-based, to determine contributory factors or potential outcomes. The latter assessment produces numerical values of the risk involved and evaluates the results against specific risk criteria. The appropriate type of risk assessment will be determined by the nature of the hazards being assessed.

1.258 Particularly where risk reduction or elimination involves high costs, companies may carry out cost benefit analysis (CBA). A CBA should clearly present its assumptions and findings so that decision-makers understand the results and uncertainties. There is an interesting discussion on the use of CBA in the *Report of the Joint Inquiry into Train Protection* chaired by Lord Cullen and Professor Uff.[1]

1 Professor John Uff QC and the Rt Hon Lord Cullen PC, *The Joint Inquiry into Train Protection* (2001).

Responsibility for risk assessment[1]

1.259 Regulation 7 of the MHSWR imposes the duty upon the employer to have someone within his or her own organisation carry out the necessary risk assessments. This person must have the appropriate expertise and training and be provided with the appropriate information. For many companies, advice from outside consultants will be the exception rather than the rule, but in complex industries consultants are more likely to be used. However, it will still be for the companies to ensure that these are the appropriate experts and that they are provided with all the relevant information to carry out the task.

1 See above.

1.260 Regulation 7(1) of the MHSWR states:

'Every employer shall ... appoint one or more competent persons to assist him in undertaking the measures he needs to take to comply with the requirements and prohibitions imposed upon him by or under the relevant statutory provisions ...'

1.261 Regulation 7(8) of the MHSWR states:

'Where there is a competent person in the employer's employment, that person shall be appointed for the purposes of paragraph (1) in preference to a competent person not in his employment.'

1.262 Regulation 7(4) of the MHSWR states:

'The employer shall ensure that–

(a) Any person appointed by him in accordance with paragraph (1) who is not in his employment–
   (i) is informed of the factors known to him to affect, or suspected by him of effecting, the health and safety of any person working in his undertaking and

(ii) has access to the information referred to in Regulation 10; and

(b) Any person appointed by him in accordance with paragraph (1) is given such information about any person working in his undertaking who is—

(i) employed by him under a fixed term contract of employment, or

(ii) employed in an employment business,

as is necessary to enable that person properly to carry out the function specified in that paragraph.'

1.263 Regulation 7 of the MHSWR makes it clear (if there was ever any doubt) that it does not matter who carries out the risk assessment; the company is responsible and liable for that risk assessment. Further, reg 21 says that a company cannot defend criminal proceeding because a person appointed to do risk assessment has failed to carry out that risk assessment adequately. See *R v HTM Ltd.*[1]

See also a series of other Regulations (such as PUWER and LOLER) that require competent people to undertake risk assessments.

1 [2006] EWCA Crim 1156, [2007] 2 All ER 665, [2006] ICR 1383.

Risk management

1.264 Risk assessment and risk management are distinct. Risk management addresses the *appraisal* of assessed risk and the *making of decisions* concerning risks, in particular safety measures and their subsequent implementation. The carrying out of a risk assessment does not remove from managers the responsibility for safety at work.

For basic information, or for people getting started in managing health and safety, the HSE has a publication called 'Health and safety made simple: The basics for your business'.

Guidance on controlling risks from specific topics can be found in the HSE publication, 'The Health and Safety Toolbox: How to control risks at work'.

A revised and online version of 'Managing for Health and Safety' (also known as HSG 65) is also available. See above.

CONSTRUCTION

1.265 Construction and issues relating to this industry are covered in **Chapter 5**. However, the industry has been subject to scrutiny for some time, investigations into the fatality rates and the introduction of the Construction (Design and Management) Regulations and the update of these Regulations and the Control of Asbestos Regulations 2006. This has been accompanied by changes to working at height and the use of work equipment. On 28 January 2014 the HSE published new guidance in relation to working at height for employers and employees. It sets out in simple terms what to do and what not to do. See also the case of *R v Willmott Dixon Construction Ltd.*[1]

1 [2012] EWCA Crim 1226, and see above.

1.266 The House of Lords in *Smith v Northamptonshire County Council*[1] reviewed various regulations, including PUWER and the Manual Handling Regulations. This is an important case and the Lords have introduced an element of common sense to the purpose of the various regulations. Their Lordships in effect found that there must be a specific link between the equipment and the employer's undertaking. This case will have far-reaching consequences.

1 [2009] UKHL 27.

HUMAN FACTORS[1]

1.267 The HSE's view is that accidents, ill health and adverse incidents are seldom random events. The immediate cause may be human or technical failure, but these in turn usually stem from organisational failures which are the responsibility of management.

1 See the amended HSG 65 above.

1.268 In March 1989 a London-bound train collided with another service outside Purley station. Five people died and more than 80 were injured in the crash. The driver, Robert Morgan, admitted two counts of manslaughter and was jailed for 18 months in September 1990, 12 of which were a suspended sentence, but the term was later cut to four months on appeal. This decision was appealed and, in December 2007, the manslaughter convictions were quashed. In the Court of Appeal,[1] Lord Justice Latham said that there were factors of 'considerable significance' which had not been considered in 1990. Following Mr Morgan's conviction, it was found that there had been four previous 'signal passed at danger' (SPAD) incidents involving the same signal in the five years before the fatal crash. Lord Justice Latham said:

> 'The history of signal 168… would have been a significant factor in any jury's evaluation of the extent to which the appellant's fault could have been said to have gone from being negligent – breach of duty – to being the sort of breach of duty which justified the imposition of criminal sanctions … Mr Morgan was placed in a position where he had no alternative other than to plead guilty.'

See also *R v Transco plc*.[2]

1 *R v Morgan* [2007] EWCA Crim 3312.
2 [2006] EWCA Crim 838, and see above.

1.269 Further, on 11 March 2011, one of the most powerful earthquakes on record hit north-east Japan. The resulting tsunami killed almost 20,000 people, and caused a meltdown at the Fukushima Daiichi nuclear plant. After 900 hours of hearings and 1,100 interviews over a six-month period, the Fukushima Nuclear Accident Independent Investigation Commission – chaired by Kiyoshi Kurokawa, an academic fellow at Tokyo's National Graduate Institute for Policy Studies – said that the accident was 'a profoundly man-made disaster that could and should have been foreseen and prevented'. Japanese Power Company TEPCO claimed that the main cause of the nuclear disaster at the Fukushima Daiichi was the 15-metre high tsunami wave that exceeded all expectations in the event of a major earthquake. This position is somewhat at odds with the results of the investigation launched by Japan's government, which claimed the main cause appears to have been the human factor element.[1] An investigation report said:

> 'We believe that the root causes were the organisational and regulatory systems that supported faulty rationales for decisions and actions, rather than issues relating to the competency of any specific individual.'

1 See **1.90** above.

1.270 On the evening of 20 April 2010, the oil rig Deepwater Horizon exploded about 50 miles off the Louisiana coast in the Gulf of Mexico, killing 11 workers and injuring 17 more. Two days later, it sank in 5,000 feet of water. A report from BP concluded that decisions made by 'multiple companies and work teams' contributed to the accident, which it says arose from 'a complex and interlinked series of mechanical failures, human judgments, engineering design, operational implementation and team interfaces'.[1]

1 See **1.89** above in relation to the trial.

REASONABLE BUSINESSES

1.271 In January 2009 the Department for Business, Enterprise and Regulatory Reform (BERR) published Sarah Anderson's review of health and safety and employment regulation, *The Good Guidance Guide*,[1] which made a variety of recommendations about how the government can best give regulatory guidance to businesses, especially SMEs. In March 2009 the government said it would take forward a number of Anderson's proposals and that it would discourage inspectors from prosecuting 'reasonable' businesses, as part of a range of commitments made in response to the guide. The government also said that it:

> 'agrees in principle that where an enforcement authority had discretion over what enforcement action is taken, they should not generally prosecute or impose punitive sanction where a business has reasonably followed their advice and there are no other factors which indicate that a prosecution would be appropriate.'

1 *The Good Guidance Guide: taking the uncertainty out of regulation*, available at www.berr.gov.uk/files/file49881.pdf.

1.272 The Government has published two reports about restoring common sense to health and safety and saving businesses millions of pounds and considerable time every year. The first, written by Professor Ragnar Löfstedt, looks at the progress made in implementing the recommendations contained in his 2011 report 'Reclaiming Health and Safety for all'. The report states that these changes will help businesses focus on growth rather than unnecessary red tape. The other report shows that the Government has already implemented 23 of the 35 recommendations in Lord Young's 2010 report 'Common Sense, Common Safety'. Lord Young called for changes to the regulatory system in order to combat the rise of a compensation culture, and address the frequently negative perceptions of health and safety legislation (see above).

REDUCING RISKS, PROTECTING PEOPLE[1]

1.273 It is very important to look at the HSE guidance on how they instruct their staff to look for evidence that duty holders have reduced risk to 'as low as reasonably practicable' ('ALARP') in certain situations. The guidance is set out on the HSE website under 'Risk Expert'.

1 See above.

1.274 The guidance is in headings:

1. Principles and Guidelines to assist HSE in its judgment that duty-holders have reduced risk as low as is reasonably practicable.

2. Assessing compliance with the law in individual cases and the use of good practice.
3. Policy and guidance on reducing risks as low as reasonably practicable in design.
4. HSE principles for cost benefit analysis.
5. Cost benefit analysis.
6. 'ALARP at a glance'.

This is a vital document for all duty-holders to consult and abide by.

IMPROVEMENT AND PROHIBITION NOTICES

1.275 To put this in context, some 13,503 notices were issued in 2012/2013 alone. They are commonly used and remain a powerful and efficient weapon for regulators.

Improvement notice

1.276 Pursuant to HSWA 1974, s 21, inspectors can issue an improvement notice when they consider health and safety legislation is being contravened. The notice can be issued whether the legislation being breached is HSWA 1974 or some other health and safety statutory provision.

1.277 The notice must specify the legal requirements that the inspector thinks are being broken and give reasons. The time allowed to put matters right cannot be less than 21 days (as this is the time limit to submit an appeal to the Employment Tribunal). How long is allowed is in the discretion of the inspector. It will depend on factors such as the seriousness of the matters involved and the ease with which action necessary to comply with the notice can be taken. In the notice, the inspector can also set out what he or she needs to be done to put matters in order.

Prohibition notice

1.278 An inspector may issue a prohibition notice pursuant to HSWA 1974, s 22 when he or she thinks that there is a risk of *serious personal injury*. The notice prohibits the carrying on of the work activity that the inspector believes is creating the risk of injury. If the inspector considers the risk is of imminent danger, the notice must take immediate effect and the work activity stopped at once. If not, the notice can be deferred, stating that the work activity must be stopped within a certain time. There does not have to be imminent danger for a notice to be issued.[1]

1 See *Tesco Stores Ltd v Kippax* (1990) COIT 7605, HSIB 180, p 8.

1.279 The case of *Railtrack v Smallwood*[1] concerned railway signal SN109 that was passed at danger on 5 October 1999, resulting in the Ladbroke Grove railway crash. On 8 October 1999 a prohibition notice was served prohibiting the use of the signal and the section of track on which it was situated. Railtrack, who controlled the infrastructure at the time, appealed the notice on the basis that, when it was issued, the track was not in use. It was held that, even though work activities were suspended at the time, 'activities' were still carried on for the purpose of HSWA 1974, s 22.

1 [2001] ICR 714.

1.280 In July 2009, in the case of *Chilcott v Thermal Transfer Ltd*[1] the High Court considered the appeal brought by Mr Simon Chilcott, HM Inspector of Health and Safety, against the decision of the Bristol Employment Tribunal to cancel a Prohibition Notice issued by him against Thermal Transfer Ltd. The facts of the case were that the main contractor had contracted with sub-contractors to carry out platform steelwork on a site for which they were the main contractor. A 'task specific risk assessment/method statement' was agreed which identified falling from height as a risk. This risk was dealt with by certain stipulations pertaining to the way in which work at height was to be carried out and the erection of handrails on a mobile platform. One of the sub-contractor's supervisors, whilst working on the mobile platform with his assistant, contrary to what was specified in the method statement, fell through a hole in the platform and broke both his ankles. The same day, the inspector made an unplanned visit to the site and was informed about the accident. Following an investigation, the inspector issued a prohibition notice identifying breaches of s 3 of the HSWA 1974 and reg 4 of the Work at Height Regulations 2005. The notice stated that the company had failed to ensure, so far as was reasonably practicable, that work at height was planned to avoid the risk of persons falling from height. The main issue for determination was whether the tribunal had applied the correct test in deciding whether to affirm or cancel the prohibition notice pursuant to an appeal under s 24 of the HSWA 1974.

1 [2009] EWHC 2086 (Admin).

1.281 The High Court adopted the provisional view expressed by Sullivan J in *Railtrack plc v Smallwood*,[1] namely that a Tribunal hearing an appeal against a notice was not limited to reviewing the genuineness and/or the reasonableness of the Inspector's opinions, but was required to form its own view, paying due regard to the Inspector's expertise.

1 [2001] ICR 714.

1.282 See also *Wandsworth v Covent Garden Market Authority*[1] regarding the strict approach to time limits taken by the courts. There also appears to be a trend of appellants being more successful when they do appeal.

1 [2011] EWHC 1245 (QB).

1.283 The prohibition notice can be issued for any activities to which the relevant statutory provisions apply. There does not have to be an actual breach of a legal duty although, if it is thought that the law is being broken, the notice must state this and detail any breach. The notice must also state the matters which, in the inspector's view, are creating the risk of serious personal injury, and may include directions on what steps should be taken to rectify matters.

Power to deal with cause of imminent danger

1.284 Where there is an article or substance that the inspector thinks creates an imminent risk of injury, he or she can seize the article or substance or 'cause it to be rendered harmless (whether by destruction or otherwise)'.[1]

1 HSWA 1974, s 25.

Service of notices

1.285 An improvement notice is served on the person responsible for the breach of legal requirements. Thus, if the legal requirement being breached imposes duties on employers, the improvement notice is served on the employer, even if employees or other people have been involved in the events that led to the issue of the notice.

1.286 A prohibition notice is served on the person carrying on or in control of the activities concerned, whether or not that person would also be responsible for any breach of legal requirements. Therefore, for example, a prohibition notice could be served on a site manager, even though the legal duty is upon his or her employers. However, a copy of the notice would also be sent to the employers as well.

1.287 The inspector usually serves the notice personally. Notices can, however, also be served by post. In the case of a corporate body (for example, a limited company or local authority) or of a partnership, it will be sent to the principal office or registered address. However, the person to be served can specify the address where the notice should be sent.

Withdrawal of notices

1.288 Once the remedial action has been taken to comply with the notice, the notice ceases to have effect and so does not have to be withdrawn. However, there are certain circumstances where the notice can be withdrawn – for example, if the situation that gave rise to the notice changes. See the HSE Enforcement Guide at section 5.5 regarding notices.

Extension of time limits

1.289 Lord Justice Underhill undertook a full review of the Employment Tribunals and the recent changes that have occurred. These came into effect on 29 July 2013.

1.290 The new rules are now Schedule 1 to the Employment Tribunals (Constitution and Rules of Procedure) Regulations 2013.[1] Rule 105 modifies the appeal procedures in relation to improvement and prohibitions notices.

Rule 105(1) states that an appellant:

> '… may appeal an improvement or a prohibition notice by presenting a claim to a tribunal office–
>
> (a) before the end of the period of 21 days beginning with the date of the service on the Appellant of the notice which is the subject of the appeal; or
>
> (b) within such further period as the Tribunal considers reasonable where it is satisfied that it was not reasonably practicable for an appeal to be presented within the time.'

1 SI 2013/1237.

OFFENCES UNDER THE HSWA 1974

1.291 HSWA 1974, s 33 sets out a number of offences that can be prosecuted under the Act. These include:

- Failing to comply with general duties imposed on employers, the self-employed, people in control of premises, manufacturers etc (ie general duties contained in HSWA 1974, ss 2–7) or failing to comply with any requirement imposed by regulations made under the Act, for example, failing to carry out a suitable and sufficient risk assessment pursuant to MHSWR, reg 3.[1]

- Obstructing or failing to comply with any requirements imposed by inspectors in the exercise of their powers.
- Failing to comply with an improvement or prohibition notice.
- Failing to supply information as required by a notice issued by the HSC.
- Failing to comply with a court order to remedy the cause of an offence.

1 SI 1999/3242, reg 3.

1.292 Individuals and corporate bodies such as limited companies and local authorities can commit offences. If an offence is committed by a corporate body was committed 'with the consent or connivance of, or to have been attributable to any neglect on the part of any director, manager, or secretary or other similar officer', he or she is also guilty of the offence and may be prosecuted as well as the corporate body.[1] See the case of *R v Chargot*.[2]

1 HSWA 1974, s 37.
2 [2008] UKHL 73, [2009] 1 WLR 1, [2009] 2 All ER 645.

1.293 If someone commits an offence under HSWA 1974 because of an 'act or default' of someone else, that person will also be guilty of an offence pursuant to HSWA 1974, whether the other person is prosecuted or not.[1] See *R v Atterby*[2] for an interesting example.

1 HSWA 1974, s 36.
2 (2009) unreported; discussed at **2.202**.

PROSECUTIONS OF EMPLOYERS FOLLOWING WORK RELATED INCIDENTS[1]

1.294 The majority of prosecutions of employers following work related incidents are on the basis of breaches of the general duties of ss 2(1) and 3(1) of the HSWA 1974 – ie a failure by the employers to conduct their business so as to ensure the health and safety of employees and non-employees (including members of the public) respectively, 'so far as is reasonably practicable'.

1 See above and **Chapter 2**.

1.295 Because of the correlation between these duties and the requirement to manage risk, it is now becoming more common for individual employers and directors also to be charged with both a failure to carry out a suitable and sufficient risk assessment pursuant to reg 3 of MHSWR and s 37 of the HSWA 1974.

1.296 Further, the employers can in certain circumstances be prosecuted with other breaches of health and safety law. For example, if an employee was injured or killed because of a failure to have a guard on a machine, the employer could also be expected to be prosecuted for a breach of reg 11 of PUWER.[1]

1 SI 1998/2306.

How bad does the failure have to be?

1.297 British Rail was prosecuted for breaches of HSWA 1974, ss 2 and 3 following the Clapham train crash of 1988. Thirty-five people died. BR pleaded guilty. The case came before Mr Justice Wright at the Old Bailey for sentencing on 14 June 1991, two and a half years after the incident. He observed that the underlying causes of the tragedy were:[1]

> '... a failure of any proper systems of preparation for work, supervision of work, inspection of work, testing and checking of work of the re-signalling work that was being carried out ... Standing instructions were not properly distributed; individual personnel were not fully trained or instructed in their responsibilities; there was no proper co-ordination of instructions or system for ensuring that those instructions were complied with.'

1 (14 June 1991, unreported).

1.298 The judge commented upon the level of failure required for an employer to be guilty of an HSWA 1974 offence. He said:[1]

> '... the charges that British Rail face today do not involve any connotation or allegation of recklessness. The allegation is no more than that; a failure to maintain and observe the high standards required by the Health and Safety at Work legislation ...'

1 (14 June 1991, unreported).

Level of exposure to risk

1.299 Crucially, in *R v Chargot*[1] the House of Lords stated the following:

> '... The prosecution must show that there is a connection between the work that the employees was doing (section 2 cases) or the conduct of the undertaking (section 3 cases) and the accident, but the fact that the cause of the accident was unknown or is irrelevant so far as the prosecution's case is concerned. This is because the duty that these provisionally lay down looks at the result, not the means of achieving it. Prima Facie a breach of section 2(1) arises where an employee is injured whilst he is at work in the workplace. That fact in itself demonstrates that the employer failed to ensure his health and safety at work. The same is true where a person not in his employment but who may be affected by the undertaking suffers injury. The effect of the reverse burden of proof must be understood against that background.'

See *R v EGS Ltd*[2] and *R v Willmott Dixon Construction Ltd*.[3]

1 [2008] UKHL 73, [2009] 1 WLR 1, [2009] 2 All ER 645, [2009] ICR 263.
2 [2009] EWCA Crim 1942.
3 [2012] EWCA Crim 1226.

Causation

1.300 It follows that it is not necessary for the prosecution to prove a causal link between any injury or death that may have occurred. The prosecution only has to prove that there was a risk of exposure to harm, so that proof of actual harm is unnecessary. However, in *Davies v Health and Safety Executive*,[1] which concerned a breach of HSWA 1974, s 3(1), the Court of Appeal made the following observation:

> 'There may be real issues about whether the defendant owes the relevant duty or whether in fact the safety standard has been breached, for example where the cause of an accident is unknown or debatable.'

1 [2002] EWCA Crim 2949.

1.301 The House of Lords in *R v Chargot*[1] dealt with this aspect and said:

> '... Nevertheless, for the other reasons that the Court of Appeal gave in *R v Davies* I would hold that placing of a legal burden of proof on the employer in the case of this legislation is not disproportionate. The penalties that may be imposed on an individual have now been increased: see para 15. But I do not think that, when account is taken of the purposes that this legislation is intended to serve, this alteration in the law renders what was previously proportionate disproportionate.'

1 [2008] UKHL 73, [2009] 1 WLR 1, [2009] 2 All ER 645, [2009] ICR 263.

Defence under HSWA 1974, s 40[1]

1.302 HWSA 1974, s 40 states:

> 'In any proceedings for an offence under any relevant statutory provisions consisting of a failure to comply with a duty or requirement to do something so far as is practicable or so far as is reasonably practicable, or to use the best means to do something, it shall be for the accused to prove (as the case may be) that it was not practicable or not reasonably practicable to do more than was in fact done to satisfy the duty or requirement, or that there was no better practicable means than was in fact used to satisfy the duty or requirement.'

1 See above.

1.303 In other words, if the prosecution proves (beyond reasonable doubt) that there has been a criminal breach of a duty, the onus then falls upon the employer to prove (on the lower standard of proof, a balance of probabilities – see *R v Carr-Briant*[1] and *R v Dunbar*)[2] that all that was reasonably practicable to do was done.

1 [1943] 2 All ER 156, CCA.
2 [1958] 1 QB 1, CCA.

1.304 The question is, when does the defence come into play?

1.305 In *R v British Steel PLC*,[1] the Court of Appeal said that a breach of HSWA 1974, s 3(1) (and thus HSWA 1974, s 2(1) as well) imposed absolute criminal liability on an employer, subject only to the defence of reasonable practicability. In the House of Lords' case of *R v Associated Octel Co Ltd*[2] it was said that s 3(1) requires the employer to conduct his undertaking in a way which, subject to reasonable practicability, does not create risks to people's health and safety.

1 [1995] 1 WLR 1356.
2 [1996] 4 All ER 846.

1.306 In *Lockhart v Kevin Oliphant*[1] it was said that, once there is a 'prima facie' case against the employer that the health, safety and welfare of employees was not ensured, the onus under HSWA 1974, s 40 is on the defendant. The prosecution 'does not have to prove that it was reasonably practicable to comply' with HSWA 1974.[2]

1 1992 SCCR 774.
2 *Redgrave's Health and Safety* (4th edn, 2002), p 99.

1.307 In *Davies v Health and Safety Executive*[1] there was an issue as to what the prosecution had to prove before the defendant was required to rely on the defence under HSWA 1974, s 40. The prosecution argued that, once it had proved an exposure to risk, the offence (under HSWA 1974, s 3(1)) was proved unless the defendant established the s 40 defence; or. as the Court of Appeal put it, 'reasonable practicability is not an essential ingredient of the offence'.

1 [2002] EWCA Crim 2949.

1.308 However, the Court of Appeal rejected that submission.[1] and went on to say:

> 'The duty cast on the defendant is a "duty ... to ensure as far as is reasonably practicable". It is a breach of a qualified duty which gives rise to the offence. [Prosecuting counsel] had to concede that but for section 40 [HSWA 1974, s 40] it would be for the [prosecution] to negative reasonable practicability.'

1 *Davies v Health and Safety Executive* [2002] EWCA Crim 2949.

1.309 Despite what is said in *Davies v Health and Safety Executive*,[1] it is suggested that, once the prosecution has proved there was a risk that personal injury could be caused, it is then for the defendant to rely upon HSWA 1974, s 40. The House of Lords appears to have confirmed that view. See also *R v EGS Ltd*.[2]

1 [2002] EWCA Crim 2949.
2 [2009] EWCA Crim 1942, and see above.

Directors and senior managers

1.310 It used to be the case that, in order to prove a breach of, for example, HSWA 1974, s 2(1) or s 3(1), it was not necessary to show that the directors or senior managers of the employers had personally failed not to expose people to the risk of harm, only that the company had failed. Further, it was no defence for the directors and senior managers to argue that they had personally done all that was reasonably practicable.[1]

1 See *R v British Steel PLC* [1995] 1 WLR 1356.

1.311 Section 47 of the HSWA 1974 has been amended,[1] so that the standard of strict liability has been removed from certain health and safety regulations. This means that no civil claim may be brought for breach of statutory duty unless a regulation expressly provides for it; this effectively reverses the old position. This is part of the Coalition Government's attempts to liberate UK businesses from what they perceived as excessive over-regulation.

1 Enterprise and Regulatory Reform Act 2013, s 69.

1.312 However, as stated earlier, if the directors or senior managers had some involvement personally in the company's breach, they can be prosecuted individually as well, pursuant to s 37 of the HSWA 1974. In a work related incident, this is most likely to be on the basis that it was 'attributable' to the director's or senior manager's 'neglect'.

1.313 The meaning of 'attributable' in relation to neglect for the purposes of s 47 of the HSWA 1974 was considered in *Wotherspoon v HM Advocate*.[1] It was said that:

> '... any degree of attributability will suffice and in that sense it is evident that the commission of a relevant offence by a body corporate may well be found to be attributable to the failure on the part of each of a number of directors, managers or other officers to take certain steps which he could or should have taken in the discharge of the particular functions of his particular office.'

1 [1978] JC 74 at 78.

1.314 In the same case,[1] the meaning of 'neglect' was also considered. It was said that neglect:

> '... in its natural meaning pre-supposes the existence of some obligation or duty on the part of the person charged with neglect.'

1 *Wotherspoon v HM Advocate* [1978] JC 74 at 78.

1.315 Further, it was said:[1]

> '... in considering in a given case whether there has been neglect ... on the part of a particular director or other particular officer charged, the search must be to discover whether the accused has failed to take some steps to prevent the commission of an offence by the corporation to which he belongs if the taking of those steps either expressly falls within or should be held to fall within the scope of the functions of the office which he holds.'

1 *Wotherspoon v HM Advocate* [1978] JC 74 at 78.

1.316 In *R v Chargot*[1] the House of Lords said

> '... In *R v P Ltd* [2008] ICR 96 Latham LJ endorsed the Lord Justice General's observation that the question, in the end of the day, will always be whether the officer in question should have been put on inquiry so as to have taken steps to determine whether or not the appropriate safety procedures were in place. I would too. The fact that the penalties that may be imposed for a breach of this section have been increased does not require any alteration in this test. On the contrary, it emphasises the importance that is attached, in the public interest, to the performance of the duty that section 37 imposes on the officer.'

1 [2008] UKHL 73, [2009] 1 WLR 1, [2009] 2 All ER 645.

1.317 In relation to the definition of 'manager' this is likely be viewed as only those responsible for deciding corporate policy and strategy (see *R v Boal (Francis)*[1] which was a prosecution in relation to the Fire Precautions Act 1971).

See also *Michael Wilson v R*.[2]

1 [1992] 3 All ER 177, CA.
2 [2013] EWCA Crim 1780.

Health and safety duties cannot be delegated

1.318 Employers cannot delegate their health and safety duties or 'contract out' of them. When a certain activity has given rise to a risk or exposure to harm, the issue is whether that activity came within the defendant's undertaking, for example, its business.

1.319 In the case of *R v Associated Octel Co Ltd*,[1] which came before the House of Lords, a specialist independent contractor using its own workers was engaged by the defendant company to repair the lining of a tank on the defendants' chlorine plant. While carrying out the maintenance, an employee of the contractors was badly burnt when a bucket of highly inflammable acetone used for cleaning ignited.

1 [1996] 4 All ER 846.

1.320 The House of Lords said it was a matter of fact whether an activity came within the defendants' undertaking. In this case, their Lordships found that any reasonable jury would find, on the facts of this case, that the activities did amount to the conduct of a defendant's undertaking. The facts were: the tank was part of the defendants' chemical plant; the contractor's employees worked on a more or less permanent basis; the defendants authorised the work to be carried out by these employees, which enabled them to impose conditions upon the way the work was carried out; and the defendants provided them with the safety equipment.

1.321 It is extremely difficult to divorce your criminal responsibilities under the Act from that of your contractors. Although you may bring in specialist contractors, you still have to check their competency, and monitor, sample and audit their safety systems to verify that they are undertaking what their systems said they would do.[1] See also the case of *R v Atterby*.[2]

1 *Managing contractors: A guide for employers* (1997 ISBN: 9780717611966), available at www.hse.gov.uk.
2 (2009) unreported; discussed at **2.202**.

1.322 In October 2013 the Supreme Court in the case of *Woodland v Essex County Council*[1] extended the doctrine of non-delegation.

Miss Woodland, then aged 10, had a serious brain injury during a swimming lesson given by an independent contractor of the council.

The court said that, in situations such as a school, duties cannot be delegated if it is in an educational context. These included swimming lessons.

Although this case turns on the facts, it is an important development in the law and may herald a further extension of the doctrine.

1 [2013] UKSC 66, [2013] 3 WLR 1227.

Failures by employees which cause employer's breach of duty

1.323 In *R v Nelson Group (Maintenance) Ltd*[1] the defendant company installed, serviced and maintained gas appliances nationally. It appealed a conviction relating to a breach of HSWA 1974, s 3(1).

1 [1998] 4 All ER 331.

1.324 One of the company's fitters had removed a defective and dangerous gas fire in a private house. The trial judge directed the jury that, if it found the fitter had been negligent by not capping the gas pipe, the company was in breach of HSWA 1974, s 3 and therefore guilty. The jury found the fitter had been negligent in this way. The fitter was also prosecuted and convicted under HSWA 1974.

1.325 The Court of Appeal considered the distinction between the duties of employers and the duties of their employees. It said it was not necessary, for the adequate protection of the public, for the company to be held criminally liable as well as the employee. It was open to the company to show it had done everything reasonably practicable to avoid the risk. The court accepted that the company had done so, because there was a good system of training and instruction. The conviction against the company was quashed.

1.326 Regulation 21 of the MHSWR is designed to reverse the effect of this case. The regulation states that an employer will not have a defence to a health and safety prosecution because of any act or default by an employee. Thus, many lawyers and commentators thought that the defence in the *Nelson* case[1] should not succeed in the future. In the case of *R v HTM Ltd*,[2] Latham LJ, when reviewing relevant authorities in relation to foreseeability, said:

> '… But it seems to us that a defendant to a charge under section 2 or indeed section 3 or 4 , in asking the jury to consider whether it has established that it has done all that is reasonably practicable, cannot be prevented from adducing evidence as to the likelihood of the incidence of the relevant risk eventuating in support of its case that it had taken all reasonable means to eliminate it.'

The law therefore appears, at first blush, to have reverted back almost to the legal situation in *Nelson* in 1998.

1 *R v Nelson Group (Maintenance) Ltd* [1998] 4 All ER 331.
2 [2006] EWCA Crim 1156.

1.327 In *R v EGS Ltd*,[1] Lord Justice Dyson said (at para 27):

> 'In any event, it is strictly inapt to speak of a risk being foreseeable. A risk is a present potential danger the existence of which may or may not be appreciated: see per Steyn LJ in *R v Board of Trustees of the Science Museum* [1993] 1 WLR 1171, 1177F, approved in *Chargot* at [20]. If the risk eventuates and an accident occurs, then a question may arise in the context of a section 40 defence as to whether the accident was foreseeable or unforeseeable: see *R v HTM Ltd* [2006] EWCA Crim 1156. But it is not relevant to the issue of whether the prosecution has proved the existence of a material risk. It may be that the judge used the word "foreseeable" inaccurately and that he used it interchangeably with "would have been appreciated".'

1 [2009] EWCA Crim 1942.

1.328 Later on, Dyson LJ critically says:

> 'Causation is not an essential ingredient of the offence. The prosecution did not have to establish that EGS *caused* the accident, although in the present case, as in most, they did in fact rely on a causal connection between EGS's acts and omissions as going to establish risk. They merely had to prove that EGS exposed persons not in its employment to risks to their health or safety. Nor do we understand the reference to remoteness. It would appear that the judge was treating the prosecution as if it were a civil claim for damages for breach of statutory duty. But it is clear from *Chargot* that this is the wrong approach.'

1.329 Further, the court said:

> '… In our judgement, judges should not normally allow a defendant to raise a section 40 defence of which reasonable notice has not been given to the prosecution.'

1.330 This specifically means that defence teams will need to cite it (and the evidence) in their defence statements. The court also said, in relation to the prosecution, that:

> 'prosecution cases should not be overloaded with evidence which is adduced solely in order to meet a technical case of which no notice has been given. That is not to say that it may not sometimes be appropriate for the prosecution to lead evidence of breach of regulations, standards and codes of practice in support of its own case: see *Chargot* at [22] and [25].'

1.331 Finally, in relation to case management, the court suggested that, in cases of fatalities or very serious injury where there was not to be an early plea of guilty, a nominated trial judge should be appointed, and that cases going to trial:

> '... should be vigorously case managed by the nominated judge, who should be astute to ensure, in advance of the trial, that the parties confine the case to the issues that really matter and that the case does not become overloaded.'

1.332 In reality, this already happened in many court centres, but this direction will ensure that, in those courts that do not already have a formalised allocation system, one will be introduced.

1.333 These practice directions were both very timely and useful. They will force both defence and prosecution teams to increasingly front-load the case preparation and management at an early stage.

1.334 This being said, it will be difficult in practical terms to 'run' this argument in many cases in front of a jury, particularly in a fatality case.

PROSECUTIONS OF EMPLOYEES FOLLOWING WORK RELATED INCIDENTS[1]

1.335 As to when an employee breaches their duty under s 7(a) of the HSWA 1974 (ie to take reasonable care for the health and safety of him or herself and of other persons who may be affected by his or her acts or omissions at work) is a somewhat grey area of law. The reason for this is that the vast majority of cases are dealt with in the magistrates' courts. As a consequence, there are very few reported cases.

1 See **Chapter 2**.

1.336 There are two types of prosecution: those of employees that can be described as 'front-line' workers; and those of employees that are at management level (normally lower or middle management).

1.337 For a front-line employee to be prosecuted, his or her actions normally have to be bordering on the reckless. An example is the case of *R v Holland*,[1] where the defendant exposed a 17-year-old colleague to hydrochloric acid as a 'practical joke'.

1 (1997, unreported), Bridgnorth Magistrates' Court.

1.338 For a manager to be prosecuted, the negligence does not have to be so bad. A manager can be prosecuted if he or she has failed to carry out their job (or has violated procedures) so that health and safety standards are significantly lowered. An example is the case of *R v Helmrich*.[1] The defendant was the health and safety manager of a national restaurant chain. At one of its restaurants, a fatal accident occurred when an employee was electrocuted. The defendant was not prosecuted in relation to this incident, but was prosecuted on the basis that he failed to implement an adequate health and safety system on behalf of his employers.

1 (2003, unreported), Lincoln Magistrates' Court. See also *R v Atterby*, discussed at **2.202**.

CONCLUSION

1.339 There have been many changes since the last edition. The future chapters will attempt to review some of these in greater detail. The health and safety regime in the UK is currently being drastically overhauled. It is hoped that, rather like the Kerplunk game ball, these changes will do nothing to deflect from the steady fall in work related deaths that we have witnessed in the last few years.

1.340 There is still no major corporation in the prosecution gunsights in relation to corporate manslaughter, but increasing numbers of smaller organisations are being prosecuted. Directors and managers who do not have health and safety at the core of their businesses run the gauntlet of prosecution in the future. Increasingly, conviction will lead to lengthy terms of imprisonment.

Chapter 2

Sentencing

Gerard Forlin QC, Barrister, Cornerstone Barristers, London; Denman Chambers, Sydney; Maxwell Chambers, Singapore

Dr Louise Smail, Risk Consultant, Ortalan, Manchester

INTRODUCTION

2.1 This chapter gives an overview to this complicated area and provides some examples of how various courts have approached the difficult task of sentencing both organisations and individuals. It looks at the sentencing powers of the courts in health and safety prosecutions and manslaughter prosecutions and gives examples of past cases for guidance. It also sets out the sentencing powers contained in the corporate killing proposals. It does not attempt to deal with all aspects of this vast area.

HEALTH AND SAFETY OFFENCES – EMPLOYERS

Sentencing powers of the courts

2.2 The courts can only impose a fine upon an employer guilty of a breach of one of the general duties under the Health and Safety at Work etc Act 1974 (HSWA 1974) (ie the duties pursuant to HSWA 1974, ss 2–6) and related regulations. On 16 October 2008 the Health and Safety (Offences) Act 2008 received Royal Assent and came into force in January 2009.This changed the maximum penalties that can be imposed.

2.3 Since the last edition there have been several major developments in the sentencing arena.

2.4 In February 2010, in accordance with s 170(9) of the Criminal Justice Act 2003, the Sentencing Guidelines Council ('SGC') issued guidelines on sentencing corporate manslaughter and health and safety offences causing death. the guidelines apply to the sentencing of organisations on or after 15 February 2010, and do not cover individuals.

It is not intended to set out all 10 pages of the guidelines, but to pick out some of the most salient factors.

2.5 Paragraph 4(c) of the guidelines makes it clear that it is only related to cases '... where it is proved that the offence was a significant cause of death, not simply that death occurred'.

Section B deals with factors likely to affect seriousness, and para 5 states:

> '... this guideline applies only to corporate manslaughter and to those health and safety offences where the offence is shown to have been a significant cause of the death. By definition, the harm involved is very serious.'

2.6 Paragraph 6 states:

> 'Beyond that, the possible range of factors affecting the seriousness of the offence will be very wide indeed. Seriousness should ordinarily be assessed first by asking:
>
> (a) How foreseeable was serious injury? ...
> (b) How far short of the applicable standard did the defendant fall?
> (c) How common is this kind of breach in this organisation? ...
> (d) How far up the organisation does the breach go? ...'

2.7 Paragraph 7 lists a non-exhaustive series of aggravating features:

> '(a) more than one death, or very grave personal injury in addition to death;
> (b) failure to heed warnings or advice, whether from officials such as the Inspectorate, or by employees (especially health and safety representatives) or other persons, or to respond appropriately to 'near misses' arising in similar circumstances;
> (c) cost-cutting at the expense of safety;
> (d) deliberate failure to obtain or comply with relevant licences, at least where the process of licensing involves some degree of control, assessment or observation by independent authorities with a health and safety responsibility;
> (e) injury to vulnerable persons.
>
> In this context, vulnerable persons would include those whose personal circumstances make them susceptible to exploitation.'

2.8 Paragraph 8 provides:

> 'Conversely, the following factors, which are similarly non-exhaustive, are likely, if present, to afford mitigation:
>
> (a) a prompt acceptance of responsibility;
> (b) a high level of co-operation with the investigation, beyond that which will always be expected;
> (c) genuine efforts to remedy the defect;
> (d) a good health and safety record;
> (e) a responsible attitude to health and safety, such as the commissioning of expert advice or the consultation of employees or others affected by the organisation's activities.'

2.9 Section C of the guidelines deals with financial information, and size and nature of the organisation in relation to sentence. Paragraph 12 states:

> '... the law must expect the same standard of behaviour from a large and a small organisation. Smallness does not by itself mitigate, and largeness does not by itself aggravate, these offences. Size may affect the approach to safety, whether because a small organisation is careless or because a large

one is bureaucratic, but these considerations affect the seriousness of the offence via the assessment set out in paragraphs 6–8 above, rather than demonstrating a direct correlation between size and culpability.'

2.10 At para 15:

'A fixed correlation between the fine and either turnover or profit is not appropriate. The circumstances of defendant organisations and the financial consequences of the fine will vary too much; similar offences committed by companies structured in differing ways ought not to attract fines which are vastly different; a fixed correlation might provide a perverse incentive to manipulation of corporate structure.'

2.11 According to para 19:

'In assessing the financial consequences of a fine, the court should consider (inter alia) the following factors:

(i) the effect on the employment of the innocent may be relevant;

(ii) any effect upon shareholders will, however, not normally be relevant, those who invest in and finance a company take the risk that its management will result in financial loss;

(iii) the effect on directors will not, likewise, normally be relevant;

(iv) nor would it ordinarily be relevant that the prices charged by the defendant might in consequence bet raised, at least unless the defendant is a monopoly supplier of public services;

(v) the effect upon the provision of services to the public will be relevant, although a public organisation such as a local authority, hospital trust or police force must be treated the same as a commercial company where the standards of behaviour to be expected are concerned and must suffer a punitive fine for breach of them, a different approach to determining the level of fine may well be justified; "the Judge has to consider how any financial penalty will be paid. If a very substantial financial penalty will inhibit the proper performance by s statutory body of the public function that it has been set up to perform, that is not something to be disregarded." [*Milford Haven Port Authority* [2000] 2 Cr App R(S) 423 per Lord Bingham CJ at 433–4] The same considerations will be likely to apply to non-statutory bodies on charities if providing public services.

(vi) the liability to pay civil compensation will ordinarily not be relevant; normally this will be provided by insurance or the resources of the defendant will be large enough to meet it from its own resources (for compensation generally see paras 27–28 below);

(vii) the cost of meeting any remedial order will not ordinarily be relevant except to the overall financial position of the defendant; such an order requires no more than should already have been done;

(viii) whether the fine will have the effect of putting the defendant out of business will be relevant; in some bad cases this may be an acceptable consequence.'

2.12 Section D of the guidelines discusses the level of fines, and para 24 states:

'The offence of corporate manslaughter, because it requires gross breach at a senior level, will ordinarily involve a level of seriousness significantly greater than a health and safety offence. The appropriate fine will seldom be less than £500,000 and may be measured in millions of pounds.'

2.13 Paragraph 25 goes on:

> 'The range of seriousness involved in health and safety offences is greater than for corporate manslaughter. However, where the offence is shown to have caused death, the appropriate fine will seldom be less than £100,000 and may be measured in hundreds of thousands of pounds or more.'

2.14 Paragraph 26 states that, 'A plea of guilty should be recognised by the appropriate reduction'.

The guidelines also deal with compensation at para 27, costs at para 29, publicity orders at para 30 and remedial orders at para 33.

2.15 In section I, para 37 helpfully sets out the normal approach to sentence:

> '(1) consider the questions at paragraph 6;
> (2) identify any particular aggravating or mitigating circumstances (paragraphs 7–11);
> (3) consider the nature, financial organisation and resources of the defendant (paragraphs 12–18);
> (4) consider the consequences of a fine (paragraphs 19–21);
> (5) consider compensation (but see paragraphs 27–28);
> (6) assess the fine in the light of the foregoing and all the circumstances of the case;
> (7) reduce as appropriate for any plea of guilty;
> (8) consider costs
> (9) consider publicity order;
> (10) consider remedial order.'

2.16 The recent case of *R v Merlin Attractions Operations Limited*[1] was discussed in **Chapter 1**. In that case, the Court of Appeal reviewed the guidelines set out above in a most concise and helpful way, and said (at para 60):

> '… The judge was clearly entitled to conclude, as he did, that there had been a serious breach of the appellant's health and safety systems in relation to the Bear and Clarence Bridge and that that had resulted in an obvious danger of at least serious injury (not so much in relation to adults, but in relation to children) to which a very large number of people had been exposed over many years, and which had been the product, in part, of a failure to react appropriately to advice in 2003. There was no question of any discount for a plea. Hence, notwithstanding the appellant's generally good health and safety systems, its good record and the other mitigating features identified in paragraphs 54 and 56 above, but given the necessary inference as to its wealth, it seems to us that the judge was entitled to conclude that the total fine (although not required to be in the realms of that imposed for corporate manslaughter) had to be measured in hundreds of thousands of pounds, and that the total actually imposed of £350,015 was within the appropriate range (albeit towards the top end of it) for these particular offences committed by this particular appellant – especially when all the purposes of sentence are borne in mind.'

1 [2012] EWCA Crim 2670.

2.17 Further, on 17 January 2014 the Court of Appeal (Criminal Division) handed down the judgment in *R v Sellafield Limited; R v Network Rail Infrastructure Limited* [2014] EWCA Crim 49. This is an important judgment and needs to be read in entirety; it is not the intention to set out all the various segments of this lengthy judgment but a number of extracts appear below. In essence, however, the court said:

> '1. These two appeals are being heard together as they raise issues of principle in relation to the level of fines to be imposed for breaches of safety and environmental protection legislation on very large companies – Sellafield Limited (Sellafield Ltd) with a turnover of £1.6bn and Network Rail Infrastructure Ltd (Network Rail) with a turnover of £6.2bn.
>
> (i) Sellafield Limited was fined £700,000 at the Crown Court at Carlisle on 7 February 2013 for offences arising out of the disposal of radioactive waste.
>
> (ii) Network Rail was fined £500,000 in the Crown Court at Ipswich on 27 June 2013 for an offence arising out of a collision at an unmanned level crossing, causing very serious injuries to a child.'

Both appeals were dismissed. The court continued:

> '5. Where a fine is to be imposed a court will therefore first consider the seriousness of the offence and then the financial circumstances of the offender. The fact that the defendant to a criminal charge is a company with a turnover in excess of £1 billion makes no difference to that basic approach.
>
> 6. The fine must be fixed to meet the statutory purposes with the objective of ensuring that the message is brought home to the directors and members of the company (usually the shareholders). The importance of the application of s 164 in relation to corporate defendants was reinforced in the Definitive Guideline of the Sentencing Guidelines Council *Corporate Manslaughter & Health and Safety offences Causing Death*, published in 2010. It has been reflected in more recent decisions of this court: see for example: *R v Tufnells Park Express Ltd* [2012] EWCA Crim 222 at para 43 (the fine after trial on a company with a turnover of £100m and profitability of £7.7m was £225,000; this represented, as the court noted, 2.8% of its operating profit).
>
> …
>
> **1. THE SERIOUSNESS OF THE OFFENDING OF SELLAFIELD LTD**
>
> **(1) The stringent standards of safety imposed by the legislative regime**
>
> 8. The processing and storage of nuclear waste is a by-product of an activity of national economic importance: the generation of electricity by nuclear power.
>
> 9. It carries with it potentially grave risks. To mitigate those risks the most stringent standards have been adopted at national and international levels. In the United Kingdom they have been laid down in licences granted under the Radioactive Substances Acts 1948, 1960 and 1993 and, more recently, by the Environmental Permitting Regulations 2010 …
>
> **(2) The offences committed**
>
> 11. Between 15 November 2008 and 19 April 2010 Sellafield Ltd breached those standards in a variety of ways in relation to the system for segregating non-radioactive waste from radioactive waste and for disposing of it.

12. A fine of £700,000 was imposed on 14 June 2013 by HH Judge Peter Hughes QC at Carlisle Crown Court, being made up of a fine of £100,000 for each of the seven offences. Five contraventions of different statutory requirements were set out in seven charges. Two pairs of charges covered the same factual allegations, but were separately laid because of change in the statutory regime which took effect on 5 April 2010. Sellafield Ltd:

...

v) On and before 12 April 2010 caused or permitted non-exempt radioactive waste to be carried in a manner which did not comply with the European Agreement for the International Carriage of Dangerous Goods by Road contrary to Regulation 5 of the Carriage of Dangerous Goods and Use of Transportable Pressure Equipment Regulations 2009.

(3) The processing of exempt waste and the failures by Sellafield Ltd

(i) Exempt waste

14. Sellafield Ltd generates substantial quantities of waste from its activities at its extensive site at Sellafield in Cumbria. Much of it is not radioactive. Such waste may be disposed of as non-toxic waste by ordinary means of waste disposal, principally in landfill sites. This is known as "exempt waste".

...

(iii) The extent of the failure in the system

19. The system was installed in May 2009 and brought into full operation in November 2009. The first batch of exempt waste was dispatched from the site on 12 April 2010 to a nearby landfill site. Four thousand further bags of supposedly exempt waste passed through the monitors with an apparent dosage of zero microsieverts per hour. In fact a small number of bags – five out of five thousand – had dosage levels in excess of 20 microsieverts per hour. The error in the setup of the devices was discovered by chance on 19 April 2010 when, during a training exercise, a bag of waste with a dosage of 41 microsieverts per hour was passed through them and wrongly classified as exempt waste.

20. The error was discovered when the operator checked the assay produced by the Exempt Waste Monitor and saw the dosage level there recorded. This led to the taking of immediate steps to identify the problem and its extent and to alert the relevant authorities and the operators of the landfill site to which the first batch of one thousand bags had been dispatched.

21. Once alerted to the problem, Sellafield Ltd did all they could to ensure that no harm came to anyone. It is as near certain as can be that none has. Five bags were identified from the assay report with dosages of between 23 and 32 microsieverts per hour. Four had been dispatched to the landfill site and one retained on site. Measurements taken at the landfill site vastly exceeded the low level permitted to be deposited at the site – 0.4 bequerels per gram (a measure of the radiation emitted by an object). Three of the bags were within the lowest category of toxic waste – low level radioactive waste – and one was within the next category – intermediate level radioactive waste.

22. There was expert evidence that exposure to the four bags by anyone who might have handled them would have been no greater than a passenger

would experience on a flight to Paris. There was agreed expert evidence that had the problem not been discovered for a long time – years – and non-exempt waste of this category had been regularly handled, there would have been a very small but perceptible increase in the risk of death from cancer to those handling it.

…

(*v*) *Culpability and harm*

24. The judge summarised the significance of the failures:

> "The mistake that was made in this case was a fundamental one in the setting up, the testing and the subsequent monitoring of the equipment. That such a basic mistake could possibly occur in what needs to be an industry managed and operated with scrupulous care for public safety and the environment is bound to be a matter of grave concern to everyone and particularly local residents in Cumbria. What adds significantly to the concern and the seriousness of the mistake is that it had been in existence and allowed to go undetected for the period of 4 months or so that the system had been in use."

25. The prosecution alleged, Sellafield Ltd accepted, and the judge found, that these failings

> "indicate basic management failures and a deeply concerning lack of procedures formally established and rigorously enforced to ensure that equipment was properly set up at the outset and regularly and routinely checked."

26. This led the judge to conclude that

> "The management failures are not confined to specific individuals or failures at certain levels to follow established procedures. They demonstrate … **a custom within the company which was too lax and** … **to a degree complacent** and senior management must bear its share of responsibility." (Our emphasis).

27. He identified three aggravating features:

i) The failure was not isolated but systemic.

ii) It potentially exposed those who handled waste off-site and the public to unnecessary risk.

iii) It was not a first offence. A prohibition notice had been served on 28 June 2008, one year before installation of the new monitors, by the Department of Transport for breach of Regulation 5 of the 2009 Regulations. Sellafield Ltd had been fined twice for incidents involving the emission of radioactive material in 2005 and 2007 – £500,000 and £75,000 respectively.

28. He also took into account the mitigating factors already mentioned. The breaches were not deliberate or reckless; no harm had been done and the actual risk of harm was relatively low; Sellafield Ltd had readily co-operated with the authorities and had pleaded guilty at the first opportunity.

…

31. The judge found that there had been a custom in the company Sellafield Ltd which was too lax and complacent; that senior management must bear a share of the responsibility. We can see no basis for criticising that finding. The failure was easily avoidable and could and should have been

detected very quickly; there was the clearest negligence. We therefore conclude that for an incident of this kind the culpability was medium.

2. THE SERIOUSNESS OF THE OFFENDING OF NETWORK RAIL

(1) The accident

...

34. On 3 July 2010 at about 8.30 in the evening, Mr Wright drove his 4x4 car to the crossing. He was accompanied by his 10 year old grandson and his dog. Mr Wright stopped at the crossing, opened the first gate looked both ways, crossed on foot and opened the second gate. He looked both ways again and then re-crossed the track back to his car. He told his grandson to get out of the driver's seat and back into the rear seats. His grandson did so after about 15–30 seconds. Mr Wright got into the car and drove towards the crossing.

...

36. The victim impact statement in respect of the grandson sets out the devastating effect that the accident had had. The hospital had given him a 5% chance of survival. He had not had a normal childhood. He would need teaching support, a scribe and a reader throughout his education; his brain's processing speed was 50% of normal. He had right side blindness, his peripheral vision as damaged and he had to avoid all contact sports. He had a titanium plate in his head. His future prognosis would not be known until he was 21 or 22.

(2) The proceedings

...

(ii) The significant failures of Network Rail

...

40. Prior to the accident there were risk assessments on 15 January 2000, 24 September 2003, 4 March 2007 and 12 April 2009. Those in 2007 and 2009 were carried out by Mobile Operations Managers for Network Rail. There was a maintenance inspection on 28 January 2010.

41. Elementary mistakes were made in the assessment. It should have been obvious to those conducting the risk inspections and to those more senior persons within Network Rail responsible for level crossings that, because of the nature of the use of the crossing and the sight lines, a telephone should have been installed so that anyone crossing the line could ring up and find out if a train was on its way. The sight lines were such that, given the time it would take to cross and the speed of the trains on the line, there was an obvious and serious risk of a collision. It is also of particular importance that in the risk assessment carried out in 2003 the assessor concluded that the crossing was not safe, but no steps were taken to remedy it over the following 6 years, particularly by installing a telephone. As we have noted, after the accident, a telephone was installed at Wright's Crossing.

42. The judge found that there was that obvious risk and it was readily reducible. He also found that the risk assessments were poorly done; there were repeated failures to follow the correct guidance. In 2007, Network Rail had installed a computer system; the risk assessments in 2007 and 2009 were inputted into it, but the programme used did not sport the inconsistencies.

Network Rail were unable to explain this failure. We consider that these findings were amply justified on the evidence.

…

(4) Our conclusion on harm and culpability

50. The actual harm caused in the present case was, as is evident from the facts we have set out, serious; much greater harm was foreseeable.

51. As to the level of culpability of Network Rail, there was no evidence of specific senior management failures. The failures, serious and persistent though they were, were at lower operational levels.

3. THE LEVEL OF FINE TO BE IMPOSED IN RESPECT OF THE SERIOUSNESS OF THE OFFENCES

…

(1) The provision of financial and corporate information to the sentencing court

53. To enable us to consider the financial circumstances of the offenders, we called for the accounts of both companies prior to the hearing of the appeal.

54. It is important that well in advance of the sentencing hearing there is provided to the sentencing court the accounts of the offending companies and any other information (including information about the corporate structure) necessary to enable the court to assess the financial circumstances of the company and the most efficacious way of giving effect to the purposes of sentencing. The provision of that information to the court and to counsel for the Crown will enable counsel for the Crown to present the information to the sentencing court so that it can carry out the analysis required.

…

(4) Our conclusion

(*a*) *Sellafield Ltd*

…

63. It is not appropriate, as was submitted on behalf of Sellafield Ltd, to consider a fine of £1 million as apposite only to a major disaster. To accept that submission would be to ignore the court's obligation under s 164 of the CJA 2003 to have regard to the financial circumstances of the offender and the approach made clear in the Sentencing Guidelines Council Guideline to which we have referred at paragraph 6 above. There is no ceiling on the amount of a fine that can be imposed.

64. In considering those financial circumstances, we have regard to its turnover of £1.6 billion (or £30.7 million per week) and its annual profit of £29m (or £560,000 per week). It is clear that viewed in the light of the financial circumstances of this company, a fine of £700,000 after a guilty plea is a fine which reflects a case where the culpability was moderate, the actual harm in effect nil and the risk of harm very low. It must be viewed against the requirement that those engaged either as directors or shareholders of companies engaged in the nuclear industry must give the highest priority to safety as Parliament has directed.

65. A fine of the size imposed, even though only a little more than a week's profit and about 2% of its weekly income, would, in our view, in the

circumstances achieve the statutory purposes of sentencing by bringing home to the directors of Sellafield Ltd and its professional shareholders the seriousness of the offences committed and provide a real incentive to the directors and shareholders to remedy the failures which the judge found existed at the site at Sellafield as we have set out at paragraphs 24–27, particularly the custom within the company which was too lax and to a degree complacent. If it does not have that effect, then, as in the case of any other offender, the sentence of a court for any further culpable failure would have to reflect that a fine of the size imposed for the current offences had not achieved some of the statutory purposes of sentencing.

66. We therefore see no grounds for in any way criticising the level of fine imposed by the judge. We dismiss the appeal.

(*b*) *Network Rail*

…

68. We reject the submission made on behalf of Network Rail that a fine of £750,000 was appropriate only where there had been a fatality. As we have explained in respect of Sellafield Ltd that would be to ignore the statutory obligation to consider the means of the offender – in the case of Network Rail a turnover of £6.2 billion (or £119m per week) and annual profitability of £750m (or £14.4m per week).

69. However, a significant fine imposed on Network Rail would, unlike the case of Sellafield, in effect inflict no direct punishment on anyone; indeed it might be said to harm the public. That is because the company's profits are invested in the rail infrastructure for the public benefit; the profits make an addition to the state funds that are otherwise provided to meet the requirements of the provision of that infrastructure. It is likely that any shortfall in the requirements as a result of a fine will have to be met from public funds or in a reduction in the investment. That is a factor which a court must take into account: see *R v Milford Haven Port Authority* [2002] 2 Cr App R 423; *R v Network Rail* [2011] Cr App R (S) 44, [2010] EWCA Crim 1225 at para 24.'

Importantly, the court added:

'70. …

iii) We have set out in paragraph 60 the statement by the non-executive directors as to how they will reduce bonuses if safety performance is poor; it appears to concentrate on a catastrophic accident and not accidents of the kind with which this appeal is concerned which has had such a devastating effect on a child. There was evidence before us (but not the judge) that the bonuses of the directors had been adjusted downwards to a minor (though inadequate) extent in part because of the poor level crossing safety record to which we have referred. Plainly the bonuses should have been very significantly reduced. For the future, it will be important for Network Rail to ensure that full information is provided to the sentencing court, as it is highly material to the assessment of the response of the board of the company to the statutory purposes of sentencing in a case where a fine inflicts no direct punishment on anyone. If, as is accepted by the board of Network Rail, a bonus incentivises an executive director to perform better, the prospect of a significant reduction of a bonus will incentivise the executive directors on the board of companies such as

Network Rail to pay the highest attention to protecting the lives of those who are at real risk from its activities. In short, it will demonstrate to the court the company's efforts, at the level of those ultimately responsible, to address its offending behaviour, to reform and rehabilitate itself and to protect the public.

...

72. Nonetheless, the fine of £500,000 imposed on a company of the size of Network Rail can only be viewed as representing a very generous discount for the mitigation advanced; we would observe that if the judge had imposed a materially greater fine, there would have been no basis for criticism of that fine. Indeed, were it not for the matters to which we have referred, a fine of the size imposed would have been very significantly below that which should be imposed for an offence committed by a company of this size where the harm was relatively serious and the culpability at local operational management was serious and persistent.'

This judgment is important as it indicates that fines and sentences should be increased even in non-fatality cases. It also reiterates the fact that individual directors should have performance based bonuses reduced in certain instances. As the court said at para 70 (quoted above):

'... the prospect of a significant reduction of a bonus will incentivise the executive directors on the board of companies such as Network Rail to pay the highest attention to protecting the lives of those who are at real risk from its activities.'

Guidance on imposing fines before the Sentencing Guidelines

2.18 The leading case on the levels of fines to be imposed upon employers for breaches of health and safety law remains *R v F Howe & Son (Engineers) Ltd.*[1] The Court of Appeal in this case arguably responded to the public's disquiet that the level of fine for health and safety offences was too low. See also *R v Balfour Beatty Rail Infrastructure Services Ltd*,[2] which indicates further guidance on sentencing large organisations, *Merlin Attractions Ltd*,[3] *R v TDG (UK) Ltd*[4] and *R v John Pointon & Sons Ltd.*[5]

The Court of Appeal acknowledged there was increasing recognition of the seriousness of such offences, and that judges and magistrates who rarely deal with these cases had found it difficult to know how to approach sentencing. It therefore took the opportunity to set out some sentencing guidelines, while emphasising that each case would turn on its own particular circumstances. The Court of Appeal said it was not possible to set any tariff or to set any specific relationship between the fine and the employer's turnover or profits. Increasingly, sentencing judges are remarking that further guidance would be helpful.

1 [1999] 2 All ER 249.
2 [2007] Bus LR 77, [2007] 1 Cr App R (S) 65, [2007] ICR 354.
3 [2012] EWCA Crim 2670.
4 [2009] ICR 127.
5 [2008] 2 Cr App R (S) 22.

2.19 Mr Justice Scott Baker (as he then was) said the law requires employers to do 'what good management and common sense requires them to do anyway, ie look at what the risks are and take sensible measures to tackle them'. He pointed out that failure to fulfil these general duties under the HSWA 1974 is particularly serious, as they are the 'foundations for protecting health and safety'. He went on to say:[1]

> 'The objective for health and safety offences in the workplace is to achieve a safe environment for those who work there and for other members of the public who may be affected. A fine need to be large enough to bring that message home where the defendant is a company not only to those who manage it but also to its shareholders.'

1 *R v F Howe & Son (Engineers) Ltd* [1999] 2 All ER 249 at 255.

2.20 Although it was accepted, in general, that a fine should not be so large as to imperil the earnings of employees or create a risk of bankruptcy, there may be cases where the offences are so serious that the defendant ought not to be in business.

2.21 Mr Justice Scott Baker also made it clear that the standard of care imposed by the health and safety legislation is the same, regardless of the size of the defendant company.

2.22 The Court of Appeal said the following factors should be taken into account.

Turnover

2.23 It is important to recall that a sentencing judge should consider only the pre-tax profits rather than the gross turnover when assessing the appropriate fine.[1] See also paras 64 and 65 of the *Sellafield* judgment (at **2.17** above).

1 *R v ESB Hotels* [2005] EWCA Crim 132 at para 39.

General factors

2.24

(1) The gravity of the offence – how far short of the appropriate standard the defendant fell in failing to take reasonably practicable steps to ensure health and safety.
(2) The degree of risk and the extent of the danger created by the offence.
(3) The extent of the breach or breaches, for example, whether it was an isolated incident or continued over a period.
(4) An important factor is the defendant's resources and the effect of the fine on its business.

Aggravating features

2.25

(1) A failure to heed warnings.
(2) Deliberately profiting financially from failure to take necessary health and safety steps or specifically running the risk to save money. It was said: 'A deliberate breach of health and safety legislation with a view to profit seriously aggravates the offence.'
(3) The breach has resulted in death. The penalty 'should reflect the public disquiet at the unnecessary loss of life'.

Mitigating factors

2.26

(1) Prompt admission of responsibility and a timely plea of guilty.

(2) Steps to remedy deficiencies after they are drawn to the defendant's attention.

(3) A good safety record.

2.27 The Court of Appeal said that any fine should reflect not only the gravity of the offence but also the means of the offender, and that this applies just as much as to corporate defendants as any other. It went on to say:[1]

> 'Difficulty is sometime found in obtaining timely and accurate information about a corporate defendant's means. The starting point is its annual accounts. If a defendant company wishes to make a submission to the court about its ability to pay a fine it should supply copies of its accounts and any other financial information on which it intends to rely in good time before the hearing both to the court and to the prosecution.
>
> This will give the prosecution the opportunity to assist the court should the court wish it. Usually accounts need to be considered with some care to avoid reaching a superficial and perhaps erroneous conclusion. Where accounts or other financial information are deliberately not supplied the court will be entitled to conclude that the company is in a position to pay any financial penalty it is minded to impose. Where the relevant information is provided late it may be desirable for sentence to be adjourned, if necessary at the defendant's expense, so as to avoid the risk of the court taking what it is told at face value and imposing an inadequate penalty.'

1 *R v F Howe & Son (Engineers) Ltd* [1999] 2 All ER 249 at 254.

2.28 It should be noted, as was pointed out by counsel for the defendant in this case, that neither the fines nor the costs imposed upon an employer are deductible against tax, and therefore the full burden falls upon the company.

2.29 The Court of Appeal also stated that, in its judgement, magistrates should always think carefully before accepting jurisdiction in health and safety at work cases where it is arguable that the fine may exceed the limit of their jurisdiction or where death or serious injury has resulted from the offence. The trend now appears for more cases to be heard in the Crown Court.

2.30 In the case of *R v Friskies Petcare Ltd*[1] the Court of Appeal recommended that, where there is a plea of guilty, the prosecution and defence should set out in advance the aggravating and mitigating features in the case. In practice, the prosecution is required to serve a schedule setting out the aggravating and mitigating features of the case for agreement. This has become known as a 'Friskies schedule'. The Court of Appeal said in this case that, if it is possible to place an agreed basis of plea before the court, that should be done. If there is a 'disagreement of substance', the presiding judge can determine whether a *Newton* hearing is required.[2]

1 [2000] Cr App R (S) 401.

2 A '*Newton* hearing' is where the defendant admits his guilt but does not accept the facts presented by the prosecution as the basis of his guilt: *R v Newton* [1983] Crim LR 198. Evidence is then called at the hearing in order to determine the facts. See also *R v Underwood* [2004] EWCA Crim 2256, where the Court of Appeal gave further direction if a *Newton* situation arose. This is now the leading case. See also *R v Martin* [2013] EWCA Crim 2565 and *Nichols v DPP* [2013] EWHC 4365 (Admin).

2.31 In this case[1] the Court of Appeal also commented that fines in excess of £500,000 tend to be reserved for cases of major public disasters. This led to defence arguments in subsequent cases that effectively there should be a ceiling on fines of £500,000 for incidents that were not major public disasters.

1 *R v Friskies Petcare Ltd* [2000] Cr App R (S) 401.

2.32 However, the Court of Appeal in the case of *R v Colthrop Board Mill* made it clear that this was not the position:[1]

> 'It appears from the authorities that financial penalties of up to around half a million pounds are appropriate for case which result in the death even of a single employee, and perhaps of the serious injury of such a single employee. We would not wish the sum of £500,000 to appear to be set in stone or to provide and sort of maximum limit for such cases. On the contrary, we anticipate that as time goes on and awareness of the importance of safety increases, that courts will uphold sums of that amount and even in excess of them in serious cases, whether or not they involve what could be described as major public disasters.'

1 [2002] EWCA Crim 520.

2.33 In *R v Rollco Screw and Rivet Co*[1] the Court of Appeal said the question was not only the level of penalty merited by the offence, but also the level that the defendants could reasonably be expected to meet. In relation to the latter, it was relevant to consider the issue of time over which the penalty should be payable. Judges are increasingly willing to allow time to pay fines. For instance, in September 2009 a Judge at Maidstone Crown Court gave two directors three and a quarter years to pay the £150,000 fine for the death of a worker.[2]

1 [1999] IRLR 439.
2 *R v North Kent Shotblasting Limited* (unreported).

2.34 In criminal proceedings against an employer following an accident at work, it is sufficient for the prosecution to prove merely a risk of injury arising from the work, without identifying and proving specific breaches of duty by the employer. Once that is done, a prima facie case of breach is established. The onus is then passed to the employer to make good the defence of reasonable practicability.

2.35 The House of Lords held this, dismissing an appeal by the defendants, Chargot Ltd (trading as Contract Services), Ruttle Contracting Ltd and George Henry Ruttle, from the dismissal by the Court of Appeal Criminal Division[1] of their appeals against conviction on 19 November 2006, in Preston Crown Court (Judge Russell QC and a jury), of contravening the HSWA 1974.

The first defendant was fined £75,000 and ordered to pay £37,500 costs; the second was fined £100,000 with £75,000 costs; the third was fined £75,000 with £103,000 costs.[2]

1 (Latham LJ, Gibbs and Lloyd Jones JJ) [2008] ICR 517.
2 *R v Chargot* [2008] UKHL 73, [2009] 1 WLR 1, [2009] 2 All ER 645.

2.36 In *R v Goodyear*,[1] the Lord Chief Justice sitting in the Court of Appeal Criminal Division issued guidelines that amount to the introduction of a formalised procedure of advance sentence indication. The principles of the judgment modify the rule of practice adopted by courts following the decision in *R v Turner*.[2] The most important case in health and safety terms is *Transco v*

R,[3] where the Lord Chief Justice gave some further guidelines. The fine was reduced from £1 million to £250,000. See also *R v Martin*.[4]

1 [2005] EWCA Crim 888.
2 [1970] 2 QB 321.
3 [2006] EWCA Crim 838.
4 [2013] EWCA Crim 2565.

Some recent cases

2.37 In 2011, Network Rail were fined £3 million and £150,000 costs following an Office of Rail Regulation prosecution arising out of the Potters Bar derailment in 2002 in which seven people died. Jarvis had gone into liquidation. In 2012, Network Rail were also fined £4 million and costs in relation to the Grayrigg crash caused by degraded points. One person died and 88 people were injured when the stretcher bars failed, causing the train to come off the tracks. The Rail Investigations branch report said that the 'immediate cause of the crash had been poor maintenance'.

2.38 In 2013, Network Rail were fined £500,000 plus £23,500 costs after a 10-year-old boy received serious injuries when the car he was in was hit by a train on a level crossing in Beaches, Suffolk in 2010. The ORR report said the crash was caused by poor visibility when people were crossing from the south side of the crossing. They were also fined £125,000 and £80,000 costs after a mobile elevated platform caused three workers to be injured. The judgment has just been handed down (see **2.17** above).

2.39 In 2012, Network Rail were fined £1 million plus costs of £60,000 after two teenage girls died after being hit by a train at Elsenham station in December 2005.

2.40 Shell UK were fined £1 million and had costs of £242,000 imposed on them in 2011 after a huge explosion and fire took place at their gas terminal in Bacton in Norfolk. No-one was killed or injured. Shell pleaded guilty to seven charges. According to the HSE:

> '… Investigations traced the cause of the explosion to a leak of highly flammable hydrocarbon liquid into a part of the plant responsible for treating waste water before discharging it into the sea. The leak was caused by the failure of a corroded metal separator vessel, which allowed water contaminated with the highly flammable condensate to enter a concrete storage tank where it was heated by an electric heater. The heater's elements were exposed, raising the surface temperature significantly causing the fire.'

2.41 In 2012, TATA Steel PLC were fined £300,000 and costs of £57,487 after an experienced worker became disoriented by a steam emission that blocked his view. He fell into molten slag that was heated to 1,500 degrees centigrade. He suffered 85 per cent burns to his body and died the next day. There had been 16 'near miss' reports filed in the previous nine months, and a further 19 reports specifying steam as a safety concern. Mr Justice Spencer said, when sentencing, that 'The warning signs were there for management to see and they did nothing about them'.

2.42 In 2012, AMEC were fined £300,000 and costs of £333,866 after a contested trial. A worker died after becoming entangled in chains and was dragged over railings, falling seven storeys to his death on a construction site.

2.43 In 2012, Svitzer Marine were fined £1.7 million after a tugboat capsized in December 2007 and three crew members died. The accident happened when the boat was towing. This was a Scottish case. No prosecution costs are awarded in Scotland.

2.44 Also in 2012, Nestlé was fined £180,000, and ordered to pay £41,826.33 in costs, after an employee was crushed in a conveyor-type machine.

2.45 In 2011, Scottish Sea Farms were fined £600,000 in Oban after two contractors died when fixing a hydraulic crane on a barge. They passed out due to very low oxygen levels when they went into the barge. Another worker also passed out. The employer was fined £40,000. There are no costs awarded in Scotland.

2.46 In 2013, Sellafield Ltd were fined £700,000 plus £72,635 costs after they sent bags of radioactive waste to a landfill site in Workington. This was a breach of the company's environmental permit and other environmental breaches. It was a joint Environmental Agency and Office for Nuclear Regulation prosecution.

In November 2013 the Court of Appeal heard an appeal against sentence in this case, and the judgment has just been handed down (see **2.17** above).[1] See also *R v Southern Water Ltd* (23 January 2014) approving *R v Sellafield* in relation to environmental fines. This appeal also was heard by the Lord Chief Justice.

1 This case is being heard together with the Network Rail case (see **2.38** above).

2.47 In 2013, Larkin Logistics were fined £350,000 plus £23,317 costs after pleading guilty to two offences under the HSWA 1974 when a lorry's brakes failed when a hitching process was occurring and the worker was killed.

2.48 In 2013, European Metal Recycling Limited, a global company, was fined £300,000 and £72,901 costs after pleading guilty to offences under ss 2 and 3 of the HSWA 1974. A temporary worker was killed after he was struck by a bucket of a loading shovel at the company's Willesden site during a shutdown in 2010.

2.49 In 2013, Adis Scaffolding was fined £300,000 and £124,468 costs. The company pleaded guilty after a skip lorry overturned and crushed the driver. The skip's locks had failed to engage on a catch bar. The HSE said:

> '… the failings by Adis Scaffolding Limited were substantial, ranging from unsuitable equipment, an inadequate risk assessment, inadequate training and instruction, and an absence of safe systems of work.'

2.50 Atec Ltd was fined £300,000 and costs of £77,500 after pleading guilty to an offence under s 2 of the HSWA 1974. An employee was killed when the main isolation valve of a casting furnace suddenly closed, trapping his head. The court found that there had been a problem with the company's isolation procedures for many years. Judge Bayliss QC said:

> '… At the time of the accident there was no robust system in place to ensure safety during maintenance. It was left to the discretion of the fitters. It was, I'm sorry to say, a shambles.'

2.51 In 2013, St George's South London Ltd (an agent for St George's, the property developers) and a sign manufacturer were fined £300,000 and £222,692 in costs after a contested trial. AE Tyler were fined £60,000 and £22,855 costs. In essence, a large sign, that had been up for nine years and decaying, hit a pedestrian whilst she walked along the pavement at Vauxhall Bridge. She was very badly injured after a blow to the head and has not been able to work ever since. The sign had been blown over by a strong gust of wind.

2.52 In 2013, the drinks manufacturer, Allied Domecq Ltd, were fined £266,667 and £10,752 costs after a contractor fell 20 feet through a fragile skylight at the company's bonded warehouse in Dover. There was no safety apparatus used. The HSE said:

> 'Allied Domecq do not just contract out their health and safety responsibilities just by contracting out a particular job. It was their duty to ensure there was a safe system of work before the job started and that their contractors followed agreed safe procedures. They failed to do so.'

2.53 BAE Systems (Operations) Ltd were fined £250,000 and £97,153 costs after pleading guilty to an offence under s 2(1) of the HSWA 1974. An employee was crushed by a 45-metre frame, used to make Hawk jet trainer components. The HSE said:

> 'Safety failings uncovered by HSE's investigation included an absence of a suitable assessment of the risk associated with the test process and a lack of engineering control measures to prevent entry by workers during testing or to stop the machine if anyone did enter a danger zone.'

2.54 Nolan Recycling Limited were fined £250,000 and costs of £53,100 after pleading guilty to one charge, after a worker, who had only been on site for about six months, was run over whilst cleaning sensors on a weighbridge. He had never done the job before. There was a blind spot, so the lorry driver could not see him.

2.55 Bison Manufacturing Ltd were fined £300,000 and £21,341 costs after pleading guilty when a worker was killed by a trailer (see Larkin Logistics Ltd above).

2.56 In 2010, SITA UK Limited were fined £210,000 and ordered to pay costs of £33,000 after a worker died trying to unblock a paper brake in the company's Swindon site. In 2012, they were also fined £200,000 and costs of £77,402 after a 21-year-old employee died when an arm of a JCB loader fell and crushed him. The company pleaded guilty in both cases.

2.57 In 2012, Hertfordshire NHS Trust were fined £150,000 and costs of £326,345 after a trial. They were found guilty of a failure to adequately risk assess a patient who stabbed a care worker, who died, and seriously wounded her colleague. The patient had been diagnosed with bipolar disorder.

2.58 A care home, New Century Care Ltd, was fined £160,000 and £18,000 costs after pleading guilty to an offence under s 3 of the HSWA 1974. A 93-year-old resident of the care home died after becoming trapped by the neck between the mattress and the safety rail on the bed.

2.59 In 2012, another care home, Your Health Ltd, was fined £110,000 and ordered to pay £26,226 costs. A diabetic resident fell down the stairs and died in his wheelchair after a fire door had been left open.

2.60 In 2012, ISS Mediclean Ltd, a national cleaning company, were fined £175,000 and costs of £42,000 after a hospital porter at the Royal Bolton Hospital died when the lid of a metal compactor slammed down on his head. According to the HSE:

'… The most likely explanation … was that he leaned against a lever whilst leaning over the compactor, causing the lid to snap down. The manufacturer's recommendations for the compactor stated that it should be loaded from the front, away from the controls, but the court was told it was standard practice for porters to load from the side.'

2.61 In 2012, Serco Plc and Birse Civils were fined £200,000 and £100,000 respectively plus approximately £200,000 costs after an employee fell off the M5 motorway near Clevedon.

2.62 The Atomic Weapons Establishment in 2013 was fined £200,000 and £80,258 costs and £2,500 compensation after pleading guilty to one count, after an employee was very badly burnt in a fireball during a manufacturing process.

2.63 Basildon and Thurrock University Hospitals NHS Foundation Trust was fined £100,000 and costs of £162,000 after pleading guilty to health and safety charges concerning Legionnaires' disease, and a patient died as a result. Five other patients were also infected. The Trust was also fined £75,000, and costs of £13,000, in relation to a separate case involving a fall from a window of an elderly patient.

There has been a series of prosecutions against various hospitals, and this trend seems to be intensifying.

2.64 Greater Manchester Police were fined £150,000 plus £90,000 costs after a police officer was killed in a firearms training exercise. A blank round hit him when he was not wearing body armour. The officer in charge of the exercise was fined £2,000 for a breach of s 7 of the HSWA 1974 and costs of £500. Very recently, the CPS announced that it was prosecuting Greater Manchester Police over an unarmed man who was shot by an officer. The CPS said, 'there is sufficient evidence to prove the force broke health and safety laws'.

2.65 In the case of *R v Merlin Attractions Operations Ltd*[1] a fine of £350,015 was appropriate in the case of a company which operated a historic building as a tourist attraction and had been convicted of health and safety failures resulting in the death of a visitor who had fallen from a bridge. Although the company had generally good health and safety systems, its failures in relation to the bridge had been such as to create a foreseeable risk of serious injury (see **Chapter 1** and above).

1 [2012] EWCA Crim 2670.

2.66 Marks & Spencer were fined £1 million for failing to protect customers, staff and workers from potential exposure to asbestos. Customers at stores in Reading and Bournemouth were put at risk of exposure to asbestos in 2006 and 2007 during refurbishments; Marks & Spencer were convicted of two charges under the HSWA 1974 of failing to ensure the health and safety of its staff and others at the Reading store in July.

They were fined £500,000 for each offence. Marks & Spencer employed contractors who removed asbestos present in ceiling tiles and elsewhere during the work at the store in Reading and also at Bournemouth. During the three-month trial at Winchester Crown Court, it was revealed that Marks & Spencer guidance on asbestos removal was not fully followed by the contractors during the refurbishments (see **Chapter 1**).

2.67 Also in 2013, three companies, namely Crown House Technologies Ltd, Kiddle Fire Precautions Ltd and Kiddle Products Ltd, were fined £117,000 and £119,393 costs, £165,000 and £59,696 costs and £165,000 and £59,696 costs respectively, after pleading guilty to a series of health and safety charges. An employee was struck and killed by one of 66 gas cylinders that was travelling at over 170 mph. One cylinder had toppled over, starting a 'chain reaction'. Other workers were also injured, and the building was severely damaged.

2.68 General Motors UK Ltd, the Vauxhall car owners, were fined £150,000 and £19,654 in costs after a worker died in the factory when he was crushed by a machine that suddenly restarted. The court was told that a risk assessment had been undertaken 10 years previously, but no modifications had been made. The HSE said:

> 'There was absolutely no point in Vauxhall carrying out a risk assessment into the dangers posed by the machines if it wasn't going to act on the recommendations … The company has now installed a new safety system on the door which means power to the machine has to cut before the door can be opened. If this had been in place in July 2010 then Mr Heard's life could have been saved.'

2.69 UK Coal Mining Limited was fined £125,000 and £175,000 costs after a locomotive driver was crushed by steel pipes that rolled on to him. The court was told that there had been four previous reports.

2.70 In October 2013 the former UK Coal Mining Company were also fined £50,000 (with no costs imposed) after a methane gas explosion at the Kellingley Colliery in Yorkshire. 218 miners were evacuated. The HSE said:

> '… if the explosion had occurred 20 minutes earlier, 10 men would have been right in it and we could have been dealing with a fatal accident. However, this happened near the end of a shift and the workers were at a safe enough distance when the gas ignited. In only slightly different circumstances, the outcome could have been very different.'

2.71 Another mining company, UK Coal, were fined £1 million plus costs in 2012 for four fatalities in collieries in Nottinghamshire and Coventry in 2006 and 2007.

2.72 SAFC Hitech Ltd was fined £120,000 and £13,328 costs after a worker was very seriously burnt after a chemical explosion. The company pleaded guilty to single breaches of the Dangerous Substances and Explosive Atmospheres Regulations 2002 and other charges.

2.73 Special Metals Wiggin Ltd were fined £120,000 and £50,000 costs after an employee was crushed and burnt by molten metal. He spent 18 days on a life support machine. He had 25% burns to his body.

2.74 Tata Chemicals, an international conglomerate, were fined £100,750 and £71,082 costs for a series of accidents at their plant in Winnington.

2.75 London Underground Ltd, Tube Lines and another contractor, Schweerbau GmbH, were fined £100,000 each and costs of £40,000 after a 39-ton train went out of control for four miles.

2.76 ThyssenKrupp Elevators UK were fined £100,000 in April 2013, and ordered to pay £25,784 costs, in respect of an employee who died when he was electrocuted whilst fixing a fault in a lift in Pentonville Prison. The same company had been fined £233,000 in 2010 arising out of a lift accident in the Broadgate Health Club in the City of London.

2.77 Aesica Pharmaceuticals Limited were fined £100,000 and £7,803 costs after a worker was seriously injured by bromine burns at a plant in Northumberland.

2.78 In December 2013, Esso Petroleum Company Limited were fined £100,000 plus £50,000 costs after a worker was killed on a docked tanker at Fawley. Another contractor was also fined £60,000 and £30,000 costs. A heavy fuel hose collapsed on Juan Romero in August 2008.

2.79 In 2013, Mapei (UK) Ltd, a glue manufacturer, were fined £173,332 and undisclosed prosecution costs in respect of an employee who died after being impaled by a forklift truck at the company's plant.

2.80 In 2013, Sheffield Forgemasters Ltd were fined £120,000 and ordered to pay £125,000 in costs after an employee died from carbon monoxide poisoning in a cellar he was working in. Four co-workers tried to rescue him, and they were almost overcome by the gas. The HSE said that the company '… had given no thought to the risks associated with the task being undertaken by Mr Wilkins, nor had they provided emergency rescue equipment'.

2.81 Also in 2013, Buccleuch Estates Limited were fined £140,000 after a worker died when cutting down trees on the estate. A 216-foot tree crushed him, after the tree was uprooted by another tree that had been felled.

2.82 Also in 2013, following a six-week jury trial, Luton Airport Operations Ltd were fined £75,000 and £197,595 costs after a passenger was crushed by a milk lorry when using a crossing between the passenger drop-off area and the terminal. Design contractor CT Aviation Solutions were also fined £70,000 and £30,000 costs.

2.83 In 2014, a care home, The Order of St John's Care Trust, was fined £140,000 and had costs of £65,000 imposed after an elderly resident with Alzheimer's went missing for eight hours and was found in the gardens on a freezing November night in 2010. She died two months later, as the hypothermia that she suffered brought on pneumonia.

2.84 In 2012, another care home, Hillcrest House, was fined £40,000 and £95,963 costs after an elderly resident with dementia fell from a window to his death. Suitable window locks had not been fitted.

2.85 In 2014, Tesco Maintenance Limited and Otis Limited were fined £115,000 and £110,000 respectively. An undisclosed sum of costs was also imposed against both companies. A Tesco employee lost all five toes from his right foot when a faulty lift trapped and crushed his foot between the lift and the lift shaft.

Legal test of risk

2.86 In *R v Porter*,[1] the Court of Appeal decision reviewed the current legal test of risk under the HSWA 1974. It states that there needs to be a real, as opposed to a theoretical, risk and considers whether the existence of a previous similar accident can be a relevant factor.

It is perhaps helpful to set out the facts: since 1975, James Porter had been head of Hillgrove School in North Wales, which educates children from ages 3 to 16. The school is housed in a 20th century house in a fairly rugged site, with a playground built in the grounds of a disused quarry. Classroom situated away from the main building are reached by brick steps. The HSE had never visited the school and there had been no complaints about these steps. In July 2004 a child at the school who was 3 years and 9 months was with a group being supervised at playtime. During the break the child moved into the area which was designated 'out of bounds'. This area contained a flight of steps. When the child reached the fourth step from the bottom, he jumped towards the base of the steps, fell forward and bumped his head. He was taken to a local hospital and then transferred to Alder Hey Hospital in Liverpool. Although the head injury was minor, there was some intracranial bleeding. He became immobile, contracted pneumonia and the MRSA virus, and died in August 2004.

The HSE visited shortly after the incident and served an improvement notice which said a gate should be provided, and failure to do so would lead to the school being closed. This was done, although there was some evidence that the existence of a fence and gate decreased the opportunity for teachers to see the playground below.

1 [2008] EWCA Crim 1271. See also **Chapter 1** in relation to developments since *R v Chargot* and *R v Porter*.

2.87 The defendant was charged, contrary to s 3(1) of the HSWA 1974, with failing to conduct his undertaking in such a way as to ensure so far as was reasonably practicable that persons not in his employment, who might be affected thereby, were not thereby exposed to risks to their health and safety. The risk identified in amended particulars of offence was a 'risk of falling down a flight of steps'. At the trial the Crown called evidence to show that other schools operated a level of supervision higher than that at the defendant's school, though in cross-examination it was found that those schools' safety records were far lower than that of the defendant's, and the defence evidence was that there had been no previous playground accidents in the 29 years during which the defendant had run the school. The judge refused an application to withdraw the case from the jury, made at the close of the prosecution case, ruling that the jury could properly conclude that the steps constituted a risk to the safety of a child if he were to descend them unsupervised, and that it would have been reasonably practicable to prevent a child from descending the steps if there had been constant supervision. The defendant was convicted.

2.88 The case went to the Court of Appeal. The Court of Appeal found that the risk that the Crown had to prove, for the purposes of a prosecution for failure to comply with s 3(1) of the HSWA 1974, should be real as opposed to fanciful or hypothetical. In determining whether the risk was real, it was important for the jury to take into account, as in this present case, the absence of any previous accidents in similar circumstances with similar levels of supervision. The evidence demonstrated that there was no real risk of the kind contemplated by the statute, and, the risk being part of the incidence of everyday life, it was less likely that it could be said that the child had been exposed to risk by the way in which the school was managed. On all the evidence, a jury properly directed could not reasonably have concluded that the child had been exposed to risk by the defendant's conduct of the school. Accordingly, while the judge had been correct to decline to stop the case at the close of the prosecution case, the case should have been withdrawn at the close of all the evidence. The appeal was allowed and the conviction quashed. See also **Chapter 1**.

The Criminal Justice Act 2003

2.89 Section 142 of the 2003 Act states that the purpose of sentencing is about:

(a) the punishment of the offender;
(b) the reduction of crime (including the reduction by deterrence);
(c) the reform and rehabilitation of offenders;
(d) the protection of the public, and
(e) the making of reparation by offenders to persons affected by their offences.

2.90 The importance of the sentencing court honouring these principles of sentencing was reaffirmed by the Lord Chief Justice in *R v Balfour Beatty Rail Infrastructure Services Ltd*[1] when reducing the £10 million fine after the Hatfield Train crash trial to one of £7.5 million. See also **2.17** above.

1 [2006] EWCA Crim 1586.

Other older fines

2.91 In July 2009, Bodycote was fined £533,000 and ordered to pay £200,000 costs after two employees died from lack of oxygen after argon leaked into a pit where people were working.

2.92 In 2009, Scottish Coal were fined £400,000 after pleading guilty to one count under s 3 of the HSWA 1974. Two men, who were employed by a sister company, were killed when their Land Rover was crushed by a 100-tonne dumper truck.

2.93 In 2008, George Wimpey were fined £300,000 plus costs after a trench collapsed, killing one worker and seriously injuring another. In 2008, Storage Industrial Products were fined £350,000 and had an order for £60,000 costs imposed after a worker was engulfed in a fireball and hospitalised for seven weeks. The HSE prosecuted the company, even though it was in administration and was not represented at the trial.

Judge Newton said it had put profit before safety and that the 'extent of the danger would have been obvious to a three-year-old child'.

2.94 In 2008, Shell UK Oil Products were fined £267,000 plus £37,000 costs for a breach of the COMAH Regulations 1999 in relation to a gas leak at its Stanlow Refinery. No one was injured, but the largish fine reflected the high risk involved. Shell were also fined £300,000 in 2009 for breaches of the Regulatory Reform (Fire Safety) Order 2005 after two small fires at the London Headquarters indicated inadequacies in their fire precautions. This is the largest fine under the legislation.

2.95 In 2009, FJ Chalcroft Ltd had the £260,000 fine, that had been imposed for the death of an employee from a fall from height, upheld in the Court of Appeal. The company had pleaded guilty to two counts.

2.96 In September 2009, ABP were fined £266,000 plus £75,000 costs after a reversing vehicle killed a worker at the Port of Ipswich. The company had pleaded guilty to an offence under s 2 of the HSWA 1974.

2.97 In 2009, Mitie Engineering Limited were fined £300,000 after being found guilty of an offence under s 2 of the HSWA 1974, following the death of a 26-year-old worker who was electrocuted in Dundee.

2.98 In June 2009, Bouygues (UK) were fined £160,000 plus costs for the death of a contractor who was killed by a reversing telehandler. They pleaded guilty.

Examples of much older fines imposed

2.99 The following are examples of fines imposed upon employers for breaches of health and safety duties. Although some pre-date the *Howe* case,[1] they are still useful as a guide.

1 *R v F Howe & Son (Engineers) Ltd* [1999] 2 All ER 249.

2.100 There are few reported cases on sentencing in health and safety cases. Details of cases can be found on the HSE website[1] or reported in health and safety magazines such as *Safety Management* and *Safety and Health Practitioner*.

1 At www.hse.gov.uk.

2.101 The figures given are for fines only and do not always include costs of the prosecution or defence.

2.102 British Railways Board was fined £250,000 on 14 June 1991 at the Old Bailey in relation to the Clapham Junction train crash of 1988 in which 35 people died.[1] British Railways Board pleaded guilty to breaches of HSWA 1974, ss 2 and 3. Immediate cause was a signal failure caused by a technician failing to isolate and remove a wire. Contributory causes included degradation of working practices, problems with training, testing quality and communication standards, poor supervision.

1 (14 June 1991, unreported), Central Criminal Court.

2.103 Mersey Docks and Harbour Company pleaded guilty to a breach of HSWA 1974, s 3. The matter went before the Court of Appeal on sentence.[1] The company had failed to ascertain whether there were any 'dangerous spaces' where flammable cases could be found after discharging the cargo of the vessel in port. Two workmen died and others were injured when a torch conflagrated gases remaining in the holds. The company was fined £250,000. Hobhouse LJ stated:

> '… nothing that we say … detracts from the duty of the Master of the vessel, or indeed those responsible for the operation of a vessel. But the duty imposed by statute upon persons in the position of the … company is that they themselves have the duty. It is a non delegable duty and they must carry out that duty. It is not mitigation for them to say that they just leave it to other people. If that is their attitude to the discharge of their statutory duty, then it is important the Courts when there has been a breach of the duty, should impose fines which leave people such as this Harbour Company in no doubt that it is their duty and that they have to discharge it. Anything less than a substantial fine will fail to bring the matter home to them.'

1 *R v Mersey Docks and Harbour Company* [1995] 6 Cr App R (S) 806.

2.104 British Railways Board was fined £200,000 at Snaresbrook Crown Court on 7 October 1996 in relation to a train collision at Wood Street in North East London in 1995. There were no casualties. British Railways Board pleaded guilty to breaches of HSWA 1974, ss 2 and 3. The immediate cause was a signal failure caused by human error when carrying out maintenance to a section of track. Underlying causes were that the work had not been properly planned and the maintenance staff had not been properly trained.

2.105 Fartygsentreprenader AB, Fartygskonstruction AB, Port Ramsgate Ltd and Lloyd's Register of Shipping had fines imposed upon them at the Central Criminal Court on 28 February 1997 totalling £1.7 million and ordered to pay £600,000 towards the HSE investigation costs and pay the trial costs of £115,000.[1] The case concerned the collapse in 1994 of a walkway suspended 30 feet above the ground, killing six people and injuring seven others. The companies were prosecuted under HSWA 1974, s 3 and Port Ramsgate was also prosecuted for a breach of reg 7 of the Docks Regulations 1988.[2]

1 28 February 1997, unreported.
2 SI 1998/1655.

2.106 South East Infrastructure Maintenance Company Ltd, Railtrack PLC and Southern Track Renewals Company received a global fine of £150,000 on 8 February 1998 following an incident at Bexley Kent in February 1997. Seven freight trains derailed at a bridge over Bexley High Street. Four members of the public who were near the bridge were injured. Each company pleaded guilty to a breach of HSWA 1974, s 3.

2.107 Sainsbury's was fined £450,000 in November 1998 at Winchester Crown Court after pleading guilty to six breaches of HSWA 1974 and other associated legislation. A warehouse worker died after a forklift truck toppled over and crushed him while he was inspecting it. The safety cut-off switch had been deliberately disconnected. Mr Justice Kay said when sentencing:

> '… The story is a picture of working procedures that date back to the dark ages.'

2.108 Milford Haven Port Authority was prosecuted by the Environmental Agency in relation to pollution resulting from an oil spillage of 72,000 tonnes of crude oil from the *Sea Empress*, a Russian ship that ran aground in 1996 because of an inexperienced pilot. Although this is not a health and safety case as such, it is important to note from a sentencing point of view. Fines in the public sector stand ready to increase (see **2.17** above).

2.109 At Cardiff Crown Court on 16 January 1999, Mr Justice Steel imposed a fine of £4 million on the Port Authority. He said: 'The accident occurred because of the "careless" navigation of a pilot who had never attempted this manoeuvre with a ship of this size at low tide.' The judge went on to say: 'The pilot was put in a position by the Port Authority where he could make an error of navigation.'

2.110 In relation to the fine the judge said this was required 'to reflect the genuine and justified public concerns'. The clean-up operation cost £60 million. The judge recognised that the Port Authority did not have the '… vast reserves of a major oil company or manufacturing company'.

2.111 F Howe & Son (Engineers) Ltd[1] pleaded guilty to a breaches of HSWA 1974, s 2, the Electricity at Work Regulations 1989, reg 4 and the Management of Health and Safety at Work Regulations 1992, regs 3 and 11. The prosecution resulted from a fatal accident in August 1996 when a 20-year-old employee was electrocuted. The company was small with limited resources. The electrical equipment in question was in a bad state. The Court of Appeal ruled that the appropriate fine in this case was £15,000.

1 *R v F Howe & Son (Engineers) Ltd* [1999] 2 All ER 249; see **2.18** above.

2.112 Balfour Beatty Civil Engineering Ltd on 15 February 1999 at the Central Criminal Court was fined £1.2 million following the collapse of three tunnels during the construction of the Heathrow Express railway link at Heathrow Airport on 21 October 1994. No one was injured in the collapse that occurred at 1 am. Balfour Beatty was the main contractor, using the New Australian Tunnelling Method (NATM), on which Geoconsult ZT GmbH were consultants.

2.113 Balfour Beatty pleaded guilty to breaches of HSWA 1974, ss 2 and 3. Geoconsult were also prosecuted for breaches of HSWA 1974, ss 2 and 3 but pleaded not guilty. The company was convicted and fined £500,000.

2.114 Great Western Trains were fined £1.5 million at the Central Criminal Court on 27 July 1999 in relation to the Southall train crash of 1997 in which seven people died.[1] The driver of the GWT train passed a signal at danger (SPAD). The automatic warning system (AWS), that was designed to help prevent SPADs and would have given the driver an audible warning of the cautionary signals, was isolated, ie inoperative. The judge, when passing sentence, said that the risk of the driver passing a danger signal was small but that, if this did happen, the consequences were likely to be 'appalling', and so took the view that the company should have ensured that the train was not running in that condition. Great Western Trains pleaded guilty to a breach of HSWA 1974, s 3.

1 (27 May 1999, unreported), Central Criminal Court.

2.115 London Underground Limited were fined £300,000 at the Central Criminal Court on 27 July 1999, having pleaded guilty to a breach of HSWA 1974, s 3 and reg 7(1) of the Railways (Safety Case) Regulations 1994.[1] This related to an incident at Eastcote Station on 31 December 1996. An elderly passenger fell between a train and the platform when alighting at the station. Apparently, she had become caught in the doors. The train moved off, killing the passenger. The driver was originally investigated for manslaughter but this eventually did not proceed. At the inquest, evidence was given that there was a 50-yard blind spot and so the driver would not have been able to see the passenger.

1 SI 1994/237.

2.116 Keltbray Ltd was fined £200,000 at Southwark Crown Court on 7 September 1999,[1] having pleaded guilty to a breach of reg 6 of the Construction (Health, Safety and Welfare) Regulations 1996.[2] *R v F Howe & Son (Engineers) Ltd*[3] was followed. The fine was upheld on appeal. The case related to the death of two men after the floor they were working on collapsed.

1 The defendants appealed the sentence which was upheld by the Court of Appeal: *R v Keltbray Ltd* [2001] 1 Cr App Rep (S) 132.
2 SI 1996/1592.
3 [1999] 2 All ER 249.

2.117 BG Exploration & Production Ltd was fined £300,000 at Kingston-upon-Hull Crown Court on 10 February 2000. This followed an incident in February 1998 when a large volume of natural gas was released from a leak in a pipework joint on the company's Rough 47/3B offshore gas platform. The gas did not ignite and there were no injuries. The company pleaded guilty to breaches of HSWA 1974, the Offshore Installations (Prevention of Fire and Explosion, and Emergency Response) Regulations 1995[1] and the Reporting of Injuries, Diseases and Dangerous Occurrences Regulations 1995.[2]

1 SI 1995/743.
2 SI 1995/3163.

2.118 Friskies Petcare (UK) Ltd had its fine reduced by the Court of Appeal in March 2000 to £250,000.[1] This related to the electrocution of an employee.

He was repairing a metal ribbon stirrer at the bottom of a silo. He was arc welding in a confined, damp, conductive environment in the silo when he was electrocuted while changing welding electrodes. The company pleaded guilty to a breach of HSWA 1974, s 2 and a breach of reg 3 of the Management of Health and Safety at Work Regulations 1999.[2]

1 [2000] Cr App R (S) 401, see **2.30–2.31** above.
2 SI 1999/3242.

2.119 Colthrop Board Mills Limited were fined £350,000 in 2001 at Reading Crown Court in relation to an accident in February 2000 in which one of its employees lost the use of his right hand.[1] The company pleaded guilty to breaches of HSWA 1974, s 2 and of reg 11 of the Provision and Use of Work Equipment Regulations 1998.[2] The accident occurred when the employee was checking to see if a roller on the mill's paper production needed cleaning by running his hand along the roller and as a consequence his hand become caught between two rollers, dragging him into the machine. The company was aware of this unsafe practice and, a month before the accident, the company had carried out a risk assessment that identified there was a high risk of a fatality and that guarding should be provided. At the time of the accident, the guarding had not been installed.

1 [2002] 2 Cr App R (S) 80.
2 SI 1998/2306.

2.120 Doncaster Metropolitan Borough Council was fined £400,000 on 20 February 2001 at Doncaster Crown Court following the electrocution of an electrician who had been called to repair a heating unit in a false ceiling. He came into contact with exposed wires. There was evidence that council managers had been aware of the danger for some time but failed to take action or notify the electrician.

2.121 On sentencing, His Honour Judge Crabtree said:

> 'The Council's performance was dismal and disgraceful. This was a death waiting to happen and nobody was bothered … they ignored the most elementary safety precautions … If the Council had been a profit-making company with £10 million annual profits I would have fined them £10 million.'

2.122 Smurfit UK Ltd was fined £100,000 in June 2001 at Burnley Crown Court after pleading guilty to a breach of HSWA 1974, s 2. This followed the death of a worker who was killed while attempting to clean a paper machine at its Burnley paper and board mill in Lancashire in January 2000. The presiding judge said the incident was 'a very serious breach of duty indeed'.

2.123 Fresha Bakeries and its owners Harvestime Ltd were fined £250,000 and £100,000 respectively for breaches of HSWA 1974, ss 2 and 3 on 18 July 2001 at Leicester Crown Court. This related to the deaths of two men who had been sent into a giant oven to retrieve a broken part and died as a result of the 100°C temperature. There were also prosecutions of individuals in relation to HSWA 1974, s 7 and HSWA 1974, s 37.[1]

1 See also **2.179** below.

2.124 Costain Ltd was fined £200,000 at Cardiff Crown Court in August 2001 after a 31-year-old labourer died when he became trapped between the end of a wall and a 13-tonne hydraulic excavator in October 1999. The labourer was employed by Aberdare Construction which had been contracted by Costain Ltd to work on the Lynfi Valley combined sewer overflow scheme in

Bridgend, Wales, where the accident happened. The company pleaded guilty to a breach of HSWA 1974, s 3.

2.125 Birse Construction was fined £80,000 with £20,000 costs at Croydon Crown Court in September 2001. The company pleaded guilty to a breach of HSWA 1974, s 3. The accident happened when an employee of S&J Chatteris, who had been contracted by Birse, slipped and fell beneath the guard-rail on the roof he was working on. The gap between the rail and the roof was too big. HSE inspectors who had visited the site two weeks before the incident had warned of the dangers of this gap.

2.126 When sentencing, the judge said he took into account the company's early plea of guilty and the fact it had taken significant steps in improving its safety management. However, he went on to say:

> 'But I have also taken into account the tragic loss of life which resulted and the fact that the company failed to act on the warning given by HSE two weeks before the incident occurred. The fines imposed should reflect the gravity of the accident and should bring home the message to Directors and Shareholders that the health and safety of its staff needs to be top priority.'

2.127 Mayer Parry (Recycling) Ltd was fined £200,000 at Blackfriars Crown Court on 27 September 2001 following the death of a fitter at one of its scrap metal processing plants. The deceased was killed during the annual overhaul of the fragmentising machine at the plant. He was lifting a 130-kg steel plate using a two-leg chain sling, when one of the legs caught on part of the machine and then released violently, striking the deceased on the head. The company pleaded guilty to a breach of HSWA 1974, s 2.

2.128 Railtrack Plc and English, Scottish and Welsh Railways were fined £50,000 and £70,000 respectively at Nottingham Crown Court on 31 October 2001 following the death of a 12-year-old boy at railway sidings. Both companies pleaded guilty to a breach of HSWA 1974, s 3. This case was the first time the HSE had prosecuted a company with respect to a failure to prevent a trespass on railway lines. However, the judge in sentencing emphasised there was no causal link between the death of the boy and the offences.

2.129 Corus UK Ltd was fined £300,000 plus costs of £11,591.38 at Cardiff Crown Court in November 2001 after pleading guilty to breaches of HSWA 1974, ss 2 and 3. The prosecution was taken following an explosion at its Basic Oxygen Steel plant at a site near Newport, South Wales, in September 2000, when slag spilled on to a floor area and came into contact with water. As a result of the incident, an employee fell from a ladder he was working on and fractured his spine.

2.130 Judge Hickinbottom when sentencing said:

> 'This was a most serious incident and was caused by Corus's failure to meet statutory safety requirements. Steel making is dangerous, but there is a regulatory scheme designed to ensure that employees are given every reasonable protection from risks and failures to comply. The objective of prosecution cases is to achieve a safe environment for workers and members of the public.
>
> Any fine need to be large enough to bring home the message about health and safety to both management and shareholders. A very substantial fine is necessary. Life was not lost, but Mr Bagnall suffered catastrophic injuries

and this was such an incident which could have led to more injuries and possible deaths.

Corus were aware of the risks of such an explosion. Repair would have cost little in direct terms. There was a gross failure by Corus to heed warnings from both employees and contractors. Advice from the HSE, contained in a letter written in 1995, also went unheeded. This is very serious breach of the regulatory scheme.'

2.131 Costain Ltd and Yarm Road Ltd, on 30 November 2001 at Bristol Crown Court, were fined a total of £500,000 for breaches of HSWA 1974, ss 2 and 3. This related to an incident at Avonmouth Bridge. Four men were working on a gantry suspended beneath the road bridge. They were removing and replacing old runway beams from the bridge. To do this, they were moving the gantry along the bridge. The gantry ran along runway beams underneath the bridge and was only restrained by rope pulling machines, called tirfors, and beam clamps.

2.132 The wind moved the gantry off the ends of the runway beams. When this happened, the gantry swung down vertically. All four men working on the gantry then plunged over 25 metres to their deaths.

2.133 Judge Owen said when sentencing:

'This is a case of great gravity. Perhaps the most tragic feature of this case is that the accident could and should have been prevented by a number of simple measures.

The failures on the part of the defendants were of a very serious nature. This is not a case of a single and isolated fault.'

2.134 The London Borough of Hammersmith and Fulham were fined £350,000 at Blackfriars Crown Court on 7 December 2001 after pleading guilty to breaches of reg 4 of the Management of Health and Safety Regulations 1992[1] (the provision is now reg 5 of the 1999 Regulations of the same name)[2] and HSWA 1974, s 3. The prosecution related to the death of two council tenants from carbon monoxide poisoning as a result of a faulty boiler that was overdue its annual safety check.

1 SI 1992/2051.
2 SI 1999/3242.

2.135 Sentencing, Judge Timothy Pontius condemned the council for 'prolonged dereliction of duty to its tenants' and said:

'... the lamentable history of failure fully to accept its responsibility dates back a considerable number of years ...

It is all the more regrettable that an earlier tragic death arose in precisely the same circumstances only five years before the two deaths in this case. That earlier tragedy provided the plainest salutary lesson imaginable, which should fully have been learned. Regrettably, it was not.'

2.136 Avon Lippiatt Hobbs (Contracting) Ltd was fined £250,000, comprising a fine of £175,000 for a breach of HSWA 1974, s 3 and £75,000 for a breach of HSWA 1974, s 2. The company pleaded guilty to the offences at Merthyr Crown Court on 13 December 2001. The prosecution related to a gas explosion that injured an employee, demolished a house and seriously damaged two others.

2.137 Sentencing the company, the Judge John Curren said:

> 'The firm was responsible for a series of failures on the day of the explosion, and inadequate steps were taken to find out whether there was a gas pipe underneath the pavement on Abercynon Road, where the incident took place.
>
> This could have very easily been a fatal accident, and firms must understand that, when dealing with gas, it is necessary to follow the most rigorous of safety standards.
>
> I believe that a penalty should be imposed to properly reflect the concerns of the public.'

2.138 BP was fined a total of £1 million at Falkirk Sheriff Court, Scotland, on 18 January 2002 following two incidents at the company's Grangemouth refinery in June 2000, one where a catalytic converter caught fire and the other where there was a steam rupture. The company pleaded guilty to breaches of HSWA 1974, ss 2 and 3. It was noted at court that only good fortune avoided fatalities and serious injuries.

2.139 London Underground Ltd (LUL) pleaded guilty to six offences for breaches of HSWA 1974, s 3 and was fined £225,000 at Blackfriars Crown Court on 10 January 2002. The court was told that, between April 1998 and January 2000, track maintenance workers were forced to carry out repairs using metal tools beside live rails on the Central Line in Loughton, Essex. The workers had been working in dark, rainy conditions, which exposed them to a risk of electrocution.

2.140 Judge Samuels said that safety was 'sacrificed' to keep trains running. He added:

> 'LUL, despite the lip service it paid to health and safety issues, fell lamentably short of the proper safety standards. A pattern of protracted disregard for basic safety procedures by senior management for so long must be marked by a heavy penalty.'

2.141 SDC Builders Ltd was fined £100,000 on 12 March 2002 for breaches of HSWA 1974, ss 2 and 3, at Cambridge Crown Court, after a contested trial. The prosecution related to the death of a sub-contractor who was knocked down and killed by a reversing delivery van on the site access road.

2.142 Sentencing, Judge Haworth said:

> 'The company, in my view, fell short of proper standards in relation to the use of the access road and neither installed nor maintained proper measures to the use of that road in relation to their work. Proper segregation was justified but no risk assessment was ever taken. Quite clear measures were necessary in regard to the access road.'

2.143 The driver of the delivery van pleaded guilty to a breach of HSWA 1974, s 7(a).[1]

1 See **2.197** below.

2.144 RMD Kwikform Ltd was fined £180,000 for three breaches of the Construction (Health, Safety and Welfare) Regulations 1996[1] and £60,000 for a breach of HSWA 1974, s 3 at Cardiff Crown Court in December 2002. In relation to the same matter, Taylor Woodrow was fined £80,000 for a breach of HSWA 1974, s 3. The prosecutions arose when poorly assembled scaffolding attached onto a 12-storey office building collapsed in high winds. Although no one was injured, the court heard how this could have been a major catastrophe if the collapse had happened in the daytime.

1 SI 1996/1592.

2.145 Klargester Environmental was fined £250,000 for two breaches of the Provision and Use of Work Equipment Regulations 1998[1] and a breach of HSWA 1974, s 2 at Aylesbury Crown Court on 22 March 2003 following the death of an employee whose head was crushed while he was setting a Hanwood vacuum-forming machine.

1 SI 1998/2306.

2.146 Passing sentence, Judge Morton Jack said:

> '… that there were multiple failings, some of which must have been present for a long period, some for many years …
>
> … the defendant was unable to demonstrate a proper system for inspection and maintenance, necessary to prevent accidents …
>
> … the situation at the factory was serious and horrifying.'

2.147 William Hare Ltd was fined £75,000 at the Old Bailey on 8 May 2003 after pleading guilty to a breach of HSWA 1974, s 2. The prosecution related to the fall of two workers who fell from height, resulting in the death of one of them, while doing construction work at the Imperial War Museum in London.

2.148 Nestle UK Ltd and Monotronic Ltd were fined £220,000 for a breach of HSWA 1974, s 3 and £25,000 for a breach of HSWA 1974, s 2 respectively at Isleworth Crown Court on 30 May 2003, having both pleaded guilty. This was in relation to the death of an employee of Monotronic who was electrocuted undertaking work at one of Nestle's factories. The incident occurred while he was pulling out redundant cables from trunking in the coffee plant of the factory.

2.149 Ford Motor Company was fined a total £300,000 for breaches of HSWA 1974, ss 2 and 3, to which it pleaded guilty, at Winchester Crown Court on 16 June 2003. An employee was killed when he fell into an emulsion paint over-spray capture tank used to collect excess paint from spray booths on a transit van production line.

2.150 Earls Court Ltd and Unusual Rigging Ltd were fined £80,000 and £20,000 respectively at the Old Bailey on 3 October 2003, having both pleaded guilty to breaches of HSWA 1974, s 3. The prosecution followed the death of a worker who fell 35 metres to his death. He fell through fragile, false ceiling tiles while dismantling mobile platforms as part of a refurbishment project. Six months earlier, another worker had been killed in a similar incident for which Earls Court Ltd was fined £70,000.

2.151 In the case of *R v Jarvis Facilities Ltd* [2005], which concerned health and safety failures of a company carrying out maintenance work on the railway, the Court of Appeal said a more serious view of defendants' breaches could be taken where there is a 'significant public element', particularly where the defendant has been entrusted to carry out work competently and efficiently which affects the public's safety. The company's fine was, however, reduced from £400,000 to £275,000.

2.152 In 2007, in *R v Balfour Beatty Rail Infrastructure Services Ltd*,[1] the company appealed against sentence in relation to its fine for its part in the Hatfield train derailment of 2000. The ground upon which it succeeded, when the sentence was reduced from £10 million to £7.5 million, was the disparity of sentence between it and its co-defendant Network Rail (formerly Railtrack). Both companies were sentenced for breaches of s 3 of the HSWA 1974.

1 [2007] Bus LR 77, [2007] 1 Cr App R (S) 65, [2007] ICR 354.

2.153 When looking at a defendant company's safety record, the size of its operation is relevant. This issue was emphasised in the pollution case of *R v Anglian Water Services Ltd*,[1] where the Court of Appeal said:

> 'The appellant's previous 65 convictions stretching over a period from November 1990 to February 2002 have to be seen in context. The appellant runs 1,075 sewerage treatment works and covers an area of 27,500 square kilometres serving nearly 5.5 million people. These, it seems to us, are the most important factors about the offence. The number of previous convictions is not a great significance when seen in the light of the ambit of the appellant's operation.'

1 [2003] EWCA Crim 2243, [2004] 1 Cr App R (S) 62.

2.154 In *R v Transco*[1] the Court of Appeal made the distinction between health and safety cases involving systematic failure and those that did not. This case concerned an incident in November 2001. Transco engineers were called to an old flour mill in Ashton-under-Lyme that had been converted into flats, after gas had entered the property from a fracture in the gas main. All the residents were evacuated and allowed back into the building once the gas main had been repaired and the flats ventilated. Unfortunately, the engineers failed to notice a two-metre wide void between the ceiling of one of flats and the one above. A resident lit a cigarette and was killed in the resultant explosion. The Lord Chief Justice pointed out that the *Howe* case[2] involved a serious systematic fault, as do most health and safety prosecutions. This case, on the other, had involved no systematic fault but a mistake on the part of the individuals managing the emergency situation. The fine of £1 million was reduced to £250,000.

1 [2006] EWCA Crim 838.
2 [1999] 2 All ER 249.

2.155 Following the death of a 17-year-old man at Davyhulme Wastewater Treatment Works in January 2004, after falling approximately 18 metres whilst working to construct a scaffold within a 20-metre high sewage digester tank, three companies were fined a total of £217,000 plus a total of £125,000 costs at Manchester Crown Court. He was employed by 3D Scaffolding Ltd and under the control of a visiting contracts manager, David Swindell. He was contracted to work for RAM Services Ltd, itself a subcontractor of the project's principal contractor, Mowlem Group PLC. 3D Scaffolding Ltd pleaded guilty to breaching s 2(1) of the HSWA 1974, in that it failed to maintain a safe system of work, and was fined £60,000 and ordered to pay £20,000 costs. David Swindell, a contracts manager employed by 3D Scaffolding Ltd, was found guilty of breaching s 7 of the HSWA 1974, in that he failed to take reasonable care for the health and safety of other persons who might be affected by his acts or omissions at work. He was fined £7,500 and ordered to pay £15,000 costs. The principal contractor, Mowlem Plc, and RAM Services Ltd, which employed 3D Scaffolding Ltd, each pleaded guilty to charges of breaching s 3(1) of the Act, in that they failed to ensure the safety of people not in their employment. Mowlem was fined £75,000 and ordered to pay £20,000 costs. RAM Services Ltd also pleaded guilty to a charge, under reg 29 of the Construction (Health and Safety at Work) Regulations 1996, in that they failed to ensure that scaffolding used by their own employees had been properly inspected to ensure that it was safe to use. RAM Services was fined a total of £75,000 and ordered to pay £70,000 costs.

2.156 In 2008, Talisman Energy UK Ltd, pleased guilt to breaching s 3(1) of the HSWA 1974, and its offshore partner firm Aker Kvaerner Offshore Partner Ltd pleaded guilty to a charge under s 2(1) of the same Act and were each fined £600,000 following the death of a worker hit by falling steel clamps. The case was heard in Scotland, and so no costs were awarded. Aberdeen Sherriff Court heard that the deceased, a pipefitter, had been working alongside two other men in the bowels of the Bleo Holm floating production, storage and offtake installation, which was operated by Talisman Energy, when he was killed. The men were working for Aker Kvaerner Offshore Partner which had been contracted by Talisman to carry out extensive work in the tank, part of which involved using a pneumatic hoist to lift loads out onto the main deck. The work relied on hand signals to signal the end of the lifting operation, and it seemed that the men below assumed that they had been given the all clear and re-entered the tank. However, clamps weighing 21.5 kg fell from the crane's lifting hook and struck two of the men, one of whom was killed at the scene. The investigation found that the operation should have been carried out using radio communication and not hand signals (and, in fact, the permit to work system required radios), but the company had neither supplied them nor enforced their use.

HEALTH AND SAFETY OFFENCES – INDIVIDUALS

Health and Safety (Offences) Act 2008

2.157 On 16 October 2008, the Health and Safety (Offences) Act 2008 received Royal Assent. The Act raises the maximum penalties that can be imposed for breaching health and safety regulations in the lower courts from £5,000 to £20,000, and the range of offences for which an individual can be imprisoned has also been broadened.

The Private Member's Bill received cross-party support, and came into force on 16 January 2009. There had been several attempts since 2000 to increase health and safety penalties. A previous attempt at a Health and Safety (Offences) Bill had been dropped after the second reading.

2.158 Before this Act, magistrates could not impose a fine higher than £20,000 if the offence related to a breach of the HSWA 1974 itself (or other similar statutes) or £5,000 if the offence related to a breach of a statutory instrument such as the Management of Health and Safety at Work Regulations 1999. When sentencing takes place in the Crown Court, there are no maximum fines.

2.159 The 2008 Act makes three changes to existing safety law: the current upper limits of penalties for most safety offences tried in the magistrates' court increase from £5,000 to £20,000; there is the option for custody in more cases for those that are especially serious; and it is possible for more cases to be tried at the Crown Court.

There are many health and safety regulations in force under the HSWA 1974 which will be affected by this legislation. One example is a failure to carry out a suitable and sufficient risk assessment which would, under the 2008 Act, carry with it a fine of up to £20,000 in the magistrates' court (or an unlimited fine in the Crown Court, which remains unchanged).

2.160 The Robens Report, which led to the HSWA 1974, recommended that criminal sanctions for health and safety breaches be reserved for offences of a 'flagrant, wilful or reckless nature'. This did not make its way into the Act.

2.161 An individual, whether a worker or manager, can be convicted of breaching s 7 of the HSWA 1974, for failing 'to take reasonable care' for their and others' health and safety, ie being criminally negligent.

2.162 A director can be convicted under s 37 of the HSWA 1974 if his/her company's breach was 'attributable to any neglect' on their part. In the case of *R v P Ltd and G*,[1] the Court of Appeal held a director could sometimes be guilty of neglect, even if he/she did not know of the unsafe practices of their company. Under the 2008 Act, both offences carry a maximum term of two years' imprisonment.[2]

1 [2007] All ER (D) 173 (Jul).
2 See *R v Chargot Ltd (t/a Contract Services)* [2008] UKHL 73, [2009] 1 WLR 1, [2009] 2 All ER 645 in **Chapter 1**.

2.163 The case of *David Janway Davies v HSE*[1] challenged the lawfulness of s 40 of the HSWA 1974, arguing that it breached the European Convention on Human Rights because it made inroads into the presumption of innocence. The Court of Appeal rejected the argument, holding that s 40 was proportionate and justified. One of the main reasons given was that the offence was not imprisonable.

1 [2003] ICR 586.

Individual fines examples[1]

2.164 Approximately 90 individuals have been sentenced to imprisonment (including suspended sentences) for health and safety at work charges and about 45 for manslaughter.

1 See also **Chapter 1**.

2.165 In 2013, a property developer, James Carlton, pleaded guilty to numerous health and safety and asbestos regulation charges. He was also found to have been in breach of a prohibition notice. He was sentenced to eight months in prison, suspended for two years for the breach of the prohibition notice. He was also fined £55,000 and ordered to pay £45,000 costs. The HSE said:

> 'Mr Carlton showed a wilful disregard for the health and safety of his employees and others. Our investigation uncovered a catalogue of serious errors, safety failings and a general ignorance of the laws around the safe and correct removal of asbestos … Workers who have been exposed to asbestos could have posed a health risk to others in the long term, even their families and loved ones, by taking home to their families and loved ones their contaminated clothing.'

2.166 In 2013, Mr Barnes, a director of a house building company, was sentenced to eight months' imprisonment, suspended for two years, disqualified from acting as a director for three years and fined £32,000 and £11,000 costs after pleading guilty to offences under s 37 of the HSWA 1974. The company, Paddle Ltd, was also fined £56,000 and £11,000 costs after pleading guilty to various health and safety at work charges. These related to two incidents, including a workman working at height in an elevated bucket, in clear view of Mr Barnes.

2.167 In 2013, Mr Lustig was sentenced to six months' imprisonment, suspended for 18 months, and disqualified from being a director for three years. There had been a jury trial. The company and Mr Lustig also admitted to a RIDDOR offence. In essence, a worker had seriously injured her leg when she slipped on debris on the factory floor. Mr Lustig was found guilty on one count of an offence under s 3(2) of the HSWA 1974.

2.168 A director, Vijay Kara, was fined £99,900 and ordered to pay £150,000 costs after being found guilty of an offence under s 37 of the HSWA 1974. The court also ordered that, if the fine and costs were not paid within six months, he would serve 22 months' imprisonment in default.

A foreign worker incurred multiple fractures after a two-tonne slab fell on him. This is the highest fine ever imposed for a breach of s 37 to date.

2.169 A director, Steven Martin, was ordered to pay £21,000 compensation to a worker's widow and daughter after pleading guilty to an offence under s 37 of the HSWA 1974. The inexperienced worker died after falling a significant height whilst working at night.

2.170 Richard Pratt was fined £4,000 after pleading guilty to an offence under s 37 of the HSWA 1974 after the death of a foreign worker who fell from height on a Dundee construction site. His company, Discovery Homes (Scotland) Limited, was also fined £5,000 but no costs, as these cannot be imposed in Scotland.

2.171 Also in Scotland, a company director, Thomas Thompson, was fined £1,800 after pleading guilty for failing to make sure his company undertook a proper risk assessment. A foreign worker was electrocuted and died. The company was also fined £9,000 by Perth Sheriff Court for a breach of s 2(1) of the HSWA 1974.

2.172 Norman Ellis was sentenced to 100 hours' community service after pleading guilty to an offence under s 37 of the HSWA 1974 after one of his employees was killed when a wall toppled over.

2.173 In September 2009, Thames Valley Police and PC Micklethwaite were fined £40,000 and £2,000 respectively after pleading guilty to breaches of the HSWA 1974 when an employee was accidently shot in a classroom demonstration. The employee survived very serious injuries and was in intensive care for two weeks.

2.174 Under the Company Directors Disqualification Act 1986, the courts may also disqualify directors who have been found guilty of health and safety offences. A total of eight directors have been disqualified since the Act came into force.[1] However, the trend is now that, if a director is convicted, the HSE requests the court to consider making a disqualification order. The maximum period for which a director can be disqualified in the magistrates' court is five years, but in the Crown Court the maximum is 15 years.[2]

1 HSE *Health and Safety Offences and Penalties 2002/2003 – A report by the Health and Safety Executive* (2003).

2 See above for some recent examples.

2.175 In 1992 the Crown Court at Lewes fined Rodney James Chapman £5,000 for breach of s 37(1) of the HSWA 1974 and disqualified him from holding office as a company director for a period of two years for contravening the terms of a prohibition notice.

2.176 A recent example of how severe this can be occurred in Scotland, when two directors were disqualified for seven years and their operating licence was revoked by the traffic commissioner. Munro and Sons (Highland) Ltd were effectively put out of business after a young woman died when a 30-tonne digger came off a low loader. The company was found to have put profit before safety.

Examples of fines imposed[1]

2.177 In October 2008, Sharaz Butt, the director of Alcon Construction Ltd of Norwich, was sentenced to 12 months' jail in Norwich Crown Court after pleading guilty to the manslaughter of Wu Zhu Weng. He also received a five-year disqualification from acting as a company director after pleading guilty to breaching s 37 of the HSWA 1974. Mr Butt was prosecuted by Norfolk police, with assistance from the HSE, over the incident on 31 January 2008. Mr Weng was working for Alcon Construction on the refurbishment of a building on Trowse, Norwich, when he fell 12 feet through a skylight, suffering fatal injuries. In his sentencing remarks, Judge Peter Jacobs said Mr Butt had shown a 'cynical disregard' for his workers, including Mr Weng, by employing them illegally and failing to ensure their safety.

1 See also **Chapter 1**.

2.178 Roger Folkes, managing director of Folkes (Great Yarmouth) Ltd, pleaded guilty of a breach of HSWA 1974, s 37 at Norwich Crown Court in June 1999 and was fined £10,000, following the death of a driver of a poorly maintained digger who drowned when his machine rolled into a lake. Both the brakes and the handbrake of the digger were not working. The company was fined £15,000.

2.179 John Bridston, the managing director of Fresha Bakeries and of Harvestime Ltd, was fined £10,000 for each company's breach of health and safety,[1] making a total fine of £20,000, on 18 July 2001, for two offences pursuant to HSWA 1974, s 37.

1 See HSE *Health and Safety Offences and Penalties 2002/2003 – A report by the Health and Safety Executive* (2003) with respect to the fines for each company.

2.180 Christopher Allot, director of Triplex Components Group, was fined £6,000 for a breach of HSWA 1974, s 37 at Cardiff Magistrates' Court in May 2002. This was in relation to an incident where an employee suffered serious injuries to his wrists while carrying out maintenance work. The company was also fined £6,000 for a breach of HSWA 1974, s 2.

2.181 Bipin Bhagani, a director of Whitefields Care Homes Ltd, was fined £7,500 after pleading guilty to a breach of HSWA 1974, s 37 at Reading Crown Court on 29 April 2003. This followed an incident at the company's care home where an 85-year-old resident sustained serious burns to her legs while asleep in her bed. Her legs had come into contact with a hot radiator and pipes. The company pleaded guilty to a breach of HSWA 1974, s 3 and was fined £10,000.

2.182 Keith Stait, managing director, and Nicholas Martin, technical director, of Mays (Pressure Diecastings) Ltd were each fined £3,000 at

Southwark Crown Court in April 2003. This was in relation to employees and members of the public being exposed to the risk of legionella bacteria because the company had failed adequately to maintain cooling towers at its site. Martin had received a report from a consultant, stating that the company needed to establish a regular cleaning and water treatment programme for the towers. The managing director was made aware of the report, but both he and the technical director failed to take action. The company was fined £30,000.

2.183 Mark Foley and Daniel O'Brian, directors of Wessex Stone Ltd, were each fined £10,000 for breaches of HSWA 1974, s 37 at Bristol Crown Court. This was as a consequence of a worker's arm being severed when it became trapped in an unguarded conveyor belt. The company was fined £15,000.

2.184 Paul Mackenzie, director of Philip Services Ltd (Europe) Ltd, and Peter Preston, service manager of the same company, were each fined £5,000 for a breach of HSWA 1974, s 37 on 16 June 2003, resulting from the death of an employee who was killed at a Ford factory when he fell into an emulsion paint over-spray capture tank (see **2.149** above). The prosecution of Philip Services (Europe) Ltd for a breach of HSWA 1974, s 2 was dropped, as the company no longer existed in the UK.

2.185 In 2000, Keith Dickson, a planning supervisor manager, was fined £5,000 and costs of £2,500 for breach of s 7 of the HSWA 1974. A property collapsed, killing one worker and injuring two others.

2.186 In 2006, at Croydon Crown Court, Lee Smith, a warehouse manager, was fined £1,000. Two workers were inspecting a forklift truck with a Man riding cages. The pair climbed inside the cage to have a closer look at it when the manager as a joke raised them up 20 feet above ground. The forklift toppled over and the two men were seriously injured.

2.187 George Graham was fined £100,000 and ordered to pay £20,000 after he pleaded guilty to an offence under s 2(1) of the HSWA 1974 after an employee died when a five-tonne vehicle toppled over, crushing Stuart Mullen.

2.188 The case of Gillian Beckingham (who was acquitted of manslaughter but convicted of a breach under s 7 of the HSWA 1974) may be an indication that fines for directors and managers convicted of health and safety offences are increasing. Barrow Council, her employers, who had pleaded guilty to a breach of s 3 of the HSWA 1974, were fined £125,000 and Ms Beckingham was fined £15,000.

Breaches of HSWA 1974, s 7

2.189 In the magistrates' court the maximum fine for a breach of HSWA 1974, s 7 is £20,000, and in the Crown Court it is unlimited. Both courts can impose a custodial sentence, limited to six months in the magistrates' court and two years in the Crown Court.

Examples of fines imposed[1]

2.190 In 2013, Kenneth Miller received a six-month suspended jail sentence after pleading guilty to an offence under s 7 of the HSWA 1974. He was also ordered to pay £600 costs. In essence, whilst operating a loading shovel, he crushed another worker who died. The HSE said that it:

'took the rather unusual decision in this case to prosecute an individual rather than a company because it was clear that Kenneth Miller had totally failed to take the care that was necessary when operating a large vehicle on a busy waste site … employees have a duty to take reasonable care for the safety of others, particularly when they are operating dangerous machinery.'

2.191 In April 2009, Peter Bacon was sentenced to £5,333 in fines and costs for a breach of s 7 of the HSWA 1974 after an employee's arm was broken in a waste screening machine.

1 See also **Chapter 1**.

2.192 Mr Paul Nolan, site foreman and employee of A&A Building Services Ltd, was found not guilty of the CPS manslaughter charge but guilty of a charge under the HSWA 1974. He was fined £5,000, to be paid within 28 days, or face three months' imprisonment. The company had already pleaded guilty, at an earlier hearing, to the health and safety breaches, so A&A Building Services Ltd was fined £55,000. Legal proceedings were brought after Mr Alexander Hayden (28) was killed when a nine-tonne dumper truck, that he was driving, overran the edge of an embankment. In trying to jump clear, Mr Hayden sustained extensive injuries when he was crushed by the toppled truck and was pronounced dead at the scene, in Fenton, Stoke-on-Trent.

2.193 Mark Holland was fined £159 for breach of HSWA 1974, s 7(a) at Bridgenorth Magistrates' Court in October 1997. He poured a small amount of floor-cleaning material, which he found in a cupboard, underneath a toilet door. A 17-year-old part-time worker was locked in the toilet. This was done as a practical joke. The liquid contained hydrochloric acid which reacted with dirt on the floor to produce toxic fumes. The 17-year-old suffered from the effects of the fumes.

2.194 Brian Watkins was fined £250 for a breach of HSWA 1974, s 7(a) at Newport Magistrates' Court in April 1999. Watkins, a school teacher, was prosecuted after pupils were splashed with molten metal when a class-room experiment went wrong.

2.195 David Alexander, a London Underground Ltd (LUL) manager, was fined £2,400 in relation to four counts under HSWA 1974, s 7(a) at Blackfriars Crown Court on 16 July 2001 because he forced track workers to carry out repairs using metal tool beside live rails, risking electrocution. LUL was later fined £225,000 in relation to the same matters in January 2003 (see **2.139** above).

2.196 Denis Masters, the Chief Engineer of Fresha Bakeries, and Brian Jones, an employee of Fresha Bakeries, were fined £2,000 and £1,000 respectively on 18 July 2001 for breaches of HSWA 1974, s 7(a). See **2.179** above with respect to the prosecution of Fresha Bakeries.

2.197 Paul Jones was fined £3,000 on 12 March 2002 at Cambridge Crown Court after pleading guilty to a breach of HSWA 1974, s 7(a). He was the driver of a delivery van reversing into an access road, which killed a sub-contracted employee to SDC Builders Ltd. The company were fined for breaches of HSWA 1974, ss 2 and 3 (see **2.141** above).

2.198 Paul Dove, a schoolteacher, was fined £2,000 at Portsmouth Crown Court in November 2002 after a nine-year-old child drowned whilst on a school sailing trip. The child was part of the group of which he was in charge. The school was also fined £25,000.

2.199 John Cullen, a site foreman employed by O'Rourke Civil and Structural Engineering Ltd, was fined £1,500 at the City of London Magistrates' Court on 26 June 2003 after pleading guilty to a breach of HSWA 1974, s 7(a). He was given the task of establishing the site, which included the erection of a site hoarding around the perimeter. Several large billboards were in place which prevented the hoarding being erected. Although it was not Cullen's job to get them removed, he came to make arrangements to have them lifted by an excavator. The project manager told him to stop, but he continued. As a result, one of the employees suffered severe injuries to his legs.

2.200 Paul Ellis, a teacher from Fleetwood High School, pleaded guilty in September 2003 to manslaughter at Manchester Crown Court. The prosecution, brought by the Crown Prosecution Service, followed the death of Max Palmer, a ten-year-old boy, in May 2002. Max was with his mother, who was helping with a trip from the school to the Lake District organised by Paul Ellis. Mr Ellis led an activity which was to jump into a natural rock pool in Glenridding Beck. Max got into difficulties in cold and turbulent water, and was swept out of the pool and down the beck. Mr Ellis was sentenced to 12 months' imprisonment in relation to the manslaughter. However, he also pleaded guilty to a breach of HSWA 1974, s 7(a) for which, according to the HSE website,[1] he was sentenced to six months' imprisonment to run concurrently. As explained at **2.189** above, the court does not have the power to impose such a sentence.

1 At www.hse.gov.uk.

2.201 In 2004, Lee Grundy was fined £1,000 for a breach of s 7 of the HSWA 1974 when he ran over an employee of another company while driving a forklift truck.

2.202 In 2009, Mr Atterby, a self-employed health and safety consultant, was fined £1,000 and ordered to pay £700 costs by Bradford Magistrates' Court. He had been found guilty of an offence under s 36(1), in that he had not undertaken an adequate risk assessment concerning exposure to Sandstone dust, which led to workers having increased exposure to the dust. He worked for a quarry company which had been prosecuted and fined. The HSE inspector is quoted as saying '… you cannot outsource your responsibilities – the duty of care remains with you as an employer and the selection and use you made of consultants is crucial' (*Health and Safety at Work*, May 2009, p 9).

2.203 In 1998, Charles Heinrich, the health and safety manager of the restaurant chain Fatty Arbuckle's, was fined £3,000 and costs for a breach of s 7 of the HSWA 1974 after an employee received a fatal electric shock from a plate warmer which had previously been reported by employees. Communication problems meant that Mr Heinrich was not aware of this. The inquest returned a verdict of unlawful killing, but there was no prosecution for manslaughter.

Section 42 of the HSWA 1974

2.204 This section provides the court with the ability to impose orders to remedy certain deficiencies within a certain period of time. This is similar to the provision in the Corporate Manslaughter and Corporate Homicide Act 2007.

Failure to obey the order can lead to an unlimited fine and two years' imprisonment in the Crown Court.

MANSLAUGHTER[1]

2.205 If an individual is convicted of manslaughter in relation to a work related incident, then he or she is likely to receive a custodial sentence in the region of two to three years. Obviously, however, each case will turn upon its own facts. In certain circumstances, the court might be prepared to suspend the sentence.

2.206 Where corporate manslaughter is proved, the company will also be fined.

1 See also **Chapter 1**.

Examples of sentencing in manslaughter

2.207 Allan Turnball was jailed in 2012 for three years for manslaughter after Mr Joyce was knocked from his cherry picker. There had been a four-week trial in Newcastle Crown Court. Another director was fined £30,000 and £50,000 costs for offence under ss 37 and 2 of the HSWA 1974.

2.208 In 2012, a gas fitter was jailed for three years for manslaughter for a botched job on a boiler. A young girl died.

2.209 In *R v Wacker*,[1] a Dutch lorry driver was sentenced to 14 years' imprisonment after being convicted of 58 counts of manslaughter after illegal immigrants suffocated in a lorry being driven by him. Two survived. The Court of Appeal interestingly found that the legal principle of *ex turpi causa* did not apply to manslaughter cases. There was an Attorney General's reference, and the Court of Appeal held that 14 years was appropriate for this '... quite dreadful case'.

1 [2003] Cr App R (S) 92.

2.210 In *Lin Liang Ren*,[1] the main perpetrator of the Cockle picking case, where at least 21 people died in Morecambe Bay, was sentenced to 12 years' imprisonment for manslaughter by Henriques J. The judge said that Mr Lin Liang Ren had cynically and callously exploited his countrymen. He also pointed out that he had waited nearly an hour before alerting the authorities. Mr Lin Liang Ren was also sentenced to two years' imprisonment for conspiracy to pervert the course of justice, which was ordered to run concurrently.

1 *Health and Safety Executive v Lin Liang Ren* (unreported, 2006) (Preston Crown Court).

2.211 In *Connolly and Kennet*[1] the two directors were sentenced after appeal to seven and two years' imprisonment. This arose out of the Tebay rail disaster, where four workers died when a rail carriage descended at high speed in the dark. Mr Connolly had also been convicted on charges of perverting the course of justice.

1 [2007] EWCA Crim 270.

2.212 Roy Dank, owner of North Eastern Roofing, was jailed for 22 months after pleading guilty on the eve of the trial for manslaughter. The case concerned the death of a worker who fell through a skylight. The director had received two previous HSE warnings about working at height.

Interestingly, the family and unions had judicially reviewed the decisions not to prosecute and the CPS then reconsidered.

2.213 In July 2009 a director, Colin Holton, was convicted of manslaughter and sentenced to three years' imprisonment for inadequate supervision and equipment after a wall collapsed, crushing the worker. The site contractor was also sentenced to nine months' imprisonment for running a company whilst disqualified.

2.214 Steven Smith was sentenced to 22 months' imprisonment after pleading guilty to manslaughter and attempting to pervert the course of justice. One of the workers fell from a height of seven metres through a skylight. He also pretended to police that the deceased had been issued with a harness, when in fact they were only bought after the accident.

2.215 David Johnson was sentenced to three and a half years' imprisonment at Norwich Crown Court after being convicted of manslaughter and of an offence under s 3 of the HSWA 1974. A client died from carbon monoxide poisoning after a gas boiler was blocked off, following chimney work that had been undertaken by Mr Johnson.

2.216 James Johnson was sentenced to 18 months' imprisonment after pleading guilty to gross negligence manslaughter at Newcastle Crown Court. He was a foreman at a recycling depot in Sunderland. He has been driving a 20-tonne vehicle 'by instinct' and could not see out of the front window when he killed an employee in the yard. The company pleaded guilty to an offence under s 2 of the HSWA 1974 and was fined £15,000 plus costs.

R v Adomako

2.217 In *R v Adomako*,[1] the leading case on gross negligence manslaughter, the defendant was sentenced to six months' imprisonment suspended for 12 months. In this case the defendant was an anaesthetist in charge during an eye operation and failed to notice that an endotracheal tube (to allow the patient to breathe normally) had become disconnected for approximately six minutes.

1 [1995] 1 AC 171.

R v OLL Ltd and Kite

2.218 In this case, OLL Ltd and Mr Kite, the managing director, were convicted of the manslaughter in November 1994 of four students who died during a canoeing trip in Lyme Regis. The group of students was accompanied by a schoolteacher and two unqualified instructors. The group was swept out to sea and the four students died. The company had not provided distress flares and had not advised the coastguard of the expedition.

2.219 Mr Kite was originally sentenced to three years' imprisonment, which was reduced to two years' on appeal. The company was fined £60,000.

2.220 When the trial judge sentenced Mr Kite, he said:

> '… beyond doubt, these matters are so serious as to demand a sentence of immediate custody and of some substance.'

2.221 At his appeal the Court of Appeal agreed with the trial judge's view but reduced the sentence. The court observed:

> 'Mr Kite is now aged 46 and he is a man of previous impeccable character. Any prison sentence imposed on a man in these circumstances is, of course

devastating to him. Nonetheless as we have said, the facts quite clearly demand a substantial sentence.'

R v Jackson Transport (Ossett) Ltd and Jackson

2.222 Alan Jackson was the sole director of the company. An employee died after cleaning chemical residues at the rear of a road tanker. The company had not provided preventive equipment, supervision or adequate training. Alan Jackson was sentenced to 12 months' imprisonment and the company was fined £15,000.

R v Litchfield

2.223 In *R v Litchfield*,[1] Mr Litchfield was both master and owner of a square-rigged schooner which foundered off the north Cornish coast. He was sentenced to 18 months' imprisonment concurrently on three counts of manslaughter.

1 [1998] Crim LR 507.

2.224 He did not appeal against sentence, but did appeal against conviction, which was unsuccessful. At that appeal, Simon Brown LJ stated:

> 'We add only this. We do not pretend to have found this an altogether easy case. Manslaughter convictions based on findings of gross negligence are always troubling. Defendants – very often, as here, of previously unblemished character – are being prosecuted for consequences they never for an instant intended or desired. It is to juries, however, that these difficult decisions are entrusted. Fidelity to the Jury system requires that [Mr Litchfield's] conviction stands.'

R v Roy Bowles Transport Ltd, Stephen Bowles and Julie Bowles

2.225 In October 1999, brother and sister Stephen and Julie Bowles were directors of the company. They were convicted of manslaughter in relation to a seven-vehicle pile-up on the M25 after a lorry driver working for them fell asleep at the wheel. The drivers at the company had been working excessive hours in the knowledge of both directors. Stephen Bowles and Julie Bowles were sentenced to 15 months' and 12 months' imprisonment respectively, suspended for two years.

R v Edward Crow and Alistair Crow

2.226 In July 2001 the two defendants, farmers, were found guilty of the manslaughter of a 16-year-old trainee. He was crushed by a seven-tonne JCB potato loader that he was operating. As a 16-year-old with limited experience, he should not have been operating the machine. Mr Alistair Crow was jailed for 15 months and his father received a one-year suspended sentence.

R v Brian Dean[1]

2.227 In April 2002, Brian Dean, former owner of Brian Dean Demolition and Civil Engineers, was convicted of the manslaughter of two employees who died when a kiln collapsed on top of them. His conviction was quashed on appeal. However, at first instance when he was sentenced he received a custodial sentence of 18 months.

1 [2002] EWCA Crim 2410.

R v Teglgaard (UK) Ltd and John Horner

2.228 Mr Horner was the managing director of the company. A worker was crushed to death when a pile of poorly stacked hardwood packs toppled over on top of him. Mr Horner and the company were convicted of manslaughter in March 2003. Mr Horner was sentenced to 15 months' imprisonment suspended for two years, and the company was fined £25,000.

R v Ian Morris

2.229 In August 2003, Ian Morris, the owner of a factory, was convicted of gross negligence manslaughter of two employees who died when they inhaled toxic fumes during an unsafe paint-stripping operation. He was sentenced to nine months' imprisonment.

R v Michael Shaw[1]

2.230 Michael Shaw, the boss of Change of Style Ltd, was jailed for 15 months for the manslaughter of a 22-year-old employee after appeal judges decided the original two-year suspended sentence was 'unduly lenient'. At a Southampton factory in May 2003, David Bail, a stone-cutting machine operator, died when his head was crushed between the machining head of a 'Bavelloni' stone-cutter and a fixed part of the unit. The prosecution argued that Michael Shaw was the directing mind of the company and a hands-on boss who knew what was going on on the factory floor. The original trial judge, in imposing the suspended sentence, took into account the fact that people would lose their jobs should Shaw be jailed. While the appeal judges agreed that this was a factor to be considered, they held that it should not be a bar to sending someone guilty of such a serious offence to prison, and that the suspended sentence was 'unduly lenient', given the nature of the offence. Shaw and Change of Style were also sentenced for a raft of health and safety offences at the original trial in July. The company pleaded not guilty to, but was found guilty of, two offences under s 2(1) of the HSWA 1974 of failing to safeguard employees; and pleaded guilty to a breach of reg 3 of the Management of Health and Safety at Work Regulations 1999, by failing to carry out a suitable and sufficient risk assessment. It was fined £10,000 on each charge. A charge of corporate manslaughter against the company was left on file.

1 [2006] EWCA Crim 2570.

COSTS

2.231 The power to award costs against a defendant is contained in the Prosecution of Offences Act 1985, s 18(1). In relation to the level of costs that the defendant is ordered to pay:

- They should be just and reasonable.
- They will include the cost of the prosecuting authority carrying out investigations with a view to prosecuting.[1]

- The defendant can be ordered to pay the whole of the prosecution costs in addition to any fine imposed.

1 See *R v Associated Octel Co Ltd* [1996] 4 All ER 846.

2.232 In the case of *Ex parte Dove, R v Northallerton Magistrates' Court*, there was an application for judicial review of the defendant's costs order. The order was challenged as being disproportionate to the fine imposed.

The decision enunciated the following principles:

- Sum of costs and fine not to exceed amount it is reasonable to order the defendant to pay in light of their means.
- Costs order not to exceed prosecution's actual and reasonable costs.
- Purpose is compensation not punishment.
- Appropriate fine should be fixed first.
- In absence of disclosure by the defendant of data regarding their financial position, reasonable inferences as to his means could be inferred from the evidence heard and all other circumstances of the case.
- The defendant must be given a fair opportunity to adduce relevant financial evidence and make submissions on the order.
- The defendant should be given advance notice of any unusual or unconventional order.

2.233 While the costs of defending a prosecution may be covered by insurance (directors' and officers' policy, employers' liability policy or public liability policy), the prosecution's costs, like any fine imposed by the court, will not usually be covered by insurance. Even if acquitted, companies and other corporate bodies cannot recover their costs since 2012.

Compensation

2.234 The courts can impose compensation orders but very rarely do so, as they can complicate civil cases, as any amount imposed by a criminal judge can be offset as part of a civil claim.

CONCLUSIONS

2.235 What can be seen from the examples, given above, of fines following breaches of the HSWA 1974 is that, since the guidance was handed down in *R v F Howe & Son (Engineers) Ltd*[1] and the SGC Guidance of 2010 (see above), the courts have on occasions been prepared to impose substantial fines. The latest *Sellafield and Network Rail* judgment (see **2.17** above) further raises the bar.

1 [1999] 2 All ER 249.

2.236 It is clear from the cases above that the size of fines for breaching health and safety has increased. Few cases have yet been decided regarding corporate manslaughter, and the fines have all remained at a similar level, and nowhere near the level of unlimited fines. Perhaps the biggest driver for looking after health and safety is the option of an individual being prosecuted and the reputation of an organisation. It will also make bidding for new contracts more difficult in the future. For a large organisation, where an employee is killed at work, the investigation process and the fact that they were being investigated for a corporate manslaughter prosecution may cause great damage to the company's value and morale. The other factor which is playing an increasing role is that of insurance companies, who are insisting that companies deal with health and safety and fire issues in specific ways, or else they will increase premiums or refuse cover. Successful corporate defendants can no longer have their costs refunded.

2.237 It should be remembered that fines and prosecution (and investigation) costs are not all covered by insurance (and are not tax deductible).

2.238 However, it is not only the cost in terms of money, but also in terms of reputation, which can have an indirect impact upon the financial well-being of the company.[1] The latest Court of Appeal judgment relating to Sellafield and Network Rail may also cause many large companies and organisations to consider, when appropriate, to enter early pleas rather than fighting cases. Only time will tell.

1 *R v F Howe & Son (Engineers) Ltd* [1999] 2 All ER 249.

Chapter 3

Practical issues

Gerard Forlin QC, Barrister, Cornerstone Barristers, London; Denman Chambers, Sydney; Maxwell Chambers, Singapore

Dr Louise Smail, Risk Consultant, Ortalan, Manchester

INTRODUCTION

3.1 There have been many changes since the last edition, and this chapter looks at a few of these.

3.2 Professor James Reason wrote:[1]

> 'The mind is prone to match like with like. It is therefore natural for us to believe that disastrous accidents must be due to equally monstrous blunders. But the close investigation of organizational catastrophes has a way of turning conventional wisdom on its head. Defences [designed to prevent accidents] can be dangerous. The best people can make mistakes. The greatest calamities can happen to conscientious and well run organizations. Most accident sequences, like the road to hell, are paved with good intentions – or with what seemed like good ideas at the time.'

1 Professor J Reason *Managing the Risks of Organizational Accidents* (1997) p 21.

3.3 When a criminal investigation is instigated, it should not be assumed that those investigating have all the necessary expertise to understand the mechanics and causes of the incident. It can be difficult, particularly in a police investigation, to explain the circumstances as a whole and why the incident may have occurred. There can be a tendency for the investigators to concentrate, to the exclusion of other factors, upon what they believe the individual or company has done wrong. This is particularly the case in complicated specialised industrial sectors.

3.4 As Professor Reason warns, the 'mind is prone to match like with like'.[1] Thus it is easy for an investigation to be swayed by the consequences of a failure or series of failures, rather than to put the incident in context in relation to each failing.

1 Reason, fn 1 to **3.2** above.

3.5 When an incident happens and an investigation begins, it is easy for panic to set in and for confusion to reign. Trying to decide the best way to proceed, at that stage, can place the company or individuals under investigation in an invidious position. Often, senior people within a company believe that they simply have to explain what has happened to those investigating and that will be an end to the matter. This can be dangerous and foolhardy. It is also fair to say that what happens in the first 48 hours after the incident, the so-called 'golden hours', will have important ramifications for the way the investigation proceeds and the direction of it. Planning, organisation and consideration of tactics are essential at an early stage – this also applies to internal reports. Obviously, the prevention of accidents should be the aim of any employer. Organisations also need to consider all aspects of emergency (including COMAH) and pandemic planning. Further, internal reports are increasingly deemed disclosable.[1]

1 See *West London Pipeline and Storage Ltd v Total UK Ltd* [2008] EWHC 1729 (Comm), particularly at para 86 onwards.

3.6 The role of the media should not be underestimated. The way in which the incident is reported can have a dramatic effect upon a company's reputation. The media will be interested in a 'good' story (see **Chapter 15**).

3.7 This chapter looks at some of the issues that will face companies, individuals and their lawyers if a criminal investigation and/or prosecution results from a work related incident. The issues covered are by no means an exhaustive list. The practical issues are increasing in both numbers and complexity.

PROGRESS OF A PROSECUTION

3.8 It is beyond the scope of this book to set out in detail how a work related incident prosecution progresses through the courts. What follows is merely an outline.

3.9 If companies are prosecuted for manslaughter or health and safety offences, they will be summonsed. Proceedings for health and safety offences must commence in the name of individual inspectors. If an individual is prosecuted for manslaughter, he or she will be charged with the offence. If an individual is prosecuted for breach of s 7 or 37 of the Health and Safety at Work etc Act 1974 (HSWA 1974), he or she will be summonsed to appear at court. Since January 2009, individuals who are convicted under certain sections of the HSWA 1974 can be imprisoned for up to two years.[1]

1 See *R v Wilson* [2013] EWCA Crim 1780 and **Chapter 2**.

3.10 Health and safety offences are 'either way' offences; in other words, they can be heard in the magistrates' court (if it accepts jurisdiction, which is increasingly rare in everything but the most minor of cases and/or injuries) or in the Crown Court. Manslaughter is an indictable offence so can only be tried in the Crown Court. If a company or individual is prosecuted for manslaughter and health and safety offences relating to the same incident, the trial of these will usually be heard together in the Crown Court. It is important to recall that magistrates should not be informed of previous convictions (or bad character) when deciding on the suitable venue.[1]

1 See also *R v Lawrence* [2014] 1 Cr App R 5 on preparatory hearings.

3.11 If the prosecution is for health and safety offences only, there will be a hearing where a plea will be entered, and it will then be determined if the matter is going to be heard before the magistrates or to go to the Crown Court (this is known as the 'mode of trial'). The magistrates may decline jurisdiction (because of the serious nature of the case), or the defendant may elect for the case to be committed to the Crown Court. Increasingly, cases are going to the Crown Court.

3.12 One also needs to consider the Criminal Procedure Rules 2013,[1] which came into effect on 7 October 2013. Lord Thomas CJ set out the new rules[2] which:

> '… had the aim to revise thoroughly the 2002 practice directions to bring it up to date with changes to the law and to best practice. The new practice directions have also been restructured with the aim of integrating them and complementing the various parts of the Criminal Procedure Rules.'[3]

See also *R v (Rahmdezfouli) v Crown Court at Wood Green*.[4]

1 SI 2013/1554.
2 [2013] EWCA Crim 1631.
3 *Archbold*, 2013.
4 [2013] EWHC 2998 (Admin).

3.13 In manslaughter cases, the first hearing will be in the magistrates' court, which will send the case direct to the Crown Court pursuant to s 51 of the Crime and Disorder Act 1998. At the first appearance in the Crown Court, a timetable will be set for the case. The prosecution will then serve the evidence upon which it relies. The defence will have to consider whether it wishes to make an application for the case to be dismissed. If no application is made, or it is unsuccessful, there will be a plea and case management hearing (PCMH). After that hearing, the matter will either proceed to trial or, if a guilty plea is entered, for sentencing. The timing of the guilty plea may greatly affect the amount of credit that the judge will ultimately allow.[1] See also *R v Martin*[2] and *Nichols v DPP*.[3]

1 See *R v Barber* [2002] 1 Cr App (S) 48 and *R v Balfour Beatty Rail Infrastructure Services Ltd* [2006] EWCA Crim 1586, [2007] Bus LR 77, [2007] 1 Cr App R (S) 65. See also the Criminal Practice Directions (preliminary proceedings) allocation guidelines (2013).
2 [2013] EWCA Crim 2565.
3 [2014] EWHC 4365 (Admin).

FUNDING

3.14 It is the professional duty of a lawyer to advise his or her client of the availability of public funding. Public funding is available to individuals in relation to criminal matters for:

- advice and assistance;
- advocacy assistance; and
- representation.

3.15 Public funding can only be provided where the solicitors hold a general criminal contract with the Legal Services Commission. If the client decides not to proceed with public funding, they may be asked to confirm that in writing. It is crucial to remember that the legal aid budget is continually decreasing, and the number of solicitors (and barristers) prepared to do legal aid criminal work is rapidly dwindling. This trend is due to the continuing changes to legal aid funding and the Jackson reforms. Further, where proceedings are commenced on or after October 2012, even an acquitted person can only, at a maximum, recover costs at a legal aid rate. This does not apply to corporates, who cannot recover any costs. This development may have an impact on cases, particularly in smaller uninsured organisations who may not want to contest the proceedings.

3.16 With a private client (ie one not in receipt of public funding), it is important that the client is sent a client care letter, not only setting out who will be dealing with the case and the complaints procedure of the firm, as required by the Solicitors' Practice Rules, r 15, but also providing details in respect of costs, as required by the Solicitors' Costs Information and Client Care Code.

3.17 In the vast majority of cases, the cost of the defence in work related incidents will be covered by insurance.

3.18 Directors' and officers' liability and company reimbursement policies (often referred to as 'D&O policies') provide insurance for senior executives against legal bills arising from their corporate responsibilities. It is not possible to obtain insurance to pay fines and penalties. D&O policies are not compulsory.

3.19 These policies are on a 'claims made' basis. That is to say, they apply not when the incident happens but when the executive needs to rely upon the policy (for example, when first interviewed by the police, which may be some time after the incident). The limit of indemnity is aggregate, and costs are included in the limit of indemnity. Thus, if more than one director being prosecuted is covered by the policy, once their costs added together have reached the limit of the policy, there will be no further cover (and so they may need to rely upon public funding for further representation). These funds are not bottomless and can run out relatively quickly, especially if there are many potential defendants or people being interviewed under caution.

3.20 As for the costs of companies in corporate manslaughter/health and safety prosecutions, these may be covered by employers' liability insurance or public liability insurance (depending upon the nature of the matter). It is important to check the terms of the policy. Some policies may state that they only cover a breach of statutory duty (which therefore will not include corporate manslaughter proceedings), and others may say they only cover proceedings in the magistrates' court. If convicted, many policies state that the money would need to be repaid. It is also crucial to remember that, if an insurer undertakes audits of an organisation, those audits can be used by the prosecution to cross-examine the organisation during a criminal or civil trial, and this can have devastating effects. Further, there appears to be less funding generally by insurers, and often public liability policies are limited in financial cover.

CONFLICTS OF INTEREST

3.21 When a work related death or serious incident occurs, it is understandable that there is a desire on the part of the employer, their insurers and legal representatives to want to investigate the matter as soon as possible. There will be a wish to interview 'front line' workers involved and no doubt also managers. However, care needs to be taken as to how and when this is done.

3.22 If there is a possibility of any criminal investigation of an individual, the first issue that a legal representative of a company needs to consider is whether there is any conflict of interest between the company and the individual. The possibility of a conflict might be more obvious if a criminal investigation (either by the police or the Health and Safety Executive (HSE)) is being dealt with in the immediate aftermath. However, a conflict might be a possibility if dealing initially with civil issues (for example, personal injury claims) or the preparation for an inquest (note that the coroner can refer the papers to the police if he or she believes that there should be a criminal investigation). See the Chief Coroner's guide to the Coroners and Justice Act 2013.

3.23 The difficulty will arise if the company and an employee have the potential to become co-defendants. The result may be that the solicitor would have to decline to act for any party. Before the solicitor could agree to continue to act for one party (for example, the company), the solicitor would need to consider carefully whether there is any information in his or her possession relating to the other 'clients' which may be relevant to the retained client.[1]

1 See *R v Ataou* [1988] QB 789.

3.24 It should also be borne in mind that conflicts of interests can arise between the company and its directors if both are being investigated. The Law Society gave specific advice relating to this in 2009.[1] In essence, however, the solicitor should err on the side of caution and offload any potential conflicts at an early stage. Increasingly, firms of solicitors have built up relationships with other specialist firms so that they are ready when a potential conflict occurs.

1 'Initial Interviews' practice note (6 October 2011), www.lawsociety.org.uk.

INTERVIEWS UNDER CAUTION

3.25 An interview under caution and a suspect's detention at a police station are governed by the provisions of the Police and Criminal Evidence Act 1984 (PACE). The tactics of how an interview under caution by the police is dealt with can have a significant bearing upon the progression of the case. It is important that any solicitor representing a suspect in this type of interview in relation to a work related incident has experience of representing clients in this position, is familiar with the Codes of Practice to PACE, and understands the legal and evidential issues involved in this type of matter. It is also vital to ascertain if the interviewee is speaking on behalf of themselves and/or the organisation.

3.26 Before the interview commences, the suspect will be cautioned. The caution is:

> 'You do not have to say anything, but it may harm your defence if you fail to mention something that you later rely on in Court. Anything you say may be used in evidence.'

3.27 Section 34 of the Criminal Justice and Public Order Act 1994 permits adverse inferences to be drawn from a defendant's failure, when interviewed by the police, to mention facts that he or she later relies on in defence at trial.

3.28 An interview with the police will often be by prior arrangement. Often the interview will take place on a voluntary basis, ie the client will not be arrested and just interviewed. He or she will, of course, still be cautioned at the start of the interview and informed of the offences for which he or she is being investigated.

3.29 However, it is becoming more common for the police to insist on arresting the client when he or she attends the police station. If this happens, then at the end of the interview, the client will normally be bailed to return to the police station at a future date. Failure to surrender to bail is a criminal offence.

3.30 It is prudent for the solicitor to discuss, with the officer who will be carrying out the interview, how he or she intends to conduct matters.

3.31 Interviews will be tape-recorded. It is now becoming common for interviews concerning manslaughter investigations carried out pursuant to the work-related deaths protocol[1] to have an HSE officer present in the interview as well. Increasingly, HSE inspectors monitor the interview remotely. This can cause the interview to become very disjointed.

1 HSE 'Work-related deaths: a protocol for liaison' (2003). The full text is also available at www.hse.gov.uk/pubns/misc491.pdf.

3.32 Before the interview takes place, the officer will normally provide disclosure of relevant documents, which set out the case against the suspect. Sometimes, this can be sent to the solicitor in advance of the interview date. The disclosure may include a case summary that the officer may be prepared to allow the legal representative to have a copy of in order to assist in taking instructions. The defence often have to write a series of letters asking for disclosure.

3.33 The decision to be taken after disclosure is how to deal with the interview. There are three options:

(1) to make no comment;
(2) to give a full interview; or
(3) for the suspect to read out a prepared statement.

3.34 Some situations where it might be appropriate for suspects to make no comment are as follows:

- the greater the complexity of the case, the more likely that a court will consider it appropriate for a suspect to give a no comment interview;[1]
- where the police have not given sufficient notice prior to disclosure to enable the suspect properly to comment; or
- the police have little or no evidence against the suspect.

1 See *R v Roble* [1997] Crim LR 449.

3.35 Often, in work related incidents the police, as lay people, have difficulty understanding the mechanics and the circumstances of the incident. The danger is that misunderstandings can arise simply because the police do not understand the explanation being given, which can lead in some cases to charges.

3.36 One way of overcoming this is for a prepared statement to be read out and then for no comment to be made in response to further questions. In the case of *R v Knight*[1] the defendant was accused of two offences of indecent assault. A prepared statement was read out at interview. At trial, the judge directed that the jury might draw adverse inferences from this. The defendant was convicted. His appeal against conviction was allowed. The Court of Appeal held that no adverse inference could be drawn under s 34 of the Criminal Justice and Public Order Act 1994 if the statement contained the matters upon which he relied on in his defence at trial (which it did). The court said that the purpose of the provision was to procure the early disclosure of a suspect's account and not the scrutiny and testing of it by the police. However, the defendant may face difficulties if something significant, which is relied upon at trial, was omitted from the prepared statement.[2] These prepared statements are increasingly being exhibited by the HSE in trials, and used in adverse inference applications. Further, see generally *Iqbal v Solicitors Regulation Authority* [2012] and *Ashiq v Bar Standards Board* [2013] on adverse inferences and the approach by regulatory bodies.

1 (29 July 2003, unreported), CA.
2 See also *R v Charisma (Louie Presence)* [2009] EWCA Crim 2345 and *R v Billingham* [2009] EWCA Crim 19.

3.37 An HSE officer can interview a suspect under caution subject to PACE. However, the important thing to remember is that an HSE officer does not have the power of arrest (therefore, if the suspect refuses to be interviewed, the HSE officer cannot force the suspect to be interviewed, unlike a police officer who can simply arrest the suspect and interview him or her). See also powers under s 20 of the HSWA 1974, where individuals can be forced to attend an interview. This information cannot be used against them individually. It is also vital to ascertain if the interviewee is speaking on behalf of themselves and/or the organisation.

DISCLOSURE

3.38 Under the Criminal Procedure and Investigations Act 1996 (CPIA 1996), on a case being committed to the Crown Court, the prosecutor is required to disclose to the accused previously undisclosed material which, in the opinion of the prosecutor, might undermine the prosecution case (this is known as primary disclosure), or else to give the accused a written statement that there is no such material (pursuant to the CPIA 1996, s 3). Once a defence statement is served, the position with respect to disclosure is considered again. If further documents are considered relevant in the light of the defence statement, secondary disclosure will be given.

Amendments made to the CPIA 1996 by the Criminal Justice Act 2003 abolished the concept of 'primary' and 'secondary' disclosure, and introduced an amalgamated test for disclosure of material that 'might reasonably be considered capable of undermining the prosecution case or assisting the case for the accused'. The 2003 Act also introduced a new Code of Practice.

3.39 In work related incident cases, because of their complex nature, the police usually gave full disclosure of documentation, as there was a fear of withholding evidence which might later be found to be relevant. Increasingly, the defence are forced to make a series of requests, both in writing and before the court under PACE, to force the disclosure. Increasingly, third party summonses are also needed.[1] See also the Criminal Procedure Rules 2013[2] on disclosure; and, since 3 December 2013, there are new guidelines from the Attorney-General replacing the 2005 Guidelines.[3]

1 See also *R (Mohamed) v Secretary of State for the Foreign and Commonwealth Affairs* [2009] EWHC 152 (Admin).
2 SI 2013/1554.
3 See *Archbold*, 2013.

DEFENCE STATEMENT

3.40 Once the case has been sent to the Crown Court, the defendant has 14 days to file a defence statement or risk an adverse inference being drawn at trial. Care should be taken in drafting these statements and, because of the complexities of these types of case, drafting will take time. Once the statutory period has expired, the trial judge cannot extend it retrospectively. It is therefore important, where necessary, to apply for and obtain an extension of time for the filing of the statement before the expiry period. It is also important to remember that a defendant can be cross-examined not only upon what he or she said in their police interview, but also upon what is contained in the defence statement. The defence need also to make sure that the prosecution have provided full disclosure (see above).

PROSECUTION CASE SUMMARY

3.41 In complex work related death cases, it is usual for the prosecution to serve a full case summary, setting out in detail the case against the defendants, fully cross-referenced to the evidence relied upon. It is essential that this document is obtained before a defence statement is served.

3.42 In health and safety prosecutions, it is said by the court that the prosecution should always be required to set out precisely what its case is. The obligation upon them to do so was emphasised in *R (on the application of Bernard) v Dudley Metropolitan Borough Council*:[1]

> 'When Informations are laid and proceedings commenced, it is manifest that there is a clear obligation upon the prosecuting authority to make it plain just what the allegation is, and upon what evidence they seek to rely to substantiate that allegation … It is not sufficient for a prosecuting authority in a Health and Safety at Work Act [HSWA 1974] case merely to serve the summonses and the statements.'

1 [2003] EWHC 147 (Admin).

The defence must carefully check to see if the prosecution are attempting to try and prosecute a defendant arising out of a similar factual scenario which they have previously dealt with. This is deemed as 'double jeopardy' and would be designated as an abuse of process.[1]

Further, the defence should review whether the prosecution have in turn reviewed their decisions to prosecute (or continue with the prosecution), if certain new evidence is uncovered by the defence or by the Crown. The prosecution have a residual duty to continually monitor and review their decision to prosecute (and investigate all reasonable lines of inquiry). If they fail that duty, it can amount to an abuse of the process.[2]

1 *R v Beedie* [1977] 2 Cr App R 167.
2 *R v Equicorp* (2008) unreported, Bristol Crown Court.

EXPERTS

3.43 In prosecutions concerning work-related incidents, so often it is not the facts of what happened that are in dispute, but the interpretation of those facts. Because of this, experts usually play a vital role. It is vital that consideration is given to instructing experts at an early stage. Finding appropriate experts can be a time-consuming exercise. In specialised areas, they are very rare and should be retained as soon as possible.

3.44 The Ministry of Justice has produced guidance on the use of expert witnesses,[1] and the Crown Prosecution Service's Guidance Booklet for Experts[2] provides guidance about how cases should proceed using expert witnesses and their evidence. Guidance on the remuneration of expert witnesses (November 2013)[3] is available from the Legal Aid Agency.

1 See www.justice.gov.uk/legal-aid/funding/using-experts.
2 See www.cps.gov.uk/legal/d_to_g/disclosure_manual/annex_k_disclosure_manual.
3 See www.justice.gov.uk/downloads/legal-aid/funding-code/remuneration-of-expert-witnesses-guidance.PDF.

3.45 The Civil Legal Aid (Remuneration) Regulations 2013[1] make provision about the payment by the Lord Chancellor to persons who provide civil legal services under arrangements made for the purposes of Part 1 of the Legal Aid, Sentencing and Punishment of Offenders Act 2012 ('the 2012 Act'); and the Criminal Legal Aid (Remuneration) Regulations 2013[2] make provisions for the funding and remuneration of advice, assistance and representation made available under ss 13, 15 and 16 of the 2012 Act.

1 SI 2013/422.
2 SI 2013/435.

3.46 It is important that experts give evidence within their own expertise. Those instructing should ensure that the experts: are provided with the appropriate evidence and material, understand their duties to the court;[1] and understand what the case is (from a legal point of view as well as a factual point of view) that the prosecution is seeking to prove. In relation to any expert report, as the law currently stands in criminal cases, litigation privilege attaches to that report.[2] See *R (on the application of Prudential plc) v Special Comr of Income Tax*[3] on the limits of privilege; it only applies to lawyers. See also the Criminal Procedure Rules on experts (and **3.5** above).

1 Note that the Practice Direction to Part 35 of the Civil Procedure Rules 1998 sets out an expert's duties to the court.
2 See *R (Morgan Grenfell) v Special Comr of Income Tax* [2002] UKHL 21, [2002] 3 All ER 1.
3 [2013] UKSC 1.

OTHER NON-CRIMINAL PROCEEDINGS

3.47 Criminal investigations and prosecutions can be lengthy. In the meantime, it is likely that claims for compensation will be made. The Jackson reforms on costs have also made a big difference to this area of the law. A company will need to ensure that this procedure does not prejudice its position in respect of the criminal proceedings. There must be liaison between the company's lawyers, the insurance company and the lawyers dealing with the civil claims (if different). There are also often other proceedings, such as disciplinary hearings and, of course, inquests.

COMPANY DOCUMENTATION

3.48 When an investigation takes place, there will be a great deal of interest in the company's documentation. Companies are well advised to have a good document management system in place so that management know what documents there are. This includes, for instance, anti-money laundering provisions needing to be robust. Very recently the UK arm of South Africa's Standard Bank was fined £7.6 million by the Financial Conduct Authority for weaknesses in its systems. The Civil Procedure Rules 1998 (CPR) define a document as 'anything in which information of any description is recorded'.[1] It is also important to recall that the giving of a false statement, or using a document to deceive, is of itself a criminal breach of the HSWA 1974.[2]

1 See CPR Part 31.
2 See s 33(1)(k), (m).

3.49 There is sanctity in the existence of documents following the commencement of an inquiry or legal proceedings. A company may have a policy for the destruction of documents. This may have to be halted, in order to fulfil the duty of preservation of relevant documents. Certain aspects of this policy may have to be specifically identified and dealt with, such as: automatic destruction of files after a period of time has elapsed; scanning of documents and destruction of originals; destruction of duplicates; and destruction of manuscript notes and inter-office memos. The destruction of documents could amount to perverting the course of justice. In the case of *R v Selvage and Morgan*,[1] it was held that the offence is made out by acts which did and were intended to interfere with pending or imminent proceedings or with investigations which might end with criminal proceedings being brought. Further, a company officer who destroys, mutilates, or falsifies a document affecting the company's property or affairs may be liable to prosecution unless the officer can prove that there was no intent to conceal or defeat the law.[2] There is a responsibility upon a solicitor to ensure his/her client understands this. It was said in *Rockwell Machine Tool Co v EP Barrus* (*Concessionaires*):[3]

> '... It seems to me necessary for solicitors to take positive steps to ensure that their clients appreciate at an early stage of the litigation, ... not only the duty of discovery and its worth but also the importance of not destroying documents which might by possibility have to be disclosed. This burden extends, in my judgment, to taking steps to ensure that in any corporate organisation knowledge of this burden is passed on to any who may be affected by it.'

See also *Erles v Barclays Bank*,[4] which restates the duty of organisations to keep documentation pursuant to the CPR.

1 [1982] 1 All ER 96.
2 Companies Act 1985, s 450(1) and (2), which was not repealed by the Companies Act 2006, makes it an offence to destroy or mutilate, falsify or make false entries, part with or alter or make omissions from company documents. This offence carries maximum penalties of seven years' imprisonment and/or fine on indictment and, on summary conviction, six months' imprisonment and/or fine up to £5,000.
3 [1968] 1 WLR 693 at 694.
4 (2009) Times, 20 October, Queen's Bench Division.

3.50 Emails are often written in a rush, are not fully considered and can be ambiguous. However, these can become evidence. It is therefore important for companies to have an email policy.

3.51 This is illustrated by the case of *United States v Microsoft*.[1] The Justice Department alleged that Microsoft used its Windows monopoly unfairly to crush the Netscape Navigator web browser. The prosecutors produced emails going back to 1993 to attack the credibility of Microsoft executives. In February 1999, the chief lawyer for the US government produced a 1996 email message from Bill Gates to the chairman of another software manufacturer, seemingly offering a bribe if the manufacturer used Microsoft Internet Explorer rather than Netscape Navigator. There have been a series of other recent examples, especially since WikiLeaks. Further, in 2012, Greater Manchester Police were fined £120,000 by the Information Commissioner's Office for serious data protection breaches.

1 US case (1999).

3.52 Note that, more so than paper documents, emails are stored. It is often assumed that deleting a record or file removes it from the hard drive. In fact, the only thing that is removed is the file name and protection overwrite, leaving data on the drive. Even using disk-erasing utilities, fragments of the file remain which can be pieced together. See also *R v Turner*[1] on professional privilege and the *Prudential* case (above).

The civil case of alleged blacklisting by major contractors was launched in 2013 in the High Court.

1 [2013] EWCA Crim 642.

INQUESTS[1]

3.53 The Coroners (Inquests) Rules 2013[2] and the Coroners (Investigations) Regulations 2013[3] came into force in July 2013. The explanatory note to the 2013 Rules provides that they are made under the Coroners and Justice Act 2009. Part 1 of the 2009 Act introduces a new regime for death investigations and inquests, which replaces the Coroners Act 1988 and the Coroners Rules 1984. Under the 2009 Act, a coroner must conduct an investigation into violent or unnatural deaths, deaths where the cause is unknown, and deaths which occur in custody or otherwise in state detention. In certain cases, this investigation will include the coroner holding an inquest.

1 See also **Chapter 1**.
2 SI 2013/1616.
3 SI 2013/1629.

3.54 Only a small proportion of deaths reported to the coroner require a public inquest, including when there is:

- a violent or an unnatural death;
- a sudden death and cause unknown; or
- when someone is being detained.

3.55 Not all inquests have a jury. Two circumstances when a jury is required are where the coroner has reason to suspect that:

- the death was caused by accident, poisoning or disease (notice of which is required to be given to the HSE under the HSWA 1974); or
- the death occurred in circumstances prejudicial to public health.

3.56 The purpose of the inquest is to determine who the deceased was, where and when the deceased died, and how they died. The means by which the deceased came by his or her death is a limited question.[1]

1 See *R v HM Coroner for North Humberside and Scunthorpe, ex p Jamieson* [1995] QB 1.

3.57 The inquest conclusions of an inquest into a work related death that might be given are:

- accidental death (misadventure);
- unlawful killing; or
- open verdict.

3.58 It is important to note that there is no statutory requirement that the verdict of the inquest be in any particular form. However, there is a prohibition on an inquest conclusion being framed in such a way as to appear to determine any question of criminal or civil liability. There is an increasing number of narrative verdicts/conclusions being recorded.

3.59 Any interested party to the inquest can be represented at the inquest. This includes any person whose act or omission, or that of his or her agent or servant, may (in the opinion of the coroner) have caused, or contributed to, the death of the deceased.

3.60 At the conclusion of the evidence, the legal representatives cannot make a speech to the jury. The coroner cannot be addressed on the facts, but can be addressed on the law in relation to each possible verdict.

3.61 An inquest will not take place until the police have confirmed that there are to be no prosecutions. However, if evidence arises in the inquest, the coroner is able to refer the matter back to the police to reconsider the matter. If there is an unlawful killing verdict, the matter is automatically referred back to the police. In relation to health and safety prosecutions, these can commence before the inquest starts.

3.62 It is now becoming more common for representatives of the family of the deceased to take a proactive role in these proceedings, with a view to raising issues in cross-examination that are relevant to criminal liability. It is therefore important to explain, to any witness who might be criticised, the rule against self-incrimination in respect of a criminal offence. With the advent of the Corporate Manslaughter and Corporate Homicide Act 2007, inquest juries are much more likely to reach unlawful killing verdicts. If, after an inquest is opened, it appears that manslaughter may have been committed, the inquest should be stopped and the matter referred to the Director of Public Prosecutions (DPP). If this occurs, the police must reopen the investigation into whether a homicide(s) may have occurred.

JUDICIAL REVIEW

3.63 The families of deceased who have died in a work related incident are more likely to make submissions to the police, the CPS and/or the HSE about who should be prosecuted and why. It should be noted that:

> 'The Crown Prosecution Service prosecutes cases on behalf of the public at large and not just in the interests of any particular individual. However, when considering the public interest test Crown Prosecutors should always take

into account the consequences for the victim of the decision whether or not to prosecute, and any views expressed by the victim or the victim's family.'[1]

1 Code for Crown Prosecutors, para 6.7.

3.64 If a decision not to prosecute is taken, it is open to the family to judicially review that decision. This is a means by which public law disputes are resolved by bringing the matter before the High Court. This form of action was described as 'a remedy invented by the judge to restrain the excess or abuse of power'.[1] The procedure for bringing a judicial review is set out in statute: s 31 of the Senior Courts Act 1981, on which CPR Part 54 elaborates.

1 Lord Templeman in *R v Secretary of State for the Home Department, ex p Bond* [1991] AC 696, HL.

3.65 There is a pre-action protocol that covers judicial reviews. A judicial review must be brought promptly and within three months of the decision or action which gives rise to the action. The defendant will be the body that has taken the decision not to prosecute. The proceedings must also be served on any interested party, which includes any likely defendant in relation to the criminal proceedings which are the subject of the judicial review.

3.66 The decision-makers subject to judicial review must not act in a way that no reasonable decision-maker would consider justifiable. The benchmark decision on this principle was made in the *Wednesbury* case,[1] which said:

> 'If a decision on a competent matter is so unreasonable that no reasonable authority could ever come to it, then the courts can interfere … but to prove a case of that kind would require something overwhelming …'

1 *Associated Provincial Picture Houses Ltd v Wednesbury Corpn Ltd* [1948] 1 KB 223, HL.

3.67 In the judicial review that considered the CPS's decision not to bring manslaughter charges in relation to the death of Simon Jones in a work related incident, the court said the relevant issues were:[1]

> '1. has the decision maker properly understood and applied the law?
> 2. has he explained the reasons for his conclusions in terms that the court can understand and act upon? And
> 3. has he taken into [account] an irrelevant matter or is there a danger that he may have done so?'

1 *R v DPP, ex p Jones* [2000] IRLR 373.

3.68 If the matter has been fully considered by the CPS, it is usually difficult for the judicial review to succeed. In *R v Director of Public Prosecutions, ex p Manning*,[1] it was said:

> 'The primary decision to prosecute or not is entrusted by Parliament to the Director [of Public Prosecutions] as head of an independent, professional prosecuting service answerable to the Attorney-General in his role as guardian of the public interest and to no-one else. It makes no difference that in practice the decision will ordinarily be taken by a senior member of the Crown Prosecution Service … In any borderline case the decision may be one of acute difficulty, since while a defendant whom a jury would be likely to convict should properly be brought to justice and tried, a defendant whom a jury would be likely to acquit should not be subjected to the trauma inherent in a criminal trial. The Director and his officials … will bring to their task of deciding whether to prosecute an experience and an expertise which most courts called upon to review their decisions could not match. In most cases the decision will turn not on an analysis of the relevant legal

principles but on the exercise of an informed judgement of how a case against a particular defendant, if brought, would be likely to fare in the context of a criminal trial before … a jury. The exercise of judgement involves an assessment of the strength by the end of the trial, of evidence against the defendant and of the likely defences. It will often be impossible to stigmatise a judgement on such matters as wrong even if one disagrees with it. So the courts will not easily find that a decision not to prosecute is bad in law, on which basis alone the court is entitled to interfere. At the same time the standard of review should not be set too high, since judicial review is the only means by which a citizen can seek redress against a decision not to prosecute and if the test were too exacting an effective remedy would be denied.'

1 [2000] 3 WLR 463.

3.69 The order that will be sought in a judicial review is that the decision not to prosecute be quashed and that there be a mandatory order that the decision be reconsidered afresh.

See the example of Roy Clarke, where the family successfully judicially reviewed the original CPS decision not to prosecute for manslaughter. This is covered in **Chapter 2**. The High Court has also said that it is 'highly probable' that it cannot extend the 21-day period to state a case.[1] There has been various curtailments to the funding of judicial review. Some very recent practice directions on costs have been issued by the High Court. See also *M v London Borough of Croydon*[2] (particularly at paras 75–77).

1 See *Chief Constable of Cleveland v Vaughan* [2009] EWHC 2831 (Admin).
2 [2012] EWCA Civ 595.

PUBLIC INQUIRIES

3.70 After a disaster or major accident has occurred, a public inquiry might be set up to look at its causes, what lessons can be learnt from the incident and to make recommendations as to how matters might be improved. There are increasing numbers of these being heard across a plethora of industries where it is felt that lessons need to be learnt.

A recent example of this is the public inquiry into the blast at the OCL Stockline plastics factory in Glasgow in 2004. The public inquiry into the nine deaths and many injured people was chaired by Lord Gill.

The HSE came under criticism for not following up a previous warning by the HSE inspector concerning the gas pipe that was at the centre of the explosion.

3.71 In 2013, we have seen a 13-week 'super-inquest' into the Lakanal House fire, in which six people died in a fire in South London in 2009, where a narrative verdict was recorded by the coroner. We have also seen the inquest into the Mid Staffordshire NHS Foundation Trust; and another inquest has recently been announced.[1]

1 See also **Chapter 1**.

3.72 Nowadays, a public inquiry is likely to be conducted pursuant to s 14(2)(b) of the HSWA 1974. Examples of such inquiries are those that followed the Southall and Ladbroke Grove train crashes. This enables the HSE (with the consent of the Secretary of State) to direct an inquiry to be held. The procedures for inquiries under this provision are regulated by the Health and Safety Inquiries (Procedure) Regulations 1975.[1]

1 SI 1975/335.

3.73 A person will be appointed to chair the inquiry and is usually a senior lawyer (although this is not always the case). The chair can appoint assessors, who will be experts in particular fields relevant to the nature of the incident, to assist him or her with understanding technical issues and evidence.

3.74 Counsel to the inquiry – an experienced advocate and, normally, Queen's Counsel – will be appointed. Counsel to the inquiry will be supported by a team of lawyers. His or her role is to advise the chair on legal issues, to arrange the evidence to be placed before the inquiry, and usually to take the witnesses called to the inquiry through their evidence. A secretary to the inquiry will also be appointed and, along with his or her team, will deal with the administration of the inquiry. Recent examples include the Duggan shooting and the new inquest into Hillsborough.

3.75 Public inquiries can be delayed as a result of criminal proceedings. This is because, if the inquiry took place before the criminal proceedings were concluded, any trial might be prejudiced. The Southall train crash inquiry was delayed for two years while criminal proceedings were taken against one of the train drivers involved and his employers, Great Western Trains.

3.76 Public inquiries are meant to be inquisitorial, and not adversarial like normal legal proceedings. However, because so much is at stake, because the outcome can influence civil and criminal liability, this is not so easy to achieve in practice.

3.77 Professor HWR Wade and CF Forsyth comment in relation to inquiries (authorised under the Tribunals of Inquiry (Evidence) Act 1921):[1]

> 'Experience of Tribunals of Inquiry has revealed the dangers to which a procedure of this kind is naturally prone. The Inquiry is inquisitorial in character, and usually takes place in a blaze of publicity. Very damaging allegations may be made against persons who may have little opportunity to defend themselves and against whom no legal charge is preferred.'

1 Professor HWR Wade and CF Forsyth *Administrative Law* (7th edn) p 1008.

3.78 Thus, any person giving evidence to an inquiry, where there is the possibility that he or she may be criticised, should consider strongly having legal representation and, if employed by a party involved in the inquiry, separate legal representation from them in case there is a possibility of a conflict of interests.

3.79 Since s 17 of the Access to Justice Act 1999 came into effect, any inquests will normally be adjourned until after the public inquiry as, effectively, the inquiry will determine the cause of the incident and thus how the deceased came to die.

HUMAN RIGHTS ACT 1998

3.80 The Human Rights Act 1998 (HRA) came into force on 2 October 2000. The Act 'gives further effect to' the European Convention on Human Rights (ECHR) in domestic law. See also *Sarjantson v Chief Constable of Humberside Police*[1] and *R (Humberstone) v Legal Services Commission and another*.[2]

1 [2013] EWCA Civ 1252.
2 (CA) Times Law Reports, 16 February 2011. See also *R v January Davies* in **Chapter 1**.

3.81 The HRA has been given partial effect since 22 May 1999, in Scotland, by virtue of the Scotland Act 1998. As a result of the HRA, ECHR rights can now be directly relied upon, argued and enforced in the UK courts and tribunals at all levels. The acts or omissions of 'public authorities' can be challenged as part of any legal proceedings, by judicial review or as an entirely new course of action under the HRA.

3.82 It should be remembered that, once all domestic remedies have been exhausted in the domestic courts, an appeal still lies to the European Court of Human Rights in Strasbourg. Decisions of the European Court of Human Rights are final and binding on member states.

3.83 The articles of the ECHR that are most relevant to criminal proceedings in relation to work related incidents are Article 2 (right to life) and Article 6 (right to a fair trial).

3.84 An offshore disaster such as Piper Alpha would undoubtedly trigger a challenge under Article 2 of the ECHR. Remember that 'life' includes 'physical integrity'. There is a clear obligation on the state to investigate deaths fully.[1]

1 See *R (on the application of Middleton) v West Somerset Coroner* [2004] UKHL 10.

3.85 Although questions of causation will always arise, public authorities such as the HSE, local authorities and health authorities may be under a duty to protect the public (and their employees) from health risks such as AIDS, BSE, E-coli, and from environmental danger and life-threatening hazards such as radiation, explosions and asbestos – for instance, Bhopal gas leak and Chernobyl.[1]

1 See *Indian Council for Enviro-Legal Action v Union of India* [1996] 2 LRC 226.

3.86 In *Guerra v Italy*,[1] illegal toxic emissions caused arsenic poisoning and possibly cancer. There was a claim under Article 2 of the ECHR. The court awarded £3,000 to the applicants to compensate them for the fact that the authorities had not provided them with sufficient information about a pesticide factory to enable them to assess the risks of living in its vicinity.

1 (1998) 26 EHRR 357.

3.87 It is clear that the HRA will have an ever-increasing importance in the areas of health and safety and corporate manslaughter. In order to properly advise and represent clients, lawyers need to be fully familiar with the provisions of the HRA and ancillary legislation, and the relevant procedures in this expanding body of ECHR law.

MEDIA

3.88 A company should have procedures in place for dealing with the media. Simply trying to hide from the media may well make matters worse. If anyone in the company is going to speak to the media, they should have the appropriate training. Obviously, care should be taken as to what is said, particularly once charges have been brought.

3.89 A company should make sure that there are systems in place to deal appropriately and sensitively with the relatives of those killed or injured as a result of an incident. A jury will give much weight to these sorts of matters. At the Southall Rail Inquiry, a great deal of criticism was made by the passengers and families of those killed and injured of how they were treated in the immediate aftermath of the accident. To act decently in such circumstances does not amount to an admission of guilt and provides good evidence of the attitude of the company.[1]

1 See **Chapter 15**.

DRIVING AT WORK

3.90 Increasingly, the police are targeting this area (which is more fully covered in **Chapter 1**).

BAD CHARACTER

3.91 Since December 2004, it has become much easier for the prosecution to apply to the court to adduce 'bad character'.

This can be very powerful evidence, especially in cases involving fatalities and/or where the prosecution is alleging that profit has been put before safety. The complexities of the Criminal Justice Act 2003 are beyond the scope of this book, but suffice to say that, despite the fact that previous convictions are not in themselves capable of indicating propensity to offend in the future, prosecutions do attempt to put this evidence before the court. The leading case is *R v Hanson and others*.[1] See also *R (Morales) v Kettering Magistrates' Court* (2014).

1 [2005] 2 Cr App R 21 (CA). See also *Johnson and others* [2009] EWCA Crim 649, *R v Hearne* [2009] EWCA Crim 103, *R v Z* [2009] EWCA Crim 20, *R v C* [2011] EWCA Crim 939, *D, P and U* [2011] EWCA Crim 1474, *R v O'Leary* [2013] EWCA Crim 1371 and *R v Speed* [2013] EWCA Crim 1650. See further the Criminal Procedure Rules 2013. See also *R v Dixon* [2014] Crim LR (2) 141–144.

CRIMINAL PROCEDURE RULES

3.92 These are vital to be evidenced by anyone acting in a criminal case. In essence, the Rules, which first came into force in April 2005, harmonise criminal practice into one set of regulations and embody the spirit of fair and efficient prosecutions.[1]

1 See www.justice.gov.uk/courts/procedure-rules/criminal; and see above for the 2013 Rules.

COSTS

3.93 The court can award costs to the prosecution, provided they are just and reasonable.[1] The costs can include the investigation costs, but not the costs of the inquest.

1 Prosecution of Offences Act 1985, s 18.

3.94 It is crucial to report and receive a schedule of costs from the prosecution as soon as possible and to carefully check it, as often a discount can be negotiated or ordered by the judge.

3.95 The costs and any fine should be looked at as a global sum 'in the round'.[1]

3.95 *Costs*

In the civil arena, the Civil Procedure Rules 1998 are being interpreted in a very strict way for any procedural irregularities.[2] However, a very recent case has shown a more lenient approach: in December 2013, Judge Oliver-Jones QC did not allow a claim to be struck out because the claimants were out of the country.[3]

1 *R v Associated Octel Ltd* (*No 2*) [1997] 1 Cr App R (S) 435; *R v Northallerton Magistrates' Court ex parte Dove* [1999] EWHC 499 (Admin), [2000] 1 Cr App R (S) 136.

2 See *Mitchell v News Group Newspapers Ltd* [2013] EWCA Civ 1537and *Re Atrium Training Services Ltd* [2013] EWHC 1562 (Ch).

3 See *Law Society Gazette*, 20 December 2013.

CONCLUSIONS

3.96 It has only been possible to touch the tip of the iceberg of the practical issues that may arise in a manslaughter and/or health and safety investigation and prosecution. Many of these issues often arise from the very nature of the case and the industry in which the incident occurred. The procedure is highly dynamic. The police and other regulators do not always fully understand the industry, and this needs to be explained to them before enforcement action is commenced, either in interview, correspondence or statement form. The proactive response will become increasingly important, especially given that successful corporate defendants cannot recover their costs. The very recent case of *R v Sellafield Ltd; R v Network Rail Infrastructure Limited* (see **2.17**) may also influence large corporates on their decisions as to how to progress future prosecutions.

Chapter 4

Work related death investigations by the police, HSE and other agencies*

Nick Kettle MBE MSc MCIEH MIFST CMIOSH, Head of the Metropolitan Police Safety and Health Risk Management Department

Mark Oliver, Detective Chief Inspector Humberside Police, PIP 3 Senior Investigating Officer, Senior Identification Manager UK DVI

Mike Elliker, Legal Director, Addleshaw Goddard LLP

* Original authors (Second Edition, 2010):
Malcolm Ross, MSc, former Detective Superintendent, West Midlands Police
Mike Elliker, Legal Director, Addleshaw Goddard, Leeds

INTRODUCTION

4.1 For a number of years, the investigation into deaths, especially those occurring in the workplace, was often seen as a matter for the Health and Safety Executive (HSE) and the coroner to determine whether any offences had been committed. If any offences were proved, they would usually be offences contrary to health and safety legislation.

4.2 In 1998 the publication of '*The protocol for liaison: work related deaths*' set out the effective liaison framework during criminal investigations of work related deaths. The protocol was last updated in September 2011.[1] From 1 April 2012, the signatories to this protocol have specific roles in investigating or prosecuting work related fatal accidents.[2] The protocol requires the police to conduct an investigation where there is an indication of the commission of a serious criminal offence (other than health and safety offences) such as manslaughter (corporate or gross negligence).

1 *Health and Safety Executive, Work-related deaths: a protocol for liaison* (England and Wales) (WRDP1 9/11). The full text is available at www.hse.gov.uk/pubns/wrdp1.pdf.

2 A work related death is a fatality resulting from an incident arising out of, or in connection with, work. This also applies to cases where the victim suffers injuries in such an incident that are so serious that there is a clear indication, according to medical opinion, of a strong likelihood of death.

4.3 This chapter considers how the police and other agencies (including the HSE) investigate work related deaths.

PROTOCOLS AND GUIDANCE

Work-related deaths – a protocol for liaison

4.4 This work-related deaths protocol was first introduced in 1998. The latest version that was published in September 2011 is the third version of this protocol. The signatories to the protocol are now:

- the Crown Prosecution Service (CPS);
- the police through their professional body, the Association of Chief Police Officers (ACPO);
- the HSE;
- local authorities through their representative bodies, the Local Government Association and the Welsh Local Government Association;
- the British Transport Police (BTP);
- the Office of Rail Regulation (ORR);
- the Maritime and Coastguard Agency (MCA);
- the Medicines and Healthcare Products Regulatory Agency (MHRA) Medical Devices Division; and
- the Fire and Rescue Services through their professional body, the Chief Fire Officers Association (CFOA).

4.5 This latest iteration reinforces the principle of multiple agencies working together to investigate thoroughly, and to prosecute appropriately, those responsible for work related deaths[1] in England and Wales. All of the signatory organisations recognise the need for investigating and prosecuting authorities to engage with each other and to share information and best practice. The protocol addresses issues concerning general liaison and is not intended to cover the operational practices of the signatory organisations.[2]

1 A 'relevant enforcing authority' is defined as the health and safety regulator with responsibility for the activity or workplace involved, and includes HSE, ORR, local authorities, MCA and the Fire and Rescue Services.

2 The Civil Aviation Authority (CAA) and the Independent Police Complaints Commission (IPCC) are not signatories to this protocol, but each has agreed to abide by the protocol's principles.

4.6 The police, CPS and relevant enforcing authorities[1] have different roles and responsibilities in relation to a work related death. The protocol defines the relevant responsibilities and is supported by a range of supplementary guides.[2]

1 There will be cases in which it is difficult to determine whether a death is work related within the application of this protocol (for example, those arising out of some road traffic incidents, or in prisons or health care institutions, or following a gas leak). Each fatality must be considered individually, on its particular facts, according to organisational internal guidance, and a decision made as to whether it should be classed as a work related death.

2 'Work-related deaths: Investigators' guide' (WRDP2 (Rev 1) 7/12); 'Work-related deaths: Guidance on the timing of criminal proceedings in a work-related death case' (WRDP3 9/11).

4.7 The police investigate serious criminal offences (other than health and safety offences) such as manslaughter (whether corporate or gross negligence), and only the CPS can decide whether such a case will proceed. The police will also have an interest in establishing the circumstances surrounding a work-related death in order to assist the coroner's inquest. Health and safety offences are usually prosecuted by the relevant enforcing authority in accordance with current enforcement policy. The CPS may also prosecute health and safety offences, but usually does so only when prosecuting other serious criminal offences, such as manslaughter, arising out of the same circumstances.

4.8 Where a work-related death from an air, marine or rail accident is being investigated, the relevant Accident Investigation Branch (AIB) of the Department for Transport has additional duties of investigation. To ensure a consistent approach is taken to enforcement across all industry sectors, agencies such as the ORR and MCA have either agreed to abide by the enforcement principles or have a separate Memorandum of Understanding (MoU). There is a separate MoU between the AIBs collectively and the CPS, and between each of the AIBs and the other relevant agencies.

4.9 When making a decision whether to prosecute, the CPS or relevant enforcing authority will review the evidence, according to the Code for Crown Prosecutors,[1] to decide if there is a realistic prospect of conviction and, if so, whether a prosecution is needed in the public interest. Enforcing authorities that have signed up to the Enforcement Concordat must follow the principles and procedures within it and also the Regulators' Compliance Code.[2]

1 'Code for Crown Prosecutors' (2013), available at www.cps.gov.uk/publications/code_for_crown_prosecutors/index.html.

2 'Regulators' Compliance Code: Statutory Code of Practice for Regulators' (2007), available at www.berr.gov.uk/files/file45019.pdf.

4.10 A police officer is to attend incidents involving a work related death to:

- identify, secure, preserve and take control of the scene, and any other relevant place;
- supervise and record all activity;
- inform a senior supervisory officer;
- enquire whether the employer or other responsible person in control of the premises or activity has informed the relevant enforcing authority; and
- contact and discuss the incident with the relevant enforcing authority, and agree arrangements for controlling the scene, for considering access to others, and for other local handling procedures to ensure the safety of the public.

4.11 A police officer of supervisory rank should also attend the scene and any other relevant place to assess the situation, review actions taken to date, and assume responsibility for the investigation. Should any other investigating or enforcing authority have staff in attendance before the police arrive, it should ensure that the police have been called, and preserve the scene.

4.12 The police and the relevant enforcing authority will also agree upon:

- how resources are to be specifically used;
- how evidence is to be disclosed between the parties;
- how the interviewing of witnesses, the instruction of experts and the forensic examination of exhibits is to be co-ordinated;
- how, and to what extent, corporate or organisational failures should be investigated;
- a strategy for keeping the bereaved, witnesses and other interested parties, such as the coroner, informed of developments in the investigation; and
- a media strategy to take account of the sensitivities of the bereaved and those involved in the incident.

4.13 Where the investigation gives rise to a suspicion that a serious criminal offence (other than a health and safety offence) may have caused the death, the police will assume primacy for the investigation and will work in partnership with the relevant enforcing authority. Where it becomes apparent during the investigation that there is insufficient evidence that a serious criminal offence (other than a health and safety offence) caused the death, the investigation should, by agreement, be taken over by the relevant enforcing authority.

4.14 There will also be rare occasions where, as a result of the coroner's inquest, judicial review or other legal proceedings, further consideration of the evidence and surrounding facts may need to be made. Where this takes place, the police, the relevant enforcing authority with primacy for the investigation and the CPS will work in partnership to ensure an early decision. There may also be a need for further investigation.

4.15 The protocol states that its underlying principles are:

- an appropriate decision concerning prosecution will be made based on a sound investigation of the circumstances surrounding work related deaths;
- the police will conduct an investigation where there is an indication of the commission of a serious criminal offence (other than a health and safety offence), and the relevant enforcing authority will investigate health and safety offences. There will usually be a joint investigation but, on the rare occasions where this would not be appropriate, there will still be liaison and co-operation between the investigating parties;
- the decision to prosecute will be co-ordinated, and made without undue delay; and
- the bereaved and witnesses will be kept suitably informed; and the parties to the protocol will maintain effective mechanisms for liaison.

Police investigation

4.16 The police will conduct an investigation into a work related death where there is an indication of a serious criminal offence, other than a health and safety offence; in the case of a health and safety offence, the matter will be investigated by the HSE, local authority or other enforcing agency. The work-related deaths protocol gives guidance as to how the police and agencies liaise, but does not detail operational practices.

It will not always be apparent, at the early stages of an investigation into the death, that a serious criminal offence has been committed. An early decision is made as to which agency maintains primacy for the investigation, based on an initial assessment as to whether there is suspicion that a serious criminal offence has been committed.

4.17 The protocol gives guidance for the police, HSE and other agencies to work together to ensure that the initial investigation is thorough and can meet the requirements of all agencies, irrespective of who eventually pursues any prosecution (ie the police, HSE or other agency).

The police are the only agency that can investigate manslaughter, and only the CPS can decide whether such a case will proceed. Police also usually carry out enquiries on behalf of the coroner into deaths. The coroner has the responsibility to hold an investigation into a death to establish whether there is reason to suspect that:

- the deceased died a violent or unnatural death;
- the cause of death is unknown, or

- the deceased died in custody or 'state detention'.

4.18 The initial assessment by the police into the circumstances of a work related death will be used to determine the primacy of the investigation, the requirements to assist the coroner and the level of resources that the police require to meet their investigative obligations and objectives.

The police, in liaison with the other agencies, will continually assess, as the investigation progresses, whether there is a suspicion that a serious criminal offence has been committed, and whether the appropriate party has primacy of the investigation. The majority of work related deaths will be investigated by the HSE and it is, therefore, necessary for police and agencies to maintain continuing liaison to discuss whether any emerging case information amounts to adequate suspicion of a serious criminal offence and the police assuming primacy.

The police will consider whether the assessment indicates that 'unlawful act manslaughter' or 'gross negligence manslaughter' attributable to any individuals has occurred. In these circumstances, it will be clear that the police will maintain or assume primacy of the investigation.

4.19 The police are responsible for investigating offences under the Corporate Manslaughter and Corporate Homicide Act 2007 (CMCHA 2007), irrespective of initial case primacy decisions in the investigations of work-related death. The evidential elements which provide an indication that the death has occurred as a result of a gross failure by an organisation in the way activities were managed or organised may not be initially apparent, and may often only be identified as the investigation by the HSE or other enforcing authority, or a joint police and enforcing authority investigation, progresses.

The case assessment will be used to identify the requisite investigative activities and the appointment of an appropriately trained and resourced investigator to lead the investigation on behalf of the police. Investigations of work related deaths can be carried out by different levels of investigator: large-scale investigations will be led by Senor Investigators setting up well-resourced Major Incident Rooms; other deaths, despite complex features, including Road Traffic Investigations, will usually be managed with less staff and administrative support.

4.20 Police investigation standards within the UK can be separated into levels of investigative activity which are each supported by national training, developed and accredited by the College of Policing under the Professionalising Investigation Programme (PIP).[1]

The four identified investigation levels are based on investigative activities:

- **PIP Level 1.** The investigation of volume crime – Police Constables, staff and their supervisors.
- **PIP Level 2.** Substantive investigation into more serious/complex offences, including road traffic deaths – Dedicated investigators (Detectives or Road Traffic Investigators).
- **PIP Level 3.** Lead investigator in cases of murder, stranger rape or kidnap – Senior Investigating Officers (SIOs) (Detective Chief Inspectors or Detective Superintendents).
- **PIP Level 4.** Critical, complex, protracted and/or linked serious crimes – SIO and Officer in Overall Command (OIOC) (Detective Superintendents or above).

4.20 *Protocols and guidance*

The aim of PIP is to develop professional investigators who can conduct investigations to a national standard based on recognised good practice.

1 College of Policing's 'Professionalising Investigation Programme'.

4.21 In October 2012, the College of Policing began replacing Manuals of Guidance, such as the Murder Investigation Manual,[1] with Authorised Professional Practice (APP), which is a body of consolidated guidance for policing. The intention is that APP will significantly reduce the amount of national guidance in circulation but bring consistency to specific police business areas including investigation, which will, under a homicide discipline, cover types of homicide including corporate manslaughter and corporate homicide and links to reference material such as the work-related deaths protocol.

At the time of writing, the Murder Investigation Manual is still used by SIOs throughout the UK whilst APP develops. In addition to the Murder Investigation Manual, the Core Investigative Doctrine[2] and the Road Death Investigation Manual[3] are used in investigative training and road death training respectively.

1 Murder Investigation Manual (Association of Chief Police Officers), introduced in November 1999 and updated in 2006.
2 Core Investigative Doctrine (Centrex/ACPO), 2005.
3 Road Death Investigation Manual (NPIA/ACPO), 2007.

Major Incident Room Standardised Administrative Procedures

4.22 Major Incident Room Standardised Administrative Procedures (MIRSAP)[1] is used with the police's computerised system for investigation called HOLMES (Home Office Large Major Inquiry System) to record investigations. The use of MIRSAP rules and conventions ensures that large investigations can be recorded and audited and, where necessary, linked with others around the UK via HOLMES.

1 Major Incident Room Standard Administrative Procedures (Association of Chief Police Officers), updated by Centrex in 2005.

PROCESS OF THE POLICE INVESTIGATION

The Murder Investigation Manual

4.23 The police investigation will follow the outline of an investigation as determined by the Murder Investigation Manual; but, in cases where corporate manslaughter is being pursued, the SIO will find that there are additional issues to consider.

Investigations and initial response

4.24 The actions taken by the first officers attending the scene of any homicide are critical to the outcome of the investigation. Police should carry out an initial assessment of the incident, providing a situation report to the control room, enabling the deployment of additional resources including the co-ordination of other emergency services.

Those making the initial response are guided by the following 'building block' principles:

- preservation of life;
- preservation of the scene;
- securing evidence;
- identifying victims; and
- identifying suspects.

Preservation of life

4.25 This is the primary responsibility of the emergency services initially deployed to the scene, assessing the condition of any victim, applying first aid and calling the ambulance service. When resources allow, the police will take action to reduce the impact of scene disturbance and destruction of evidence, by taking an early record of the scene, and establishing common approach and exit paths.

Preservation of the scene

4.26 The police aim to identify, secure, preserve and take control of each crime scene. Police establish the parameters of all identified crime scenes at the earliest stage, preventing unauthorised access to the scene and removal of potential evidence. Scenes include routes taken in and out of the scene by suspects and victims, including casualties and witnesses taken from the scene to additional locations such as hospitals. The deceased will, for instance, be considered as a scene at any additional location. A cordon, with monitored access to and from each scene, via single entry and exit points, recorded via a crime scene log, is essential to maintain the integrity of the scene. Once the cordon is established, entry will be controlled to those with an investigative requirement, and wearing protective clothing.

Securing the evidence

4.27 In addition to securing physical evidence at the scene, police will identify available witness evidence by speaking to people who are present at the scene. This is especially important when dealing with a death or deaths, where issues may arise concerning the accountability of others, eg when attempting to establish a causation chain (those in an organisation who are responsible for making decisions that may have had a bearing on the death). Initial inquiries at scenes may determine the identity of the victim and the identity of potential witnesses. Once an SIO is informed of the incident, their role is to review what has been done and bring additional enquiries under their command and control.

Identifying the victim

4.28 The victim should be identified quickly, from apparent investigative leads. The available indicators will be case specific, often (in the case of work related deaths) from witnesses at the scene. If the identity of the victim is unknown, examination of them or their personal effects at the scene by Crime Scene Examiners (or, during the forensic post mortem, by a Forensic Pathologist and Crime Scene Investigators) may provide putative identifiers, or samples suitable for scientific comparison against ante mortem records, such as fingerprints, DNA or dental records. In the case of mass fatalities, a Senior Identification Manager may be appointed to lead the identification process, according to Disaster Victim Identification principles. For example, these principles were applied at the Ladbroke Grove rail crash (also known as the Paddington train crash) which occurred on 5 October 1999 at Ladbroke Grove, London.

There will arise many 'fast track actions' that officers will need to complete, such as the deployment of family liaison officers (FLOs) to act as a single conduit between the investigation and the family. This, in turn, may lead to the early identification of suspects or additional witnesses, further scenes or evidence.

Identifying suspects

4.29 When a death occurs in the workplace (or, indeed, elsewhere) which is deemed to be suspicious, the identification of possible suspects must be at the forefront of the mind of the officer attending the scene. Inquiries must be made by the police to establish if the suspect poses a continued threat to others. The identification of suspects is an ongoing responsibility for the SIO.

The investigation

4.30 Once the scene has been secured, the investigation will continue under the command of an appointed SIO (usually PIP3) or Roads Policing SIO, according to the complexity of the incident. The case will take the form of a homicide investigation. The SIO will need to comply with the guidelines set out in the work-related deaths protocol.

The protocol places the emphasis on the first officer at the scene securing the scene, supervising and recording activity, and informing a senior supervisory officer. This is in accordance with standard force guidelines regarding unexplained death, when a detective of supervisory rank attends the scene of a death in the workplace and makes an initial assessment as to whether the circumstances may amount to homicide or other serious criminal offence. As a general rule, the police will investigate where there is evidence of, or a suspicion of, deliberate intent or gross negligence or recklessness on the part of an individual or company, rather than human error or carelessness. It may take some time before the police are sure that there have been no serious criminal offences committed, so the investigation needs to commence and continue with homicide (gross negligence manslaughter or corporate manslaughter) in mind from the beginning.

The protocol further states that the investigation will be a joint investigation, with the police and the HSE (or other enforcing authority) working closely together throughout. If the police reach a stage where they are satisfied that no serious criminal offences have been committed, the case will be passed wholly to the HSE or other enforcing authority to continue in the prosecution of any health and safety offences. Clearly, this indicates that the preservation of the scene and the initial actions of the police are paramount, as failure in this respect may affect or compromise subsequent safety prosecutions.

4.31 Where the relevant enforcing authority is investigating a death, and new information is discovered which may assist the police in considering whether a serious criminal offence (other than a health and safety offence) has been committed, the enforcing authority will pass that new information to the police. An enforcing authority inspector may do this, but it may also be from the enforcing authority's solicitors via the CPS. The police should then consider whether to resume primacy for the investigation.

With the protocol and the Murder Investigation Manual in mind, the SIO commences the investigation. If the incident is of sufficient size in terms of casualties, deaths, the number of witnesses and potential defendants, or due to its complexity, the SIO may well consider opening a major incident room adopting MIRSAP principles.

4.32 Officers will be appointed to lead teams of staff to discharge the strategies set by the SIO, including: forensic management; exhibits management; disclosure; media; witness and suspect interviews; family liaison delivered via family liaison officers (FLOs); intelligence accessing passive data enquiries such as data communications; search; house to house; and covert, including surveillance.

4.33 The protocol determines that resourcing, disclosure, witness, family liaison and media strategies should be agreed between the police and the HSE or other enforcing authority.

Working to the structure outlined in the Murder Investigation Manual, the SIO will commence 'fast track actions' or those tasks that need to be considered at a very early stage of the investigation.

Interviewing suspects and witnesses

4.34 It is not unusual for HSE inspectors and police officers to interview witnesses jointly, although separate statements may be obtained by each investigating body. As and when potential responsibility has been considered, it may be necessary for 'PACE interviews' (ie interviews to which the Police and Criminal Evidence Act 1984 applies) to be conducted and, again, it is increasingly common for those interviews to be conducted jointly by the police and the HSE or other enforcing authority inspectors.

4.35 On a practical basis, it is similarly not uncommon for inspectors to listen in and monitor the interviews from a separate room, with the consent of all concerned, and provide technical guidance and assistance to the officers conducting the questioning. The interviewing of witnesses is now considered a much more crucial role of the police within investigations, and guidance is given especially when dealing with vulnerable, intimidated and significant witnesses.[1] Witnesses in work related deaths may, by the nature of their involvement, be significant and will require being dealt with in a specific manner by properly trained officers. Significant witnesses, sometimes referred to as 'key' witnesses, are those who:

'• have or claim to have witnessed, visually or otherwise, an indictable offence, part of such an offence or events closely connected with it (including any incriminating comments made by the suspected offender either before or after the offence); and/or
• have a particular relationship to the victim or have a central position in an investigation into an indictable offence.'

1 'Achieving Best Evidence in Criminal Proceedings: Guidance on interviewing victims and witnesses, and guidance on using special measures' (March 2011).

4.36 The SIO will stipulate which witnesses should be treated as key witnesses and the method of achieving best evidence in each case, such as video recording prior to converting their statement into a witness statement.

Post mortems

4.37 In the case of deaths which are deemed to be suspicious, the usual practice is for the local coroner to be informed and a decision made as to whether a post-mortem examination is required. Provisions under the Coroners and Justice Act 2009 ('the 2009 Act') allow for coroners to hold investigations if the deceased has died a violent or unexpected death, the cause of death is unknown, or the deceased died in custody or 'state detention'.[1]

If, however, the death is deemed suspicious (ie the SIO considers that there is sufficient evidence to commence a manslaughter investigation), the coroner may agree to a forensically qualified pathologist to perform the post-mortem examination, as would be the case for a murder or manslaughter arising from any other cause.

1 The Coroners (Inquests) Rules 2013, SI 2013/1616, regulate the practice and procedure at, or in connection with, inquests which form part of an investigation into a death held under Part 1 of the 2009 Act.

4.38 From the post-mortem examination, the SIO would hope to establish the cause of death and recover any evidence or indication to support charges against an individual under common law or a corporation under the CMCHA 2007. It would be determined whether the deceased had had a natural illness that contributed to his or her death, or whether the death was caused as a direct result of the incident under investigation. It may be the case that the death had been caused as a result of an industrial disease, which is something the SIO would expect the pathologist to determine during the post mortem and from evidence gained during subsequent tests.

4.39 The coroner would open an inquest, and would hear evidence of identification, the time of death, the place of death and the cause of death, if known at that stage. The inquest would normally be adjourned to a date to be arranged, when the coroner would hear additional evidence regarding the reasons for the death and whether any individual or corporation is to be charged in connection with the death.

Victim identification

4.40 This brings the SIO to the first major line of inquiry that he or she would have to consider: that of the formal positive identification of the deceased person(s).

4.41 The coroner has statutory responsibility for the identification of the body. The standard of proof for identification is set by the coroner on a case-by-case basis, and the requisite standard of proof may be reached on the balance of probabilities. In a case anticipating criminal proceedings (or, particularly, involving multiple fatalities), the coroner is likely to specify that the higher criminal standard of proof should be used to determine an established identification, and avoid chances of misidentification.

4.42 The coroner may provide a set of identification criteria as to how the standard of proof can be met using established categories, or an incident-specific combination of criteria:

Primary Identification criteria, using comparative scientific techniques:

- DNA comparison;
- forensic odontology comparison;
- fingerprint comparison.

Secondary Identification criteria, using:

- unique physical match (such as a serial number on a body implant);
- marks;
- scars;
- tattoos;
- medical deformity/condition;
- unique & identifiable jewellery;
- unique clothing;
- personal effects.

Assistance Factors:

- photographs;
- clothing;
- jewellery;
- medical deformity/condition;
- location of deceased person;
- witness description;
- visual identification.

4.43 Visual identification is used to assist in identification in individual death cases, but its reliability should be regarded with circumspection. In mass fatality cases, visual identification has proved unreliable, and viewing by family members should only take place for social and grieving purposes after identification is confirmed by the coroner at the Identification Commission.[1]

1 'Public Inquiry into the Identification of Victims following Major Transport Accidents Report', Lord Justice Clarke, 2001.

Family liaison

4.44 The SIO is responsible for the delivery of the family liaison strategy via family liaison officers (FLOs) and family liaison coordinators (FLCs). FLOs act as a single conduit between the families of the deceased and the SIO and or other investigating agencies. A record is made of all interaction with family members by the FLOs

Training of investigators as FLOs, within all police forces, is designed to ensure that relatives are treated sympathetically and provided with honest and, as far as possible, accurate information at all times and at every stage throughout the investigation. The SIO's investigative requirements from the families are established via the FLOs. In cases of deaths where the protocol is involved, a joint family liaison strategy between the police and enforcing authority is essential in order to keep the families of those involved informed with up-to-date information. In the instances where primacy transfers from the police to other agencies, the family liaison strategy set by the police should be continued by the HSE or other investigative agency.

4.45 On 27 January 2010, the Justice Secretary announced the creation of the Victims Support National Homicide Service. The service is funded by the government to deliver a national service to support people who have been bereaved by murder and manslaughter in England and Wales. Once a homicide has occurred, Victim Support receive a referral from the relevant FLO (acting on behalf of the SIO). The Homicide Service delivers access to a range of support for families, including counselling, general financial support and access to services.

Relatives of victims of major disasters, or even work-related deaths, form themselves into international, well-organised, well-financed and legally well-briefed support groups, such as Disaster Action and SAMM (Support after Murder and Manslaughter) and place high information expectations on SIOs. The police should consider communication with groups at an early stage.

Lines of inquiry

4.46 In any major investigation, the SIO will determine 'lines of inquiry' or those inquiries that will achieve the overall aim of the investigation. Lines of inquiry are an essential element in the investigation, as they will assist in prioritising the actions raised for officers to follow.

The fundamental purpose of the investigation would be to:

- determine how and why the person died or was seriously injured;
- prove or disprove that a criminal offence has been committed (including any Health and Safety at Work etc Act 1974 (HSWA 1974) offences);
- determine who is responsible if a criminal offence has been committed; and
- determine whether there is sufficient evidence to prosecute any individual or organisation for any criminal offence (including HSWA 1974 offences).

4.47 There may be human offenders under common law and corporation offenders under the CMCHA 2007. Issues will arise regarding the identification of 'senior management' in order to prove an offence of criminal gross negligent manslaughter.

The incident may involve multiple victims – far more victims, either dead or injured, than the SIO has had experience of before. When large numbers are involved, a Senior Identification Manager (SIM) will be appointed to have overall management of the identification matters, including family liaison. This leaves the SIO to concentrate on the causes of the incident and identifying those culpable.

4.48 The process required to manage the identification of the dead following mass fatalities is known as Disaster Victim Identification (DVI).[1] The processes have been agreed internationally via INTERPOL agreements.[2] The investigative requirements include obtaining ante mortem information from families who may be spread throughout the UK or the world, and comparing the information with the post mortem information obtained in the mortuary through a reconciliation process. The coroner will chair an Identification Commission, prior to any inquest, to confirm proposed identifications.

The UK police and UK Forensic Medical Practitioners are now trained and co-ordinated via UK DVI and the National Police Coordination Centre (NPoCC), with proportionate contribution from police forces throughout the UK, to form a large mutual aid capability available for UK incidents or to support the FCO in mass fatality incidents overseas, such as the Tsunami in December 2006.

1 For a useful source of information, see Black, Walker, Hackman, Brooks (eds), *Disaster Victim Identification: The Practitioner's Guide* Dundee University Press Ltd, 2009.
2 See www.interpol.int/INTERPOL-expertise/Forensics/DVI-Pages/DVI-guide.

4.49 If a judicial or public inquiry commences during the time the investigation is live, this may present challenges for the SIO, in that the judicial inquiry may be hearing evidence before any manslaughter investigation is in possession of it. Whilst that process is a matter for the chair of the judicial inquiry and for the government, experience indicates that it causes problems with the investigation tracing of witnesses and subsequently taking statements from them, especially if the witness has already given verbal evidence to the inquiry. Previous similar situations have arisen and witnesses can often be reluctant to provide the police with statements, on the understanding that they have already given evidence before the inquiry and decline to repeat the process. In such circumstances, the SIO will consider tracing significant witnesses and arranging to interview them early in the investigation process.

4.50 The SIO will consider that the co-ordination and sequencing of the interviews is necessary, in order to ensure that the application of different powers by the relevant agencies does not lead to unfairness allegations at trial. The SIO and enforcing authorities must agree the sequencing of interviews, and powers used in interviews, ensuring that clear and lawful objectives are set to achieve each agency's lawful purpose.

The SIO will give early consideration to obtaining technical expert assistance, usually from the HSE, but also from other industry sectors, as appropriate, and from universities both in the UK and abroad.

The company profile

4.51 The SIO will consider examining the company's safety strategy, culture and performance. This may provide evidence to support a charge against a company or an organisation and will portray an image of the company's attitude towards safety. The SIO may well turn to the HSE or other enforcing authority for assistance and use their expertise.

The SIO will quickly commission officers to create a profile of the company, concentrating on the management and structure of the company. The investigation will need to determine who the management are (including job description and role in the organisation) and whether they have come to the notice of the police, the HSE or any other investigative body before, either within this company/corporation or another. Where organisations have been taken over, as such ceasing in legal terms, the new organisation will assume the liabilities of the old via a transfer of undertakings. Police may object to a voluntary liquidation request by a company or seek an injunction to prevent the liquidation taking place.

4.52 The SIO will acquire a financial profile of the organisation to determine whether spending on safety was appropriate, given the amount of gross profit. The object of this line of inquiry would be to ascertain if the company took safety seriously. This may require the employment of forensic accountants.

The decision-making process within an organisation will also be of great interest to the SIO. Senior management will be asked to justify decisions made regarding safety management. With the need to determine the functions of the senior management, the SIO will be looking for documentation which relates to these decisions in the form of policies, memoranda and meeting minutes (for example, maintenance procedures, response to concerns expressed by the workforce, and comments made by union or safety representatives).

Disclosure and the legal strategy

4.53 Each investigative agency under the work-related death protocol will be responsible for scheduling their documentation. The SIO will appoint disclosure officers to schedule the police material and any third party material, which will be presented to the CPS. Early liaison by the police, HSE or other enforcing authority with the CPS is required for consultation about the nature of charges, and legal and evidential issues surrounding the investigation, including advice about expert evidence.

Dealing with the media

4.54 It is the responsibility of the SIO to deal with the media at the scene and as the investigation progresses. To prevent loss of evidence and to ensure the release of crucial information pertinent to the investigation, it is important that the media are managed correctly and joint information requests are agreed by both the police and the HSE or other investigative authorities.

HEALTH AND SAFETY ENFORCEMENT

4.55 The Health and Safety (Enforcing Authority) Regulations 1989[1] allocate investigation and enforcement responsibilities. The Health and Safety Commission (HSC), HSE or local authorities have an agreed enforcement approach. Other enforcement agencies (such as the ORR and MCA) are not necessarily bound by these agreements.

1 SI 1989/1903.

4.56 In addition, there are other authorities and agencies with responsibilities for the investigation and enforcement of the HSWA 1974 or other similar legislation. These include:

- the Care Quality Commission (health and social care premises) in England;
- the Care and Social Services Inspectorate Wales and Health Inspectorate for Wales;
- the Environment Agency;
- the Civil Aviation Authority;
- Trading Standards;
- the Department for Business, Innovation and Skills (BIS);
- the Marine Accident Investigation Branch (MAIB);
- the Rail Accident Investigation Board (RAIB); and
- the Air Accident Investigation Board (AAIB).

4.57 The main thrust of health and safety enforcement activity resides with the HSE and local authorities. As a general rule, the HSE is responsible for enforcement activity in factories, construction sites and agricultural locations, and local authorities are responsible for enforcement activity in wholesale and retail premises, offices and catering premises. When assessing enforcement responsibility, consideration will need to be made of the detailed regulations and relevant enforcing authority guidance[1] if the question of jurisdiction needs to be considered. It may sometimes be necessary for enforcement action to be transferred between the HSE and local authorities, and procedures exist for this to be done.

1 HSE Operational Guidance OC 124/7 gives detailed guidance on the Health and Safety (Enforcing Authority) Regulations 1989; and HSE Operational Guidance OC 124/9 gives detailed guidance on considerations when considering the main activity in relation to enforcement responsibility.

4.58 The overall investigation should be pursued in such a way that will best serve the public interest and enable there to be a sound investigation of all the relevant circumstances. The investigation aims will include:

- establishing whether prosecution is warranted and, if so, whether for manslaughter (gross negligence or corporate), other serious criminal offence, or HSWA 1974 offences, or a combination of these; and
- gathering admissible evidence where prosecution is warranted.

4.59 Therefore the investigation of any health and safety offences should be undertaken by the relevant enforcing authority in parallel with the police investigation of manslaughter.[1] The investigation powers under s 20 of the HSWA 1974 include:

- the power to enter premises;
- the power to examine equipment;
- the power to take photographs;
- the power to seize items of plant and equipment;
- the power to copy documentation;
- the power to require those who have information relating to the incident to provide that information to the inspector and then sign a written statement setting out such information; and
- any other power required to give effect to the preceding powers.

1 'Work-related deaths: liaison with police, prosecuting authorities, local authorities and other interested authorities including consideration of individual and corporate manslaughter / homicide', HSE publication OC 165/10.

4.60 The investigative powers under s 20 of the HSWA 1974 are more extensive and wide-ranging than those possessed by police officers. In addition, the power to require individuals to provide information and an authorised investigator can begin an interview conducted pursuant to s 20 with the words, 'I require you to answer my questions', and can continue by insisting that the person questioned should then sign a written statement containing the information provided under such compulsory questioning. The safeguard in the legislation, however, is that, if any individual is compelled pursuant to s 20 to provide information, the information so provided cannot be used in the course of any prosecution of that individual. In practical terms, it means that employees can be required to give any information which is required relating to the employer's business and the circumstances leading up to and relating to the actual death. It also follows that, if there is any suggestion that the individual concerned might have personal responsibility for what has occurred, the powers under s 20 cannot be used; in such circumstances, the individual must be cautioned and an interview conducted in accordance with the requirements of the Police and Criminal Evidence Act 1984.

4.61 Once a decision has been taken by the CPS that no charge of manslaughter should be pursued, further conduct of the investigation is passed to the HSE. It is the usual practice for all police material to be transferred to the HSE, and this will then form part of the prosecution's evidence and either be relied upon or be disclosed as unused material. Practitioners should be alert to that fact when considering evidence disclosed to them.

DEATH IN POLICE CONTACT

4.62 Police forces in England and Wales have a statutory duty to refer all deaths following police contact, including those that occur in or following police custody, to the Independent Police Complaints Commission (IPCC).[1] The IPCC will, subject to the provisions of its legal obligations, have regard to the work-related deaths protocol obligations of the police, to the extent that they have not already been fulfilled; and the HSE will deal with the IPCC in the same way as it deals with the police under the protocol.

1 The IPCC was established by the Police Reform Act 2002 and became operational in April 2004.

CLOSING THOUGHTS

4.63 The work-related deaths protocol has had a big impact upon the way in which work related deaths are investigated by the police and multiple enforcement agencies. In recent years, the protocol has been refreshed with additional signatories, further defining responsibilities and providing an effective framework for joint working.

4.64 The latest edition of the protocol will continue to drive joint working practices and, as these continue to mature, it is likely that the number of prosecution for offences under the CMCHA 2007 will increase.

Chapter 5

Construction

Dr Louise Smail, Risk Consultant, Ortalan, Manchester

INTRODUCTION

5.1 Construction is one of the major contributors to the UK economy and, with the steady decline and loss of the traditional 'heavy industries' over the last 50 years, construction is now one of the highest risk industries and accounts for about 31 per cent of UK reported workplace fatalities. Although recent years have seen a decline in injuries and fatalities, there has also been a decline in the amount of construction taking place. There were 39 fatal injuries to workers in construction, 26% lower than the average figure of 53. The latest rate of fatal injury is 1.9 per 100 000 workers, compared to a five-year average of 2.3.[1]

1 This is based on the Health and Safety Executive (HSE) statistics for 2012/13.

5.2 Traditional tendering procedures have historically placed the emphasis on the lowest bidder, sometimes to the detriment of other factors such as safe working practices. However, the introduction of recent legislation, such as the Construction (Design and Management) Regulations 2007[1] (CDM Regulations) and the Corporate Manslaughter and Corporate Homicide Act 2007 (CMCHA 2007), has caused clients to take greater account of factors such as competence, compliance with safe working practices, and allowance for adequate resourcing of projects, as well as cost considerations.

1 SI 2007/320.

5.3 The CDM Regulations ensure that the client has a key role in ensuring that there are adequate management arrangements for construction projects and that the client is involved in key decisions about the commencement of the construction works. The client's role has well-defined duties, and these regulations place needs for the client's positive actions throughout the construction process and in compliance with their duties.

5.4 The CDM Regulations are due for review during 2014. The HSE has not yet issued the consultative document. It is expected that the revision with be based on a 'copy out' of Directive 92/57/EEC, and the HSE has indicated that this may be a radical overhaul of the current system.

5.5 The 1960s saw major reforms in health and safety legislation with the introduction of the Factories Act in 1961, the Offices, Shops and Railway Premises Act in 1963 (updating and reforming previous legislation) and various sets of regulations specific to the construction industry, including:

- the Construction (General Provisions) Regulations 1961;
- the Construction (Lifting Operations) Regulations 1961;
- the Construction (Working Places) Regulations 1966; and
- the Construction (Health and Welfare) Regulations 1966.

The Construction (General Provisions) Regulations 1961 remained in place until 1996, when they were replaced by the Construction (Health, Safety and Welfare) Regulations 1996 (which have now been incorporated into the CDM Regulations).

EC DIRECTIVES

5.6 In the UK, EC Directives are introduced in the form of regulations, or statutory instruments which are made in accordance with the provisions of the Health and Safety at Work etc Act 1974 (HSWA 1974), which is the primary legislation that provides the framework under which health and safety regulations are enacted.

The two main Directives affecting the management of construction activities are the Temporary or Mobile Construction Sites Directive[1] (TMCS Directive) and the Management Directive 1992. The TMCS Directive dealt with the implementation of minimum safety and health requirements at temporary or mobile construction sites, and was part-implemented in the UK as the CDM Regulations 1994.

1 92/57/EEC of 24 June 1992.

Management Directive

5.7 The Management Directive stemmed from the Framework Directive 89/391/EEC, which was adopted in 1989. It was implemented in the UK as the Management of Health and Safety at Work Regulations 1992, which aim to direct the principles of health and safety management and, in particular, risk assessment. These were revoked and replaced in 1999 by the Management of Health and Safety at Work Regulations 1999 (Management Regulations 1999).

Workplace Directive

5.8 The Workplace Directive (89/654/EEC) is mostly implemented by the Workplace (Health, Safety and Welfare) Regulations 1992 (Workplace Regulations 1992), which address welfare facilities and conditions in the workplace.

5.9 The CDM Regulations make specific reference to the 'workplace', and there is now a specific requirement for designers to take account of the 'workplace' when preparing a design of a structure which is to be used as a 'workplace'.

Work at Height Regulations

5.10 The Work at Height Regulations 2005 (SI 2005/735) implement EC Directive 2001/45/EC of 27 June 2001 amending EC Directive 89/655/EEC concerning the minimum health and safety requirements for the use of work equipment by workers at height (The Temporary Work at Height Directive). This consolidates previous legislation relating to the use and inspection of scaffolds and ladders, fragile roofs and edge protection which were previously included under the Construction (Health, Safety and Welfare Regulations 1996 and in the Workplace Regulations 1992.

CORPORATE LIABILITY AND CONSTRUCTION REGULATIONS

5.11 The introduction in April 2007 of the Construction (Design and Management) Regulations 2007 was an update on the previous CDM Regulations which, when they were introduced, provided a significant change and set out the framework for managing the construction process through the design and construction phases. As in the previous version, the regulations demand that consideration be given as to how buildings and structures will be maintained, cleaned and repaired throughout their life cycle, as well as their eventual demolition. They also require all those involved in the design process to take account of the building or structure in use where the structure is to be used as a 'workplace'.

5.12 An architect's practice and a construction company involved in a Somerset development have been fined a total of £195,000 following a fatality on the site. Express Park Construction Company Limited (EPCC), of Harley Street, London, pleaded guilty to breaching s 3(1) of the HSWA 1974 for failing to safely manage subcontractors working for it. The architects involved, Oxford Architects Partnership, of Bagley Croft, Hinksey Hill, Oxford, pleaded guilty to breaching regs 13 and 14 of the CDM Regulations 1994, which require designers to take safety considerations into account. EPCC was fined £75,000 and ordered to pay costs of £68,000, and Oxford Architects Partnership was fined £120,000 and ordered to pay costs of £60,000. The worker was on the air conditioning plant, which was built on a platform accessed via a ladder at the edge of a flat roof. The roof only had a low parapet, which was not high enough to prevent him falling nine metres to the ground.

5.13 The CMCHA 2007 makes specific mention of construction in s 2(1)(c)(ii), where it refers to the '… carrying on by the organisation of any construction or maintenance operations'. The CDM Regulations set out clear duties and responsibilities for all those companies or organisations involved in construction work or in a construction project. All of these duty holders are equally liable to find themselves subject to prosecution in the event of a fatality occurring on a construction project where a breach of health and safety legislation is established. The guidelines that are in place allow the various duty holders to document and manage the way in which they carry out construction work and to demonstrate that they have done all that is necessary to achieve compliance with the law. The CMCHA 2007 sets out quite clearly, at s 8, the matters which a jury should take into account when considering whether or not the prosecution has established that significant breaches have taken place that amount to gross negligence on the part of the accused parties.

HEALTH AND SAFETY GUIDANCE

5.14 Section 8(3)(b) of the CMCHA 2007 states that the jury may also have regard to:

> '... any health and safety guidance that relates to the alleged breach.'

Section 8(5) provides some further clarification, stating:

> '... In this section "health and safety guidance" means any code, guidance, manual or similar publication that is concerned with health and safety matters and is made or issued (under a statutory provision or otherwise) by an authority responsible for the enforcement of any health and safety legislation.'

5.15 There is extensive 'health and safety guidance' available to the construction sector, including:

- Approved Codes of Practice (ACOPs), issued by the HSE, which are accorded special status by virtue of their approval by the HSE with the consent of the Secretary of State; following the Löfstedt review, the HSE is reviewing many of the ACOPs;
- Codes of Practice;
- guidance including the HSE/IOD Guidance – Leading Health and Safety at Work,[1] which provides guidance to directors and board members;
- British Standards published by the British Standards Institution; and
- guidance issued by professional organisations and trade bodies.

1 Health and Safety Executive, 'Leading Health and Safety at Work' INDG417 10/07 C2000, available at www.hse.gov.uk/leadership and at www.iod.com/hsguide.

5.16 The construction industry is subject to a number of regulations and supporting guidance which would be relevant in the investigation and prosecution of a manslaughter case. The CDM Regulations, ACOPs and industry guidance may also be used by prosecutors to identify shortcomings by the accused in their management and health and safety arrangements for construction work, including subsequent repairs and maintenance.

5.17 Although more emphasis seems to be placed upon major projects and new build, a significant part of construction comprises minor works (including repairs and maintenance), extensions and refurbishments to existing buildings, minor alterations, and internal and external redecorations. Construction forms a major part of the UK economy, and the definition of 'construction or maintenance operations' provided by s 2(7) of the CMCHA 2007 means that virtually every construction activity is covered by both the Act and the CDM Regulations.

5.18 The definition provided at s 2(7) of the CMCHA 2007 states:

> ' "construction or maintenance operations" means operations of any of the following descriptions–
>
> (a) construction, installation, alteration, extension, improvement, repair, maintenance, decoration, cleaning, demolition or dismantling of–
>   (i) any building or structure,
>   (ii) anything else that forms, or is to form, part of the land, or
>   (iii) any plant, vehicle or other thing;
>
> (b) operations that form an integral part of, or are preparatory to, or are for rendering complete, any operations within paragraph (a).'

5.19 The refurbishment sector was targeted by the HSE, following a 61 per cent increase in fatalities on sites where refurbishment works were being undertaken – over half of the workers who died on construction sites in 2006/07 worked in refurbishment.

5.20 In March 2013 the HSE carried out an initiative to visit construction sites. Nearly one in five construction sites visited across Britain has been subject to enforcement action after failing safety checks. In the month-long initiative, inspectors from the HSE visited a total of 2,363 sites where refurbishment or repair work was taking place and saw 2,976 contractors. A total of 631 enforcement notices were served across 433 sites for poor practices that could put workers at risk, with 451 notices ordering that work stop immediately until the situation was put right.

THE CDM REGULATIONS – THE ROLES AND RESPONSIBILITIES OF DUTY HOLDERS

5.21 The CDM Regulations place duties upon clients and designers as well as contractors, and any investigation is likely to include each of these duty holders and to examine the part played by each of them in the events leading up to the incident (their roles are defined by the CDM Regulations), and any failures by the client project team during the design phases which may have contributed to an incident.

5.22 Individuals fulfilling these roles may be regarded by the investigating body (the police) as undertaking these roles at a senior management level and responsible for their own actions and capable of acting on behalf of the organisation or body that they represent. The prosecution will be seeking to demonstrate that failures have occurred at a senior management level. Where there is any evidence to suggest that practices within the organisation tolerated or accepted failures, the prosecution can use such evidence to convince the jury that the failures are symptomatic and amount to a gross breach of the duty of care.

COMPETENCE

5.23 The CDM Regulations require that clients appoint competent companies and individuals to the various duty holder roles – a client should ensure, under reg 4, that 'he has taken reasonable steps to ensure that the person to be appointed or engaged is competent'. Regulation 4 also applies to any person appointing or engaging another duty holder, such as a designer subcontracting out design work, or a principal contractor engaging other contractors. There is also a duty on any person accepting an appointment or engagement to accept this only if they are competent to do so.

5.24 Where the construction work involves safety-critical activities, it is especially important that clients ensure that those appointed or engaged to undertake such work are competent and that the checks made are sufficient to demonstrate that all reasonable steps have been taken to establish that the management and supervision of such work will in fact be undertaken by suitably qualified and experienced people.

5.25 The ACOP[1] provides relevant information as to the standards required in respect of competence and how the client can carry out an assessment of competence.

Appendix 5 to the ACOP deals with the competence for CDM coordinators involved with larger or more complex projects, but fails to adequately define what a larger or more complex project might be, although reference is made elsewhere in the ACOP (for example, at para 21) to more complex construction work in the context of non-notifiable projects and the requirement for documentation more akin to the construction phase plan is required for notifiable projects.

1 Approved Code of Practice, 'Managing Health and Safety in Construction' ISBN 978 0 7176 6223 4.

5.26 A list of construction activities where the risk is higher is included at para 21 of the ACOP L144, which is loosely based upon Annex II of the Directive (Council Directive 92/57/EEC). These activities include:

(a) structural alterations;
(b) deep excavations, and those in unstable or contaminated ground;
(c) unusual working methods or safeguards;
(d) ionising radiation or other significant health hazards;
(e) nearby high-voltage powerlines;
(f) a risk of falling into water which is, or may become, fast flowing;
(g) diving;
(h) explosives; and
(i) heavy or complex lifting operations.

5.27 The requirements under reg 4 of the CDM Regulations relating to competence are not restricted to appointments made by the client: they apply to any appointment made by any of the duty holders, including designers, contractors and principal contractors.

5.28 Duty holders will need to measure their knowledge and experience against the core criteria contained within the ACOP and should only take on projects where they have the requisite knowledge or experience or can bring in the additional expertise necessary where these skills are not already available within the company.

5.29 The Work at Height Regulations 2005 require that persons involved in any activity where work at height is to be undertaken are competent, as set out in reg 5, which states:

> 'Every employer shall ensure that no person engages in any activity, including organisation, planning, and supervision, in relation to work at

height or equipment for use in such work unless competent to do so or, if being trained, is supervised by a competent person.'

Those engaging contractors to undertake works on their premises should ensure that contractors are able to achieve compliance with reg 5 of the Work at Height Regulations 2005 – a similar duty is placed upon clients under reg 4 of the CDM Regulations.

5.30 In July 2013, Lion Steel Equipment Limited became the third company in the UK to be convicted of corporate manslaughter and were fined £480,000 and ordered to pay prosecution costs of £84,000. Lion Steel were charged following the death of an employee who died after falling through a fragile roof at its site in Hyde, Cheshire in May 2008. The company admitted the offence, part way through a trial, on the basis that all charges against its directors would be dropped (three men had been charged with gross negligence manslaughter and health and safety charges).

APPROVED CODE OF PRACTICE

5.31 The broad principles for the assessment of competence found at Appendices 4 and 5 to the Approved Code of Practice (ACOP) L144 largely reflect the research carried out by John Carpenter, in a report commissioned by the HSE, in which he examined how competence was assessed and dealt with under the original CDM Regulations 1994.[1]

1 RR422 – Developing guidelines for the selection of designers and contractors under the Construction (Design and Management) Regulations 1994', available from the HSE website.

5.32 Competence is regarded as a measure of skills, experience and knowledge, and Appendix 4 to the ACOP sets out, in tabulated form, how clients can make an assessment of these components for each of the main duty holders.

5.33 Appendix 4 is set out in three columns: the first column sets out the Core Criteria; the second the standards to be achieved; and the third column gives examples of the evidence that could be used to demonstrate how the required standards have been met.

5.34 The section dealing with individual qualifications and experience refers to health and safety training and qualifications and, although general qualifications such as the NEBOSH Construction Certificate and the Site Management Safety Training Scheme are mentioned, it is important that clients making appointments are able to make a judgement as to the relevance of such qualifications in relation to the complexity of the project to be undertaken: these qualifications are likely to be taken to provide a minimum standard, and additional expertise may be required when dealing with specific activities, such as the design or construction of tunnels or bridges.

SAFETY-CRITICAL ACTIVITIES

5.35 In the case of safety-critical work, the standards applied to both the management and the organisation and procedures whilst the work is being undertaken must reflect the fact that failures are likely to lead to major injuries or to fatalities, and that any failures in applying such standards are likely to be construed in the courts as a gross breach of the duty of care owed to the injured parties.

5.36 The CDM Regulations place duties upon all of the parties involved in a construction project, and each duty holder should be able to identify safety-critical activities and consider the nature of the risks that they are introducing into the project. This includes both clients procuring the work and designers who are translating the client's requirements and incorporating them in their design.

5.37 Where client organisations or designers fail to recognise the impact of their requirements or designs on the safety of construction workers or the end user, they are unlikely to be able to demonstrate due diligence in fulfilling their role as duty holders under the CDM Regulations.

5.38 It is important that 'safety-critical activities' are supervised by competent persons, and that the actual work is undertaken by competent operatives.

5.39 The types of activities that would come within this category include work at height, lifting operations and work in 'confined spaces', where the task-specific activities are covered by their own sets of regulations, such as the Work at Height Regulations 2005, the Lifting Operations and Lifting Equipment Regulations 1998 (LOLER) and the Confined Spaces Regulations 1997.

5.40 An examination of past incidents involving construction site plant or equipment suggests that a major factor in a number of these cases has been directly attributable to the competence of those supervising or operating plant or equipment, with in many cases the operatives not having the necessary certification for the plant or equipment that they have been using and, in a number of cases, for which they have not received any training or instruction.

5.41 In October 2008, Carmarthenshire County Council was fined a total of £45,000 under LOLER and ss 3 and 2 of the HSWA 1974. A volunteer with the Urdd Eisteddfod fell into an unguarded lift shaft at the Lyric Theatre.

5.42 A worker died at the site run by Harris Calnan Construction Co in Hampstead when a skip full of liquid concrete overturned as the crane was lifting it onto the site. The crane's boom fell onto Page, causing fatal crush injuries to his upper body. Harris Calnan was fined £80,000,[1] with costs of £66,244, after it had earlier pleaded guilty to breaching reg 8(1)(c) of LOLER, which required it to ensure lifting operations were carried out safely. Harris Calnan admitted failing to ensure that every lifting operation was properly planned by a competent person, contrary to reg 8(1)(a) of LOLER. The judge fined him £7,500 plus costs of £25,000.

1 *Harris Calnan Construction Co Ltd v Ridgewood (Kensington) Ltd* [2007] EWHC 2738 (TCC).

MANAGING CONSTRUCTION PROJECTS

5.43 Clients have a major role to play in ensuring that arrangements for managing health and safety for construction projects are dealt with effectively throughout the whole of the construction project, both during the design stages and when the construction project is underway on site. Clients are not required to manage the project themselves, but they do need to ensure that the management arrangements that are in place are adequate: this includes the arrangements made by each of the duty holders, including designers, contractors and others who might be involved in the project.

CLIENT'S DUTY IN RELATION TO ARRANGEMENTS FOR MANAGING PROJECTS

5.44 A major change to the CDM Regulations is the way in which information flows are managed. The CDM Regulations 1994 concentrated on the pre-construction health and safety plan, whereas the new regulations place more emphasis on managing the flow of information between the respective duty holders, with clients taking on the responsibility for obtaining and passing on any information which they might already have in their possession (including any existing health and safety file) and which can be obtained by survey or other investigation.

5.45 Information needs to be passed on by clients, initially to designers and then to any contractors who are tendering for the work. A number of prosecutions have been taken under the CDM Regulations for the failure by duty holders to pass on information, for example, about fragile surfaces (such as asbestos cement roofing sheets and fragile roof lights), which has led to construction or maintenance workers sustaining serious or fatal injuries after falling through the fragile roof surface or roof light.

5.46 In February 2009, IC Roofing Limited's director Colin Cooper was found guilty of manslaughter after one of his employees was killed in a six-metre fall and was sentenced to 12 months in jail. He was also fined £10,000. The company were fined £20,000 and Mr Cooper was also disqualified from holding a director's position for three years. The employee died whilst carrying out roofing repairs at a unit at the Bellbrook Industrial Estate in Uckfield. He had not been wearing a harness and there was no safety net in place to catch him when he fell through a skylight on to the concrete floor. He died in hospital the day after the fall. Mr Cooper had failed to carry out the necessary risk assessment procedures for the work. The employee who died was a trained electrical engineer, but had only received limited training during the six months he had been employed by the company.

SAFETY CULTURE

5.47 In recent years, effort has been concentrated on the safety culture of the organisation and its workforce. This has led to the implementation of procedures and policies (which have been driven by both the client and the main contractors) to implement safe working practices throughout the supply chain and achieve changes in behaviour and attitude at all levels in the workforce.

NOTIFICATION AND REPORTING OF INJURIES AND DANGEROUS OCCURRENCES

5.48 The central reporting system which is provided under the Reporting of Injuries, Diseases and Dangerous Occurrences Regulations (RIDDOR) requires employers to report certain designated incidents, including major injuries and fatalities, and dangerous occurrences to the relevant authorities. The system, which makes reporting of incidents compulsory, also provides an opportunity for management to use the information to monitor performance, identify trends and introduce appropriate measures, either to correct poor behaviour and performance or to provide additional training or schemes that are designed to improve performance. The inclusion of 'near misses' as part of the internal reporting system is generally recognised as a useful tool in the identification of poor practices within the organisation.

5.49 These employers' reports are the source of the accident statistics which are produced by the HSE on an annual basis.

5.50 In the event of a fatality on a construction site, the 'responsible person' is required to notify the HSE by the quickest possible means, and an investigation will be carried out jointly by the police and the HSE in accordance with the Work-related Deaths Protocol.

5.51 It is important to note that the recession in late 2008 and in 2009 had a large impact on the amount of construction work being carried out, and this may partially explain the decrease in the accident figures for the industry. The construction industry accounts for about 5% of the employees in Britain, but it accounts for 27% of fatal injuries to employees and 10% of reported major injuries.

The latest statistics show that there were 39 fatal injuries to workers in 2012/13, of which 12 fatalities were to the self-employed. This compares with an average of 53 over the previous five years, including an average of 18 to the self-employed.

According to the Office for National Statistics,[1] output in the construction sector grew steadily (with the exception of a mild downturn in 2004) from 2000 until early 2008, when a financial market shock affected the UK and global economic growth. The deterioration in economic conditions that followed had a marked effect on the three main sectors of the economy and, in particular, construction. Between Q1 2008 and Q2 2009, construction output fell by 17.1%, while output in the production and services industries fell by 11.7% and 5.4% respectively. Quarterly estimates for Q2 2013 show that output in the construction industry remains 14.7% below its pre-downturn peak in Q1 2008.

It should be remembered that any decrease may be due to the downturn in work and not necessarily attributable to an increased safe working culture.

1 Statistical Bulletin, 'Construction Output' (August 2013), available at www.ons.gov.uk/ons/dcp171778_330068.pdf.

CONSTRUCTION RISK MANAGEMENT

5.52 The nature of construction work, which can change during the course of a project, can make risk management difficult on a day-to-day basis, compared to a factory or even the rail infrastructure, where the assets are fixed. This is recognised at European level, where the EC Directive refers to 'temporary or mobile' construction sites.

5.53 The issues that make construction projects more difficult to manage and control include the following:

- Construction involves projects where the design is not standardised to any extent: many projects are unique in their combination of site and location, type of structure and mix of components and finishes.
- Project teams tend to change on a project-to-project basis and also during the course of a project, although some clients recognise the benefits of collaborative working with the same teams.
- The timescales for planning and mobilisation can be more concentrated in the construction sector at all levels and stages of the project.

- The initial design stages can be slowed because they are subject to external influences (whilst planning consent and any other approvals are obtained), but historically the timescales allowed to both designers and contractors have often been too short. This has historically been the case in the time allowed to contractors between winning a tender and being required to commence the construction works on site. Such issues are partly addressed by the requirements under the CDM Regulations for clients to state the time that will be allowed to contractors for preparation and planning.
- The transient nature of the workforce.
- The use of innovative materials and components and the move away from a craft-based industry.

It is therefore important that procedures are introduced to ensure that safety-critical issues are properly addressed and control measures implemented and monitored in order to overcome the issues outlined above.

5.54 Construction work does involve a number of activities that are intrinsically dangerous, and Parliament has introduced legislation, especially over the last 10 years or so in response to EC Directives, to ensure that organisations recognise and understand the nature of the risks involved and implement control measures that are designed to prevent major injuries and fatalities from occurring.

5.55 Some of the legislation is specific to construction but, because construction is so widely based, and encompasses such a broad range of activities, much of the legislation (for example, that relating to work at height, lifting operations, use of work equipment, noise and vibration, work in confined spaces, and ionising radiation) applies across all industries but does have specific relevance to construction activities.

5.56 A common feature of much of this legislation is the requirement for the work to be properly planned, organised and supervised, for the works to be undertaken and supervised by 'competent persons', for a written plan or method statement including emergency or evacuation procedures, and for permit systems to be in place.

5.57 The manner in which the new CDM Regulations have been set out, with Pt 4 looking at the construction site activities and the emphasis on the roles played by the various duty holders, provides a framework for providing such procedures.

SUB-CONTRACTOR MANAGEMENT

5.58 A contractor has been fined for safety failings after two subcontractors were unwittingly exposed to asbestos fibres at Reading University. Gardner Mechanical Services Ltd (GMS) had been contracted to undertake a mechanical services upgrade in a room at the University. They subcontracted the work to another company, which in turn hired two self-employed men to carry out the work.

The two men had not been made aware of the presence of asbestos in the room and believed that all asbestos-containing material had been removed by specialist contractors prior to their work commencing. They drilled through a sprayed asbestos ceiling coating, releasing asbestos fibres into the air.

5.58 *Sub-contractor management*

The HSE found that, as the principal contractor, GMS was aware that some asbestos-containing materials were to be left on the site but had failed to pass on this information to the subcontractors. The company pleaded guilty to breaching reg 11(1) of the Control of Asbestos Regulations 2006 and reg 23(1)(a) of the CDM Regulations and was fined a total of £28,000 and ordered to pay £22,631 in costs.

5.59 Bethell Construction Limited, of Preston, had been hired to perform maintenance work in a large underground structure forming part of the local sewerage system and subcontracted the work out to another contractor, who in turn took on additional subcontracted labour. The prosecution arose following the death of a subcontracted labourer who fell five metres onto a concrete platform.

5.60 Following a trial at Liverpool Crown Court, both contractors were found guilty of breaches of s 3 of the HSWA 1974 and reg 3(1) of the Management Regulations 1999.

5.61 The judge commented:

> 'the subcontracting arrangements had been very loose and should have been properly managed so that each party had understood its roles and responsibilities.'

Concerns were also raised that 'Method statements were crossing in the post with the contract', and that method statements and risk assessments were rushed to and fro 'without an opportunity for paused reflection'.

5.62 Under the CDM Regulations 2007, clients need to give more thought as to whether such arrangements are adequate. As a result of the successful prosecution, Bethells were fined £15,000 plus £49,110 costs.

DUTY OF CARE

5.63 A duty of care is owed to various groups of people who may be affected by construction activities, either directly as workers on a site or indirectly as visitors to the site including the client and consultants, as well as those who are unauthorised, such as trespassers, to whom a duty of care is owed under the Occupiers' Liability Act 1957.

5.64 Marks and Spencer plc and three of its contractors were fined for putting members of the public, staff and construction workers at risk of exposure to asbestos-containing materials during the refurbishment of two stores in Reading and Bournemouth: Marks and Spencer plc were fined £1 million and ordered to pay costs of £600,000; PA Realisations Ltd were fined £200 for contravening reg 15 of the Control of Asbestos at Work Regulations; and Styles & Wood Limited were fined £100,000 and ordered to pay costs of £40,000; all for breaches that took place at the Marks and Spencer plc store in Broad Street, Reading. Willmott Dixon Construction Ltd was fined £50,000 and ordered to pay costs of £75,000, for contravening ss 2(1) and 3(1) of the HSWA 1974 at the Marks and Spencer plc store in Commercial Road, Bournemouth. Willmott Dixon Construction Ltd is applying for permission to appeal against conviction.

As a result of a prosecution brought by the HSE, Marks and Spencer plc, Willmott Dixon Construction Ltd and PA Realisations Ltd (formerly Pectel Ltd) were found guilty in July 2011. Styles & Wood Limited pleaded guilty at an earlier hearing in January 2010. The work was carried out between 2006 and 2007 on shops in Reading and Bournemouth.

The court heard that the client, Marks and Spencer plc, did not allocate sufficient time and space for the removal of asbestos-containing materials at the Reading store. The contractors had to work overnight in enclosures on the shop floor, with the aim of completing small areas of asbestos removal before the shop opened to the public each day.

The contractor, PA Realisations Ltd, failed to reduce to a minimum the spread of asbestos to the Reading shop floor. Witnesses said that areas cleaned by the company were re-contaminated by air moving through the void between the ceiling tiles and the floor above, and by poor standards of work.

Styles & Wood Limited, the principal contractor at the Reading store, admitted that it should not have permitted a method of asbestos removal which did not allow for adequate sealing of the ceiling void, which resulted in risks to contractors on site.

The principal contractor at the Bournemouth store, Willmott Dixon Construction Ltd, failed to plan, manage and monitor removal of asbestos-containing materials. It did not prevent the possibility of asbestos being disturbed by its workers in areas that had not been surveyed extensively.

5.65 Mowlem (which was bought by Carillion in 2006) was one of three companies and a manager fined a total of £217,500 and £125,000 in costs following the death of a 17-year-old trainee scaffolder in 2004. The scaffolder died at Davyhulme Wastewater Treatment Works in January 2004 after falling approximately 18 metres while working to construct a scaffold within a 20-metre-high sewage digester tank. The scaffolder, who was employed by 3D Scaffolding, was under the supervision of a visiting contracts manager. 3D Scaffolding was contracted to work for RAM Services, a subcontractor for principal contractor, Mowlem. 3D Scaffolding pleaded guilty to breaching s 2(1) of the HSWA 1974, in that it failed to maintain a safe system of work, and was fined £60,000 and ordered to pay £20,000 costs.

5.66 In 2008, Alfred McAlpine Capital Projects Ltd was fined £250,000 and ordered to pay £5,859 in costs for breaching s 3(1) of the HSWA 1974 at Maidstone Crown Court. This followed an HSE investigation into the death of a motorcyclist at a roadworks site in August 2005. HSE Inspector John Underwood said:

> 'This was a wholly avoidable incident which led to unnecessary loss of life. The "Road Closed" signs and the traffic cones had been missing for more than eight weeks before the incident – yet the contractor was working in the area almost every day. The temporary traffic management system should have been checked at least once a day but this was never done properly. The barriers were only 450mm high and from a distance they could have been mistaken for a shadow or a change in the colour of the tarmac. It is clear the motorcyclist braked hard and tried to avoid the barriers but could not stop in time.'

5.67 In August 2005, a fatal incident occurred on the former A228 old Ratcliffe Highway, which is a single carriageway road near Hoo on the Isle of Grain, Kent. The motorcyclist hit temporary concrete crash barriers, weighing over two tonnes each, that had been laid in a line across the road. There were no warning signs or traffic cones along the route to warn drivers of the closure. The old A228 was being downgraded to a local through-road following the opening of a new section of dual carriageway built by Alfred McAlpine Capital Projects Ltd running parallel to it. When the old A228 was closed, a series of 'Road Closed Ahead' signs and traffic cones were put in place along the route. However, by late June 2005 these had disappeared and there was nothing to warn drivers about the concrete barriers.

5.68 In 2004, four construction companies were fined a total of £37,000 for breaches of health and safety legislation at Medway Magistrates' Court. The prosecution followed an investigation by Kent Police and the HSE into an incident in which two vehicles collided on the A229 at Buckmore Park near Chatham in November 2001. The driver of one of the vehicles was killed and his wife was injured. A slip road onto the A229 had earlier become contaminated with mud, from a construction site exit. Costain Ltd, Skanska JV Projects Ltd, Mowlem plc and Ovendens Earthmoving Co Ltd all pleaded guilty to a breach of duties under s 3(1) of the HSWA 1974, in that they failed to ensure that persons not in their employment were not exposed to risk to their safety.

5.69 The driver's employer, Elmsgold Haulage Ltd, and the managing director of Elmsgold Haulage Ltd pleaded guilty to two charges under s 2(1) of the HSWA 1974, in that they failed to provide a safe system of work and failed to ensure that people working on site were properly trained and supervised, and a third charge under reg 9(3) of LOLER, in that they failed to ensure that lifting equipment was properly examined and inspected.

5.70 Demolition contractor, Excavation & Contracting (UK) Ltd of Manchester, the principal contractor for the Chesford Grange project, and the company's former managing director, pleaded guilty to a charge under s 3(1) of the HSWA 1974, in that they each failed to ensure that risks to non-employees were adequately controlled.

5.71 The foreman on Elmsgold Haulage's site pleaded guilty to a charge under s 7 of the HSWA 1974 in that he failed to ensure the safety of other employees. He was fined £2,500 and ordered to pay £2,500 costs.

5.72 At an earlier hearing in January 2006, John Edge of Knight Frank, a property management company acting for the owner of Chesford Grange and planning supervisor for the project, pleaded guilty to two charges under reg 15 of the CDM Regulations 1994, for which Knight Frank was fined a total of £7,000 plus full prosecution costs of £4,500.

5.73 In this case, which was heard before the CMCHA 2007 came into place, Knight Frank operated as a partnership, which did not constitute a legal entity for the purposes of prosecution, and the case was taken against one of the partners, Mr John Edge.

5.74 Section 1(2)(d) of the CMCHA 2007 makes specific reference to partnerships, and future cases involving gross negligence manslaughter can be taken against the partnership.

FALLS FROM HEIGHT

5.75 In August 2008, Sharaz Butt,[1] Company Director of Alcon Construction, Norwich pleaded guilty at Norwich Crown Court to a charge of manslaughter arising from the death of Wu Zhu Weng, a Chinese construction worker, who fell to his death whilst working at the Norfolk and Norwich

University Hospital on the refurbishment of The Panary, a bakery and cafe within the hospital site. Butt was sentenced to 12 months' imprisonment and a nominal fine of £100 after admitting the offence.

1 HSE Reference 4132437.

5.76 In January 2005, Lee Harper,[1] the sole Director of Harper Building Contractors Limited of Cannock, Staffordshire pleaded guilty to charges of manslaughter at Manchester Crown Court following the death of Daryl Arnold at a site on the Lynton Industrial Estate in Salford. Daryl Arnold had been employed by Harper, alongside several other workers, to remove and replace the roof of a warehouse on the industrial estate, although he had never undertaken this type of work before. Whilst working on the roof, he stepped backwards onto a fragile roof light and fell approximately 6.75 metres to his death when the roof light gave way. Harper was jailed for 16 months.

1 HSE Reference 2013478.

STABILITY OF STRUCTURES

5.77 RIDDOR requires that all incidents of unplanned collapse of structures be reported to the enforcing authority and may lead to an investigation in order to ascertain the cause of the collapse. The collapse in 1968 of Ronan Point, a 1960s system-built tower block, that suffered progressive collapse following a gas explosion, led to changes in design practice and the introduction of a Code of Practice which deals with the progressive collapse of buildings and the changes required in the design and construction of such structures in order to prevent this type of collapse.

5.78 A more recent case, involving the collapse of a walkway at Port Ramsgate in 1994, led to the prosecution of the designers, the manufacturers/suppliers, and the contractor, as well as a third party accreditation body. The summary to the investigation into the Port Ramsgate collapse emphasises the fact that the failure by the client 'to provide vital information to the designer, eg a design brief, specification for the project and environmental data' was a major contributory factor leading to the collapse. This failure was further compounded by the designers producing design calculations which were 'based upon inadequate assumptions, were inaccurate and failed to provide a safe design'. In this case, the client was prosecuted along with the designers, contractors and a third party accreditation organisation (who failed to adequately check the design).

5.79 Both of the above cases involved buildings or structures after they had been constructed, commissioned and handed over to the client for occupation or use: both cases involved issues relating to their design and resulted in fatalities.

5.80 Fatalities during construction arising from building collapses are relatively rare, as in most cases there is normally some visual or audible indication or other warning of the impending collapse, and the building or structure can be successfully vacated. However, in those cases where there is no warning, the number of casualties and fatalities can be significant and also expose the rescuers and the emergency services to risk whilst casualties are removed from the scene.

COLLAPSE OF BUILDING OR STRUCTURE

5.81 The collapse of a building or structure is reportable under Sch 2(18) of RIDDOR as a 'dangerous occurrence'. A case in August 1995 raised a number of issues following a coroner's inquest heard at Chertsey Coroner's Court in the following year, at which a verdict of unlawful killing was reached by an inquest jury. The unlawful killing verdict related to the death of four construction workers who were killed when a building collapsed in Ashford, Surrey, in August 1995 during works to add an extension to a three-storey building that had originally been constructed as a single-storey building in 1968 and extended with two additional floors in 1969. The incorporation of lightweight blocks into the structure, subsequently covered up, caused a sudden collapse of the structure when the four men, who were working on the property 27 years later, removed panels between the piers which had provided an alternative path for the load. This led to the men being buried under the debris when the building collapsed. Although the inquest jury gave a verdict of unlawful killing, the elapse of 27 years since the defective work was carried out made it difficult to bring those involved before the courts: the contractor and site foreman involved in the original extension works had both died, as had several other potential witnesses.

5.82 Clients and designers have a duty under the CDM Regulations to provide information relating to issues concerning the stability of structures, including the requirements for temporary propping or support, the sequence of work, and the nature and physical characteristics of the ground and sub-soils and/or the nature and condition of any existing structures, where such information is required in order to safeguard and prevent any injury to any persons involved in the work or construction activity.

5.83 In 2005, following the death of a worker during work on the refurbishment of St Mary's Church in London in 2001, fines totalling £95,000, plus costs of £60,000, were made against Lindsay Barr,[1] David O'Keefe & Co Ltd and Britin Construction Ltd. The worker had been working in an excavation beneath a section of wall at St Mary's Church. The refurbishment work included lowering the crypt floor, which required the foundations to be reformed below their existing level using underpinning. This involved excavating beneath the existing foundations in short sections and casting concrete 'pins' underneath the wall. Whilst working in such an excavation, a 1.5 tonne section fell onto him from the underside of an unsupported wall.

1 HSE Reference 2016437.

5.84 Following sentencing, HSE Inspector Barry Mullen, who investigated the case, said:

> 'The untimely death of Mr Simionica came about through the failure to take appropriate action in relation to a potential risk in the underpinning work that had been brought to the attention of both the structural engineer and the contractors. The possible risks should have been addressed by uncomplicated measures including a detailed structural investigation, suitable and sufficient risk assessments and adequate protective measures, such as propping of the foundations.'

5.85 The structural consultant engineer, Lindsay Barr of Lindsay Barr Associates, Buckinghamshire, was found guilty of breaching s 3(2) of the HSWA 1974, in that he failed to ensure that persons not in his employment were not exposed to risks to their health and safety. He was fined £45,000 and ordered to pay costs of £30,000 to the HSE. The principal contractor, David O'Keefe & Co Ltd of Prince George's Road, London, SW19 were fined £25,000 and ordered to pay costs of £15,000 to HSE, having been found guilty of charges of breaching s 3(1) of the HSWA 1974, reg 16(1)(a) of the CDM Regulations 1994, and reg 9(1) of the Construction (Health, Safety and Welfare) Regulations 1996.

5.86 Britin Construction Ltd[1] of Aylesford, Kent, were fined £25,000 and ordered to pay costs of £15,000 to HSE, having been found guilty of the charges of breaching s 2(1) of the HSWA 1974, reg 3(1) of the Management of Health and Safety at Work Regulations 1999, and reg 9(1) of the Construction (Health, Safety and Welfare) Regulations 1996.

1 HSE Reference 2016457.

DEMOLITION

5.87 Demolition works are regarded as intrinsically dangerous, although the utilisation of innovative techniques and modern plant and equipment has led to significant reductions in major injuries by reducing the need for manual demolition of buildings and structures. However, demolition work does need to be properly planned, controlled and supervised, and it remains a high-risk or safety-critical activity.

5.88 The main issues relating to demolition works are:

- unplanned or uncontrolled collapse of the structure or part of the structure;
- damage to plant, machinery or adjoining buildings or structures;
- demolition workers being buried under debris, trapped or crushed;
- operatives and plant falling through holes or gaps in floor slabs or other parts of the structure;
- demolition workers or others (including members of the public) being struck by flying or falling objects;
- gas explosions where supplies are not disconnected; and
- a significant risk involved in the rescue of injured workers, as further collapse of unstable structures is likely to occur.

The demolition of structures is covered by reg 29 of the CDM Regulations 2007.

5.89 AA Construction (London) Ltd of Wimbledon were ordered to pay over £45,000 for endangering workers and the public with unsafe demolition work in south west London during February 2011.

Local residents had raised concerns that asbestos materials were being 'smashed up and littering the site', that debris was dropping from height onto the road and footpath, and that the site was insecure despite its close proximity to a local school. HSE was notified and served three enforcement notices relating to unsafe practices that forced the site to be closed until urgent improvements were made. Westminster Magistrates' Court heard (on 24 October 2012) that HSE investigators found numerous precautions were lacking to make the site safe.

The company engaged inexperienced labourers to carry out demolition and asbestos removal, and failed to provide sufficient instruction, training, or supervision. The buildings could have been demolished remotely (for instance, using long-reach demolition plant).

The company pleaded guilty to breaching ss 2(1) and 3(1) of the HSWA 1974 and was fined £36,000 plus a £15 surcharge and ordered to pay £9,159 in costs.

5.90 In May 2003 at Liverpool Crown Court, KDP Capita Property Services Limited[1] were fined a total of £30,000, with additional costs of £37,860 awarded to the HSE, for breaches of s 3(1) of the HSWA 1974. A demolition worker was killed and two other site operatives seriously injured following the collapse of two properties at the end of a three-storey brick-walled Victorian terrace undergoing demolition works in windy conditions. The defendant company was employed as the planning supervisor for the project.

1 HSE Reference 2017927.

EXCAVATIONS

5.91 The high-risk nature of this work is recognised in reg 31 of the CDM Regulations 2007. Regulation 31(4) requires that construction work should not be carried out in an excavation where it has been necessary to provide supports or battering (formation of sloped sides or steps that reduces the risk of the sides collapsing into the excavation), unless the excavation has been inspected by a competent person prior to works commencing.

5.92 This type of work does need to be properly planned, and any necessary safety measures made available from the outset, with appropriate details included in the method statement.

5.93 Additional control measures that might be introduced include the use of permit systems, including:

- permit to dig – necessary where electrical cables, gas or water pipes are likely to be present; and
- permit to enter – necessary where explosive concentrations of gases or asphyxiating fumes may be present in the excavation, or the atmosphere is oxygen deficient.

Where it is likely that gases are present in an excavation, procedures should also be in place to ensure that air and gas monitoring is undertaken.

5.94 Adequate measures should also be provided to enable workers to escape from any excavation in the event that part of the excavation collapses or is flooded, or in any other circumstances that require emergency evacuation procedures to be implemented – these measures all need to be considered and included in the initial planning for the work.

R V HATTON TRAFFIC MANAGEMENT LTD

5.95 The Court of Appeal made an important ruling in this case[1] that allows employers to put forward a defence that, by providing safe systems of work, training and instructions, they have followed the requirements of the law, even where employees have not followed the correct procedures.

1 *R v Hatton Traffic Management Ltd* [2006] EWCA Crim 1156.

GAS INSTALLATIONS AND PIPELINES

5.96 A major gas explosion, in which a family of four were killed in Larkhall, South Lanarkshire in December 1999, with considerable damage to neighbouring properties, resulted in charges being brought against Transco Plc[1] and a record £15 million fine.

1 *Transco plc v HM Advocate* [2005] BCC 296, 2004 JC 29, 2004 SLT 41.

5.97 The case also made Scottish legal history, with Transco being the first company to be prosecuted for the common law crime of culpable homicide in Scotland. The indictment against Transco plc cited the company's failure between August 1986 and December 1999 to 'devise, institute, implement and maintain any adequate or effective safety policy, or strategy, for the use, maintenance, inspection, repair or replacement of ductile iron pipes', such failings being a cause of the subsequent gas explosion that occurred on 22 December 1999.

5.98 The culpable homicide charge was dismissed at a Scottish Appeal Court hearing in June 2003, but the HSE was successful in bringing charges under s 31 of the HSWA 1974. In evidence, it was heard that Transco had failed to keep accurate records of its gas pipelines, and information on the pipes (such as what they were made from) was often inaccurate.

TRAFFIC ROUTES AND VEHICLES

5.99 Although it is left to the contractor to organise the construction site in such a way that both pedestrians and vehicles can move around the site safely (CDM Regulations 2007, reg 36 – traffic routes), these issues should be considered at the design stage. Where possible, designers should ensure that a one-way traffic operation can be provided in the workplace – such traffic operations can then be planned into the construction phase of the project.

5.100 Accidents involving vehicles reversing on construction sites, or reversing into a site or back out onto the highway, are a major cause of injury and fatalities.

5.101 Regulation 37 of the CDM Regulations 2007 deals specifically with the use and operation of vehicles on construction sites.

GAS INSTALLATIONS AND CARBON MONOXIDE

5.102 Fatalities have occurred due to gas explosions and exposure to carbon monoxide, and any failure on the part of the occupier to comply with statutory requirements in respect of inspections or maintaining plant or equipment in a safe condition can be taken as evidence of a failure to carry out a duty of care.

5.103 David Johnson,[1] a roofing contractor, was convicted of gross negligence manslaughter at Norwich Crown Court in April 2008 and sentenced to three and a half years' imprisonment in May 2008 following the death in May 2006 of Robert Schenker who died of carbon monoxide poisoning following the blockage of a flue serving a gas boiler by mortar droppings, which prevented gases escaping from the boiler flue.

1 *R v David Johnson* [2008] EWCA Crim 2976.

5.104 A case was heard at the Old Bailey involving the deaths of a woman and her two young grandchildren due to the faulty installation of a boiler by Atalokhia Omo-Bare at a home in East London in April 2006. Atalokhia Omo-Bare pleaded guilty to three counts of manslaughter, when it was disclosed that he was not qualified to carry out gas installations and failed to fit a flue to the appliance in order to extract noxious fumes – this failure led to a build-up of carbon monoxide fumes and the death of three people. In March 2007, Atalokhia Omo-Bare was jailed for 20 months.

5.105 Scott Lee Stuart,[1] a gas fitter, was jailed for two years in October 2006 following sentence at Cardiff Crown Court after pleading guilty to a charge of manslaughter. He had incorrectly installed a flueless gas fire (he had failed to check that the gas pressure was set at the correct level, which led to a leak), which resulted in the death of a 14-year-old school girl at her family home. Stuart had received training and was qualified and registered to install gas central heating boilers, but was not registered to fit flueless gas fires.

1 HSE Reference 2029173.

LIFTING OPERATIONS AND LIFTING EQUIPMENT REGULATIONS 1998 (LOLER)

5.106 All lifting operations are governed by the Lifting Operations and Lifting Equipment Regulations 1998[1] (LOLER). These regulations cover both the equipment to be used, ie the fixed or mobile crane or lifting device, as well as the chains, shackles or strops and the actual lifting operation itself. All lifting equipment is subject to regular inspections.

1 SI 1998/2307.

5.107 In recent years, there have been a number of major incidents in Canary Wharf, Worthing, Battersea and Liverpool, all of which have involved multiple fatalities due to incidents involving tower cranes. There have also been a number of incidents involving mobile cranes which have toppled over or the load has shifted, resulting in a fatality.

CONCLUSION

5.108 Construction industry accidents continue to be a source of concern. The Department for Work and Pensions (DWP) has published the long-awaited report[1] into the underlying causes of fatal accidents in the construction industry. Published on 8 July 2009, the report, resulting from the inquiry led by former ACAS chair Rita Donaghy, recommends extending the Building Regulations to include health and safety processes when building control applications are being considered.

1 'One Death is too Many – Inquiry into the Underlying Causes of Construction Fatal Accidents' (July 2009) CM 7657, ISBN 978–0-10–176572-5.

5.109 The report also calls for the Gangmasters (Licensing) Act 2004 to be extended to cover construction, and favours the introduction of statutory duties on directors to ensure good health and safety management. The appointment of a full-time minister for construction is also proposed, to give the industry higher status within government.

5.110 Whether these recommendations will be taken up remains to be seen and also, if they are, whether this will make a difference to the safety of the construction industry.

Chapter 6

Transport

Michael Appleby, Housemans Solicitors

INTRODUCTION

6.1 A look at both health and safety prosecutions and those for manslaughter related to transport incidents show that these often reflect general trends in the way safety and culpability are viewed.

6.2 The prosecution of British Rail under the Health and Safety at Work etc Act 1974 (HSWA 1974) in relation to the Clapham Junction train crash of 1988 was a defining moment for the Act in setting out the level of negligent conduct required to prosecute an employer.[1] Mr Justice Wright when sentencing made it clear that the charges against British Rail did not involve any connotation or allegation of recklessness: the allegation was no more than a failure to observe its high standards of safety. This case was the first time that the HSWA 1974 had been used to prosecute an organisation following a disaster where members of the public had died.[2]

1 *HSE v British Rail*, 14 June 1991, unreported, Central Criminal Court.
2 This case is described in more detail at **6.51** onwards.

6.3 In November 1994, OLL Ltd became the first company in English legal history to be convicted of manslaughter, and Peter Kite also the first director to be given an immediate custodial sentence arising out of the operation of a business.[1] The defendants were prosecuted following a canoeing accident in 1993 in which four students died at Lyme Bay.[2]

1 *R v OLL Ltd and Kite* (1994) unreported.
2 See **6.27** and **6.28** below.

6.4 Earlier, in 1990, P&O Ferries had been prosecuted in connection with the sinking of *Herald of Free Enterprise* in 1987, killing 192 passengers. This case confirmed it was possible to prosecute a company for manslaughter under the common law.[1] The company and seven individuals, including directors and some of the crew, were charged with manslaughter based upon recklessness. The trial collapsed when Mr Justice Turner directed the jury that, as a matter of law, there was no evidence upon which they could properly convict the company of manslaughter. He said that, in order to convict the company of manslaughter, one of the individual defendants who could be 'identified' within the company as the directing mind would have to himself be guilty of manslaughter. Since there was insufficient evidence on which to convict any of the individual defendants, the case against the company had to fail.[2] After this case, there were calls for a change in the law from the media, public, trades unions and MPs.

1 *R v P&O European Ferries (Dover) Ltd* (1991) 93 Cr App R 72.
2 See **6.19** to **6.22** below for more details of the case.

6.5 Other high-profile manslaughter cases followed against companies in relation to transport disasters which failed, each time renewing the call for a change in the law. In 1999 the prosecution of Great Western Trains (GWT) failed in respect of the Southall train crash of 1997, which killed seven passengers.[1] In prosecuting GWT, the prosecutor took the view (in the light of the prosecution of an anaesthetist for gross negligence manslaughter: *R v Adomako* [1995] 1 AC 171) that it was no longer necessary to prosecute a directing mind to secure a conviction of a company for manslaughter. The trial judge disagreed. In 2005 the prosecution of the maintenance arm of Balfour Beatty in relation to the Hatfield derailment of 2000, killing four passengers, also failed.[2] Although, when dismissing the case, the trial judge, Mr Justice Mackay, called for a change in the law, the case against Balfour Beatty actually failed because the prosecution could not prove that the directing mind, the head civil engineer of the maintenance company, was guilty of manslaughter. The prosecution of the engineer for an offence pursuant to s 7 of the HSWA 1974 also failed.

1 *R v GWT* (1999, unreported), Central Criminal Court, described in **6.56** to **6.58** below.
2 *R v Balfour Beatty Rail Infrastructure Services Ltd* (2005, unreported), Central Criminal Court. The manslaughter case against Network Rail had been dismissed at an earlier hearing in 2004 by the trial judge. This case is described in more detail at **6.65** to **6.77** below.

6.6 This chapter considers the types of investigation which follow accidents in the sea, railways, aviation and road industries, along with some of the significant prosecutions in these areas.

SEA

Investigations

6.7 The Marine Accident Investigation Branch (MAIB) investigates all types of marine accidents to or on board UK ships worldwide, and other ships in UK territorial waters.

6.8 The powers of MAIB inspectors, and the framework for reporting and investigating accidents, are contained in the Merchant Shipping Act 1995 (MSA 1995). The Merchant Shipping (Accident Reporting and Investigation) Regulations 2012 put the framework in the MSA into effect.

6.9 The MAIB is a separate branch within the Department for Transport (DfT). It is not part of the Maritime and Coastguard Agency (MCA): see below.

6.10 The MAIB does not enforce law or prosecuted criminal prosecutions. These are carried out by other agencies. The MAIB's objective is to determine the circumstances and causes of an accident, with the aim of improving safety. Its purpose is not to apportion blame.

6.11 However, reports by the MAIB have been used as the basis of prosecutions.

6.12 The MCA is also an agency of the DfT. It is responsible throughout the UK for implementing the Government's maritime safety policy. The MCA's statutory powers and responsibilities derive primarily from the Coastguard Act 1925, the MSA 1995 and the Merchant Shipping and Maritime Security Act 1997.

6.13 The MCA is responsible for enforcing all merchant shipping regulations in respect of occupational health and safety, the safety of vessels, safe navigation and operation (including manning levels and crew competency). It can issue improvement and prohibition notices in addition to prosecuting offences.

6.14 The Health and Safety Executive (HSE) is, however, responsible for enforcing the HSWA 1974 in respect of land-based and offshore work activities, including the loading and unloading of ships, and for all work activities carried out in a dry dock.

6.15 Manslaughter and an offence under the Corporate Manslaughter and Corporate Homicide Act 2007 (CMCHA 2007) are investigated by the police and prosecuted by the Crown Prosecution Service (CPS). However, the jurisdiction of the CMCHA 2007 is limited in terms of ships. By s 28(3) of the CMCHA 2007 the Act applies within the seaward limits of the territorial sea adjacent to the UK and on a ship registered under Pt 2 of the MSA 1995. It should be noted that the police and CPS can also prosecute offences under the MSA 1995.

SOLAS

6.16 SOLAS stands for 'Safety of Life at Sea' and is the most important treaty protecting the safety of merchant ships. The first version of the treaty was in 1914 and was in response to the sinking of the *Titanic*. The latest version is 1974 as amended.

6.17 Chapter IX of SOLAS concerns the 'Management for the Safe Operation of Ships'. It is highly significant because, for the first time internationally, it sets a benchmark against which a corporation's criminal responsibility for deaths at sea can be judged. Chapter IX makes mandatory the International Safety Management Code (ISM Code).

6.18 In order to comply with the ISM Code, each ship class must have a working Safety Management System (SMS). Each SMS comprises:

(a) commitment from top management;
(b) a Top Tier Policy Manual;
(c) a Procedures Manual that documents what is done on board the ship;
(d) procedures for conducting both internal and external audits to ensure the ship is doing what is documented in the Procedures Manual;
(e) a Designated Person to serve as the link between the ships and shore staff;
(f) a system for identifying where actual practices do not meet those that are documented and for implementing associated corrective action; and
(g) regular management reviews.

Herald of Free Enterprise

6.19 On 6 March 1987 the Ro-Ro ferry *Herald of Free Enterprise* sailed from the inner harbour at Zeebrugge bound for Dover. Approximately 20 minutes after leaving the harbour, the ferry capsized.

6.20 The Secretary of State for Transport ordered a formal investigation (pursuant to the MSA 1970), which was chaired by Mr Justice Sheen. Mr Justice Sheen in his report[1] concluded that the ship capsized because she went to sea with her inner and outer bow doors open. The central problem was that there was no system to monitor whether the bow doors were closed. He found that the root cause of the disaster was the management of the company that owned the ship – Townsend Thoresen (later acquired by P&O Ferries). He observed there was a 'disease of sloppiness' and negligence at every level of the corporation's hierarchy.

1 Department for Transport, 'M/V *Herald of Free Enterprise* Report of the Court no 8074' (1987).

6.21 Following the public inquiry an inquest was held which returned a verdict of 'unlawful killing'. On the basis of this finding, a summons for manslaughter was issued against P&O Ferries. The company sought to have the summons set aside on the basis that it was not a natural person and so could not be convicted of manslaughter. This was rejected by the trial judge, Mr Justice Turner, who found that it was possible. He said:[1]

> 'Since the 19th century there has been a huge increase in numbers and activities of corporations whether nationalised, municipal or commercial which enter the private lives of all or most of "men and subjects" in a diversity of ways. A clear case can be made for imputing to such corporations social duties including duty not to offend all relevant parts of the criminal law. By tracing the history of the cases decided by the English courts over the period of the last 150 years, it can be seen how first tentatively and finally confidently the courts have been able to ascribe to corporations a "mind" which is generally one of the essential ingredients of common law and statutory offences. Indeed, it can be seen that in many acts of Parliament the same concept has been embraced … once a state of mind could be effectively attributed to a corporation, all that remained was to determine the means by which that state of mind could be ascribed and imputed to a non natural person. That done, the obstacle to the acceptance of general criminal liability of a corporation was overcome … there is nothing essentially incongruous in the notion that a corporation, through the controlling mind of one of its agents, does an act which fulfils the pre-requisites of the crime of manslaughter, it is properly indictable for the crime of manslaughter.'

1 *R v P&O European Ferries (Dover) Ltd* (1991) 93 Cr App R 72.

6.22 Turner J emphasised the need to prove that the person prosecuted as the 'directing mind' had sufficient responsibility for safety to establish guilt. It was not permissible to 'aggregate' the faults of a number of different individuals. The case failed because the prosecution could not establish that those directors in charge of the company's senior management were at fault.

Marchioness

6.23 On 20 August 1989 the pleasure boat the *Marchioness* collided with the dredger *Bowbelle* near Southwark Bridge in the Thames. Fifty-one passengers were drowned. The master of the *Bowbelle* was charged with breaching s 32 of the MSA 1988 for failing to keep a proper lookout. However, two juries failed to agree a verdict, and so no evidence was eventually offered against him, with a verdict of 'not guilty' being entered.[1]

1 Lloyd's List, 12 April 1995.

6.24 After the criminal proceeding had been concluded, the MAIB published its report in August 1991. The report did not identify any one party as being to blame but concluded that the disaster was caused by poor look-outs on both ships. It also criticised the Department for Transport, which it said was aware of previous collisions on the Thames between large vessels and smaller passenger ships due to visibility problems but had failed to take action.

6.25 In late 1991 a private prosecution was brought against company officials and the managing company of the *Bowbelle*. In June 1992 the Chief Stipendiary Magistrate discontinued these proceedings due to lack of evidence.[1]

1 Lloyd's List, 12 April 1995.

6.26 An inquest into the incident took place in 1995 and returned a verdict of 'unlawful killing'. In October 2000, 11 years later and after much lobbying, a public inquiry was held into the incident before Lord Justice Clarke.[1] This inquiry again attributed responsibility upon the poor look-outs on both ships but, in addition, also criticised the owners and managers for failing to instruct their crews adequately.

1 The two reports of Lord Justice Clarke, the Formal Investigation and Non-Statutory Inquiry, were published on 23 March 2001 by HMSO.

Lyme Bay

6.27 On 22 March 1993 a teacher, a group of eight students and two instructors from OLL Ltd tried to canoe across the sea from Lyme Regis to Charmouth, a distance of about one-and-a-half miles. The teacher got into difficulties, and one of the instructors went over to assist. The other instructor proceeded with the students. The group drifted out to sea where the students drowned.

6.28 Mr Kite was one of two directors of the company. He was 'hands-on' and had the primary responsibility for devising, instituting, enforcing and maintaining an appropriate safety policy. The jury found that he failed in his responsibility and this was a substantial cause of the deaths. Mr Kite was sentenced to three years,[1] which was later reduced to two years on appeal.[2] At the appeal, Lord Justice Swinton Thomas noted that Mr Kite had been convicted on the basis of negligence and that there had been no criminal intent on his part. In the circumstances, the original sentence was too severe. OLL Ltd was fined £60,000, which represented its entire assets.

1 *R v OLL Ltd and Kite* (1994) unreported.
2 *R v Kite* [1996] Cr App Rep (S) 295.

Maria Asumpta

6.29 The *Maria Asumpta* was a two-masted sailing ship built in the 1850s which had been fitted with engines in the 1930s. On 30 May 1995, on her first voyage following a refit, the ship was preparing to enter Padstow harbour in Cornwall. The captain and owner, Mark Litchfield, decided to take the *Maria Asumpta* between the Mouls and Pentire Point, which was not a route recommended by the Admiralty.

6.30 The engines suddenly stopped and more sail was raised. Look-outs on the ship failed to spot submerged rocks, and the ship foundered. Three of the crew drowned after abandoning ship.

6.31 Following an MAIB investigation, Mr Litchfield was prosecuted for manslaughter based on his gross negligence. The case against him was that he followed an unsafe course and relied too heavily on his engines, even though he knew the fuel was contaminated. He was convicted by majority verdict and sentenced to 18 months' imprisonment.

6.32 Mr Litchfield's appeal against conviction was dismissed.[1]

1 *R v Litchfield* [1998] Crim LR 507, CA.

6.33 The following direction to the jury on gross negligence was given in this case and approved by the Court of Appeal:[1]

> 'Before you could convict this defendant of manslaughter, the negligence established must go way beyond the mere matter of compensation between parties. It must be more than just some degree of fault, or mistake, or error of judgement, or carelessness even though that led to death. It must be such as to demonstrate a reckless disregard for the lives of others of such a nature, and to such an extent, that in your judgement the negligence is so bad that it can properly amount to a criminal act.'

1 *R v Litchfield* [1998] Crim LR 507, CA.

Sea Empress

6.34 The grounding of the oil tanker *Sea Empress* in February 1996 caused one of the largest and most environmentally damaging oil spills in European history. About 72,000 tonnes of crude oil were released into the seas around the coast of South West Wales.

6.35 The MAIB in its report found that the tanker ran aground because an inexperienced pilot failed to keep it in the deepest part of the navigation channel. The MAIB also criticised Milford Haven Port Authority's disaster planning.

6.36 Before the publication of MAIB's report the Environmental Agency began a prosecution of the Port Authority, which subsequently pleaded guilty to one charge under the Water Resources Act 1991 of causing polluting matter to enter controlled waters. In passing sentence of a fine of £4 million, Mr Justice Steel said: 'The pilot was put in a position by the Port Authority where he could make an error of navigation'. On appeal the fine was reduced to £750,000.[1] This is an important case in respect of the sentencing of a public authority. The Court of Appeal stated that, where a defendant is a public body, it is not immune from criminal penalties because it has no shareholders or directors 'in receipt of handsome annual bonuses'. However, the court should take into account the effect of a substantial fine upon a public body if it will affect its ability to perform the public duty that it was set up to do.

1 *R v Milford Haven Port Authority* [2000] 2 Cr App R (S) 423.

Death of Simon Jones

6.37 Simon Jones was killed in May 1998 on his first day of employment with Euromin Ltd as a stevedore at Shoreham Harbour, West Sussex. The jaws of a clam-shaped grab, used for moving large quantities of slag and aggregate, accidentally closed over his head, fracturing his skull and severing his head while he was working in the hold of a ship.

6.38 Mr Martell, the managing director, had designed a new system for the use of the grab. This involved the addition of two hooks which were welded to the centre columns of the grab. The hooks could then be attached to chains and those chains could be used to lift bags of cargo. This system meant that the lifting operation had to be carried out with the grabs open. Workers would have to stand very close to the bucket so as to attach the chains.

6.39 The CPS determined that there was insufficient evidence to bring a manslaughter prosecution. This decision was challenged by the deceased's family through judicial review.[1] The court ruled that the CPS had failed to apply the law of manslaughter correctly and that part of the reasons given by the CPS were 'irrational'.[2] In November 2001 at the Old Bailey, the company and Mr Martell were acquitted of manslaughter. The company was convicted of breaching the HSWA 1974 and fined £50,000.

1 A judicial review is a form of court proceedings, in the Administrative Court in central London, in which a judge reviews the lawfulness of a decision or action made by a public body. It is a challenge to the way the decision has been made, not to the decision itself.

2 *R v DPP, ex p Jones* [2000] IRLR 373.

6.40 Following the case, there were calls for a change in the law. But notably, when the incident occurred, there was confusion as to who had responsibility for investigating the accident and so the investigation was delayed. It was eventually investigated by the HSE. There are those who take the view that the case failed not because of the state of the law but because of the quality of the evidence that was available. Due to the delay in the investigation and deciding which agency should take the lead, valuable evidence was lost. Today it is unlikely that such a situation would arise because of the protocol *Work Related Deaths: A protocol for liaison.*[1]

1 This can be downloaded at www.hse.gov.uk/pubns/wrdp1.pdf.

Costa Concordia

6.41 The most recent sea disaster, killing over 30 passengers, has been the partial sinking on 13 January 2012 off the Tuscany cost of the Costa Concordia, a cruise ship operated by Costa Cruises, a unit of the world's largest cruise operator, Carnival Corp. Prosecutors say the ship's captain, Francesco Schettino, deviated from the ship's computer-programmed route and, as a consequence, hit a reef off Isola del Giglio. The ship started to take in water, flooding the engine room and generators, causing the ship to drift for more than an hour before running aground.

6.42 Coverage of the shipwreck dominated international media in the days after the disaster. Costa Cruises were prompt to issue a press statement expressing their great sorrow for what had happened. In July 2012 the captain gave a television interview where he acknowledged his responsibility as captain and said he thought constantly about the victims of the disaster.

6.43 In April 2013 the cruise operator agreed to pay a fine of $1.3m (£860,000) to settle possible criminal charges. In July 2013 a court in Italy, following plea bargains, convicted five people in relation to the disaster. Two officers, the helmsman, the head of cabin services and the head of the crisis team (who was ashore co-ordinating the company's response) were given custodial sentences of between 18 months and two years and 10 months for multiple manslaughter, negligence and shipwreck. In Italy, sentences of less than two years do not generally have to be served in prison, except in some categories of homicide (of which manslaughter is not one).

6.44 The captain is charged with manslaughter and abandoning ship and has been tried separately. The prosecutors rejected his request for a plea bargain. The prosecutors are calling over 350 witnesses, and the trial is expected to last over a year.

6.45 Various safety reviews were announced following the incident, including a Cruise Industry Operation Safety Review by the Cruise Lines International Association, Inc and a regulator's Passenger Ship Safety Review by the International Maritime Organisation (IMO). The IMO called upon Member States to ensure that their national safety regulations and procedures were implemented fully and effectively.

RAILWAYS

Investigations

6.46 The Rail Accident Investigation Branch (RAIB) was set up following recommendations by Lord Cullen in his public inquiry into railway safety following the Ladbroke Grove train crash of 1999 in which 31 people died. It is similar in its remit to the MAIB and it, too, is part of the Department for Transport.

6.47 The RAIB was established by the Railways and Transport Safety Act 2003. The Railways (Accident and Investigation) Regulations 2005 define the RAIB's powers and duties, the scope of its work and dealings with other people and organisations that are involved in rail accidents. Its investigations are focused on safety issues and are not intended to apportion blame. The RAIB does not enforce law or carry out prosecutions.

6.48 It should be noted there is also the Railway Safety and Standards Board (RSSB), a not-for-profit company established in April 2003. Its prime objective is to lead and facilitate the railway industry's work to achieve continuous improvement in the health and safety performance of the railways.

6.49 The Office of Rail Regulation (ORR) was established on 5 July 2004 under the Railways and Transport Safety Act 2003. The ORR is an independent statutory body led by a Board. It is part of the DfT. On 1 April 2006, in addition to its role as economic regulator, the ORR took on the role of safety regulator when Her Majesty's Railway Inspectorate (HMRI) moved from the HSE to the ORR. HMRI enforces the HSWA 1974 on the railways. The HMRI dates back to 1840 when Inspecting Officers of Railways were first appointed by the Board of Trade.

6.50 The British Transport Police (BTP), which can also trace its history back to the 1800s, is responsible for investigating crime on the railways, including manslaughter. The BTP covers the National Rail network across England, Scotland and Wales, the London Underground system, Docklands Light Railway, the Glasgow Subway, the Midland Metro tram system and Croydon Tramlink. The BTP also police international services operated by Eurostar. Cases investigated by the BTP are prosecuted by the CPS.

Clapham Junction rail crash

6.51 At 8.10 am on 12 December 1988, a northbound commuter train ran into the back of a stationary train in a cutting just south of Clapham Junction station. A third train, going south, ran into the wreckage of the first train. Thirty-five people died.

6.52 The immediate cause of the crash was a signal that failed because it was showing a green light instead of a red one, thus concealing the presence of a stationary train from the driver of the northbound commuter train. By the time the driver observed the stationary train, because of the speed he was travelling (as he believed he had a clear track), there was insufficient time and distance to stop the train.

6.53 The signal failure was directly due to the working practices of a technician engaged in rewiring the signal on the previous day. Rather than cutting off or tying back the old wires, he merely pushed them aside. It was also his practice to re-use old insulating tape, though on this occasion no tape at all was wrapped around the bare ends of the wire. As a result, the wire came into contact with nearby equipment causing a 'wrongside' signal failure.

6.54 At the Public Inquiry chaired by Mr Anthony Hidden QC (as he was then), the following comments were made in relation to the way British Rail (which ran the railways before they were privatised in the 1990s) managed safety:[1]

> '[British Rail] is responsible for an industry where concern for safety should be at the forefront of the minds of everyone, from the Board itself at the top to the newest beginner at the bottom. The concept of absolute safety must be a gospel spread across the whole workforce and paramount in the minds of management. The vital importance of this concept of absolute safety was acknowledged time and time again in the evidence which the Court heard. This was perfectly understandable because it is so self-evident.
>
> The problem with such concern for safety was that the remainder of the evidence demonstrated beyond dispute two things:
>
> (1) there was total sincerity on the part of all who spoke of safety in this way, but nevertheless
> (2) there was failure to carry those beliefs through from thought into deed.
>
> The appearance was not the reality. The concern for safety was permitted to co-exist with working practices … Were positively dangerous. This unhappy co-existence was never detected by management and so the bad practices were never eradicated. The best of intentions regarding safe working practices was permitted to go hand in hand with the worst of inaction in ensuring that such practices were put into effect.
>
> The evidence therefore showed the sincerity of the concern for safety. Sadly, however, it also showed the reality of the failure to carry that concern through into action. It has to be said that a concern for safety which is sincerely held and repeatedly expressed but, nevertheless, is not carried through into action, is as much protection from danger as no concern at all.'

1 A Hidden QC, *Investigation into the Clapham Junction Railway Accident* (Cm 820, 1989).

6.55 No individual was prosecuted for manslaughter or health and safety offences. British Rail pleaded guilty to breaches of ss 2 and 3 of the HSWA 1974.[1] Mr Justice Wright, when sentencing, made it clear there was no question of recklessness on the part of British Rail. He said the case against BR was 'a failure to maintain and observe the high standards required by the Health and Safety at Work legislation …'. British Rail was fined £250,000.

1 *R v BRB* (14 June 1991, unreported), Central Criminal Court.

Southall train crash

6.56 At approximately 1.15 pm on 19 September 1997, the 10.32 am Swansea to Paddington, a high-speed train (HST) driven by Driver Harrison passed a red signal (known as a SPAD – signal passed at danger) and collided with a goods train routed across its path. Driver Harrison applied the brakes upon seeing the red signal when his train was travelling at line speed of 125 mph, but this was too late to be able to stop the train and avoid the collision. Miraculously, he survived. Seven passengers lost their lives. The Automatic Warning System (AWS) on the train – which gives an auditory warning to a driver of a cautionary signal – was not working. This had been reported as faulty on the outward journey, by the previous driver of the train, earlier in the day, but had not been rectified.

6.57 Mr Harrison was charged with gross negligence manslaughter and breaching s 7 of the HSWA 1974. At a later stage, proceedings were commenced against his employers, Great Western Trains (GWT), for manslaughter and breaching s 3 of the HSWA 1974. No director or senior manager of GWT was prosecuted as the 'directing mind'.

6.58 At a preliminary hearing in June 1999, the trial judge, Mr Justice Scott-Baker (as he was then), dismissed the manslaughter case against the company because no one as directing mind had been charged. This ruling was later confirmed by the Court of Appeal.[1] Following this ruling, GWT pleaded guilty to breaching s 3 of the HSWA 1974 and was fined £1.5 million. The trial judge stated that the risk of a driver passing a danger signal was small, but that, if it did happen, the consequences were likely to be 'appalling' and GWT should have not allowed the train to run without an operational AWS.[2] No evidence was offered against the driver and he was acquitted on all counts.

1 *A-G's Reference (No 2 of 1999)* [2000] 2 Cr App Rep 207.
2 *R v GWT* (1999) unreported, Central Criminal Court.

Ladbroke Grove train crash

6.59 The Ladbroke Grove train crash occurred on 5 October 1999, killing both drivers of the trains involved and 29 passengers. A Turbo of Thames Trains, driven by Michael Hodder, left Paddington Station at approximately 8.06 am bound for Bedwyn in Wiltshire. At 8.25 am the Turbo passed signal SN109 on gantry 8 at red. It had passed the previous signal SN87 that was showing a single yellow. Because of the way the points had been set, it meant that, if the Thames train failed to stop at SN109, it would be heading for the Up main line. Meanwhile, an HST of First Great Western was approaching the Up main line. Both trains collided almost head on as the Thames train entered onto the main line.

6.60 Signal SN109 had been passed on nine occasions in just under six years. Immediately after the accident, a prohibition notice was served preventing the use of signal SN109.

6.61 A public inquiry chaired by Lord Cullen took place in 2000. Lord Cullen was satisfied that Driver Hodder thought he had a proceed signal. He wrote in his report:[1]

> 'It is more probable than not that the poor sighting of SN109, both in itself and in comparison with the other signals on and at gantry 8, allied to the effect of the bright sunlight at a low angle, were factors which led [Driver Hodder] to believe that he had a proceed aspect and so that it was appropriate for him to accelerate as he did after passing SN87 … The unusual configuration of SN109 … not only impaired initial sighting of the red aspect, but also might well have misled an inexperienced driver … into thinking it was not showing a red but a proceed aspect.'

1 The Rt Hon Lord WD Cullen PC, *The Ladbroke Grove Rail Inquiry, Part 1 Report* (2001).

6.62 Prior to the crash, there had been problems with SPADs in the Paddington area. Lord Cullen concluded:[1]

> 'There was a lamentable failure on the part of Railtrack (the then infrastructure owner) to respond to recommendations of inquiries into two serious incidents, namely the accident at Royal Oak on 10 November 1995 and the serious SPAD at SN109 on 4 February 1998. The recognition of the problem of SPADs in the Paddington area led to the formation of a number of groups to consider the problem. However, this activity was so disjointed and ineffective that little was achieved. The problem was not dealt with in a prompt, proactive and effective manner. In 1998 Railtrack dispensed with the services of a significant number of senior Great Western Zone personnel or moved them on to other work. In the course of what was correctly described as a "devastating critique" of their deficiencies and other management failures, the incoming Zone Director said: "The culture of the place had gone seriously adrift over many years".'

1 The Rt Hon Lord WD Cullen PC, *The Ladbroke Grove Rail Inquiry, Part 1 Report* (2001).

6.63 Lord Cullen also had criticisms of Thames Trains in respect of training deficiencies. The company was later prosecuted for breaching the HSWA 1974 and fined £2 million.[1]

1 *R v Thames Trains* (2004) unreported, Central Criminal Court.

6.64 Shortly before the public inquiry started in May 2000, the police announced there would be no manslaughter prosecution of any company involved in the crash. However, following the publication of the findings of Lord Cullen, the families of the victims campaigned for the police to reopen its investigation into Railtrack. In 2005 the CPS announced that it had concluded again there was insufficient evidence to prosecute Network Rail (the successor of Railtrack) for manslaughter. The company was eventually prosecuted for breaching s 3 of the HSWA 1974 and fined £4 million following a 'guilty' plea (the judge did consider the appropriate fine was £5 million but concluded that, if he fined the company that amount, it would have been out of step with the fine imposed upon Thames Trains and so reduced the fine accordingly). Mr Justice Bean in his sentencing remarks observed:[1]

> 'If a signal has been passed at danger on several occasions by different drivers the common factor is the signal, not the driver.'

1 *R v Network Rail* (2007) unreported, Blackfriars Crown Court.

Hatfield derailment

6.65 On 19 October 2000 a section of rail on a bend at Hatfield failed, shattered into a number of pieces, causing a high-speed train to derail and kill four passengers. The rail was suffering from a metal fatigue condition known as 'gauge corner cracking' (GCC) that manifests itself as hairline cracks on the surface of the rail.

6.66 Balfour Beatty Rail Infrastructure Services Ltd was responsible for the maintenance of the track. The rail in question had been reported to Railtrack, owner of the infrastructure and the predecessor of Network Rail, and was due to be replaced by another contractor.

6.67 Fifty-five suspects were interviewed under caution, directors and managers of varying seniority of both companies. Eventually, in 2003, both companies were summonsed for manslaughter, six individuals were prosecuted for manslaughter in their own right (four of them being identified as directing minds – two for each company) and breaching the HSWA 1974, and a further six, including the former CEO of Railtrack, Gerard Corbett, were prosecuted for breaching the HSWA 1974 only.

6.68 The case against Balfour Beatty was that it had failed to apply a speed restriction when inspecting the track (through visual inspection and ultrasonic inspection) prior to the derailment. The case against Network Rail was that it had failed to manage Balfour Beatty properly.

6.69 At the time of the derailment, GCC was a growing problem on the railway network nationally. Railtrack was formulating new instructions, known as 'Line Standards', for its maintenance contractors to follow, setting out what action to take when certain conditions occurred. Some ten months before the derailment, Railtrack had sent a letter to infrastructure maintenance contractors detailing best practice with respect to the management of GCC. Much of the prosecution centred upon the effect this letter should have had on ultrasonic inspection and whether, if followed, it would have resulted in a speed restriction being applied to the line.

6.70 At a preliminary hearing, the trial judge, Mr Justice Mackay, ruled that there was not a case to answer for one of the alleged 'directing mind' defendants, and that two other defendants were not directing minds. As a consequence, there were no defendants that were left in the trial as directing minds for Network Rail (and so the manslaughter failed against the company) and only one directing mind for Balfour Beatty, Mr Nicholas Jeffries, who had been the civil engineer for the maintenance company.[1]

1 Preliminary Ruling in *R v Network Rail*, 1 September 2004, Central Criminal Court. It should also be noted that the charges under s 37 of the HSWA 1974 against Gerald Corbett and another director were dismissed at this hearing. However, the trial judge's interpretation and application of s 37 was later questioned by the Court of Appeal in the leading case upon directors' health and safety duties: *R v P Ltd & G* [2007] All ER (D) 173 (Jul) (CA).

6.71 Mr Jeffries was not a director of the maintenance arm of Balfour Beatty, and neither did he have any management responsibility for the maintenance contract of the East Coast Main Line that covered the Hatfield area. However, he was responsible for advising the company with respect to new line standards that were issued by Railtrack and to be applied to the various maintenance contracts operated by the company on the network. At the pre-trial hearing, although the judge was concerned by Mr Jeffries's level of seniority, he accepted the prosecution's arguments that, in relation to line standards, Mr Jeffries 'reigned supreme'.

6.72 At the end of the prosecution case at trial, the trial judge dismissed all the manslaughter charges, including those against Mr Jeffries, on the basis that there was insufficient evidence. As a consequence the case against Balfour Beatty for manslaughter failed. Subsequently, on 18 July 2005, Balfour Beatty entered a plea of guilty to breaching s 3 of the HSWA 1974. The trial continued against the remaining defendants on health and safety charges.

6.73 On 6 September 2005 (there having been a three-week break in August), the jury returned verdicts of 'not guilty' against the individual defendants in relation to the health and safety offences and 'guilty' against Network Rail for breaching s 3 of the HSWA 1974.

6.74 On sentencing Balfour Beatty for its failure to carry out visual and ultrasonic inspection adequately over a 21-month period, the trial judge described this as 'the worst example of sustained, industrial negligence in a high risk industry I have seen' and 'lying at the top of the scale'. Network Rail was sentenced upon its failure to adequately mange the maintenance contractor. The judge said the company's failures 'were lamentable, but of a lower order by a clear margin'. Balfour Beatty was fined £10 million and Network Rail £3.5 million.

6.75 On appeal, the maintenance contractor's fine was reduced to £7.5 million because of the disparity of sentence between its fine and that of Network Rail.[1]

1 *R v Balfour Beatty Rail Infrastructure Services Ltd* [2007] 1 Bus LR 77.

6.76 The Court of Appeal said that the disparity between the sentences imposed on two defendants is not an automatic reason for reducing a sentence, but relied upon the test set out in *R v Fawcett* (1983) 5 Cr App R (S) 158 which it said had been satisfied:

> '… would right thinking members of the public, with knowledge of the relevant facts and circumstances, learning of this sentence consider that something had gone wrong with the administration of justice.'

6.77 The Court of Appeal went on to explain its reduction in sentence:

> 'To restore appropriate proportionality between the two fines would require Balfour Beatty's fine to be reduced to a level at which it failed to give proper effect to the [*Howe* principles]. We do not consider that this would be right. Those principles do, however, provide more assistance in identifying the lower limit of an appropriate range of fine than the upper limit. They leave the sentencing judge a wide discretion as to the level at which to pitch the fine. The fine of £10 million on Balfour Beatty was severe. We consider that there is scope for a reduction in the interest of proportionality which will still do justice to the applicable principles and, in particular, to the victims of the Hatfield disaster. We have decided that Balfour Beatty's fine should be reduced to £7.5 million, thereby reducing the disparity between its sentence and that of [Network Rail].'

Purley train crash

6.78 On Saturday, 4 March 1989 the 12.50 pm Electrical Multiple Unit from Horsham to London Victoria made its usual stop on the slow line platform in the direction of London at Purley. On departure, it crossed from the slow line to the fast line in the direction of London. A non-stopping train was closely approaching Purley on the fast line heading for London, driven by Mr Robert Morgan. Mr Morgan failed to stop at signal T168 which was showing red. As a consequence, his train collided with the rear of the Horsham to London train, killing five people.

6.79 The AWS was working on the train. Mr Morgan saw signal T168 at red and made an emergency brake application, but was going too fast to stop. He could not recollect the previous cautionary yellow signals but must have cancelled the horn of the AWS; otherwise, the train would have been automatically stopped. Like the signal at Ladbroke Grove, T168 had been passed on a number of occasions when red.

6.80 Mr Morgan was unable to give an explanation for his error. In September 1990 he pleaded guilty to manslaughter and was sentenced to imprisonment for 18 months, 12 of them suspended.[1] This was later reduced on appeal to a sentence of four months' imprisonment.[2] There were no other prosecutions.

1 *R v Morgan* (1990), unreported, Central Criminal Court.
2 *R v Morgan* (1990) unreported, CA.

6.81 Seventeen years after his conviction, Mr Morgan was given leave to appeal against his conviction by the full court of the Court of Appeal. The prosecution did not contest his appeal and his conviction was finally quashed. Lord Justice Latham stated:

> '... something about the infrastructure of this particular junction was causing mistakes to be made. Had a jury known that, it is, at the very least, impossible for us to conclude that the jury would inevitably have nonetheless convicted the appellant of manslaughter. Those facts would have all been matters which the jury would have taken into account when assessing the level of fault of Mr Morgan. As a result, there is no way that we can say this conviction is safe. The position, accordingly, is that we allow the appeal against conviction.'

1 *R v Morgan* [2007] EWCA Crim 3313.

Potters Bar derailment

6.82 On 10 May 2002, at Potters Bar station a northbound high-speed train was derailed, killing seven passengers. Part of the train ended up wedged between the station platforms and building structures. At the time of the incident, Railtrack contracted the inspection and maintenance of the track to Jarvis Rail Ltd. It was found that the points at a crossing were poorly maintained. The bolts that held the stretcher bars that keep the rails apart had come loose or gone missing, resulting in the points moving while the train passed over them. In the aftermath, Jarvis originally claimed that the points had been sabotaged. On 17 October 2005 the CPS announced that it had concluded there was no realistic prospect of conviction for an offence of gross negligence manslaughter against any individual or corporation.

6.83 In February 2007, following a fatal derailment of a train at Grayrigg (see below), the inquest into the fatalities at Potters Bar was adjourned, pending the decision of the Secretary of State for Transport on whether a public inquiry or joint inquest should be held into the Potters Bar and Grayrigg incidents. In June 2009 the Secretary of State for Transport decided that separate inquests should be held into the Potters Bar and Grayrigg incidents. The inquest into the deaths at Potters Bar took place during June and July 2010. The jury returned seven verdicts of accidental death. In October 2010 the CPS informed the ORR that it had decided that there were no grounds for it to reconsider its decision of October 2005. The ORR announced its decision to prosecute Network Rail, the successor to Railtrack, and Jarvis Rail for breaches of s 3 of the HSWA 1974 on 10 November 2010. Network Rail pleaded guilty. The case against Jarvis Rail did not proceed because the company had gone into administration and the ORR concluded that to continue against the company was not in the public interest. On 14 May 2011, nine years after the incident, Network Rail was fined £3 million and ordered to pay costs of £150,000 at St Albans Crown Court.

Grayrigg derailment

6.84 A high-speed train was derailed 23 February 2007, just to the south of Grayrigg, Cumbria causing the death of one passenger. The investigation concluded that the derailment was caused by a faulty set of points. Prior to the incident, the maintenance of the railways had been brought 'in-house' by Network Rail. Because of the apparent similarities with the Potters Bar derailment, consideration was given to the inquests of both incidents being heard together (see above).

6.85 The inquest into the death of Mrs Masson, who died in the incident, finished on 4 November 2011. The inquest concluded that the derailment was caused by the degradation of a set of points, which had been poorly maintained. Network Rail was prosecuted by the ORR. On 4 April 2012, having pleaded guilty to a breach of s 3 of the HSWA 1974, Network Rail was fined £4 million and ordered to pay costs of £118,052 at Preston Crown Court.

AVIATION

Investigations

6.86 The Air Accidents Investigation Branch (AAIB) is part of the DfT and is responsible for the investigation of civil aircraft accidents in the UK. Its purpose is to investigate accidents to determine the circumstances and causes and not to apportion blame.

6.87 The Civil Aviation Authority (CAA) is the UK's independent specialist aviation regulator. Its activities include economic regulation, airspace policy, safety regulation and consumer protection. It is part of DfT and is tasked with investigating and prosecuting breaches of aviation safety rules and some aviation-related consumer protection and health and safety requirements. It enforces the requirements of the Air Navigation Order 2005.

6.88 The CMCHA 2007 applies on a British-controlled aircraft, as defined in s 92 of the Civil Aviation Act 1982. Such incidents will be investigated by the police.

6.89 Over the years, the industry at international level has raised concerns about the criminalisation of aviation accidents because of prosecutions of airlines and individuals.

6.90 There have, to date, been no manslaughter prosecutions arising out of aviation accidents in the UK. However, there have been some in other jurisdictions, possibly the most famous one being the prosecution relating to the ValuJet crash in the US.

ValuJet crash

6.91 This case brought about the first criminal indictment in the US of an aviation company for manslaughter.

6.92 On 11 May 1996, ValuJet flight 592 took off from Miami International Airport bound for Atlanta. The aircraft was carrying 110 passengers. Shortly after take-off, the aeroplane crashed when oxygen generators in its cargo hold caught fire.

6.93 The National Transportation Safety Board (NTSB) published its report in August 1997.[1] ValuJet had subcontracted the maintenance of its aircraft to a company called SabreTech. It was found that SabreTech's employees had loaded more than 100 armed oxygen generators into the cargo hold. These had been improperly labelled as 'empty'.

1 NTSB/ARR-96/03 (PB96–910404).

6.94 These canisters came from three used jets that SabreTech was renovating for ValuJet. The canisters, about the size of soft drink cans, provide oxygen to passengers when aircraft cabin pressure drops. These 'empty' generators were being shipped to ValuJet's stores in Atlanta.

6.95 As a safety precaution, SabreTech's employees should have fitted small plastic caps on the old canisters to prevent inadvertent activation. The total cost of sufficient caps for all the canisters that were removed was in the region of $9. Instead, the employees concerned disconnected the canisters, tied them off and falsely labelled them as 'empty' and consigned them for shipment.

6.96 The NTSB found that the correct course of action would have been for SabreTech's employees to have taken the canisters to a safe place, intentionally activate them and dispose of the empty cans. In addition, the two SabreTech employees involved also signed work cards that falsely indicated that the shipping caps had been installed on the oxygen canisters.

6.97 SabreTech was prosecuted with 110 counts of murder and 110 counts of manslaughter. In addition, 24 charges were brought against SabreTech and three of SabreTech's employees: Daniel Gonzales, Maintenance Director, and Eugene Florence and Mauro Valenzuela, SabreTech mechanics. The wide-ranging charges against the employees were of conspiring to falsify aircraft records, falsifying aircraft records, violating hazardous materials regulations and placing a destructive device on board the aircraft. It was alleged that the maintenance director pressured employees to skip prescribed work steps and then falsely sign documents indicating work had been completed, a practice known in the industry as 'pencil whipping'. It was alleged that the mechanics signed the work cards.

6.98 A key allegation in the indictment was that SabreTech had put profit before safety.

6.99 At the trial, two of the three employees were acquitted of the charges against them. The third absconded before the trial. The acquittals were on the basis that there was insufficient evidence of any intention to commit an offence.

6.100 The jury found SabreTech guilty of eight counts of recklessly causing hazardous materials to be transported, and one count of failing to train employees in handling hazardous materials. SabreTech was fined $2 million and ordered to pay $9 million in compensation to the families of 16 passengers. By this stage, ValuJet's insurers had already paid over $262 million in damages to the victims' families and SabreTech was a bankrupt company.

6.101 SabreTech appealed and, in October 2001, the Federal Appeals Court dismissed (on technical legal grounds) eight of the nine federal convictions against SabreTech. The remaining conviction was for failing to train employees in hazardous materials. The fines were reduced to $500,000.[1]

1 *United States v SabreTech Inc*, 271 F 3d 1018 US App LEXIS 23595 (11th Cir Fla, 2001).

6.102 On 7 December 2001, the district court in Miami formally dismissed the 110 murder and 110 manslaughter charges against SabreTech, in exchange for a no-contest plea to the one count of unlawfully causing the transportation of hazardous waste.

Other notable prosecutions

6.103 Greek prosecutors brought charges of negligent manslaughter, negligent bodily injury, and disrupting the safety of air services against the captain and first officer in connection with the 1979 Swissair crash in Athens, and the pilots received sentences of four years' imprisonment. This was later converted to a fine.

6.104 In Italy, in 2006, the convictions for manslaughter of an air traffic controller, a director of Milano Linate airport and the chief executive and a former director-general of ENAV (the Italian air traffic control agency) were upheld. This prosecution arose from the runway collision in Milan between an SAS aircraft and Cessna jet. It was found that an inoperative ground radar system was an underlying cause of the crash.

6.105 In 2007, four middle managers of a Swiss air traffic control company were convicted by a Swiss court of negligent homicide in relation to the 2002 mid-air collision of a passenger plane and a cargo jet that killed 71 people. Three of the convicted were given 12 months' suspended prison sentences, and the fourth a fine. Four other defendants were acquitted. Handing down the verdicts, the court said that the collision could have been avoided if two controllers had been on duty at the time of the incident. Prosecutors said that a culture of negligence and lack of risk awareness at the company contributed to the crash. The three managers who received the suspended sentences were criticised by the court for tolerating the single controller policy, which it said was incompatible with air safety rules.

6.106 In 2008, in New Zealand, the former owner of an aviation workshop and a licensed aircraft engineer whom he employed were convicted of manslaughter, after a helicopter that they were responsible for servicing crashed, killing the pilot. The engineer was meant to approve the assembly of a gearbox, which had been put together by unlicensed engineers, before it went inside the tail boom.

6.107 Instead, the gearbox was installed and checked only through a small hole, so small that it was impossible to see that the work had been carried out incorrectly. The mistake caused the driveshaft for the rear rotor to snap in flight. The engineer admitted he would have seen the misassembled component if he had viewed the tail rotor driveshaft before it was placed in the tail boom.

6.108 The trial judge sentenced each man to 300 hours' community service, saying to the defendants, 'You are both basically decent men who have made a serious criminal blunder'.

6.109 In July 2008 a French judge ordered Continental Airlines and five individuals to stand trial on charges of manslaughter for the crash in July 2000 of a Concorde jet that killed 113 people.

6.110 The Air France Concorde crashed shortly after take-off from Charles de Gaulle Airport, Paris. French investigators said that a metal strip from a Continental Airlines DC-10, lying on the runway when the Concorde took off, caused one of the supersonic jet's tyres to burst. Debris from the force of the explosion punctured the jet's fuel tanks. The French judicial inquiry also concluded that the tanks lacked sufficient protection from shock and that the manufacturers of Concorde had been aware of this problem since 1979. The crash marked the beginning of the end for Concorde, which ceased flying in 2003.

6.111 The director of Aerospatiale's Concorde programme between 1978 to 1994, a former chief engineer of Aerospatiale, an ex-director of the French civil aviation authority, Continental Airlines and two Continental employees were all charged in relation to the incident. The principal charges were manslaughter, which carries a maximum sentence in France of five years' imprisonment and a fine 75,000 Euros for individuals and 375,000 Euros for companies.

6.112 The trial began on 23 February 2010 at Pontoise Criminal Court, with judgment being given on 6 December 2010. All were acquitted except Continental Airlines and one of its employees, a welder. The welder received a 15-month suspended jail sentence, and the airline was fined 200,000 Euros.

6.113 In 2012, the appeals court in Versailles overturned the verdict against Continental Airlines, absolving it of 'criminal responsibility', and quashed the conviction of the welder.

ROAD

Work-related road risk

6.114 It is thought that a third of all accidents on UK road are work related. According to the Royal Society for the Prevention of Accidents (RoSPA):

> 'Employees who drive more than 25,000 miles a year have at least a 1 in 8,000 chance of dying behind the wheel of their company vehicle – a risk similar to that of miners dying at the coal face.'

6.115 It is not surprising, therefore, that work-related road risk (WRRR) is a risk rapidly moving up the health and safety agenda. It is relevant to all organisations where employees drive on its business, whether they are a lorry driver in a haulage company or a salesman visiting a client.

Investigations

6.116 To date, the HSE has been reluctant to become involved in regulating WRRR. This may, in part, be due to lack of resource and also because the HSE sees incidents on the road as being the remit of the police alone. It has, however, produced guidance with respect to the risk, entitled *Driving at Work –*

Managing work-related road safety (which can be downloaded at www.hse.gov.uk/pubs/indg382.pdf).

6.117 The guidance applies to 'any employer, manager or supervisor with staff who drive, or ride a motorcycle or bicycle at work, and in particular those with responsibility for fleet management'. It also states that:

> 'some employers believe, incorrectly, that provided that they comply with certain road traffic requirements, eg company vehicles have a valid MOT certificate, and that drivers hold a valid licence, this is enough to ensure the safety of their employees, and others, when they are on the road'.

6.118 The guidance divides the risk into three elements:

(1) the vehicle;
(2) the driver; and
(3) the journey (activity).

6.119 The guidance should be seen as the benchmark of defining a company's duty of care in respect of WRRR.

6.120 The police are responsible for investigating road traffic accidents. Since 2004, police road accident investigations have been specifically instructed to consider corporate liability through its Road Death Investigation Manual (RDIM). This can be downloaded at www.npia.police.uk/en/docs/Road_Death_Investigation_Manual_2007_PA.pdf.

6.121 The police will consider corporate responsibility where:

(a) There is sufficient indication that *failures in safety management* by the employer have *significantly contributed* to the incident; and
  (i) these failures cannot be addressed by the 'cause and permit' provisions in the road traffic legislation; and
  (ii) the risks are foreseeable and beyond the direct control of the driver.
And/or
(b) There is a *serious continuing risk* (ie one that could result in a similar incident occurring in similar circumstances) which cannot be addressed by the police using road traffic legislation, or by another appropriate enforcing authority (eg the Vehicle and Operator Services Agency (VOSA)).

6.122 The RDIM specifically refers to the HSE guidance and includes the following examples:

- **Driver Competency** – the employer has failed to ensure that drivers are competent and capable of doing their work in a way that is safe for them and other people (for example, has the employer considered whether the driver has the necessary driving licence and, if so, whether further training is required)?
- **Fitness and Health** – the employer has ignored obvious signs that an employee is unfit to drive (for example, from the effects of drink or drugs or through some medical condition).
- **Vehicle Suitability** – vehicles are being used for a purpose for which they were not intended (for example, saloon cars used to transport heavy or bulky goods without appropriate means to secure the load safely, or vehicles being too powerful for the competency of the driver).

6.123 In respect of commercial vehicles, VOSA is also relevant. VOSA was formed on 1 April 2003 following the merger of the Vehicle Inspectorate and the

Traffic Area Network division of the Department for Transport. VOSA provides a range of licensing, testing and enforcement services and supports the independent Traffic Commissioners.

Prosecutions

6.124 There have been relatively few prosecutions of companies and managers in relation to WRRR. However, with the increase in corporate manslaughter investigations, this may well change, as the police are more used to investigating road incidents than accidents at work. If there are prosecutions, then as the followings cases suggest, these are likely to be related to driver fatigue or vehicle defect.

6.125 *R v Stephen Bowles and Julie Bowles* (1999) concerned the directors of a family haulage company who were convicted of manslaughter and received custodial sentences of 15 months and 12 months, suspended for two years. A driver employed by their company fell asleep at the wheel of his lorry, having regularly spent more than 60 hours per week on the road. He was involved in a seven-vehicle pile-up on the M25, killing two people.

6.126 The prosecution argued that the defendants should have known about the driver's excessive hours and the effects this would have on his driving.

6.127 In *R v Melvyn Spree* (2004) the defendant was a director of Keymark Services, a road haulage company. He was convicted of manslaughter and conspiracy to falsifying driving records and sentenced to seven years in prison. Two deaths occurred after a lorry careered across the M1 during a driver's 18-hour shift. The police discovered a widespread tachograph fraud in the company. The company was also fined £50,000 for manslaughter.

6.128 In *R v Yaqoob* (2005),[1] the defendant was prosecuted for manslaughter arising from his failure, as a partner/manager of a business, to inspect the tyres of a minibus involved in a fatal accident. The Court of Appeal said it was entirely open to a jury to find that there was a duty to inspect and maintain a vehicle beyond the standard required for an MOT and other duties imposed by legislation. The Court of Appeal reduced his sentence of imprisonment to two and a half years and reduced his period of disqualification to three years.

1 [2005] EWCA Crim 2169.

6.129 *R v Produce Connection* (2006) concerned a rare prosecution by the HSE in respect of WRRR. The defendant company was fined £30,000 for health and safety breaches concerning driver fatigue and ordered to pay costs of £24,000. An employee was killed driving home from work when his van drifted into the path of an oncoming lorry. He had worked an average of 17 hours a day for 11 days without a day off.

6.130 In *R v Translact and others* (2009) the company and its directors were prosecuted for manslaughter following a fatal road crash involving one of its lorries. The lorry driver was imprisoned, having pleaded guilty to causing death by dangerous driving.

6.131 At trial, the jury heard evidence that the driver was pressured into working long hours and that the company, instead of employing more drivers, expected their drivers to cover the work by driving longer hours. The jury acquitted the defendants of manslaughter but convicted them of health and safety offences. Translact was fined £40,000 and two directors were fined £7,200 and £4,500 respectively.

6.132 In *R v Oughton and Lote* (2009), coach driver Robert Oughton was sentenced to five years and three months in prison after he admitted charges of causing death by dangerous driving and gross negligence manslaughter. His business partner in travel firm 1.4.You Coaches, John Lote, who pleaded guilty to gross negligence manslaughter, was jailed for three years.

The coach driven by Oughton was involved in a fatal car crash. The police investigation by North Yorkshire Police, working with the Vehicle and Operator Services Agency, concluded that the cause of the crash was acute brake failure due to poor maintenance. The coach's maintenance history showed that the two men had repeatedly failed to act on warnings to fix faults with the braking system. Officers also discovered that Oughton had ignored passenger complaints about the smell of burning during the journey.

THE FUTURE

6.133 Transport disasters that result in loss of life are, thankfully, rare. However, when they do occur, they attract much publicity. When a public inquiry follows, it often points to senior management failures and poor safety culture within organisations. Most of the incidents relied upon in the campaign for corporate manslaughter legislation cited disasters on the railways and at sea, where there has been a perceived inadequate response from the criminal justice system.

6.134 The cases set out in this chapter give an indication as to how future prosecutions for corporate manslaughter might be brought and the issues that are likely to be relevant. There are a number of lessons that can be learnt.

6.135 The development in transport prosecutions demonstrates that prosecutors and courts are now more willing to look beyond the initial causes of an incident, often the failures of front-line workers, to underlying causes which often involve failures by organisations and their management. Those organisations that believe they can simply place the blame on those at the coalface, or delegate the responsibility for risk to junior management, in order to escape prosecution for corporate manslaughter are likely to find themselves disappointed.

6.136 Criminal prosecutions arising out of a workplace death tend to take a long time, often two to three years, before proceedings are instituted. It is important, therefore, that, when organisations investigate incidents, they gain an understanding at an early stage of the causes, both initial and underlying. This will be essential for tactical and strategic decisions if prosecuted. Some companies put their heads in the sand, hoping the investigation will go away and, when a prosecution does arise, find themselves trying to investigate an incident properly years down the line when evidence and an understanding of the incident are more difficult to come by, and key figures within the organisation may have moved on.

6.137 Cases can be won or lost by the way in which an organisation responds to an investigation within the first 72 hours. Organisations ought to consider putting together an 'investigation protocol' before any incident occurs, so that they have a plan as to how they will respond to an investigation in the event of the worst case scenario. As part of this process, it is important for organisations to understand the powers of investigators in respect of disclosure of documents and the interviewing of individuals. It is also important that organisations consult with their insurers to ascertain the exact nature of its cover, eg some insurance policies do not provide cover for legal advice and representation until there is a prosecution, thus the important investigation process will not be covered.

6.138 There are those who believe that, when an organisation is prosecuted for corporate manslaughter, there will be no individual prosecutions. However, without prosecuting individuals, from a practical point of view, it may be difficult for a prosecutor to establish senior management failure in the minds of the jury. While there may be insufficient evidence to prosecute a senior manager of manslaughter, it may be possible to succeed in a health and safety prosecution. In the Hatfield case (see **6.65** to **6.77** above), of the 12 individual defendants, six were prosecuted for health and safety offences only.

6.139 With the passing of the Health and Safety (Offences) Act 2008, as from 16 January 2009 courts have the power to impose custodial sentences of up to two years for defendants convicted in the Crown Court of offences pursuant to s 7 or s 37 of the HSWA 1974. The impact of s 143 of the Criminal Justice Act 2003 requires the outcome of offending to be taken into account when sentencing. This is likely to lead to individuals convicted of health and safety offences that have contributed to a fatality being sent to prison. Further custodial terms for work related gross negligence manslaughter will also increase in the future with the application of s 143.

6.140 If the approach of the trial judge in the Hatfield case is taken to determining who is senior management, the fact that a manager is, in many respects, junior may not be sufficient protection against the prosecution establishing senior management failure if, for certain aspects of his/her job, that manager 'reigns supreme' in relation to the management of risk that is proved as being a cause of the accident.

6.141 As demonstrated by some of the transport cases, just because the police may decide not to prosecute does not mean that this is an end to the matter. Inquests are now, in many respects, becoming the new corporate battleground. The evidence that emerges may be crucial to whether a manslaughter investigation is reopened or, indeed, how the HSE presents its case if it decides to prosecute for health and safety breaches. There is also the possibility that the family (or trades union) of the deceased may decide to institute judicial review proceedings.

6.142 These cases also show that any prosecution can expect to receive a great deal of media coverage, which in turn will have an impact on an organisation's reputation. An organisation should have procedures in place for dealing with the media. Simply trying to hide from the media may well make matters worse. It should also be appreciated that dealing with the media in relation to criminal proceedings is very different to dealing with it on other company issues.

6.143 However, it is not just the media with whom an organisation will need to communicate but also with its staff (and their trades unions if they represent sections of the workforce), management, its investors and financial analysts. What may be said to these various stakeholders may vary depending upon the group and its interests.

6.144 Since the CMCHA 2007 came into effect in April 2008, there have been a handful of convictions although more trials are scheduled for 2014. In March 2012, questions were asked in Parliament about the lack of prosecutions. The written answer to this question revealed that, at that time, there were in the region of 50 cases under review by the CPS where one of the charges under consideration was corporate manslaughter. So, while the Act may not have resulted in more prosecutions, it has certainly been a catalyst for more corporate manslaughter investigations.

6.145 There are those who believe that high-profile prosecutions of companies in relation to transport accidents have failed because the prosecutors have not understood the nature of the incident and its complex causes. If this is correct, a change in the law alone will not rectify the position. It will be interesting to discover whether any of the cases under consideration by the CPS concern transport incidents and, if prosecuted, whether the Act makes any difference.

Chapter 7

The chemical industry*

Rob Elvin, Partner and European Head of the Environmental, Safety & Health Group, Squire Sanders (UK) LLP, Manchester

Dr David Gordon, Partner in the Environmental, Safety & Health Group, Squire Sanders (UK) LLP, Birmingham

Philip Bonner, Associate, Squire Sanders (UK) LLP, Manchester

* Material for the second edition (2010) contributed by Dr John Bond, Retired Safety Advisor and Auditor, BP Chemicals.

INTRODUCTION

7.1 The use of oil and chemicals has been known from the earliest times of recorded history. The Assyrians[1] used fire-pots of burning pitch as incendiaries in the siege of towns in the ninth century BC, and Pliny refers to a pool on the Euphrates discharging a flammable mud. Gerald of Wales (c 1188) quoted Julius Caesar's reference to the Britons who painted their faces with a 'nitrous ointment' which gave them such a frightening appearance that they unnerved their enemies. On the other side of the world, China was developing a mixture of sulphur, graphite and potassium nitrate, subsequently called black powder.

1 JR Partington, *A History of Greek Fire and Gunpowder* (1960).

7.2 The modern oil and chemical industries have developed many of the essentials for present-day living. There is insufficient wool, cotton and linen to clothe the world's population, and synthetic fibres have been developed to fill the gap. Many diseases have already been eliminated or controlled by drugs made from a variety of chemicals.

7.3 *Introduction*

7.3 With the wide variety of chemicals in use today, there are risks in their manufacture and use. These risks are generally well understood by industry although there have been a few well-publicised cases where the risks were not at first appreciated. These include asbestos, cigarette smoke, thalidomide for pregnant mothers, and methyl isocyanate used at Bhopal. Industry has learnt some lessons from these mistakes but there can be accidents in the manufacture of many chemicals which affect employees and the community adjacent to the plant. There have been numerous disasters. Flixborough in 1974 (production of caprolactam), Piper Alpha in the North Sea in 1988 (oil extraction) and Milford Haven in 1994 (explosion at an oil refinery) are well known in the UK. In Europe there have been disasters at Feyzin in France in 1966 (propane fire), Beek in The Netherlands in 1975 (vapour explosion), Seveso in Italy in 1976 (oil extraction) and Los Alfaques in Spain in 1978 (propylene explosion). In the rest of the world there have been similar disasters at Mexico and at Bhopal. In more recent years, there have been explosions on-site. The 1994 explosion at the Texaco Milford Haven Refinery injured 26 people and caused damage of £48 million and significant production loss. In 2001, another explosion killed 30 people and left almost 2,000 injured at AZF's fertiliser plant in France. The explosion at ICL Plastics' plant in Glasgow killed nine workers and left 33 injured in 2004. In December 2005, a number of explosions occurred at Buncefield Oil Storage depot in Hemel Hempstead, Hertfordshire. Fortunately there were no fatalities but over 40 people were injured, and significant damage was caused to both commercial and residential properties in the vicinity.[1] There were two incidents of note in the UK in 2010: an explosion, and subsequent fire, at a waste management facility in Burscough, West Lancashire, caused three people to be severely burnt, whilst another explosion at a chemical plant in Gloucestershire, led to three people being injured and another being killed. More recently, in 2012, a large explosion at a chemical specialist in Hazelwood, Derbyshire resulted in over 300 people being evacuated from the area; whilst a caustic soda explosion in West Thurrock, Essex in October 2013 led to two injuries and earthquake-like tremors felt several miles away.

1 The prosecutions arising from this incident are dealt with in detail at **7.164**.

7.4 Perhaps the most noteworthy disaster in recent memory is the Deepwater Horizon oil spill (commonly referred to as the 'BP oil spill') which occurred in the Gulf of Mexico off the coast of Louisiana in April 2010. During the final phase of the exploratory drilling of the Macondo oil well, an explosion led to the sinking of the Deepwater Horizon oil rig, resulting in 11 deaths and giving rise to the largest accidental spill of oil into the deep ocean the petroleum industry has ever seen. The blowout continued to release oil into the Gulf of Mexico for nearly three months until, following several failed attempts to stem the flow, the well was finally capped on 15 July 2010. The subsequent internal and external investigations into the incident made numerous recommendations and imposed fines on BP of over $4 billion to be paid to various organisations over a period of five years.[1] The National Commission on the BP Deepwater Horizon Oil Spill and Offshore Drilling produced its final report to President Obama in January 2011.[2] One of the key findings of the report was that human error, engineering mistakes and management failures all played a key part in the disaster, which was then exacerbated by the industry's and the government's lack of foresight and forward-planning, in order to ensure that they were adequately prepared to respond to a deep-water blowout of such scale. In response to the events in the Gulf of Mexico, the HSE's Energy Division (formerly known as the Offshore Safety Division) set up an internal Deepwater Horizon Incident Review Group in order to assess the implications of the incident and how it would affect the UK Continental Shelf. This led to a number of new priorities being established for those members of the offshore oil and gas industry operating on the UK Continental Shelf.[3]

1 http://www.bp.com/en/global/corporate/gulf-of-mexico-restoration/investigations-and-legal-proceedings.html.
2 http://cybercemetery.unt.edu/archive/oilspill/20121210200431/http://www.oilspillcommission.gov/final-report.
3 Further information in relation to these priorities is outlined at **7.157**.

7.5 Despite these disasters, the oil and chemical industries, with a high risk potential, have good records for safety in the UK. Close scrutiny by the Health and Safety Executive (HSE), coupled with the stringent regulatory regime surrounding the chemical industry, has meant that the number of accidents is not as high as other manufacturing industries.

Number and rate of reported major injuries within manufacturing[1]

Average number of employee major injuries 07/08 to 11/12p	Manufacture of:	Average rate of major injuries per 100,000 employees 07/08 to 11/12p
751	food products	250.7
677	fabricated metal products	253.4
265	rubber & plastic products	175.9
227	machinery & equipment nec	82.9
221	basic metals	215.2
216	other manufacturing	230.5
203	non-metallic mineral products	225.3
191	wood products	321.6
189	vehicles	96.4
149	chemicals & chemical prod	132.3
143	other transport equipment	76.7
132	printing & recorded media	98.7
129	repair & installation	126.4
120	furniture	146.3
102	paper & paper products	149.6
74	beverages	135.7
72	textiles	120
69	electrical equipment	86.8
61	computer, electronic & optical	30.6
40	pharmaceutical	35.8
12	clothes	30.7
12	coke & refined petrol	35.4
3	leather & related products	36.1
1	tobacco products	25.4

1 Table taken from HSE report, *Health and safety in manufacturing in Great Britain, 2013*, available at www.hse.gov.uk/statistics/industry/manufacturing/manufacturing.pdf.

Chemical industry and health and safety

7.6 Health and safety have been important issues in the chemical industry due to the toxicity and flammability characteristics of many chemicals. The containment of these chemicals is therefore an important factor in chemical plant operation. Release of any chemical is avoided by careful and detailed control of all operations.

7.7 During the 1950s and 1960s, chemical plants had increased rapidly in size and, at the same time, the pressures and temperatures required for their operation also increased.[1] This resulted in many serious fires, explosions and releases of materials from plants.

Figure 7.1 ICI's fatal accident rates, five-year moving average, for the period 1960–1982

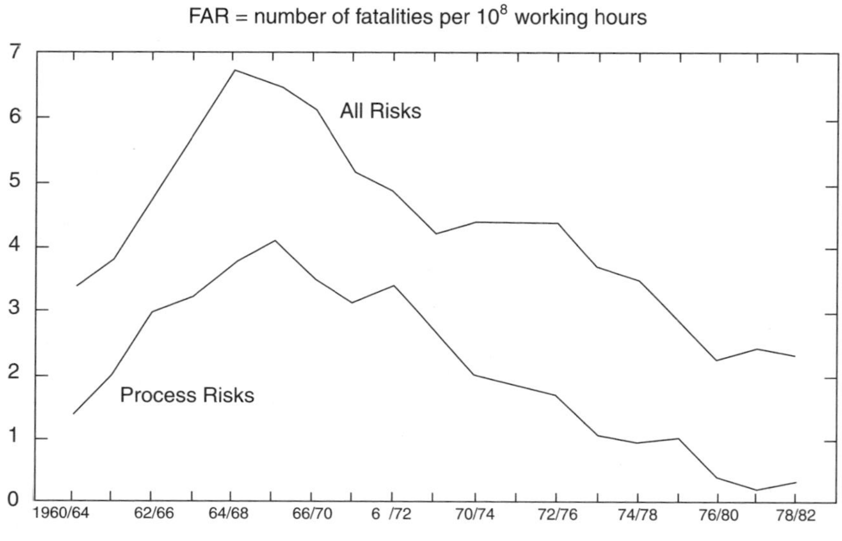

1 T Kletz (1999) 77 Trans IChemE, Part B, May, at 109.

7.8 The cost of these incidents escalated to such an extent that much more effort was given to building safer plants. The effort was mainly in the design and operational field and started both in the US and UK.

7.9 The increase in fatalities was attributed to the more risky conditions of chemical processing being used, and this had expanded both the potential for loss and the financial value of any single loss. The need for a systematic approach to technical safety became a necessity, and this led to the concept of loss prevention.

7.10 Prior to this period, safety had been treated as a reactive approach to accidents, after the event had been examined, and usually only the basic causes were found. Loss prevention was a totally different approach which sought the causes of potential accidents and introduced control systems to prevent the undesired event from happening. This proactive approach to safety became the basis of loss prevention, inspired by the following principle:[1]

'Those who cannot remember the past are condemned to repeat it.'

1 G Santayana 'The Life of Reason; or, The phases of human progress' (1905) New York: C. Scribner's Sons.

7.11 A major step forward in 1964 was the introduction of Hazard and Operability Studies (HAZOP).[1] The procedure was a systematic methodology for identifying most hazards at the design stage and hence designing out problems before the plant was built.

1 T Kletz 'HAZOP and HAZAN. Identifying and Assessing Process Industry Hazards' (1999) IChemE.

7.12 Improvements in safety in the chemical industry came about as a result of voluntary effort by the leading companies. Despite these efforts, there were a number of major incidents.

7.13 The Flixborough accident of June 1974 focused the attention of the new Health and Safety Executive on the chemical industry, which came under the Control of Industrial Major Accident Hazards Regulations 1984[1] (CIMAH), followed by the Control of Major Accident Hazards Regulations 1999[2] (COMAH) as amended. The HSE initially enforced the Health and Safety at Work etc Act 1974 (HSWA 1974) in the shore-based industries, while the Department of Industry enforced it in the pipeline and offshore industry. After the Piper Alpha incident of July 1988, responsibility for the offshore industry was transferred to the HSE.

1 SI 1984/1902, as amended by SI 1990/2325 and SI 1994/118.
2 SI 1999/743, as amended by SI 2005/1088 and SI 2009/1595.

7.14 As the 20th century was ending, it became clear that HSWA 1974, which had been in force for 25 years, was an effective regulatory means of controlling industry, although the self-regulating nature of the original intent was mainly forgotten. The European Directives, particularly those known as Seveso I[1] and Seveso II,[2] were brought in by regulations under the Act. Seveso I came about in the aftermath of the Seveso disaster in 1976 where 3,000kg of chemicals were released into the air from a factory near the town of Seveso in Italy. The Seveso II Directive was updated in 2003 by Directive 2003/105/EC, in the light of more recent industrial accidents such as the cyanide spill at Baia Mare in Romania in 2000, the explosion at a fertiliser plant in Toulouse in September 2001 and in the light of studies into carcinogens and substances dangerous to the environment. It is important to note that the Seveso III[3] Directive will repeal and replace Seveso II by 1 June 2015. The driving force behind the introduction of the Seveso III Directive was the changes that had been made to the classification of dangerous substances, by the EC Regulation on Classification, Labelling and Packaging of Substances and Mixtures (CLP Regulation),[4] and Seveso III will re-define what is classified as a 'dangerous' substance for the purposes of the Seveso regime.

1 Council Directive 82/501/EEC (now repealed), was implemented in the UK by the Control of Industrial Major Accidents Hazards Regulations 1984, SI 1984/1902, now amended by SI 1990/2325 and SI 1994/118.
2 Council Directive 96/82/EC, amended by Council Directive 2003/105/EC, implemented in the UK by the Control of Major Accident Hazards Regulations 1999, SI 1999/743, now amended by SI 2005/1088 and SI 2009/1595.
3 Council Directive 2012/18/EU on the control of major-accident hazards involving dangerous substances, amending and subsequently repealing Council Directive 96/82/EC.
4 Regulation (EC) 1272/2008 (as amended by Regulation (EU) 286/2011 to bring it in line with the 3rd edition of the GHS – see **7.58** below), amending and repealing Directives 67/548/EEC and 1999/45/EC, and amending Regulation (EC) 1907/2006. More information is available at http://www.hse.gov.uk/seveso/

7.15 Further regulations were brought in, although the chemical and oil industries had built up a self-regulating and monitoring culture, with extensive auditing systems based mainly on the International Safety Rating System.[1]

1 FE Bird and GL Germain, *Practical Loss Control Leadership* (1985).

7.16 The ideas of loss prevention expanded rapidly in the last part of the 20th century, with particular advances in:

- risk assessment;
- accident investigation;
- auditing of the management system;
- learning lessons from past accidents;
- detailed management systems;
- safety performance indicators; and
- inherent safety.

7.17 The new millennium has introduced fresh approaches to reducing accidents but still within the concept of loss prevention. Further development in the chemical industry of earlier systems, particularly inherent safety and safety performance indicators, will take place. However, the frequency of serious accidents in the chemical industry continues to be a cause for concern.

7.18 These safety areas are discussed below, but the management of risk involved in chemical plant operations is a very large subject, and this book merely provides an overview of the key pieces of legislation covering the regulation of the industry.

THE LEGAL FRAMEWORK FOR HEALTH AND SAFETY IN THE CHEMICAL INDUSTRY

Introduction

7.19 As in most areas of health and safety law, there are two principal levels of regulation in the chemical industry: national law and international law. International law is drafted by both the UN and the EU. For example, in relation to the safe transportation of chemicals, the UN prescribes regulations and guidelines which the EU then adopts. EU law overrides national law, and EU Member States cannot introduce valid contradictory legislation. It is important to understand what the key EU laws and national laws are, as non-compliance leaves a company open to potential prosecutions.

7.20 A number of legislative vehicles are available with which to pass EU law, and it is important to understand what they are:

(1) *Regulations* 'shall have general application' and 'shall be binding in [their] entirety and directly applicable in all Member States'.
(2) *Directives* 'shall be binding, as to the result to be achieved, upon each Member State to which it is addressed, but shall leave to the national authorities the choice of form and methods'.
(3) *Decisions* 'shall be binding in [their] entirety upon those to whom [they are] addressed'.
(4) *Recommendations* and *Opinions* 'shall have no binding force'.

7.21 Regulations are 'directly applicable' in the sense that they become part of a Member State's law without the requirement for any implementation. In other words, Parliament need not pass any legislation to bring such a regulation into force. Furthermore, once a piece of EU legislation is transposed into national law, it must be interpreted in accordance with the Directive or Regulation from which it originates.

7.22 EU Regulations should not be confused with regulations passed by Parliament, which form part of national law.

7.23 It should be noted that, in the instance of the carriage of dangerous goods legislation in the UK, domestic regulations have gradually converged with international legislation.

7.24 However, the main way in which EU law impacts on the chemical industry is through Regulations and Directives, such as the Chemical Agents Directive[1] and the Regulation on Registration, Evaluation, Authorisation and Restriction of Chemicals (REACH).[2] Directives must be transposed into national law within the prescribed period. If this does not occur, the Directive may take on the same direct effect features as a Regulation.[3] In addition, once the implementation deadline has expired, an individual can enforce the provisions of a Directive against the government of the Member State, and so Directives are capable of vertical effect.[4] Directives are not, however, capable of horizontal effect[5] in this way.

1 Council Directive 98/24/EC.
2 Regulation (EC) 1907/2006, as amended by Regulation (EU) 252/2011.
3 (Case 41/74) *Van Duyn v Home Office* – it is possible for Directives to be directly effective in this way but it will depend on the facts of each case and the Directive in question. It is to be questioned 'whether the nature, general scheme and wording of the provision in question are capable of having direct effects'.
4 *Francovich v Italy* (Joined Cases C-6/90 and C-9/90) [1991] ECR I-5357; [1993] CMLR 66.
5 Ie between two individuals.

Key legislation

7.25 We can therefore divide the relevant legislation into two groups:

- EU Directives and Regulations; and
- key national statutes and regulations.

7.26 There are many pieces of national legislation, and indeed EU legislation, which impact not only on the chemical industry, but more generally on all places of work.

7.27 Of particular importance are HSWA 1974 and the Management of Health and Safety at Work Regulations 1999 (MHSWR).[1] These impose both general and specific duties on employers to maintain a safe place and system of work, in addition to a number of other duties. This chapter reviews the key elements of chemical specific legislation and briefly touches on its interface with the Corporate Manslaughter and Corporate Homicide Act 2007.

1 SI 1999/3242, as amended by the Management of Health and Safety at Work (Amendment) Regulations 2006, SI 2006/438.

Chemical Agents Directive

7.28 The Chemical Agents Directive[1] (CAD) establishes minimum requirements for the protection of workers from risks arising (or likely to arise) out of the presence or use of hazardous chemical agents. Most of the health requirements under CAD have been implemented through the Control of Substances Hazardous to Health Regulations 2002,[2] with some changes to other existing legislation on asbestos and lead. Much of CAD follows the well-accepted principles previously present in UK legislation, in particular:

- Employers are required to assess the risks to the health and safety of workers from the presence of chemical agents in the workplace.[3]
- Employers must take both general and specific protection and prevention measures to eliminate or reduce the risks to workers' health and safety arising from chemical agents.[4]
- Employers are required to monitor the exposure of workers to chemical agents in certain circumstances[5] and are obliged to place them under health surveillance where such an assessment discloses a risk to their health.[6]
- The Directive imposes a duty on employers to establish arrangements to deal with accidents and incidents, including emergency situations.[7]
- Employers must provide workers with appropriate information and training relating to chemical agents. This information must be updated to take account of changing circumstances.[8]
- The production, manufacture or use at work of certain specified chemical agents, or activities involving their use, is prohibited.[9]
- Employers are placed under a duty to consult workers or their representatives on, and allow them to participate in, the matters covered by the Directive.[10]

1 Council Directive 98/24/EC.
2 SI 2002/2677 (as amended).
3 Council Directive 98/24/EC, Art 4.
4 Council Directive 98/24/EC, Arts 5 and 6.
5 Council Directive 98/24/EC, Art 6(4).
6 Council Directive 98/24/EC, Art 10.
7 Council Directive 98/24/EC, Art 7.
8 Council Directive 98/24/EC, Art 8.
9 Council Directive 98/24/EC, Art 9 and Annex III.
10 Council Directive 98/24/EC, Art 11. An employer already has general duties under the Health and Safety (Consultation with Employees) Regulations 1996, SI 1996/1513, and the Safety Representatives and Safety Committees Regulations 1977, SI 1977/500, to consult with employees or their representatives and to provide them with information. However, the CAD imposes much more specific requirements than these.

CONTROL OF SUBSTANCES HAZARDOUS TO HEALTH REGULATIONS 2002

7.29 Statistics show that, in 2011/12 in Great Britain, approximately 1.1 million[1] people suffered from an illness that they believe was caused or made worse by their current or past work. In 2010, 2,347[2] people died of mesothelioma, and thousands more died of occupational cancers and lung diseases. The Control of Substances Hazardous to Health Regulations 2002 (COSHH)[3] aims to protect the health of both workers and others from exposure to hazardous substances during work activities, by requiring employers to introduce controls on such exposure.

1 *Health and Safety Executive Statistics*, published in the HSE's Annual Statistics Report for 2011/12, available at www.hse.gov.uk/statistics/overall/hssh1112.pdf.
2 *Health and Safety Statistics,* published by the HSE, available at www.hse.gov.uk/statistics/causdis/asbestos.htm.
3 SI 2002/2677 (as amended).

7.30 The original version of COSHH came into force in October 1989[1] but has been amended and replaced several times since. The 2002 version of

COSHH[2] implements the Chemical Agents Directive (CAD)[3] and is supported by an Approved Code of Practice (ACOP)[4] which gives detailed practical advice on compliance. It should be noted that all ACOPs are under review by the HSE.[5] As of Autumn 2013, the HSE has conducted an initial review of 32 of its 52 ACOPs. The remaining 20 are still to be reviewed due to their association with ongoing sector-specific consolidations or other regulatory amendments, and the HSE plans to review them in the course of the delivery of those processes.

1 SI 1988/1657, as amended.
2 SI 2002/2677 (as amended).
3 Council Directive 98/24/EC.
4 L5 (5th Edition) 2005.
5 More information on the HSE's review of its ACOPs is available at www.hse.gov.uk/consult/condocs/cd241-acops.htm.

7.31 COSHH imposes a duty on employers to assess the risks to the health of their employees arising from exposure to hazardous substances, and to prevent such exposure. Where this is not reasonably practicable, exposure must be adequately controlled. In some instances, employers must monitor employees' exposure to a hazardous substance and place them under health surveillance.

7.32 The most recent amendments to COSHH[1] came into force in April 2005. There were three significant amendments. The main differences is the extension of reg 7, relating to the prevention or control of exposure to substances hazardous to health. COSHH used to apply solely to Maximum Exposure Limits (MELs) and Occupational Exposure Standards (OESs), but it now covers *all* hazardous substances and describes a new single limit, Workplace Exposure Limits (WELs).[2] The other major change is in relation to reg 9. Originally under COSHH there was a duty to maintain any control measure 'in an efficient state, efficient working order, in good repair and in a clean condition'. This has now been extended so that, as well as ensuring that all physical risk control measures meet the necessary requirements,[3] there is also an obligation to ensure that management arrangements (ie the provision of systems of work and supervision of any other measure) undergo reviews 'at suitable intervals and are revised if necessary'. The HSE suggest that the frequency with which this should be done will depend on the likelihood of the deterioration of that particular element of the control measure and its importance.[4]

1 SI 2002/2677 (as amended).
2 The requirements are contained in reg 7 of SI 2002/2677 (as amended).
3 This is due to change: see **7.40** below.
4 More information and HSE guidance in relation to the amendments are available at www.hse.gov.uk/aboutus/meetings/iacs/acts/watch/091106/p6annex1.pdf.

Hazardous substances

7.33 Hazardous substances are those that can harm the health of employees working with them if they are not properly controlled, for example, by using adequate ventilation.[1] As such, they are found in nearly all workplaces within the chemicals industry. For the great majority of commercial chemicals, whether a warning label is present or not will indicate whether COSHH[2] is relevant. For example, liquid soap for household use does not contain a warning label, but bleach does, so COSHH applies to bleach but not to the liquid soap when used at work.

1 A 'substance hazardous to health' is defined under SI 2002/2677 (as amended), reg 2.
2 SI 2002/2677 (as amended).

Approaching COSHH

7.34 The Health Directorate, part of the HSE, summarises that there are eight steps, which must be followed, in order to comply with COSHH[1].

1 SI 2002/2677 (as amended), and HSE, *COSHH: A Brief guide to the Regulations* (available at /www.materials.ox.ac.uk/uploads/file/COSHHRegulations.pdf) and *COSHH Essentials: Easy Steps to Control Chemicals* (available at www.coshh-essentials.org.uk/assets/live/CETB.pdf).

Step 1 – assessing the risks

7.35 The first step is to carry out a risk assessment.[1] This should:

(a) identify the hazardous substances present in the workplace; and
(b) consider the risks the substances present to peoples' health.

1 SI 2002/2677 (as amended), reg 6.

7.36 It is necessary to consider substances that have been supplied to the company, those produced by its work activity and those naturally or incidentally present in the workplace. There must then be an assessment of what risks these substances present to people's health.

7.37 This involves making a judgement on how likely it is that a hazardous substance will affect someone's health. The HSE has developed a generic risk assessment guide for supplied substances.[1] Key questions to be considered are:

1 How much of the substance is in use or produced by the work activity and how could people be exposed to it?
2 Who could be exposed to the substance and how often? This involves a consideration of all the groups of people who could come into contact with the substance, ie visitors, members of the public, contractors etc, as well as the company's employees.
3 Is there a possibility of substances being absorbed through the skin or swallowed?
4 Are there risks to employees working at locations other than the main workplace?

1 HSE, *COSHH Essentials: Easy Steps to Control Chemicals* (available at www.coshh-essentials.org.uk/assets/live/CETB.pdf).

7.38 The legal responsibility for the assessment under COSHH[1] is on the employer. However, outside parties can be engaged to carry out some or all of the work in preparing the assessment. Employees or their safety representatives must be involved in the preparation of assessments and informed of the results of the assessment.

1 SI 2002/2677 (as amended).

Step 2 – decide what precautions are needed

7.39 If significant risks are identified, companies must decide on the action they need to take to remove or reduce them to acceptable levels. Pre-existing controls may be compared with:

- *COSHH Essentials: Easy Steps to Control Chemicals.*
- Occupational Exposure Limits (OELs) and Workplace Exposure Limits[1] (WELs).
- Good work practices and standards endorsed for the company's industry sector, for example, the Chemicals Industry Association.

1 Inserted by SI 2004/3386.

7.40 WELs are Occupational Exposure Limits set under the COSHH (Amendment) Regulations 2004.[1] They replace OESs and MELs. WELs are concentrations of hazardous substances in the air, averaged over a specified period of time. It is an offence to exceed WELs. In addition, there is an overriding obligation to reduce exposure, as far as is practicable, below the WEL. Harmful substances and their WELs are listed in a publication called EH40/2005, *Workplace Exposure Limits.*[2] The introduction of WELs does not drastically change the way in which COSHH risk assessments are carried out, as most WELs are equivalent to the old OESs and MELs. Therefore, previous risk assessments under COSHH will still be regarded as valid. However, COSHH risk assessment should be reviewed on an annual basis so that they now refer to the appropriate WEL. In December 2009, the European Commission adopted a further Directive, establishing a third list of indicative occupational exposure limit values (IOELVs).[3] This required Member States to introduce domestic occupational exposure limits for the 19 substances listed in the Annex to the Directive, whilst also taking into account the IOELVs. Implementation in the UK was secured in December 2011, with the HSE approving a revised set of WELs, set out in EH40 (see above). The revised list included new WELs for five hazardous workplace chemicals, whilst nine substances required reductions to the existing WELs. In addition, one substance had its WEL increased, as the IOELV in the Directive was higher than the limit that had formerly been in place.

1 SI 2004/3386.
2 2nd edition available at www.hse.gov.uk/pubns/priced/eh40.pdf.
3 Commission Directive 2009/161/EU (implementing Council Directive 98/24/EC and amending Commission Directive 2000/39/EC).

7.41 It is essential for a company to check that its control systems are effective. If risks to health are identified, the company must take action to protect the health of both its employees and others. If the company has five or more employees, it must make and keep a record of the main findings of the risk assessment, either in writing or in computerised form.[1] This record must be made as soon as practicable after the assessment and contain sufficient information to explain the decisions taken about whether risks are significant and the need for any control measures.[2]

1 SI 2002/2677 (as amended), reg 6(4).
2 See the COSHH ACOP, which provides information on what the record of the main findings of the assessment should contain.

7.42 The HSE emphasises that an assessment should be a 'living' document, which is revised regularly and changed when necessary. Assessments must be reviewed when:

- there is reason to suspect that the assessment is no longer valid;
- there has been a significant change in the work carried out; and/or
- the results of monitoring employees' exposure show that the revision is necessary.[1]

Indeed, each assessment should state when the next review is expected to take place.

1 SI 2002/2677 (as amended), reg 6(3).

Step 3 – prevent or control exposure sufficiently

7.43 COSHH[1] imposes a requirement on companies to prevent exposure to substances which are hazardous to health, if it is reasonable to do so. As such, the company might be expected to use the substance in a safer form, change the process or activity to remove the need for the hazardous substance, or replace the substance with a safer alternative.

1 SI 2002/2677 (as amended).

7.44 Exposure must be adequately controlled if prevention is not reasonably practicable, and the measures that are put in place should be both appropriate and consistent with the risk assessment. They may involve:

- use of appropriate work processes;
- control of exposure at source (for example, a ventilation system at source); and
- provision of personal protective equipment.

7.45 As guidance as to what 'adequate control' constitutes, COSHH states that this means reducing exposure to a level that most workers could be exposed to every day at work without any adverse effects on their health.

Step 4 – ensure that control measures are used and maintained

7.46 COSHH, reg 9[1] requires employees to make proper use of control measures and to report defects. It is the duty of the employer to take all reasonable steps to ensure that they do so. The specific duties that COSHH imposes aim to ensure that controls are kept in efficient working order. For example, respiratory protective equipment must be examined and, where appropriate, tested at suitable intervals. Specific intervals are laid down by COSHH between examinations for local exhaust ventilation equipment. In addition, companies must retain records (or a summary of them) for examinations and tests carried out for at least five years.

1 SI 2002/2677 (as amended), reg 9.

Step 5 – monitoring exposure

7.47 Where a risk assessment concludes that:

- there could be serious risks to health if control measures failed; or
- exposure limits might be exceeded; or
- control measures might not be functioning fully,

an employer must monitor the exposure of employees to hazardous substances in the air that they breathe.

7.48 This need not be done if the employer can show that another method of control is in place adequately to prevent employees' exposure to hazardous substances (for example, an alarm system which activates when hazardous substances are detected).

Step 6 – carrying out appropriate health surveillance

7.49 Health surveillance[1] must be carried out in the following circumstances:

(a) (i) Where an employee is exposed to one of the substances listed in COSHH, Sch 6[2] and is engaged in a related process; and
(ii) there is a reasonable likelihood that an identifiable disease or adverse health effect will result from that exposure; *or*

(b) (i) Where employees are exposed to a substance linked to a particular disease or adverse health effect; and
(ii) there is a reasonable likelihood under the condition of work of that disease or effect occurring; and
(iii) it is possible to detect a disease or health effect.

In addition, any technique used must pose a low risk to the employee.

1 SI 2002/2677 (as amended), reg 11.
2 SI 2002/2677 (as amended), Sch 6.

7.50 Health surveillance could involve examination by a doctor, trained nurse or, in some cases, a trained supervisor. A basic record ('health record') should be kept of any health surveillance carried out. Health records should be kept for at least 40 years.[1]

1 Further information is available in HSE, *Understanding health surveillance at work: An introduction for employers* (available at www.hse.gov.uk/pubns/indg304.pdf) and *Health Surveillance under work* (available at www.hse.gov.uk/pubns/priced/hsg61.pdf).

Step 7 – preparation of plans and procedures to deal with accidents, incidents and emergencies

7.51 Plans and procedures must be established in workplaces where there is a risk of an accident, incident or emergency involving exposure to a hazardous substance which goes well beyond the risks associated with normal day-to-day work. The objective is to have a response in place before any such incident takes place. It will involve the preparation of procedures and the setting up of warning and communication systems, which can be activated immediately should any such incident occur. Safety drills should be practised at regular intervals.

7.52 Emergency procedures need not be introduced if:

- the quantities of substances hazardous to health that are present in the workplace are such that they present only a slight risk to the health of employees; and
- the measures taken under Step 3 are sufficient to control that risk.

Step 8 – ensuring that employees are properly informed, trained and supervised

7.53 COSHH, reg 12[1] requires employers to provide their employees with suitable and sufficient information, instruction and training, which must include:[2]

(a) the names of substances that the employee could be exposed to, the risks created by such exposure and access to any safety data sheets that apply to those substances;
(b) the significant findings of the risk assessment;

(c) the precautions that should be taken to protect themselves and other employees;
(d) how to use the personal protective equipment provided;
(e) the results of any exposure monitoring and health surveillance (whilst maintaining the anonymity of individual employees); and
(f) the emergency procedures which must be followed.

1 SI 2002/2677 (as amended).
2 SI 2002/2677 (as amended), reg 12(2).

7.54 As with many health and safety legal obligations, employers need to ensure that they update and adapt the information, instruction and training to take account of significant changes in the nature of work carried out or work methods used. The HSE regards it as vitally important that the information provided should be appropriate to the level of risk identified by the assessment and be presented in a way which will be understood by employees.

CONTROL OF MAJOR ACCIDENT HAZARDS REGULATIONS 1999

7.55 The main HSE guidance to the Control of Major Accident Hazards Regulations 1999[1] (COMAH) is contained in the *Guide to the Control of Major Accident Hazards Regulations 1999*.[2] COMAH currently implements the requirements of the Seveso II Directive,[3] which covers the control of major accident hazards involving dangerous substances such as chlorine and explosives.

1 SI 1999/743 (as amended).
2 L111 (Second Edition) 2006 (available at www.hse.gov.uk/pubns/priced/l111.pdf).
3 Council Directive 96/82/EC, as amended by Council Directive 2003/105/EC.

7.56 COMAH was amended in June 2005, to broaden its scope and to take account of research into carcinogens and those substances that are harmful to the environment as well as certain amendments to the Seveso II Directive.

7.57 The key revisions to COMAH were amendments to the lists of dangerous substances and general categories of substances that are used to determine whether COMAH applies or not.[1]

1 The amendments to the lists of substances are outlined below:
- redefinition of ammonium nitrate to cover lower percentage composition, and new classes covering self-sustaining decomposition and reject material;
- new named category for potassium nitrate fertilisers;
- specification of seven new carcinogens, and raised threshold limits for all carcinogens;
- new category for petroleum products to include gas oils such as diesel, naphtha and kerosene, with thresholds that are half those of the previous automotive petrol category;
- redefinition of the classes for explosives;
- lower qualifying thresholds for substances dangerous for the environment; and
- change to the aggregation rule that is applied to all substances classified as toxic, dangerous for the environment, flammable or oxidising.

7.58 However, in July 2012, the Seveso III Directive[1] was published. The Seveso III Directive will repeal and replace its predecessor, Seveso II, with changes transposed into national legislation by 1 June 2015. This was made necessary by changes to the EU's classification of dangerous substances, which were motivated by the United Nations' Globally Harmonised System of Classification and Labelling of Chemicals (GHS).[2] The GHS seeks to ensure worldwide continuity in the classification and labelling of chemicals. Consequently, the Regulation on Classification, Labelling and Packaging of

Substances and Mixtures (CLP Regulation)[3] came into force on 20 January 2009, with the aim of aligning the EU system with the GHS. The CLP Regulation will gradually replace the two Directives by which the Seveso II Directive currently defines dangerous substances.[4] Consequently, amendments will need to be made to how dangerous substances are classified under the Seveso regime. The COMAH Competent Authority in the UK (CA)[5] plans to consult on the proposals to implement Seveso III prior to the 1 June 2015 deadline, with draft regulations due to be published for consultation in Spring 2014.[6] Transitional arrangements are being considered for the post-Seveso III system in order to minimise any additional burden on operators.

1 Council Directive 2012/18/EU on the control of major-accident hazards involving dangerous substances, amending and subsequently repealing Council Directive 96/82/EC.
2 More information on the GHC is available at www.unece.org/trans/danger/publi/ghs/ghs_welcome_e.html.
3 Regulation (EC) 1272/2008 (as amended by Regulation (EU) 286/2011 to bring it in line with the 3rd edition of the GHS), amending and repealing Directives 67/548/EEC and 1999/45/EC, and amending Regulation (EC) 1907/2006.
4 Council Directive 67/548/EEC on approximation of laws, regulations and administrative provisions relating to the classification, packaging and labelling of dangerous substances, and Directive 1999/45/EC concerning the approximation of the laws, regulations and the administrative provisions of the Member States relating to the classification, packaging and labelling of dangerous preparations.
5 The COMAH Competent Authority consists of the Health and Safety Executive and the Environment Agency in England and Wales, and the HSE and the Scottish Environment Protection Agency in Scotland.
6 See www.hse.gov.uk/seveso/faqs.htm.

7.59 The principal aim of COMAH is to prevent and mitigate the effects of those major accidents involving dangerous substances which can cause serious harm to people and the environment. COMAH mainly affects the chemical industry, but also some storage activities, explosives and nuclear sites and other industries, where threshold levels of specified 'dangerous substances' are used.[1]

1 COMAH, SI 1999/743 (as amended).

7.60 COMAH imposes two categories of duties on operators: 'lower tier' and 'top tier'. The lower-tier duties apply to all operators, whereas the top-tier duties apply only to certain operators.

7.61 An incident is a major accident if:

- it results from uncontrolled developments[1] (ie unplanned and unexpected) in the course of the operation of an establishment to which COMAH[2] applies;
- it results in a serious danger to people or the environment, whether on or off the site in question and whether immediate or delayed; and
- it involves one or more of the dangerous substances identified by COMAH, irrespective of the quantity involved.

1 'Loss of control' is not limited to sudden, unplanned events but also includes expected, planned or permitted discharges.
2 SI 1999/743 (as amended).

7.62 COMAH, reg 2(2)[1] defines an 'operator' as a person who is in control of the operation of an establishment or installation.[2] This can include a company or partnership or, arguably, an insolvency practitioner. Where the establishment or installation is to be constructed or operated, the 'operator' is the person who proposes to control its operation. If that person is not known, it will be the person who has commissioned its design and construction. The

Court of Appeal has also given further guidance on how 'operator' is defined for the purposes of COMAH. It held that the 'operator' is the person who has identified itself to the COMAH CA as the operator of the installation and is treated accordingly by the CA. Once the person has identified itself to the CA, it cannot seek to abrogate responsibility by claiming they were not 'in control of the operation'.[3]

1 SI 1999/743 (as amended), reg 2(2).

2 COMAH, SI 1999/743 (as amended), applies to an 'establishment' where:

- dangerous substances are present; or
- the presence of dangerous substances is anticipated; or
- it is reasonable to believe that they may be generated during the loss of control of an industrial chemical process.

'Establishment' is defined as the whole area under the control of the same person where dangerous substances are present in one or more installations.

A substance is a 'dangerous substance' for the purposes of COMAH if it is:

- specified at the appropriate threshold; or
- falls within a generic category at the appropriate threshold.

'Industrial chemical process' means that premises with no such chemical process do not come within the scope of COMAH solely because dangerous substances are generated during an accident.

3 See *Hertfordshire Oil Storage Ltd v R* [2010] EWCA Crim 493.

The thresholds are split between lower tier and top tier: lower-tier duties

7.63 Under COMAH, reg 4,[1] every operator has a general duty to take all measures necessary to prevent major accidents and limit their consequences to persons and the environment.

1 SI 1999/743 (as amended), reg 4.

7.64 COMAH recognises that eliminating a risk entirely will not always be possible. However, prevention measures should aim to reduce risk to as low a level as is reasonably practicable. When considering what is reasonably practicable, the balance between cost and benefit, as well as what is technically achievable, will be considered. Where hazard levels are high, high standards will be required to ensure that risks are sufficiently low. Consequently, operators are expected to demonstrate that they have put in place control measures that are sufficient for the risks identified.

7.65 If the lower-tier threshold is equalled or exceeded, an operator must notify the HSE,[1] ensuring that it provides the details outlined in Sch 3. Lower-tier operators are under a duty to take 'all measures necessary' to prevent major accidents. Any major accidents must be reported.

1 SI 1999/743 (as amended), reg 6(1).

7.66 COMAH, reg 5(1)[1] requires every operator to prepare and keep a major accident prevention policy (MAPP) document, detailing its policy with respect to the prevention of major accidents. This duty highlights the important role that management systems play as an accident prevention mechanism. Schedule 2 outlines the key elements of a safety management system, which must be addressed by the MAPP. Safety policies and risk assessments undertaken by industry to comply with the requirements of other health and safety legislation, such as the HSWA 1974 and the MHSWR, can be used as part of the paperwork for MAPP.[2] Operators who come into the scope of COMAH must prepare their MAPP without delay, and in any event within three months of becoming subject to COMAH.

1 SI 1999/743 (as amended), reg 5(1).
2 For further details, refer to SI 1999/743, reg 5, Sch 2 as amended, ACOP L111 and the HSE's chemical industry information sheet No 3, 'Major accident prevention policies for lower-tier COMAH establishments' (available at www.hse.gov.uk/pubns/chis3.pdf).

7.67 The MAPP document needs to be in writing and include sufficient particulars to demonstrate that the operator has established an appropriate safety management system. The principles that must be taken into account are specified in COMAH, Sch 2, paras 1–4.[1] The MAPP has to be reviewed and updated when required, so that it is kept up to date.[2] Furthermore, it is the operator's duty to implement the policy set out in the MAPP.[3]

1 SI 1999/743 (as amended), Sch 2, paras 1–4.
2 SI 1999/743 (as amended), reg 5(4).
3 SI 1999/743 (as amended), reg 5(5).

7.68 An establishment should therefore apply the following tests:

- Does COMAH apply?[1]
- If so, the MAPP should be drawn up in accordance with COMAH, reg 5.[2]

The establishment should then consider the following questions, which appear in the guidance to COMAH:[3]

- Does the MAPP meet the requirements of COMAH?
- Will it deliver a high level of protection for people and the environment?
- Are there management systems in place, which achieve the objectives set out in the policy?
- Are the policy, management systems, risk control systems and workplace precautions kept under review to ensure that they are implemented and that they are relevant?

An operator should also bear in mind the provisions of the Seveso III Directive with regard to MAPPs (which will come into effect on 1 June 2015, when the new COMAH Regulations come into force).[4] Article 8(1) of the Directive stipulates that a MAPP needs to include a commitment towards continuously improving the control of major-accident hazards and improving safety.

1 See COMAH, SI 1999/743 (as amended), reg 3, Sch 1 and check existing and likely inventory of dangerous substances.
2 SI 1999/743 (as amended), reg 5.
3 L111 Second Edition 2006, section 134.
4 Council Directive 2012/18/EU on the control of major-accident hazards involving dangerous substances, amending and subsequently repealing Council Directive 96/82/EC.

Notifications

7.69 COMAH, reg 6[1] contains the notification requirements that must be observed by establishments. Notifications should be in writing; this can include e-mail or by such other means as the recipient may allow.[2]

1 SI 1999/743 (as amended), reg 6.
2 SI 1999/743 (as amended), reg 2.

7.70 Within a reasonable period of time prior to the start of the operation of an establishment, the operator of the establishment must send to the CA a notification containing the information specified in COMAH, Sch 3.[1] The information required includes, amongst other information, the address of the establishment concerned and the quantity and physical form of the dangerous substances present.[2]

1 See COMAH, SI 1999/743 (as amended), Sch 3.
2 Although there is no prescribed form for notifying the CA under reg 6, template notification forms are available on the HSE website (www.hse.gov.uk/comah/notification/notif2.htm), and operators are encouraged to use the forms where possible.

7.71 For the purposes of COMAH,[1] the CA is a joint inspection and enforcement body that comprises the HSE and the Environment Agency for England and Wales and the Scottish Environment Protection Agency for Scotland. It should be noted that the COMAH audit of one UK-based chemical company resulted in many days of scrutiny by the regulatory authorities and numerous improvement and enforcement notices being issued. It is clear that COMAH is being used as a vehicle to examine health and safety and environmental systems in some significant detail, and the chemical industry is reporting an increase in the intervention activities of the CA.

1 SI 1999/743 (as amended) and Council Directive 2012/18/EU, art 20.

7.72 This will only be heightened when the Seveso III Directive comes into effect, with article 20 specifying a more detailed and frequent system of inspections.[1] Top-tier sites will be inspected once a year, whilst lower-tier sites must be inspected every three years. Article 9 of Seveso III introduces the concept of the potential 'domino effect' from a number of sites being in close proximity. Using the information garnered from notifications by operators, inspections of establishments and the safety reports produced by the operators of top-tier establishments,[2] the CA must identify those establishments whose geographical proximity increases the risk of a major incident. In addition, the CA must ensure that these operators share information, so that their emergency planning (including any MAPP and safety report) adequately reflects the heightened risks, as well as helping to inform the public of the potential hazards. Article 15 of Seveso III will also mean more extensive and earlier public consultation in relation to the siting of new establishments and modifications to, or the establishment of new developments close to, existing sites. This requirement ensures that the Seveso regime complies with the Aarhus Convention.[3] Article 15 of Seveso III ensures that the public are afforded an early opportunity to give their opinions to the CA before the relevant decision is taken, and the results of this consultation must be taken into account when the final decision is made.

1 Council Directive 2012/18/EU, art 20.
2 See **7.75** below for more information.
3 Convention on access to information, public participation in decision-making and access to justice in environmental matters (available at www.unece.org/fileadmin/DAM/env/pp/documents/cep43e.pdf).

7.73 Once the initial notification has been made, a continuing duty remains to notify the CA in the event of:[1]

- any significant increase in the quantity of dangerous substances previously notified;
- any significant change in the nature or physical form of the substances previously notified, the processes employing them or any other information notified to the CA in respect of the establishment;
- COMAH, reg 7[2] ceasing to apply to the establishment by virtue of a change in the quantity of dangerous substances present there; or
- permanent closure of an existing installation in the establishment.

1 SI 1999/743 (as amended), reg 6(4).
2 SI 1999/743 (as amended), reg 7.

Notification of a major accident

7.74 Where a major accident[1] occurs at an establishment, the operator is under a duty to inform the CA of the accident immediately.[2] The CA is then required to conduct a thorough investigation into the accident.

1 A major accident means an occurrence (including in particular, a major emission, fire or explosion) resulting from uncontrolled developments in the course of the operation of any establishment and leading to serious danger to human health or the environment – immediate or delayed, inside or outside the establishment, and involving one or more dangerous substances (SI 1999/743 (as amended), reg 2).
2 SI 1999/743 (as amended), reg 15(3). NB: the duty to notify will be satisfied if the operator notifies a major accident to the HSE pursuant to the requirements of the Reporting of Injuries, Diseases and Dangerous Occurrences Regulations 1995, SI 1995/3163, as amended (SI 1999/743 (as amended), reg 15(4)).

Top-tier duties

Safety reports

7.75 If any top-tier thresholds are equalled or exceeded, an operator must comply with additional requirements under COMAH.[1] All operators of top-tier establishments must produce a safety report,[2] the key purpose of which is to demonstrate that they have taken all necessary measures to prevent major accidents. In addition, they must show that all steps have been taken to limit the consequences to people and the environment of any such accidents which do occur. Regulations 9–13 of COMAH cover emergency planning requirements,[3] including the need to test both on- and off-site emergency plans and arrangements for charging by local authorities for preparing and testing off-site plans. Regulation 14 of COMAH covers the information that the operators must give to the public. Public registers are used by CAs to make safety reports available to the public.

1 SI 1999/743 (as amended), regs 7–13.
2 SI 1999/743 (as amended), reg 7.
3 Additional guidance is available on emergency planning for major accidents (HSG191).

7.76 The operator is required to submit a safety report to the CA, within a reasonable period of time both before construction commences and before start-up, in order to ensure that safety is considered at the design stage.[1] The information that these must contain is specified in COMAH, Sch 4,[2] in order to demonstrate to the CA that all the necessary measures for the prevention and mitigation of major accidents are in place. However, to avoid duplication, any information already contained in the pre-construction report need not be repeated in the pre-start-up report. In addition, an operator may limit the information that a safety report contains if he can show that certain dangerous substances are in a state incapable of creating a major accident hazard.[3]

1 SI 1999/743 (as amended), reg 7(1) and (5).
2 SI 1999/743 (as amended), Sch 4.
3 SI 1999/743 (as amended), reg 7(12).

7.77 Neither the construction of a new establishment nor its operation may be begun before the operator has received the CA's conclusions on the report.[1] However, the CA must ensure that these are communicated within a reasonable period of time of receiving the safety report.[2] Should the CA reasonably request further information in writing following its review of the safety report, the operator must provide such information within the time period specified in the CA's request.[3]

1 SI 1999/743 (as amended), reg 7(1) and (4).
2 SI 1999/743 (as amended), reg 17(1)(a).
3 SI 1999/743 (as amended), reg 7(13).

7.78 All of the information that is required to be included under COMAH, Sch 2[1] can be done so by reference to the content of another notification or report made pursuant to statutory requirements. The information in that notification must be up to date and sufficiently detailed.

1 SI 1999/743 (as amended), Sch 2.

7.79 It is important for companies to realise that the operator must review the safety report:[1]

- fully, at least every five years;
- whenever such a review is necessary because of new facts or to take account of new technical knowledge about safety matters; and
- whenever the operator makes a change to the safety management system which could have significant repercussions with respect to the prevention of major accidents to the persons and the environment.

1 SI 1999/743 (as amended), reg 8(1).

7.80 Where a review determines that the report needs revision, the operator must carry that revision out immediately and inform the CA of the details.[1] The operator must also inform the CA if the five-year review does not lead to a revision of the report.[2]

1 SI 1999/743 (as amended), reg 8(1).
2 SI 1999/743 (as amended), reg 8(2).

7.81 The HSE publication *Preparing Safety Reports*[1] provides practical guidance to operators on the preparation of COMAH safety reports.[2] In addition, the *Review and Revision of COMAH Safety Reports – Guidance for Operators from the Competent Authority*[3] gives non-statutory guidance on the requirements relating to the review and revision of safety reports as outlined in reg 8 of COMAH.[4] The guidance stresses how important it is to continually review and revise safety reports to ensure that they remain accurate and up to date. In addition, the revised version of the guidance, which was published in 2010, re-modelled the CA's preferred approach to the five-year review process. It introduced a meeting between the operator and the CA six months before the revision is due to take place, in order to discuss its scope, breadth and depth, as well as signalling the move to more on-site and fewer desktop assessments of the five-year revision.[5]

1 HSG 190 (2nd Edition) 2005, available at www.hse.gov.uk/pubns/priced/hsg190.pdf.
2 SI 1999/743 (as amended), reg 8(1).
3 COMAH RO1 '*Revised guidance for operators of top tier COMAH establishments*' (January 2010), available at www.hse.gov.uk/comah/report-review.pdf.
4 SI 1999/743 (as amended), reg 8.
5 COMAH RO1 '*Revised guidance for operators of top tier COMAH establishments*' (January 2010), paras 4.10 and 4.11, available at www.hse.gov.uk/comah/report-review.pdf.

Emergency plans

7.82 COMAH imposes a requirement[1] on top-tier establishments to draw up a written on-site emergency plan,[2] which is sufficient to achieve the following objectives:

(1) containing and controlling incidents so as to minimise the effects, and to limit damage to persons, the environment and property;

(2) implementing the measures necessary to protect persons and the environment from the effects of major accidents;

(3) communicating the necessary information to the public and to the emergency services and authorities concerned in the area; and

(4) providing for the restoration and clean-up of the environment following a major accident.

1 SI 1999/743 (as amended), reg 9.
2 SI 1999/743 (as amended), Sch 5, Pt 1.

7.83 The information that must be included in this plan is specified in Sch 5, Pt 2 of COMAH[1] and includes details such as the emergency contact details of the persons authorised to set emergency procedures in motion, and the arrangements for providing early warning of the incident to the local authority responsible for implementing the off-site emergency plan.

1 As specified in SI 1999/743 (as amended), Sch 5, Pt 2.

7.84 An adequate off-site emergency plan must be drawn up by the local authority (LA) for the area where a top-tier establishment is located.[1] The same objectives apply to this plan as to the on-site plan[2] and it must also be in writing. COMAH provides a list of required information that plans must contain,[3] and an operator must supply all necessary information to the LA to enable the plan to be prepared.[4] In addition, the LA must consult:

- the operator;
- the CA;
- Environment Agency (EA)/Scottish Environment Protection Agency (SEPA);
- the emergency services;
- each health authority for the area in the vicinity of the establishment; and
- such members of the public as it considers appropriate.[5]

1 SI 1999/743 (as amended), reg 10(1).
2 SI 1999/743 (as amended), Sch 5, Pt 1.
3 SI 1999/743 (as amended), Sch 5, Pt 3.
4 SI 1999/743 (as amended), reg 10(3).
5 SI 1999/743 (as amended), reg 10(6).

7.85 The CA may, in light of a safety report, exempt an LA from the requirement to prepare an off-site emergency plan in respect of an establishment. Any such exemption must be in written form and state the reasons for granting it.[1]

1 SI 1999/743 (as amended), reg 10(7).

7.86 Once emergency plans have been put in place, it is obligatory that these are reviewed (and, if necessary, revised) at least every three years.[1] COMAH, reg 11(1) specifies certain factors that should be taken into account when carrying out such a review, such as new technical knowledge, knowledge concerning the response to major incidents and, in the case of an off-site emergency plan, required consultation by the LA of such members of the public as it considers appropriate.

1 SI 1999/743 (as amended), reg 11.

7.87 The LA is entitled to charge the operator a fee[1] to cover costs that are reasonably incurred in performing its functions of preparing, reviewing and testing off-site emergency plans.[2] When requesting the fee from the operator, the LA must provide a detailed statement of the work carried out and the costs incurred.

1 SI 1999/743 (as amended), reg 13.
2 Under SI 1999/743 (as amended), regs 10 and 11.

7.88 Where the CA believes that the measures taken by an operator for the prevention and mitigation of major accidents are seriously deficient, it must prohibit the operation or bringing into operation[1] of that establishment or installation, or any part of it. The CA also has the discretion to prohibit the operation or bringing into operation of any establishment or installation if the operator has failed to submit any notification, safety report or other information required under COMAH within the required time. The CA must serve a notice on the operator giving reasons for any prohibition and the date when it is to take effect.[2]

1 SI 1999/743 (as amended), reg 18(1).
2 SI 1999/743 (as amended), reg 18(2) and (3).

7.89 A final obligation which operators should take into consideration is the duty to supply information to the public within an area without their having to request it. The CA will notify the area to the operator as being one in which people are liable to be affected by a major accident occurring at the establishment.[1] This information must be supplied within a reasonable period of time after the off-site emergency plan has been prepared.[2]

1 SI 1999/743 (as amended), reg 14(1) and (2).
2 SI 1999/743 (as amended), Sch 6 specifies the minimum information to be supplied to the public.

Pollution Prevention Guidelines

7.90 The Environment Agency, together with other UK environmental regulators, last published Pollution Prevention Guidelines in April 2011 (PPG 22).[1] PPG 22 are aimed at:

- site operators of industrial and commercial premises;
- vehicle operators;
- other organisations who store or handle polluting materials;
- sewage treatment providers; and
- those responsible for responding to spills and transporting or storing waste from spills (for example, the Fire and Rescue Services and local authorities).

PPG 22 provide good practice guidelines to aid the identification of measures to prevent, limit or reduce the damage to the environment and the risk to public health in the event of a spill. It includes information and advice about a site operators' Pollution Incident Response Plan, pollution control methods and site-specific pollution control options.

1 'Pollution Prevention Guidelines Incident Response – Dealing with spills: PPG 22 (April 2011)' (available at http://cdn.environment-agency.gov.uk/pmho0411btez-e-e.pdf).

7.91 Notably, PPG 22 recommend that they are read in conjunction with previous Pollution Prevention Guidelines, published in March 2009 (PPG 21).[1]

PPG 21 intend to:

- stop an incident occurring, or help the site operator act quickly if it does occur;

- provide a template for the basic information that a site operator will need to prepare a plan; and
- give advice on how to train staff.

PPG 21 are aimed at site operators that do not have a legal obligation to produce a plan, in contrast to operators of high-risk sites that are regulated under the COMAH and Environmental Permitting (EP) regimes. However, COMAH and EP operators can also use the guidelines as supplemental guidance for preparing their plans. In particular, PPG 21 are targeted at:

- operators of industrial and commercial premises that store or use toxic or potentially polluting substances, such as chemicals, oil and food;
- the Fire and Rescue Service, HSE and other regulators involved in the production of plans; and
- insurers and underwriters of operators.

There has been considerable work on incident response planning in the light of the recommendations of the investigation into the Buncefield explosion.

1 'Pollution Prevention Guidelines Incident Response Planning: PPG 21 (March 2009)' (available at http://cdn.environment-agency.gov.uk/pmho0309bpna-e-e.pdf).

TRANSPORT AND CARRIAGE OF DANGEROUS CHEMICALS

7.92 The law on transportation of dangerous goods, including radioactive material, is a complex evolution of global, European and domestic legislation, as illustrated in the table below at **7.94**. In the UK, the interpretation and enforcement of the law can be carried through either the criminal or the civil courts. This is particularly relevant for producers of dangerous goods, transport operators and designers of transport containers who may have to deal with either system. For example, a prosecution under the criminal law may open the way for a large civil claim.

7.93 In addition to the extra safety constraints, operators must also consider handling and vehicle construction, driver training and security plans. Particular attention must be paid to road use regulations in respect of maximum weights permissible. The need for security and shielding of some goods means 'special types' gross weight regulations apply.

7.94 Operators may also need to comply with physical security regimes and special measures applied to ensure that certain materials are protected from interference by terrorist groups.

LIABILITIES **Who is held responsible for breaches, and what are the consequences in criminal law?**	**LIABILITIES** **Who is held responsible for breaches, and what are the consequences in civil law?** **Is insurance available?**	**OPERATIONAL REQUIREMENTS** **What are industry rules?**
Global laws and standards	International treaties Civil liability	United Nations (UN) regulations
↓ EU codes and legislation	↓ Implemented into EU legislation	↓ Implemented into EU legislation
↓ Domestic legislation	↓ Implemented into domestic legislation	↓ Implemented into domestic legislation

At UN level

7.95 The United Nations Economic Commission for Europe (UNECE) provides analysis, policy advice and assistance to governments. The UNECE also sets out norms, standards and conventions to facilitate international cooperation within and outside the region. The EU, as part of this organisation, adopts those standards and reports of the UNECE expert bodies. The UNECE produces model regulation and published guidance known as:

- The European Agreement concerning the International Carriage of Dangerous Goods by Inland Waterways (ADN);[1]
- The European Agreement concerning the International Carriage of Dangerous Goods by Road (ADR);[2] and
- The European Agreement concerning the International Carriage of Dangerous Goods by Rail (RID).

1 ECE/Trans/2003 (Vol I).
2 ECE/Trans/2002 (Vols I and II).

7.96 This produces a cycle of updates, which are then followed by implementation in the EU by Directives. Typically, the EU is slightly behind the advances made by the UN, and implementation into national legislation then follows at a further delay.

7.97 EU Directives[1] require the UK to implement the requirements of RID and ADR for domestic as well as for international carriage of dangerous goods by rail and road.

1 Specifically, Directive 2008/68/EC on the inland transport of dangerous goods, available at http://eur-lex.europa.eu/LexUriServ/LexUriServ.do?uri=OJ:L:2008:260:0013:0059:EN:PDF.

7.98 The technical annexes to the UN documents provide extensive detail of what should be done in any circumstance: for instance, section 5.3 of ADR sets out placarding requirements for loads, with detailed descriptions and diagrams of the placards required.

At European Union level

7.99 The Dangerous Goods (DG) Directive is the latest in a succession of Directives applying RID and ADR; it has been amended primarily to take into account technical progress and ensure the continued safe and secure transport of dangerous goods.[1]

1 Directive 2008/68/EC on the inland transport of dangerous goods, available at http://eur-lex.europa.eu/LexUriServ/LexUriServ.do?uri=OJ:L:2008:260:0013:0059:EN:PDF.

7.100 To simplify the legislation, the Commission has decided to consolidate five Directives into one. It is the DG Directive that Member States are required to implement into domestic legislation. It pulls together regulation for three different transport modes: road, rail and inland waterway.

7.101 The five EU Directives being consolidated and replaced by the DG Directive are:

- Council Directive 94/55/EC on the transport of dangerous goods by road;
- Council Directive 96/49/EC on the transport of dangerous goods by rail;
- Council Directive 96/35/EC on the appointment and vocational qualification of safety advisers for the transport of dangerous goods by road, rail and inland waterway;

- Council Directive 2000/18/EC on the minimum examination requirements for safety advisers for the transport of dangerous goods by road, rail and inland waterway; and
- Council Directive 82/714/EEC that lays down the technical requirements for inland waterway vessels.

7.102 The annexes to the DG Directive contain lists of national derogations, which allow specific national circumstances to be taken into account in relation to the transportation of dangerous goods within Member States' territories.[1] These lists are regularly updated, with the most recent revision being made in May 2013.[2]

1 Directive 2008/68/EC on the inland transport of dangerous goods, Annex I, Section I.3, Annex II, Section II.3 and Annex III, Section III.3, available at http://eur-lex.europa.eu/LexUriServ/LexUriServ.do?uri=OJ:L:2008:260:0013:0059:EN:PDF.
2 http://eur-lex.europa.eu/LexUriServ/LexUriServ.do?uri=OJ:L:2013:130:0026:0059:EN:PDF

Options for implementation delivery in the UK

7.103 The DG Directive required Member States to introduce the necessary transposition measures by 1 July 2009. The Department for Transport (DfT), having produced draft regulations and consulted on the implementation of this legislation, published the final version of the regulations, the Carriage of Dangerous Goods and Use of Transportable Pressure Equipment Regulations 2009 (CDG 2009),[1] which came into force on 1 July 2009. CDG 2009 made a number of proposals to improve UK regulation of this area as part of an ongoing process, which started in 2007 with the CDG 2007.[2]

1 SI 2009/1348, as amended by SI 2011/1885.
2 SI 2007/1573.

7.104 CDG 2009 consolidated previous regulations which separately covered different classes of dangerous goods, following CDG 2007, which resulted in one set of regulations covering all classes of dangerous goods.

7.105 In line with industry wishes, CDG 2007 also made more use of referencing the technical annexes of RID and ADR. As a result of the pre-consultation and stakeholder feedback to the DfT, CDG 2009 went further by including simpler direct referencing of RID and ADR. The idea was that this would eliminate the 'time lag' for domestic regulations to catch up with the UN provisions, avoiding the biannual cycle of new regulation, as the UN recommendations pass through the EU and UK legislative process.

7.106 The introduction of CDG 2009 marked a major change, in that the new regime does not specify responsibilities for specific duty holders (eg a consignor) in relation to RID and ADR requirements. Instead, the emphasis is on all parties involved in the transport of dangerous goods to comply with the obligations and responsibilities set out in RID and ADR that are applicable to them. This is coupled with a new approach to enforcement, which looks at causes and culpability rather than prescriptive regulatory duties. Enforcement is focused on the most important requirements, being those that 'contribute significantly to the safety of the public';[1] these include the integrity of packages and tanks, documentation, equipment including fire extinguishers, driver training, package marking and labelling, and other matters of obvious concern (eg insecure loads). The HSE maintains that enforcement should 'be in accordance with the HSE policy',[2] although failure to comply with the regulations may often be judged to 'represent a serious risk' for the purposes of Prohibition Notices. The enforcement advice was developed by the HSE, jointly with the police and the Vehicle and Operator Services Agency, and is reproduced on the HSE website.[3]

1 See www.hse.gov.uk/cdg/manual/opstratenforce.htm#enforcement at 11.
2 See www.hse.gov.uk/cdg/manual/opstratenforce.htm#enforcement at 14.
3 See www.hse.gov.uk/cdg/manual/opstratenforce.htm#annex3–1.

7.107 In October 2011, the Carriage of Dangerous Goods and Use of Transportable Equipment (Amendment) Regulations 2011[1] (CDG 2011) were published. These amended the CDG 2009, with the driving force behind their introduction being the need to transpose the Transportable Pressure Equipment (TPE) Directive[2] in Great Britain, which made changes to the EU regime governing transportable pressure equipment. CDG 2011 also provide for the classification of class 1 goods (explosives) and further consolidate previous regulations, by revoking the Classification and Labelling of Explosives Regulations 1983,[3] which had been deemed by the HSE as redundant and an unnecessary burden on the industry. In addition, the Secretary of State for Energy and Climate Change will now undertake the competent authority and enforcement functions in relation to class 7 (radioactive material), whilst the Secretary of State for Defence has been given enforcement responsibilities for the transportation of military-related goods.

1 SI 2011/1885, amending SI 2009/1348. Explanatory memorandum available at www.legislation.gov.uk/uksi/2011/1885/pdfs/uksiem_20111885_en.pdf.
2 Directive 2010/35/EU, repealing, amongst others, Directive 1999/36/EC (which had previously been implemented by CDG 2009).
3 SI 1983/1140.

Secure transportation

7.108 In response to the events of 11 September 2001, the United Nations agreed enhancements to the security of transporting dangerous goods. The European Commission adopted the new road and rail security measures. The requirements are split into two levels: a general level applicable to the carriage of all dangerous goods; and a higher level for the carriage of high-consequence dangerous goods. These are defined as those which have the potential for misuse in a terrorist incident and which may, as a result, produce serious consequences such as mass casualties or mass destruction.

7.109 The measures for road and rail are currently in place through CDG 2009 (as amended by CDG 2011) and are supported by a comprehensive set of guidance. The guidance is not intended to be a prescriptive document, and organisations are free to consider other ways of meeting the requirements of the regulations. However, there is a requirement for carriers, consignors and others (including infrastructure managers) engaged in the transport of high-consequence dangerous goods to adopt, implement and comply with a security plan. It is difficult to see how this can be met by other measures, and operators should now address themselves to the issues of preventing deliberate misuse of their goods.

7.110 As noted at **7.107**, the CDG 2011[1] made the Secretary of State for Energy and Climate Change the competent authority for the purposes of the carriage of radioactive material. This was part of the government's wider plan to restructure the UK's civil nuclear industry, by consolidating the former regulatory and security responsibilities of the HSE and DfT into a single body. Consequently, the Office for Nuclear Regulation (ONR) was formed on 1 April 2011. The transport of all civil nuclear material, as defined by the Nuclear Industries Security Regulations 2003 (NISR),[2] was previously regulated by the

Office for Civil Nuclear Security (OCNS), but the OCNS has now been subsumed within the ONR. The ONR's Civil Nuclear Security Programme has taken on the responsibility of regulating the movement of civil nuclear material, in accordance with the NISR. Whilst the ONR is currently an agency of the HSE, the long-term plan is for it to be an independent statutory corporation. As of 24 October 2011, the Radioactive Materials Transport Team (RMTT), which was formerly based at the DfT, moved to the ONR. RMTT inspectors will continue to undertake inspections to enforce the security regulations of the CDG 2009 (as amended by CDG 2011) for radioactive material.

1 SI 2011/1885, amending SI 2009/1348.
2 SI 2003/403, as amended by SI 2006/2815.

Department for Transport's (DfT) approach to regulation

7.111 The DfT followed up on its previously stated aim – to continue the consolidation after the CDG 2009 to simplify regulations for all classes of dangerous goods, in line with industry wishes, by introducing further amendments via the CDG 2011.

7.112 The DfT has also established an Industry Advisory Group,[1] to aid with the monitoring and implementation of the security requirements and the supporting guidance.

1 More information on the Industry Advisory Group for the Security of Carriage of Dangerous Goods by Road and Rail Industry is available at www.dft.gov.uk/publications/iag-security-of-carriage-of-dangerous-goods-by-road-and-rail.

7.113 The DfT aims to apply derogations under the DG Directive (Art 6(2) to (4)) which allow Member States to be exempt from certain requirements of the Dangerous Goods Directive,[1] which may change or become redundant over a fairly short space of time. For example, in April 2012, the DfT published a list of circumstances under which particular types of carriage or carriage in particular circumstances will be exempt from the requirements of the CDG 2009 (as amended by CDG 2011).[2] This contained an extensive list of derogations for both transport by road and transport by rail.

1 Directive 2008/68/EC, on the inland transport of dangerous goods available at http://eur-lex.europa.eu/LexUriServ/LexUriServ.do?uri=OJ:L:2008:260:0013:0059:EN:PDF. See **7.102** above for more detail.
2 'Carriage of Dangerous Goods: Approved Derogations and Transitional Provisions (April 2012)', available at http://assets.dft.gov.uk/publications/carriage-of-dangerous-goods-approved-derogations-transitional-provisions/approved-derogations-transitional-provisions.pdf.

7.114 The CDG 2009 (as amended by CDG 2011) also remove some of the additional domestic requirements (which were in CDG 2007) that are over and above RID and ADR. These are the:

- miscellaneous security requirements for carriage of class 1 goods by rail (reg 84);
- carriage of class 1 goods in vehicles used to carry passengers for hire or reward (reg 86);
- carriage of class 1 goods by road in motor vehicles (reg 87);
- marshalling and formation of trains (reg 88); and
- removal of the due diligence defence (reg 93).

7.115 The pattern of regulation is as follows:

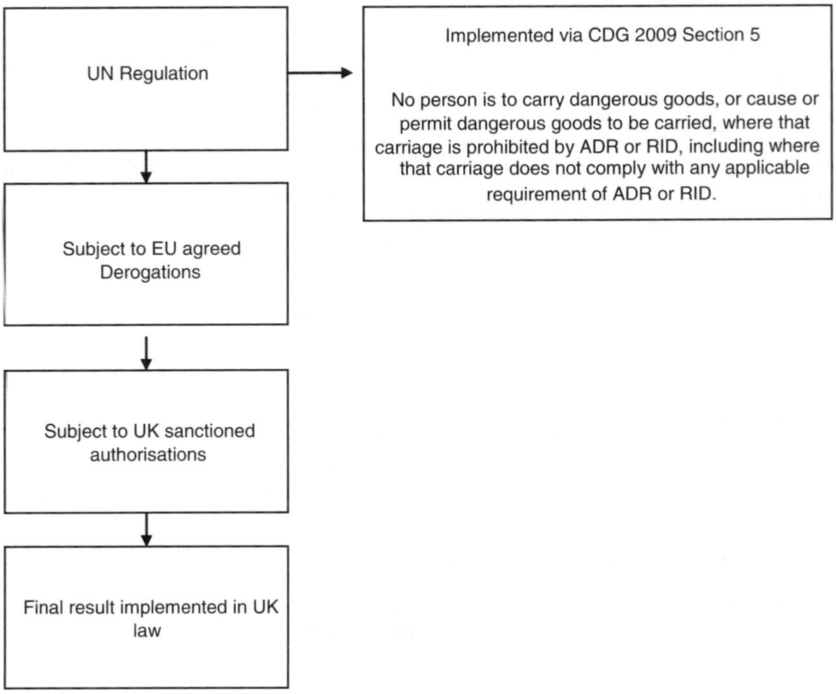

CHEMICALS (HAZARD INFORMATION AND PACKAGING FOR SUPPLY) REGULATIONS 2009

7.116 The Chemicals (Hazard Information and Packaging for Supply) Regulations 2002[1] (CHIP) requires suppliers of chemicals to classify, package and label dangerous chemicals appropriately and to supply information for their safe use. CHIP was amended in October 2005. The main change related to the introduction of a new approved supply list, implementing in full Commission Directive 2004/73/EC.[2] CHIP was further amended in October 2008; these changes did not affect the main legal duties under CHIP but did adjust the rules and procedures for classifying and labelling certain chemical preparations. CHIP was, however, revoked and replaced by the Chemicals (Hazard Information and Packaging for Supply) Regulations 2009[3] (CHIP4), which came into force in April 2009.

1 SI 2002/1689, as amended by Chemicals (Hazard Information and Packaging for Supply) (Amendment) Regulations 2005 (SI 2005/2571) and Chemicals (Hazard Information and Packaging for Supply) (Amendment) Regulations 2008 (SI 2008/2337).

2 Available at http://eur-lex.europa.eu/LexUriServ/LexUriServ.do?uri=OJ:L:2004:216:0003:0310:EN:PDF.

3 SI 2009/716.

7.117 The EC is in the process of updating the current EU classification and labelling system for hazardous chemicals by aligning it with the United Nations' Globally Harmonised System of Classification and Labelling of Chemicals (GHS).[1] This new system will complement the Regulation on Registration, Evaluation and Restriction of Chemicals (REACH)[2] and take effect in all Member States through the Regulation on Classification, Labelling and Packaging of Substances and Mixtures (CLP Regulation),[3] and amending Directive 67/548/EEC; and, over time, it will replace the CHIP regime. CHIP4 dovetails the requirements of CHIP with CLP, allowing both to be enforced in

Great Britain.[4] CHIP4 also consolidates all amendments to the CHIP regime since 2002, and repeals the requirements of CHIP at the end of the transitional measures provided for in the CLP Regulation, when the CLP regime comes fully into force on 1 June 2015. In the meantime, CHIP and its supporting guidance will be amended, as the transitional process set out in the CLP Regulation progresses. Duty holders are already required to apply the CLP Regulation provisions for substances placed on the market on or after 20 January 2010; whilst, in relation to mixtures, duty holders can voluntarily apply the CLP Regulation during the transitional period, prior to the mandatory compliance deadline for mixtures of 1 June 2015.

1 More information on the GHS is available at www.unece.org/trans/danger/publi/ghs/ghs_welcome_e.html.
2 Regulation (EC) 1907/2006, as amended by Regulation (EU) 252/2011, Regulation (EU) 847/2012 and Regulation (EU) 848/2012.
3 Regulation (EC) 1272/2008 (as amended by Regulation (EU) 286/2011 to bring it in line with the 3rd edition of the GHS), amending and repealing Directives 67/548/EEC and 1999/45/EC, and amending Regulation (EC) 1907/2006.
4 The explanatory memorandum to CHIP4 is available at www.legislation.gov.uk/uksi/2009/716/pdfs/uksiem_20090716_en.pdf.

7.118 The aim of CHIP4 is to ensure that people who are supplied with chemicals receive the information that they need to protect themselves, others and the environment. To achieve this aim, the CHIP regime requires suppliers of chemicals to identify their hazards (for example, reactivity, toxicity) and to supply this information, together with advice on safe use, to the people to whom they supply the chemicals. Usually the most suitable means of doing this will be through package labels.[1] It also requires suppliers to ensure chemicals are packaged safely.

1 Safety Data Sheets (SDS) are no longer covered by CHIP4. The laws requiring SDSs to be provided have been transferred to the REACH Regulation (as amended), Art 31.

7.119 Whilst CHIP4 applies to most chemicals, it does not apply to all; cosmetics and medicines are outside the scope and are regulated by separate legislation. CHIP4 applies to the supply of chemicals – this is distinct from controlling dangerous chemicals in the workplace, or their transportation. 'Supply' means making the substance or preparation available to another person, including importation of the substance or preparation to Great Britain. 'Supplier' is construed accordingly.[1]

1 SI 2009/716, reg 2.

7.120 In a nutshell, CHIP4 is intended to protect both people and the environment from the harmful effects of dangerous chemicals, by ensuring that their users are supplied with adequate information about the dangers.[1] Changes to the classification of a chemical may make it subject to the regime under COMAH[2] or could trigger specific workplace controls under COSHH.[3] However, it should be remembered that CHIP4 deals with the packaging or labelling of chemicals, not with deciding if new chemicals on the market are dangerous. The REACH Regulation[4] is designed to deal with the effects of dangerous chemicals on human health and the environment, and is addressed at **7.133** below.

1 See HSE's CHIP homepage at www.hse.gov.uk/chip/index.htm.
2 SI 1999/743 (as amended).
3 SI 2002/2677 (as amended).
4 Regulation (EC) 1907/2006 (as amended).

Key goals of CHIP

Classification

7.121 The fundamental requirement under CHIP4 is to decide whether a chemical that is being supplied is hazardous. No dangerous substance or dangerous preparation may be supplied unless it has been classified in accordance with CHIP4, reg 4 and Sch 3.[1]

1 SI 2009/716, reg 4 and Sch 3.

Provisions of hazard information

7.122 Once the classification of the substance or preparation has been established, customers must be informed about the hazards and how they can use the chemicals safely through labelling,[1] and provision of a safety data sheet (SDS).

1 SI 2009/716, reg 7.

7.123 Labelling is governed by CHIP4, regs 7–10.[1] If a dangerous chemical is supplied in a package that package has to be labelled. Clearly, where chemicals are not supplied in a package but, for example, via a pipeline, labelling is not possible.

1 SI 2009/716, regs 7–10.

7.124 The purpose of the label is to inform anyone handling the chemicals about their hazards and to give brief advice on what precautions should be taken. CHIP4[1] gives specific instructions as to what has to go on the label and how packages should be labelled.

1 SI 2009/716.

Safety data sheets

7.125 In the past, manufacturers, importers, downstream users and distributors supplying substances meeting the criteria for classification as dangerous would have been required to complete an SDS under CHIP. Whilst CHIP[1] did not provide exact instructions as to what should go into an SDS, it did set a minimum standard which needed to be attained. In addition, it provided headings under which the information should be provided. REACH has now taken over the SDS system and introduced some changes which are detailed below. A supplier would need to provide an SDS if:

(1) it supplies a substance or a mixture that is either:
  (a) classified as dangerous under the Dangerous Substances Directive[2] or Dangerous Preparations Directive;[3] or
  (b) persistent, bioaccumulative and toxic (PBT), or very persistent and very bioaccumulative (vPvB), as defined in Annex XIII of REACH;[4] or
  (c) included in the European Chemicals Agency's 'Candidate List' of substances of very high concern for reasons other than (a) and (b) given here; or

(2) a supplier's customer requests an SDS for a mixture that is not classified as dangerous under Directive 1999/45/EC, but contains either:
  (a) a substance posing human health or environmental hazards in an

individual concentration of ≥ 1% by weight for non-gaseous mixtures or ≥ 0.2% by volume for gaseous mixtures; or

(b) a substance that is persistent, bioaccumulative and toxic (PBTs), or very persistent and very bioaccumulative (vPvBs) as defined in Annex XIII of REACH[5] in an individual concentration of ≥ 0.1% by weight for non-gaseous mixtures; or

(c) a substance on the 'Candidate List' of substances of very high concern (for reasons other than those listed above), in an individual concentration of ≥ 0.1% by weight for non-gaseous mixtures; or

(d) a substance for which there are Europe-wide workplace exposure limits, e.g. a substance that has indicative occupational exposure limit value (IOELV).

Although not required by REACH, suppliers to EU countries other than the UK may need to supply a SDS for mixtures that are not classified as dangerous but that contain substances with national workplace exposure limit values in other EU countries.

Suppliers do not need to provide a SDS:

(1) if they offer or sell dangerous substances or mixtures to the general public and they provide sufficient information to enable users to take the necessary measures as regards safety and the protection of human health and the environment, unless an SDS is requested by a downstream user or distributor;

(2) if the substances/mixtures are supplied in the UK and are not classified as dangerous; and

(3) for certain products intended for the final user, eg medicinal products or cosmetics.

Additional obligations apply for any member in the supply chain that deals with dangerous substances, PBTs or vPvBs, and which are produced or imported in volumes of ten tonnes or more a year. In accordance with the registration requirements of REACH,[6] exposure scenarios must be included in the chemical safety reports for these substances. Annex II to REACH now ensures that the relevant exposure scenarios must be annexed to the SDS for an applicable substance, forming extended safety data sheets (eSDS).

1 SI 2002/1689, as amended by SI 2005/2571 and SI 2008/2337.
2 67/548/EEC.
3 1999/45/EC.
4 Regulation (EC) 1907/2006 (as amended).
5 Regulation (EC) 1907/2006 (as amended).
6 See **7.136** below for more detail.

7.126 An important point to remember is that, whilst provision of an SDS is now covered by REACH,[1] an employer's use of SDS to carry out risk assessment and management is covered by workplace control law, such as COSHH.[2]

1 Regulation (EC) 1907/2006 (as amended).
2 SI 2002/2677 (as amended).

Packaging

7.127 The packaging requirements in CHIP are contained in regs 7–10.[1] The package must be suitable and there are particular labelling provisions for different preparations. There are requirements imposed in respect of the particulars to be shown on the labels for dangerous substances and dangerous preparations.[2] Special labelling requirements are also imposed in relation to labelling of single receptacles and receptacles in outer packages, as well as particular labelling requirements for certain preparations.[3] Furthermore, CHIP4 also imposes requirements in respect of the methods of marking and labelling packages that contain dangerous substances or dangerous preparations.[4]

1 SI 2009/716.
2 SI 2009/716, reg 7.
3 SI 2009/716, regs 8 and 9.
4 SI 2009/716, reg 10.

Consumer protection measures

7.128 CHIP4, reg 11[1] sets out certain consumer protection measures relating to the packaging of specified chemicals that are sold to the public. Packaging may not, for example, be of a shape or design that might mislead consumers or arouse the active curiosity of children. The packaging of some dangerous substances sold to the general public must have child-resistant fastenings, to prevent young children from opening the package. In addition, some must have a tactile danger warning to alert blind and partially sighted people that they are handling a dangerous product.

1 SI 2009/716, reg 11.

REGULATION IN THE CHEMICALS INDUSTRY AND THE CORPORATE MANSLAUGHTER AND CORPORATE HOMICIDE ACT 2007

7.129 The regulations that govern the use of chemicals in the workplace and the chemicals industry itself form part of a much wider framework of legislation covering health and safety at work. In many cases, compliance with the industry's specific regulations may help companies fulfil obligations imposed by regulations applicable to all workplaces.

7.130 Compliance with industry-specific legislation is increasingly important, following the introduction of the Corporate Manslaughter and Corporate Homicide Act 2007 (CMCHA 2007). As the CMCHA 2007 is dealt with elsewhere in this book, it is only briefly touched upon in this chapter.

7.131 The CMCHA 2007 statutory criteria[1] that a jury is required to consider, when deciding whether an organisation's conduct is a gross breach[2] of a relevant duty of care, mean it is even more important for chemical companies to ensure that they have robust health and safety management systems in place and that those systems are updated and implemented correctly. This is because the offence of corporate manslaughter is concerned not only with there being a system for managing activities within an organisation but also with the way in which those activities are carried out in practice. Juries are required to consider breaches of health and safety legislation in determining liability of companies and other corporate bodies for corporate manslaughter/homicide. Juries may also be asked to consider whether a company or organisation has taken account of any appropriate health and safety guidance and the extent to which the evidence shows that there were attitudes, policies, systems or accepted practices within the organisation that were likely to have encouraged any such serious management failure or have produced tolerance of it.

1 Corporate Manslaughter and Corporate Homicide Act 2007, s 8. This includes:
a) The extent to which the evidence shows there were attitudes, policies, systems or accepted practices within the organisation that were likely to encourage 'the relevant failure' or to have produced 'tolerance' of it;
b) any health and safety guidance that relates to the alleged breach; and
c) any other matter the jury considers relevant.
2 Corporate Manslaughter and Corporate Homicide Act 2007, s 1(1)(b).

7.132 Whilst CMCHA 2007 does not require chemical companies to comply with any new regulatory standards, it sharpens the focus on whether senior management in chemical companies are taking adequate steps to ensure that their organisation is complying with existing legal obligations in respect of health and safety. The CMCHA 2007 will inevitably prompt chemical companies to renew their existing systems, and it should ensure that they comply with industry-specific and more general workplace regulations and statutory duties.

REACH REGULATION: THE REVISED CHEMICAL STRATEGY

7.133 The European Commission made a commitment in 1998 to assess the operation of the four main legal instruments governing chemicals in the EC,[1] due to the failure of the previous system to ascertain sufficient information on chemical substances and also in response to concerns that legislation for chemicals did not provide sufficient protection in relation to the potential impact of chemicals on health and the environment. This culminated in the naming of the REACH[2] regime, which came into force on 1 June 2007. Under REACH, there is a gradual increase in obligations on the chemical industry up to 2018. The REACH Enforcement Regulations 2008 (RER)[3] help to implement REACH in the UK. RER came into force on 1 December 2008 and create the enforcement regime that is designed to fulfil the UK's obligations under art 126 of REACH (that is, to provide effective enforcement, a list of offences and set out the penalties for non-compliance).

1 Council Directive 67/548/EEC relating to the classification, packaging and labelling of dangerous substances; Council Directive 76/769/EEC on the marketing and use of certain dangerous substances and preparations; Council Directive 88/379/EEC relating to the classification, packaging and labelling of dangerous preparations; and Council Regulation 793/93/EEC on evaluation and control of risks of existing substances. These instruments have been repealed or superseded.
2 Regulation (EC) 1907/2006 (as amended).
3 SI 2008/2852.

7.134 Under REACH, legislation for new and existing chemicals is merged into one, with all chemicals manufactured at over one tonne being registered. Higher tonnage of manufacture will attract an increasing degree of testing. Chemicals of high concern will have to be positively listed, with manufacturer and use approved prior to marketing. Whilst there is a clear improvement for new substances, the proposal has the potential to be extremely burdensome for all other substances, due to proposed time scales imposed by the Regulation.

7.135 The three elements of the REACH system, to be applied to new and existing substances, are: registration; evaluation; and authorisation.

Registration

7.136 The threshold for registration of substances is the manufacture or import into the EU of one tonne per year (for manufacturers and importers) – which covers approximately 30,000 substances.

7.137 Manufacturers or importers must complete a registration dossier for each individual substance that is required to be registered under REACH. The registration dossier for each substance should include a technical dossier, which provides information such as the identity of the manufacturer/importer, the identity of the substance and guidance on its safe use. If more than 10 tonnes of that substance is manufactured or imported, the registration dossier should also include a chemical safety report including detailed information on the potential effects that the substance may have on the environment and the harm that it may have to human health.

7.138 The definition of 'substances' is purposefully wide, covering organic and inorganic substances and both consumer and industrial applications. Limited exemptions are set out in the legislation – for example, many radioactive substances are exempted; essentially, they concern substances already covered by equivalent legislation (eg EURATOM) elsewhere or low-risk substances that occur naturally.

7.139 There was a period for pre-registration of certain 'phase-in substances' that ran from 1 June 2008 to 1 December 2008. The substances were pre-registered at the European Chemicals Agency (ECHA). Phase-in substances are substances listed on the European Inventory of Existing Commercial Chemical Substances (EINECS), or that were manufactured in the EU between 1 January 1995 and 1 May 2004, that were not placed on the market by the manufacturer or that are 'no longer polymers'. The benefit of pre-registration was that companies that complied with pre-registration could benefit from extended registration deadlines. Companies that start manufacturing or importing substances after 1 December 2008 will still be able to pre-register from 5 January 2009 until 1 June 2018 (at the latest) under art 28(6) of REACH,[1] which states that first-time importers or manufacturers of phase-in substances can still pre-register up to 12 months before the relevant deadlines for registration as set out in art 23 of REACH[2] (these vary according to type and amount of substance that is manufactured or imported; substances amounting to 1,000 tonnes or more per year were required to be registered by 1 December 2010, 100 tonnes or more by 1 June 2013, and substances of 1 tonne or more need to be registered before 1 June 2018). Companies that already manufacture or import substances, which did not pre-register by 1 December 2008, are now required to submit a full registration dossier and cannot import or manufacture the substances until they do.

1 Regulation (EC) 1907/2006 (as amended), reg 26.
2 Regulation (EC) 1907/2006 (as amended), reg 23.

7.140 The first deadline for registration under REACH closed in 2010,[1] with 3,472 phase-in substances registered across the EU and almost 25,000 registration dossiers having been submitted. The UK made up 12% of these registrations, putting it in second place across the EU.[2] The closing date for the second stage of registration passed on 31 May 2013, with registration required for phase-in substances manufactured and imported between 100 and 1,000 tonnes per year. For a second stage running, the UK accounted for the second highest number of registrations.[3]

1 Regulation (EC) 1907/2006 (as amended), reg 23(1).
2 Information as at 28 February 2011. Figures taken from the Chemicals Industries Association (CIA) Associate Membership Briefing Seminar – REACH & CLP update (8 March 2011), by Silvia Segna (CIA Chemical Policy Adviser).
3 Information as at 4 June 2013. Figures taken from www.cia.org.uk/Newsroom/PressReleases/PressRelease/tabid/114/pwnid/148/Default.aspx.

Evaluation

7.141 The information provided by the registration process will allow the authorities to carry out more in-depth evaluations, assessing substances whose registration information suggests a potentially high risk. The evaluation process has two methods: dossier evaluation; and substance evaluation.

7.142 Dossier evaluation is the process by which all testing proposals that are submitted with a registration dossier are checked – the main purpose of this process has been cited as ensuring that no unnecessary animal testing is conducted. Five per cent of all registration dossiers will be subject to a full compliance check by the authorities.

7.143 Each year, EU Member States and the Commission will agree a list of substances which will require in-depth assessment. This substance evaluation will be carried out by competent authorities (such as the HSE). The outcome of this evaluation will be, if necessary, new control measures.

Authorisation

7.144 Authorisation will be required for all substances of 'very high concern'. This will cover approximately 1,500 substances. Substances of very high concern will include those that can be categorised as:

- CMRs: carcinogens, mutagens and those toxic to reproduction;
- BTs: persistent, bioaccumulative and toxic; and
- vPvBs: very persistent and very bioaccumulative.

For authorisation to be granted, the risks of a substance will have to be demonstrated as either 'under adequate control' or, if there is no suitable (and safer) alternative, on socio-economic grounds. The list of substances that should be considered for the authorisation process is published by the ECHA on its website[1] and, at the time of writing, contains 144 substances.

1 See http://echa.europa.eu/candidate-list-table.

7.145 This requirement demonstrates the tension in the new legislation between achieving a safer environment and maintaining a sustainable chemicals industry in the EU. Companies are being asked to find safer substitutes as part of the authorisation process, with any substitute having a lower risk level but also being sustainable from a technical and economic perspective.

7.146 The system is managed centrally through the ECHA, which is based in Helsinki. The ECHA evaluates the information provided by companies and the most hazardous substances.

7.147 The main objective of the revised chemicals strategy is to protect health and the environment and to stimulate innovation and the competitiveness of the chemical industry. It is intended to anticipate and avoid risks as far as possible, particularly with regard to chemicals that are not degradable and which accumulate in the human body and the environment.

7.148 The chemicals industry has more responsibility to prove that a substance is harmless and they will be responsible for the testing and risk assessment of chemicals. Eleven years after the legislation comes into force, all existing chemicals produced in volumes of more than one tonne per annum, approximately 30,000 substances, will be registered in a central database.

REACH Enforcement

7.149 On 1 December 2008, RER came into force in the UK. RER creates an enforcement regime for the UK, as requested by art 126 of REACH. RER sets out who the enforcing authorities are, what offences exist and the penalties for non-compliance under REACH.

7.150 In the UK, the principal enforcing authorities are as follows:

- the HSE (UK REACH CA);
- the EA, SEPA and NIA;
- local authorities (LAs)/Trading Standards (for consumer safety and health and safety); and
- the Department for Energy and Climate Change (DECC) (for offshore facilities).

7.151 Offences are set out in Pt 5 of the RER,[1] and include breaching a 'listed REACH provision' (these are listed in Sch 1 to RER) or causing or permitting another person to do so.[2] The maximum penalty for breaching a REACH provision, or breaching a provision under reg 13 or other offences, is a £5,000 fine and/or up to three months' imprisonment (in the magistrates' court) and an unlimited fine and/or up to two years' imprisonment (in the Crown Court). There are a number of other offences and penalties outlined in reg 13, which include providing false or misleading statements in purported compliance with a listed REACH provision, and not complying with the request of an enforcing authority.[3] Regulation 15 establishes that, where an offence is committed with the consent or connivance of an 'officer'[4] of that company relating to a listed REACH provision, or where the offence is 'attributable to any neglect on the officer's part', the officer, as well as the company, will be liable for the offence and could be punished accordingly. The term 'officer' includes directors, managers, secretaries or other persons acting in such a capacity.

1 SI 2008/2852.
2 SI 2008/2852, reg 11.
3 SI 2008/2852, reg 13.
4 SI 2008/2852, reg 15.

7.152 The enforcement data that is available from the HSE and EA to date provides the most accurate indication as to where the main breaches of the REACH regime have occurred in the UK. The HSE has thus far issued 121 improvement notices, the vast majority of which have related to the failure of duty holders to pre-register or register substances in accordance with reg 5 of REACH. In addition, 15 enforcement notices have been issued, along with one prohibition notice.[1] The EA has only issued one enforcement notice, but has made extensive use of information notices, with 57 being issued in the period 2011 to 2013 alone.[2] These enable the EA to obtain further information from the recipient, with the information provided then allowing the EA to decide whether further measures need to be taken.

1 Enforcement data accurate to 4 October 2013 and available at www.hse.gov.uk/notices/default.asp.

2 Enforcement data taken from the Environment Agency's Chemical Compliance Team Annual Enforcement Reports for 2011/12 and 2012/13, both of which are available at http://publications.environment-agency.gov.uk/.

Concerns about how REACH will affect the chemical industry

7.153 Virtually all manufactured articles depend on man-made chemicals, and there is the fear that REACH and the costs associated with it may adversely affect the EU manufacturing industry. EU finished products may cost more. Given that the obligations of REACH are being gradually phased in up to 2018, the real impact of REACH will not be apparent for a number of years. However, supporters of REACH argue that Europe will stand to gain a competitive advantage in the long term, for being the first to develop and market the new safer substitute products that REACH will encourage.

7.154 A 2012 baseline study on REACH, published by the European Commission, has confirmed that progress on two of the main goals of REACH is being made. The study confirmed that there had been a marked increase in the quality of publicly available data for the assessment of chemicals, as well as a better control of the risks posed by chemicals to both humans and the environment.[1] Many chemicals may be taken off the market because their turnover is insufficient to bear the costs. However, the proposal for reduced testing requirements for new substances at lower volumes may alleviate a barrier to innovation. The wide definition of 'downstream users'[2] in REACH means that the EU manufacturing industry will be heavily affected by REACH. UK manufacturers who use chemicals in the course of their 'industrial or professional activities' will need to be certain, whether by virtue of contractual guarantees or otherwise, that their suppliers have fulfilled the requirements of REACH. Otherwise, they face the prospect of enforcement action and the attendant costs.

1 'The REACH baseline study 5 years update' (Eurostat) (2012 edition) is available at http://epp.eurostat.ec.europa.eu/cache/ITY_OFFPUB/KS-RA-12–024/EN/KS-RA-12–024-EN.PDF.

2 Defined in Regulation (EC) 1907/2006 (as amended), reg 1 as 'any natural or legal person established within the Community, other than the manufacturer or the importer, who uses a substance, either on its own or in a preparation, in the course of his industrial or professional activities. A distributor or a consumer is not a downstream user'.

7.155 Following a recent review that was undertaken in order to identify ways in which the impact and burden of the REACH regime on SMEs could be reduced, the European Commission published a report in June 2013 recording its findings.[1] The report makes several recommendations to improve REACH implementation, including improving the quality of registration dossiers, encouraging companies to enhance the use of safety data sheets as a central risk management tool, and addressing issues related to the transparency of cost sharing within the Substance Information Exchange Forums (SIEFs). The report also noted that the reduction of the financial and administrative burden on SMEs would ensure the proportionality of legislation and the ongoing compliance with REACH obligations. In terms of enforcement, the report acknowledged that improvements could be made, and that this is the responsibility of the Member States to reinforce coordination among them. Although the report identifies the need for certain adjustments to current legislation, the Commission made it clear that it wants to guarantee legislative stability for European businesses. However, on 28 October 2013 the Prime Minister David Cameron made a speech to the House of Commons pledging to continue the deregulation of industries more generally in reaction to a report by the government's business taskforce which found that 'EU red tape' was costing UK businesses billions of pounds.[2] REACH was one of the regimes mentioned in the House of Commons, but concrete plans as to deregulation are not yet in place and, as such, no substantive alterations to REACH's main terms are proposed at the time of writing.

1 http://eur-lex.europa.eu/LexUriServ/LexUriServ.do?uri=CELEX:52013DC0049:EN:NOT.
2 See www.gov.uk/government/speeches/pm-statement-on-european-council-october-2013.

ENFORCEMENT TRENDS

General

7.156 Investigation and enforcement of the chemical industry's compliance with these laws are mostly carried out by the Hazardous Installations Directorate (HID), a specialist department of the HSE. This consists of three divisions: Specialised Industries; the Offshore Industry; and the Chemicals Industry.[1] The Chemicals Industry Division is responsible for the monitoring and regulation of health and safety standards for chemicals and also for inspecting relevant sites, the investigation of accidents or incidents, enforcement of health and safety law and assessing safety reports under COMAH[2] obligations. The Specialised Industries Division consists of four distinct business units, one of which is the Gas and Pipelines Unit (GPU). The GPU regulates the natural gas supply industry and industries transporting hazardous substances via pipelines, as well as other major hazards, such as gas explosives, biological agents and mines.

1 More information available at www.hse.gov.uk/hid.
2 SI 1999/743 (as amended).

7.157 The Energy Division (which subsumed the Offshore Safety Division (OSD) in April 2013, as part the HSE's reorganisation of its major hazard directorate) monitors the offshore oil and gas industry on the UK Continental Shelf (UKCS).[1] The creation of the Energy Division, which coincided with the 25th anniversary of the Piper Alpha disaster, attracted criticism from the National Union of Rail, Maritime and Transport Workers, who feared that the disappearance of the OSD would weaken offshore regulation. These concerns led to public reassurances being issued by the HSE, which stated that the creation of the Energy Division would lead to the existing offshore expertise being supplemented with knowledge acquired from other major hazard sectors.[2] The division's reaction to the Deepwater Horizon spill in April 2010 was notable, with the frequency of well control issue assessments being immediately increased for those members of the offshore oil and gas industry operating mobile drilling rigs on the UKCS. In addition, the Energy Division's priorities for 2012/13 included maintaining its focused inspection programme to ensure compliance with its verification scheme for well control equipment[3] and the creation of a blow-out prevention integrity management system, with structured inspections to test compliance. This emphasis on blow-out prevention will include inspecting:

(i) duty holders disconnect criteria for blow-out prevention;
(ii) duty holders blow-out prevention control systems; and
(iii) the competency systems for wells personnel to ensure that all well operators conducting well operations in the UKCS have adopted the

'Well Life Cycle Practices Forum competency guidelines' (or have an equivalent competence framework in place).[4]

The Central Division works in tandem with the Offshore and Land Divisions by providing support on matters such as policy, technical advice and finance.

1 More information is available at www.hse.gov.uk/offshore/index.htm.
2 The HSE's full response is available at www.hse.gov.uk/press/record/2013/pandjenergy020413.htm.
3 Pursuant to reg 19 of the Offshore Installations (Safety Case) Regulations 2005, SI 2005/3117. A review of the HSE database (as at 14 October 2013) revealed no prosecutions under the Offshore Installations (Safety Case) Regulations 2005.
4 Further information is available at www.hse.gov.uk/offshore/priorities.htm.

7.158 One piece of legislation proving to be problematic for the chemical industry is COMAH.[1] The majority of improvement and prohibition notices issued for failure to comply with COMAH have been due to the poor standard of MAPP documents, which must conclusively show that health and safety systems are adequate to limit the number and extent of accidents. There have been 216 improvement notices and 24 prohibition notices issued for COMAH-related offences between 2008 and June 2013.[2] Of these notices, many were attributable to breaches of regulation 4 of COMAH, which provides for a general duty on operators to 'take all measures necessary to prevent major accidents and limit their consequences to persons and the environment'. This was also illustrated by the average fine per conviction for the five proceedings brought in 2010/11 and 2011/12, which was £246,764.[3] In contrast, the average fine per conviction for the 20 proceedings brought for breaches of COSSH[4] in the same period was £3,573.

1 SI 1999/743, as amended by SI 2005/1088.
2 Enforcement data accurate to 4 October 2013 and available at www.hse.gov.uk/notices/default.asp.
3 Data available at www.hse.gov.uk/statistics/tables/index.htm#enforcement. It should be noted that this figure was largely influenced by the convictions in relation to the Buncefield incident which occurred in 2005, including the £1 million fine handed to Hertfordshire Oil Storage Limited, which is dealt with in more detail at **7.164** below.
4 SI 2002/2677 (as amended).

7.159 During 2011, there were 699 prohibition notices[1] issued for breaches of the Carriage of Dangerous Goods legislation. These were predominantly for failures to carry the requisite type of fire extinguishers, not having the necessary equipment to be used in the event of an emergency, or carrying chemicals that were inadequately marked. All relate to endangering safety in some way.

1 Data available at www.hse.gov.uk/enforce/cdg-notices.htm.

Prosecution

7.160 In line with the principle of proportionality,[1] the HID must use its discretion in deciding whether incidents need to be investigated, and the investigation resources tend to be focused on the most serious events. The HID may decide to prosecute without using any alternative measures, such as a formal caution.

1 The HSE's enforcement policy statement is available at www.hse.gov.uk/pubns/hse41.pdf.

7.161 An analysis of the HSE database for successful prosecutions reveals that there was only one prosecution of a chemical company in 2012, down from eight in each of the previous two years. The available data reveals that the more hefty fines over the past three years have been imposed predominantly for failure to control hazardous substances and inadequate risk assessment of employee practices, particularly in relation to the maintenance of machinery.

7.162 Whilst it appears that, in more recent years, the number of prosecutions brought by the HSE against companies in the chemical sector has gone down,[1] the average fines have remained relatively high.[2]

1 Prosecutions in 2010, 2011 and 2012 were 8, 8 and 1 respectively.

2 Highest fines in 2010, 2011 and 2012 were £330,000, £150,000 and £120,000 respectively, whilst the average fines were £51,623.13, £27,200 and £120,000 (for the single prosecution in 2012) respectively.

7.163 By way of example, in June 2005, Conoco Philips Ltd were fined £895,000 and ordered to pay £218,854 in costs after pleading guilty to breaching s 2(2) of the HSWA. The fine related to two incidents: a fire and an explosion that occurred in 2001 at the Humber Refinery in North Lincolnshire. More recently, Chevron and Total were prosecuted following the explosion at Buncefield in 2005 (see below).

7.164 The HSE and EA, as the 'Competent Authority' responsible for regulating non-nuclear major hazardous industrial sites in England and Wales under COMAH,[1] were in charge of the investigation into the Buncefield incident. After a lengthy investigation, which included the publishing of a report by the Buncefield Major Incident Investigation Board chaired by Lord Braintree,[2] the HSE and EA successfully prosecuted five companies for their roles in the incident. Notably, Hertfordshire Oil Storage Limited (owned by Total and Chevron) was fined £1,000,000 for breaching regulation 4 of COMAH, whilst the British Pipeline Agency Ltd (jointly owned by BP and Shell) were fined £150,000 for the same offence. In total, the fines and costs imposed came to £9.5 million.[3] One of the key findings of the joint investigation by the HSE and EA was that a culture had developed in which maintaining operations was seen as more important than safe processes and management, which did not get the attention, resources or priority that they required.[4] The report set out a number of important process management principles that had been reinforced by the incident:

- there should be a clear understanding of the major accident risks and the safety critical equipment and systems designed to control them;
- there should be systems and a culture in place to detect signals of failure in safety critical equipment and to respond quickly and effectively to them;
- time and resources for process safety should be made available;
- once the above is in place, there should be effective auditing systems in place which test the quality of management systems and ensure that these systems are actually being used on the ground and are effective; and
- at the core of managing a major hazard business should be clear and positive process safety leadership with board-level involvement and competence to ensure that major hazard risks are being properly managed.

1 SI 1999/743 (as amended).

2 Available at www.buncefieldinvestigation.gov.uk/reports/index.htm#final.

3 See www.hse.gov.uk/news/buncefield/index.htm.

4 'Buncefield: Why did it happen?', available at www.hse.gov.uk/comah/buncefield/buncefield-report.pdf.

Inadequate control of hazardous substance

7.165 The HSE report into the three incidents that occurred at BP Grangemouth in 2001 was published in 2003 and stated that the underlying failures were attributable to a number of weaknesses in the safety management systems on site.

7.166 A 1998 prosecution of ExxonMobil Chemical Limited resulted in a £30,000 fine and a heavy bill for costs imposed upon the organic basic chemicals manufacturer when inadequate isolation procedures resulted in 17 tonnes of flammable liquid being released unintentionally.[1]

1 See www.hse.gov.uk/enforce/prosecutions.htm.

7.167 In 1997 a £300,000 fine was imposed on chemicals giant BOC Gases Ltd after an investigation into a fatal explosion during the filling of compressed gas cylinders revealed that there was a lack of adequate risk assessments, unsafe plant, unsafe systems of work and inadequate training and supervision. Shell was fined £900,000 in April 2005 after two workers died in a gas escape on the Brent Bravo platform in September 2003.

7.168 More recently, in 2009 Millennium Inorganic Chemicals were fined £65,000 for a 2006 incident, in which titanium tetrachloride was released into the atmosphere during routine plant maintenance and formed a toxic vapour cloud.[1] In addition, Euticals Limited were fined £100,000 for the failure to comply with three improvement notices which had been issued by the HSE. After repeated visits by the HSE and an enforcement notice twice being extended to allow for compliance, the company was found to have failed to demonstrate that it had a sufficient understanding of the measures necessary to prevent major incidents and limit their consequences to employees and the wider environment, which included a failure to produce a suitable safety report, as required by COMAH.[2] The most recent chemicals industry prosecution led to the fine of an SAFC subsidiary, SAFC Hitech Limited, in February 2012. A mixture of Trimethylindium (TMI) waste in Isooctane originating from the TMI purification process was released from a glass container, leading to it spontaneously combusting on contact with an employee who sustained severe burn injuries. The company was found to have failed to complete a Dangerous Substances and Explosive Atmospheres Regulations 2002 (DSEAR) risk assessment and had endangered the safety of employees by having inadequate control measures in place to minimise the risk of exposure to TMI. Additionally, the investigation found that the company had inadequate safe systems of work, risk assessments, training, instruction and supervision.[3]

1 See www.hse.gov.uk/enforce/prosecutions.htm.
2 See www.hse.gov.uk/press/2012/rnn-w-archimica.htm.
3 See www.hse.gov.uk/prosecutions/case/case_details.asp?SF=CN&SV=4305566.

Accidents during machinery maintenance

7.169 The lack of adequate risk assessment identified in the BOC and SAFC Hitech prosecutions[1] is a common complaint against the chemical industry. The majority of prosecutions follow as a result of accidents in the workplace during machinery and plant maintenance.

1 See **7.167** and **7.168** above.

7.170 This is because such incidents attract a large amount of investigation resources and often expose great risks and breaches of health and safety legislation. The HID is under an obligation to conduct a full site investigation of a work related death in nearly all circumstances. The circumstances of the case may even justify a charge of manslaughter, which the police will be responsible for pursuing.

7.171 The number of incidents where limbs or digits have been severed – due to a combination of unsafe working practices and insufficient training of employees – are high, yet these incidents attract a lower fine upon successful prosecution. For example, in March 2013, London-based chemicals company Nuplex Resins Limited were fined £20,000 after an employee was run over by a forklift truck and had to have part of his leg amputated.[1] In 2009, BASF Construction Chemicals were fined £12,000 following an accident which led to one of their employees requiring his thumb and fingers to be amputated. In 2007, Epichem Ltd were fined £20,000 following two explosions at their premises in Bromborough which injured two employees. Investigations after employee deaths, however, have tended to reveal gross health and safety breaches deserving of the larger fines. An example of this occurred in 2002, where Klargester Environmental were fined £250,000 after a worker was killed when his head was crushed in a piece of specialist machinery. The judge in the case said 'that there were multiple failings, some of which had been present for a long time, some for many years'.[2] This example shows that companies should not be complacent about their health and safety duties just because they have not had any major incidents. The courts have made it clear that they do not look kindly upon long periods of non-compliance with health and safety requirements resulting in accidents.

1 See www.hse.gov.uk/press/2013/rnn-ldn-6913.htm.
2 See www.hse.gov.uk/press/2002/e02062.htm.

The effect of enforcement

7.172 The Chemical Industry Association (CIA) has adopted the Responsible Care programme[1] to encourage improvement in the area of safety, health and environment. The accident statistics gathered by the CIA show a 40 per cent drop in accidents between 1990 and 2005, and the International Council of Chemical Associations' (ICCA) 2012 Responsible Care Progress Report shows a downward trend in RIDDOR major injuries, dangerous occurrences and 'over three day' lost time accidents within the chemicals industry, from 2000 to 2010.[2]

1 See www.cia.org.uk/ResponsibleCareRoot/ResponsibleCare.aspx.
2 ICCA Responsible Care Progress Report 2012, available at www.icca-chem.org/ICCADocs/RC%20annual%20report.pdf.

7.173 Other statistics gathered by the CIA indicate that the number of RIDDOR[1] reportable major injuries decreased from 59 in 2004 to 28 in 2005. Thus, the number of health and safety incidents in the chemicals industry are on the decline.

1 SI 1995/3163.

7.174 Whilst HSE investigations and enforcement action can result in large fines and imprisonment, they have ensured that companies take the health and safety of their employees seriously and place it high on a company's agenda. Investigations may be prompted by what an inspector finds on a routine visit, by complaints or by a reported incident. However an investigation originates, there is no escaping the fact that a health and safety offence is a criminal offence which can cause a stain on a company's reputation and may also result in an individual being prosecuted.

RISK ASSESSMENT – ANALYSIS AND RISK MINIMISATION

Management systems, learning lessons and the corporate memory

7.175 A safety management system[1] is necessary to ensure that lessons are learnt from accidents that occur not only within a chemical company but from those that occur in other companies operating the same equipment or process. Any management system must cover procedures for learning lessons when developing the concept of the chemical plant, through design, operations, maintenance and inspection to emergency operations and demolition.

1 B Mellin and J Bond, 'Learning Lessons from Accidents – The Problems facing an Organisation' (2000) Hazards XV, The Institution of Chemical Engineers Symposium Series no 147.

7.176 Learning the lessons from accidents with chemicals or chemical equipment, from within an organisation and from others outside the organisation, is a key requirement for achieving continuous improvements in safety. Unless such a commitment is forthcoming, the enhancement of corporate knowledge in the chemical industry will be limited.

7.177 To be effective, health and safety policies need to demonstrate that health and safety is an integral part of business performance and efficiency. The policy statement must reflect a commitment to learn the lessons from accidents and hence produce safer processes, safer designs, safer operations, safer systems of work and a more informed, competent and knowledgeable workforce.

7.178 Corporate memory can be captured and maintained within an establishment, but it needs to be augmented by the experience of others, gleaned from an accident database. Of course, in light of the new corporate manslaughter offence, failure by a corporate to act upon learning of a failure in its health and safety obligations may render a management failure a gross one.

7.179 To ensure that this enlarged experience is regularly tapped, a management system has to be devised and used. As with all safety matters, commitment starts at the top, and the management system described above is essential for maintenance of the corporate memory.

Management of change

7.180 The modification of a plant may well introduce new hazards and, unless these are identified, problems may well be introduced. At Flixborough, a modification was made to an existing plant in order to overcome a production problem. The result was a vapour cloud explosion that killed 28 people, destroyed the plant, and damaged many houses surrounding the works.

7.181 A management system for any change is required to control the change, so that the authorisation for the change or modification is not given until all possible hazards are considered and the risk assessed. This includes a check against the original design intent.

7.182 Changes require a careful consideration of all circumstances to identify any new hazard and the risk established. The change/modification must only be authorised when the risk has been fully examined and reduced to an acceptable level. All documents must be kept with the details of the design.

7.183 A management system to control a change/modification is essential and should be part of the safety manual.

Inspection system

7.184 Inspection of equipment, both internally and externally, is an essential part of maintaining the integrity of the process. A management system describing and laying down the frequency and conditions for the inspection is necessary to obtain a quality system.

7.185 Inspection of equipment internally and pressure testing is a vital part of ensuring the integrity, and hence safety, of the equipment. Some of these inspections are a requirement of statutory regulation, and some a requirement of company or industry standards.

7.186 A comprehensive list of equipment should be maintained and detailed internal inspection records kept. From the results of the inspection, the period to the next inspection can be set (unless covered by a statutory inspection requirement).

7.187 It is important to ensure the competency of those who are carrying out the inspection system and to see that records are properly kept for all equipment and their state at each inspection. Recommendations from the inspectors should be recorded. Signatures for each inspection and recommendations must be obtained, and the reports passed to a senior person in the organisation who can authorise the necessary repair or maintenance work.

7.188 A follow-up procedure for all recommendations should be made to ensure that no recommendation is forgotten.

Emergency plans and crisis management

7.189 An emergency starts the moment a person discovers something significantly unusual, whether it be a fire on the plant, a problem on a ship or a drum at the roadside. The actions that follow the start of the emergency will determine how soon the problem is cleared up or whether it escalates. Emergency instructions local to the event will be activated to deal with the situation and, if these are followed, there is a good chance that that will be the end of the problem. However, in a number of cases it is not and, as the local emergency starts to escalate, the problem becomes greater than the local team can manage. It is important at this stage that the company representatives can handle the emergency so that it does not get out of hand.

7.190 Crisis management builds upon the philosophy of emergency management, to address the full range of crises that can strike at all organisational levels of a company, and to manage the consequences of those crises. A crisis management programme will apply to those plant emergencies that might create a crisis for the company, as well as the range of other types of potential crises (such as financial, legal or communication problems, company image and reputation, natural disasters, product hazards and changes in political or regulatory climate).

7.191 Crisis management is a broader strategic effort, of which the tactical approach of emergency management is an element.

7.192 It is important to recognise that the scale of an incident is not the sole determinant of its potential to develop into a crisis. Other factors could come into play (for example, communications, political environments and recent similar incidents).

7.193 Crisis management can generally be defined as those activities undertaken to anticipate or prevent, prepare for, respond to and recover from any incident that has the potential greatly to affect the way a company conducts its business. Rather than being a reactionary process, crisis management encompasses several proactive activities. It is important that the management system is detailed and training carried out to ensure its effectiveness. Dummy runs should also be organised.

PROFESSIONAL RESPONSIBILITY IN THE CHEMICAL INDUSTRY

General duty of care

7.194 Every person employed in the chemical industry has a duty of care, under s 7 of the HSWA 1974, to:

> '(a) take reasonable care for the health and safety of himself and other persons who may be affected by his acts or omissions at work, and
> (b) as regards any duty or requirement imposed on his employer or any other person by or under any of the relevant statutory provisions, to co-operate with him so far as is necessary to enable that duty or requirement to be performed or complied with.'

7.195 Chartered members of the chemicals industry not only have this duty of care under the HSWA 1974 but also have duties laid down by virtue of membership of their professional body, such as the Institution of Chemical Engineers (IChemE). These professional bodies encourage professional development and usually operate a recording system to ensure that professional development includes safety matters. They also provide safety information, and it could be argued that a professionally qualified person should be a member of their professional body in order to obtain this information to discharge their duty of care.

7.196 The chemicals industry professional must use all resources that he or she considers appropriate in order to reduce the risks to him or herself, to others at his or her place of work and to the public at large. If these resources are withheld or not available, this should be drawn to the attention of his or her manager.

Company culture

7.197 The prime concerns of any organisation are:

- to produce a quality product or service that satisfies the customer;
- to operate efficiently so that there is an adequate return on the capital involved; and
- to provide a safe system of work that minimises the impact on the environment.

7.198 The emphasis on each of these concerns will be dependent upon various factors, including the potential to cause harm. The amount of time, money and effort spent on these primary areas of concern has to be judged by the management. The operations in the chemical industry by their nature require a high level of money and effort to maintain a satisfactory level of safety and environmental protection. The successful organisation is the one that gets the blend of these correct. Organisations have to remember that, as a result of their operations:

- at risk are the employees, the community and the customers; and
- the audience is the authorities, the pressure groups, the media, other companies and the trades unions.

7.199 The clear responsibility for all this lies with the management of the organisation, and the chief operating executive as the head of the management team. Safety, occupational health and environmental issues have, in the past, been based on reacting to events, but the more modern concept of loss prevention is concerned with preventing the loss of people or their services, of equipment or of material. It incorporates safety, occupational health and environmental issues and is concerned with preventing an unwanted event from happening. Loss prevention is thus a proactive approach to accident prevention and must be part of the policy of the organisation.

7.200 An organisation will reflect its culture in its commitment:

- to the customer, the employee and the community by the senior management team;
- to the importance of human factors, covering the job, the individual and the organisation;
- to measuring the performance of individuals against standards; and
- to sharing accident information to prevent a recurrence.

This is increasingly important, given the fact that juries can make reference to an organisation's 'attitudes, notices, systems or accepted practices' when looking at relevant failures for the purposes of CMCHA 2007 (see **7.129** above). The importance of board-level awareness and engagement in both a company's safety culture and operational safety was heightened by the findings of the investigation into the Buncefield disaster (see **7.164** above). The report into the disaster specifically stated that board-level visibility and promotion of process safety leadership was essential to enshrining a positive safety culture throughout an organisation.[1]

1 'Buncefield: Why did it happen?', available at www.hse.gov.uk/comah/buncefield/buncefield-report.pdf.

7.201 The Institute of Directors (IOD) and HSE have set out, in their guidance note 'Leading Health and Safety at Work',[1] to clarify what directors should do, both individually and collectively. The guidance aims to ensure that directors are aware of their health and safety responsibilities and how best to ensure good health and safety performance in these organisations. The guidance comprises:

- an introduction to health and safety at board level;
- a four-step agenda (care actions/good practice);
- lay questions for the board; and
- an outline of legal liabilities.

1 IOD/HSE Guidance, available at www.hse.gov.uk/pubns/indg417.pdf.

7.202 The guidance is likely to be regarded as highly authoritative by the courts when a director or manager is charged with a breach of s 37 of the HSWA 1974 or where an organisation is charged with corporate manslaughter or other serious safety offences. The guidance should therefore be given serious attention at board level.

7.203 Historically, there have been very few successful prosecutions under s 37 of the HSWA 1974. This is thought to be because of the difficulty in prosecuting the offence and proving that the director knew of reasonably practicable steps that he could have taken, but chose not to take. The Court of Appeal case of *R v E* (2007)[1] held that the standard used to determine neglect has been too high and sets a new test to determine the correct standard of proof found on the following:

- What was the state of fact that required action to be taken?
- Was the individual director aware or ought he/she to have been aware of those facts?
- What appropriate investigation ought the director to have made to establish those facts?

It is believed that this case will make it easier to bring prosecutions under s 37 of the HSWA 1974 against directors and senior managers.

1 *R v E* [2007] EWCA Crim 1937.

7.204 Furthermore, with the advent of the Health and Safety (Offences) Act 2008 (HSOA), which came into force on 16 January 2009 and which brings increased levels of fines and prison sentences for health and safety offences in both the magistrates' and crown courts, attention to health and safety issues is likely to climb higher on the boardroom agenda.

CONCLUSION

7.205 While management systems and auditing of chemical plants are well-established, it is important that these are maintained at all times. Other important areas that must be established and recognised are learning lessons from other accidents and recognising the responsibilities of engineers and scientists. This is especially important, given the advent of the CMCHA 2007 and the matters which can be considered by a jury when deciding whether there is enough evidence to find an organisation guilty of corporate manslaughter.

7.206 As the 21st century advances and the pace of technological innovation quickens, it can be expected that fresh hazards will present themselves. There will consequently be a greater onus on the professional to keep abreast of new developments in his or her sphere of expertise and to investigate the safety implications of them.

Chapter 8

Healthcare

David Firth, Partner, Capsticks Solicitors LLP

INTRODUCTION

8.1 The provision of healthcare, whether by National Health Service (NHS) bodies or within the private and voluntary sectors, is an inherently risky undertaking in which unexpected deaths can and do occur. Criminal sanctions have been imposed in a number of instances, either against individuals for the common law offence of gross negligence manslaughter, or against undertakings for offences under the Health and Safety at Work etc Act 1974 (HSWA 1974). No convictions were secured against healthcare providers for the common law manslaughter offence but, with the change in law with effect from 6 April 2008, the prospects of a conviction under the Corporate Manslaughter and Corporate Homicide Act 2007 (CMCHA 2007) have increased significantly, if only in contrast to the corporate manslaughter law as it stood before that date.

8.2 The way in which the corporate manslaughter offence has been defined in the CMCHA 2007, and the potential difficulties for the prosecutor in establishing, variously, senior management involvement to the required level, causation, and the absence of significant intervening events in the lead-up to any fatal incident, means that corporate manslaughter convictions will not be easily won by the prosecutor. That has not, however, prevented a 40 per cent increase in the number of corporate manslaughter cases opened by the Crown Prosecution Service (CPS) in the last 12 months (63 in 2012, up from 45 in 2011).[1] Any such prosecution is likely to be vigorously contested: not only does the CMCHA 2007 afford, by contrast with HSWA 1974 prosecutions, a greater prospect of achieving a successful defence, but also the indirect consequences of conviction – such as the damage to reputation in an increasingly competitive healthcare market – will provide added impetus to defending cases brought under the CMCHA 2007.

1 'Corporate manslaughter cases soar', (2013) Daily Telegraph, 28 January.

8.3 Before any prosecution has been commenced, however, the principal impact of the CMCHA 2007 has been to prompt an urgent review of the way in which risk is identified and managed. Where such reviews have not been undertaken, or completed, healthcare providers and commissioners of healthcare services should do so as a matter of priority. Poorly managed organisations which fail to address the risks associated with their activities, and whose management of risk falls far below the standard that can be expected in the circumstances, increase the prospect of a corporate manslaughter conviction, alongside existing criminal sanctions under health and safety legislation, in the event of an unexpected fatality. This increased focus on the management of risk has been evidenced by the corresponding increase in the number of HSWA 1974 prosecutions against healthcare bodies, including where the circumstances of the fatal incident do not merit a corporate manslaughter prosecution.

THE NATIONAL HEALTH SERVICE – STRUCTURE AND ORGANISATION

8.4 The process of identifying and managing risk is a particularly hard task for the NHS, which must be seen more as an ecosystem than an organisation due to the myriad individual bodies it comprises. Its scale, complexity, and the nature of work it undertakes and the near-constant evolution it has undergone since its inception on 5 July 1948, mean that risk within the NHS is difficult to quantify and manage. From the outset, as a system the NHS has been responsible for a huge workforce (1.4 million in England alone) and budget (£106 billion), while facing increasing challenges in organising and managing the demands and expectations of patients, their carers and staff.

8.5 The National Health Service and Community Care Act 1990 became law on 1 April 1991 and created the 'purchaser–provider' split. This division has been amplified over the past 20 years, culminating in the most significant evolution of this division being recently introduced by the Health and Social Care Act 2012 (the H&SCA). The split was designed to ensure that those planning, paying for and monitoring the care needed by particular populations were separate from those tasked with delivering that care.

The governance arrangements for the National Health Service changed after the H&SCA came fully into force on 1 April 2013. Up to that date, the Department of Health was directly responsible for the NHS, with the Secretary of State for Health as the head of that Department and reporting to the prime minister. Formerly, the Department of Health controlled England's 10 Strategic Health Authorities (SHAs), themselves further 'clustered' (effectively, amalgamated) prior to abolition, which oversaw all NHS activities in England. In turn, each SHA supervised all the NHS trusts in its area (these being Primary Care Trusts, and both mental health and secondary care NHS Trusts). A distinction must been drawn for NHS Foundation Trusts (which may provide community, mental health or acute services), which, having gained a degree of independence from the above management structure, are authorised and regulated from a governance and financial perspective by Monitor (see below), the independent regulator.

However, in terms of the quality and safety of care across the NHS, this is regulated by the Care Quality Commission (CQC), a non-departmental public body.

The devolved administrations of Scotland, Wales and Northern Ireland run their local NHS services separately.

8.6 From 1 April 2013, NHS England has taken on many of the functions of the former primary care trusts with regard to the commissioning of primary care health services, as well as some nationally based functions previously undertaken by the Department of Health. It oversees all care commissioned either by itself or through 2012 GP-Led Clinical Commissioning Groups organised geographically to cover all the population within England. NHS England's powers derive from its 'mandate' from the Department of Health, but it remains a non-departmental government body maintaining a distance from the Department of Health and ministerial intervention.

NHS England is responsible for primary care contracts and has a duty to commission primary care services in ways that improve quality, reduce inequalities, promote patient involvement, and promote more integrated care.

Monitor's role has been extended from 1 April 2013 to become the regulator of all NHS-funded care, including: regulating prices; enabling services to be provided in an integrated way; safeguarding choice and competition; and supporting commissioners so that they can ensure essential health services continue to run if a provider gets into financial difficulties. Monitor will continue to ensure that the boards of NHS foundation trusts focus on good leadership and governance, in line with their duty to be effective, efficient and economic. It is anticipated that Monitor will work far more hand in glove with the CQC going forward.

8.7 Local authorities are already playing an increasing role in the planning and purchasing of health care services and, in particular, in the preventive public health arena. The H&SCA created a new body, Public Health England, and new Directors of Public Health within local Authorities, and transferred public health functions to local authorities on 1 April 2013.

8.8 The 2006 report by the Chief Medical Officer[1] identified the greater emphasis by the NHS on the quality and safety of care since the turn of the century. The summary to the report[2] quoted the creation of a legal duty of quality for all NHS organisations, clear national standards, clinical governance arrangements, inspection against national standards in hospitals and primary care services, the establishment of a national patient safety programme, including adverse event and near miss reporting, and a range of measures to empower patients and their representatives. A specific service to support the NHS in assessing and dealing with concerns about the performance of doctors, the National Clinical Assessment Service, has also been established.

1 'Good doctors, safer patients: Proposals to strengthen the system to assure and improve the performance of doctors and to protect the safety of patients'.

2 Report, para 11, 'the NHS Quality Landscape' (DH Publications).

8.9 In summary, healthcare organisations need to appreciate the role(s) they perform in order to identify and manage the risks that their activities create. The experience of frequent and significant restructuring certainly does not provide immunity from prosecution, and recent history demonstrates that prosecutions of NHS organisations, principally by the Health and Safety Executive (HSE) but also by the CPS, are becoming increasingly common. The introduction of the corporate manslaughter legislation may see that trend continue, but with more severe consequences, in the event of a conviction for the corporate manslaughter offence under the CMCHA 2007.

CRIMINAL SANCTIONS FOLLOWING UNEXPECTED DEATHS IN THE HEALTHCARE SECTOR PRE-2008

Manslaughter prosecutions in the sector restricted to individuals

8.10 In the context of work related deaths in the healthcare sector, criminal liability to date has focused on the person(s) directly involved in the incident leading to death. Prosecutions against individual clinicians for the common law offence of gross negligence manslaughter have increased substantially since 1989, albeit with a significantly lower conviction rate than in cases not involving clinicians.[1] If the public truly wish to punish doctors who make fatal errors, and there is scant evidence that this is what they want, the gross negligent manslaughter charge has arguably proved to be a blunt and inefficient way of achieving that aim.

1 Of the six doctors prosecuted between 2000 and mid-2002, for example, one was found guilty. By contrast, there were 278 prosecutions for the same involuntary manslaughter offence outside the healthcare sector, with 238 convictions and only 40 acquittals.

Corporate criminal charge – health and safety offences

8.11 A preliminary hearing in the trial following the 1997 Southall Rail Crash established, in principle, that an organisation can commit manslaughter[1] although no such charge was alleged, still less proven, against an NHS body under the law in place before the CMCHA 2007 came into force. Public sector bodies were certainly not immune from prosecution under the former corporate manslaughter law and where poor management has been suspected as a contributory cause of fatalities, as the (failed) manslaughter prosecution brought against Barrow in Furness Council[2] demonstrates. However, perhaps due in large part to the shortcomings of the involuntary manslaughter offence, and the well-publicised difficulties in securing a conviction against larger undertakings, corporate criminal sanctions have mostly been limited to charges under the HSWA 1974, typically prosecuted by the HSE but, on occasion, by the CPS.[3]

1 *R v Great Western Trains* – Scott Baker J's ruling approved by the Court of Appeal in *Attorney General's Reference (No 2 of 1999)* [2000] QB 796.
2 *R v Beckingham* [2006] EWCA Crim 773.
3 The HSE is the authority that enforces health and safety law in the UK in relation to health services and other authorities. Whilst the police would lead any investigation into a major incident involving the NHS and decide whether or not to bring charges against the organisation or an individual, the HSE will be involved in the investigations with a view to determining the root cause of the incident.

8.12 Numerous fatal incidents over recent years, involving allegations of gross carelessness by individual doctors, help to highlight two main facts. First, that clinical intervention can be very complex and inherently dangerous. Secondly, questions are being asked more frequently about the structures and systems in place to address the risk of error by the clinician.

Involuntary manslaughter to 1989

8.13 Up until the 1990s, it was rare for charges of manslaughter to be brought against doctors. Cases such as *R v Doherty*[1] and the 1925 case of *R v Bateman*,[2] who was convicted of manslaughter (but pardoned on appeal) after an obstetric patient died, were based on the principle that, in order for an act of clinical negligence to be 'gross' and meriting punishment, a doctor had to show a total disregard for the life and safety of others..

1 (1887) Cox CC 306, 14 Digest (Rep) 71, 327.
2 (1925) Cr App R 8.

Since 1990

8.14 Things have certainly changed. Writing in the *British Medical Journal* in November 2000,[1] Dr Robin Ferner found only seven prosecutions against doctors for gross negligence manslaughter in the 120 years to 1989. The position altered radically after that date and, since 1990, there have been in excess of 30 prosecutions against doctors in the UK for manslaughter. This major change has contributed to further debate on what constitutes 'gross negligence' and has highlighted that those at one step removed from the actions or omissions of the doctors – typically, their employing organisations – may properly bear criminal liability. However, due to the problems in proving the offence of gross negligence manslaughter against any but the smallest undertakings, the shortcomings of the then manslaughter law were brought into sharp focus.

1 BMJ 2000;321:1212–1216 (11 November).

8.15 The reality is that, where death does occur at the hands of a healthcare professional, the death is not necessarily as a consequence of any 'disregard' on the part of that individual, but may be due to inadequate or failed systems within the organisation employing that professional or otherwise responsible for his or her work. Where 'systems failures' in particular have been shown to be the cause of the death, it has been very difficult to pursue a successful prosecution for gross negligence manslaughter, as was demonstrated in the case described below.

Involuntary manslaughter cases – not always the individual's fault

8.16 In 1999, a 12-year-old boy with T cell non-Hodgkin's lymphoma was admitted to a general paediatric ward instead of a paediatric oncology ward because the paediatric oncology ward was full. His lumbar puncture procedure had been postponed and was consequently to be undertaken by a registrar in paediatric anaesthesia who had never given intrathecal cytotoxic treatment before, although he had discussed the procedure by telephone with his haematological colleague. Unaware that vincristine should not be taken into theatre, a nurse provided the paediatric anaesthetist with two syringes, one of which contained vincristine and bore a label indicating that the drug was for intravenous use only. Without reading the label, the paediatric anaesthetist administered the drug intrathecally, in consequence of which the boy died some days later. The doctors were formally cleared of manslaughter charges after the CPS were unable to offer any evidence of gross negligence on their part, and the evidence showed that significant systems failures within the hospital administration were significant factors in the boy's death.[1]

1 Report of an expert group learning from adverse incidents in the NHS, chaired by the Chief Medical Officer, 'An Organisation with a Memory' (2000, The Stationery Office).

Convictions against individuals for gross negligence manslaughter

8.17 Systemic failures within an organisation have not, however, precluded successful prosecutions against individuals. In *R v Adomako,*[1] a case involving anaesthesia, an endotracheal tube (ET tube) was inserted to enable a patient to breathe by mechanical means. Approximately one hour later, the ET tube became disconnected, depriving the patient of oxygen and causing cardiac arrest some ten minutes later. During this period, Mr Adomako, the anaesthetist, failed to notice or remedy the disconnection, only becoming aware that any problem had arisen when a blood pressure alarm sounded. On the evidence, it appeared that some four-and-a-half minutes would have elapsed between the disconnection and the sounding of the blood pressure alarm. On hearing the alarm, Mr Adomako checked the equipment and administered atropine to raise the patient's pulse, but at no stage before the cardiac arrest did he check the integrity of the ET tube connections. The disconnection itself was not discovered until after resuscitation had been commenced.

1 [1995] 1 AC 171.

8.18 The prosecution's case was that the anaesthetist was guilty of gross negligence arising from the failure to notice or respond appropriately to what were obvious signs of a disconnection.[1] The prosecution also alleged that he had failed to notice at various stages during the period, after the disconnection and before the arrest either occurred or became inevitable, that the patient's chest was not moving and the dials on the mechanical ventilating machine were not operating. It was accepted that the anaesthetist had been negligent, but the question then arose as to whether his actions constituted a crime.

1 *R v Adomako* [1995] 1 AC 171.

8.19 The jury felt they did, and found him guilty of manslaughter; a six-month prison sentence was imposed, suspended for 12 months. In dismissing his subsequent appeal, the House of Lords held that: the ordinary principles of the law of negligence applied to ascertain whether the defendant had been in breach of a duty of care towards the victim; on the establishment of such breach of duty, the next question to ask was whether it caused the death of the victim and, if so, whether it should be characterised as 'gross negligence' and therefore a crime.[1] It was up to the jury to decide whether, having regard to the risk of death involved, the conduct was so bad in all the circumstances as to amount to a criminal act or omission. They clearly thought so in this case.

1 *R v Adomako* [1995] 1 AC 171.

8.20 Arrogance was identified by the judge in *R v Ramnath*[1] as the defining error by an NHS doctor when she injected a patient with adrenaline at Stafford District General Hospital in July 1998, hastening the death of a 51-year-old patient. Dr Priya Ramnath, who was found guilty of manslaughter by gross negligence, had chosen not to listen to the advice of a senior nurse working alongside her, as well as two other colleagues, and had administered the adrenaline without speaking to a consultant anaesthetist.

1 Birmingham Crown Court, reference T20087026.

Convictions outside the acute sector

8.21 Unexpected deaths and ensuing prosecutions have not been confined to incidents arising at acute hospitals. A general practitioner was sentenced to

12 months in prison, suspended for 12 months, after he pleaded guilty to manslaughter following the death of a nine-year-old boy who remained unconscious and suffered irreversible brain following a circumcision, at which time diamorphine was administered as a sedative. The general practitioner admitted to using 'excessive amounts of … drugs, and it was accepted that they were wholly inappropriate as sedatives'.[1]

1 'Doctor admits killing' (1995) Guardian, 7 March.

8.22 Three years later, another general practitioner was convicted and sentenced to 12 months' imprisonment, suspended for two years, following the death of a 41-year-old woman.[1] The GP in this instance had been charged with administering an amount of drug (pethidine) which amounted to gross negligence. The patient in question had been seen by the GP for severe migraine. Whilst prochlorperazine and diazepam had stopped her vomiting, they had not relieved her pain. At the suggestion of the patient's husband (a consultant surgeon), the GP sought to administer pethidine. As the pharmacy had no pethidine, the GP obtained an ampoule of diamorphine containing 100 mg and administered the whole dose intramuscularly. Whilst the dosage was reasonable for pethidine, it was ten times too high for diamorphine and the woman went to sleep and died within the hour.

1 Stokes P, 'Suspended jail term for doctor in lethal injection case' (1998) Daily Telegraph, 28 November.

Systems failures

8.23 Arguably, and whether or not gross negligence was established, the cases referred to above properly confine criminal liability to the clinician whose actions or omissions were a substantial contributing factor to the unexpected death. '*Vincristine* cases' (such as that following the tragic death of Wayne Jowett) may be distinguished, in the sense that systems failures are more evident, if not the principal cause of death.

8.24 In September 2003, Dr Mulhem, who had mistakenly ordered the wrong chemotherapy drug to be injected into a cancer patient's spine, pleaded guilty to manslaughter at Nottingham Crown Court and was subsequently sentenced to eight months' imprisonment.

8.25 The facts are relatively straightforward. Wayne Jowett,[1] a teenager, had been diagnosed with leukaemia at the age of 16 but was in remission at the time of the incident. Dr Mulhem, who was only three days into his first specialist registrar appointment, supervised the injection, which was given by a junior doctor who was also new to the ward. The prosecution submitted that it was the specialist registrar's job to check the route of administration and the syringe, and had he done so he would have realised that the drug was to be injected into a vein. The prosecution also submitted that his conduct fell far below that which could be expected of a competent doctor in his field, and that the doctor had failed in a number of respects which were absolutely basic, and that these failures led to Wayne Jowett's death.

1 (2003) BMJ, 27 September.

8.26 It is significant that an independent report[1] into the circumstances leading up to Wayne Jowett's death criticised procedures at the Queen's Medical Centre in Nottingham and highlighted design faults in syringes. The report concluded that Wayne Jowett died as a result of a 'complex amalgam of human organisational, technical and social interactions' at the hospital where he received the injection. Furthermore, 'the safety culture surrounding Ward E17's patient chemotherapy supply and administration system [did] not appear to be all that it should'. It was said that the inquiry undertaken identified 'classic systems failures'. Such failures can attract criminal sanctions to the employing organisation, under health and safety law (see below), and are likely to be subject to very close scrutiny with the CMCHA 2007 now in force.

1 External inquiry into the adverse incident that occurred at Queen's Medical Centre, Nottingham, 4 January 2001 by Professor Brian Toft.

THE CURRENT AND ONGOING CONNECTION BETWEEN HEALTH AND SAFETY LEGISLATION AND DEATHS IN THE HEALTHCARE SECTOR

8.27 It has been frequently stated that the CMCHA 2007 imposes no new duties and merely introduces a new offence. Of the existing duties, the requirement to have proper regard for health and safety is fundamentally connected to criminal liability (or lack of it) for corporate manslaughter in its new form. For healthcare bodies, this duty extends widely, to include, among other matters, estates management, stress at work and violence towards staff.

8.28 Prior to the introduction of the CMCHA 2007, the close association of health and safety duties and death caused by clinical oversight and error was brought to the attention of NHS organisations following the death of Sean Phillips at Southampton University Hospital in 2000. Mr Phillips, aged 31, died of toxic shock syndrome following routine knee surgery. Two clinicians, Drs Misra and Srivastava, were convicted of gross negligence manslaughter after they had failed to monitor Mr Phillips' abnormal temperature and pulse, despite having checked on him at least four times. Both doctors were sentenced to 18 months' imprisonment, suspended for two years, as a result of death caused by gross negligence in co-operative care.

8.29 In the context of the cases outlined above, there is little unusual about the prosecution that followed this tragic incident. What was novel, however, was that the CPS then prosecuted the Hospital Trust, securing a conviction against it under s 3 of the HSWA 1974 for a failure to manage clinical staff. Despite narrowing down the charges against it, and limiting its plea to management failures over an eight-week period, and furthermore with the prosecution agreeing that that failure did not cause Mr Phillips' death, the Trust was fined £100,000 (reduced on appeal to £40,000) and ordered to pay £10,000 in costs.

8.30 Health and safety prosecutions against NHS bodies, including those relating to patient safety (rather than, for example, estates maintenance issues), are increasing. In its written evidence to the House of Commons Select Committee on Health in September 2008,[1] the HSE considered the interaction between its own regulatory regime and that of other inspection bodies and regulators (including, from April 2009, the Care Quality Commission). The HSE also referred to a number of prosecutions, including the following:

- In 2008, Avon and Wiltshire Mental Health Partnership NHS Trust was prosecuted after a 77-year-old patient fell 5.5 metres from a first-floor window sustaining major injuries. Suitable window-opening restrictors should have been in place to prevent this accident. The Trust was fined £20,000 and ordered to pay costs of £12,502.30.

- The previous year, Heart of England NHS Foundation Trust was prosecuted because they did not have preventive maintenance systems for bed use or bed rails, and did not provide suitable information on patient transfer from ward to ward, leading to patients being nursed on inappropriate beds. One patient fell from a bed when the bed rail collapsed and suffered a fracture of the right hip. The patient later fell again when on a bed without bed rails but suffered no further injury. The Trust was fined £25,000 and ordered to pay costs of £30,000.
- In 2006, Mid Essex Hospital Services NHS Trust was prosecuted after the death of a child during a minor operation, where the tube providing oxygen to the child was blocked by a foreign object. The Trust was fined £30,000 and ordered to pay costs of £10,000.
- Two years earlier, Basildon & Thurrock University Hospitals NHS Trust was prosecuted following a confirmed case of legionnaires' disease and the identification of widespread failure to manage microbiological risks in hot water services, which led to the proliferation of legionella bacteria. The Trust was fined £25,000 and ordered to pay costs of £12,225.
- Lastly, in 2000, Maidstone & Tunbridge Wells NHS Trust was prosecuted when an elderly patient died after being given an incorrect blood transfusion. The systems of work for ordering and collecting blood, and for checking that it is given to the right patient, were confused, and staff had not been trained in safe working procedures. The Trust was fined £7,000.

1 Memorandum by the Health and Safety Executive (HSE) (PS 07).

8.31 The HSE's enforcement action against NHS bodies certainly did not halt in 2008, and prosecutions have continued since then. In October 2013, Mid Staffordshire NHS Trust pleaded guilty to safety breaches relating to the death of Gillian Astbury, on 11 April 2007, when she was an in-patient at the hospital. The immediate cause of death was the failure to administer insulin to a known diabetic patient. The HSE's case was that the Trust failed to devise, implement or properly manage structured and effective systems of communication for sharing patient information, including in relation to shift handovers and record-keeping.[1]

1 BBC News website, 9 October 2013.

8.32 The majority of deaths that occur at work are a consequence not of any particular act or omission on the part of an individual but due to omissions on the part of the employer in ensuring a safe working environment. Where a death at work occurs, there is a need (and, for NHS bodies in particular, a requirement) to investigate the incident in order to determine the causes and to try and prevent similar events from arising in the future. This internal investigation process is explored in more detail below.

APPLYING THE CORPORATE MANSLAUGHTER LEGISLATION TO HEALTHCARE BODIES

8.33 Subject to certain exemptions, as outlined below, the corporate manslaughter offence applies to healthcare bodies, including NHS organisations, in that the majority of their activities fall within the ambit of a 'relevant duty of care' as defined in s 2 of the CMCHA 2007.

8.34 Whilst the proposed application of the CMCHA 2007 to NHS bodies was not universally welcomed, it does appear logical that a national health service body should owe the same duty – and suffer the same consequences if that duty is breached – as a private sector body undertaking, in some circumstances, identical activities. Differently put, the NHS should not be exempt from the offence created by the Act simply by reason of its status as a public service organisation.

8.35 As noted elsewhere, the penalty on conviction for corporate manslaughter is an unlimited fine. In relation to NHS providers of healthcare, the Sentencing Guidelines Council notes that the effect upon the provision of services to the public will be relevant; although a public organisation such as a hospital trust must be treated the same as a commercial company where the standards of behaviour to be expected are concerned, and must suffer a punitive fine for breach of them, a different approach to determining the level of fine may well be justified:[1]

> 'The Judge has to consider how any financial penalty will be paid. If a very substantial financial penalty will inhibit the proper performance by a statutory body of the public function that it has been set up to perform, that is not something to be disregarded.'

1 'Corporate Manslaughter & Health and Safety Offences Causing Death' (Sentencing Guidelines Council, 2010), para 19(v).

Comprehensive exemptions

8.36 Situations do arise, however, where exemptions should and do apply. Section 3 of the CMCHA 2007 sets out exemptions in relation to public policy decisions, exclusively public functions and statutory inspections. The exemption most likely to apply to NHS bodies relates to public policy decisions (set out below), but it should be noted that the provision of healthcare services generally will engage the Act in the event of gross breach by the organisation of its duty of care to those affected by its activities.

Public policy decisions

8.37 Section 3(1) provides:

> 'Any duty of care owed by a public authority in respect of a decision as to matters of public policy (including in particular the allocation of public resources or the weighing of competing public interests) is not a "relevant duty of care".'

'Public Authority' has the same definition, with limited exceptions, as is contained in s 6 of the Human Rights Act 1998, and includes within its ambit National Health Service bodies. The public policy exemption covers, for example, decisions by a PCT to fund (or not) medical treatments or prescription drugs. It does not, however, exempt decisions as to how resources are allocated.[1]

1 *A Guide to the Corporate Manslaughter and Corporate Homicide Act* (Ministry of Justice, 2007), p 9.

Partial exemption – emergencies

8.38 The partial exemption in s 6 of the Act is of particular note to NHS Acute Trusts and Ambulance Trusts in emergency situations, and also in the same circumstances to organisations providing organ and blood transport services. The offence under the CMCHA 2007 does not apply unless the death relates to the organisation's responsibilities as employer or as an occupier of the premises.

8.39 In emergency circumstances,[1] triage decisions (determining the order in which injured people are treated) fall within the partial exemption. However, the duty of care relating to the subsequent medical treatment, once the triage decision has been made, is not exempted, and the potential for a corporate manslaughter case, assuming the other elements of the offence are met, may potentially arise from that point onwards. For the purposes of the CMCHA 2007, 'emergency circumstances' are defined as being: life threatening or causing, or threatening to cause, serious injury or illness or serious harm to the environment or buildings or other property. Healthcare bodies should note, however, that emergency circumstances do not cover the routine activities of an Accident and Emergency Department in the normal conduct of its business; predictable fluctuations in the department's activity do not meet the definition.

1 Defined at CMCHA 2007, s 7(7).

Detained patients/prison healthcare

8.40 An impasse had been reached in the days and weeks before the Corporate Manslaughter and Corporate Homicide Bill gained Royal Assent in July 2007 in relation to the application of the proposed new law to deaths in custody. The Home Office strongly argued that deaths in prison and in police custody cells should fall outside the ambit of the new law, whereas equally strong views, expressed in the House of Lords in particular, held that it should. That argument was not merely technical in nature, and the Prison Service in particular might, without further time to prepare, have felt the impact of the new Act at an early stage. In 2009, the first full year of the Act, there were more than 150 deaths in prison, with a number of inquests returning verdicts of unlawful killing, and cases such as the death of Seni Lewis in September 2010.

8.41 Mr Lewis, a 23-year-old postgraduate from south London, died following prolonged restraint by police officers while he was an in-patient in the Bethlem Royal Hospital. It appears that he was held face down on the floor for a total of at least 40 minutes in the course of two successive episodes of restraint, altogether involving some 11 police officers.

8.42 Whilst clearly concerned to address risks associated with detained patients in particular, the position taken by the Department of Health was more sanguine. It assessed that the NHS and independent mental health services sector were in a position to deal with the impact of the CMCHA 2007 without significant delay. In the five-year period from 2000 to 2004, 1,672 deaths of detained patients were notified to the Mental Health Act Commission, 75 per cent of which were ascribed to natural causes. Inquests found 16 per cent to be suicide, with the remaining 8 per cent largely being ascribed to accidental death or misadventure.[1]

1 Mental Health Act Commission 11th biennial report/DH letter, 9 January 2008.

8.43 The agreed solution is that the Act does apply to the so-called 'custody duty' (owed to a person who, by reason of being in custody, is someone for whose safety the organisation is responsible) but, unlike the majority of its provisions, this specific duty did not come into force on 6 April 2008. Little over three years later, in September 2011, however, the position changed, and corporate manslaughter now applies to organisations detaining individuals in custody, including patients detained under the Mental Health Act.

8.44 For the providers of healthcare to detained patients or otherwise in a custodial setting, the impact of the CMCHA 2007 was arguably likely to be less than for the Prison Service. Providers of mental health services have been subject to existing law (common law in the case of corporate manslaughter, notwithstanding its evident inadequacies, and offences under the HSWA 1974). Moreover, the majority of activities undertaken by these providers in relation to detained patients are covered by the other sections of the Act already in force from April 2008. In relation to these activities, the duty is engaged as supplier of goods or services or as occupier of premises, and would almost certainly extend to cover the provision of healthcare and hotel services to detained patients, like any other patients. The exercise of identifying duties in relation to detained patients owed solely by reason of their detention is not an easy one but the potential consequences of serious mismanagement are clear enough. Providers of healthcare (as well as the police and Prison Service) can be at risk of prosecution for corporate manslaughter when the organisation is responsible for the safety of a person who is detained in a custodial institution, living in secure accommodation, or where they are a patient detained under the Mental Health Act. Senior managers need to be aware of the provisions, to have a realistic approach to what this means for their own organisation, and to understand the steps that can be taken to minimise corporate risk.

MANAGEMENT OF HEALTH AND SAFETY RISKS

8.45 In the same way as any other employer, healthcare organisations, including NHS bodies, are required to manage their business to minimise the health and safety risks to their employees and those affected by their undertakings. If they fail to do so, they risk prosecution for health and safety offences. Moreover, whilst a corporate charge of involuntary manslaughter was technically a possibility, albeit a distant one, under the former regime, there is now a very real prospect of a corporate manslaughter investigation in certain circumstances where serious management failure is identified as a potential cause of death. The process of investigation alone, irrespective of the decision whether or not to prosecute, will be disruptive, demoralising and very time-consuming and reason enough, if one were needed, to ensure that risks arising from an organisation's activities are identified and effectively managed.

8.46 The duties associated with the identification of risk and, more specifically, the steps necessary to address it, were examined on appeal to the House of Lords in *R v Chargot.*[1] The prosecution's assertion that it was only required to identify and prove a risk of injury from a state of affairs at work was upheld by the Law Lords, who rejected the defendants' assertion that the prosecution should instead identify the specific acts and omissions by which it was alleged that there were failures to comply with the HSWA 1974.

1 [2009] 1 WLR 1.

8.47 *Chargot* is additionally significant because it was an appeal by three individual defendants against conviction under s 37 of HSWA 1974. Individuals may be exposed to prosecution as employees under s 7 of HSWA 1974, for failing to take reasonable care for health and safety, or as directors, managers or officers under s 37 of the same Act, if an offence by the organisation is committed with their consent, connivance, or attributable to their neglect. The question of individual liability for a corporate health and safety offence has become even more acute from 16 January 2009 when the Health and Safety (Offences) Act 2008 came into force. Whilst the CMCHA 2007 pre-dates the events in *Chargot* and does not have retrospective effect, the defendants would have been concerned to note the introduction of a penalty of up to six months' imprisonment upon conviction of the individual offence.

8.48 It is necessary for healthcare bodies' clinical and corporate governance and control assurance systems to encompass the management of health and safety risks to their employees, contractors, visitors and patients. Indeed, clinical and health and safety risks should not be viewed in isolation; there can be an overlap between clinical and health and safety risks that emphasises the need for authorities and trusts to co-ordinate their management and assessment of these risks.

An example of this overlap arose in 2007 (ie before the CMCHA 2007 came into force) in the mental healthcare context when a service user was placed in a care home, satisfactory from the viewpoint of both the CQC and the local authority which had itself purchased services there, but which (because of local issues, including inadequate staff numbers) was wholly unsatisfactory to the service user. Tragically, a member of staff was killed and another seriously injured. The prosecution against the Trust as one of two defendants was undertaken due to failures in assessment of the type of care required and the training of employees at the home. Criticism was directed to the interface between management, administration and clinical expertise involved in the placement of persons accessing mental health services. Following a seven-week Crown Court trial, both the care home's owner and the Trust were convicted for breaches of duty under the HSWA 1974 and fined £75,000 and £150,000 respectively and each required to pay prosecution costs in excess of £300,000.

1 *R (HSE) v (1) Hertfordshire Partnership NHS Foundation Trust and* (2) *Chelvanayagam Menna* (Luton Crown Court, June 2012).

8.49 The role and responsibilities of the boards of healthcare bodies to ensure that the management of health and safety risks are addressed are entirely analogous to those of directors in the private sector and include the risk assessment responsibilities of the Management of Health and Safety at Work Regulations 1999.[1] Like the HSWA 1974, they apply to every work activity, and impose the requirements: to carry out and record the findings of risk assessments in relation to those activities and to make arrangements for implementing the health and safety measures identified as necessary by the risk assessment; to appoint competent people to help implement the arrangements; to set up emergency procedures; and to provide clear information and training to employees.

1 SI 1999/3242.

8.50 Healthcare organisations are required to manage a variety of risks, some general and associated with any business and some specific to their industries. By way of example, general risks will include those relating to the control of contractors and those associated with sharing their sites with organisations contracted to supply services, such as catering and laundry services and also managing the risks of vehicular and pedestrian traffic on site. The HSE's imposition in 2009 of an improvement notice on United Lincolnshire Hospitals NHS Trust for failings surrounding its management of work related stress demonstrates the importance of NHS and private sector healthcare bodies addressing risks common to the current provision of healthcare. There is no question that enforcement action by the HSE is solely reactive (ie prompted by a serious incident, an adverse incident or a near miss). Improvement Notices and, more infrequently, Prohibition Notices have been served on healthcare bodies following routine surveillance visits and with the aim of addressing, before any incident arises, a very wide range of risks, including falls from height, manual handling, scalding, legionella and construction-related risks such as insecure scaffolding.

8.51 Moreover, the increasing focus within the NHS on the role of governance in delivering high-quality care in a safe environment is also the subject of external scrutiny. The health care industry is closely regulated, arguably over-regulated, with more than one body overseeing compliance with similar or identical standards. To place this comment in context, an NHS Trust, in the final stages of its progression to Foundation status, may be subject to review and/or performance management from a very wide range of organisations including its Strategic Health Authority (until April 2013), Monitor, commissioning bodies such as PCTs and their replacements, the Care Quality Commission, the HSE and the police.

8.52 The publication of the report by Robert Francis QC into failings at Mid-Staffordshire NHS Foundation Trust has profound implications for all those involved in the commissioning, provision and regulation of healthcare services. It calls for a seismic shift in the way organisations operate so that patients' wellbeing is prioritised demonstrably and without exception. Some of the changes proposed by the Francis report are subject to government acceptance or legislation, but many are not, and acute trusts must show by the end of 2013 how they intend to respond to the report.

A full examination of the recommendations of the Francis report is beyond the scope of this chapter, but one of the most significant cultural changes, if implemented, is in relation to transparency, openness and candour. The need for (and the benefits of) openness are clear enough and include maintaining trust in the service, properly involving patients in their own care and identifying concerns early. However, the effect of the proposed statutory duty of candour, if adopted, could prove controversial in relation to enforcement action that may follow a serious incident involving the death of a patient. That controversy arises because of the tension between candour and self-incrimination.

8.53 Candour, as proposed in this context, would require that, where death or serious harm has been (or may have been) caused to a patient by an act or omission of the organisation or its staff, the patient (or any lawfully entitled personal representative or authorised person) should be informed of the incident, given full disclosure of the surrounding circumstances and be offered an appropriate level of support, whether or not the patient or their representative has asked for this information. Crucially, under the report's proposals, individual health professionals as well as organisations could face criminal sanctions should they wilfully obstruct anyone in the performance of this statutory duty.

That is not to say that the existing framework does not promote openness. The NHS Constitution makes clear that, when mistakes happen, they should be acknowledged and an explanation provided, along with an apology. The GMC's Good Medical Practice as well as the Code of Conduct of the Nursing and Midwifery Council (NMC) both place an ethical duty on the health professionals to be open and honest with patients if things go wrong and to offer full explanations promptly when a patient has suffered harm. Lastly, the NHS Litigation Authority Risk Management Standards 2012–13 require organisations to have 'an appropriate documented process for open and honest communications following an incident, complaint or claim'.

8.54 There are also many specific risks relevant to the health sector that require priority management, some of which have been the subject of HSE guidance, such as manual handling risks associated with risk groups, such as ambulance personnel and nursing staff,[1] slips and trips[2] and violence and aggression towards staff.[3] However, an element of realism is nonetheless required in assessing whether there is a risk of injury within the meaning of the Manual Handling Operations Regulations 2002 (for example, when assessing the risk of injury to a domestic care worker in a children's home associated with pulling out a bed to make it).[4] The lesson here is that a degree of discretion and common sense judgement need to be applied by those responsible for managing risk. Where guidance is available, and is relevant, it makes sense to follow it, but doing so without a wider consideration of risk will not always discharge liability if injury or loss arises.

1 HSC 03/98 'Manual Handling in the Health Services' (1998).
2 HSE Information sheet ' "Slips and trips" in the Health Services' (2003).
3 HSC Health Services Advisory Committee 'Violence and aggression to staff in Health Services, guidance on assessment and management' (1997).
4 *Koonjul v Thameslink Healthcare Services* [2000] PIQR P123.

8.55 Compliance with the Provision and Use of Work Equipment Regulations 1998[1] (PUWER) in relation to medical devices and other work equipment, the Personal Protective Equipment at Work Regulations 1992[2] (PPE) and the Control of Substances Hazardous to Health Regulations 1999[3] (COSHH) remain priority areas in the NHS. For example, in the case of *Dugmore v Swansea NHS Trust and Another*,[4] the Trust was found liable to an employee for a latex allergy from latex gloves under its absolute duty under the COSHH Regulations. The HSE has issued a number of information sheets by way of guidance to health service providers on the safe use of particular equipment and handling of drugs to comply with the COSHH Regulations.[5]

1 SI 1998/2306.
2 SI 1992/2966.
3 SI 1999/437.
4 [2002] EWCA Civ 1689.
5 HSE information sheets 09/99 'Safe use of pneumatic air tube transport systems for pathology specimens' (1999); 09/03 'Safe handling of cytotoxic drugs' (2003).

8.56 Failure to manage these issues may lead to enforcement steps by the HSE, including prosecution under the HSWA 1974. Not all of the risks associated with the delivery of healthcare relate to serious injury or death and prosecutions have taken place where cumulative shortcomings in relation to, for example, traffic segregation, risk of scalding by radiators, storage and disposal of clinical waste and manual handling training and equipment have been deemed insufficient by the HSE but no injury or loss has, in fact, arisen. In theory, however, each of these shortcomings could lead to a fatality, and the subsequent investigation of the role played by senior management in identifying and managing the risk, as well as the organisation's general approach to health and safety management, could be closely examined as part of a corporate manslaughter investigation.

RISK MANAGEMENT CHECKLIST

8.57 As part of a rolling review, and to assist healthcare bodies in countering the very damaging perception of cultural laxity towards potentially fatal risks, the following questions should be asked, with answers recorded and, where necessary, acted upon.

Identifying risk

8.58

- Which of the organisation's activities are inherently dangerous?
- What other activities have the potential to cause death, and in which circumstances?
- How are different activities ranked in terms of risk?

Identifying the management team and its roles

8.59

- Who in the organisation falls within the definition of senior manager? Do they understand what the title means, particularly in the context of the CMCHA 2007?
- Can senior managers explain their role in managing risk and do they have sufficient autonomy to discharge their duties properly? Would they agree with the analysis of a board representative on these issues?
- Which board director is responsible for the management of risk? How are shortcomings and suggested improvements brought to his or her attention? Are these acted upon?

Risk management and related documentation

8.60

- Are policies, procedures and risk assessments specific, comprehensive, relevant and up to date? Who updates them? Do they take account of all available relevant information, including any Approved Codes of Practice?
- Are the organisation's reporting systems robust? Are adverse incidents and near misses reported and recorded? Have the recommendations in any serious incident report been carried into effect?
- Does theory and practice differ substantially? There is no value in written policies and procedures if they are not followed. How is the documentation disseminated, audited and checked?

Recruitment and training

8.61

- Does the organisation recruit competent, appropriately qualified staff (including clinical staff)? How is their suitability for promotion and additional responsibility assessed? Are they adequately trained and supervised? Could the organisation prove, if challenged, that each employee works within his level of skill and competence unless closely supervised? Do employees have the expertise to match their job descriptions, and how is that expertise kept up to date?

- How are staff and contractors trained in the management of risk?
- Could the organisation defend an allegation that attitudes, policies, systems, or accepted practices exist and these are likely to encourage health and safety failures, or produce tolerance of them?

8.62 In summary, does the organisation have suitable policies in place to ensure that its activities are conducted safely? Is evidence available to show that the policies are effectively implemented, and would all members of staff agree? The review of risk and the means for addressing it is a live process and so policies and related risk management procedures should be kept updated, taking into account new industry practice and guidance from a number of sources including the Department of Health and the HSE. Risk management is seldom perfect but the consequences of gross negligence in the way senior managers address the risks associated with the activities undertaken by their organisation may be very severe indeed.

INFECTION CONTROL AND REGISTRATION

8.63 The prevention of healthcare-associated infections (HCAIs), and in particular failures to do so, have been the subject of intense public disquiet and recent media attention. Repeat outbreaks of *clostridium difficile* at Stoke Mandeville Hospital in 2004 and 2005, leading to the death of 30 patients, and more recently at Maidstone Hospital, involving 90 deaths, both pre-dated the coming into force of the CMCHA 2007.

8.64 As with all potential prosecutions for corporate manslaughter, however, there are a number of obstacles for the CPS to overcome, particularly in relation to deaths allegedly caused by HCAIs. It may be assumed that any breach of its duty of care to patients by the organisation would constitute a breach of a relevant duty of care and also, given the fundamental importance of infection control to the safe delivery of healthcare, that senior management had substantial input into that breach.

8.65 The issue of causation, namely that the death(s) were caused by the breach of duty, may prove more problematic for the prosecution. Leaving aside the question of whether or not any breach in these circumstances was gross,[1] it may be difficult to link the death by MRSA/*clostridium difficile* to the failure to control the risk of infection properly and effectively. In particular, the prosecution must establish beyond reasonable doubt that the patient acquired the HCAI after admission to hospital and/or that its acquisition was the result of very serious failures by senior management, and further that the HCAI, rather than the condition for which the patient was first admitted, caused death. It is possible, for instance, that death from *clostridium difficile* infection may occur even where the patient had received the best possible care. Given the severe penalties, direct and indirect, following a corporate manslaughter conviction, admissions on these points are likely to be in short supply, with the prosecutor being put to strict proof on each issue.

1 Within the meaning given in CMCHA 2007, s 8.

8.66 The vital importance of infection control in itself would suggest that any failure to manage the associated risks could, depending on the circumstances, constitute a gross failure on the part of those responsible. A substantial volume of guidance is available from the Department of Health[1] reflecting an accumulating body of evidence that takes account of current clinical practices. From 1 April 2009, as a pre-cursor to 'full' registration the following year in relation to all of their activities, all NHS Trusts providing healthcare directly to patients were required to register with the newly created Care Quality Commission (CQC) for the purposes of infection control.

Registration is dependent upon compliance with regulations[2] aimed at ensuring that patients, workers and others are protected against the identifiable risks of acquiring HCAIs, so far as is reasonably practicable.[3] In the absence of registration, the full sanctions available to the CQC under the Health and Social Care Act 2008 are available to the CQC, and these now apply across the full range of healthcare services, including those provided by NHS organisations (ie not just in relation to infection control). Typically, but not exclusively, following announced or unannounced inspections of premises from which healthcare services are delivered, the sanctions range from requiring actions to be taken in order to comply with required standards, imposition or variation of conditions of registration, to suspension or cancellation of registration. For example, following an unannounced inspection to the Royal Blackburn Hospital in July 2013, the CQC issued formal warnings to East Lancashire Hospitals NHS Trust requiring improvements to care and welfare of patients, and assessing and monitoring the quality of service provision (SI 2009/660, regs 9 and 10 respectively).[4]

Although not evidently contemplated in the above instance, cancellation of registration for the services of an acute hospital, for example, would not be undertaken lightly since it would, in effect, constitute a reconfiguration of services without due consultation. Well before any such extreme action were taken, however, a number of agencies (potentially including Monitor, NHS England, Clinical Commissioning Groups, as well as the CQC itself) would almost certainly be exerting influence to improve services. Criminal sanctions including penalty notices and/or prosecution are also available to the CQC and are explained in detail in their periodically updated Enforcement Policy document.[5]

1 Department of Health Guidance includes the following: 'Getting Ahead of the Curve: A strategy for combating infectious diseases (including other aspects of health protection)' (2002); 'Winning Ways: Working together to reduce Healthcare Associated Infection in England' (2003); 'Towards cleaner hospitals and lower rates of infection: A summary of action' (2004); and 'Clean, Safe Care – reducing infections and saving lives' (2008).

2 Health and Social Care Act 2008 (Registration of Regulated Activities) Regulations 2009, SI 2009/660.

3 The Health and Social Care Act 2008 Code of Practice on the prevention and control of infections and related guidance, Department of Health, Gateway Reference 14808.

4 'CQC warns East Lancashire Hospitals NHS Trust that they must make improvements', CQC website, 16 October 2013.

5 Care Quality Commission, Enforcement Policy, June 2013, pages 12–19.

8.67 The intention of guidance from, among others, the National Institute for Health and Care Excellence (NICE)[1] is to help NHS bodies plan and implement how they can prevent and control HCAIs. It sets out criteria by which managers of NHS organisations are to ensure that patients are cared for in a clean environment, where the risk of HCAIs is kept as low as possible.

1 See http://publications.nice.org.uk/prevention-and-control-of-healthcare-associated-infections-ph36.

8.68 That is not to suggest that an absence of imposed remedial steps will denote compliance with the Code, nor will that situation necessarily offer a defence to prosecutions under the CMCHA 2007 or the HSWA 1974. Conversely, however, adopting best industry practice and ongoing compliance with the Code should provide a robust response to allegations that any breach of the organisation's duty of care in this context was a gross one, ie that the conduct by senior managers fell far below what could reasonably be expected of it in the circumstances.

8.69 Management of the high-profile risks associated with infectious diseases in the healthcare sector is, however, just one aspect of the wider regime by which risks are assessed and managed in the NHS across the full spectrum of its activities, both clinical and non-clinical, as examined in the following section.

8.70 Additionally, special measures may be imposed in order to bring about compliance. Monitor publishes a list of actions it is undertaking at NHS Foundation Trusts, ranging from 'special measures' to 'other regulatory action', where Foundation Trusts have failed to comply with the conditions set out in the NHS provider licence, and the regulator also names those organisations 'currently under investigation'. For NHS provider organisations seeking Foundation Trust status, the NHS Trust Development Agency carries out a similar role, in which special measures are imposed where serious and systemic failings in relation to quality of care have been identified, and the persons responsible for leading and managing the organisation are unable to resolve the problems without intensive support. Special measures include: publication of an Improvement Plan, listing the steps to be taken by the organisation to improve its services; review (and, potentially, change) of its leadership; intense external scrutiny; and extensive support, for example from commissioners or advice from a partner in a high-performing Trust.

INTEGRATED GOVERNANCE

8.71 Managing and minimising risks has been a high priority in the NHS for many years now. This is achieved through clinical and corporate governance, which now come together as 'integrated governance' to ensure that NHS organisations coordinate all strands of governance and to help NHS Boards direct and control organisations more effectively.

8.72 'Integrated governance' is defined as:[1]

> 'systems, processes and behaviours by which Trusts lead, direct and control their functions in order to achieve organisational objectives, safety and quality of service and in which they relate to patients and carers, the wider community and partner organisations.'

1 Integrated Governance Handbook (2006).

8.73 There have been a number of initiatives in the NHS to support good governance and patient safety. The now repealed s 18 of the Health Act 1999 brought in the statutory duty of quality. Under s 45 of the Health and Social Care (Community Health and Standards) Act 2003, this covers all NHS bodies. This section imposes a duty upon each NHS body to have arrangements for monitoring and improving the quality of healthcare provided by it. Although not governance specific, the Health and Social Care Act 2012 has introduced obligations on both the Secretary of State and the NHS Commissioning Board Authority to improve the quality of services and quality standards which, pursuant to s 234 of the 2012 Act, are to be developed by NICE in relation to NHS services, public health services and/or social care in England.

8.74 The National Patient Safety Agency (NPSA) was set up in 2001 to promote patient safety and to learn from patient safety incidents. From 1 June

2012, the patient safety remit of the NPSA has moved to the NHS Commissioning Board (NHS CB). The NHS CB will utilise the information collected by the National Reporting and Learning System (NRLS) to review and address patient safety issues at the root cause. The transferred functions include the administration of a mandatory reporting scheme for adverse health events and 'near misses', as well as the issuing of patient safety alerts on specific matters from time to time. The NHS CB also releases a report concerning Organisation Patient Safety Incident Reports, although this is populated from NRLS data, which is a voluntary reporting tool.

8.75 As previously mentioned, the Care Quality Commission (CQC) was created pursuant to the Health and Social Care Act 2008 to carry out the work previously undertaken by the Healthcare Commission, the Commission for Social Care Inspection and the Mental Health Act Commission. Chapter 3, Part 1 of the 2008 Act requires the CQC to review health care provided by PCTs and NHS Trusts, and adult social care services provided by local authorities. The CQC may also undertake special investigations and reviews into the provision of NHS care, adult social services or the exercise of functions by English Health Authorities, either with the permission of the Secretary of State, or they must do so if instructed by the Secretary of State. The Health and Social Care Act 2012 will also provide the CQC with the ability to conduct an investigation without the permission of the Secretary of State if it believes that the health, safety or welfare of persons receiving health or social care is at risk. Where such investigations are carried out, the CQC must prepare reports and recommendations to improve patient safety. The Health and Social Care Act 2012 also provides a duty for some of the CQC's functions to be undertaken by Healthwatch England. These functions relate primarily to the provision of advice.

8.76 The Secretary of State retains the ability to order a public enquiry where a service failure results in serious harm to patients or where there is serious national concern or where a major ethics or policy issue is raised.

8.77 The National Health Service Litigation Authority was established in 1995 and also has a risk management function in its Framework Document. It has an extensive risk management programme for both clinical and non-clinical risks, primarily achieved through a range of standards that are regularly assessed. There are standards for each type of NHS body, and also for the independent sector if they provide NHS services, and separate maternity standards. The standards are divided into three levels, and it is an indication of good governance for organisations that achieve the highest level. The standards also provide essential guidance for NHS bodies in controlling their risk exposure and include useful preventive measures.

8.78 National standards are essentially a method for ensuring that NHS organisations are well managed and are implementing good practice locally. Quality Accounts are annual reports by providers of NHS health care, and addressed primarily to the public, relating to the service quality. These are designed to encourage boards of healthcare organisations to properly assess the quality of all of the healthcare services that they offer, and they help the organisations and those involved to demonstrate a commitment to continuous, evidence-based quality improvement. All NHS organisations are required to have an Assurance Framework to underpin their Statements of Internal Control. This provides a simple method to manage the principal risks that arise. The Board Governance Assurance Framework has been implemented for aspirant

Foundation Trusts. This is designed to assist NHS Trusts in ensuring that their board governance is appropriately skilled and prepared for Foundation Trust status. NHS Foundation Trusts themselves have to comply with Monitor's Quality Governance Framework, which is now incorporated into the Compliance Framework. This assesses processes and structures that a Foundation Trust has in place, which assist the board in monitoring and ensuring that it provides the appropriate quality of care.

8.79 There have been a number of helpful publications providing guidance on governance matters.[1] Monitor published its 'Quality Governance Framework' and this is now included in the 'Compliance Framework 2012/13' published in March 2012. Monitor first published the 'Code of Governance for NHS Foundation Trusts' (the Code) in 2006; the updated version has been in place since 1 April 2010.

1 These include: 'Quality governance in the NHS – a guide for provider boards' (2011), 'Quality Accounts Toolkit' and 'Board Governance Assurance Framework for Aspirant Foundation Trusts: Board Governance Memorandum', all published by the Department of Health.

8.80 However, the effectiveness of all these standards and whether good governance is achieved depends on the implementation in the individual NHS body. There needs to be a comprehensive programme of quality improvement for clinical and corporate matters, and poor performance must be identified and remedied. If this does not happen, it leaves organisations open to possible corporate and, on rare occasions, personal liability.

MANAGEMENT IN THE EVENT OF AN ADVERSE INCIDENT RESULTING IN DEATH

8.81 The delivery of healthcare, as noted above, is an inherently risky business, and unexpected deaths do occur. Of these, only a small proportion are pursued by the investigating bodies as criminal prosecutions, and an even smaller number result in the prosecution (typically under the HSWA 1974) of the undertaking whose activities caused death.

8.82 However, a small but significant number of corporate prosecutions under the HSWA 1974 do occur, and evidence suggests that the trend to pursue criminal enforcement steps is rising. Recent cases within the NHS itself have involved health and safety offences in a wide variety of theatres, not solely deriving from clinical systems failures, and these include estates-related fatalities/serious personal injury (falls from height; steam scalds), lone working (death of mental health nurse working alone in a ward environment) and prosecutions contemplated following high-profile and widely publicised failures to manage infection control appropriately.

8.83 The way in which an organisation responds in the initial stages of a police or HSE investigation is crucial, as it speaks volumes for the culture of management of health and safety risks specifically and of the organisation as a whole. A well-handled, co-operative and efficient response from the outset could make the difference between a prosecution and a decision by the investigator and prosecutor not to proceed with a criminal prosecution. The reverse situation could demonstrate a poor approach to risk management, potentially suggesting that any failure by the organisation's senior managers is 'gross', within the meaning given in s 8 of the CMCHA 2007.

8.84 In the event of an adverse incident, it is essential that the organisation is able to deal with the incident as quickly and efficiently as possible and to ensure that the incident is managed and reported to a designated officer within the organisation in accordance with its internal protocols. The learning points may be equally important, whether or not the incident falls within the SIRI (Serious Incident Requiring Investigation) category, an Adverse Incident or a Near Miss. The categorisation derives from the national framework developed by the National Patient Safety Agency relating to serious incidents in the NHS. Completion of the framework followed consultation and collaboration with key NHS stakeholders including the CQC, the Department of Health, the Medicines and Healthcare Products Regulatory Agency, the NHS Litigation Authority, Monitor and the Independent Advisory Service.

8.85 The incident should be graded according to its seriousness and the impact on the victim, their family and the individuals concerned. Seriousness is assessing by reference to the level of harm that has arisen (from Level 1 – No harm: impact prevented – to Level 5 – Death)[1] and the context within which it arose. Evidence should be gathered and preserved (for example, medical records and defective equipment). An investigation should be undertaken to understand the events and circumstances leading up to the incident and a root cause analysis undertaken. In certain cases, as where the incident results in the death of the patient, the NHS organisation will need to notify the Department of Health and other stakeholders of the incident, including, for example, the coroner, so that they can undertake their own inquiries.

1 'Seven steps to patient safety', National Patient Safety Agency, July 2004.

8.86 Where obvious errors have been made, consideration should be given to early admissions, with approval from those stakeholders who will ultimately be responsible financially for any civil claims (such as the NHSLA). However, where both a corporate manslaughter offence and an offence under the HSWA 1974 are prosecuted in parallel, an admission to the health and safety offence could be wholly prejudicial to any defence offered to the corporate manslaughter offence. Such decisions need to be carefully thought through by the defence team.

8.87 Organisations should be open and keep those individuals affected by the incident informed and up to date as well as agreeing timescales for further action. To a large extent, these considerations fall within the Memorandum of Understanding in place between the NHS, the Association of Chief Police Officers and the HSE[1] following patient safety incidents and specifically the Incident Coordination Group between those bodies, usually convened at the instigation of the NHS party, but it may be called by either of the other agencies.

1 'Memorandum of Understanding: Investigating patient safety incidents involving unexpected death or serious untoward harm' (February 2006), accessible via Department of Health website.

8.88 There may be occasions where the mistake made is such that the NHS organisation itself would wish to take disciplinary action against the individual healthcare professional; and it is imperative that, if this is the case, the individual be notified of any such disciplinary action at the earliest opportunity and, if necessary, steps be taken in order to ensure patient safety, to relieve the individual of his or her duties until such time as the investigation into the incident is complete and an informed decision can be made as to that individual's future employment.

8.89 Staff should be interviewed, site inspections undertaken, and policies and procedures reviewed. It is inevitable that an adverse incident will become public knowledge and, with this in mind, consideration should be given at an early stage to press statements and media handling. Where an incident is likely to have an impact on other patients, consideration should be given to setting up a 24-hour helpline so that patients can be provided with information and, where appropriate, counselling.

8.90 Once an investigation is complete and a final report is to hand, it is important that the organisation study the report to understand the root cause of the incident and determine what lessons can be learned so as to prevent a similar occurrence in the future. This may necessitate a change in systems, policies, procedures and/or working practices, but the ability to learn lessons is fundamentally important in demonstrating an organisation's responsible attitude to risk management. More immediately, if a corporate prosecution and conviction were to follow the incident, progress made since that time, including close monitoring of how risk is managed, forms a powerful part of the organisation's points in mitigation following a guilty verdict or plea.

CONCLUSION

8.91 The disruptive effect of a manslaughter investigation upon any healthcare organisation will be widely felt, even if it does not lead to prosecution. The consequences of a conviction are clear enough, not simply in terms of the fine imposed (which, in the case of a NHS Foundation Trust in particular, could lead to it ceasing to trade, since it is unable to fall back on the National Health Service (Residual Liabilities) Act 1996) but also in the lasting damage to corporate reputation. In addition to the penalty imposed by the court, including remedial or publicity orders, any NHS body whose management practice falls far below the threshold of reasonableness will have problems with staff recruitment and retention. Moreover, with commissioning decisions far less certain than they once were, convicted provider organisations, such as Acute Trusts and Foundation Trusts, could begin to lose out to local competition and ultimately fail as autonomous healthcare bodies.

8.92 The identification and management of risk needs to be kept under constant review and updated to take account of developing best practice in the healthcare sector. The assessments reached must be borne out in practice, without which the most well-informed suite of risk management documentation is worthless. Communication is also vital so that, when interviewed, those who were involved in, or otherwise witnessed, a fatal incident will be able to provide an accurate summary of risk management procedures and, more importantly, demonstrate that those procedures were adhered to. Consideration of how an organisation would withstand close scrutiny review by an investigating body, be it the police, the HSE or the CQC, is a useful and revealing starting point of any health and safety review.

8.93 Securing a corporate manslaughter conviction will, without question, be a tough task for the prosecutor, and the defendant arguably has greater scope to defend such cases by contrast with the more limited opportunity to defend a health and safety prosecution brought on the same facts. However, the focus has to be in preventing charges being brought – by actively treating patient safety and the safety of all those potentially affected by an organisation's activities as an absolute priority. The process of investigation and the steps leading to trial will be time-consuming and stressful, and, with none of the safeguards afforded to individual defendants, senior managers giving evidence may find their reputation and career prospects irretrievably damaged, even if a conviction is not secured against their organisation.

8.94 *Conclusion*

8.94 The public appetite to hold poorly managed organisations to account in the criminal court, particularly where the consequences of management failure are fatal, shows no sign of abating. Just as the number of prosecutions against individual clinicians for involuntary manslaughter has increased substantially since 1990, so it would appear has the volume of prosecutions against healthcare providers, including NHS organisations. That trend of stronger, more frequent intervention by the HSE looks set to continue, and there is no reason to suppose that police/CPS-initiated prosecutions for corporate manslaughter will not do the same.

8.95 The increased number of prosecutions of NHS bodies may be attributable to a wide range of factors, including the fact that, prior to 1 April 1991, onerous legislation such as the HSWA 1974 did not apply to NHS bodies, by reason of Crown immunity. Even with that immunity removed, and the activities of NHS bodies exposed to the full force of criminal law, there was inevitably a delay in criminal proceedings coming to court and, arguably, uncertainty as to the occasions on which it is appropriate to prosecute such bodies. Part of the answer may lie in the guidelines for the CPS as to the decision whether or not to prosecute. The two-part test includes, first, an analysis of the strength of evidence and the prospect of obtaining a conviction. The second limb relates to whether or not to prosecute would be in the public interest. The utility and public benefit of the work carried out by NHS bodies is not in doubt and, to that extent, public interest arguably militates against prosecution; but, placed in full context, the reality is different. Public confidence and trust cannot reside in an organisation where poor management of risk could potentially cause injury or death to persons in that organisation's care. In the most serious of cases, where grossly negligent management contributes to one or more fatalities, the public interest in a prosecution will be overwhelming.

8.96 While organisations that meet, and continue to meet, the duties they owed prior to 6 April 2008 have little to fear from the CMCHA 2007 in force from that date onward, they should ensure that they understand the risks that their activities generate, and demonstrate best practice in mitigating those risks as their organisation and their sector continue to evolve.

Chapter 9

Waste management

Paul Rice, Pinsent Masons LLP, London

INTRODUCTION

9.1 The twin disciplines of environment and health and safety (EHS) law are amongst the most heavily regulated areas under both national and European legislation. As a specific subset of environmental law, the law and policy on waste and waste management is particularly sophisticated, which presents a challenge to all involved in the waste management chain. It is relevant to waste producers, carriers, reprocessors, waste collection and disposal authorities and all involved in the waste industry, not to mention related industries such as the emerging Energy Recovery Facilities or Energy from Waste markets and, of course, the general public.

9.2 The industry is currently facing increased demands from EU legislation for increases in recycling and recovery of waste, as well as diversion from landfill,[1] confusion over the application of certain carbon-saving legislation (such as the Carbon Reduction Commitment Energy Efficiency Scheme) and new punitive systems and the problems of fraud within specific parts of the industry.

1 Landfill of Waste Directive, Council Directive 99/31/EC.

9.3 In recent years, fines for waste-related offences have increased dramatically, while the courts have also indicated a greater willingness to award both fines and custodial sentences to directors and officers of offending companies, in addition to exercising powers in relation to the disqualification of directors and confiscation of vehicles. We have set out below a new approach to sanctions within the industry and the use of civil penalties.

9.4 This chapter looks at the corporate liability aspects of the following:

- the general regulatory framework for waste management in the UK, including assimilation into law and permits of the EU 'waste hierarchy';
- permitting of waste management activities;
- the recently revised statutory 'duty of care';
- site waste management plans;

- transfrontier movements of waste;
- producer responsibility for work;
- waste from electrical and electronic equipment;
- waste batteries and accumulators; and
- asbestos waste.

9.5 Most relevant of all the recent changes is the introduction of provisions to implement the revised Waste Framework Directive.[1] The revised Directive implements a number of changes, including amendments to the definition of 'waste', new definitions of various terms (such as 'by-products', 'recovery' and 'recycling') and the repeal of various Directives including the Waste Oils Directive,[2] the Hazardous Waste Directive[3] and the current Waste Framework Directive. The revised Directive also introduces a refined waste hierarchy (prevention; preparing for re-use; recycling; other recovery, eg energy recovery; and disposal) as well as setting various targets for Member States to comply with (such as a target by 2020 of 50 per cent re-use and recycling of waste materials such as paper, metal, plastic and glass from households).

1 Directive 2008/98/EC.
2 Directive 75/439/EEC.
3 Directive 91/689/EEC.

THE WASTE REGULATORY FRAMEWORK

9.6 Before the introduction of the modern age of waste regulation in the UK under Pt II of the Environmental Protection Act 1990 (EPA 1990), the deposit or disposal of waste was regulated through 'waste disposal' licences pursuant to the Control of Pollution Act 1974 (COPA 1974).

9.7 Even in 1974, the maximum penalty for waste-related offences under COPA 1974 was an unlimited fine and/or a prison term of up to five years,[1] although the range of possible offences was rather limited. However, Pt II of the EPA 1990 established a much more comprehensive system of waste management licensing, introducing new offences and also requiring applicants for waste management licences to prove their competence, both financial and technical, in order to hold a licence. We deal with new approaches to waste-related offences for breach of licence or permit conditions, as well as breach of waste law pursuant to the Regulatory Enforcement and Sanctions Act 2008, further below.

1 COPA 1974, s 3(3).

9.8 The EPA 1990 also established:

- a statutory duty of care[1] (known as the 'waste duty of care') on all persons who produce, import, handle, store, keep, treat or dispose of waste to take all appropriate measures to ensure that it does not cause harm; and
- the outsourcing of local authority waste management functions to local authority waste disposal companies[2] (LAWDCs) which, in turn, led to new regulatory provisions in relation to the collection, disposal and treatment of controlled waste.

1 EPA 1990, s 34 (as amended), and Environmental Protection (Duty of Care) Regulations 1991, SI 1991/2839 (as amended).
2 EPA 1990, s 34.

WASTE PERMITTING

9.9 The introduction of waste permitting is now regulated by the Environmental Permitting (England and Wales) Regulations 2010 (the EPR 2010).[1] The Regulations consolidate the Environmental Permitting (England and Wales) Regulations 2007 (the EPR 2007),[2] which previously carried out a major overhaul of pre-existing waste management licensing and the pollution prevention and control (PPC) permitting regime. The EPR 2007 revoked and amended a large number of environmental regulations and other legislation (including the Waste Management Licensing Regulations 1994[3] and the Pollution Prevention and Control (England and Wales) Regulations 2000).[4] The EPR 2010 repealed and consolidated existing environmental permitting legislation and extended the regime. Defra is also looking to incorporate water abstraction and impoundment licensing into the single environmental permitting regime, with new regulations expected imminently. In addition, consultation on proposed new and revised rules and risk assessments in relation to activities that will become installations under the Industrial Emissions Directive closed on 4 September 2012. Defra set up a consultation on charges for LAPPC and LA-IPPC processes for 2014/15, which invited views on the proposed fees and charges for 2014/15.[5] On 20 December 2013, Defra published 13 revised Process Guidance Notes for local authority regulated industries, including nine with simplified model permits.[6]

1 SI 2010/675 (corrected by SI 2010/676).
2 SI 2007/3538.
3 SI 1994/1056.
4 SI 2000/1973.
5 See www.defra.gov.uk/industrial-emissions/2013/consultation-on-charges-for-lappc-and-la-ippc-processes-for-201415.
6 See www.defra.gov.uk/industrial-emissions/las-regulations/guidance.

9.10 The EPR 2007 came into force in England and Wales on 6 April 2008, and their aim was to provide a new unified and streamlined environmental permitting system for activities regulated under the PPC regime and the waste management licensing regime by the introduction of a single integrated 'environmental permit'.

9.11 The EPR 2010 came into force in England and Wales on 6 April 2010. Their scope is wider than the previous EPR 2007, and now includes water discharge consents, groundwater activities and radioactive substances authorisations, in addition to covering the requirements of a number of EU Directives, including the Mining Waste Directive and Batteries Directive. The EPR 2010 also reflect the amendments to waste exemptions made in the Environmental Permitting (England and Wales) (Amendment) (No 2) Regulations 2009 (the No 2 2009 Regs).[1]

1 SI 2009/3381.

9.12 The No 2 2009 Regs simplified the regulation of low-risk waste recovery and disposal operations and so amended the existing exemptions from the requirements to hold an environmental permit for certain types of waste activities. In particular, these Regulations removed the availability of exemptions for higher-risk activities so that they need to obtain an environmental permit. These Regulations came into force on 6 April 2010, immediately before the EPR 2010.

9.13 The environmental permitting regime deals with the mechanics of permitting (for example, the application for an environmental permit and the varying, transferring, surrendering or revocation of a permit). There are no changes to the activities that are regulated or to the standards required of a permit holder. The EPR 2010 regime continues to provide operators with increased flexibility, including the ability to:

- transfer or surrender a permit, either in full or in part;
- extend the area covered by a permit by applying for a variation to their permit;
- change the terms of a permit to take into account changes to activities or the commencement of new activities; and
- demonstrate operator competence through a choice of schemes (see **9.37** onwards below).

The environmental permitting regime forms part of Defra's wider Environmental Permitting Programme, and this in turn forms part of the Government's Better Regulation Initiative.

9.14 Environmental permits are issued by the regulator (this will either be the Environment Agency in England and Wales or the local authority, depending on the activity to be permitted) in the form of either a bespoke or a standard permit. The type of permit required depends on the complexity and environmental risk of the operation. A standard permit is a permit which sets out standard conditions in the form of a set of fixed rules for common activities with which the permit holder is required to comply. Currently, standard permits are available for 52 low- to medium-risk waste activities. Bespoke permits are written specifically for the particular activity concerned.

9.15 An operator can apply for a standard permit if the activity meets all of the requirements of the standard rules set out in the EPR 2010. Standard permits are normally quicker to obtain (in particular, there is no need for public consultation) and easier to apply for than bespoke permits, and are subject to a fixed charge. In contrast, bespoke permits are written specifically for a particular facility (ie they contain site-specific conditions). Bespoke permits should be applied for when there are no suitable standard permits (for example, if an operator is using a novel technology, the process is complex, or the process will produce high volumes of material or waste). If an operator is unsure as to whether to apply for a standard or a bespoke permit, the EPR 2012 Environmental Permitting Core Guidance (the Core Guidance) recommends requesting a 'pre-application' discussion with the regulator.

9.16 An environmental permit for an operation that generates waste will include a condition concerning the application of the 'waste hierarchy'. The Waste (England and Wales) Regulations 2011 apply the waste hierarchy which was introduced by the revised Waste Framework Directive,[1] ie waste prevention, preparing for re-use, recycling, other recovery and disposal. Operators of existing sites will have the condition attached to their permit when they are next reviewed as part of the EA's periodic review process.

1 Directive 2008/98/EC.

9.17 In addition to the environmental permit, if an operator produces or holds hazardous waste at any premises they must also register as a hazardous waste producer unless the total quantity produced or held is less than 500 kg a year. If the quantity is below this threshold, the Environment Agency does not need to be informed.

9.18 If two or more operators run different parts of a single facility, they will each require an environmental permit for their part of the facility. A single permit can only be granted for more than one regulated facility where:

- the regulator is the same for each facility;
- the operator is the same for each facility; and
- all the facilities are on the same site (subject to limited exceptions).

When is an environmental permit required?

9.19 The EPR 2010 requires that an environmental permit is held for any of the following activities:

- an installation which carries out the activities listed in Sch 1 to the EPR 2010 and any activities that are technically linked;
- a waste operation (as defined in reg 2 of the EPR 2010 by reference to the recovery and disposal operations in the Waste Framework Directive). Any recovery or disposal of waste is a waste operation and will require an environmental permit, unless the waste operation is specifically excluded or exempt from the need to hold a permit;
- a mobile plant (carrying out either a Sch 1 activity or a waste operation);
- a mining waste operation (carrying out the activities set out in para 2(1) of Sch 20); and
- a radioactive substances activity (one of the activities set out in para 11 of Pt 2 of Sch 23).

The Environment Agency's website includes a number of guidance documents on a variety of activities to assist an operator in England and Wales to decide whether or not it needs to apply for a permit.

9.20 The collective term used in the EPR 2010 for installations, waste operations and mobile plant is 'regulated facility'. In 2012, the Environment Agency published updated regulatory guidance on the meaning of 'regulated facility'. There may be more than one regulated facility on the same site. In such a case, there are arrangements in the EPR 2010 to allow for all such facilities to be regulated by the same regulator and to allow, in many cases, for a single permit to be issued (see **9.19** above). Certain waste operations regulated under other regimes are described as 'excluded' waste operations and are not regulated facilities.

9.21 It is an offence to operate a regulated facility without an environmental permit.[1]

1 EPR 2010, reg 12.

Exemptions

9.22 The EPR 2010 introduced significant changes to the waste exemption regime, with 60 new waste exemptions applying immediately. Operators who held a waste exemption under the old system had between 18 months and 3½ years (from 1 April 2010) to register under the new system pursuant to transitional arrangements. Most of the exemptions only relate to the recovery of waste and cannot be relied upon for the disposal of waste.

9.23 Certain waste management activities that pose a sufficiently low risk are exempted from the need to hold an environmental permit. A waste operation must meet certain criteria in order to benefit from such an exemption. There will be no requirement for an environmental permit for an 'exempt facility' if the following requirements are met:[1]

- the requirements in the relevant paragraph of Sch 2 are satisfied;
- it is registered as an exempt operation;
- an establishment or undertaking is registered in relation to it; and
- the type and quantity of waste submitted to the waste operation, and the method of disposal or recovery, are consistent with the need to attain the objectives set out in Art 4(1) of the Waste Framework Directive. (These objectives state that an operation should be carried out without endangering human health and without using processes or methods likely to harm the environment.)

1 EPR 2010, reg 5.

9.24 Defra has published specific guidance on exempt waste operations called 'Environmental Permitting Guidance, Exempt Waste Operations for the Environmental Permitting (England and Wales) Regulations 2010, March 2010'.

The 'low risk' initiative

9.25 The low risk waste position is an enforcement position that the Environment Agency takes when they are informed or come across waste management activities that are low risk but do not benefit from a waste exemption. The Agency recognises that the regulation of the waste sector should be proportionate to risk. As a consequence, its low risk initiative identifies low-risk waste operations that are exempt from the requirement to hold an environmental permit and which do not justify enforcement action.

9.26 An exempt facility can only be exempt from the need to hold a permit where an establishment or an undertaking has been registered. The exemption registration authority (ie the regulator) for each type of exempt waste operation is identified in para 2 of Sch 2 to the EPR 2010 and the No 2 2009 Regs; it is generally the Environment Agency in England and Wales but, for some activities, it is the local authority or Animal Health and Veterinary Laboratories Agency.

9.27 The regulator is required to maintain a register of establishments and undertakings carrying out exempt waste operations. Some exemptions require a greater level of regulatory scrutiny where the level of environmental risk is higher, and these have specific registration requirements; in particular, a notice must be served on the relevant regulator containing:

- the relevant particulars of the establishment or undertaking;
- a description of the waste operation;
- the place where the waste operation takes place; and
- if the waste operation is a WEEE operation, the type and quantity of waste subject to the operation.

9.28 An establishment or undertaking seeking to be registered in relation to an exempt waste operation, or seeking to renew such a registration, must notify the exemption registration authority of the name and contact details of an individual officer or employee designated by the establishment or undertaking to be the primary contact for the purposes of registration.

9.29 An exempt registration is valid for three years and then must be renewed. No fee is chargeable other than for the repair and refurbishment of WEEE.

9.30 Operators of exempt operations will not need to show technical competence (see **9.37–9.40** below) under this new waste exemption regime. Technical competence is only relevant to the process for environmental permits.

9.31 In 2011,[1] further amendments to the EPR 2010 simplified the regulation of very low-risk radioactive substances, including the definition of radioactive waste. Following this, amendments to the EPR 2010 in 2012[2] placed a requirement on the Secretary of State to review the legislation implementing the EPR 2010 by 2017 and then every five years thereafter. It was also confirmed that there would be less complex regulation of combustion of waste-derived fuel where it has ceased to be waste before being burned as a fuel and meets three tests set out in the Court of Appeal decision in *R* (*OSS Group Ltd*) *v EA and others*.[3]

1 SI 2011/2043.
2 SI 2012/630.
3 [2007] EWCA Civ 611. The Court of Appeal set out a test to determine when lubricating oil ceases to be waste. The Environment Agency has since applied the test to any waste-derived fuels. The test sets out that a product ceases to be waste when: 1) the waste has been converted into a distinct and marketable product; 2) the processed substance can be used in exactly the same way as non-waste; and 3) the processed substance can be stored and used with no worse environmental effects when compared to the raw material it is intended to replace.

Temporary storage and waste

9.32 The temporary storage of waste, either at the place of production, a place controlled by the producer or at a collection point before it is sent for disposal or recovery (subject to it being stored for no longer than 12 months and being stored in a secure place), does not need to be registered.

Who must apply for an environmental permit?

9.33 The only person who can obtain or hold an environmental permit is the 'operator', which, in relation to a regulated activity, means:

(a) the person who has control over the operation of the regulated facility,
(b) if the regulated facility has not yet been put into operation, the person who will have control over the regulated facility when it is put into operation, or
(c) if a regulated facility authorised by an environmental permit has ceased to be in operation, the person who holds the environmental permit.

9.34 This is a major difference from the pre-EPR 2007 waste management licensing regime, where the holder of a waste management licence did not necessarily have to be the operator. In order to assist the industry in adjusting to the new regime, transitional provisions in the EPR 2007 allowed the holder of a waste management licence on 6 April 2008 to be treated as the operator for the purposes of the EPR 2007. Permits had to be transferred over time to the operator. The transitional provisions set out in the EPR 2007 were revoked by the EPR 2010, which introduced similar transitional arrangements for existing permits.

The operator

9.35 As noted at **9.33** above, para (b) of the definition of 'operator' provides that a regulated facility need not be in operation for there to be an operator. Legal obligations may also be imposed on an operator during pre- and post-operational phases. The operator must demonstrably have the authority and the ability to ensure that the environmental permit is complied with. The Environment Agency has published guidance entitled 'Understanding the meaning of Operator'. This guidance provides that effective regulation under the EPR 2010 requires both an initial correct identification of the operator of a regulated facility and also continued scrutiny that an operator remains in control over such operations. A permit will not be granted if the regulator considers that the applicant will not be the operator. Similarly, a regulator must not transfer a permit if it considers that the proposed transfer will not be the operator of the regulated facility covered by the proposed transfer.

9.36 The operator guidance provides that the regulator, when considering whether an operator or proposed operator has 'control' over the operation of the regulated facility, will consider, amongst other factors, whether the operator or proposed operator has the authority and ability to:

- manage site operations through having day-to-day control of the plant including the manner and rate of operation;
- ensure that permit conditions that will be imposed or which apply are effectively complied with;
- decide who holds key staff positions and have incompetent staff removed;
- make investment and/or other financial decisions affecting performance of the facility; and
- ensure that regulated activities are suitably controlled in an emergency.

Technical competence

9.37 Under the previous EPA 1990 regime, in order to hold a waste management licence for a specified waste management activity, an applicant was required to demonstrate that they were a 'fit and proper person'. The EPR 2010 do not replicate this test. Instead, the regulator must be satisfied that the applicant is able to operate the facility in accordance with the environmental permit and is "competent".[1]

1 EPR 2010, Sch 5, para 13 (identity and competence of the operator).

9.38 The regulator, when deciding whether or not to grant or transfer an environmental permit, is under an obligation to consider whether the operator or proposed operator is 'competent'. Operator competence supports the objectives of examining and maintaining the operator's ability to carry out the relevant activities and fulfil the obligations of an operator. The regulator must not issue or transfer an environmental permit if it considers that the operator will not operate the facility in accordance with the permit. Defra's Core Guidance[AQ is this the same as the Guidance referred to in para 9.31? if so, include 'Core' there?] provides that the regulator might doubt whether the operator could or is likely to comply with the permit conditions, if, for example:

- the operator's management system is inadequate;
- the operator's technical competence is inadequate;
- the operator has a poor record of compliance; or
- the operator's financial competence is inadequate.

9.39 Operators should be technically competent to operate their facility. The Core Guidance provides that the operator's wider management system should contain mechanisms for assessing and maintaining technical competence. The Environment Agency encourages the development of industry-led competence schemes. Certificates of Technical Competence (CoTCs) are still an appropriate means of demonstrating technical competence for relevant waste operations. The Core Guidance prescribes the specific criteria that waste operations must satisfy in addition to a CoTC.

9.40 There are currently two approved schemes for waste operations. These are:

- the CIWM/WAMITAB scheme developed jointly by the Chartered Institution of Wastes Management (CIWM) and the Waste Management Industry Training and Advisory Board (WAMITAB); and
- the ESA/EU Skills scheme developed jointly by the Environmental Services Association (ESA) and Energy and Utility Skills (EU Skills).

When should an operator apply for a permit?

9.41 The Core Guidance provides that operators should normally make an application for an environmental permit when they have drawn up full designs but before construction work commences (whether on a new regulated facility or when making changes to an existing one). There are no provisions in the EPR 2010 preventing an operator from beginning construction before a permit has been issued; however, regulators may not agree with the design and infrastructure put in place. Therefore, in order to avoid any expensive and timely delays, it is in the operator's best interest to submit applications at the design stages where possible. If the regulated facility also needs planning permission, the Core Guidance recommends that the operator should make both applications in parallel, whenever possible, on the basis that 'this will allow the pollution control regulator to start its formal consideration early on, thus allowing it to have a more informed input to the planning process'. For certain waste operations and certain mining waste facilities, planning permission must be in force before an environmental permit can be granted. Planning permission may also be required for exempt facilities.

What is the procedure for making and deciding applications?

9.42 As set out above, the EPR 2010 provide for two types of environmental permit (ie standard permits and bespoke permits), depending on the nature and risks associated with the regulated facilities.

9.43 Operators must use the application forms provided by the regulator and provide all the information that the regulator needs to make a determination. If an operator fails to do this, the application may not be 'duly made' and the regulator will request additional information, thereby lengthening the application process and the period within which they can determine the application.

9.44 The determination period begins on the date the regulator receives an application which is subsequently determined to be 'duly made'. The period for determining an application for a bespoke permit is four months (and three months for a standard permit). It is always open to the regulator and the applicant to agree a longer period if this is necessary.

9.45 Where the regulator has not determined the application within the required timeframe, the applicant can notify the regulator that it considers the application to be refused. The applicant may then appeal the refusal to the Secretary of State/the Welsh Ministers.

9.46 Generally, public consultation is not required for applications for a standard permit. In relation to applications for bespoke permits, the regulator has a duty to consult and consider representations made pursuant to the determination process during the allowed consultation period, usually 30 days. These representations may be received from members of the public or interested bodies, persons with rights to land, anyone who the regulator considers will be or is likely to be affected by the application, and, if relevant, other Member States.

9.47 If the regulator decides to grant a permit, a person with sufficient interest in the facility may apply to have the decision judicially reviewed. Any application must be filed within three months of the date upon which the grounds to make the challenge first arose (for example, the date of the decision to grant the permit). If the application is successful, the case will be heard in the High Court.

Appeals

9.48 An appeal can be made to the Secretary of State/Welsh Ministers when the regulator has refused an application for a permit, an application is deemed to have been refused, or the regulator has deemed an application to be withdrawn. Time limits for making appeals vary according to the type of appeal.

9.49 The Secretary of State or Welsh Ministers have the power to extend some of the limits, but will only do so in the most compelling circumstances. In summary, an appeal must be made:

- no later than 15 working days after receiving notice that the application is deemed to be withdrawn; or
- in all other cases, within six months of the date of the decision or deemed decision.

9.50 The appeal may be conducted by written representations, or through a hearing or inquiry. The time period for an appeal will depend on the complexities of the case and whether the written representations (which are likely to be quicker) or the hearing procedure is used. It is not therefore possible to set out a precise time period for the appeal process; however, this could take six months or more.

9.51 If an appeal is unsuccessful, an operator may apply to have the decision judicially reviewed. Any claim form would have to be filed promptly and, in any event, not later than three months after the grounds to make a claim first arose. Chapter 5 of the Core Guidance provides further information on the procedures.

Transfer and surrender of environmental permits

9.52 The EPR 2010 allow for permits to be transferred[1] or surrendered.[2] A permit can be transferred or surrendered completely or partially (if the original operator retains control of some of the facility and, in the case of a transfer, another operator takes over the operation of the transferred part of the facility).

The 2012 Amendment Regulations[3] make it easier to transfer an environmental permit if a permit holder cannot be found.

1 Regulation 21.
2 Regulations 24 and 25.
3 SI 2012/630.

9.53 If an operator wishes to transfer all, or part, of a permit to another party, the parties must make a joint application. For a partial transfer, where the original operator retains part of the permit, the application must include a plan identifying which parts of the site and which regulated facility (or facilities) the operator proposes to transfer. Regulators may need to vary permit conditions as a result of a partial transfer, to reflect any shared operations.

9.54 An environmental permit remains in force until it is surrendered, revoked or consolidated. Where a permit is held by personal representatives following the death of an individual permit holder, this will cease if no transfer application or notification is made within six months of the death of the permit holder. Until that time, the operator remains subject to its conditions. When a regulated facility ceases to operate, an operator should (but is not compelled to) seek surrender of the permit, so as to end regulation under the environmental permitting regime and the requirement to pay the associated annual charges for its environmental permit.

9.55 The EPR 2010 set out two separate methods for surrender of a permit. Operators of certain facilities (ie operators of Part B installations, except to the extent that they relate to a waste operation and mobile plants) may simply notify the regulator of a surrender. All other operations require an application to be made to the regulator to surrender the permit.

9.56 The general requirement for permit surrender[1] is that the regulator must accept the surrender of the environmental permit if it is satisfied that the necessary measures have been taken:

- to avoid any pollution risk resulting from the operation of the regulated facility (para 7.22 of the Core Guidance), and
- to return the site of the regulated facility to a satisfactory state, having regard to the state of the site before the facility was put into operation (paras 7.23 to 7.33 of the Core Guidance).

1 EPR 2010, Sch 5, Pt 1, para 14.

Offences

9.57 Proposals to introduce the use of civil sanctions for compliance with the EPR 2010 is ongoing but have still not been brought into play. The Government is currently undertaking work on whether and when such sanctions may be brought into force for offences under the EPR 2010.

9.58 The main offences under the EPR 2010 are:

- operating a regulated facility without an environmental permit;
- operating a regulated facility in breach of any conditions attached to an environmental permit; and
- failure to comply with an enforcement notice or suspension notice.

The regulator must place details of any convictions or formal cautions on a public register. Formal cautions must be removed after five years.

9.59 The offences are punishable by a fine and/or imprisonment.[1] A person found guilty of such offences in the magistrates' court will be subject to a fine of up to £50,000 and/or up to 12 months' imprisonment. If convicted on indictment in the Crown Court, the penalty is an unlimited fine and/or up to five years' imprisonment. Furthermore, where a corporate body is shown to have committed an offence with the consent or connivance of an officer or that is attributable to any neglect on his part, the officer as well as the body corporate is liable to be proceeded against and punished accordingly.[2]

1 EPR 2010, reg 39.
2 EPR 2010, reg 41.

Waste offences

9.60 Almost all corporate liability for waste offences arises either out of a failure to have, or to act in accordance with, an environmental permit[1] or to observe the statutory waste duty of care.[2]

1 EPA 1990, s 33.
2 EPA 1990, s 34.

9.61 All operations where waste is stored, transported or finally disposed of must be carried out pursuant to and in accordance with an environmental permit issued by the Environment Agency. Sections 33 and 34 of the EPA 1990 detail various offences in relation to the unauthorised handling and disposal of waste.

9.62 EPA 1990, s 33 requires that a person shall not:

'(a) deposit controlled waste,[1] or knowingly cause or knowingly permit controlled waste to be deposited in or on any land unless an environmental permit authorising the deposit is in force and the deposit is in accordance with the permit;

(b) submit controlled waste, or knowingly cause or knowingly permit controlled waste to be submitted, to any listed operation (other than an operation within subs 33(1)(a)) that:
  (i) is carried out in or on any land, or by means of any mobile plant, and
  (ii) is not carried out under and in accordance with an environmental permit;

(c) treat, keep or dispose of controlled waste in a manner likely to cause pollution of the environment or harm to human health.'

1 'Controlled waste' is rather vaguely defined in s 75 of the EPA 1990 as '... household, industrial and commercial waste or any such waste'.

UNAUTHORISED DISPOSAL AND HANDLING OF WASTE

The offences

9.63 Whatever the type of waste involved, a person committing an offence under EPA 1990, s 33 will be liable:

(a) on conviction in the Crown Court, to imprisonment for a term not exceeding twelve months or a fine not exceeding £50,000 or both; and

(b) on conviction in the Crown Court, to imprisonment for a term not exceeding five years or a fine or both.

9.64 In an attempt to deal with the growing problem of the fly tipping of waste at unlicensed sites, EPA 1990, s 33(5) places the onus on the owner or manager of vehicles used to transport controlled waste to ensure that their employees, and those to whom they might lend their vehicles, do not fly tip waste. It provides that the person who controls or who is in a position to control the use of the vehicle is to be treated as 'knowingly causing the deposit', whether or not he or she has given any instructions for this to be done. The main rationale would seem to be to attempt to redress the common claim by companies that they cannot be responsible for the unlawful and unauthorised use of their vehicles by employees.

9.65 In *Environment Agency v Melland*[1] the court held that ownership of a vehicle was sufficient to establish evidence of control, or of a position to control, the use of a vehicle for the purposes of an EPA 1990, s 33(5) offence. So, it would seem that businesses need also to monitor, or have written policies in place relating to, the use of vehicles by employees and/or other third parties. The court did, however, recognise that ownership would not point to control in situations involving a vehicle that has been stolen, loaned or hired.

In *Mountpace Ltd v Haringey LBC*[2] the court held that a commercial freeholder who verbally contracted out renovation works on its property, including the removal and disposal of any waste, to an independent contractor had remained a producer of waste and in control of it, with a duty of care under EPA 1990, s 34. It was held that it had been foreseeable that the waste would have to be taken off the site by a transfer to a third party and the freeholder had been criminally liable for its disposal to a 'fly tipper'.

1 [2002] EWHC 904 (Admin).
2 [2012] EWHC 698 (Admin).

9.66 The Clean Neighbourhoods and Environment Act 2005 inserted various new sections, including a new EPA 1990, s 33A, which provides that a person convicted under s 33 offences may be ordered by the court to pay the costs of the Environment Agency in investigating and bringing about the conviction. New s 33B further provides that a person charged under a s 33 offence may be required to pay cleaning-up costs for the land concerned. Lastly, it is now possible under s 33 for the vehicle in which the offender has rights, and which was used in the commission of the offence, to be ordered by the court to be forfeit.

9.67 A significant recent development has been the categorisation of offences under s 33 as 'serious crimes' in the Serious Crime Act 2007 (SCA 2007), Sch 1, Pt 1, para 13. This means that a person involved in committing a s 33 offence, whether by committing, facilitating or being likely to facilitate the serious crime, and:

- where this person is convicted of the serious crime; and
- where there are grounds to indicate that a protection order would protect the public by disrupting involvement,

may be subject to a serious crime prevention order (SCPO).

9.68 An SCPO can be issued by the High Court against those involved in the serious crime. Breach of an order is a criminal offence and can result in a prison sentence. The SCPO includes (amongst other matters) prohibitions, restrictions or requirements that may be imposed on individuals in relation to premises to which an individual has access, an individual's finances, property or business dealings, the use of any premises, or an individual's travel. The effect of the SCA 2007 is to bring fly tipping and other waste offences in line with crimes such as money laundering and drugs trafficking. An SCPO will allow the Environment Agency to act against suspects without taking them to court.

9.69 Under the Proceeds of Crime Act 2002 (POCA), a company and its directors can be prosecuted in respect of any unlawful benefit received when operating without a waste permit. POCA allows a court to deprive convicted offenders of the assets they have gained from crime. Once a court or jury has found someone guilty of a crime, the Environment Agency can confiscate the unlawful proceeds. The Environment Agency may also investigate an offender's finances to locate profits, obtain an order to prevent an offender disposing of assets while investigators identify their illegal profits, obtain a confiscation order for assets to the value of the amount the offender has to pay, and enforce a default prison sentence against the offender if they fail to pay fines. The Environment Agency may also be able to recover its costs from the offender.

9.70 The highest confiscation order secured under POCA for waste crimes in 2011 was over £800,000 in relation to an illegal waste transfer station and scrap metal yard in Slough. In addition to the fine, the owner of the business in question also received a two-year community order. The total amount confiscated in 2011 under POCA was £2,246,000, with 16 confiscation orders being made.

9.71 The Environment Agency launched a task force in December 2011 to crack down on serious and organised waste crime, and published its first waste crime report for 2011–2012 in July 2012. The Environment Agency pursued more than 400 waste-related prosecutions in 2010/2011, and 335 in 2011/2012. A total of 759 illegal waste sites were closed, and 262 incidents of illegal dumping were dealt with in 2011/2012. Total fines in 2011 amounted to £1,700,000, with the highest fine being £170,000, and 16 custodial sentences were imposed, the longest being 27 months. The equivalent figures in 2010 were £942,000, £50,000, five and 16 months respectively.

Defences

9.72 It is a defence under the EPR 2010 that an act alleged to constitute an offence under regulation 38(1), (2) or (3)[1] was done in an emergency in order to avoid danger to human health in a case where:

- the party concerned took all such steps as were reasonably practicable in the circumstances for minimising pollution; and
- particulars of the acts were furnished to the regulator as soon as reasonably practicable after they were done.[2]

1 Regulation 38(1), (2) and (3) makes it an offence: to contravene, or knowingly cause or knowingly permit the contravention of, reg 12 (requirement for an environmental permit); to fail to comply with or to contravene an environmental permit condition; and to fail to comply with the requirements of an enforcement notice, a suspension notice or a landfill closure notice.

2 EPR 2010, reg 40.

9.73 There are similar defences to EPA 1990, s 33 proceedings, and these are set out below.

9.74 There are two statutory defences available to companies or individuals charged with s 33 offences:[1]

- that the company or individual took all reasonable precautions and exercised all due diligence to avoid the commission of an offence; or
- that the act alleged to constitute the contravention was done in an emergency in order to avoid danger to the public and that, as soon as reasonably practicable thereafter, full details were given to the local Environment Agency offices.

1 EPA 1990, s 33(7).

9.75 It is important to note that the defence of acting on an employer's instructions has been removed,[1] although it still exists for offences committed before 7 June 2005. This will have a significant impact on potential liability of companies where management structures necessitate the delegation of responsibilities. Most commonly invoked is the 'reasonable precautions'/'due diligence' defence. Companies that can show that they had adequate management systems in place may be able to prove a defence. Co-operation at the early stages of any investigation by the Environment Agency is recommended where any company seeks to rely on such a defence.

1 Clean Neighbourhoods and Environment Act 2005, s 40.

9.76 The Environment Agency has extensive powers to investigate and bring proceedings in relation to waste offences, and the courts are willing to award considerable fines and custodial sentences. In the 1997 case of *R v Hertfordshire County Council, ex p Green Environmental Industries*,[1] the director of Green Environmental Industries was sentenced to 18 months' imprisonment in relation to the unauthorised disposal of clinical waste and a further nine months' imprisonment in relation to fraudulent activity.

1 [1998] Env LR 153.

9.77 In July 1998, John Hawksworth was prosecuted for running an illegal landfill operation.[1] He was sentenced to 21 months' imprisonment. Mr Hawksworth had a long history of waste-related offences but continued accepting waste onto his site and ignored an Environment Agency waste removal notice (served pursuant to EPA 1990, s 59) requiring him to remove the waste on his land. Thereafter, the Environment Agency obtained evidence that up to 7,000 tonnes of industrial waste had been deposited at the site over a period of two years, which had raised the land by a height of eight metres. Mr Hawksworth pleaded guilty to 19 charges of depositing, keeping and burning waste on unlicensed land, contrary to EPA 1990, s 33.

1 ENDS Report, July 1998, No 282.

9.78 EPA 1990, s 59 empowers authorities to serve land clearance notices in relation to unlawfully deposited waste. It is an offence to fail to comply with an s 59 notice and, in certain circumstances, the Environment Agency may carry out the clearance itself and seek to recover its costs from the person(s) served with the notice.

9.79 The longest ever prison sentence for waste crime offences was given in June 2011 to organised crime boss Hugh O'Donnell, after a three-year investigation by the Environment Agency and New Scotland Yard showed he had laundered millions of pounds in illegal profits.[1] Mr O'Donnell ran an illegal waste site near Reading, spread over land the size of five football pitches. He was ordered to repay more than £800,000 under POCA and warned he could face up to four and a half years in prison if he did not pay.

1 Environment Agency 2012.

9.80 In June 2012, the owner of a haulage and waste management company was ordered to pay more than £100,000 in fines and costs after pleading guilty to illegally disposing of thousands of tonnes of waste at two sites in Cornwall, including the illegal use of unsuitable waste to create an earth bund.[1]

1 Environment Agency 2012.

9.81 In June 2011, a West Yorkshire man was given a 12-month prison sentence at Selby Magistrates' Court after the Environment Agency found thousands of tyres dumped illegally at a York airfield and other locations around the city.[1] The man admitted to the offences and was found guilty after a trial on a charge relating to fly tipping at Goodmanham near Market Weighton.

1 Environment Agency 2012.

9.82 In Derby in 2012, a composting firm was fined £75,000 and ordered to pay over £13,000 costs after pleading guilty to three charges of depositing controlled waste on land without a permit. The firm had supplied a farmer with compost that was found to contain a level of plastic contamination between six and ten times the permitted limit, and therefore should have been classified as waste.

9.83 In Scunthorpe, three directors of a waste recycling company were ordered to do 200 hours of unpaid work for operating a regulated waste facility without a permit. They were also ordered to pay over £4,000 costs and were disqualified from being directors of a company for five years.

In December 2011, the owner of a Devon waste company was fined £21,000 and costs for storing illegal amounts of food waste at three farms and causing odour problems.

9.84 In April 2012, a gang of waste bosses in Preston were successfully prosecuted for illegally storing tonnes of dangerous chemical waste, some of which ended up fly-tipped across the North of England. They were given prison sentences of 18 months, 12 months and 9 months respectively.[1]

1 Environment Agency 2012.

9.85 As detailed above, it is an offence to fail to comply with permit/licence conditions. In April 2007 a waste management company was fined £10,000 with £1,292 costs for contravening a condition of its licence, contrary to EPA 1990, s 33(6). The waste management licence required clean-up of any litter by the end of each day, and the Environment Agency officer found that this had not been carried out on the day of their inspection.[1]

A scrap merchant, Easco (Wheelers) Ltd, was ordered to pay £160,000 for operating a facility outside the hours stated in its waste management licence, resulting in noise exposure to local residents.[2]

1 ENDS Report, August 2007, No 391.
2 ENDS Report, April 2007, No 387.

DUTY OF CARE

EPA 1990, s 34

9.86 The statutory duty of care was introduced by EPA 1990, s 34. It applies to all persons in the waste management chain and requires all persons involved in the 'cradle to grave' handling of waste (ie from producers to final disposers) to take all reasonable and appropriate measures to ensure the following:

- Waste is only kept, treated, deposited or disposed of in accordance with an environmental permit or other authorisation. This will require that the original producer or holder of the waste checks that all subsequent holders of the waste within the chain are duly authorised and that the waste has been disposed of correctly. Technically, transfer to a waste contractor will not relieve a waste producer of his duty.
- Waste does not escape from the control of the holder. This requires that the waste is stored or packaged properly (for example, in drums or secure containers) and lorries are covered to prevent waste becoming windborne.
- Waste is only transferred to authorised persons, such as registered waste carriers or licensed disposal operations permitted to accept that type of waste. As above, this can extend to checking authorisations, including confirming that they are still valid before the waste is transferred and, for more specialist operations, can even extend to site visits.
- All transfers/movements of the waste are accompanied by an adequate written description (a 'waste transfer note') which will allow the waste to be identified and handled correctly. Following the enactment of the Waste (England and Wales) Regulations 2011,[1] from 28 September 2011 all waste transfer notes must include a declaration from the transferor of the waste that the waste management hierarchy has been applied (ie the transferor must have considered reusing or recycling its waste before deciding to dispose of it).[2]

1 SI 2011/988.
2 Note that, unlike in England and Wales, in Scotland a new section 34(2A) has been inserted into the EPA 1990 to amend the duty of care to require 'any person who produces, keeps or manages controlled waste, or as a broker or dealer has control of such waste, to take all such measures available to that person as are reasonable in the circumstances to apply the waste hierarchy'.

9.87 The Waste (England and Wales) Regulations 2011 are supplemented by a statutory Code of Practice which was originally issued in 1991 and was re-released by the Department of the Environment in March 1996. An updated version of the Code was released for consultation in April 2009, but a new Code was not subsequently published. The Government has indicated that it intends to consult on a new Code of Practice in 2014 to reflect the most recent changes in law.

Duty of care offences

9.88 Failure to comply with the duty of care (as set out in the EPA 1990, s 34 and its accompanying regulations) is amongst the most commonly prosecuted of environmental offences. Moreover, while it is not an offence to fail to follow the Code of Practice, the Code can be taken into account by the court when considering any failure to comply with the s 34 duty itself and the penalty to be imposed.[1] The Code of Practice is an extremely useful document, with advice on waste labelling, secure storage and transport, as well as completion of the appropriate waste tracking documents.

1 EPA 1990, s 34(10).

Waste transfer notes

9.89 The requirement in the EPA 1990 for an adequate description of the waste has been extended, in the supporting regulations, to require a waste transfer note (WTN) to accompany each movement of waste, with paper copies being retained for future inspection.[1] The WTN itself has no definitive form, although the Environment Agency has provided a style which parties may use. As a minimum, the WTN must record certain information, including the type and quantity of the waste including the type of container, details of the transferor and transferee, as well as the time and place of transfer. As noted previously, from 28 September 2011 all WTNs must also include a declaration from the transferor of the waste that the waste management hierarchy has been applied, which means the transferor of the waste must consider reusing or recycling its waste before deciding to dispose of it. WTNs must be retained for at least two years, and three years if they relate to hazardous waste.[2]

1 In March 2012 the Government announced plans to streamline environmental legislation. One of the proposals was to use other forms of evidence (for example, invoices) to record information covering the transfer of waste instead of requiring separate WTNs.

2 The Environment Agency, in partnership with related governmental and non-governmental bodies, is developing an electronic duty of care ('edoc') programme to replace the current paper-based system of WTNs. The edoc programme will create a national web-based system to capture information from production to collection, transportation, treatment and disposal of waste. It is expected to go live in 2014.

9.90 While hazardous wastes are also subject to the duty of care, they must comply with a different transfer note system. Any premises in England and Wales which produce or hold hazardous waste must be registered on an annual basis with the Environment Agency, unless the total quantity of hazardous waste produced or held is less than 500kg each year. It is an offence to remove hazardous waste from a site unless it is registered or exempt.

Duty of care penalties

9.91 It is an offence not to comply with the duty of care or any regulations made under s 34 of the EPA 1990. Duty of care offences are punishable with a fine of up to £5,000 in the magistrates' court, rising to an unlimited fine if prosecuted upon indictment in the Crown Court. Proposals contained in Defra's 2008 consultation on the duty of care were to increase the level of the fine to £10,000. This would require a change to primary legislation. It is not known when or if this amendment will be introduced.

Following enactment of the Regulatory Enforcement and Sanctions Act 2008, the Environment Agency in England and Wales may also impose a fixed penalty notice (currently set at £300) should a party fail to comply with the obligation under s 34(5) of the EPA 1990 to retain and provide copies of WTNs. In addition, the Environment Agency, Waste Collection Authorities and the police have powers to search and seize vehicles which they reasonably believe have been, are being or are about to be, used to commit an offence under s 34 of the EPA 1990.[1]

1 EPA 1990, ss 34B and 34C.

Duty of care case law

9.92 The case of *London Borough of Camden v Mortgage Times Group Ltd*[1] has clarified an element of the duty of care. The cleaners of a company office had, on several occasions, placed bags of shredded paper by the side of the road, hours before the time at which business waste could be put out for collection. The rubbish was 'controlled waste', and therefore the duty of care under s 34(1)(b) imposed a duty on the company to 'take all such measures … as are reasonable in the circumstances … to prevent the escape of the waste from his control or that of any other person'.

1 [2006] EWHC 1615 (Admin).

9.93 Previous case law[1] had held that the use of the word 'escape' precluded the act of depositing bags of waste for collection from being caught by the duty. *London Borough of Camden v Mortgage Times Group Ltd* has now made it clear that s 34(1)(b) actually imposes a duty to take reasonable measures to prevent an escape of waste. It is the failure to take reasonable steps which constitutes the offence; an escape of waste is not a prerequisite of liability.

1 *Gateway Professional Services (Management) Ltd v Kingston upon Hull City Council* [2004] EWHC 597 (Admin).

9.94 The case of *Milton Keynes Council v Leisure Connection Ltd*[1] also provides a reminder of the need for businesses to take clear and demonstrable steps to manage their waste and evidence compliance with the duty of care. Here, the High Court noted that the duty under s 34(1)(b) of the EPA 1990 is to take *all* reasonable measures to prevent the escape of waste. It was therefore determined that simply providing facilities for waste disposal may not be enough to discharge that duty. Rather, businesses should additionally ensure they set up management systems and/or provide training to ensure that employees and others use the waste disposal facilities correctly and record these steps to evidence the efforts made to discharge their duty.

Similarly, in *Mountpace Ltd v Haringey LBC*,[2] the Divisional Court determined that a property owner who verbally engaged a contractor to remove and dispose of waste during renovation was liable under s 34(1)(c) of the EPA 1990 when the contractor subsequently fly tipped some of that waste. Whilst the owner was not present on site when the waste was transferred to the contractor, the court was satisfied both that the owner remained in legal control of the waste and that it was reasonably foreseeable that the waste would have to be taken off the site by a transfer to a third party. The court also noted that the owner had not put in place any mechanism to comply with its duty of care under s 34(1)(c) on the occasion of any transfer of waste – rather, it had chosen to trust the contractor to comply with its statutory requirements on waste disposal, particularly since the contractor had worked for the owner before. It was determined that this assumption as to the reliability of the contractor was insufficient to discharge the owner's duty.

1 [2009] EWHC 1541 (Admin).
2 [2012] EWHC 698 (Admin).

9.95 The 2009 prosecution of a director of a firm dealing in electrical waste contaminated with various chemicals demonstrates the impact that a breach of the duty of care under s 34 can have. A director of Eurotech Waste Management (now in liquidation) was sentenced to 240 hours of unpaid work for the community as a result of the prosecution and was disqualified from holding a company directorship or being involved in the management of a company for a period of two years. No fine was imposed, due to the limited means of the director in question. The judge described the management of the site as 'lamentable'. The serious outcome of the director in question being 'out of his depth' reinforces the need for directors of any company involved in waste management to ensure that they are aware of, and comply with, their duty of care under s 34 of the EPA 1990.

SITE WASTE MANAGEMENT PLANS

9.96 The government has introduced a mandatory requirement for Site Waste Management Plans (SWMPs) for new construction projects of a value greater than £300,000 (excluding VAT). The aim is to reduce construction waste and fly tipping and improve resource efficiency. SWMPs are currently only required for construction projects in England, although SWMPs are promoted as a means of reducing the production of construction waste in Scotland, Northern Ireland and Wales. Recent legislation[1] has given Welsh Ministers the power to introduce regulations requiring SWMPs; however, thus far, they have chosen not to use these powers. This may be because the UK government is currently reviewing SWMP Regulations with a view to abolishing them by October 2013.

1 Waste (Wales) Measure 2010, s 12.

9.97 The Site Waste Management Plans Regulations 2008 (the 'SWMP Regulations')[1] came into force on 6 April 2008, and a SWMP must be in place for all qualifying construction projects before construction work begins. Projects planned on or before 6 April 2008, or projects where construction has commenced before 1 July 2008, are not affected by the SWMP Regulations. A SWMP is not needed if the construction project has a Part A environmental permit.

1 SI 2008/314.

9.98 SWMPs are plans which identify the different types and amounts of waste expected to be produced by a project and the waste management action proposed for the different types of waste. They will also detail the nature of the project along with design and materials specification.[1] They will detail how waste is going to be disposed of and the recycling or re-use options available for the different waste streams produced by the project. There is a requirement for construction companies to declare that they will take all reasonable steps to ensure duty of care compliance and efficient handling of waste.[2] Waste quantities produced also need to be recorded. Responsibility for writing, implementing and updating the SWMP falls to one individual, who must be identified on the SWMP, along with the client and principal contractor.[3] The author of the initial SWMP will be the client, who will then pass the SWMP to the principal contractor, who will keep the SWMP on site during construction (and for two years after the project completes). SWMPs must be updated throughout the course of a project (with differing requirements depending upon whether the construction value of a project is between £300,000 and £500,000 (excluding VAT), or greater than £500,000 (excluding VAT)).

1 SWMP Regulations, reg 6(3).
2 SWMP Regulations, reg 6(5).
3 SWMP Regulations, reg 6(1).

9.99 Where the project total value of materials and labour is between £300,000 and £500,000 (excluding VAT), the SWMP is required to contain information on the types of waste removed from the site, the party removing the waste and where the waste was delivered to. Projects costing more than £500,000 (excluding VAT) will require an even more detailed SWMP. In addition to all the information required for lower-value projects, such SWMPs also need to include waste carrier registration details, description of the waste and environmental permits or exemptions held. SWMPs should be prepared as early as possible in the project planning process, and certainly before commencement of construction.

9.100 Enforcement falls to local authorities and the Environment Agency, with fixed penalties for failure to make, keep or produce a SWMP. Offences include failure to update SWMPs,[1] failure to make the SWMP available to every contractor carrying out the work detailed in the SWMP,[2] and knowingly or recklessly making a false or misleading statement in a SWMP. A person found guilty of an offence under the Regulations will be liable on conviction in the magistrates' court to a fine of up to £50,000. Conviction in the Crown Court will result in a penalty or an unlimited fine. Where an offence is shown to be committed with the consent or connivance of, or as a result of the neglect of, a company director/manager/secretary or person purporting to be of that capacity, that person may be guilty of an offence under the Regulations.

1 SWMP Regulations, regs 7(3) and 8(5).
2 SWMP Regulations, reg 9(3).

9.101 In March 2012, Defra announced its intention to abolish the SWMP Regulations. Defra described the Regulations as unnecessary and also stated that the burdens they impose do not help achieve environmental goals. Defra ran a consultation between 16 June and 18 July 2013[1] to ensure that it understood the implications of the proposed repeal of the SWMP Regulations. Defra's original intention had been to abolish the Regulations by October 2013.[2] Defra published a summary of the responses that it received,[3] which stated in its conclusion:

> 'To summarise the key findings above, opinions on the proposal to repeal were quite mixed with an even split between those in favour and those against. A slight majority indicated that they were in agreement with the impacts identified in the impact assessment that had been produced, and significantly, respondents were unable to provide sufficient evidence that there would be any additional impacts and no evidence that impacts would be very different to those expected by Government. The majority of respondents also indicated that they would continue to use SWMPs or a similar process if the Regulations were repealed, which gives more substance to that expectation.
>
> A common theme running through responses was that further strides can be made in reducing waste, particularly with a focus further up the hierarchy to resource efficiency and waste prevention. Waste prevention is a complex issue and will require many different actions in order to deliver significant change. Resource efficiency and waste prevention can deliver real financial savings to business, and therefore the incentives to take action already exist. There are many alternative levers to legislation to prevent waste and Government will consider which interventions are most appropriate to support action by business.'

1 DEFRA Site Waste Management Plans Consultation, available online at https://consult.defra.gov.uk/waste/site_waste_management/consult_view.
2 Defra, 'Red Tape Challenge – Environment Theme Implementation Plan', September 2012, at p 21 (www.defra.gov.uk/publications/files/pb13819-red-tape-environment.pdf).
3 'Proposed repeal of construction Site Waste Management Plan Regulations (2008) Summary of responses and Government response', available online at www.gov.uk/government/uploads/system/uploads/attachment_data/file/237398/site-waste-manage-consult-sum-resp-20130830.pdf.

TRANSFRONTIER SHIPMENT OF WASTE

9.102 When waste crosses national borders, international, EU and national regulations will apply. In the UK, this will be of primary concern to movements of waste for recovery and recycling in other countries.

9.103 The1989 Basel Convention on the Control of Transboundary Movements of Hazardous Wastes and their Disposal came into force in May 1992, in response to international concern regarding the threat that the unregulated transportation of hazardous waste was posing to the environment and human health. The European Community ('EC') became a party to the Basel Convention in 1994, by virtue of Council Decision 93/98/EEC, and it was the Convention that formed the basis for the 1993 EC Regulation on Transfrontier Shipments of Waste.[1]

In light of the Basel Convention and the economic value of recovering materials from waste, the Organisation for Economic Co-operation and Development ('OECD') adopted Decision C(2001)107/final.[2] This further developed the framework for OECD countries to control the transboundary movement of waste destined for recovery operations within the OECD area. This decision established a classification of wastes comprising 'Green List', 'Amber List' and 'Red List' materials, based on their perceived environmental risk, with greater controls on the more hazardous 'Red List' wastes. This regime has now been incorporated into EC legislation by virtue of the EC Waste Shipment Regulation ('WSR'),[3] which subsequently updated the1993 EC Regulation.

1 Council Regulation (EEC) 259/93.
2 As amended by Decision C(2004)20.
3 Regulation (EC) 1013/2006.

9.104 The WSR applies to all international waste shipments for all purposes, and therefore includes waste shipments under the Basel Convention and OECD control system. Again, it classifies waste into 'green', 'amber' and 'red' categories, and the movement of waste is regulated based upon:

- the type of waste;
- whether the waste is being shipped for disposal or recovery purposes; and
- the destination country of the waste (OECD, non-OECD, EU Member State or a signatory to the Basel Convention).

In accordance with the WSR, a notifier (producer, collector, dealer, broker or holder of waste) must submit a prior written notification through the relevant authority in the country of dispatch. This notification must be accompanied by a movement document and a concluded contract with the consignee of the waste. Additionally, the notifier must provide evidence that he has in place a form of financial surety, bond or guarantee to cover the cost in the event that the waste is returned to the country of despatch.

9.105 The WSR directly transposes the Basel Convention and the OECD Decision into UK domestic law by virtue of its direct effect. However, national legislation in the guise of the Transfrontier Shipment of Waste Regulations 2007 (' the TSWR')[1] was enacted to set out offences and penalties for breaches and to specify the competent authorities to deal with such breaches. The TSWR also require the Secretary of State to prepare a UK policy on waste shipment for disposal into and out of the UK, namely the UK Plan for Shipments of Waste ('the UK Plan').[2]

1 SI 2007/1711.
2 Department of the Environment (May 2012).

9.106 The UK Plan is a legally binding document and implements the government's waste management plan by strictly limiting when waste may be shipped to or from the UK for disposal. However, there are exceptions allowing for the transportation of certain types of waste in an emergency, for trial waste disposal purposes, and where a specified method of disposal is not available in the country producing the waste. The UK Plan also provides guidance to the UK competent authorities when making decisions on proposed waste shipments.

9.107 The Environment Agency is the competent authority for transfrontier movements of waste in England and Wales;[1] in Scotland, it is the Scottish Environment Protection Agency (SEPA); and, in Northern Ireland, the Northern Ireland Environment Agency. As well as checking consent applications, financial security provisions and shipment documents prior to granting approval to movements, the competent authority will also prosecute breaches of the TSWR.

1 TSWR, reg 6.

9.108 The Court of Appeal has recently considered elements of the transfrontier shipment of waste legislation.

KV & Others v R[1] involved prosecutions for transporting waste destined for recovery in a non-OECD, contrary to reg 23 of the TSWR. Regulation 23 makes it a criminal offence to breach article 36 of the WSR which, in turn, prohibits exports of hazardous and other specified waste destined for recovery to non-OECD countries.

1 [2011] EWCA Crim 2342.

9.109 It was the defendant's argument that, in accordance with the definition in article 2(31) of the WSR,[1] an export does not occur until the waste actually leaves the EC. This construction of 'export' was rejected, and the court held that waste can be destined for recovery in a non-OECD country before it reaches the actual point of leaving the EC. It was held that the action of waste leaving the EC has a transactional nature and it is a process commencing at the 'point of origin, and continuing until the waste reaches its ultimate destination in the foreign country'.[2] The court based this interpretation on the plain meaning of the words in article 36.

The court also resisted arguments that reg 23 of the TSWR was incongruent with article 36 of the 1993 EC Regulation, in that it created a much broader offence. Regulation 23 makes it an offence to 'transport' waste destined for recovery in a non-OECD country, and the definition of 'transport' includes the notifier of waste, any transporter, any freight forwarder and any other person involved in the shipment of waste.[3]

1 Article 2(31) of the WSR states that ' "export" means the action of waste leaving the Community but excluding transit through the Community'.
2 [2011] EWCA Crim 2342 at 29.
3 TSWR, reg 5.

9.110 In *KV & Others v R*, the defendants submitted that the wide definition of 'transport' within the TSWR has the effect of criminalising a wider range of participants in the export shipment chain than article 36 of the 1993 EC Regulation. It was therefore incompatible with EU law and *ultra vires*. Additionally, it was argued that this created a strict liability offence (punishable by imprisonment) that could be committed by a greater range of individuals and organisations and that this was disproportionate.

9.111 Given that the court had found that the proper construction of 'export' under article 36 encompassed all stages in the export chain (from the point of origin in the country of dispatch onwards) and, moreover, because article 50[1] of the WSR requires Member States to impose penalties for breaches, the court also concluded that reg 23 successfully transposes the UK's obligations and was therefore not incompatible with EU law or disproportionate.

It is worth noting that, at the time of writing, the defendants have been given leave to appeal to the Court of Appeal as to whether reg 23 is a strict liability offence.

The case has wide ramifications in terms of the scope of the transfrontier shipment of waste regime in relation to criminal culpability. There are considerable duties and obligations imposed by the legislation that are applicable throughout all stages in the export chain.

1 Article 50 of WSR states that 'Member States shall lay down the rules on penalties applicable for infringement of the provisions of this regulation and shall take all measures necessary to ensure that they are implemented. The penalties provided for must be effective, proportionate and dissuasive. Member States shall notify the Commission of their national legislation relating to prevention and detection of illegal shipments and penalties for such shipments'.

PRODUCER RESPONSIBILITY

Generally

9.112 The Environment Act 1995 introduced the concept that a producer bears responsibility for the environmental consequences of its products. These new provisions were to facilitate implementation of the EU Packaging and Packaging Waste Directive,[1] but went much wider by giving the Secretary of State a power to introduce regulations governing '… such products or materials, as may be prescribed'.

1 Council Directive 94/62/EC.

9.113 Since 1995, an ever-increasing number of EU laws have been introduced to force producers to take responsibility for the environmental impact of their products, especially at 'end of life' and when they become waste. In the UK, a regulatory scheme in respect of packaging and packaging waste was introduced in 1997. This has been followed by the introduction of regulations on waste electrical and electronic equipment (WEEE), end of life vehicles (ELVs), and batteries and accumulators.[1] Within the current trend towards lighter-touch regulation and 'red tape' cutting, it remains to be seen if further producer responsibility legislation in relation to other products (for example, newspapers and magazines) will follow.

1 Council Directive 2006/66/EC on batteries and accumulators and waste batteries and accumulators.

Packaging waste

9.114 The Producer Responsibility Obligations (Packaging Waste) Regulations 1997[1] (the 1997 Regulations) implemented (in part) the provisions of the 1994 EC Directive on Packaging and Packaging Waste. The 1997 Regulations were revoked and replaced by the Producer Responsibility

Obligations (Packaging Waste) Regulations 2005[2] (the 2005 Regulations). In short, they impose obligations, on producers of both packaging and packaging waste, to recover and recycle the waste packaging for which they are deemed responsible (ie the packaging they make or in which they place their products). As such, the EC obligations on the UK are a shared obligation, spread across the parties in the packaging supply chain (from producers of packaging through to manufacturers and retailers).

1 SI 1997/648.
2 SI 2005/3468.

9.115 The 2005 Regulations (further amended in 2007, 2008 and 2010) set out how producers must comply with the packaging waste regime and meet targets for recovery and recycling; and they sit alongside the Packaging (Essential Requirements) Regulations 2003,[1] which establish minimum standards and requirements for packing of products.

1 SI 2003/1941.

9.116 Packaging waste recovery and recycling targets (which obligated parties must meet) increase over time. Each waste stream has its own targets. The overall recovery targets are set out below:

2008 – 72%
2009 – 73%
2010 – 74%
2011 – 74%
2012 – 74%
2013 – 75%

9.117 A new Packaging Directive is expected in 2014. The draft directive plans to expand the types of packaging covered by the regime. For example, the definition of 'packaging' may, in the future, expressly include CD and video cases, flower pots, tubes and cylinders around which flexible material is wound, clothes hangers, plastic film, and refillable steel cylinders used for various kinds of gas, excluding fire extinguishers. This could increase the types of waste that may be included in the waste recycling stream, the quantities of waste recovered, and the quality of recycling.

9.118 Since the introduction of the 1997 Regulations, packaging recycling has improved significantly, from just 27 per cent in 1997 to over 57 per cent in 2008 and 61 per cent in 2010.[1] However, the recession has seen a dramatic drop in prices for recyclable materials, and there is uncertainty as to whether the UK will continue to meet its packaging recycling targets, due to the decline. There is a particular concern with regard to whether enough steel will be recycled to maintain current levels. According to Defra, local authority-collected waste sent to landfill decreased by 2.7 per cent to 25.6 million tonnes, while household waste sent to recycling, composting or reuse increased by 1.3 per cent between 2010/11 and 2011/12.

1 Defra News Release Ref: 362/07 dated 11 October 2007.

9.119 In order not to impact unduly on small businesses, the UK scheme only applies to those packaging waste producers who have an annual turnover of above £2 million and who handle more than 50 tonnes of packaging a year. Groups of companies will need to consider, in aggregate, the total amount of packaging handled and the turnover of all its subsidiaries.

9.120 All obligated businesses must either register with the Environment Agency and be allocated an annual recovery and recycling target or join a registered compliance scheme. If obligated, it is an offence to fail to register with the Agency or a scheme, punishable on summary conviction by a fine not exceeding the statutory maximum (currently £5,000) and, on conviction on indictment, to an unlimited fine.

9.121 Fines for committing waste packaging offences have been relatively modest. On 22 January 2009, a Guildford-based wine distributor (Les Caves de Pyrene) was ordered to pay £33,626 for packaging offences. The firm was charged with failing to register, recycle and produce evidence of recycling between 2000 and 2007.[1]

1 ENDS Report 407, December 2008.

9.122 In January 2008, a magistrates' court fined Western Wines Ltd £225,000 for failing to comply with the regime. The fine is the largest recorded for a packaging waste prosecution to date, and is in contrast to the fines in the early days of packaging waste prosecutions. Investigations showed that the company had failed to comply with the requirements of the regime over a three-year period, and as a result had avoided costs of around £187,000.

9.123 Despite these levels of fines, they may not be high enough, as companies continue to profit from failing to comply with the regulations. For example, wine importer Berkmann was fined more than £48,000 on 20 July 2007, despite the Environment Agency estimating that the firm saved over £72,000 by avoiding its obligations.[1]

From January 2011, a certain number of offences became punishable by the Environment Agency by way of a civil sanction.[2] The Environment Agency may now impose non-criminal or civil sanctions, including compliance notices, restoration notices, stop notices and monetary penalties (fixed/variable) for packaging waste offenders. These civil sanctions provide an alternative to criminal prosecution.

1 ENDS Report, 391, August 2007.
2 Regulatory Enforcement and Sanctions Act 2008.

Registered compliance schemes

9.124 Compliance schemes need to register with the Environment Agency, Scottish Environment Protection Agency (SEPA) or the EHS in Northern Ireland, and they are then responsible for their members' recovery and recycling obligations. The scheme contracts with those facilities undertaking the recovery/recycling, and provides the regulators with evidence of compliance on behalf of its members.

9.125 Accredited recovery/recycling processes issue packaging recovery notes (PRNs) for the packaging waste they process. Where packaging waste is exported for recovery, accredited exporters are also entitled to issue packaging waste export recovery notes (PERNs). Both PRNs and PERNs are accepted as evidence of compliance with recovery and recycling obligations by the relevant Regulator. An online system, the National Packaging Waste Database (NPWD), started operating on 1 February 2007 and allows obligated businesses to provide packaging data to the Environment Agency online.

End of Life Vehicles (ELVs)

9.126 Every year, ELVs generate between eight and nine million tonnes of waste in the European Union. ELVs are vehicles which fall within the definition of waste as set out in the Waste Framework Directive. In the UK alone, around two million vehicles a year become waste. As a result, in 1997 the European Commission adopted a proposal for a Directive which aimed at making vehicle dismantling and recycling more environmentally friendly, setting clear quantified targets for re-use, recycling and recovery of vehicles and their components, and pushing producers to manufacture new vehicles with a view to their recyclability. The EU Directive on End of Life Vehicles (the 'ELV Directive') came into force in October 2000. The ELV Directive requires Member States to take measures to:

- ensure that ELVs can only be scrapped ('treated') by authorised dismantlers or shredders, who must meet high environmental treatment standards;
- require 'economic operators' (this term includes manufacturers, dismantlers and shredders among others) to establish adequate systems for the collection of ELVs from the outset;
- ensure that last-owners must be able to return vehicles free of charge to these systems from January 2007;
- require vehicle manufacturers or importers to pay 'all or a significant part' of the costs of take back and treatment from January 2007;
- set rising re-use, recycling and recovery targets which must be met by January 2006 and 2015. Article 7 (Re-Use and Recovery) sets re-use and recycling targets for post-1980 cars of 85 per cent, and re-use and recycling targets of at least 80 per cent by weight per vehicle (model) per year from 2006 to 31 December 2014; and
- restrict the use of certain heavy metals in new vehicles from July 2003.

9.127 The UK transposed the ELV Directive through the End-of-Life Vehicles Regulations 2003[1] and the End-of-Life Vehicles (Producer Responsibility) Regulations 2005.[2] The Environmental Permitting (England and Wales) Regulations 2010[3] implement the provisions of the ELV Directive concerning the permitting requirements for the storage and treatment of ELVs in England and Wales.

1 SI 2003/2635 (amended by SI 2010/1094).
2 SI 2005/263 (amended by SI 2010/1095).
3 SI 2010/675 (corrected by SI 2010/676).

9.128 The 2003 Regulations put in place most of the requirements of the Directive, including:

- restrictions on the use of certain heavy metals in vehicle manufacture (Pt III);
- marking of certain rubber and plastics vehicle components, and publication of design and dismantling information (Pt IV);
- the introduction of a Certificate of Destruction (CoD) system (Pt V). The CoD system was established to prove that scrap cars are recycled to the required environmental standards;
- free take-back of vehicles put on the market from 1 July 2002, if having no value when scrapped (Pt VI); and
- licensing of authorised treatment facilities, and the site and operating standards with which they must comply (Pt VII).

9.129 The remaining provisions of the ELV Directive were implemented into UK legislation through the 2005 Regulations that came into force on 3 March 2005. The 2005 Regulations transpose those aspects of the ELV Directive relating to producer responsibility for establishing collection systems to take back end-of-life vehicles from 2007 and the arrangements for meeting re-use, recycling and recovery targets from 2006.

9.130 A public register, showing whether vehicle producers have met their 2006 re-use, recovery and recycling target of at least 85 per cent for their brand of vehicles, is published on the BIS website.

9.131 If a business manufactures or imports certain motor vehicles (essentially, cars and vans below 3.5 tonnes), it will have a number of responsibilities, including:

- the need to register with BIS;
- a free take-back system for the ELVs that it is responsible for. Such a system must be approved by BIS. If a business sells or transfers its business to another person, it can still retain legal liability for its ELVs. It will be important to include provisions in the sale contract if a producer does not wish to retain such liability;
- design and information requirements. Vehicles must comply with certain design requirements, including a restriction on the use of certain heavy metals in vehicle components (cadmium, lead, mercury, and hexavalent chromium); and
- record keeping. Technical documents must be kept for four years from the date the vehicles are placed on the market.

The enforcement authority is the Vehicle Certification Agency.

9.132 The 2005 Regulations were amended in 2010[1] in relation to compliance notices and certificates.

1 SI 2010/1095.

WASTE FROM ELECTRICAL AND ELECTRONIC EQUIPMENT

9.133 In order to address the impact of the increasing volumes of Waste from Electrical and Electronic Equipment (WEEE) in the European Community, the Waste Electrical and Electronic Equipment Directive 2002[1] was brought in, together with the related Directive on the Restrictions on the Use of Certain Hazardous Substances in Electrical and Electronic Equipment (RoHS).[2] In summary, the WEEE Directive:

- requires that producers finance the collection, treatment, recovery and disposal of their allocated market share of domestic WEEE and the treatment, recovery and disposal of replaced equipment where new equivalent equipment has been supplied to a business user;
- makes distributors and retailers responsible for taking back WEEE free of charge;
- sets targets for the amount of household WEEE to be separately collected;
- requires electrical and electronic equipment (EEE) to be marked with a 'crossed out wheeled bin symbol';
- requires all separately collected WEEE to be treated; and

- introduces recycling and recovery targets for the various categories of WEEE.

1 Council Directive 2002/96/EC.
2 Council Directive 2002/95/EC.

9.134 The implementation of the WEEE Directive in the UK was a long process. After nearly four years of consultations, the WEEE Directive was finally implemented into UK law by the Waste Electrical and Electronic Equipment Regulations 2006[1] (the WEEE Regulations).

1 SI 2006/3289 and subsequently amended in 2007 (SI 2007/7454), 2009 (SI 2009/2957), 2009 (SI 2009/3216) and 2010 (SI 2010/1155).

9.135 A producer of EEE has obligations in relation to the EEE it places on the market, together with responsibilities for financing the treatment, reprocessing and environmentally sound disposal of EEE when it becomes waste. A 'producer' of EEE includes any person, irrespective of the selling technique used (for example, whether by internet sales or distance selling), who:

- manufactures and sells EEE under his own brand;
- resells (under his own brand in the UK) EEE manufactured by another person; and/or
- imports or exports EEE on a professional basis into a Member State.

9.136 The WEEE Regulations obligate producers to join a Producer Compliance Scheme (PCS) to discharge their financial responsibilities in respect of the treatment, re-use, recovery, recycling and environmentally sound disposal of EEE they have placed on the market when it becomes waste. A network of approved authorised treatment facilities (AATFs) and approved exporters (AEs) provide evidence to the PCS on the amounts of WEEE received for treatment. An end of year settlement period ensures that each PCS is able to meet the obligations of its members and evidence can be traded between one PCS and another.

9.137 The full producer take back and financing obligations came into force on 1 July 2007. Due to the complexities of the WEEE Directive, combined with the UK waste market, the process of implementing a WEEE regime that worked within the existing structure of waste sector and the licensing system was not an easy task. BERR issued non-statutory government guidance in August 2007,[1] and updated this in December 2007 in order to provide additional information to producers, distributors and PCSs on how the WEEE regime works.

1 WEEE Regulations – Government Guidance Notes – August 2007 (URN 07/1303).

9.138 On 1 January 2008, the WEEE (Amendment) Regulations 2007[1] came into force (the 'Amendment Regulations'). The Amendment Regulations made minor amendments to the WEEE Regulations. Amendments included:

- the deadline for submitting evidence relating to a compliance period was extended from 15 February to 30 April after the end of each compliance period (on 31 December);
- the introduction of a standard form of reporting for distributors;
- a requirement to record WEEE to the nearest kilogram, rather than tonne, on evidence notes;
- an offence of disclosing commercially sensitive information received in the course of regulating WEEE operators and schemes;
- powers for the regulating authority to refuse an application for, or

withdraw existing approval of, a WEEE scheme where the operator has knowingly or recklessly supplied misleading information; and

- a new right for the final holder of household WEEE to return it free of charge to the WEEE system.

1 SI 2007/3454.

9.139 There was generally a good uptake of Producer Compliance Schemes but, due to the complexities of the WEEE regime, the waste industry found it difficult to fully understand and comply with the WEEE Regulations. As a result, in 2007, BERR established the WEEE Advisory Board and the WEEE Settlement Centre to record evidence and monitor WEEE treatment.

9.140 A number of deficiencies have been identified with the WEEE Directive, and the obligatory registration and reporting requirements in Member States have resulted in considerable difficulties for companies operating across the EU.[1] As a result, in December 2008, the European Commission proposed revising the WEEE Directive 2002. Almost four years after the proposal, and ten years after the WEEE Directive 2002 came into force, the recast WEEE Directive[2] was adopted in June 2012.

The main purpose of the recast WEEE Directive is to increase the amount of WEEE that is appropriately treated and to reduce the volume of WEEE which requires disposal. The Commission have reported that only an estimated 30 to 40 per cent of WEEE disposed of in the EU is actually captured through the WEEE regime and is properly accounted for; and, in the EU, almost two thirds of WEEE ends up in landfills or sub-standard treatment sites.[3]

1 'Europe-wide concerns over proposed WEEE collection goals', www.letsrecycle.com, 16 June 2009.
2 Directive 2012/19/EU.
3 'The challenge of transposing WEEE II into national law' (September 2012), WEEE Forum.

9.141 The recast WEEE Directive came into force on 13 August 2012, and repeals and replaces the WEEE Directive 2002 from 15 February 2014. Member States must transpose the recast WEEE Directive into national law by 14 February 2014.

The recast WEEE Directive broadens producer responsibilities and aims to improve collection, treatment and recycling. It extends the scope of the regime to cover all WEEE (including photovoltaic panels, small lighting equipment, electronic and electrical tools, fluorescent lamps containing mercury, and any temperature exchange equipment).

Other major changes will be the requirement that each Member State collects annually 45 per cent of the average weight of WEEE placed on their national markets by 2016, and 65 per cent by 2019.

9.142 The UK Government has until 14 February 2014 to implement the provisions of the recast WEEE Directive. It is likely to combine any implementing regulations with proposals to amend the WEEE regime to cut compliance costs.

9.143 Boots (Retail) Ireland Ltd was the first EU company to be prosecuted under the Irish WEEE Regulations. On 23 January 2006, Boots pleaded guilty to two charges, admitting to failing to provide a specified notice in-store alerting customers to the fact that prices included a contribution to a 'Producer Recycling Fund'. In May 2006, a second Irish prosecution was made against Spectra Photo Limited, again, for failing to alert customers that prices included a contribution to recovery and recycling costs. Although similar 'visible fee' obligations have not been included in the UK's WEEE Regulations, these cases serve to illustrate that the enforcing authorities will be taking a tough stance in enforcing the WEEE Regulations. This is in line with the enforcement of the Packaging Waste Regulations, where prosecutions continue to be taken. UK companies must comply with the WEEE Regulations or risk prosecution.

9.144 In order to prevent the generation of hazardous waste (particularly WEEE), the RoHS Directive[1] on the restriction of the use of certain hazardous substances requires the substitution of various heavy metals and brominated flame-retardants in new EEE.

1 Directive 2002/95/EC.

9.145 The Restrictions on the Use of certain Hazardous Substances in Electrical and Electronic Equipment Regulations 2008[1] (the RoHS Regulations) implemented the RoHS Directive into UK law. The RoHS Regulations prohibit producers of EEE from marketing new EEE in the EU which contains more than the prescribed levels of six hazardous substances from 1 July 2006.[2] The intention is that reducing hazardous EEE through the RoHS Directive should make it easier to handle and dispose of EEE under the WEEE Directive (as referred to above).

On 21 July 2011, a recast RoHS Directive[3] (RoHS 2) entered into force which repeals and replaces the RoHS Directive. Member States had until 2 January 2013 to implement the new Directive into national law.

All categories of EEE will be subject to the ban on the use of certain hazardous substances (including cables and spare parts). A transitional period will apply in respect of medical devices, in vitro medical devices, industrial control appliances, and monitoring and control devices. The six hazardous substances will remain the same, although RoHS 2 will allow the Commission to more effectively review and amend the list of restricted substances.

In September 2013 the government issued guidance notes on RoHS 2.

The RoHS Regulations 2012[4] came into force on 2 January 2013, replacing the original Regulations that came into force on 1 February 2008. The RoHS Regulations 2012 impose obligations on operators throughout the supply chain (as defined in the Regulations) in relation to the placing and making available of EEE on the market. The key restriction is that economic operators may not place, or make available, EEE containing the six hazardous substances (see above) in amounts that exceed the established maximum concentration values, on the market. There are also requirements for finished EEE to be CE marked. The Regulations outline what EEE are in scope, exclusions (EEE to which these Regulations do not apply) and makes reference to the specific exemptions to the restricted substances. With the exception of some categories (where certain parts of the Regulations apply on special dates), EEE that was outside the scope of the 2008 Regulations does not need to comply with the new Regulations until 22 July 2019.

Operators must be able to demonstrate compliance by submitting an EU Declaration of Conformity and technical documentation or other information to the market surveillance authority on request, and must retain such documentation for a period of ten years after the EEE is placed on the market.

1 SI 2008/37, revoking and replacing the RoHS Regulations 2006, SI 2006/1463.
2 The six hazardous substances are lead, cadmium, mercury, hexavalent chromium, polybrominated biphenyls (PBB) and polybrominated diphenyl ethers (PBDE).
3 Directive 2011/65/EU.
4 SI 2012/3032.

WASTE BATTERIES AND ACCUMULATORS

9.146 The legislative regime applying to the disposal of batteries and accumulators changed with the arrival of the Directive on batteries and accumulators and waste batteries and accumulators (the 'Batteries Directive').[1] The Batteries Directive focuses on all batteries and accumulators (regardless of shape, volume, weight, composition or use) and requires the separate collection and recycling of spent batteries and accumulators and the reduction of battery and accumulator disposal in municipal waste.

1 Directive 2006/66/EC.

9.147 Key provisions in the Batteries Directive include:

- collection targets for spent batteries;
- bans on disposal of untreated automotive and industrial batteries to landfill or incineration;
- prohibitions on mercury and cadmium (where more than 0.002 per cent weight of battery);
- recycling aids, such as labelling with a crossed out wheelie bin for batteries and accumulators or a chemical symbol for certain batteries or button cells (where more than 0.005 per cent weight of battery);
- capacity labelling; and
- requirement for the ability to remove waste batteries from certain appliances.

9.148 The Batteries and Accumulators (Containing Dangerous Substances) Regulations 1994[1] (the 1994 Regulations) did not have a particularly significant impact on companies dealing with the recycling and disposal of batteries, as they merely introduced restrictions on the use of mercury and cadmium in batteries. The 1994 Regulations were revoked by the Batteries and Accumulators (Placing on the Market) Regulations 2008 (the 2008 Regulations),[2] which partially implement the Batteries Directive.

1 SI 1994/232.
2 SI 2008/2164.

9.149 The 2008 Regulations implemented the internal market provisions of the Batteries Directive and set out technical requirements which apply to any person wishing to place new batteries or battery-powered appliances on the UK market from 26 September 2008. Batteries placed on the market on or after 26 September 2008 which do not comply with these regulations will be prohibited from sale or will be required to be withdrawn from the market.

9.150 Enforcement of the 2008 Regulations falls to the Secretary of State for Business, Innovation and Skills, who has appointed the National Measurement Office (NMO), an executive agency of the Department, to act on his behalf. Offences introduced by the 2008 Regulations include:

- failure to comply with an enforcement notice;
- putting batteries on the market which exceed the maximum allowed percentage by weight of mercury or cadmium;
- putting unlabelled/incorrectly labelled batteries on the market; or
- putting appliances on the market which are not designed in such a way that waste batteries can be removed easily.

9.151 Following the consultation on implementation of the provisions in the Batteries Directive dealing with waste battery collection and recycling, the Waste Batteries and Accumulators Regulations 2009[1] (the 2009 Regulations) have now been introduced. The 2009 Regulations implement further parts of the Batteries Directive, and largely came into force on 5 May 2009. Various requirements are set out in the 2009 Regulations for the treatment and recycling of waste automotive, industrial and portable batteries. These Regulations apply to all batteries (except those used for military or space purposes). All companies that place batteries on the market will have to register as a producer of batteries.

1 SI 2009/890.

9.152 Key responsibilities in the 2009 Regulations in relation to portable batteries include:

- Producers (other than those placing less than one tonne of portable batteries onto the market, so-called 'small producers') will each finance the costs of collection, treatment and recycling of a share of all portable batteries placed onto the market.
- Producers (other than small producers) are required to register with a battery compliance scheme, with the first compliance period running from 1 January 2010.
- Where producers place less than 1 tonne of portable batteries on the market, they will register with the Environment Agency but will not have to fund collection, treatment and recycling. Small producers may, however, be subject to a charge imposed by the relevant authority.
- Producers must keep records of the total number of batteries in tonnes, and chemistry type, of portable batteries which they have placed onto the UK market during 2009, and each subsequent compliance period. The reporting obligations are slightly less rigorous for small producers.
- Retailers of household batteries are required to collect in-store waste household batteries from February 2010.
- Shops or distributors that sell over 32 kilograms of portable batteries are required to accept back waste batteries from 1 February 2010. Shops or distributors selling less than 32 kilograms are not obliged to comply with this take-back duty.

9.153 Key responsibilities in relation to waste industrial and automotive batteries include:

- Producers of these batteries are required to arrange separate collection and recycling of waste industrial and automotive batteries, where necessary, including take-back responsibilities from the end-user.
- A ban on the disposal of these batteries by landfill or by incineration from 1 January 2010.
- Producers must keep records of the total amount of automotive and industrial batteries in tonnes that they have placed onto the market in 2009, and any subsequent compliance period. They must also keep records of the total amount of automotive and industrial batteries in tonnes that they have been responsible for taking back.
- Producers have certain reporting obligations. The required information must be provided to the Secretary of State, in the required format, by 31 March 2010.
- From 16 October 2009, all producers of these types of batteries must be registered with the Secretary of State, with certain limited exceptions.

9.154 The environment agencies (the Environment Agency, the Scottish Environment Protection Agency and the Northern Ireland Environment Agency) will be responsible for enforcement of portable battery provisions. The Secretary of State for Business, Innovation and Skills will be responsible for enforcing the waste industrial and automotive battery provisions. Offences for producers under the 2009 Regulations include failure to join a compliance scheme, and failure to keep records, and a person found guilty of these offences can be found liable, on summary conviction, to a fine not exceeding the statutory maximum or, on conviction on indictment, to a fine. Defra have appointed VCA, an Executive Agency of the Department of Transport, to enforce the requirements on distributors with effect from 1 February 2010.

9.155 The 2009 Regulations set out a requirement for treatment operators who are approved under these Regulations (Approved Battery Treatment Operators (ABTOs) and Approved Battery Exporters (ABEs)) to meet certain recycling efficiencies for lead acid batteries, nickel cadmium batteries and other waste batteries. These requirements have now been superseded,[1] and the Environment Agency has stated that they will not take action against such recyclers until the EU provisions take effect.[2] The Regulation applies to the recycling processes carried out to waste batteries and accumulators from 1 January 2014.[3] Recyclers must calculate the recycling efficiency of waste lead acid, nickel-cadmium and other batteries and accumulators, including the rate of recycled lead and recycled cadmium, using formulas in the Regulation.[4] Recyclers must then report that information on an annual basis and send it to the relevant authorities (yet to be specified by government, but likely to reflect the enforcement authorities above) in due course. Recyclers must submit their first annual reports no later than 30 April 2015.[5]

1 Regulation (EU) 493/2012.
2 Environment Agency, 'Regulatory Position Statement – Battery recycling efficiencies', March 2012, at p 1.
3 Regulation (EU) 493/2012, art 1.
4 Article 3.
5 Article 4.

ASBESTOS WASTE

9.156 The prohibition and control of asbestos, as well as asbestos licensing, is dealt with under the Control of Asbestos Regulations 2012 (COAR 2012),[1] which came into force on 6 April 2012. The COAR 2012 revoke and replace the Control of Asbestos Regulations 2006,[2] which in turn revoked and replaced the Control of Asbestos at Work Regulations 2002.[3] COAR 2012 prohibit the importation, supply and new use of all forms of asbestos.

1 SI 2012/632.
2 SI 2006/2739.
3 SI 2002/2675.

9.157 Regulation 4 of COAR 2012 contains a duty to manage asbestos in non-domestic properties, and this duty applies to the person or organisation obligated to maintain and repair non-domestic properties, whether by virtue of contract or tenancy, or where the person has 'control' of the non-domestic property. There is a range of guidance available from the HSE advising on the duty to manage asbestos.[1]

1 For example, (i) the 'Approved Code of Practice: The Management of Asbestos in Non-Domestic Premises' L127, and (ii) the 'Approved Code of Practice and Guidance: Work with Materials Containing Asbestos' L143. An HSE consultation on revising a number of Approved Codes of Practice (including L127 and L143) closed in September 2012. The consultation proposed the consolidation of L127 and L143 into a single Approved Code of Practice by the end of 2013 to reflect the introduction of COAR 2012.

9.158 The persons to whom the duty applies are required to inspect their premises for the existence of asbestos and assess the risk attached to any asbestos present. Although it is a criminal offence to fail to carry out the survey, it is likely that insurers will require surveys to be completed before buildings and/or employers cover is reissued. Moreover, failure to undertake the survey will also provide powerful evidence in any claim for loss or death following exposure to asbestos by employees or visitors. It is therefore likely to be insisted upon by all subsequent tenants, licensees or purchasers of property where asbestos or asbestos-containing materials are (or are likely to be) present.

9.159 Whilst existing asbestos-containing materials (in good condition) can be left in place and monitored, there is nonetheless an obligation to draw up a maintenance plan setting out details for monitoring the condition of the asbestos, setting out the plan for maintenance and safe removal, and also identifying the location and condition of asbestos for those likely to disturb it. There is a presumption in favour of asbestos being present.[1]

1 COAR 2012, reg 5.

9.160 The asbestos management plan must be regularly reviewed, particularly where it is intended to significantly change the asbestos-containing premises in some way. Any work proposed on the premises by an employer requires that an employer carry out a risk assessment, which shall be reviewed regularly.[1]

1 COAR 2012, reg 6.

9.161 Employers must ensure that employees at risk from exposure to asbestos fibres at work receive training,[1] whether they are removing or disturbing asbestos or simply coming into contact with it. Medical surveillance must be paid for by the employer. Employers have a duty to prevent employee exposure to asbestos where reasonably practicable,[2] failing which, they have a duty to take measures necessary to reduce exposure, providing information to any employees concerned where control limits are exceeded and providing protective equipment, as well as washing and changing facilities, where employees are exposed to any level of asbestos.

1 COAR 2012, reg 10.
2 COAR 2012, reg 11.

9.162 COAR 2012 now include a definition of 'licensable work with asbestos',[1] which relates to work which is neither sporadic nor of low intensity or where there is risk of dangerous exposure to asbestos. In practice, this may include nearly all work with asbestos insulation, asbestos coatings and asbestos insulating board. No licensable work with asbestos can be carried out unless a licence is held by an employer for such work. Licences must be obtained from the HSE and are granted subject to a maximum duration of three years.[2] Notification of the relevant work must be given to the enforcing authority at least 14 days before commencing the work.[3] In addition, the employer must maintain registers of work (health records) for each employee exposed to asbestos and ensure any such employee is under adequate medical surveillance as stipulated in COAR 2012.[4]

1 COAR 2012, reg 2(1).
2 COAR 2012, reg 8.
3 COAR 2012, reg 9.
4 COAR 2012, reg 22.

9.163 In addition to licensable work with asbestos, there are two other categories of work referred to in COAR 2012: 'non-licensed work'; and (introduced for the first time by COAR 2012) 'notifiable non-licensed work'. 'Non-licensed work' is that work with asbestos which is exempt from COAR 2012 by reg 3(2) (eg low-risk, sporadic work). 'Notifiable non-licensed work' is work which is neither exempt nor licensable. Where 'notifiable non-licensed work' is to be undertaken, the employer must:

- notify the enforcing authority before commencing the work (there is no minimum period);[1]
- by 30 April 2015, ensure all workers carrying out the work have had medical examinations;[2] and
- maintain registers of work (health records) of 'notifiable non-licensed work' for each employee exposed to asbestos.[3]

1 COAR 2012, reg 9(2).
2 COAR 2012, reg 22(3).
3 COAR 2012, reg 22(1).

9.164 Employers working with raw asbestos or asbestos-containing waste must ensure it is stored or handled in a sealed receptacle and clearly marked.[1] The spread of asbestos must be prevented. COAR 2012 set out requirements for labelling raw asbestos, and state that waste containing asbestos shall be labelled where it is transported in accordance with the Carriage of Dangerous Goods and Use of Transportable Pressure Equipment Regulations 2009.[2]

1 COAR 2012, reg 24.
2 SI 2009/1348.

9.165 The HSE is entitled to exempt parties from certain requirements under COAR 2012 subject to conditions, and may produce a certificate in writing to that effect,[1] provided the health and safety of persons who are likely to be affected by the exemption will not be prejudiced in consequence of it. The fact that no licence is required to deal with the asbestos-containing material does not preclude the requirement for adherence to the other COAR 2012 obligations, where applicable.

1 COAR 2012, reg 29.

9.166 Where work with asbestos has been carried out, a certificate for site clearance may only be issued by a body accredited to do so (presently the United Kingdom Accreditation Service), and only where the asbestos has been thoroughly cleared and the premises are suitable for re-occupation.[1]

1 COAR 2012, reg 20.

9.167 Materials containing or contaminated with asbestos must be disposed of as hazardous waste. This includes clothing and footwear that may have come into contact with asbestos-containing materials, as well as respiratory equipment.

9.168 As COAR 2012 are made under the HSWA 1974, failure to comply with the requirements of COAR 2012 constitutes a criminal offence. This means that (i) for summary conviction, the penalty for each offence is imprisonment for up to 12 months and/or a fine not exceeding £20,000, and (ii) for conviction on indictment, the penalty for each offence is imprisonment for up to two years and/or an unlimited fine. Please note that s 85 of the Legal Aid, Sentencing and Punishment of Offenders Act 2012, once in force, will mean that, when an offence is triable either way, magistrates will have the ability to commit the case to the Crown Court for sentence, where fines will be unlimited. However, the Secretary of State will have the power to disapply these changes and set new caps by secondary legislation.

9.169 In September 2012 the director of a Bromley firm was sentenced for removing asbestos without a licence contrary to the Control of Asbestos Regulations 2006, failing to properly clean and decontaminate the relevant property, and thereafter for deceiving the householders by providing a doctored air test saying the relevant room was safe to re-enter. The director had pleaded guilty to three charges and, as a result, was (i) given a six-month prison sentence for each charge, to run concurrently and suspended for two years, (ii) required to undertake 300 hours' unpaid community service, (iii) given an electronic curfew between 9pm and 6am for three months, and (iv) ordered to pay £11,340 to the affected Residents' Association and £10,160 costs.

9.170 In July 2012, Stockton-on-Tees Borough Council was fined for putting workers at risk of exposure to asbestos during clearance work of premises in Stockton prior to demolition. The council was fined a total of £20,000 (£10,000 for each offence) and ordered to pay £5,555.60 costs after it pleaded guilty to two breaches of the Control of Asbestos Regulations 2006.

In September 2011, Marks and Spencer plc and three of its contractors were prosecuted and fined under the HSWA 1974 for mishandling asbestos during removal from two of its stores during refurbishment works. Marks and Spencer was fined £1m and ordered to pay £600,000 in costs.[1] The court ruled that it had failed to protect customers, staff and workers from potential exposure.

1 ENDS Report, October 2011, 441, p 62.

9.171 Where a party is found guilty of a criminal act, a referral can be made to the Serious Organised Crime Agency (SOCA) under the Proceeds of Crime Act 2002, asking SOCA to confiscate assets equal to the financial benefit gained from the criminal activity. This power has been used in relation to asbestos-related crimes.

Chapter 10

Environmental liability including nuclear

Caroline May, Partner and Head of the Environment, Safety and Planning Group, Norton Rose LLP

INTRODUCTION

10.1 As has been discussed elsewhere, environmental regulation is increasing. Many laws applicable in the UK arise as a direct result of European Directives. There is increasing pressure on enforcement standards to minimise emissions to the environment as a result of the government's pledge to reduce greenhouse gas emissions by 80 per cent below 1990 levels by 2050 in an attempt to meet the UK's targets under the Kyoto Protocol.

10.2 The interrelationship between environmental and health and safety issues is highlighted in this chapter, by recognising that what is discharged into the environment also has implications for human health and other life forms. This interrelationship means that there is often an overlap between the enforcement authorities such as the Environment Agency (EA) and the Health and Safety Executive (HSE) in the work they do to ensure that health and safety and environmental legislation is complied with. A single incident may involve an investigation by both the EA and the HSE.

10.3 The law has evolved from the basic concept of recovery of damages for harm to property, through the torts of nuisance and negligence, to today's criminal regulatory regime. Many companies are surprised to find that, if they are in breach of environmental and health and safety permits, the enforcement remedies include not only criminal prosecution for the corporate vehicle, but potentially also its individual directors/senior management. This can often be an unpleasant surprise for senior management. The potential for individual liability (as discussed elsewhere in this chapter) also impresses upon directors and other senior officers the importance of demonstrably discharging their environmental and health and safety obligations to minimise risks. An individual, – company and/or its directors and senior officers with a criminal record will find that its operations will come under increasing scrutiny, not only by the regulators and its employees and shareholders, but also its financiers, analysts, competitors and potential clients. Public authority tenders often require details of any criminal convictions to be disclosed, which (if disclosed) will seriously inhibit a company's ability to secure future work. Clearly, the risks are to be minimised or avoided at all costs.

10.4 In this chapter, we examine some key environmental and public health-related areas, which are the subject of specific regulations and specific criminal penalties. Topics covered in this chapter are not intended to be exhaustive, but rather to highlight the principal areas of likely concern.

ENVIRONMENTAL LIABILITY DIRECTIVE

10.5 The Environmental Liability Directive (the 'EL Directive')[1] was adopted on 21 April 2004 and came into force on 30 April 2007. In England and Wales, the Environmental Damage (Prevention and Remediation) Regulations 2009 (the 'ED Regulations'),[2] implementing the EL Directive into national law, came into force on 1 March 2009. The ED Regulations apply to incidents which occur after such regulations came into force. The EL Directive is aimed at the prevention and remedying of environmental damage, specifically damage to habitats and species protected by EC law, damage to water resources, and land contamination which presents a threat to human health.

1 Directive 2004/35/EC.
2 SI 2009/153 (as amended).

10.6 The EL Directive is based on the 'polluter pays' principle. It requires operators to inform the regulatory authority of environmental damage that has occurred and to take all reasonable steps to deal with it. Polluters will meet their liability by remedying the damaged environment directly, or by undertaking 'complementary' or 'compensatory' remediation. Regulatory authorities will be responsible for enforcing the regime in the public interest, including determining remediation standards, or taking action to remediate or prevent damage and recover the costs from the operator. Individuals and others who may be directly affected by actual or possible damage, and NGOs in certain circumstances, may request action by a regulatory competent authority and seek judicial review of the regulatory authority's action or inaction if there are grounds for believing that the regulator is failing in its public duty to act.

10.7 The Department for the Environment, Food and Rural Affairs (DEFRA) has indicated that the EL Directive is aimed at the most significant cases of environmental damage. It estimates that fewer than one per cent of approximately 30,000 cases of environmental damage reported to enforcement authorities in the UK each year will fall within the scope of the ED Regulations (in 2011, for example, they were only used in three incidents in England and Wales, Scotland and Northern Ireland).

ENVIRONMENTAL PERMITTING

Introduction

10.8 The environmental permitting regime came into force on 6 April 2008 under the Environmental Permitting (England and Wales) Regulations 2007 (the 'EP Regulations 2007').[1] This created a single permitting regime incorporating (1) the Integrated Pollution Prevention and Control (IPPC) permitting regime (authorising the operation of certain industrial installations); and (2) the Waste Management (WM) licensing regime (authorising the treatment, keeping or disposal of any waste) into a single environmental permit ('EP').

1 SI 2007/3538 (as amended).

10.9 The EP Regulations 2007 were revoked and replaced by the Environmental Permitting (England and Wales) Regulations 2010 (the 'Consolidated EP Regulations')[1] which came into force on 6 April 2010. The Consolidated EP Regulations revoked over 40 sets of environmental regulations (in particular, regulations relating to waste, mining and batteries), and extended the EP regime to include:

(a) Water Discharge Consents (permits to control certain discharges to surface water);
(b) Groundwater Authorisations (permits to control the disposal of specific substances into groundwater); and
(c) Radioactive Substances Authorisations (permits for keeping and use of radioactive materials and for accumulation and disposal of radioactive waste).

There have been a number of amendments to the Consolidated EP Regulations, including the Environmental Permitting (England and Wales) (Amendment) Regulations 2013 (the 'EP Amendment Regulations 2013'),[2] which implement the Industrial Emissions Directive.[3] This came into force for the most part on 27 February 2013. The Industrial Emissions Directive replaces and consolidates the IPPC Directive[4] and six sector-specific Directives into a single Directive.

In addition, there are plans to include into the Consolidated EP Regulations (at a later date):

(a) civil sanctions (to give the EA power to use civil sanctions under the Regulatory Enforcement and Sanctions Act 2008 to enforce breaches of the EP regime);
(b) the regulation of waste batteries;
(c) Water Abstraction and Impoundment Licences;
(d) fish pass and flood defence consents; and
(e) clarification of the division of responsibilities between the EP regime and the marine licensing regime under the Marine and Coastal Access Act 2009 (especially for activities carried out partly on land and partly below mean high water springs (high tide) such as ship dismantling).

1 SI 2010/675 (as amended).
2 SI 2013/390.
3 Directive 2010/75/EU on industrial emissions (recast).
4 Directive 2008/1/EC concerning integrated pollution prevention and control (codified version).

The environmental permitting regime

Regulated facilities

10.10 Regulated facilities, as defined under the Consolidated EP Regulations, include:

(a) **installations** or **mobile plant** where specified controlled activities within the energy, chemicals, metal processing and waste management industries are carried out;
(b) facilities (not including installations and mobile plant) carrying out **waste operations**;
(c) facilities dealing with the management of waste resulting directly from prospecting, extraction, treatment and storage of mineral resources and working quarries;
(d) **water discharge activities**;
(e) **groundwater activities**; and
(f) **radioactive substance activities** (including various activities relating to the keeping, using, disposing and accumulating of radioactive material or waste on premises).

The general permitting procedure

Who needs to apply for an environmental permit?

10.11 Under the Consolidated EP Regulations, an Operator is required to obtain or hold an EP before its Regulated Facility can be lawfully operated.

10.12 An 'Operator' is defined under the Consolidated EP Regulations as being a person who is 'in control' of the Regulated Facility. It is worth noting that the Regulated Facility is not required to be in operation for there to be an Operator. Rather, a person may be deemed to be an Operator during the pre- and post-operational phases of a Regulated Facility. In addition, where two or more Operators run different parts of a single Regulated Facility, they will each need to obtain (and comply with) a separate EP.

The relevant regulator

10.13 The Regulator under the Consolidated EP Regulations is either the EA or the appropriate local authority (LA) in whose area the Regulated Facility will be operated. The EA will take responsibility for Regulated Facilities undertaking activities which have the potential to cause significant harm to the environment and/or human health (eg the disposal of waste in a landfill with a total capacity of more than 25,000 tonnes). LAs, on the other hand, will be responsible for Regulated Facilities carrying out lower-risk polluting processes (eg the storage of chemicals and manufacturing processes at relatively low capacities).

Standard and bespoke EPs

10.14 The Consolidated EP Regulations provide Operators with a choice of applying for a standard or bespoke EP, depending on the nature and risks associated with the relevant Regulated Facility.

10.15 A standard EP is not site-specific and contains only one condition. The condition refers to a fixed package of standard rules (SRs) to which a standard EP holder needs to adhere. SRs define how the Operator must carry out the activity (for example, by limiting the types of waste that can be brought onto the site). There is no right of appeal against SRs, and SRs may not be amended (SR packages are developed for each type of Regulated Facility in consultation with industry and are therefore deemed to be suitable). It is intended that standard EPs will generally be suitable for Regulated Facilities carrying on activities that are thought to be of low risk to the environment and/or human health. There are two types of SRs: one for fixed sites, and one for mobile plant.

10.16 Bespoke EPs have conditions that are set by the EA for a specific site/facility. Conditions may include any conditions which the Regulator sees fit, as long as they are both necessary (ie justifiable) and enforceable (eg they must state their objective, standard or desired outcome, so that the operator can understand what is required). EP conditions may comprise some or all of the following:

(a) conditions stipulating objectives or outcomes;
(b) standards to mitigate a particular hazard/risk;
(c) conditions addressing particular legislative requirements; and/or
(d) conditions setting out steps to be taken before, during and after the operation of the Regulated Facility.

It is intended that bespoke EPs will be suitable for Regulated Facilities that present a medium to high risk to the environment and/or human health.

10.17 It is possible for any Operator holding a standard EP to apply for a bespoke EP where a change in permit conditions (from standard to facility-specific) is considered to be desirable (eg resulting from a change in the way the Regulated Facility is operated).

10.18 Applications for EPs need to be made using the forms provided by the Regulator. The appropriate fee will also have to be paid.

10.19 An Operator of two or more Regulated Facilities may apply for a single EP where certain conditions are met. Generally speaking, a single EP may be granted to an Operator of two or more Regulated Facilities where the Regulated Facilities are found on the same site and have the same Regulator.

Assessment by the Regulator

10.20 The Consolidated EP Regulations require the Regulator to assess whether the Operator is 'competent' (ie whether the operator will be able to operate the facility in accordance with the terms of an EP).

10.21 In assessing competency, the Regulator may need to consider certain factors, including whether the Operator:

(a) has adequate systems in place;
(b) is technically competent to operate the facility;
(c) has been convicted of an environmental offence before; and
(d) has sufficient financial means to operate the facility in accordance with the terms of an EP (NB financial solvency should only be considered (except for landfill) in cases where running costs are high relative to the profitability of the activity).

10.22 In making a decision on applications, Regulators are likely to be subject to prescribed timeframes. The guidance to the Consolidated EP Regulations provides that applications for a new EP should be decided within four months of an application being submitted for a bespoke EP, or three months for a standard EP. The Regulator and Operator can agree longer periods, if appropriate, and the clock is temporarily suspended when a Regulator requests further information to be provided. If the Regulator fails to reach a decision within the set timeframes, the application is deemed to have been refused and the Operator may then appeal against the deemed refusal.

Historic permits

10.23 On 6 April 2010, all existing permits under any of the previous regimes which the Consolidated EP Regulations replaced automatically became EPs.

Enforcement

Enforcement tools

10.24 The core guidance to the Consolidated EP Regulations provides that the nature and extent of the regulatory effort (permitting and enforcement) under the EP regime should be appropriate and proportionate to:

(a) the risk posed by the Regulated Facility;
(b) the impact of the activities carried out by the Regulated Facility; and
(c) the Operator's performance in mitigating the risk and impacts of the Regulated Facility.

10.25 The Regulator will be able to enforce the requirements of the Consolidated EP Regulations in the following ways:

- Serving an **Enforcement Notice** – if it believes the Operator has breached, is breaching or is likely to breach a condition in an EP.
- Serving a **Suspension Notice** – if it believes there is a risk of serious pollution, regardless of whether or not the Operator has breached any of the conditions in its EP.
- Serving a **Revocation Notice** (revoking an EP in whole or in part) – if it considers that other enforcement tools have failed to sufficiently protect the environment.
- Carrying out the necessary works itself and recovering the costs from the operator.
- Prosecuting.

Appeals

10.26 An Operator has the right to appeal to the Secretary of State for Environment, Food and Rural Affairs (or, in Wales, to the Welsh Assembly Government) if, for example:

(a) the Regulator has refused to grant, vary, transfer or accept the surrender of an EP;
(b) the Operator disagrees with any conditions imposed in a bespoke EP; or
(c) the Regulator has served an Enforcement, Suspension or Revocation Notice.

Offences and penalties

10.27 The principal offences under the Consolidated EP Regulations include:

- operating a Regulated Facility without an EP; or
- operating a Regulated Facility in breach of EP conditions; and/or
- failure to comply with an Enforcement or Suspension Notice.

10.28 All offences (with the exception of an offence under reg 38(5) which is heard in a magistrates' court only) under the Consolidated EP Regulations[1] are criminal and are triable either way, with the more serious offences having a maximum term of imprisonment of 12 months and/or a maximum fine of £50,000 in a magistrates' court, and a maximum term of imprisonment of five years and/or an unlimited fine in the Crown Court. The court may also (where appropriate) order that the offence be remedied.

1 Regulations 12, 38 and 39. Regulation 40 contains a list of defences.

10.29 Where an offence under the Consolidated EP Regulations[1] committed by a body corporate is proved to have been committed with the consent or connivance of, or is attributable to any neglect on the part of, any director, manager, secretary or similar officer, he or she and the body corporate and/or the officers in their personal capacity[2] may all be guilty of an offence and liable to criminal prosecution.

1 Regulation 4.
2 Regulation 41.

WASTE

Introduction

10.30 Under Part 2 of the Environmental Protection Act 1990 (EPA 1990), there is a wide-ranging system for controlling waste. All waste producers are subject to a duty of care under s 34 of the EPA 1990 to take reasonable and appropriate steps in the handling of waste. This duty involves preventing any escape of waste, and ensuring that the waste is transferred only to an authorised and competent person, and that an adequate written description of the waste is given to the transferee. Waste that is particularly hazardous is known as 'hazardous waste' and is subject to further controls. Hazardous waste requires greater care by those producing and handling it. For example, hazardous waste may be an issue where asbestos is being removed from a site or where contaminated soil is removed because of building works. By way of guidance, in 2010 the government published a 'Strategy for Hazardous Waste Management in England'.

Revisions to the Waste Framework Directive,[1] which is the primary European legislation for the management of waste, have been implemented in England and Wales through the Waste (England and Wales) Regulations 2011.[2]

1 Directive 2008/98/EC.
2 SI 2011/988.

Permit to manage waste

10.31 Pursuant to the Consolidated EP Regulations (see **10.9**), an Operator of either a waste operation (ie a facility operated to recover or dispose of waste) or a mobile plant used to carry on waste operations is required to obtain an EP. The EA monitors compliance with the permit conditions as part of its role as the waste regulation authority in England and Wales.

Offences

10.32 It is a criminal offence not to comply with the requirements of s 34 of the EPA 1990, and new powers have been given to regulators to impose fixed penalties to search and seize vehicles under amendments introduced by the

Consolidated EP Regulations. In 2011, the EA set up a specialist environmental crime taskforce to deal with illegal waste sites, shutting down 1,135 of these in 2012. In 2011 the courts issued £1.7 million in fines for serious waste offences, including a single fine for £170,000.[1]

1 EA, Waste Report 2011–2012.

10.33 In 2013, three Kent waste firms (Countrystyle Recycling Limited, FGS AGRI Limited and Mark Luck Limited), as well as a director, were fined £233,670 for illegally depositing waste on golf courses and farms in the area. Also in 2013, Biffa Waste Services Ltd, which manages a Tyne and Wear landfill site, was fined £105,000 for failing to manage pollution from the site in line with environmental legislation.

10.34 It should be noted that higher fines may apply to waste offences relating to packaging and recycling or hazardous waste.

WATER POLLUTION

Introduction

10.35 Water pollution incidents are costly to companies. Licences may be revoked, penalties imposed, clean-up, investigation and legal costs will be incurred (all of these directly affecting the company's profit and loss accounts). Further cost considerations include the damage done to a company's image, brand and public relations by press coverage of water pollution incidents and the EA's policy to 'name and shame' offenders.

10.36 In terms of criminal liability, the EA's Enforcement Policy[1] contains a mandate to prosecute not only companies, but also individuals where they can be shown to have 'turned a blind eye' to the circumstances which caused the offence.

1 In force from 1 November 1998.

Brief legislative history

10.37 The first recorded law prohibiting water pollution dates from 1388. However, modern water pollution legislation derives from the Consolidation Act.[1] Of particular importance to this chapter are the Consolidated EP Regulations (which have replaced the provisions of Pt III of the Water Resources Act 1991 (WRA 1991)), and the Water Industry Act 1991 which regulates the privatised water industry.

1 Namely the Water Consolidation (Consequential Provisions) Act 1991, which seeks to consolidate and repeal a number of Acts, including the Water Resources Act 1991, the Water Industry Act 1991, the Land Drainage Act 1991, and the Statutory Water Companies Act 1991.

The regulators

10.38 Pursuant to the Environment Act 1995 (EA 1995), the EA has general responsibility for water quality in England and Wales.

10.39 Section 4 of the EA 1995 requires the EA, in discharging its functions, to protect or enhance the environment (taken as a whole) in order to achieve its objective of sustainable development. Section 6 of the EA 1995 furthermore obliges the EA to promote the conservation and enhancement of the natural beauty and amenity of inland and coastal waters and of land associated with such waters, the conservation of flora and fauna which are dependent on an aquatic environment, and the use of such waters and land for recreational purposes.

10.40 The Water Services Regulatory Authority (Ofwat) is the economic regulator of the water and sewage industry in England and Wales and may take action against a water/sewage undertaker for failing to achieve performance standards (eg targets in relation to leakage/failure to prevent pollution).

Liability under the Environmental Permitting Regulations 2010

10.41 The Consolidated EP Regulations have replaced Part III of the WRA 1991 that contained the main offences, although the two regimes are very similar in practice.[1]

The principal offence relating to water pollution is contained in regulation 12 of the Consolidated EP Regulations, which makes it a criminal offence for a person *to cause or knowingly permit* a water discharge activity or groundwater activity unless authorised to do so by an EP.

The meaning of 'water discharge activity' is widely construed to include any discharge of poisonous, noxious or polluting matter, any waste matter or any trade effluent or sewage effluent[2] into fresh water, coastal waters or relevant continental waters. It may also include causing deposits to be carried away on water, or cutting or uprooting a substantial amount of vegetation in or near inland fresh waters, if the debris falls into the water and is not retrieved.

'Groundwater activities' include the discharge of a polluting substance that results in or might lead to a direct or indirect impact to groundwater.[3] Polluting substances controlled under the Consolidated EP Regulations include toxic, persistent and biocumulative substances (hazardous substances) and any other non-hazardous pollutant.

1 The Consolidated EP Regulations cover activities that require an EP, and all unlawful discharges to water, whether deliberate or accidental.
2 Consolidated EP Regulations, Sch 21, para 3(i)(a).
3 The Consolidated EP Regulations, Sch 22 para 3)(1).

To 'cause' or 'knowingly permit'

10.42 The leading authorities on 'to cause' for water pollution cases are the *Alphacell* case[1] and the *Empress Car* case.[2] In both cases, the House of Lords distinguished between the two limbs 'to cause' and 'to knowingly permit'. In the *Alphacell* case, a paper manufacturing company had a settling tank on the bank of a river. The tank was controlled by two pumps which prevented it from overflowing into the river. Despite regular inspections, the tank had become blocked with leaves within 45 minutes, causing the contents of the tank to pour into the river. Alphacell was convicted of causing pollution and appealed on the basis that the company had no knowledge of the pollution, nor had it acted negligently. The House of Lords held that there was no need to show intent on the part of the offender, and that 'to cause' had a common sense meaning which did not require knowledge of the offender's action or inaction. It was enough to show that the tank was the company's tank, and under its control, and that the overflow of the contents of the tank had caused the pollution.

1 *Alphacell v Woodward* [1972] AC 824.
2 *Environment Agency v Empress Car Co (Abertillery) Ltd* [1998] 1 All ER 481.

10.43 In the *Empress Car* case, the defendants were held liable for the actions of an unknown third party who opened an unlocked tap on the defendants' oil tank, causing pollution of the river. The House of Lords gave useful guidance on the meaning of 'to cause':

- the defendants must cause the pollution;
- the defendants' actions do not have to be the immediate cause of the pollution – in this case, maintaining the oil tank is sufficient; and
- whether an act by a third party or a natural event that caused the oil to escape is an ordinary event and fact of life (as in this case) or an abnormal or extraordinary event (eg terrorism) is a question of fact and degree.

10.44 To 'knowingly permit' involves a failure to act to prevent pollution accompanied by knowledge. However, knowledge can be either implied or express.

The trigger for prosecution

10.45 A prosecution can be brought when polluting, noxious or poisonous matter or solid waste or trade or sewerage effluent enters controlled waters or where a person fails to comply with a notice issued by the EA in respect of the activity.

10.46 'Poisonous, noxious or polluting matter' is not defined by statute, but the courts have interpreted it as substances that are likely to 'harm' the environment.

10.47 In *Express Ltd (t/a Express Dairies Distribution) v Environment Agency*[1] (a case concerning a spillage of dairy cream which subsequently entered into a nearby water course), the Divisional Court held that it is not necessary to cause harm to controlled waters before a substance can be found to have caused pollution under s 85(1) of the WRA 1991. All that is needed to prove the offence is the potential for harm to be caused.

1 [2005] 1 WLR 223.

EA enforcement policy

10.48 At the core of the EA's Enforcement Statement (the 'Statement') are five governing principles: proportionality, consistency, transparency, targeted enforcement action, and accountability for the enforcement action taken. What is 'proportional' depends on the balance between the risks taken, the harm caused to the environment and the cost of bringing a prosecution. 'Consistency' means that the EA should be taking a similar approach in similar circumstances. 'Transparency' is translated into a need to inform those who are regulated what is expected of them. 'Targeted action' means targeting prosecution of those activities which give rise to the most serious effects on the environment. Accountability means that the EA will take responsibility for its decision and justify it, where appropriate, including providing details of the right of appeal.

10.49 In all cases, the EA will not prosecute until it is satisfied that there is sufficient admissible and reliable evidence to bring a prosecution, there is a realistic prospect of securing a conviction, and it is not against the public interest to bring a prosecution. The EA will look at factors such as the seriousness of the offence, the foreseeability of the harm, the intention of the offender, and the offender's action/inaction following the incident.

10.50 The EA has published Guidance[1] to accompany the Statement. This clarifies the Statement and confirms that prosecution or a formal caution may result following a finding that a company has failed to comply with an EP where, for example, the non-compliance is serious or has a potentially serious environmental outcome. For incidents which cause minor impact, a warning will normally be issued.

1 *EA Enforcement and Sanctions – Guidance* (4 January 2011).

10.51 The Statement encourages the EA to prosecute senior company officers individually if they believe that the company officers have 'turned a blind eye' to the circumstances which caused the pollution incident. The senior company officers will be charged with the same offence as the company.[1]

1 Consolidated EP Regulations, reg 41.

Penalties

10.52 Water pollution offences are criminal offences and can be tried by a magistrates' court or the Crown Court.[1]

1 Consolidated EP Regulations, Regulation 38 (offence of polluting controlled waters).

10.53 The penalties are as set out in the Consolidated EP Regulations.[1]

If convicted of a water pollution offence under Regulation 38 and it appears that remediation is within the offender's power, the Court can order the offender to remediate the pollution within a specified period[2]. This may be instead of, or in addition to, any fines/imprisonment under Regulation 39.

1 See **10.32** above.
2 Consolidated EP Regulations, reg 44.

10.54 It is unusual for custodial sentences to be imposed by the courts. However, directors, managers, company secretaries and other officers of a company can be prosecuted individually if the company is found to have committed a water pollution offence with their consent, connivance or neglect.[1] With today's best practice requiring companies to report on by whom and how environmental management decisions are taken from site to board level, a paper trail is being left behind which may in future allow more prosecutions to be brought against individual company officers. Disqualification of directors is also possible following a conviction and, in the case of indictable offences, the maximum disqualification period is five years in a magistrates' court and 15 years in the Crown Court.[2]

1 Consolidated EP Regulations, reg 41.
2 Company Directors Disqualification Act 1986, s 2.

10.55 There is increasing public pressure for larger fines to be levied but this is often a fraction of the true cost. When the *Sea Empress*[1] tanker was grounded at the entrance to the Milford Haven port due to the negligent navigation of the pilot, causing 72,000 tonnes of crude oil to be released, the Milford Haven Port Authority was fined £4 million (reduced on appeal to £750,000). The clean-up operation was estimated at £60 million.

1 *Environment Agency v Milford Haven Port Authority* (*The Sea Empress*) [1999] 1 Lloyds Rep 673.

10.56 Despite the level of public concern, the courts have indicated a reluctance to impose large fines for single offences. In *Hart v Anglian Water Services Ltd*,[1] the Court of Appeal reduced the fine imposed on Anglian Water Services Limited for water pollution under s 85(3) of the WRA 1991 to £60,000 (from £200,000). The court considered that the higher fine was excessive for a single offence.

1 [2003] EWCA Crim 2243.

10.57 In 2011, Bromley Magistrates' Court required Thames Water Utilities Ltd to pay a total of £345,000 in fines and costs for offences relating to the release of sewage under s 85 of the WRA 1991 and s 33 of the EPA 1990.[1] In 2012, South West Water was ordered to pay £32,697 in fines and costs for illegal sewage discharges into Salcombe harbour, in breach of the Consolidated EP Regulations (discussed more fully in **10.9** above).

1 *R v Thames Water Utilities Limited* (unreported), 8 March 2001, Bromley Magistrates' Court).

10.58 The Sentencing Council, an independent body that develops sentencing guidelines for English courts, closed a consultation in June 2013 on proposals for new guidelines for the sentencing of environmental offences. If implemented, the level of fines imposed for environmental offences is likely to increase. Publication of the guidelines is awaited at the time of publication.

Environmental Liability Directive (2004/35/EC)

10.59 The EL Directive, which is discussed in detail elsewhere in this chapter, provides that any natural or legal, private or public person operating or controlling occupational activities which result, whether directly or indirectly, in water damage (eg damage which significantly affects the ecological, chemical and/or ecological potential of water) may be liable under one of the EL Directive's two liability regimes. The EL Directive was implemented in England by the Environmental Damage (Prevention and Remediation) Regulations 2009.[1]

1 SI 2009/153.

10.60 The first regime imposes strict liability where environmental damage is caused by certain occupational activities (listed in Annex III to the EL Directive), including discharges into the inland surface water and groundwater. The second regime imposes fault-based liability where:

(a) activities (other than those falling under the strict liability regime) cause damage to protected species and natural habitats; and
(b) the operator is found to have been at fault or negligent.

Both liability regimes also apply where there is an imminent threat of damage occurring by reason of the activities carried out.

10.61 Where an operator falls under one of the EL Directive's liability regimes, it will be required to take either preventative or remedial action, and to provide certain information to the competent authority (ie the EA) notifying the situation and (if applicable) whether a threat of damage persists.

Liability under the Salmon and Freshwater Fisheries Act 1975

10.62 Section 4 of the Salmon and Freshwater Fisheries Act 1975 ('the 1975 Act') provides that any person who causes or knowingly permits to flow, or puts or knowingly permits to be put, into any waters containing fish or into any tributaries of water containing fish, any liquid or solid matter to such an extent as to cause the waters to be poisonous or injurious for fish or the spawning ground, spawn or food of fish, shall be guilty of an offence. Schedule 4, Pt 1 to the 1975 Act provides that a person guilty of an offence under s 4 is liable on summary conviction to a fine not exceeding £2,500 (and £40 for each day for which the offence continues after a conviction thereof) and, on conviction on indictment, to two years' imprisonment and/or an unlimited fine. For some of the minor offences, such as fishing without a rod licence, there are powers to issue a fixed penalty notice as an alternative to prosecution.

Relationship with contaminated land

10.63 Water pollution and contaminated land can be directly linked. The enforcing authority may choose to prosecute under WRA 1991 or EPA 1990. Contaminated land is dealt with elsewhere in this chapter, but here it is sufficient to point out that, under the contaminated land regime, the enforcing authority can serve a remediation notice specifying exactly what work is to be done and the degree of remediation to be carried out, potentially increasing the offender's clean-up costs.

Coastal pollution

10.64 Regulations 12 and 13 of the Merchant Shipping (Prevention of Oil Pollution) Regulations 1996 ('the 1996 Regulations') prohibit the discharge into the sea of chemicals or other substances in quantities or concentrations which are hazardous to the marine environment and discharges containing chemicals or other substances introduced for the purposes of circumventing the conditions of discharge prescribed by regs 12 and 13 retrospectively.

10.65 Section 131 of the Merchant Shipping Act 1995 ('MSA 1995') deals with the discharge of oil from ships. Section 131(1) of the MSA 1995 provides that, if any oil or mixture containing oil is discharged from a ship into the UK National Waters (including most waters within the jurisdiction of a harbour authority) which are navigable by seagoing ships, the owner or master of the ship shall be guilty of an offence. Section 131(3) provides that a person guilty of an offence under s 131 shall be liable on summary conviction to a fine not exceeding £250,000 and on conviction on indictment to an unlimited fine.

10.66 Under reg 36(2) of the 1996 Regulations, if a ship fails to comply with reg 12 or 13, the owner and master are each guilty of an offence under s 131(3) of the MSA 1995 (see above).

10.67 Under s 2(1) of the Prevention of Oil Pollution Act 1971 ('the 1971 Act'), the discharge of oil, or a mixture containing oil, from a place on land into waters (including territorial and inland waters) is an offence on the part of the occupier of the place in question, unless he proves that the discharge was caused by a person who was in that place without his permission, in which case that person is guilty of the offence. Under s 2(4) of the 1971 Act, the person guilty of an offence under s 2 is liable on summary conviction to a fine not exceeding £50,000 and, on conviction on indictment, to an unlimited fine.

CONTAMINATED LAND

Introduction

10.68 Part 2A of the EPA 1990 introduced the regulatory regime for dealing with contaminated land. A proposal for public registers of contaminated land (under s 143 of the EPA 1990) was shelved amid fears of its potential impact upon an already depressed property market. Following further lengthy consultation, s 57 of the EA 1995 inserted more detailed proposals of the current statutory regime which did not come into force in England and Wales until 1 April 2000.[1] The statutory scheme has since been extended to address land that is contaminated by virtue of radioactivity and is currently set out in the Contaminated Land (England) Regulations 2006[2] and further augmented by guidance.[3]

1 SI 2000/340.
2 SI 2006/1380.
3 DEFRA, 'EPA 1990: Part 2A Contaminated Land Statutory Guidance', April 2012, and also EA guidance for local authorities regarding 'Detailed inspection of radioactive contaminated land under Part 2A EPA 1990'.

10.69 The contaminated land regime, commonly referred to as 'Part 2A',[1] introduces a complex scheme of strict and retrospective liability for the remediation of contaminated land. This raises the prospect that innocent owners or occupiers of land may find themselves liable to criminal fines and penalties (and/or civil damages) for contamination liabilities pre-dating their period of ownership or occupation of the land. This is a landmark piece of legislation in environmental law and has far-reaching consequences.

1 EPA 1990, Pt 2A.

Definitions

10.70 The provisions of Part 2A envisage a sequence of steps starting with the identification of contaminated land, following a process of consultation which leads to voluntary or mandatory remediation of the land. The statutory definition of 'contaminated land' is as follows:

> 'Contaminated land is any land which, by reason of substances in, on or under the land, appears to be in such a condition that either of the following apply–
>
> (a) significant harm is being caused or there is a significant possibility of such harm being caused; or
>
> (b) significant pollution of controlled waters is being caused, or there is significant possibility of such harm being caused; …'[1]

This definition has been amended by s 86 of the Water Act 2003 in respect of pollution of controlled waters, but this provision has not yet been brought into force.

1 EPA 1990, s 78A(2).

10.71 Part 2A places responsibility upon LAs to apply the statutory definition to sites within their area to see whether they meet the definition provided in the EPA 1990. The definition itself relies upon the concept of 'harm' which is difficult to define and capable of subjective judgement. The following definition is offered by way of assistance:

> ' "Harm" means harm to the health of living organisms or other interference with the ecological systems of which they form part and, in the case of man, includes harm to his property.'[1]

While DEFRA's Statutory Guidance[2] ('the Statutory Guidance') provides some assistance as to how these concepts are to be dealt with, it still poses problems for LAs in determining what harm is to be regarded as significant and at what level to apply any *de minimis* concept with regard to pollution of controlled waters.

1 EPA 1990, s 78A(4).
2 'EPA 1990: Part 2A Contaminated Land Statutory Guidance', April 2012.

Pollution linkages

10.72 The LA must consider and identify the pollution linkages which it considers may exist on the land which are sufficient to render it capable of classification as contaminated. In this, the LA must seek to identify a source, a receptor and a pathway linking the two. It is not necessary to have an offsite receptor; onsite contamination is sufficient in itself to trigger liability.

Appropriate persons

10.73 If the LA is satisfied that such linkages exist and that the site does meet the definition under the EPA 1990, it must identify and notify all potentially 'appropriate persons', giving full details of each and every pollution linkage and why it considers that each appropriate person may be liable for some or all of them. If a site is badly contaminated and affects a major aquifer or controlled waters used for the supply of drinking water, it may be further notified as a special site, in which case it will pass to the EA who will become the appropriate regulator. Together the LA and EA are referred to as the 'Regulator'.

The EPA 1990 offers the following definition of the 'appropriate person':

> 'any person or any of the persons, who caused or knowingly permitted the substances, or any of the substances, by reason of which the contaminated land in question is such land to be in, on or under that land is an appropriate person'[1]

1 EPA 1990, s 78F(2).

10.74 Appropriate persons fall into two potential liability categories. The greatest liability category, known as 'Class A', identifies as appropriate persons those who have caused or knowingly permitted the contamination to have occurred (ie the 'polluter pays' principle). It has recently been established that this category does not extend to successors of long-defunct polluting companies which have been privatised under statutory schemes.[1] The unusual aspect of the regime is that there is a secondary class of appropriate person, known as 'Class B', which can be an innocent owner or occupier of the land who may not have even known of the contamination and not done anything to contribute to it.

1 *R (on the application of National Grid Plc (formerly Transco Plc)) v Environment Agency* [2007] UKHL 30.

10.75 EPA 1990, Part 2A envisages that, having identified a site as potentially contaminated, the LA will then seek to identify the appropriate person(s) (either Class A or Class B) and notify them of the identified pollution linkages. There should then follow a consultation period (the Statutory Guidance suggests three months, but this is not definitive) during which time the parties are intended to agree upon an appropriate programme of remediation for the land. The LA, in consultation with the appropriate persons and in line with the Statutory Guidance, will seek to apportion liability for those remediation works identified.

1 DEFRA, 'EPA 1990: Part 2A Contaminated Land Statutory Guidance', April 2012.

10.76 The Statutory Guidance sets out a system of complex rules for apportioning liability between Class A and Class B persons. Class A persons will rank in priority. If a Class A person is identified, a Class B person will not be pursued. If more than one Class A person is found liable, responsibility for remediation costs etc are apportioned between the group of Class A persons according to their perceived culpability (to be determined by the Regulator). If Class A persons cannot be found or identified, liability will rest with Class B persons irrespective of fault. Here, the Statutory Guidance suggests that liability should be apportioned between Class B persons according to their period of ownership or occupation of the land in question. Innocent occupiers or owners of sites with latent or previously unidentified pollution may, therefore, find themselves exposed to potential liabilities if they are identified as appropriate persons and could become involved in a lengthy consultation process which ends with a substantive costs bill for remediation.

Enforcement

10.77 Part 2A envisages an entirely voluntary system of consultation with clean-up standards imposed on a 'suitable for current use' basis. However, Part 2A also provides for mandatory controls in circumstances where appropriate persons do not cooperate or volunteer to meet remediation costs. Part 2A provides in those circumstances that the Regulator may use enforcement powers similar to those discussed elsewhere in this chapter relating to statutory nuisance. It provides that the Regulator has a duty, in circumstances where it has followed the regime set out under Part 2A, to serve on each appropriate person a remediation notice specifying what the person is to do by way of remediation and the periods within which he or she is required to do each of the specified things.

10.78 Further details are given in the Statutory Guidance of the form and content of such a notice and what may be reasonably required. The duty to serve a notice is one that may be enforced by judicial review proceedings. It is not a matter for the general discretion of the Regulator. Remediation notices will be placed on a public register and are therefore available for public consultation. If satisfied, the Regulator may serve more than one remediation notice upon more than one appropriate person, as circumstances dictate, and may also serve consecutive notices where phased remediation of the land is appropriate.

Offences

10.79 Section 78M(1) of the EPA 1990 provides that a person served with a remediation notice who fails, without reasonable excuse, to comply with any of its requirements shall be guilty of a criminal offence. It is open to the Regulator to commence a prosecution in a magistrates' court (the Statutory Guidance only envisages summary conviction – s 78M(3) of the EPA 1990), which must be brought within six months of the time when the offence was deemed to be committed.[1] The six-month period will run from the compliance deadline as specified in the remediation notice.

1 Magistrates' Courts Act 1980, s 127.

10.80 In the event of a successful prosecution, the potential maximum penalty would be level 5 on the standard scale (ie currently a maximum of £5,000 for individuals and £20,000 where the notice relates to industrial, trade or business premises), with the potential for continuing daily fines of a further tenth of the maximum sum for each day the offence continues. The Secretary of State can order an increase in the upper limit of the penalty for business premises. It should also be noted that a company so prosecuted will have a criminal conviction which will remain on the record.[1]

1 Note that the Rehabilitation of Offenders Act 1974 does not apply to companies.

10.81 The question of whether to prosecute (unlike whether to serve a notice) is a discretionary matter for the Regulator who will usually have regard to their own local strategy (in the case of a LA) or published prosecution policy (in the case of the EA or the Scottish Environmental Protection Agency (SEPA)). However, in most cases where a notice has been disregarded, prosecution will follow. Once a matter is before the magistrates' court, the usual evidential rules will apply, and it will be for the prosecution to prove its case beyond all reasonable doubt.[1]

1 The proceedings will be governed by the Magistrates' Courts Act 1980.

Civil remedies

10.82 Alternatively, if the Regulator feels that criminal proceedings under s 78M(1) of the EPA 1990 would be an ineffectual remedy, it may commence proceedings in the civil courts (High Court in England and Wales and a court of competent jurisdiction in Scotland) for the purposes of securing compliance with the remediation notice by means of a mandatory injunction (a specific implement in Scotland). This would secure compliance with the remediation notice, whereas criminal prosecution would only effect a criminal penalty for non-compliance.

Appeals

10.83 Section 78L of the EPA 1990 provides a 21-day period in which to bring an appeal against a remediation notice, beginning with the day on which the remediation notice is served. Appeals against notices served by LAs will go to a magistrates' court in England and Wales, or in Scotland to the Sheriff by way of summary application. Appeals against notices served by the EA or SEPA will go to the Secretary of State or Scottish ministers. While appeals to a magistrates' court follow the tradition of statutory nuisance (and other similar legislation), the Statutory Guidance provides only scant detail of the rules which should govern appeals, although the grounds of appeal are set out in the Contaminated Land (England) Regulations 2006, reg 7.[1] These Regulations provide no less than 19 separate grounds of appeal. Again, concerns have been expressed about the complexity (and lack of particularity) of the appeals procedure and of the difficulties for the lower criminal courts in having to deal with such technical and specialist legislation.

1 SI 2006/1380, reg 7.

10.84 Appeals to the Secretary of State are even less adequately specified. It appears that the Secretary of State may, in practice, pass such appeals to the Planning Inspectorate for determination. This complexity and lack of uniformity has hastened calls for an environmental court or tribunal (perhaps along similar lines to the Planning Inspectorate or the recently published proposals by the Ministry of Justice for a specialist planning court) as an appropriate forum in which such appeals and prosecutions could be heard. As yet, there is no indication that such proposals are likely to be implemented in the short to medium term. Clearly, there is also scope for alternative dispute resolution, particularly with regard to the question of the appeal procedure.

10.85 There is also the possibility to challenge a remediation notice by way of judicial review, proceeding on the basis that the Regulator has acted outside the grounds of its authority (ie its actions were *ultra vires*), or that there has been some procedural defect or failure to take relevant considerations into account. An applicant may seek judicial review in the High Court in order to quash the notice. However, leave to seek judicial review is only given in the most extreme of circumstances and is a very limited remedy which is used sparingly by the courts and only for the clearest examples of error. Although Part 2A does not specifically exclude judicial review, the fact that there is an appeal procedure under the regime may, of itself, preclude the grant of leave to bring a judicial review.

Conclusion

10.86 Part 2A of the EPA 1990 is an extremely complex and technical piece of legislation, which relies heavily upon the wording of the Statutory Guidance for its implementation. Progress has been slow in getting the practicalities of the legislation up and running and, after seven years since the legislation came into force, precedent has still not been established and there have been few challenges. At the time of writing, only two challenges under Part 2A have been brought through the court system.[1] However, as the regime 'beds in', it is inevitable that appeals, challenges and prosecutions will emerge which may assist in further defining the scope of the legislation. It is clear, however, that the concepts of criminalising liability for contamination of land or water have been firmly established and that, in the case of contaminated land, principles of strict (ie no fault) liability can be applied retrospectively.

1 *Circular Facilities (London) Ltd v Sevenoaks DC* [2005] EWHC 865 (Admin) and *R (on the application of National Grid Gas Plc (formerly Transco Plc)) v Environment Agency* [2007] UKHL 30.

10.87 As at December 2008, the EA estimated that approximately 325,000 sites (300,000 hectares) in England and Wales have had some form of current or previous use that could have led to contamination. Approximately 33,500 sites have been identified by the EA as contaminated to some extent, and 21,000 sites have been treated (predominantly through the planning regime).

AIR EMISSIONS

10.88 Air pollution legislation remains focused on reducing the adverse human effects and environmental damage caused by air pollutants. The levels of air pollutants measured today can still give rise to significant health impacts.

10.89 The regulatory framework for air pollution consists of the following:

- the permitting system under the Consolidated EP Regulations (see **10.8** onwards); and
- the common law of nuisance, which enables a claim for damages where emissions unreasonably interfere with the enjoyment of private rights or property. Common law nuisance is outside the scope of this chapter, although the Court of Appeal held that it was not a defence to a common law claim of nuisance to show that the activities that were the subject of a claim were carried out in accordance with a permit.[1] Further discussion of statutory nuisance can be seen elsewhere in this chapter, but the ability to take action under the statutory nuisance provisions is restricted where other statutory controls apply.

1 *Barr & Others v Biffa Waste Services Ltd* [2012] EWCA Civ 312.

10.90 In addition, the European Commission has adopted the Directive on Ambient Air Quality and Cleaner Air for Europe. It sets standards and target dates for reducing concentrations of fine particles (ie PM2.5) in urban areas which, together with coarser particles known as PM10 (already subject to legislation), are among the most dangerous pollutants for human health. The Directive came into force in June 2008 and was transposed into English law by the Air Quality Standards Regulations 2010,[1] which came into force in June 2010 ('the Air Regulations'). Similar instruments covering Wales, Scotland and Northern Ireland were published by the devolved administrations. The Air Regulations contain options for the Government to secure additional time to meet the limit values for particulate matter, subject to conditions and assessment by the European Commission. The obligations under the Air Regulations fall on DEFRA, but LAs have statutory duties for local air quality management under EA 1995. In March 2012, DEFRA announced that it will use the European Commission's review of the Directive, expected in 2013, to seek amendments to reduce the risk that the UK breaches EU legislation on air quality, as well as simplification of the EU air quality framework.

1 SI 2010/1001.

10.91 Whilst the Directive applies directly to the UK Government, the Government may seek to fulfil its requirements by passing on obligations to corporate bodies to reduce air pollution in urban areas (for example, by complying with Air Emission Zones aimed at reducing emissions from vehicles, or imposing conditions in environmental permits/consents or planning permissions to implement legislation governing air emissions which implement EU legislation).

10.92 In 2011, the European Commission consulted on its current policy on air quality as part of a comprehensive review of EU air quality policy. Between December 2012 and March 2013, the European Commission consulted on improving implementation of existing air quality legislation and any further regulation that may be required, as part of its review of the EU Thematic Strategy on Air Pollution. The results of the consultation will be considered in a comprehensive review, due to take place in 2013, of EU air policies and the European Commission's proposals for legislation.

STATUTORY NUISANCE

Introduction

10.93 Statutory nuisance provides the basis of public health protection against a range of environmental threats, such as emissions to air, dust, vibration, odour and noise pollution, as well as providing indirect protection for the environment.

10.94 The primary purpose of the statutory nuisance provisions in the EPA 1990 is to provide a prompt local remedy to prevent a nuisance from recurring. Statutory nuisance provides a basis for requiring the abatement, non-recurrence or prohibition of certain activities or states of affairs where they are deemed to be a nuisance or prejudicial to health. The LA, through its environmental health officer, is the regulatory authority responsible for enforcing statutory nuisances.

The basis of statutory nuisance

10.95 Nuisance is a common law concept which is not defined in the EPA 1990. An actionable nuisance at common law arises where the use or enjoyment of land, or some right over or in connection with land, is interfered with by the action of another and, as a result, an owner or occupier of the land may sue for an injunction prohibiting the recurrence of the nuisance or damages in lieu of an injunction.

10.96 Section 79 of the EPA 1990 lists a number of activities which amount to a statutory nuisance if 'prejudicial to health or a nuisance'. These are as follows:

- state of premises;
- smoke emitted from premises;
- fumes or gases emitted from premises;
- dust, steam, smell or other effluvia from industrial, trade or business premises;
- accumulation or deposits;
- animals (kept in such a place or manner so as to cause a nuisance);
- insects emanating from relevant industrial, trade or business premises;
- artificial light emitted from premises;
- noise emitted from premises;
- noise emitted from or caused by a vehicle, machinery or equipment in a street; and
- other matters declared by any enactment.

10.97 An activity or matter will not amount to a statutory nuisance where it consists of, or is caused by, any land being in a contaminated state. Land is in a contaminated state if, and only if, it is in such a condition, by reason of substances in, on or under the land, that harm is being caused or there is a possibility of harm being caused or pollution of controlled water is being, or likely to be, caused.[1]

1 The terms 'harm', 'pollution of controlled waters' and 'substance' have the same meaning as applied under the contaminated land regime in Part 2A of the EPA 1990.

10.98 The expression 'prejudicial to health' is defined in s 79(7) of the EPA 1990 as 'injurious, or likely to cause injury, to health'. In *Coventry City Council v Cartwright*,[1] the council owned a vacant site within a residential area on which people dumped inert rubbish. It was held in the Divisional Court that such accumulations or deposits were prejudicial to health if they were likely to cause a threat of disease or attract vermin but not if, as in this case, they were inert matter that could cause physical injury only to people who walked on it.

1 [1975] 1 WLR 845.

10.99 In *R v Bristol City Council, ex p Everett*,[1] the Court of Appeal confirmed that the phrase 'injury to health' in s 79(7) could not be applied to include the risk of accidental physical injury arising from a steep staircase.

1 [1999] 1 WLR 1170.

10.100 This raises the problem of particularly vulnerable groups suffering adverse effects. In the noise nuisance case of *Heath v Brighton Corpn*,[1] the claimant failed in his action in respect of noise from a generator as it was established that he had hypersensitive hearing. It was held that the test for determining whether a statutory nuisance existed was an objective one and should be judged according to the standards of an average man. The courts should not have regard to abnormal sensitiveness of particular occupiers. This was demonstrated again in *Cunningham v Birmingham City Council*,[2] where claims that a property was hazardous to an autistic child failed, since the property was not a statutory nuisance for an ordinary child.

1 (1908) 98 LT 718.
2 [1998] Env LR 1.

10.101 The court in *R (on the application of Anne) v Test Valley BC*,[1] confirmed that the test for statutory nuisance was an objective one, ie not whether the claimant's health subjectively had been prejudiced by the statutory nuisance but whether the average person would be prejudiced.

1 [2001] EWHC 1019 (Admin).

Regulation of statutory nuisance

10.102 LAs have a duty to inspect their areas from time to time in order to detect statutory nuisances which ought to be dealt with. Most cases are prompted by public complaint and, where a complaint of a statutory nuisance is made to an LA by a person living within its area, the LA is required under the EPA 1990 to take such steps as are reasonably practicable to investigate the complaint. An environmental health officer may visit the premises to make an objective decision as to whether or not the state of the premises, or anything on those premises, amounts to a nuisance. When considering the common law definition of 'nuisance', it is up to the environmental health officer to balance the various factors used when deciding on whether or not a common law nuisance exists. The most important of these factors are:

(a) the nature and location of the nuisance;
(b) the time and duration of the nuisance; and
(c) the utility of the activity concerned.

10.103 For an LA to act, the statutory nuisance must exist or be likely to occur or recur.[1] The LA must be satisfied, only on the balance of probabilities, that the statutory nuisance exists and, in the case of anticipated nuisances, there must be evidence that the forthcoming activity is likely to give rise to a statutory nuisance.

1 EPA 1990, s 80(1).

Person responsible

10.104 The enforcement of a statutory nuisance is normally against the person responsible for the nuisance, defined in s 79(7) of the EPA 1990 as being 'the person to whose act, default or sufferance the nuisance is attributable'. This is a very wide definition and can include an LA or a landlord who has allowed a tenant to carry on offensive activities, and can also include those who fail to abate nuisances which arise naturally, for example, occupiers of land where tree roots cause damage. In addition, it can include a tenant who denies access to a landlord who wishes to carry out works to abate a nuisance. The courts will look to see whether there has been some failure to meet acceptable standards (for example, sound insulation standards, at the time of construction).

10.105 In relation to a vehicle, the person responsible will also include the registered owner of the vehicle and any person who is for the time being the driver of the vehicle – and, in relation to machinery or equipment, will include any person who is for the time being the operator of that machinery or equipment.

10.106 If there are any difficulties in locating the person responsible for the statutory nuisance, section 80(2) of the EPA 1990 states that the definition of the person responsible can be extended to include the owner or occupier of the premises in question.

Enforcement scheme

The abatement notice regime

10.107 Where the LA is satisfied that a statutory nuisance exists or is likely to occur or recur in its area, it serves an abatement notice as required under s 80(1) of the EPA 1990. The notice may either require the abatement of the nuisance or it may require steps to be taken or works to be executed to abate the nuisance. An appropriate time period for complying with the requirements of the notice must be specified in the body of the notice.

10.108 If a statutory nuisance is occurring as a result of noise being emitted from a premises, prior to serving an abatement notice, the LA has the option to take such steps as it thinks appropriate in order to persuade the appropriate person to abate the nuisance or prohibit or restrict its occurrence or recurrence. If, by the end of seven days from the date the LA identified the nuisance, the steps taken by the LA have not been successful or the LA is satisfied that the nuisance continues to exist, or is likely to occur or recur, the LA shall serve an abatement notice.

10.109 The recipient of a notice may appeal to a magistrates' court within 21 days from the date on which he/she was served with the notice. If the notice involves expenditure and/or relates to noise caused in performance of a legal duty, the recipient of the notice can seek a stay in the performance of the notice, but otherwise failing to comply with the abatement notice is a criminal offence. Appealing an abatement notice within 21 days of it being served may be preferable to challenging the notice upon prosecution, because the appeal is subject to civil (rather than criminal) law, and because of the grounds available under the Statutory Nuisance (Appeals) Regulations 1995.[1] These include:

- the abatement notice is not justified under s 80 of the EPA 1990;

- substantive error or informality in the notice or procedure error in its service;
- the requirements of the notice or the period of compliance is unreasonable and the LA refuses to accept alternative arrangements;
- best practicable means were used to counteract the effects of a nuisance from an industrial, business or trade premises; and
- the notice was not served on the appropriate person or should have been served on another person in addition to the appellant.[2]

1 SI 1995/2644.
2 Regulation 3(1)(b).

10.110 In general, an appeal will have the advantage of suspending the operation of the notice. However, when the nuisance is prejudicial to human health, or where the timescales demand immediate action, the LA can insist that the notice is not suspended. In making this decision, the LA should also consider weighing costs of compliance against the public benefit expected from compliance.

10.111 An individual may also bring a private action in the magistrates' court to issue an abatement notice. Such a notice so issued would have the same consequences as a notice issued by the environmental health officer.

Contents of notices

10.112 In general, the notice must clearly and precisely identify the nuisance complained of, and provide information as to what is required to avoid recurrence or, indeed, occurrence.

10.113 To what extent the notice can give directions in relation to day-to-day business operations is arguable. A prohibition may be sufficient, as in *McGillibay v Stephenson*,[1] in which there was a requirement to desist from using premises for keeping pigs. The Divisional Court held that this was a good notice, as a statement of steps which might be taken would in these circumstances be surplus to requirements.

1 [1950] 1 All ER 924.

10.114 In some situations, the person responsible may have the freedom to abate the nuisance in a manner of their choice. However, they may need to know precisely what will satisfy the LA to abate the nuisance. A great deal would depend on the content of the notice, including the type of nuisance, the technical competence of the person responsible and the possible means of abatement. The notice should be 'sufficiently precise to enable the respondent to know what was wrong and what he had to do about it'.[1] It must be borne in mind that, in general, it may not be reasonable to demand a complete cessation of an activity[2] and that the contravention of the notice will constitute a criminal offence under s 80(4) of the EPA 1990. The courts may demand that the LA specifies the plan of work required to alleviate the statutory nuisance.

1 *Stanley v Ealing* [2000] Env LR D18.
2 *R v Falmouth and Truro Port Authority, ex p SW Water* [2000] 3 All ER 306.

Service of the notice

10.115 Notices should be served on the person responsible for the nuisance, as defined at **10.104** above.

Consequences of non-compliance

10.116 Failure to comply with a notice without reasonable excuse is an offence under s 80(4) of the EPA 1990. If the nuisance arises on industrial, trade or business premises, failure to comply with a notice in respect of that nuisance carries a maximum penalty of £20,000 in accordance with s 80(6) of the EPA 1990. In respect of all other types of premises, failure to comply with a notice will result in a maximum penalty of £5,000 being imposed, plus £500 for each day the offence continues after conviction. In addition to these sanctions, the magistrates have discretion to award compensation, but only up to a maximum of £5,000.

10.117 A company can be guilty of breaching a notice and hence can commit a criminal offence, as can an officer of the company under s 157 of the EPA 1990. Section 157(2) provides that, where the affairs of the corporation are managed by its members, a member fulfilling a managerial function shall be liable where it can be proved that the officer was a decision-maker in the company and he was at least partly responsible for the breach.

10.118 A company in the hands of administrators may be prosecuted, but the requirements of the Insolvency Act 1986, s 11 will apply, so the consent of the company's administrator must be given or leave obtained from the court.[1] It will depend upon the circumstances of the case, as the company is unlikely to have funds available to pay any fine or restoration costs. The EA may restore a company under s 1029 of the Companies Act 2006 to enable it to commence a prosecution in the public interest. Insolvency practitioners may disclaim a licence as 'onerous property' under s 178 of the Insolvency Act 1986,[2] and they can also be prosecuted in their personal capacity if they commit an offence.

1 *Re Rhondda Waste Disposal Co Ltd* [2001] Ch 57. The leave of the High Court is required for the prosecution to be commenced or continued in the magistrates' court.

2 *Re Celtic Extraction Ltd* [2001] Ch 475. The Court of Appeal confirmed that a waste management licence was 'property' within the meaning of s 436 of the Insolvency Act 1986, and could disclaim it as 'onerous property' under s 178.

Defences

10.119 The usual defences on a matter of fact and/or law may apply, but there are also specific defences written into the legislation. The first of these is the use of 'best practicable means' to prevent or counteract the effects of the statutory nuisance under s 80(7) and (8) of the EPA 1990.

10.120 Certain elements of 'best practicable means' are considered in EPA 1990, s 79(9) but this stops short of a full definition. For example, 'practicable' is defined as meaning reasonably practicable, having regard, among other things, to local conditions and circumstances, the current state of technical knowledge and the financial implications. The 'means' to be employed include design, installation, maintenance, manner and periods of operation of plant and machinery, as well as the design, construction, and maintenance of buildings and structures. It is not necessary to show that the means employed brought the nuisance to an end, but it is probably enough if they were adequate to prevent or to counteract the effects of the nuisance.

10.121 As this is a statutory defence, the burden of proof lies on the defendant. The defence is available only where the nuisance arises on industrial, trade or business premises. Furthermore, the defence is not available when considering certain specific nuisances, eg smoke being emitted from premises (unless emitted from a chimney) and fumes or gases emitted from premises and nuisances under other statutes.

10.122 The other available defence is 'reasonable excuse' under s 80(4) of the EPA 1990, which provides that a person who is served with an abatement notice will be guilty of an offence if he/she, 'without reasonable excuse', fails to comply with the requirements of the notice. The defendant who wishes to rely on such a defence must discharge the evidential burden by specifying the excuse, and the prosecutor must prove that the excuse is not a reasonable one beyond all reasonable doubt.

Interaction between nuisance and statutory consents

10.123 In the parallel field of private nuisance, two recent cases have considered the interaction between nuisance and the environmental permitting and planning regimes.

10.124 In *Barr v Biffa*,[1] the Court of Appeal confirmed that, when dealing with allegations of nuisance against sites regulated by an EP, nuisance would continue to be assessed against traditional nuisance principles (based on the character of the neighbourhood where the site is situated), rather than the modern approach (taken by the High Court at first instance) of considering allegations of nuisance against the statutory regulation of the site by the Regulator.

1 *Barr & others v Biffa Waste Management Services Limited* [2012] EWCA Civ 312.

10.125 *Coventry v Lawrence*[1] concerned residents who purchased a house near a racetrack where motor races were held. The races had been authorised by the Local Planning Authority (LPA) before the residents moved in. The residents alleged that, notwithstanding the grant of planning permission, the races constituted a private nuisance. At first instance, the High Court had granted the residents an injunction and awarded damages on the basis that the planning permission granted by the LPA had not altered the locality permanently. The Court of Appeal overturned this decision, holding that the grant of planning permission by the LPA did not amount to a licence to commit nuisance, but had changed the character of the neighbourhood. The case has gone to the Supreme Court, whose decision is likely to provide much-needed clarity on the relevance of planning permission to nuisance.

1 *Coventry (t/a RDC Promotions) v Lawrence* [2012] 1 WLR 2127.

ASBESTOS

10.126 Asbestos is a naturally occurring fibrous material that was widely used in the UK until 1999, when its use was finally banned. It had a number of uses, such as lagging for pipe work, fire protection and insulation. There are three main types of asbestos. These are:

- Crocidolite (blue);
- Amosite (brown); and
- Chrysotile (white).

All are dangerous, but brown and blue asbestos is generally recognised as being more dangerous than white. Exposure to asbestos can cause asbestosis, mesothelioma, diffuse pleural thickening and lung cancer, and is the largest cause of work related death in the UK. Although the actual number of deaths caused by exposure to asbestos is difficult to accurately determine because of the long latency period, which can be between 15 and 60 years, it is estimated that these numbers could be somewhere between 4,000 and 20,000 deaths per year. The number of deaths is predicted to continue to rise for some years to come, and currently stands at around 4,500 per year.

10.127 There are estimated to be around 1.5 million commercial properties in the UK still containing some form of asbestos-containing materials (ACMs). The possibility of exposure to ACMs during the maintenance or repair of these properties led to the introduction of the Control of Asbestos Regulations 2006. These regulations were updated in April 2012 by the Control of Asbestos Regulations 2012 ('CAR 2012')[1] to fully implement the EU Directive on exposure to asbestos.[2] These regulations require the 'duty holder' to manage asbestos in non-domestic premises. The duty holder is defined as every person who by virtue of a contract or tenancy has obligations for the maintenance or repair of those premises or where that person has control of the premises. In a shared office block, there might be more than one duty holder. Responsibilities for the duty holder, under the CAR 2012, include making a recorded assessment, which should be reviewed as necessary, of asbestos on the premises and to ascertain its location and condition. There is no specific duty on the duty holder to remove asbestos, but there is a requirement to compile a plan detailing how the asbestos is to be managed and its condition monitored. Crucially, the duty to manage includes a requirement that anyone liable to disturb it is told of its location and the content of the material.

The CAR 2012 introduced a new category of asbestos work, in addition to the two already in existence which were 'licensed asbestos work' and 'non-licensed asbestos work'. The third category is 'non-notifiable asbestos work' and, whilst not requiring a licence to carry out the work, will involve the duty holder providing notification of the works to the relevant enforcement authority, carrying out medical surveillance and keeping registers of work done by employees exposed to asbestos.

1 SI 2012/632.
2 Directive 2009/148/EC.

10.128 The costs of compliance with the CAR 2012 are determined by the extent of each duty holder's obligations. In practice, owners and landlords of buildings will need to negotiate with all the other parties involved and may agree to undertake works and recharge the costs to the tenants in accordance with the service charge provisions in the lease.

10.129 Architects, surveyors, contractors and construction and maintenance companies may also be subject to a duty to cooperate with the primary duty holder if they have information about the presence or otherwise of asbestos in a building. If such persons fail to pass on relevant information, and workers are exposed, those third parties can be expected to be prosecuted as well as the primary duty holder. A breach of the CAR 2012 is a criminal offence.

10.130 There have been a number of civil cases coming before the courts concerning claimants who have been exposed to asbestos whilst at work and the extent to which they are entitled to damages. In *Fairchild v Glenhaven Funeral Services Ltd*[1] it was decided that an employer was liable if it had materially and negligently contributed to the risk of causing mesothelioma, without need to prove direct causation of the condition. Any unlawful exposure of the employee to airborne asbestos would ordinarily constitute such a contribution if mesothelioma develops. This adequately dealt with cases where the employee had worked for a number of employers throughout his working life, and been exposed to asbestos at a number of those employers, but because of the length of time between the exposure and the diagnosis of the disease was unable to pinpoint which exposure caused the development of the disease. However, many commentators, especially from within the insurance industry, felt that this could leave employers (or, more accurately, their insurers) potentially liable for large claims where their insured was not actually primarily responsible for the damage.

1 [2003] UKHL 22, [2003] 2 All ER 689, [2003] 1 WLR 983.

10.131 In *Barker v Corus (UK) plc*[1] the House of Lords sought to address this apparent inequity by ruling that, under *Fairchild*, a defendant is treated as having materially contributed to the 'risk' that the disease would occur and not deemed to have actually caused the injury. Therefore, liability was several but not joint. In effect, this meant that, if there was more than one employer who contributed to the risk of injury, each would only be liable to the extent (or percentage) to which they had contributed to that risk. This resulted in circumstances where claimants were unable to recover fully if the employer who was the main contributor to the onset of the disease had gone out of business or whose insurer could not be traced.

1 [2006] UKHL 20.

10.132 The government swiftly reversed *Barker* with the introduction of the Compensation Act 2006. Section 3 of this Act allowed employees who had contracted mesothelioma to claim from a single employer, so long as the employer had negligently or in breach of statutory duty exposed that employee to asbestos, whether or not they had been exposed elsewhere. The Act allowed for the circumstances considered in *Barker* by stating that employers could seek a contribution from other parties and that the court would take into account the length of periods of exposure in each term of employment in considering the extent to which each party was liable.

In *Sienkiewicz v Grief (UK) Ltd*,[1] the Supreme Court upheld the principle, established in *Fairchild* and codified in section 3 of the Compensation Act 2006, that a defendant could be liable even if it had caused only a 'single exposure' to asbestos. This appeal involves two linked mesothelioma appeals, which were joined in the Supreme Court. In single exposure cases, the *Fairchild* exception applies and a claimant succeeds if he proves, on the balance of probability, that the defendant's breach of duty materially increased the risk that he would develop mesothelioma.

1 *Sienkiewicz v Grief (UK) Ltd and Knowsley MBC v Willmore* [2011] UKSC 10.

10.133 Another asbestos case of major importance to come before the courts was in the House of Lords on 17 October 2007. *Rothwell*[1] was a test case for claimants who had contracted pleural plaques as a result of exposure to asbestos dust. Pleural plaques are areas of fibrous thickening of the pleural membrane which surrounds the lungs. There are no symptoms and it is thought that many 'sufferers' are not actually aware that they have contracted pleural plaques. However, contracting pleural plaques is a sign that exposure to asbestos has taken place and presupposes that mesothelioma may, but not 'will', develop sometime in the future. Naturally this state of affairs causes distress and anxiety. These 'sufferers' were quickly named by commentators as the 'worried well'.

1 *Rothwell v Chemical and Insulating Co Ltd* [2007] UKHL 39.

10.134 The medical evidence in the cases heard before the lower courts was that pleural plaques would never cause any symptoms, did not increase the susceptibility to other diseases, including mesothelioma, or shorten the expectation of life. They had no effect upon health at all. Further, it was not the plaques themselves that caused anxiety or distress; it was the evidence of the internal presence of asbestos that led to concern. However, the extent to which this distress would be any worse than someone who had spent many years working in conditions in which they will have been exposed to asbestos is questionable.

10.135 In any case, the House of Lords ruled in *Rothwell* that damages for plural plaques, caused by negligent exposure to asbestos, were not recoverable and neither were damages for distress, even where this has been diagnosed as a psychiatric illness, as this would have been unlikely to have been foreseeable in a person of 'ordinary fortitude'.

10.136 Whilst closing the door on claims for tortious exposure to asbestos resulting in pleural plaques, the House of Lords has left the door, at least, slightly ajar so far as claims for breach of contract are concerned. However, whether this route will prove worthwhile, given the likelihood that only nominal damages will be recoverable in the absence of loss suffered by the claimant, remains to be seen.

10.137 Another recent case of note is where a parent company was found liable to an employee of its subsidiary under tort law. In *Chandler v Cape plc*,[1] the Court of Appeal found that a parent company owed a direct duty of care to employees of its subsidiary. Here, the employee had contracted asbestosis through exposure to asbestos dust. The court considered the fact that the parent company was aware of the works going on and had superior knowledge to the subsidiary about the risks and management requirements when dealing with asbestos. The parent company therefore had a duty to either educate the subsidiary on how to provide its employees with a safe system of work or to ensure that such a system was in place.

Parent companies and former parent companies may find that they face an increase in claims as a result of this decision.

1 [2012] EWCA Civ 525.

10.138 Recently, the Supreme Court decided how asbestos-related disease claims affect employers' liability policies. In *Employers' Liability Insurance 'Trigger' Litigation*,[1] the court found that mesothelioma is contracted or sustained at the time of exposure to asbestos, not at the time the disease manifests itself. This means that an insurer will be liable to indemnify an employer under an employers' liability policy that was in place at the time of the asbestos inhalation. The Supreme Court also examined the issue of whether such a liability on the insurer would apply in cases where the employer had become liable under the *Fairchild* rule. The court considered that the purpose of an employers' liability policy was to cover the employer for any liability that had arisen from its employees. Therefore, once it has been established that the employer is liable to the employee, under the rule in *Fairchild* or otherwise, the insurer is liable to pay out to the employer.

This approach was followed by the Court of Appeal in *International Energy Group Ltd v Zurich Insurance plc UK Branch*.[2] Here the court found that the insurer was liable to pay out for the entire compensation amount paid by the employer to an employee who had contracted mesothelioma at some point during her 27 years in employment there. This was despite the insurer only insuring the employer for six out of those 27 years.

1 [2012] UKSC 14.
2 [2013] EWCA Civ 39.

NUCLEAR

10.139 Nuclear energy continues to be a hot topic in the UK with the end of life of the existing stations, the development of new stations and the proposals to build a deep geological facility. Although corporate liability for nuclear incidents is not of general application to all corporate bodies, it was felt that the subject should be included briefly in this chapter. Following the Fukushima incident in Japan, the EU Energy Ministers introduced stress tests for nuclear power plants in the EU. The results of the independent tests carried out by the national authorities were published in October 2012. Member States have produced national action plans which have to be implemented by 2014.

10.140 There have been many recent developments in the nuclear sector and in the regulation of and framework for nuclear new build in the UK. However, in the context of corporate environmental liability we limit our consideration of the nuclear industry to a brief explanation of liability under the Nuclear Installations Act 1965 (the NIA 1965), the proposed amendments to the NIA 1965 and the extension of the contaminated land regime to radioactive contamination. These principles will be relevant not only to companies which operate nuclear installations, carry out activities involving radioactive substances and those to which radioactive contamination is a by-product of their activities, but also to those who have an interest in land or who carry out activities in the vicinity of those nuclear operations. In the absence of any further amendment, these principles would be equally applicable to those corporate bodies which operate or which are affected by the new nuclear facilities intended to be built in the UK over the next ten years, as well as those operating and affected by existing facilities.

10.141 Specific issues of environmental liability for companies involved in the transport or disposal of radioactive materials, and which provide support services to nuclear installations and the nuclear sector, are outside the scope of this book.

Nuclear Installations Act 1965 (NIA 1965)

10.142 The NIA 1965 governs the licensing and regulation of nuclear-licensed sites used in connection with civil nuclear purposes. Under that Act, a site cannot be used for the purpose of installing or operating a nuclear installation unless a licence has been granted by the Office for Nuclear Regulation (ONR)[1] and is in force.[2] A licence can only be held by a body corporate[3] (the licensee). In terms of corporate liability, the NIA 1965 set up a statutory regime which imposes strict liability on the licensee of a nuclear licensed site for personal injury or property damage caused to third parties by an occurrence involving nuclear matter and which:

> 'arises from or results from the radioactive properties, or a combination of those, and any toxic, explosive or other hazardous properties, of that nuclear matter.'[4]

1 At the time of going to press, the Energy Bill is progressing through Parliament under which the ONR will be established as an independent statutory body outside the HSE to regulate the

nuclear power industry. The ONR will be responsible for civil nuclear, radioactive transport safety and security registration. The costs are intended to be covered by operators in the nuclear industry. Pending adoption of this legislation, a non-statutory agency within the HSE was established in 2011.
2 NIA 1965, s 1.
3 NIA 1965, s 3.
4 NIA 1965, s 7.

10.143 The licensee is also under a duty to ensure that ionising radiation emitted during the period of the licensee's responsibility from non-nuclear matter on site or from any waste discharged (in whatever form) on or from the site does not cause injury to any person or damage to any property of any person other than the licensee.[1] Liability for this damage is also strict.

1 NIA 1965, s 7(2).

10.144 The liability of the licensee is currently capped at £140 million. However, under s 19 of the NIA 1965, the licensee is under an obligation to ensure that sufficient funds are available for the duration of the licensee's responsibility, by taking out insurance through the UK nuclear insurance pool or by some alternative approved means. Above the caps, the Government bears liability for damage in accordance with tiers 2 and 3 of the Brussels Convention.

10.145 Any corporate bodies which hold a nuclear site licence for their operations will be subject to this liability regime.

The 2004 Amending Protocols to the Paris and Brussels Conventions

10.146 In January 2011, the UK Government issued a public consultation on the implementation of the 2004 Protocols to Amend the Paris and Brussels Conventions.

Under the consultation, among other changes, it is proposed that:

- the financial liability levels for Operators/ Licensees will increase from £140 million to €1200 million over a period of five years starting at €700 million from the date the legislation is in force and increasing to €100 million per annum (thereby leaving the UK Government with only its tier 3 Brussels liability); and
- the definition of 'nuclear damage' will be changed to include environmental damage and economic loss. Therefore the definition of 'nuclear damage' will be extended to include not only personal injury and property damage but also certain types of economic loss, the cost of measures to reinstate a significantly impaired environment, loss of income resulting from that impaired environment and the cost of preventive measures, including loss or damage caused by such measures.

Contaminated land regime

10.147 The contaminated land regime contained in Part 2A of the EPA 1990 is discussed elsewhere in this chapter in relation to non-radioactive contamination. Prior to 2006, radioactive contamination was not covered by that regime.

10.148 The Radioactive Contaminated Land (Modification of Enactments) (England) Regulations 2006,[1] which came into force on 4 August 2006, brought some, but not all, radioactive contaminated land within the definition of 'contaminated land' for the purpose of Part 2A of the EPA 1990. The 2006 Regulations do not apply to and, therefore, Part 2A was not extended to cover, radioactive contamination which was caused by the holder of a nuclear site licence and who is liable under the provisions of the NIA 1965[2] described above, nor were they extended to cover natural background radioactivity or radon gas.

1 SI 2006/1379 (equivalent regulations have also been introduced in Scotland, Wales and Northern Ireland).
2 See para 1 of the explanatory note to the regulations.

10.149 The way in which Part 2A applies to radioactive contaminated land is slightly different from the way in which it applies to non-radioactive contaminated land. Non-radioactive contaminated land will be classed as 'contaminated land' by the enforcing authority, when:

'(a) significant harm is being caused or there is a significant possibility of such harm being caused; or
(b) pollution of controlled waters is being, or is likely to be, caused.'

By contrast, the definition of 'radioactive contaminated land' requires that the contamination is causing or is likely to cause harm to humans. It does not cover harm to the wider environment or water pollution.[1]

1 Section 78A(2) and (4) of the EPA 1990, as amended by SI 2006/1379.

10.150 Responsibility for the remediation of radioactive contamination under Part 2A, as amended by the 2006 Regulations, is wider than that imposed by the NIA 1965. It rests with the 'appropriate person' who 'caused' or 'knowingly permitted' the substances to be in, on or under the land (a Class A person).[1] If the Class A person cannot be found, liability may pass to the current owner or occupier of the land (a Class B person). The class of persons who may be liable for the remediation of radioactive contamination is therefore much wider than under the NIA 1965 (and holders of a nuclear site licence who are liable under the NIA 1965 are excluded from its application). Liability for harm caused by radioactive contamination under Part 2A will therefore extend to corporate bodies other than those which hold a nuclear site licence.

1 EPA 1990, s 78F.

10.151 In 2007, further legislation – the Radioactive Contaminated Land (Modification of Enactments) (England) (Amendment) Regulations 2007[1] (the RCL Regulations) – was introduced, which extended the Part 2A contaminated land regime to include 'land contaminated by a nuclear occurrence'. Prior to the EP Regulations 2007 coming into force on 10 December 2007, any such contamination was dealt with exclusively under the NIA 1965.

1 SI 2007/3245 (equivalent regulations have also been introduced in Wales).

10.152 The RCL Regulations were introduced to comply with the remainder of the UK's obligations under the Basic Safety Standards Directive[1] in the wake of the threat of infraction proceedings from the European Commission. However, in practical terms, the circumstances in which a third party claim would be made under Part 2A, where land is contaminated by a nuclear occurrence, rather than under the NIA 1965, are few and far between.

1 96/29/EURATOM.

10.153 Where a claim is made under Part 2A, in the event of land being contaminated by a nuclear occurrence, the 'appropriate person' liable for remediation of that contamination is deemed to be the Secretary of State, who will fund remediation works.[1]

1 EPA 1990, s 78F(1A).

Conclusion

10.154 While it is often assumed that the nuclear industry only affects a very small number of corporate bodies, the recent focus on the future of that industry has drawn attention to the fact that the impact and influence of the industry is felt much more widely than commonly thought.

10.155 In light of the Government's proposals for the future of nuclear power and the approach to the management of radioactive waste, the development of the environmental liability regime in the nuclear sector should be watched closely over the next ten years and beyond, as the industry expands further.

Chapter 11

Local authorities

Dr Louise Smail, Risk Consultant, Ortalan, Manchester

Gerard Forlin QC, Barrister, Cornerstone Barristers, London; Denman Chambers, Sydney; Maxwell Chambers, Singapore

Piero Ionta, Senior Solicitor (Litigation), Royal Borough of Kensington & Chelsea and the London Borough of Hammersmith and Fulham

INTRODUCTION

11.1 Historically, a local authority (being a creature of statute and, as such, could only do that which statute permitted or required of it) had no power to act other than where it was expressly authorised by law to do so. The Local Government Act 1972 set about lifting some of these restrictions and, over recent decades, authorities have been empowered to do anything (whether or not involving the expenditure, borrowing or lending of money or the acquisition or disposal of any property or rights) which is calculated to facilitate, or is conducive or incidental to, the discharge of any of their functions.

11.2 The journey towards lifting the remaining restrictions upon the powers of local authorities was aided by the Localism Act 2011. There is a now a 'general power' of competence for county councils in England, district councils, London borough councils, the Common Council of the City of London, the Council of the Isles of Scilly, and parish councils that meet criteria set out in regulations.

This general power replaces the 'well-being powers' in section 2 of the Local Government Act 2000. It is a power 'to do anything that individuals generally may do', even if it is unlike anything that the authority, or other public bodies, may do:

- The power is not limited by the existence of any other overlapping power, and vice versa.
- The power can be exercised anywhere in the United Kingdom or elsewhere, for a commercial purpose or otherwise, for a charge, or without charge, and for, or otherwise than for, the benefit of the authority, its area or persons resident or present in its area.
- The power is subject to any 'prohibition, restriction or other limitation expressly imposed by a statutory provision' contained in existing legislation or contained in future legislation expressed in appropriate terms.
- The power is subject to any restrictions on the exercise of any existing power insofar as it overlaps with the new power.

11.3 The general law in relation to offences relating to property, people and public order applies as much to local government as to others. There are a number of offences, arising under statute, which exist to support and preserve the efficient administration at elections. Here, the criminal law performs an administrative function which complements the imposition of public law and civil liabilities. Except where an offence or the available sentence is confined to an individual, local authorities are liable under the general law.

11.4 The structure of local authorities and their decision-making processes and control are an important issue when looking at prosecutions for work-related deaths.

LEGAL ENTITY OF A LOCAL AUTHORITY

11.5 Local authorities are clearly within the scope of the Corporate Manslaughter and Corporate Homicide Act 2007 (CMCHA 2007), as they are corporate bodies by virtue of s 2 of the Local Government Act 1972. They also owe a duty of care to employees as employers, and to clients and service users as occupiers of land and suppliers of services. Therefore, their exposure is significant. Where there is a dispute as to whether a 'relevant duty of care' is owed, the judge will decide the point on the facts.

It is also important to recall that, in 2012, the London Liaison Group was formed, arising out of the National Code for Local Authorities. This looks, inter alia, at ways of targeting the highest-risk sectors inside the local authority. In essence, there are three methods: advisory visits; targeted inspections; and reactive investigations and prosecutions. This regime will almost inevitably lead to more prosecutions in the future. See further **1.11**.

PARTNERSHIPS

11.6 In recent times, local authorities have formed partnerships to carry out their work. These take on many forms and can be of a contractual nature or arrangements which are much looser. It is important that the procurement strategy addresses the liability issues when such partnerships are formed.

11.7 It has not been possible in the past to prosecute an unincorporated association, such as a partnership, which has no legal personality. A partnership can now be prosecuted, in the name of the partnership only, and any fine is payable out of the funds held in the partnership name and not by the individual partners.[1] This does not prevent an individual partner from also being prosecuted at common law. A limited liability partnership, however, is a body corporate.[2] Note, however, that a partnership may only be prosecuted for corporate manslaughter if it is an employer. If the partnership does employ anyone, its relevant duties of care are not limited to its employees. Enforcement agencies tend to pursue parties who they see as having the resources and knowledge to 'know better'. For example, if a council partners with a small charity for them to carry out a council service, and they (the charity) injure a member of the public, the council would be asked what they did to manage the service/risk.

1 CMCHA 2007, s 14.
2 Limited Liability Partnerships Act 2000, s 1(2).

SUB-CONTRACTORS

11.8 Where formal partnerships have not been adopted, it is not uncommon to find that local authorities contract out work. Whilst the work itself may be contracted out, responsibility for health and safety cannot be contracted out.

In October 2013, the Supreme Court extended the doctrine of non-delegation in the important case of *Woodland v Essex County Council*. Miss Woodland, then aged 10, sustained a serious brain injury during a swimming lesson given by an independent contractor of the council. The Supreme Court stated that, in an educational context, such duties cannot be delegated. This is an important extension of the doctrine and will be of major concern to local authorities (and the public sector) and their insurers.

In March 2013, Carrickfergus Borough Council in Northern Ireland was fined for health and safety breaches. In September 2011, the council had contracted Patrick Buckler, trading as Water Management Services Ltd, to carry out a risk assessment of the water systems at the Carrickfergus Amphitheatre. Mr Buckler had no specific training in legionella management or health and safety.

The HSENI carried out a routine visit to the site in November 2011 and found the risk assessment for the water system to be insufficient. An Improvement Notice was issued, which ordered the council to carry out a full and proper risk assessment. The council subsequently employed a different water management company, which found that the risk of legionella was high.

The council voluntarily closed the leisure centre until remedial work on the water system was completed. The most recent risk assessment referred to a build-up of slime and scale, which are known hazards associated with legionella.

The council pleaded guilty to breaching articles 4 and 5 of the Health and Safety at Work Order (Northern Ireland) 1978.[1] It was fined £1,500 and ordered to pay

£61 in costs. Buckler also pleaded guilty to the same breaches of the 1978 Order, as well as a breach of regulation 6 of the Control of Substances Hazardous to Health Regulations (NI) 2003.[2] He was fined £750, plus £61 in costs.

It is also important to look at the three-month inquest in 2013 into Lakanal House, where six people died in a South London block of flats (see **Chapter 1**).

1 SI 1978/1039.
2 SI 2003/34.

11.9 In relation to corporate manslaughter, the offence applies to all companies and employing partnerships, including those who are in a contracting relationship. However, whether a particular contractor might be liable for the offence will depend, in the first instance, on whether it owed a relevant duty of care to the victim. The CMCHA 2007 does not impose new duties of care, but the offence will apply in respect of existing obligations on the main contractor and sub-contractors for the safety of worksites, employees and other workers whom they supervise. The council has a duty of care to confirm that contractors have arrangements in place.

LOCAL AUTHORITY AS EMPLOYER

11.10 The local authority is clearly an employer when there are full-time employees (in which case, there is probably a contract of employment), but there may be times when the relationship is less formal, either in terms of the time taken to complete a particular project or the arrangement between the parties (for example, where the worker works from home). Councils may also be the 'employer' for some agency staff if they are directing and managing their work.

11.11 Local authorities do, in some respects, differ from other employers in that, along with government undertakings or bodies charged by the state with some of its functions, they are considered for the purposes of European Community law as 'emanations of the state'.[1] The consequence of this is that claims may be made against them for breaches of statutory duty.

1 *Fratelli Constanzo v Commune di Milano* [1990] 3 CMLR 239 and *Marshall v Southampton and South West Hampshire Area Health Authority* [1986] ICR 335.

11.12 It is particularly important that, when the employer/employee role is considered in the areas of partnership arrangements and outsourcing, the authority is clear as to where it stands in relation to its duty of care. There are a number of principles that can be considered.

11.13 The control test is:

- who lays down what has to be done and the way in which it is to be done;
- who provides (ie hires and fires) the people by which it is done; and
- who provides the plant and machinery that is to be used.

There is also the consideration of the skilled employees and who has discretion to decide how work should be done – here is it important to consider whose business is being carried out, and that of the injured worker or that of the employer. All of the above needs to be considered in the context of who is responsible for the overall safety of the people carrying out the work.

MANSLAUGHTER – PRE-CORPORATE MANSLAUGHTER AND CORPORATE HOMICIDE ACT 2007

11.14 In 2005, Gillian Beckingham, a council architect who was head of the design services group at Barrow Borough Council, was prosecuted as the directing mind of the council. This was the first prosecution of a local authority for manslaughter. The case related to Britain's biggest outbreak of Legionnaires' disease, which occurred in the summer of 2002. It was traced to an air-conditioning unit in the arts centre, Forum 28, run by the council. It killed seven people, and another 180 were infected. The case focused on Ms Beckingham's responsibilities for the maintenance of the air-conditioning unit at Forum 28. The prosecution alleged that she was instrumental in cancelling this maintenance contract and renegotiating a new contract that did not provide a water treatment regime.

11.15 Mr Justice Poole ruled:

> '[Ms Beckingham] occupied a third-tier post in the organisation. Her role was essentially one of consultant … Cost centre heads could consult with her and then decide not to follow her advice …
>
> No reasonable jury could conclude that she was at the material time, and in discharge of her functions in forming or coordinating the contract or in supervision of its performance, the council's [directing] mind.'

11.16 Gillian Beckingham was acquitted of manslaughter but convicted, by a majority of 11 to 1, of a breach of s 7 of the Health and Safety at Work etc Act 1974 (HSWA 1974) and fined £15,000. The charge of manslaughter against the council was dropped. The judge, Mr Justice Poole, dismissed the manslaughter case against Barrow Borough Council on the basis that Ms Beckingham was not at a sufficiently senior level within the council to embody the corporation and was not one of the controlling minds of the council. It was on that legal point that the manslaughter case against Barrow Borough Council was concluded.

This was the second time that Ms Beckingham had faced a manslaughter trial over the outbreak. At the original trial, in 2005, Barrow Borough Council was charged with manslaughter and Ms Beckingham had been identified as the 'controlling mind' of the council. This was the first case where a council had been prosecuted for corporate manslaughter.

11.17 The law required a 'controlling mind', someone who was instrumental in directing the strategy and actions of the 'corporate' body that can be 'identified' with the body. In *Tesco Supermarkets Ltd v Nattrass* [1972] AC 153, which concerned a prosecution under the Trade Descriptions Act 1968, the House of Lords defined individuals who are controlling minds as '… the Board of Directors, the Managing Director, and perhaps other superior officers of the company … [who] … carry out the functions of management and speak and act as the company'. This requires one individual whose acts or omissions can be identified with the corporate entity. As no local authority had ever been charged before with the corporate manslaughter offence, it had never been necessary for a court to address the issue of who such an individual might be in the local authority context. This issue is crucial when looking at any cases under this law.

11.18 To succeed, the prosecution had to show that Gillian Beckingham was the 'controlling mind', that she owed the victims a duty of care, and that she acted in a grossly negligent manner, with a direct link to the fatalities that occurred. A 'controlling mind' has to be convicted of manslaughter in order for a company (or council) to be convicted. The council's defence argued that it was impossible, from the evidence before the court, to show that she was the person in control of the affairs of Barrow to such an extent that her actions and intent were the actions and intent of the council. They argued that, even the chief executive, let alone any officer, had no right to decide issues of policy. These issues were decided by members and the democratic process.

11.19 Ms Beckingham was a third-tier officer, and was considered not to be high enough within the council to be considered the embodiment of the corporate entity – the council – itself. In evidence, it was shown that she had a line manager to whom she reported and a budget that she could not exceed. She also had to report to, and seek the approval of, the senior management team with routine items.

11.20 These submissions were accepted by the court. The judge concluded:

> 'There are, as it seems to me, in the present state of the law relating to corporate manslaughter, considerable difficulties facing those who contemplate the prosecution of a local authority … A local authority is not, in all material particulars, to be equated with a commercial enterprise of any size; still less perhaps with a very small corporation with few directors or a sole director … It is far from clear to me that even the chief executive officer could properly be described as the controlling mind of a council of elected members, but that is not something I have to decide.'

The judge, when sentencing Barrow Borough Council, said its failings were grave 'in the extreme' and that 'the failings were not only at the lowest levels, or at the levels of Ms Beckingham. These failings were all the way to the top of the Council'. Ms Beckingham was the only one from the council prosecuted for any criminal offence.

11.21 This was, for example, in contrast to the decision of the trial judge in the prosecution of Balfour Beatty for manslaughter in relation to the Hatfield train derailment, where he ruled that the company's civil engineer could be a controlling mind for certain functions. The civil engineer was not a director of the company, and neither did he have any management responsibility for the maintenance contract for the East Coast Main Line that covered the Hatfield area. However, he was responsible for advising the company in respect to the new line standards that were issued by Railtrack and to be applied to various maintenance contracts operated by the company on the network. At the end of the trial, the judge dismissed the manslaughter case against Balfour Beatty Infrastructure Services Ltd and also against the civil engineer, because he ruled there was insufficient evidence to prove that the engineer had been grossly negligent.

11.22 It is worth noting that, although in the Barrow case Ms Beckingham was not at a level sufficient to embody the mind of the council, it may not be the case in other authorities. The way in which authorities are structured often depends on the make-up of the elected council, and the decision-making processes and responsibilities can be very different. In some cases, elected members take responsibility for decisions on expenditure related to issues of health and safety (for instance, in the allocation of highway maintenance budgets). In other circumstances, it might well be argued that a senior officer of an authority would meet the test of the controlling mind. However, what it does illustrate is the possibility that, under new forms of executive local government, portfolio-holding cabinet members or elected mayors may be seen as the directing mind, and they would then be liable for an individual prosecution.

11.23 The CMCHA 2007 removes the need to find a controlling mind but looks for management failure.

IMPACT OF THE CORPORATE MANSLAUGHTER AND CORPORATE HOMICIDE ACT 2007

11.24 Under the Corporate Manslaughter and Corporate Homicide Act 2007 (CMCHA 2007), it is likely that there will be a greater number of police investigations into work related deaths for possible corporate manslaughter charges. There may also be an expectation that, where a charge of corporate manslaughter is not laid, health and safety charges will be. This may place greater pressure on the Health and Safety Executive (HSE). It is likely that there will be more convictions, as they will be easier to obtain.

AREAS NOT COVERED BY THE CMCHA 2007

11.25 The Act specifically refers to public policy decisions that do not carry with them a 'relevant duty of care'. The specific example used is the 'allocation of public resources or the weighting of competing public interests', but it does not exempt decisions about how resources were managed. These may include decisions by Primary Care Trusts about the funding of particular treatments.

11.26 Also, undertaking statutory inspections does not carry with it a 'relevant duty of care'.

PUBLIC POLICY DECISIONS

11.27 Section 3 of CMCHA 2007 makes provision specifically to exclude certain matters from the offence. Section 3(1) deals with decisions of public policy taken by public authorities. At present, the law of negligence recognises that some decisions taken by public bodies are not susceptible to review in the courts, such as decisions involving competing public priorities or other questions of public policy. These may include decisions by Primary Care Trusts about the funding of particular treatments. A recent example, in which the courts declined to find a duty of care on this basis, related to whether the Department of Health owed a duty of care to issue interim advice about the safety of a particular drug.

11.28 Section 3(2) of CMCHA 2007 grants an exemption in respect of intrinsically public functions. An organisation will not be liable for a breach of any duty of care owed in respect of things done in the exercise of 'exclusively public functions', unless the organisation owes the duty in its capacity as an employer or as an occupier of premises. Often, functions like this will not be covered by the categories of duty set out in s 2. However, it is possible that some such functions will amount to the supply of goods or services or be performed commercially, particularly if performed by the private sector on behalf of the State.

11.29 In other circumstances, the exercise of such a function will involve the use of equipment or vehicles. This test is not confined to Crown or other public bodies, but also excludes any organisation (public or otherwise) performing that particular type of function. This does not affect questions of individual liability, and prosecutions for gross negligence manslaughter and other offences will remain possible against individuals performing these functions who are themselves culpable. The management of these functions will continue to be subject to other forms of accountability, such as independent investigations, public inquiries and the accountability of Ministers through Parliament.

11.30 A definition of 'exclusively public functions' is found in s 3(4). The test covers both functions falling within the prerogative of the Crown – for instance, where the Government provides services in a civil emergency – and types of activity that, by their nature, require a statutory or prerogative basis (in other words, cannot be independently performed by private bodies).

11.31 A private company that carries out public functions is broadly in the same position as a public body. A number of exemptions are written in a general way, to exclude a particular activity regardless of what sort of organisation is carrying it out. In some instances, the Act makes specific provision for organisations in both the public and private sectors. The intention of the Act is to ensure a broadly level playing field, for public and private sector bodies, when they are in a comparable situation.

PARTIAL EXEMPTIONS

11.32 In such circumstances, the new offence does not apply unless the death relates to the organisation's responsibility as employer (or to others working for the organisation) or as an occupier of premises. These include:

- the emergency response of fire and rescue authorities, relevant NHS bodies, ambulance services (but this does not exempt duties of care relating to medical treatment in an emergency, other than decisions which determine the order in which injured people are treated), organ carriers and the armed forces;
- carrying out statutory inspection work (CMCHA 2007, s 3(3)), child-protection functions or probation activities (s 7);
- care and supervision orders made under Pt 4 of the Children Act 1989 (removal and protection of children) or under Pt 5 (local authority's duty to investigate whether to take action to protect a child's welfare);
- the exercise by a local probation board or other authority made under the Criminal Justice and Court Services Act 2000 (s 2(2)(a) gives one of the aims of the service as 'the protection of the public'); and
- functions carried out by the government using prerogative powers, such as acting in a civil emergency, and functions that, by their nature, require statutory authority. This does not exempt an activity simply because statute provides an organisation with the power to carry it out (as is the case, for example, with legislation relating to NHS bodies and local authorities). Nor does it exempt an activity because it requires a licence (such as selling alcohol). It is the activity which must be of a sort that cannot be independently performed by a private body. The type of activity involved must intrinsically require statutory authority, such as licensing drugs or conducting international diplomacy.

DIFFERENCE BETWEEN CORPORATE MANSLAUGHTER AND OFFENCES UNDER THE HSWA 1974

11.33 Essentially, for corporate manslaughter to be proved, a person to whom a duty of care is owed has to be killed, and that death must be due to management failures which amount to a gross breach of the 'relevant duty of care'. Even if someone is very seriously injured and there are gross management failings, a charge of corporate manslaughter cannot be brought, because no-one was killed.

11.34 In these circumstances, inspectors from the HSE would investigate and may prosecute under the HSWA 1974. The fines for health and safety breaches can be unlimited in the Crown Court, and custodial sentences can also be imposed for failure to comply with notices. The Sentencing Guidelines Council have issued guidelines dealing with offences covered by the CMCHA 2007. These are discussed in more detail in **Chapter 2**.

11.35 Conversely, if a person is killed and the organisation has failed to comply fully with their duties under the HSWA 1974, it does not automatically mean that there will be a charge of corporate manslaughter. This offence is reserved for the most serious breaches that lead to a death. That is not to say that the police and the HSE inspectors would not initiate a corporate manslaughter investigation under the death at work protocol.

MANAGEMENT FAILURE

11.36 Under the CMCHA 2007, the offence relates to senior management failures that led to a death – what does that mean?

11.37 The Act refers to senior managers as persons who play a significant role in:

(a) making decisions about how the whole or a substantial part of its activities are to be managed or organised; or
(b) the actual managing or organising of the whole or a substantial part of those activities.

For local authorities, this appears to suggest that senior managers would be the corporate management team, who set in place the strategic agenda in terms of the organisation's management, and operational managers who are engaged in actually managing the authority's activities.

11.38 Further, the definition of senior managers does not exclude elected members; they may, therefore, be considered senior managers within the authority. They must understand that they may become part of a corporate manslaughter investigation if a death occurs.

11.39 It has been suggested that senior managers can avoid liability for actions of the organisation by delegating their role to junior managers. This is not practical, as strategic decisions about how the organisation or substantial parts of it are managed or organised cannot be delegated. This would paralyse the organisation. Further, strategic managers, line managers and individual employees have health and safety responsibilities that cannot be delegated.

WHO DECIDES, AND HOW, IF THE LOCAL AUTHORITY HAS COMMITTED A GROSS BREACH OF THE DUTY OF CARE OWED?

11.40 A gross breach of the duty means falling far below what can reasonably be expected of the organisation in the circumstances. The breach could be as a result of an act or omission.

11.41 In deciding what is a gross breach of the 'relevant duty of care', the jury will consider amongst other things whether the organisation failed to comply with health and safety legislation, how serious any breach was and how much of a risk of death it posed. The jury may also consider the prevailing culture within the organisation (which may encourage failure, or tolerance of it), and failure to apply any health and safety guidance (which may be relevant to the alleged breach). The HSE issues Approved Codes of Practice (ACOPs) that, although not legally binding, are the standard against which a company's actions would be judged. The Institute of Directors and the HSE have produced a document called 'Leading Health and Safety at Work' which contains guidelines about how directors of companies should ensure that health and safety issues are communicated and promoted throughout the company.

The Corby Case and a breach of duty of care

11.42 The case of *Corby Group Litigation v Corby District Council*[1] highlights the need for local authorities to be aware of their duties in relation to contaminated land. In a group litigation, the claimants alleged negligence, breach of statutory duty and public nuisance on the part of the defendant local authority in connection with the reclamation of an extensive industrial site to the east of the town of Corby. The claim related to birth defects said to have been caused to a group of children born between 1986 and 1999 consisting of shortened or missing arms, legs and fingers. It was the claimant's case that the birth defects had been caused as the result of their pregnant mothers' ingestion or inhalation of harmful substances generated by the reclamation works and spread in various ways through many parts of Corby. The court had to address specific issues, which were generic and common to claimants to the determination of liability, to any individual claimant. Judgment was made for the claimants. The judgment found that, through the management and execution of land reclamation contracts which involved toxic waste management, the local authority had owed a duty of care to the claimants to take reasonable care to prevent the airborne exposure of the claimants' mothers to such toxic waste before or during the embryonic stage of pregnancy. In practice, that duty involved taking reasonable care to prevent the dispersal of mud and dust containing contaminants from the sites, which the district council owned or operated. The judgment also found that the local authority had been in breach of that duty from 1985 until August 1997. The contaminants included cadmium, chromium, nickel, polycyclic aromatic hydrocarbons and dioxins. They had been present in mud and dust disturbed at the reclamation sites, which was spread either by the wind or by lorries and vehicles and deposited on public roads in the Corby area. The local authority had failed to carry out the reclamation safely and in accordance with best practices at the time for the management and disposal of contaminated waste. As a result of that breach of duty, the claimants' mothers had been exposed to relevant teratogenic substances. Those contaminants had the ability to cause the limb defects suffered by all but two of the claimants. The conclusion was supported by the evidence of a statistically significant cluster of birth defects to children born of mothers living in Corby in the period 1989 to 1998. It was therefore reasonably foreseeable that the local population might be exposed to hazardous or contaminated substances as a result of the land reclamation programme, and that pregnant mothers could inhale or ingest sufficient quantities of the relevant contaminants that could lead to birth defects of the type complained of. The local authority was liable for public nuisance in causing, allowing or permitting the dispersal of dangerous or noxious contaminants. It was also in breach of its statutory duty under s 34 of the Environmental Protection Act 1990 to the same extent as its breaches of its duty of care in tort, as from April 1992.

1 *Corby Group Litigation v Corby District Council* [2009] EWHC 2109 (TCC).

11.43 Corby Borough Council admitted that it owed a duty of care to the claimants and their mothers. The court found that the standard of care was that of an 'ordinary careful local authority' that was undertaking this type of reclamation works, and the duty was assessed by reference to the standards known (or reasonably ascertainable) and knowledge available at the time. It was not necessary to reasonably foresee the precise type of birth defect as suffered by the claimants, rather the harm or damage that might be caused to embryos or foetuses being carried by mothers at the material time.

Public nuisance

11.44 Corby District Council had previously attempted unsuccessfully to argue that claims for personal injury were not covered by the law of 'public nuisance'. In this judgment, it was noted that the person commits a public nuisance if, by an unlawful act, he or she endangers the life, health or safety of the public.

WHAT WOULD IT MEAN IF A LOCAL AUTHORITY WERE CONVICTED OF CORPORATE MANSLAUGHTER?

11.45 If any organisation is convicted, it will be subject to an unlimited fine. There have already been huge fines for private companies for offences under the HSWA 1974. For example, Transco were fined £15 million after an explosion which killed four people in their own home. Therefore, it is likely that fines imposed upon conviction for corporate manslaughter will be very high.

11.46 However, judges have hitherto been sympathetic when imposing sentence on local authorities for health and safety offences. Fines have been smaller than those imposed on private companies. This point was made by the judge in the Legionnaires' disease outbreak prosecution of Barrow Borough Council, who said that the fine would have been much higher had it not been a local authority. See *R v Sellafield Limited*; *R v Network Rail Infrastructure Limited* [2014] EWCA Crim 49 at para 69, where the court deals with sentencing the public sector (see further at **2.17**).

11.47 It makes little sense to impose a huge fine that essentially recycles public money back to the Treasury and away from local services. However, upon conviction there will be significant image, reputation and political issues which will be explored by council taxpayers and the local and national media. In addition, although not an outgoing cash expense, being charged with and defending a corporate manslaughter charge will occupy significant amounts of senior manager, legal and officer time. It is, therefore, better not to get convicted.

11.48 The Sentencing Guidelines Council have produced guidelines for the sentencing for corporate manslaughter. The guidelines mention the following in relation to publicly funded bodies:

> 'Where the organisation is a public body the issue of price rises will not generally arise, but a large fine may force it to reduce its services, which will have an adverse impact on users … In these circumstances, the court may contemplate a reduction in fine, although it should first consider whether spreading the payment of the fine would be a more appropriate course of action.'

11.49 Fines of publicly funded bodies have been increasing. In 2007 the Metropolitan Police were fined a total of £125,000 (£75,000 fine and £50,000 costs) over the deaths of two boys who drowned at the police swimming pool in Hendon. Barnet Council was also fined £16,500, with £10,000 costs, after it earlier admitted failing to carry out a proper risk assessment of the use of the pool. Coventry City Council was fined £125,000 and ordered to pay costs of £40,000 at Coventry Crown Court after pleading guilty to breaching s 3(1) of the HSWA 1974. On the morning of 4 April 2006, Amy Robinson, 11, was on her way to school when she was struck and killed by a reversing City Council collection vehicle at the junction of Longfellow Road and Coleridge Road in the Stoke area of Coventry.

11.50 The London Borough of Newham was fined £125,000 and ordered to pay costs of £6,000 after pleading guilty to a charge brought by the HSE of failing to maintain adequately the communal areas of Walter Hurford Parade.

11.51 More recently, in 2012, Bassetlaw District Council was fined following the death of a member of the public who was struck by a reversing bin lorry:

- In July 2008, Derrick Baines, aged 76, was returning to his home in Langold, Nottinghamshire, on his mobility scooter when he was struck by a reversing bin lorry. He suffered fatal multiple injuries.
- The lorry was on a missed bin collection. It had a one-man crew. The fatal incident could have been prevented if there had been a reversing assistant at the back of the vehicle.
- The driver became aware that something was wrong when he noticed shopping spilled in the road behind his vehicle.

Bassetlaw District Council was fined £25,000 plus £12,900 costs for a breach of s 3 of the HSWA 1974, for failing to ensure the health and safety of non-employees.

CAN INDIVIDUALS BE PROSECUTED FOR CORPORATE MANSLAUGHTER?

11.52 This offence can only be committed by organisations. At present, there is no provision for secondary liability within the CMCHA 2007, so individuals cannot be prosecuted for aiding and abetting corporate manslaughter. However, a separate charge of gross negligence individual manslaughter remains possible if the evidence is considered sufficient. In addition, sanctions against individuals may be considered, such as prosecution by the HSE under health and safety legislation, and possible disciplinary action.

STEPS TO TAKE TO AVOID PROSECUTION

11.53 The simple answer is to do all that is reasonably possible to ensure that no-one is killed as a result of the way that a local authority manages or organises its activities. Therefore, authorities should not be distracted from their primary aim of compliance with the HSWA 1974.

11.54 There are a number of simple steps to take. These should not be expensive. The CMCHA 2007 does not impose new duties, but introduces a new offence and authorities should, therefore, be doing the following already:

- Ensure that the authority has in place an effective health and safety management system: see HSG 65, 'Successful Health and Safety Management', and 'Leading Health and Safety at Work' (Institute of Directors/HSE).
- As part of the management system, ensure that a robust health and safety policy is in place. Within that policy, all health and safety roles and responsibilities should be articulated.
- Job descriptions of staff should reflect their role in health and safety management, particularly at a senior level.
- Reflect upon the competencies of existing senior managers with respect to health and safety management. Provide additional developmental opportunities to address any deficiencies.
- Strategic managers should be trained to ensure they understand their role in the effective management of health and safety.
- Line managers should receive training to enable them to manage health and safety within the part of the organisation for which they are responsible.
- Elected members should receive training on their role and responsibilities within the health and safety system. This is particularly important for portfolio holders/cabinet members and chairs of scrutiny committees.
- Ensure that health and safety performance is regularly considered at board level. To ensure adequate representation of these issues, appoint a director responsible for health and safety.
- Mainstream health and safety into decision-making processes and ensure proper scrutiny of the health and safety implications of policy decisions.
- In short, if the authority complies with its duties under the HSWA 1974, it should have little to fear from the CMCHA 2007 offence.
- Yes, the Act specifically refers to public policy decisions that do not carry with them a 'relevant duty of care', such as operations dealing with terrorism and violent disorder. This also covers, for instance, DEFRA who could not be prosecuted for any deaths from flooding based on its funding decisions for the Environment Agency's flood defence programme.

OFFICERS' AND MEMBERS' INDEMNITIES

11.55 The power of local authorities to provide indemnities and insurance has been the subject of some debate over the years. In October 1996, Judge

Neuberger in the Chancery Division ruled on the case of *Burgoine and another v Waltham Forest London Borough Council*, which indicated to local government the limits of council indemnities when it is held that the indemnity in question did not cover participation in ultra vires activities.

11.56 However, on 23 November 2004, the Local Authorities (Indemnities for Members and Officers) Order 2004[1] came into force and gives local authorities broadly similar powers to provide indemnities as those available to private sector companies. This includes a limited power to provide an indemnity in respect of acts or omissions which are outside the legal powers of the authority or the relevant member or officer, but where the person indemnified reasonably believed that the matter in question was not outside those powers or, where an untrue statement has been made as to the authority's powers, or as to the steps taken or requirements fulfilled, reasonably believed that the statement was true when it was issued or authorised. On the other hand, the provision of an indemnity in relation to criminal acts, any other wrongdoing, fraud, recklessness, or in relation to the bringing of any action in defamation, is expressly prevented.

1 SI 2004/3082.

11.57 This Order enables authorities to provide indemnities for their officers and members in relation to any act or omission by the person in question which is:

(a) authorised by the authority; and
(b) forms part of or arises from any powers conferred or duties placed upon that person in consequence of the exercise of a function (whether or not the function is exercised as member or officer of the council) at the request of, with the approval or for the purpose of the authority (art 5).

11.58 Article 6 prohibits the provision of an indemnity order in relation to any matter constituting a criminal offence or which 'is the result of fraud, or other deliberate wrongdoing or recklessness on the part of that member or officer'. Nevertheless (subject to art 8), an indemnity may be provided in relation to the defence of criminal proceedings brought against the officer or member for civil liability arising from circumstances which also constitute a criminal offence.

11.59 Article 8 makes it clear that, if there is a conviction (which is not overturned on appeal), the member or officer is under a duty to reimburse the authority (or insurer, if appropriate) for any money expended in relation to the proceedings. Article 8 also covers proceedings under Pt 3 of the Local Government Act 2000, where the requirement to reimburse will apply if there is a finding made in the proceedings (not overturned on appeal) that the members failed to comply with the Code of Conduct adopted by the authority, or if the member admits that they failed to comply with the Code of Conduct.

11.60 In art 7 of the Order, the problem of dealing with matters that exceed the powers of the authority, member or officer is addressed. Despite the limitation of the powers of the authority which grants an indemnity, the authority may provide an indemnity to the extent that the member or officer in question:

- believed the act or omission in question was within the powers of the authority; or
- where the act or omission involved the issuing or authorisation of any document containing any statement as to the powers of the authority (or

any statement that certain steps have been taken or requirements fulfilled), believed the contents to be true.

It must be reasonable for the officer or member to hold that belief at the time of the act or omission in question.

11.61 If an act or omission is subsequently found to be beyond the powers of the member or officer, the authority may still provide an indemnity in such circumstances, but only to the extent that the member or officer reasonably believed that the act or omission in question was within their powers at the time at which they acted.

11.62 The terms of any indemnity, including any insurance secured subject to administration law constraints (ie fiduciary duty and reasonableness), 'may be such as the authority in question shall agree'.

11.63 In *Burgoine*, Neuberger J referred to the first part of s 265 of the Public Health Act 1875 as being concerned with excluding liability, the second part as being concerned with an indemnity, and the third part being the proviso. He then stated as follows:

> 'The purpose of the first part of section 265 is to confer immunity from suit from the persons therein mentioned in the circumstances therein mentioned. This would strongly suggest that the "expenses" against which such persons are to be indemnified under the second part of section 265 are not intended to be substantive sums for which they are sued, because the first part of section 265 renders them immune from liability for such sums. This reinforces the view that the reference to "expenses" is to the expenses incurred by the relevant persons in connection with the claim in respect of which they are rendered exempt by the first part of section 265. It appears to me that this is consistent with what was said by Wightman J giving the judgment of the court in *Ward v Lee* (1857) 7 E&B 426, 430, where he said: "The clause at the end of the 128th section is not for the repayment of 'damages' recovered against a person acting bona fide in the execution of the Act, but for the repayment of his [expenses]; which may well be construed, consistently with our view of the meaning of the section, to be repayment of the [expenses] he may have been put to in defending an action brought against him personally, and in which he may have been successful on the ground that he was acting bona fide in the execution of the Act, and therefore not liable. [The section there referred to being a provision of similar effect to section 265 of the Public Health Act 1875]".'

FIRE SERVICE

11.64 As has already been mentioned, there are three elements which need to be present in order to establish negligence. These principles apply equally to the liability of the emergency services, although the establishment of the duty of care needs to be carefully considered.

11.65 The position of cases against the emergency services was summed up in *Capital Counties plc v Hampshire County Council*. This case involves duties owed by a fire brigade. The case in question was when a fire began in the premises of Digital at about 10.00. The sprinklers began to operate at 10.23, which was about the same time that the fire brigade arrived. On the instructions of the fire brigade, the sprinkler system was shut down at about 10.50. The judge in this case found that the fire brigade had not found the seat of the fire at that point and were, therefore, not effectively fighting the fire, and held that the sprinklers were the only way of fighting the fire at that time. At 10.55 the fire brigade found the seat of the fire, at 11.10 the roof collapsed and the sprinklers were reactivated. At 12.10 the total building was lost. The council was found to be negligent because the fire officers turned off a sprinkler system without first of all having control of the fire, or identifying the seat of the fire, and as a result the fire spread out of control and destroyed a number of office blocks. Damages (amounting to £15.9 million, plus £5.5 million interest) were awarded against the Fire Brigade, which had to be met by the county council.

11.66 The judge said:

> 'The peculiarity of fire brigades, together with other rescue services, such as ambulance or coastal rescue and protective services such as the police, is that they do not as a rule create the danger which causes injury to the plaintiff or loss to his property. For the most part they act in the context of a danger already created and damage already caused, whether by the forces of nature, or the acts of some third party or even of the plaintiff himself, and whether those acts are criminal, negligent or non-culpable.'

The court also acknowledged that the Fire Brigade are not under a common law duty to answer a 999 call for assistance and are not under a duty to take care to do so. Therefore, if they failed to turn up, or failed to turn up in time, because they had carelessly misunderstood the message or got lost on the way, the Fire Brigade are not liable. Neither do the Fire Brigade have a liability for failing to ensure an adequate water supply.

HIGHWAYS

11.67 Claims against local authorities, in their role as the highway authority, are high on the list in terms of liability claims overall. The statutory duty to maintain the highway is in s 41(1) of the Highways Act 1980, as follows:

> 'The authority who are for the time being the Highway Authority for a highway maintainable at the public expense are under a duty … to maintain the highway.'

Although claims for damages, from chipped windscreens to damaged tyres, are commonplace, recently a trend towards looking at the highway as a cause of accidents where death has occurred has led to several councils facing possible charges for manslaughter. The police have a manual called the 'Road Death Investigation Manual'. This specifically says that road death should be treated as homicide, and this manual should be used in conjunction with the ACPO murder investigation manual.

11.68 The response from some local authorities has been to assign an individual or number of individuals to liaise directly with the people in the event of a fatality on the roads that the authority is responsible for. In some cases, the individuals attend the scene of a fatal road accident with the police – this is to ensure that the authority is aware of the condition of its road, signage and any other factors at the time of the accident.

11.69 In the document on Guidance for Local Transport Plans (LTPs) from the Department for Transport (DfT), there are a number of priorities outlined which local authorities should take into account when formulating their transport plans. These priorities include raising the standards across schools, transforming the local environment and meeting local transport needs more effectively, and they are a focus for the efforts of Government and councils for improving public services. The shared priority for transport includes improving accessibility and public transport and reducing the problems of congestion, pollution and safety. A number of other quality of life issues are also related to transport and covered under the sustainable communities shared priority. The DfT says it will look for evidence that the aim of delivering the shared priorities is at the heart of all local transport strategies and LTPs, but it is for authorities to decide the relative importance of each of the shared priority themes in their area. There is no set formula, however, for deciding the priority for maintenance and improvement of highways, which are the responsibility of a local authority. In metropolitan areas with heavy traffic, accident statistics can be used to identify areas for priority work and, indeed, can then be statistically analysed later to see how effective measures have been. This becomes more difficult in rural areas, where fewer accidents may occur. Spending related to safety issues then becomes more contentious. It is accidents that result from a failure of the highway, signage, or road surface that can lead to the local authority being prosecuted for contributing to an accident; and, if the accident results in a fatality, there is the possibility that a manslaughter charge could be brought.

PENALTIES

11.70 There has been a trend of escalating fines for breaches of health and safety law, and the sentencing guidelines from the Sentencing Guidelines Council indicate that this could be further increased for penalties resulting in successful convictions for manslaughter. In the case of a local authority, where income is difficult to generate and savings are being asked for year on year, the level of fine is difficult to establish without a direct effect on the public served by the local authority.

11.71 In *R v Essex County Council*, Essex County Council pleaded guilty to a breach of s 2 of the HSWA 1974 following the death of an employee. On 27 March 2003 a 30-foot high maple tree at the edge of Chalkney Wood, near Earls Colne, needed to be felled as it was leaning dangerously towards a cottage. The owner of the cottage had expressed his concern to a senior park ranger, who did not have a current chainsaw certificate. Another senior park ranger, Hadrian Robinson, was called in to help fell the tree. Although Mr Robinson had a chainsaw certificate, he only used such saws occasionally.

The men decided to use a hand-winch system, with a cable attached to an anchor tree to secure a stem of the maple that was to be removed with the chainsaw. When the tree started to fall, Mr Robinson tried to step back, but the tree fell wrongly and struck him.

In mitigation, the council said that it did have procedures, albeit not the correct ones. It highlighted a number of safety measures that it had taken since the incident to prevent a similar recurrence, such as restricting chainsaw work and ensuring that chainsaw operators undertook a refresher course every two years; method statements were now done for all complex tree work, and employees taken on to use chainsaws were assessed before they started using them.

The employee had been removing trees, using a chainsaw, when he slipped and was fatally injured with the chainsaw. The judge considered the factors in the following paragraph when deciding on the level of the fine and, in particular, how to apply the guidance given to a local authority.

11.72 He referred to the case of *R v Howe*, where the Court of Appeal gave some guidance for sentences in HSWA 1974 cases. The key points here are:

- It is impossible to lay down any tariff or to say that the fine should bear any specific relationship to the turnover or net profit of the defendant.
- Each case must be dealt with according to its own particular circumstances.
- In assessing the gravity of the breach, it is often helpful to look at how far short of the appropriate standard the defendant fell in failing to meet the reasonably practicable test.
- Generally, where death is the consequence of a criminal act, it is regarded as an aggravating feature of the offence.
- The penalty should reflect public disquiet at the unnecessary loss of life.
- A deliberate breach of the health and safety legislation with a view to profit would seriously aggravate the offence.

11.73 In this case, the judge decided that the last point was not relevant, as a local authority operates without a profit motive, and nothing suggested that the defendants were deliberately cutting corners. However, in the current funding environment for local authorities, it is likely that the issue of budget cuts would be considered and whether, when these were made, sufficient consideration was given to the safety of employees and others.

11.74 In this case, the notable aggravating features which the judge took into account were:

- the death of the worker must be reflected, to some extent at least, in the fine; and
- some improvement notices issued to the defendants in 2002, which raised questions about the systems that should have been in place, which the experts in this case commented on.

11.75 In mitigation, it was said that the defendants had a relatively good record and they are still in a position to say they have, bearing in mind that they employ over 30,000 people in a vast spectrum of different tasks, many of them dangerous, many of them involving schools and hazardous activities, including building works, and have only been the subject of these four improvement notices, this conviction, and about four previous and relatively minor convictions since the HSWA 1974 was first implemented.

11.76 As to the extent to which this breach fell below the appropriate standards, the judge decided that it '… fell significantly, but not very far, below the standard required':

- the defendants cooperated readily with the investigation;
- they gave full access to all the necessary information without being compelled to do so;
- they entered their plea of guilty at the earliest point they could; and
- they carried out an extensive review and revision of their safety procedures, not limited to those involving chainsaw activities and tree felling, but extending to all other hazardous activities under their control.

11.77 It was noted that, although this accident occurred in March 2003 and the sentencing took place in May 2005, the employees (numbering in excess of 30,000) working for the defendants continued to do their work during those two years, and there is no suggestion of any further breach of any kind, nor any need for any improvement notice. The judge decided that that suggested an organisation that had improved and streamlined its safety procedures.

The judge then looked at how to decide the level of fine in relation to a local authority:

- All the guidelines (for sentencing) deal with business or commercial corporations. A local authority is not a business or a commercial organisation. It may have a huge income or revenue (in this case, in the order of £1.25 billion).
- A local authority has no shareholders who will suffer a loss of actual or notional dividend or capital value if their corporation is fined.
- A local authority has no shareholders who could directly influence the conduct or policies of the managers of the corporation.
- A local authority has no profits or capital value which might be reduced by a fine.
- A local authority has only its funding by central government, by council tax and business rate payers.

11.78 The judge also considered the role of political officers:

> 'It does have political officers over whom the central government and taxpayers exercise an intermittent and very indirect control, particularly at elections, in matters of policy and culture. The extent to which those political officers can influence and control the sort of minutiae of everyday micro management of which the court has been hearing today is, in practical reality, limited, and the employed officers who control such matters will not normally have a real financial stake in the capital or income of the authority. They may be expected, however, to respond to public disquiet about their activities and to peer group pressures on their reputation. They are themselves, I bear in mind, a prosecuting and enforcement authority in, for example, environmental law, public health, planning and trading standards matters.'

11.79 A fine of £200,000 was imposed, to which costs of the prosecuting authorities should be added, making a total (excluding their costs, which will have to be paid by the defendants) of £215,013.

11.80 Where a prosecution is brought against an individual who is found guilty and fined, this fine and their costs are for them personally. Where insurance has paid for a defence, on conviction this will also have to be repaid.

VICARIOUS LIABILITY

11.81 The law of vicarious liability and local authorities has been affected by cases such as *The Catholic Child Welfare Society & Ors v Various Claimants, The Institute of the Brothers of the Christian Schools & Ors*[1] concerning the abuse of pupils at a Catholic Residential School for boys. This case raises the question of how far that liability could extend, and potentially affect local authorities. The case came from a claim that abuse took place at St William's Community Home, which was founded in 1865 by benefactors as a reformatory school for boys.

1 [2012] UKSC 56, [2013] 1 All ER 670; [2013] IRLR 219.

11.82 The benefactors had entrusted the day-to-day running of the school to the Institute of the Brothers of the Christian Schools, whose purpose it was to 'teach children, especially poor children, those things that pertain to a good and Christian life'. All members of the Institute are bound by rules laid down by the Pope in 1724 and later adapted. The rules specifically prohibit the touching of children through familiarity or playfulness. The Institute appointed the head teacher and staff at St Williams, and the staff team comprised Brother teachers and lay teachers, in varying proportions. School inspections were carried out by the Institute annually.

Government legislation in the 1960s and 1970s, however, led to St William's School becoming an 'assisted community home for children in the care of the local authority'. Management responsibility was initially transferred from the Institute to the Middlesbrough Diocesan Rescue Society and, in 1982, it was transferred again to the Catholic Child Welfare Society.

The Institute selected the head teacher at St William's School, although their contract of employment was with the Catholic Child Welfare Society. The arrangement was such that any wages paid to the head teacher by the Society were returned to the Institute. In the 1990s, the head teacher of St William's School was dismissed and later expelled from the Institute following a finding of 'systemic sexual abuse of boys in his care'. The school closed in 1994. The head teacher was tried in 1994, and again in 2004, for acts of sexual abuse on pupils in his care. He was sentenced on both occasions to lengthy terms of imprisonment. Other staff were also implicated. More than 170 former residents of St William's School have been seeking damages since 2004.

The Catholic Child Welfare Society did not seek to challenge the finding of vicarious liability made against it in earlier proceedings. It accepted that, as manager of the school, it had a responsibility towards those attending it. It did, however, seek to allege that the Institute should share the liability.

This was challenged by the Institute. The Institute was told by the Supreme Court that it must share the cost of paying an estimated £7 million compensation to boys who were sexually and physically abused, the largest compensation claim arising from abuse at a single children's home.

The Supreme Court held that the Institute was not a contemplative Order but had, from inception, existed to give a Christian education to boys. To perform that activity, it sent out Brother teachers, to teach in schools managed by other bodies, and it made Trust funds available to fulfil this purpose.

The Supreme Court found that the Institute had elected the head teacher and directed his teaching activity. The manner in which the Brother teachers were obliged to conduct themselves was also directed by the Institute. This amounted to 'a strong connection'.

The activity of providing Christian teaching also created an opportunity, namely an opportunity to have access to boys. In turn, this created a potential risk; indeed, the risk of sexual abuse was judged to be high, as boys and teachers resided together within the school grounds, and the boys themselves were vulnerable.

The Supreme Court found that the relationship between the Institute and the Brothers was akin to employment. Therefore, both the Catholic Child Welfare Society and the Institute of Christian Brothers were deemed be jointly liable.

Vicarious liability involves a two-stage test. First, it is necessary to consider the relationship between the defendant and the tortfeasor to see whether it was one that was capable of giving rise to vicarious liability. Second, there should be a connection that links the relationship between the defendant and the tortfeasor to the act or omission of the latter.

The test for vicarious liability is less concerned with whether there exists an employer/employee relationship, but rather whether the relationship between abuser and defendant is sufficiently close to render it just, fair and reasonable that the defendant should be held to account for the abuser's actions (namely, 'akin to employment').

It is now possible for a local authority to be held vicariously liable for the tortious act of, for example, a foster carer. The courts may take a different view of the tortious acts of casual workers or volunteers, such as Youth or Scout and Guide leaders, and indeed the charitable organisations, companies or public service providers behind them, who permit them to access the public.

CONCLUSION

11.83 Until only a few years ago, the trend with prosecutions from the HSE was to apply charges for breaches of health and safety at work to companies and organisations. The consequence was that the conviction was of the company or organisation, which was then liable to pay a fine. Now it is not difficult to find a prosecution under the same legislation which, although it may be targeted at a company or organisation, is also aimed at an individual.

11.84 In view of the high-risk areas that local authorities have responsibility for, such as highways, housing, leisure and social care, if deaths do occur, the CMCHA 2007 and issues of liability are a real concern. The issue of fines is a particular problem when the offending organisation is a publicly funded body such as a local authority. This is because any fine imposed may be considered as only mainly affecting public services, which suffer as a result. It is important that local authorities are clear where issues relating to health and safety are managed, in particular within partnership and sub-contracted arrangements.

Chapter 12

Ireland

Aisling Butler, William Fry Solicitors, Dublin

Niav O'Higgins, Arthur Cox Solicitors, Dublin

NORTHERN IRELAND

12.1 Most of the law in Northern Ireland originates from the UK. Insofar as health and safety legislation is concerned, the law is substantively the same – where there are Acts in the UK, they are enacted in Northern Ireland by way of equivalent Orders. In addition, Northern Ireland is subject to the same 'six pack' regulations in respect of health and safety. For example, the most important primary legislation with regard to health and safety in Northern Ireland is the Health and Safety at Work (Northern Ireland) Order 1978. This Order is akin to the 1974 Act in the UK. In addition, the Corporate Manslaughter and Corporate Homicide Act 2007, enforced in the UK from 6 April 2008, will also apply in Northern Ireland.[1]

1 June 2012 saw the first corporate manslaughter conviction in Northern Ireland under the 2007 Act: *R v JMW Farm Limited* (Belfast Crown Court). In this case, an employee died when he fell while washing the inside of a large metal bin which was positioned on the forks of a forklift truck. The forklift truck was a replacement for the normal truck which had gone for servicing some weeks earlier. When the employee jumped on the side of the bin, it toppled. Given that he was standing on an unsecured piece of equipment, the judge held that it was 'an inherent and foreseeable danger' and imposed a fine of £187,000.

12.2 The Health and Safety Executive for Northern Ireland (HSENI) is the body responsible for the promotion and enforcement of health and safety at work standards in Northern Ireland. This responsibility is shared with 26 District Councils in Northern Ireland. However, there are specific divisions of enforcement responsibilities between the HSENI and the District Councils.[1]

Effectively, the HSENI and the District Councils cover all work situations that are subject to the Health and Safety at Work (Northern Ireland) Order 1978 and its subordinate regulations.[2]

1 The HSENI is responsible for factories, building sites, farms, motor vehicle repairs, mines and quarries, chemical plants, schools and universities, leisure and entertainment facilities (owned

by District Councils), fairgrounds, hospitals and nursing homes, District Councils, fire and police, government departments, railways, and any other works places not listed under District Councils. The District Councils, in turn, are responsible for offices, retail and wholesale shops, tyre and exhaust fitters, restaurants (including take-away food shops, mobile snack bars and catering services), hotels (including guest houses and residential homes), wholesale and retail warehouses, leisure and entertainment facilities (privately owned), exhibitions, religious activities, undertakers, the practice or presentation or the arts, sports, games, entertainment or other cultural or recreational activities, therapeutic and beauty services, and animal care.

2 The most recent secondary legislation in Northern Ireland includes: the Biocidal Products (Fees and Charges) Regulations (Northern Ireland) 2013 (SR 2013/207); the Biocidal Products and Chemicals (Appointment of Authorities and Enforcement) Regulations (Northern Ireland) 2013 (SR 2013/206); the Employer's Liability (Compulsory Insurance) (Amendment) Regulations (Northern Ireland) 2013 (SR 2013/199); and the Health and Safety (Sharp Instruments in Healthcare) Regulations (Northern Ireland) 2013 (SR 2013/108).

12.3 In light of the fact that Great Britain and Northern Ireland are governed by very similar legislation, there appears to be a common approach adopted on the enforcement and practice/procedure in relation to health and safety as adopted by the HSENI and the HSE in the UK. However, in the recent past, there have been a couple of initiatives between the HSENI and the Health and Safety Authority (HSA) (in the Republic) – in relation to the promotion of awareness with regard to stress in the workplace, and health and safety in engineering practice – which could be seen by some to signify a move to closer cross-border relations insofar as safety, health and welfare at work is concerned.

IRELAND

INTRODUCTION

12.4 Irish law is based on a system of common law, inherited, historically, from the English legal system, with which it shares many similarities. Whilst decisions of the English courts have no formal status, they, together with decisions of other common law jurisdictions,[1] may be referred to in the Irish courts. Given the close similarities between the two statutory frameworks for the regulation of health and safety,[2] it is not unusual for English decisions to be referred to on a fairly regular basis in the context of Irish judicial determinations on health and safety matters.

1 Principally, Australia, New Zealand, Canada and Hong Kong.
2 Although, with the advent of the Safety, Health and Welfare at Work Act 2005 (see further **12.22** below), the legislative structures are increasingly divergent.

12.5 The court structure differs somewhat from the English court system. In civil matters, there are three courts with different jurisdictions: the District Court,[1] the Circuit Court[2] and the High Court.[3] The Safety, Health and Welfare at Work Act 2005 ('the 2005 Act') has also given a limited jurisdiction to tribunals[4] responsible for employment matters through its provisions on 'penalisation'.[5]

1 The District Court has jurisdiction over claims of up to €6,348.69 (formerly £IR5,000). There are 23 District Court areas across Ireland.
2 The Circuit Court has jurisdiction over claims of up to €38,092.14 (formerly £IR30,000).
3 The High Court has jurisdiction over claims above €38,092.14 (formerly £IR30,000).
4 The Rights Commissioners Service at the Labour Relations Commission.
5 See **12.42** to **12.48** below.

12.6 A single judge in the District Court deals with criminal matters on a summary basis, and any appeal is made to the Circuit Criminal Court (CCC).

12.7 Prosecutions brought on indictment will be before a judge and jury in the CCC, with appeals going to the Court of Criminal Appeal (CCA), where three judges sit.

12.8 The Supreme Court, consisting of the Chief Justice and seven judges (with usually three judges sitting at any one hearing), is the court of final appeal for both civil and criminal cases.

HEALTH AND SAFETY REGULATION

12.9 Health and safety regulation in Ireland is based on many of the same principles, and contains many of the same duties, as in the UK. However, whilst in the UK the 1974 Act sets a statutory framework for the regulation of health and safety at work, it was not until 1989 that Ireland saw its first consolidated statute governing all workplaces. Prior to this, there were a number of statutes and regulations governing health and safety in specific industries, such as factories and workshops,[1] mines and quarries[2] and construction.[3] This piecemeal regulation of health and safety was inadequate as it omitted a number of workplaces from its ambit and therefore left a significant percentage of the workforce without statutory protection.

1 'An Act for the Preservation of the Health and Morals of Apprentices and Others Employed in Cotton and Other Mills and Cotton and Other Factories' 1802, Factories and Workshop Acts 1878 and 1901, Factories Act 1955, Factories (Report of Examination of Hoists and Lifts) Regulations 1956 (SI No 249/1956).

2 Mines and Quarries Act 1965, Mines (Electricity) Regulations 1972 (SI No 51/1972), Mines (Electricity) Regulations 1972 (SI No 50/1972), Mines (Electricity) Regulations 1974 (SI No 146/1974) and Mines (Electricity) (Amendment) Regulations 1979 (SI No 125/1979).

3 Safety in Industry Act 1980, 1989 Framework Directive on Health and Safety, Construction (Safety, Health and Welfare) (Amendment) Regulations 1988 (SI No 270/1988) and Construction (Safety, Health and Welfare) (Amendment) Regulations 1975 (SI No 282/1975).

12.10 Prior to the enactment of the Safety, Health and Welfare at Work Act 1989 ('the 1989 Act'), the Irish Government had established the Barrington Commission to undertake a comprehensive review of occupational health and safety.[1] Many of the subsequent recommendations of the Barrington Report[2] were similar to those contained in the UK Robens report[3] and included general obligations to be placed on both employers and employees in all workplaces (based, in many instances, on the existing common law duties), with specific statutory duties imposed upon manufacturers, designers and suppliers. In addition, the Report recommended that statutory protection be extended to third parties such as independent contractors and other visitors attending at a place of work. The Barrington Report also recognised that new structures would be needed to ensure improved protection, including an appropriate consultation structure between employers and employees on health and safety issues.[4]

1 The Barrington Commission was established in December 1980 following the Government's publication of the Safety in Industry Bill in 1978. This became the Safety in Industry Act 1980 and incorporated many of the provisions of the UK's 1974 Act, although its application was (as with other existing health and safety statutes) restricted to factory workers.

2 The Barrington Report was published on 14 July 1983.

3 Lord Robens' Report on 'Safety and Health at Work', published in June 1972.

4 An excellent overview of the Barrington Report and background to health and safety regulation in Ireland is provided by Raymond Byrne BL in his annotated version of the Safety, Health and Welfare at Work Act 2005, published by Round Hall (Dublin, 2006).

12.11 The Barrington Report eventually led to the enactment of the1989 Act incorporating many of the Report's recommendations. There were many similarities between the 1989 Act and the UK 1974 Act and, indeed, many of its provisions were identical.

12.12 The 1989 Act also established a statutory body, officially known as the National Authority for Occupational Safety and Health (which colloquially became known as 'the Health and Safety Authority' (HSA) – which name has now been given statutory effect in the 2005 Act).

12.13 The 1989 Act was supplemented over time, by a number of subsidiary regulations. The first, and most important of these, was the Safety, Health and Welfare at Work (General Application) Regulations 1993[1] ('the 1993 General Application Regulations'),[2] which implemented both the provisions of the European Framework Directive on Health and Safety[3] and subsequent supporting European Directives.[4] Other significant regulations included the Safety, Health and Welfare at Work (Construction) Regulations 1995, 2001, 2003 and 2006,[5] various regulations controlling the use of chemical and biological substances,[6] Asbestos Regulations[7] and Quarries Regulations,[8] to name but a few.

1 The 1989 Act was enacted on 19 April 1989 and came into effect on 1 November 1989 (SI No 7/1989).

2 The 1993 General Application Regulations have been substantially revoked and repealed by the Safety, Health and Welfare at Work (General Application) Regulations 2007 to 2012 (see **12.15** below).

3 Council Directive 89/391/EEC of 12 June 1989 on the introduction of measures to encourage improvements of the health and safety of employees at the workplace ('the Framework Directive').

4 The principal Directives issued under the Framework Directive were the Workplace Directive (89/654/EEC), the Work Equipment Directive (89/655/EEC), the Personal Protective Equipment Directive (89/656/EEC), the Handling of Loads Directive (90/269/EEC), the Visual Display Screens Directive (90/270/EEC), and the Directive on Temporary and Fixed-Term Employees (91/383/EEC). Note also that the 1993 General Application Regulations cover the 'Six Pack' Regulations 1992 to 1999 in the UK.

5 These have been revoked and replaced by the Safety, Health and Welfare at Work (Construction) Regulations 2013 (see **12.16** below).

6 Dangerous Substances (Retail and Private Petroleum Stores) (Amendment) Regulations 2001 (SI No 584/2001), European Communities (Carriage of Dangerous Goods by Road) (ADR Miscellaneous Provisions) Regulations 2006 (SI No 406/2006), European Communities (Carriage of Dangerous Goods by Road) (ADR Miscellaneous Provisions) Regulations 2007 (SI No 289/2007), European Communities (Classification, Packaging and Labelling of Dangerous Preparations) (Amendment) Regulations 2007 (SI No 76/2007), European Communities (Classification, Packaging and Labelling of Dangerous Preparations) (Amendment) Regulations 2004 (SI No 62/2004), European Communities (Classification, Packaging, Labelling and Notification of Dangerous Substances Regulations 2006 (SI No 25/2006), European Communities (Control of Major Accident Hazards involving Dangerous Substances) Regulations 2000 (SI No 476/2000), European Communities (Control of Major Accident Hazards involving Dangerous Substances) (Amendment) Regulations 2003 (SI No 402/2003), European Communities (Control of Major Accident Hazards involving Dangerous Substances) Regulations 2006 (SI No 74/2006), and European Communities (Dangerous Substances and Preparations) (Marketing and Use) (Amendment) Regulations 2006 (SI No 364/2006).

7 European Communities (Protection of Workers) (Exposure to Asbestos) Regulations 1989 (SI No 34/1989), European Communities (Protection of Workers) (Exposure to Asbestos) (Amendment) Regulations 1993 (SI No 273/1993), European Communities (Protection of Workers) (Exposure to Asbestos) (Amendment) Regulations 2000 (SI No 74/2000), Safety, Health and Welfare at Work (Exposure to Asbestos) Regulations 2006 (SI No 386/2006), European Communities (Asbestos Waste) Regulations 1994 (SI No 90/1994), European Communities (Control of Water Pollution by Asbestos) Regulations 1990 (SI No 31/1990), European Communities (Asbestos Waste) Regulations 1990 (SI No 30/1990), Air Pollution Act 1987 (Emission Limit Value for Use of Asbestos) Regulations 1990 (SI No 28/1990), Factories (Asbestos Processes) Regulations 1975 (SI No 238/1975), and Factories (Asbestos Processes) Regulations 1972 (SI No 188/1972).

8 Safety, Health and Welfare at Work (Quarries) Regulations 2008 (SI No 28/2008) and Safety, Health and Welfare at Work (Quarries) (Amendment) Regulations 2013 (SI No 9/2013).

12.14 Health and safety legislation in Ireland has seen many significant changes and advancements over the past few years, commencing primarily with the enactment of the Safety, Health and Welfare at Work Act 2005 ('the 2005 Act'). The 2005 Act entirely revoked and repealed the 1989 Act. It not only expanded the duties and responsibility placed upon employers, employees and others, but it also incorporated many of the provisions of the 1993 General Application Regulations, thereby placing the provisions of the underlying European Directives onto a firmer statutory footing.

12.15 The Safety, Health and Welfare at Work (General Application) Regulations 2007, 2010 and 2012 ('the General Application Regulations 2007 to 2012'), which revoked and replaced the General Application Regulations 1993–2003 (save in respect of reporting requirements)[1] were introduced with a view to simplifying the core health and safety framework by consolidating various regulations into the General Application Regulations 2007. These Regulations are central to the framework of health and safety compliance in all workplaces in Ireland, as they encompass detailed duties and responsibilities, primarily on employers, relating to requirements for the workplace and use of equipment, specific activities, specific risks and sensitive risk groups of employees.[2]

1 Contained in Pt X of the General Application Regulations 1993. See **12.20** below.
2 The General Application Regulations 2007 to 2012 now cover workplace and work equipment, electricity, work at height, physical agents (eg noise, vibrations), sensitive risk groups (such as children and young persons, pregnant, post-natal and breastfeeding employees, night workers, shift workers), safety signs and first aid, explosive atmospheres at places of work, artificial optical radiation and pressure systems.

Other Regulations

12.16 The construction sector has attracted particular attention for specific health and safety regulation.[1] In 1995, the Safety, Health and Welfare at Work (Construction) Regulations 1995 ('the 1995 Construction Regulations')[2] were enacted in Ireland, implementing the European Construction Directive of 1993.[3] The implementing regulations in Ireland have undergone a number of changes, and the 1995 Construction Regulations were revoked and repealed by the Safety, Health and Welfare at Work (Construction) Regulations 2001,[4] subsequently amended by the Safety, Health and Welfare at Work (Construction) (Amendment) Regulations 2003,[5] which again were both ultimately revoked and repealed by the Safety, Health and Welfare at Work (Construction) Regulations 2006 ('the 2006 Construction Regulations').[6] Thereafter, the 2006 Construction Regulations were further amended by: the Safety, Health and Welfare at Work (Construction) (Amendment) Regulations 2008, which introduced additional skills certification requirements;[7] the Safety, Health and Welfare at Work (Construction) (Amendment) (No 2) Regulations 2008, which introduced additional requirements relating to construction works undertaken in the vicinity of roadways;[8] and the Safety, Health and Welfare at Work (Construction) (Amendment) Regulations 2010, which introduced a new definition of 'confined spaces', and expanded and clarified information to be included in the notification to the HSA given by the Project Supervisor for the Construction Stage (PSCS).[9]

1 Even prior to the enactment of the 1989 Act, specific regulation of health and safety in construction could be found in the Construction (Health, Safety and Welfare) Regulations 1975

(SI No 282/1975) and the Construction (Safety, Health and Welfare) (Amendment) Regulations 1988 (SI No 270/1988).
2 SI No 138/1995.
3 Council Directive 92/57/EEC on the minimum safety and health requirements at temporary or mobile construction sites.
4 SI No 481/2001.
5 SI No 277/2003.
6 SI No 504/2006. When first brought into force on 6 November 2006, regulations 80 to 123 of the 2001 and 2003 Construction Regulations (dealing with Lifting Appliance, Chains, Ropes and Lifting Gear, Hoists and Carriage of Persons and Secureness of Loads) remained in force, but have since been revoked and repealed by the General Application Regulations 2007 and are now contained, with some amendment, in regulations 42 to 58. The 2006 Regulations have since been amended in 2008 and 2010.
7 SI No 130/2008.
8 SI No 423/2008.
9 SI No 523/2010.

12.17

The Safety, Health and Welfare at Work (Construction) Regulations 2013 ('the Construction Regulations 2013'),[1] which are effective from 1 August 2013, amend and revoke all previous regulations and bring the Irish regulations governing health and safety in construction into line with Council Directive 92/57/EEC. The Construction Regulations 2013 also bring into one place the various amending regulations that were issued since 2006. A significant change is the application of the construction regulations to both commercial and domestic 'clients' (a previous carve-out for clients procuring residential projects for their intended occupation as a dwelling has been removed). There are some relaxations in the requirements for domestic clients.[2] The Construction Regulations 2013 also ease the requirement to appoint a Project Supervisor Design Process (PSDP) and PSCS for short and routine projects involving construction. They now provide that project supervisors are only required to be appointed where: the project is likely to exceed 30 working days or 500 person days; a particular risk is involved; and/or there is more than one contractor.

Unlike the CDM Regulations 2007 in the UK, the Construction Regulations 2013 (as with the previous implementing legislation) go beyond the requirements of the Construction Directive. The Construction Directive creates specific duty-holders, who are responsible for the planning, design and management of health and safety on site. The Construction Regulations 2013 not only impose responsibilities on the specific duty-holders, but also go further, by including detailed requirements for protecting the health and safety of employees and others engaged in specific operations and activities on site. The Safety, Health and Welfare at Work (General Application) Regulations 2007 to 2012 also contain various regulations relevant to the construction sector.[3]

1 SI No 291/2013.
2 For example, the preliminary safety and health plan only has to be provided to the PSCS appointed on the project (and not all those tendering for the PSCS role, as remains the case for commercial projects). It is for contractors and designers appointed by a domestic client to demonstrate their competence and that they have adequate resources, rather than requiring the client to satisfy itself of this.
3 For example, work at height, lifting appliances, mobile work platforms, work equipment, personal protective equipment, manual handling, noise, vibration, electricity and explosive atmospheres.

ASBESTOS

12.18 Regulation to provide protection to those exposed, or at risk of exposure, to asbestos has existed in Ireland for some time,[1] reflecting the very specific risks that may be presented by exposure to asbestos dust. The Safety, Health and Welfare at Work (Exposure to Asbestos) Regulations 2006[2] were signed into law on 21 July 2006, and revoked and repealed all previous legislation. These Regulations were subsequently amended by way of the Safety, Health and Welfare at Work (Exposure to Asbestos) Regulations 2010[3] (together known as the Safety, Health and Welfare at Work (Exposure to Asbestos) Regulations 2006 to 2010). The Asbestos Regulations apply to all activities in which employees are or are likely to be exposed to dust arising from either or both asbestos and asbestos-containing materials during their work. There is an absolute prohibition on employees being engaged in activities in contravention of the Asbestos Regulations. In line with the ethos of the 2005 Act, an employer also owes duties to third parties, such as the employees of contractors undertaking refurbishment or repair and maintenance works.[4]

1 European Communities (Protection of Workers) (Exposure to Asbestos) Regulations 1989, the European Communities (Protection of Workers) (Exposure to Asbestos) (Amendment) Regulations 1993 and the European Communities (Protection of Workers) (Exposure to Asbestos) (Amendment) Regulations 2000). The Asbestos Regulations give effect to Directive 2003/18/EC (amending Council Directive 83/477/EEC, as previously amended by Council Directive 91/382/EEC) and, with the latest amending Regulations, Council Directive 2009/148/EC on the protection of workers from the risks related to exposure to asbestos at work.

2 SI No 386/2006.

3 SI No 589/2010.

4 In August 2013, the HSA issued new guidelines for the management of asbestos-containing materials: 'Asbestos-Containing Materials (ACMs) in Workplace – Practical Guidelines on ACM Management and Abatement'.

Reporting of accidents and dangerous occurrences

12.19 The reporting of accidents and dangerous occurrences is contained in Pt X of the 1993 General Application Regulations. Despite the revocation of the 1993 General Application Regulations, Pt X is the only part which was not transposed into the 2007 to 2012 General Application Regulations. The HSA, in February 2012, issued draft amendment Regulations for consultation. The principal amendments proposed included a new requirement to report certain occupational illnesses and also included a proposed draft list of illnesses which would fall to be reported. The draft Regulations also sought to clarify the timeframes for the reporting of fatalities and certain other injuries. However, since finalisation of the consultation process in March 2012, no confirmation or further publication has yet resulted. Part X of the 1993 General Application Regulations therefore remains in force for the time being.[1]

1 The anomaly of retaining only Part X of the 1993 General Application Regulations was not altogether clear, but this may now be explained by the apparent intention of the HSA to include the requirement to report occupational illnesses (as evidenced by way of the 2012 consultation document).

12.20 Part X of the 1993 General Application Regulations places a requirement on employers and other persons in control of a place of a work to report accidents resulting in death or injury and 'dangerous occurrences'[1] (whether or not resulting in death/injury) to the HSA as soon as practicable.

1 A list of what constitutes 'dangerous occurrences' is contained in the Twelfth Schedule to the 1993 General Application Regulations.

12.21 Where, in the course of carrying out work, a person is killed, or sustains an injury which results in their being absent from work for three or more consecutive days after the date of the accident,[1] or where a person visiting the place of work is killed or injured such that they require medical treatment (from either a GP or hospital), a written report (in prescribed form)[2] must be submitted to the HSA. In the case of a fatality, the requirement is to notify the HSA by the quickest practicable means (generally by telephone), followed by a written report, again in the prescribed form.

1 Weekend and bank holidays will be included in the computation of the three consecutive days.
2 To include the name of the injured person/deceased and brief particulars of the cause and location of the accident. If death occurs within one year of an accident, this is also required to be notified to the HSA. Reporting can now be done online via the HSA website using forms IR1 (for general accidents) or AF2 (for construction reports).

THE SAFETY, HEALTH AND WELFARE AT WORK ACT 2005

12.22 The Safety, Health and Welfare at Work Act 2005 ('the 2005 Act') included many of the provisions of the 1989 Act but incorporated many important new provisions. For example, the 2005 Act contains a number of new statutory definitions, notably 'competent person'[1] and 'reasonably practicable'.[2] It also expanded the duties and responsibilities of employers and employees, as well as introducing specific duties for other persons, such as those who commission, procure, design or construct places of work. For the first time, the 2005 Act provides for Joint Health and Safety Agreements[3] and new dispute resolution mechanisms for issues arising between employers and employees.[4] The impact of the 2005 Act has been greatly reinforced due to the expanded responsibilities placed upon directors and others occupying managerial positions, and the increased potential for personal liability of such individuals arising from breaches of the legislation.[5]

1 See **12.49** below.
2 See **12.37** to **12.41** below.
3 Section 24 provides for those types of agreements whereby unions separately representing employers and employees can enter into or vary agreements providing practical guidance on health and safety issues. This section of the 2005 Act sets out the criteria which must apply for such agreement to be validly approved and finalised.
4 See **12.46** below on 'legislation'.
5 See **12.60** to **12.61** below.

Employers' duties

Employers' duties to employees

12.23 In the UK, employers' duties are provided for in both the 1974 Act and the relevant subsidiary regulations.[1] The 2005 Act has the advantage of bringing together, in a comprehensive way, the more general duties and responsibilities on employers with regard to safety, health and welfare at work for effective safety management. There are, of course, further duties, which will apply to specific hazards, activities or industries, provided for in relevant regulations applicable to such hazards, activities or industries.

1 See reference to this in UK section above.

12.24 Employers' general duties are set out in s 8 of the 2005 Act. That section commences with a 'catch-all' provision[1] that employers must do whatever is 'reasonably practicable'[2] to ensure the safety, health and welfare at work of his or her employees. It then goes on[3] to provide for certain specific duties such as:

(a) managing and conducting work activities to ensure the safety and health of employees;[4]

(b) preventing any improper conduct or behaviour likely to put health and safety at risk;[5]

(c) with regard to the place of work itself, ensuring it is designed, provided and maintained in a condition that is safe, with safe access and egress to and from the place of work. This same provision applies to plant and machinery and the use of any article or substance (to include exposure to noise, vibration or ionising or other radiations);[6]

(d) systems of work must be provided in a manner that is planned, organised, performed, maintained and revised so as to be safe and without risk;[7]

(e) providing and maintaining facilities and arrangements for the welfare of his or her employees at work;[8]

(f) providing information, instruction, training and supervision necessary to ensure health and safety;[9]

(g) determining and implementing the safety, health and welfare measures necessary for the protection of the safety, health and welfare of employees when identifying hazards and carrying out risk assessments or when preparing a safety statement and ensuring measures take account of changing circumstances and the general principles of prevention;[10]

(h) where risks cannot be eliminated entirely, and having regard to the general principles of prevention,[11] to provide and maintain suitable protective clothing and equipment relevant to the risks;[12]

(i) adequate plans and procedures must be in place and revised as appropriate in respect of measures to be taken in the case of emergency or serious and/or imminent danger[13] and accidents and dangerous occurrences must be reported as prescribed;[14] and

(j) where it is necessary to do so, employers must obtain the services of a 'competent'[15] person to ensure health and safety.[16]

Other employer duties that feature throughout the 2005 Act are summarised in the following paragraphs.

1 At s 8(1).
2 See **12.37** to **12.41** below.
3 At s 8(2).
4 Section 8(2)(a): this is an entirely new statutory provision.
5 Section 8(2)(b): this is an entirely new statutory provision, which was intended to cover behaviour such as horseplay and bullying, for example.
6 Section 8(2)(c)(i)–(iii) and (d).
7 Section 8(2)(e).
8 Section 8(2)(f).
9 Section 8(2)(g).
10 Section 8(2)(h). Note also that the fact of an employer having to 'implement' improvements identified is a new (and important) feature of the health and safety legislation.
11 As set out in Sch 3 to the 2005 Act.
12 Section 8(2)(i).
13 Section 8(2)(j) and see also s 11.
14 Section 8(2)(k); detailed requirements for reporting are contained in Part X of the 1993 General Application Regulations.
15 Note the definition of 'competent' in s 2(2) of the 2005 Act, covered at **12.49** below.
16 Section 8(2)(l). See also s 18 in relation to the appointment of sufficient 'competent' persons.

12.25 Specific duties are imposed upon employers in relation to information, instruction and training.[1] Notably, these duties take account of the potential difficulties with language, arising from literacy issues and also the cultural array of workers in the Irish workforce. Any information given, whether generally or for the purposes of instruction, training and/or supervision, must be given in a form, manner and, as appropriate, language that is reasonably likely to be understood by the employee concerned.[2]

1 Sections 9 and 10.
2 This provision is indicative of the Government's will to include *all* workers within the scope of the protections afforded by the health and safety legislation.

12.26 This information must include advising of the hazards and risks identified, protective and preventive measures which the employer will take to control those risks, and the names of any person(s) assigned to implement its plans and procedures for the purposes of emergencies and/or serious and imminent dangers, together with the names of the safety representatives selected to consult with the employer for the purposes of health and safety.[1]

1 See s 9.

12.27 In addition, employers who share a place of work must cooperate towards compliance and implementation of statutory duties and coordinate their actions to ensure the safety and health of all.[1] This may include, for example, exchanging some or all of their safety statements.

1 See s 21.

12.28 Where the work carried out involves some risk to the health of employees, employers must provide appropriate health surveillance. In addition, where the work involves serious risk to health, the employer must ensure that employees assigned to such work are fit for such performance, by having the employee reviewed by a registered medical practitioner.[1]

1 See ss 22 and 23.

12.29 Another important duty of the employer is to ensure that he or she appoints one or more 'competent' person(s) to perform health and safety tasks relating to the protection from and prevention of risks.[1]

Employees must be protected against the dangers that specifically affect them. Therefore:

- upon recruitment;[2]
- where employees are transferred to new or other duties;
- where any person, even if that person is the employee of another employer, is carrying out work at the place of work and may be affected by an existing risk;[3] or
- where new work equipment, systems or new technology is introduced,

training must be provided, taking into account new or changed risks to health and safety. Where appropriate, training should be repeated periodically.[4]

1 Section 18.
2 The requirement for health and safety induction upon commencement of employment is a new feature of the legislation and reflects the statistics showing that many recorded accidents occur within the first week of starting employment.
3 Per s 10(5). See also the general duty in s 12, which relates to employer's duties to persons other than their employees who are present at the place of work and who might be affected by any risk.
4 See s 10. It should be noted that what constitutes the appropriate interval for repeat training will very much depend on the type of activity. It is usual that certificates of competency issued (for

example, certificates for first aid, manual handling, safe pass or CSCS) will highlight an expiry date before which refresher training is required.

12.30 Employers have specific duties in relation to emergencies and serious and imminent dangers and, in particular, must prepare and revise adequate plans and procedures in the event thereof. The adequacy of plans and procedures will be dependent on the size of the place of work and the type of work being carried on. It is not sufficient to simply put such plans and procedures in place; rather, employers have a statutory obligation to ensure their effective implementation. Clearly, therefore, and by way of example, any person assigned to a task such as first aid or evacuation must be 'competent' to do so. Employers must also arrange any necessary contacts with the appropriate emergency services. For the purposes of implementing the plans and procedures, the employer must designate employees to implement them and to ensure that the number of employees so appointed are properly trained and have the appropriate equipment available to them. Once trained, these employees must be the only ones with access to the areas at the place of work where a serious or specific danger exists.[1]

1 See s 11(1)–(5).

Employers' duties to others

12.31 In addition to the duties owed by the employer to his or her employee, the 2005 Act imposes duties on the employer with respect to third parties. Employers must control and manage the potential risks to which any third party might be exposed at the place of work, in such a way so as to ensure, so far as is reasonably practicable, the safety, health and welfare of that third party.[1] Allied to this specific duty is the duty on an employer to ensure that any employee of another undertaking (such as a subcontractor on the premises) who is engaged to carry out work, and may be affected by a hazard at the place of work, must be informed and instructed on the hazard(s) and the protective and preventive measures in place (in the same way as the employer's own employees).[2]

1 Section 12. Section 12 was invoked by the DPP for the HSA (in one of the longest-running prosecutions in Ireland) as a result of an accident where a 15-year-old schoolboy was killed and other schoolchildren were injured in April 2006 when the bus they were travelling in lost control and flipped over when the back axle of the bus came away. The DPP prosecuted the bus owner, Raymond McKeown, and a company that had tested the bus, O'Reilly Commercials Ltd, for breaches of health and safety legislation. The trials took place in the Dublin Circuit Criminal Court in July 2013.
The bus owner was charged under section 12 of the 2005 Act for failing, insofar as reasonably practicable, to manage and conduct his undertaking by, in particular, failing to maintain a bus so that persons not being his employees were exposed to risks to their health and safety. The bus owner changed his plea to guilty during the trial, and the judge imposed a 12-month suspended sentence. The judge stated that there was 'little point imposing a fine'.
The vehicle testing company was charged with four breaches of health and safety legislation, in particular for failing to conduct its undertaking to ensure that persons not in its employment, who may be affected, were not exposed to risks to their safety. The testing company had failed to note the modified rear suspension on the bus while carrying out a statutory Department of Environment test prior to the crash. This modification had earlier been the subject of a recall.
After a 23-day trial, the testing company was found guilty of the charge and was fined €25,000 (*DPP for HSA v Raymond McKeown; DPP for HSA v O'Reilly Commercials Limited* (Dublin Circuit Criminal Court, July 2013)).

2 Sections 9(2) and 10(5).

Persons in control of a place of work

12.32 The 2005 Act has provided further for workplaces that are controlled wholly or in part by someone other than the employer. In this regard, it provides[1] that, where a person has control to any extent of a non-domestic place of work, or has control over the access or egress to that place, or any article or substance provided for use, that person is the person responsible for health and safety to the extent of his or her obligations. These obligations extend to any article or substance provided for use by persons other than employees of the person who has control of the article or substance. Landlords are specifically included,[2] as the provision refers to obligations by way of contract, tenancy, licence or other interest.[3]

1 At s 15.
2 By virtue of s 12(2).
3 A further interesting development, with regard to duties on persons in control of a place of work, is seen in the 2007 General Application Regulations (in particular, reg 26 which relates to 'Agreements as to Premises Use as a Place of Work'). The provisions of this regulation (which reflects ss 115 and 116 of the Factories Act 1955) allow for applications to be made to the Circuit Courts to either vary or set aside the 'Agreement for places used as a place of work' (such as a lease) in certain circumstances, whereby one or other of the parties to the agreement prevents structural or other alterations to the premises which would allow compliance with the Regulations.

12.33 The HSA has been striving to extend the application of the duty to persons in control of places of work. For instance, in 2007, the HSA issued improvement notices to various local councils throughout the Republic of Ireland on the basis that the HSA identified areas where roadworks were being undertaken as a place of work to which third parties have access. In the aftermath of a number of serious road accidents on these temporary surfaces, the HSA put forward the proposition that the local councils failed in their duty to properly risk assess the hazards associated with use of vehicles on temporary road surfaces and/or implement adequate and safe traffic management systems. However, the local councils had refused to accept that the 2005 Act can be applied in this way, and so they appealed the improvement notices to their local District Courts. Subsequently, it was decided that one of the cases[1] would lead the way.

1 *Cork County Council v HSA* (High Court), 7 October 2008.

12.34 The District Court in Cork referred a number of questions of law to the High Court. This case came before Mr Justice Hedigan at the end of November 2007 and he handed down his judgment on 7 October 2008. In that judgment, Hedigan J confirmed that, where roadworks are ongoing, this constitutes a place of work[1] and therefore falls within the jurisdiction of the HSA. However, this is not the case where the site has been demobilised and the road has been opened up to unimpeded flow of traffic. It is therefore clear that, where there are ongoing roadworks, this will constitute a place of work, and the duties imposed by the 2005 Act will apply.[2]

1 As subsequently confirmed in the case of *Donegal County Council v HSA* [2010] IEHS 286. In April 2013, the High Court heard an application by *Kerry County Council* to resist investigation by the HSA of a fatal accident (which occurred over a weekend) on the basis that the area of road in question was not a 'place of work'. The High Court held that, where works were suspended over the weekend but were not yet complete, that part of the road remained a place of work, despite temporary suspension of the works.
2 This decision may, in part, have prompted the enactment of the Safety, Health and Welfare at Work (Construction) (Amendment) (No 2) Regulations 2008 (SI No 423/2008) relating to works in the vicinity of roadways.

12.35 A further point made by the court was that the Improvement Notice should be addressed to the person in control of the activity in question which, in this case, was the contractor and not the local authority.

12.36 Similarly, the HSA is also seeking to extend the application of s 12 such that persons (such as engineers or architects) who are contractually authorised to issue instructions in respect of work carried out on a construction site are seen as the person in control to an extent of that place of work. It remains to be seen how this will unfold.

'Reasonably practicable'

12.37 Health and safety duties on employers and others have historically been to the standard of that which is 'reasonably practicable' to ensure the health and safety of those who might be affected by their activities at a place of work. The meaning given to 'reasonably practicable' was not based on any statutory definition but had developed over time through a series of court decisions. In this regard, the Irish courts had adopted the same position as the UK courts. For example, the Supreme Court decision in *Boyle v Marathon Petroleum (Irl) Ltd*[1] expressly approved the judgment of Asquith LJ in *Edwards v The National Coal Board.*[2]

1 [1999] 2 IR 460.
2 [1949] 1 KB 704.

12.38 In the absence of any statutory definition, the European Commission took the view that, by using the term 'reasonably practicable', the UK and Ireland were restricting the duty upon employers, to ensure the safety and health of workers in all aspects related to work to a duty to do this only 'so far as is reasonably practicable', and, therefore, were failing to fulfil their obligations under Art 5(1) and (4) of the Framework Directive.[1] Prior to the Commission's case coming before the ECJ, Ireland enacted the 2005 Act, which incorporated a specific statutory definition on the term 'reasonably practicable'. This definition came directly from the Framework Directive. In light of the new statutory definition, the Commission was satisfied that Ireland no longer restricted the duty upon employers and was, therefore, no longer in breach of the Framework Directive. The case against Ireland was, therefore, dropped. Ultimately, however, the Commission's case against the UK was unsuccessful.[2]

1 Article 5(1) of the Framework Directive provides that the employer is obliged to ensure the safety and health of workers, while Article 5(4) allows Member States the option of limiting or excluding employers' responsibility 'where occurrences are due to unusual and unforeseeable circumstances, beyond the employers' control, or to exceptional events, the consequences of which could not have been avoided despite the exercise of all due care'.
2 See reference to this in UK section above for details concerning the basis of the ECJ's decision on the case against the UK.

12.39 Consequently, a principal innovation of the 2005 Act was the inclusion of a statutory definition of 'reasonably practicable', as follows:

> 'For the purposes of the relevant statutory provisions, "reasonably practicable", in relation to the duties of an employer, means that an employer has exercised all due care by putting in place the necessary protective and preventive measures, having identified the hazards and assessed the risks to safety and health likely to result in accidents or injury to health at the place of work concerned and where the putting in place of any further measures is grossly disproportionate having regard to the unusual, unforeseeable and exceptional nature of any circumstance or occurrence that may result in an accident at work or injury to health at that place of work'.[1]

1 Section 2(6).

12.40 Interestingly, by its wording, the application of the statutory meaning of 'reasonably practicable' has been restricted to the duties only of the employer for the purposes of the 2005 Act. Where the 2005 Act provides for duties on others (such as employees, persons in control of a place of work, persons designing, manufacturing, importing and supplying articles and substances for use at a place of work, and any person who procures, commissions, designs or constructs a place of work), insofar as such duties are subject to doing all that is 'reasonably practicable', the common law meaning will apply.[1]

1 The common law meaning given to 'reasonably practicable' in Ireland is the same as has been applied in the UK by way of, for example, *Edwards v The National Coal Board* (see reference to this in the UK section above).

12.41 In order to discharge the duty to do all that is reasonably practicable, an employer must first identify the hazards and assess the risks at the place of work, and then put in place the appropriate measures and procedures required to prevent accidents occurring and to protect the health and safety of employees. In addition, an employer must be able to show that the putting in place of further measures would be 'grossly disproportionate', given the 'unusual, unforeseeable or exceptional' circumstances giving rise to an accident. There has yet to be any judicial decision on what this may mean in practice, and whilst the time, cost and inconvenience of such additional measures may be balanced against the risk of harm (as under the common law definition), where an accident occurs, it is likely to be difficult in many instances to argue successfully that it was 'unforeseeable'. That is to say that quite often, after an incident has occurred, and upon investigating the causes of that incident, it becomes evident that the chain of events giving rise to the incident was, in fact, 'foreseeable'.

Penalisation

12.42 Section 27(1) provides that penalisation:

> 'includes any act or omission by an employer or a person acting on behalf of an employer that affects, to his or her detriment, an employee with respect to any term or condition of his or her employment.'

Furthermore, s 27(2) states that:

> 'penalisation includes –
>
> (a) suspension, layoff or dismissal (including a dismissal within the meaning of the Unfair Dismissals Acts 1977 to 2001), or the threat of suspension, layoff or dismissal,
> (b) demotion or loss of opportunity for promotion,
> (c) transfer of duties, change of location of place of work, reduction in wages or change in working hours,
> (d) imposition of any discipline, reprimand or other penalty (including a financial penalty), and coercion or intimidation.'

12.43 Section 27(3)(a)–(f) goes on to list the activities for which an employee is protected. It was the EU Framework Directive on health and safety[1] that inspired the introduction of a statutory provision protecting employees against 'penalisation' by their employer.[2] In essence, this provision prohibits employers from placing their employees at a disadvantage because they have exercised their entitlements or otherwise made complaints in respect of health and safety. This protection mirrors the equivalent provisions in the UK's Employment Rights Act 1996.

1 89/391/EEC.
2 This statutory protection does not affect the employee's existing common law right to take an action for breach of contract against his employer.

12.44 The definition of 'penalisation' is widely drawn, as it covers both persons appointed to health and safety roles and also others who bring to the attention of the employer, or make a complaint to the HSA relating to, any matter which they reasonably believe may affect their health and safety at work.[1] Indeed, the definition is also likely to include psychological injury arising from behaviour such as bullying in the workplace.[2] It appears, therefore, that this provision may protect a wider class of situations than its English counterpart.[3]

1 This entitlement was recognised by the Employment Appeals Tribunal in the case of *Doody v Denis J Downey Limited* EAT UD2426/2010 (September 2012) in awarding the employee €35,000 for having been unfairly dismissed following the employee making a complaint to the HSA.
2 After the introduction of the 2005 Act, there was a flurry of claims to the Rights Commissioner arising from bullying in the workplace brought under s 27 of the 2005 Act. For example, in *A Complainant v A Hospital* (29 January 2007) the allegation related to the employer's inordinate and inexcusable delay in dealing with allegations relating to dealing with a complaint of bullying at work. In that case, the Rights Commissioner held in favour of the employee and ordered the respondent to pay substantial compensation, as well as making a direction to the respondent to resolve the issue within a prescribed period, in accordance with the power bestowed by s 28(3)(b) of the 2005 Act.
3 It is noted that the Employment Rights Act 1996 protects employees from detriment where: a complaint is made directly to the employer; the employee can show that there was no safety representative or safety committee; or it was not reasonably practicable for the employee to raise the matter through the safety representative or the safety committee.

12.45 Having regard to the definition in s 27(1), penalisation can be defined as any act or omission that affects 'to his or her detriment' an employee with respect to any term or condition of employment. The term 'detriment' appears to have its origin in the UK's Employment Rights Act 1996, as it does not appear in any employment rights legislation in Ireland.[1] With regard to the use of the term, the English courts have held that it has to be given an objective meaning.[2]

1 'Penalisation under the Safety, Health and Welfare at Work Act 2005', paper presented by Marcus Dowling BL at the June 2007 Health and Safety Lawyers Association of Ireland Conference, in Dublin.
2 *Shamoon v Chief Constable of the Royal Ulster Constabulary* [2003] UKHL 11, [2003] IRLR 286.

12.46 The detriment alleged by the employee to have been suffered must relate to a term or condition of employment. Difficulties may arise in distinguishing what amounts to an employer's right to make operational changes, and what amounts to a change in a term or condition of employment. Some of the provisions set out in s 27(2), such as the 'transfer of duties', 'change of location of place of work', and/or 'a change of working hours', are regularly reserved in the employment contract as conditions of employment subject to change as the employer sees fit. However, if any of these changes occur following a protected activity,[1] it is likely to sway the Rights Commissioner.[2]

1 Listed in s 27(3).
2 The Rights Commissioner service at the Labour Relations Commission is the forum specified in the 2005 Act as having jurisdiction to hear any claim of 'penalisation' under the Act.

12.47 What certainly seems clear is that, in order to succeed, an employee must demonstrate three things, namely that:

(1) he has suffered a detriment within the meaning of s 27(1) or (2);

(2) he has acted in accordance with s 27(3)(a)–(f); and
(3) the reason why the employer imposed the detriment was due to the employee engaging in a protected activity.[1]

In saying this, where the claim arises from a dismissal, if an employer can show that the steps taken by the employee were so negligent that it was reasonable for the employer to dismiss the employee, the employee's claim will fail.[2]

1 As listed in s 27(3).
2 If the alleged act of penalisation constitutes a 'dismissal', an employee is entitled to bring a claim to the Employment Appeals Tribunal under the Unfair Dismissals Act 1977 to 2007. However, if brought under that legislation, an employee cannot also bring a claim to the Rights Commissioners under the 2005 Act.

12.48 Under the English legislation, it has been specifically provided that the onus is on the employer to justify detrimental treatment where it has been established. Although there is no equivalent provision in the 2005 Act, it can be said that this would be an appropriate approach to adopt in Ireland.

Safety management

'*Competence*'

12.49 The General Application (Amendment) (No 2) Regulations 2003[1] included a definition of 'competent person'. However, by including the definition in the 2005 Act, it has been placed on a firmer statutory footing. The definition has been expanded, and the concept of 'competency' is now central to effective compliance with health and safety legislation in Ireland. The statutory definition provides:

> 'For the purposes of the relevant statutory provisions, a person is deemed to be a competent person where, having regard to the task he or she is required to perform and taking account of the size or hazards (or both of them) of the undertaking or establishment in which he or she undertakes work, the person possesses sufficient training, experience and knowledge appropriate to the nature of the work to be undertaken.'[2]

In this regard, any such competent person (for example, a safety representative,[3] if appointed) shall have access to all relevant information for the purposes of performing his or her functions. This will include access to risk assessments, information relating to accidents and dangerous occurrences, and information arising from protective and preventive measures put in place.

1 SI No 53/2003.
2 Section 2(2)(a). Section 2(2)(b) goes on to state that account should be taken of the framework of qualifications referred to in the Qualifications (Education and Training) Act 1999. The 1999 Act sets up standards for certain qualifications and establishes relevant Certifying Bodies. The definition in the 2005 Act makes it clear that a person assigned to undertake certain tasks must be trained to the appropriate level. For example, Fetac certification pursuant to the 1999 Act is now required to establish competency in manual handling and first aid.
3 Section 25 deals with appointment of safety representatives.

Risk assessment

12.50 While the requirement for carrying out risk assessments is contained in the Management of Health and Safety at Work Regulations in the UK, the corresponding requirement in Ireland is set out in the 2005 Act.[1] One of the key elements to the carrying out of a risk assessment is that, although it is the employer's duty to ensure that this is carried out, the employer need not do this him or herself but may (or indeed, in certain circumstances, must) employ a competent person to fulfil this duty.[2] However, the statutory duty for ensuring that appropriate risk assessments are carried out remains with the employer, who cannot, in line with the common law duty of care,[3] delegate this duty.

1 Section 19. In addition to the s 19 general requirements, there are also requirements in subsidiary regulations which require the carrying out of risk assessments in specific circumstances.

2 See **12.49** above on who is a 'competent person'.

3 *Wilson and Clyde Coal Co Ltd v English* [1937] 3 All ER 628, in which it was reinforced that the duty of care owed by the employer to his employees was personal and was not capable of being discharged merely by delegating its performance to another apparently competent person.

12.51 All risk assessments are underpinned by the 'principles of prevention', which now form part of the 2005 Act.[1] Implementation of the improvements identified by the risk assessments is a key element of the employer's statutory duties in this regard, and must take account of all activities and all levels of the place of work.[2] The duty to carry out risk assessments must take account of employers' duties to persons other than their own employees and also extends to persons in control of a place of work.

1 At Sch 3. The principles of prevention originate from the Framework Directive and were previously contained in the 1993 General Application Regulations. They are also incorporated by the Management of Health and Safety at Work Regulations in the UK.

2 See s 19(4).

12.52 It is noteworthy that, whilst there is an Approved Code of Practice in relation to the carrying out of a suitable and sufficient risk assessment in the UK, there is no equivalent guidance document in Ireland. It is considered by the HSA to be within the employer's knowledge as to what will be sufficient and appropriate for the specific industry in which the employer is carrying out his activities, and the HSA appears reluctant to go further than restating the statutory duties on a general basis.

12.53 With regard to the management of risks, the HSA has a firm view that responsibility rests at the top of an organisation and has issued guidance for directors and managers to assist in fulfilling their responsibilities.[1]

1 This is discussed further at **12.61** below.

Safety statements[1]

12.54 Following the risk assessment, every employer who employs three or more employees must ensure that a safety statement is, or is caused to be, prepared.[2] The duty to undertake risk assessments, and the putting in place of a safety statement, is central to the operation of the 2005 Act and the ethos of ensuring effective safety management by employers.

1 In the UK, there does not appear to be provision for this name. UK organisations generally have more detailed health and safety policies in manuals. For UK organisations coming into Ireland, it is vital that they are aware of these references and that they 'Irish-ise' their safety documents. Otherwise, they are open to being served with an improvement notice or being prosecuted by the HSA.

2 When an employer has less than three employees, it will be sufficient compliance for that employer to ensure that, insofar as health and safety is concerned, he or she is fully compliant with the codes of practice relevant to the business (see s 20(8)). It should also be noted that the HSA has recently published a Code of Practice entitled 'Construction Safety Code of Practice for Contractors with Three or Less Employees'.

12.55 The safety statement is based on the hazards identified in the risk assessment and sets out in detail how the employer intends to manage and control the risks at the place of work.[1] The 2005 Act, in particular, provides that every employer or person who may have control over a place of work, or may have duties to persons other than his or her own employees, must also have a written safety statement, setting out the measures put in place for protecting those persons.[2]

1 See s 20. Safety statements had previously been addressed in s 12 in the 1989 Act, but the provision of the 1989 Act has been substantially amended under s 20 of the 2005 Act.

2 In addition, where an employer contracts with another for the provision of services, the first employer must ensure that the second employer is in possession of an up-to-date safety statement: see s 20(6).

12.56 The 2005 Act also provides for what should be included in the safety statement. In particular, it should specify:

- the hazards identified and risks assessed;
- the protective and preventive measures to be taken and resources provided for same;
- the plans and procedures to be followed in the event of an emergency and/or serious and imminent danger;
- the duties of employees with regard to health and safety;
- the names and, where applicable, the job title or position of each person responsible for performing a safety task; and
- any arrangements made regarding the appointment of safety representatives and arrangements for consultation with employees and safety representatives on health and safety issues.[1]

1 Section 20(2)(a)–(f).

12.57 It is incumbent on every employer to bring the safety statement to the attention of all employees or any other person to whom it might relate, and this should be done in a form, manner and, as appropriate, language that is reasonably likely to be understood by the relevant person.[1] The employer must also ensure that his employees' attention is drawn to the document on at least an annual basis or at any time following an amendment to it. Prior to the commencement of work by any newly recruited employee, or when any other person attends at the place of work that might be affected by any risk that has been identified, their attention must also be brought to the safety statement.

1 The same statutory duty arises in relation to the furnishing of all health and safety information, instruction and/or training.

12.58 The safety statement is a living document[1] and must be amended where there has been a significant change to the matters to which it refers or where there is any other reason for believing it is no longer valid. In addition, an HSA Inspector may direct that a safety statement be amended. A copy of the safety statement (or a relevant extract of it) must be kept available for inspection at or near every place of work to which it relates whilst work is being carried out there.

1 As is the risk assessment.

12.59 Given that the safety statement is for the purposes of notifying the relevant persons of how the employer proposes to manage and control the hazards and risks identified, part of that process will be to identify the individuals who have been assigned tasks which assist in the management of health and safety and employees who may be exposed to any risks identified. Clearly, not only is it common sense that these individuals should be appropriately and properly trained to carry out those functions, but there is also a statutory duty upon the employer to ensure that any person assigned a safety task must be 'competent', as defined under the legislation.

Directors' responsibilities for health and safety

12.60 Section 48(19) of the 1989 Act[1] was the existing provision for liability against directors and managers arising out of health and safety breaches prior to the introduction of the 2005 Act. That section of the 1989 Act was substantively repeated in the 2005 Act, as follows:

> 'Where an offence under any of the relevant statutory provisions has been committed by an undertaking and the doing of the acts that constituted the offence has been authorised, or consented to by, or is attributable to connivance or neglect on the part of, a person, being a director, manager or other similar officer of the undertaking, or a person who purports to act in any such capacity that person as well as the undertaking shall be guilty of an offence and shall be liable to be proceeded against and punished as if he or she were guilty of the first-mentioned offence.'[2]

However, the 2005 Act went much further with regard to personal liability for directors and managers (or others purporting to act in that capacity) by providing that, where a company is found guilty of an offence under the 2005 Act, there is now a presumption that a person occupying a managerial position whose duties could have affected the management of the undertaking to a significant extent, is also guilty of an offence under the 2005 Act. It will then be for that person to prove otherwise. This is tantamount to a reversal of the usual principle applicable to criminal matters that one is 'innocent until proven guilty'. Whilst the constitutionality of this reversal may be somewhat questionable, the principle in the 2005 Act has not yet undergone legal challenge. This may be because, in practice, the prosecutor in any criminal proceedings must discharge the evidential burden of proof and establish the guilt of any accused 'beyond reasonable doubt'. The 'statutory presumption' referred to above cannot, and does not, alter the onus of proof on any prosecutor.

Section 81 of the 2005 Act further provides:

> 'In any proceedings for an offence under any of the relevant statutory provisions consisting of a failure to comply with a duty or requirement to do something so far as is practicable or so far as is reasonably practicable, or to use the best practicable means to do something, it shall be for the accused to prove (as the case may be) that it was not practicable or not reasonably practicable to do more than was in fact done to satisfy the duty or requirement, or that there was no better practicable means than was in fact used to satisfy the duty or requirement.'

The shifting of the evidential burden contained in this section was at the heart of the appeal against conviction in the case of *The People* (*DPP*) *v PJ Carey* (*Contractors*) *Ltd*.[3] In this case, the Court of Criminal Appeal found that, before the evidential burden can properly pass to the accused, the prosecution must first establish a *prima facie* case in respect of the breach(es) alleged.[4]

1 Which corresponded with s 37(1) of the UK's 1974 Act.
2 Section 80(1).
3 [2011] IECCA 63 (Court of Criminal Appeal, 17 October 2011). This case related to a conviction under s 50 of the Safety, Health and Welfare at Work Act 1989 (identical to s 81 of the 2005 Act). This provision also corresponds with s 40 of the UK's 1974 Act.

4. This principle was successfully applied in the case of *DPP v SIAC Construction Ltd and Ferrovial Agroman (Ireland) Ltd* (Trim Circuit Court, 9 March 2012).

12.61 In the meantime, the provision in the 2005 Act states:

> 'Where the person is proceeded against as aforesaid for such an offence and it is proved that, at the material time, he or she was a director of the undertaking concerned or a person employed by it whose duties included making decisions that, to a significant extent, could have affected the management of the undertaking, or a person who purported to act in any such capacity, it shall be presumed, until the contrary is proved, that the doing of the acts by the undertaking which constituted the commission by it of the offence concerned under any of the relevant statutory provisions was authorised, consented to or attributable to connivance or neglect on the part of that person.'[1]

This significant provision of the 2005 Act has attracted, and continues to attract, the attention of directors and managers in many organisations based in Ireland. Indeed, the HSA has made no secret of the fact that it is actively seeking out an appropriate case for the imposition of a custodial sentence on an individual manager or director under this provision.[2] It is indicative of the weight the HSA attaches to this provision that it issued the 'Guidance for Directors and Senior Managers on their Responsibilities for Workplace Safety and Health'[3] in order to remind directors and managers of their statutory responsibilities and assist them in understanding their obligations. It also sets out steps which may (and sometimes must) be taken to achieve the most effective safety management.[4]

1 Section 80(2). In addition, s 80(3) (which repeats s 48(19)(b) of the 1989 Act) goes on to provide that, where the affairs of a body corporate are managed by its members, s 80(1) and (2) shall apply in relation to the acts or defaults of a member in connection with his or her functions of management as if he or she were a director of the body corporate.

2 To date, there has been no case of a director, manager or other company officer having received a *custodial* sentence for any breach of health and safety legislation, although there have been a number of cases where such company officers have received fines and suspended sentences. Previously, such sentences tended to arise pursuant to s 13 of the Non-Fatal Offences Against the Person Act 1997, in the absence of sufficient provision in the 1989 Act. However, given the penalties now provided for in the 2005 Act, the more recent suspended sentences have arisen directly from prosecution under that Act: see **12.81** below.

3 Published October 2007 and formally launched by the HSA on 6 November 2007 (see www.hsa.ie).

4 See **12.64** to **12.66** below in relation to the legal status of guidance documents and codes of practice.

ROLE OF THE HEALTH AND SAFETY AUTHORITY

12.62 As in the UK (since the amalgamation of the Commission and the HSE in 2008), there is only one body (the HSA) which fulfils the roles of advisor, monitor and enforcer of health and safety legislation in Ireland.[1] In the main, these roles are all discharged through the HSA Inspectors. The involvement of the HSA in the roles of advisor and enforcer may at times give rise to conflict and, in fact, there is a reluctance in individual cases from the HSA to furnish advices with a greater leaning towards the fulfilment of its enforcement duties.

1 See s 34 of the 2005 Act for the functions of the HSA.

Advisory

12.63 The HSA has been assigned[1] a number of functions for the promotion and encouragement of health and safety at work including providing education and training, fostering cooperation between bodies representative of employers and employees, as well as supporting and commissioning studies into general health and safety compliance. As part of its functions, the HSA issues guidance notes on various aspects of health and safety legislation[2] to assist employers and others in discharging their statutory duties. It also has the power to publish or approve Codes of Practice to support the statutory provision.[3]

1 See s 34.
2 See www.hsa.ie.
3 Section 60 provides that Codes of Practice are issued for the purposes 'of providing practical guidance to employers, employees and any other persons … with respect to safety, health and welfare at work, or the requirements or prohibitions of any of the relevant statutory provisions'.

Codes of Practice and other guidance issued by the HSA

12.64 Although the HSA has the power to issue Codes of Practice, it may only do so having first obtained the consent of the Minister for Enterprise, Trade and Employment. The draft Code must first be published for the purposes of allowing a 28-day consultation period for interested parties to make submissions in respect of the terms of the draft.[1]

1 Section 60(2).

12.65 Codes of Practice published or approved in accordance with the 2005 Act are afforded a particular status in criminal proceedings,[1] and s 61 of the 2005 Act provides:

> '(1) Where in proceedings for an offence … relating to an alleged contravention or any requirement or prohibition imposed by or under a relevant statutory provision, being a provision for which a code of practice had been published or approved … at the time of the alleged contravention, subsection (2) shall have effect with respect to that code of practice in relation to those proceedings.
>
> (2) (a) Where a code of practice … appears to the Court to give practical guidance as to the observance of the requirement or prohibition alleged to have been contravened, the code of practice shall be admissible in evidence.
>
> (b) Where it is proved that any act or omission of the defendant alleged to constitute the contravention (i) is a failure to observe a code of practice …, or (ii) is a compliance with that code of practice,
>
> then such failure or compliance is admissible in evidence.'

Therefore, whilst complying with a Code of Practice is not compulsory, to do so will be persuasive evidence in any proceedings arising from an alleged statutory breach.[2]

1 As in the UK.
2 Examples of current Codes of Practice issued by the HSA are: Code of Practice for Employers and Employees on the Prevention and Resolution of Bullying at Work (2007), Code of Practice on Access and Working Scaffolds (2008), Construction Safety Code of Practice for Contractors with Three or Less Employees (2008), Code of Practice for Working in Confined Spaces (2010), Code of Practice for the Design and Installation of Anchors (2010), Code of Practice for Avoiding Danger from Underground Services (2010), Code of Practice for Safety in Roofwork (2011), Code of Practice for Operators of Quarry Vehicles (2012). The National Safety Authority of Ireland, in 2013, also published a Code of Practice on the 'Safe Use of Tower Cranes and Self-Erecting Cranes' (IS 361.2013). This is not an exhaustive list.

12.66 In addition to formal Codes of Practice, the HSA has also issued several Guidance Notes on practical implementation of certain regulations or other statutory duties. HSA Guidance Notes do not have to comply with any formal procedure (such as consultation with interested bodies and/or seeking Ministerial consent required for Codes of Practice) prior to publication.[1] Whilst the 2005 Act has given specific statutory weight, however limited, to Codes of Practice, the same, however, cannot be said of the Guidance Notes. That being said, it is interesting that the HSA has, in the past, suggested a similar standing for its Guidance Notes by including the following wording:[2]

> 'This guidance is issued by the Health and Safety Authority. Following the guidance is not compulsory and you are free to take other actions to achieve compliance. But if you do follow the guidance, you will normally be doing enough to comply with the law. Health and Safety Authority inspectors seek to secure compliance with the law and may refer to this guidance as illustrating good practice and compliance.'

It is clear from the above that, in the event of a breach of the 2005 Act, where relevant guidance exists and has not been adhered to, the HSA is likely to form its own view based on that guidance and will no doubt refer to it in any subsequent prosecution.[3]

1 Examples of current guidance documents are: Health and Safety Authority Guide to the Safety, Health and Welfare at Work Act 2005 (2005); Guidelines on Risk Assessments and Safety Statements (January 2006); Workplace Safety and Health Management (January 2006); Summary of Key Duties under the Procurement, Design and Site Management Requirements of the Safety, Health and Welfare at Work (Construction) Regulations 2006 (September 2006); Guide to the Safety, Health and Welfare at Work (General Application) Regulations 2007 (2007); Guidance for Directors and Senior Managers on their Responsibilities for Workplace Safety and Health (2007); Guidelines on the Preparation of a Safety Statement for a Farm (2008); Guidance on the Manual Handling Training System (2010); Guidance on Lone Working in the Healthcare Sector (2011); Guidance on the Management of Manual Handling in Healthcare (2011); Guide for Contractors and Project Supervisors (2013); and Guide for Homeowners (2013). This is not an exhaustive list.

2 Taken from the 'Guidelines on the Procurement, Design and Management Requirements of the Safety, Health & Welfare at Work (Construction) Regulations 2006' – HSA October 2006 (introduction).

3 Similar to the use of a Code of Practice in criminal proceedings (see **12.65** above).

Enforcement

Inspectors' powers

12.67 HSA Inspectors have extremely broad powers. The powers prescribed[1] apply to all persons authorised by the HSA[2] and include the power to:

- enter a place of work to inquire into, search, examine and inspect work activities, installations, processes or procedures, articles, substances or records;
- direct that the place of work or part thereof be left undisturbed;
- direct that a safety statement be amended;[3]
- require the production of records in legible form;
- inspect and take copies of or extracts from records (including electronic records for which the Inspector may seek assistance from the person on whose behalf a computer is or has been used);
- remove and retain records;
- require records to be maintained;
- require the furnishing of relevant information;
- require assistance to enable him or her to carry out the exercise of his or her functions;

- summon, by written notice, the person concerned to furnish any relevant information;[4]
- subject to the right of a person not to incriminate him or herself, examine any such person who may be in a position to furnish relevant information;
- require that a procedure be carried out or an article be operated where deemed to be relevant to any search, examination, investigation, inspection or inquiry;[5]
- take measurements, photographs or make tape (or other) recordings;
- install, use and maintain in, at or on the place of work, monitoring instruments, systems or seals;[6]
- carry out, or have carried out, testing of article(s) or substance(s);
- cause any article or substance which may pose a risk to be dismantled or tested and/or remove and retain any article or substance posing a risk for examination purposes; and
- take samples of the atmosphere.

There is also a general power to the extent that the Inspector may 'exercise such other powers as may be necessary for carrying out his or her functions'.[7]

1 At s 64.
2 Pursuant to ss 33 and 62.
3 This is a new power which did not previously exist in the 1989 Act.
4 This is a significant new power and the person summoned will be required to give information which the Inspector may reasonably require in relation to the place, article or substance, work activity, installation or procedure. That person may also be required to produce relevant records in their power or control.
5 This is a new power which did not previously exist in the 1989 Act.
6 This is a new power which did not previously exist in the 1989 Act.
7 Section 64(1)(t).

12.68 HSA Inspectors also have various enforcement options as part of their available armoury for ensuring compliance with health and safety legislation.

Directions for improvement plan

12.69 Where an Inspector considers that a work activity involves, or is likely to involve, a risk to safety and health, he or she may issue a written direction to the employer requiring the submission of an 'improvement plan'. The written direction[1] must include certain information, and a copy must be provided by the Inspector to the employee and to the safety representative (if any) at the relevant place of work. Once the employer submits the improvement plan, the Inspector must, within one month of receipt, confirm whether he or she is satisfied with it. Where this is the case, the employer must implement the plan. Alternatively, the Inspector may request that the plan be revised.

1 As specified at s 65(2).

Improvement Notice[1]

12.70 Where the Inspector is of the opinion that the improvement (or revised) plan has not been implemented or that the employer is in contravention of its statutory duties, the Inspector may serve an 'Improvement Notice'[2] setting out the basis for his or her belief that a breach is occurring, or has occurred, and detailing the alleged breach. The Notice will also direct what the employer needs to do to remedy the situation and will inform the employer of the right to appeal the Notice. Once the matters have been remedied in accordance with the Notice, the person upon whom the Improvement Notice was served must advise the Inspector in writing. Within one month, the Inspector will confirm whether he or she is satisfied that the Notice has been fully complied with.[3]

1 Section 66, as amended by section 12 of the Chemicals (Amendment) Act 2010.
2 A copy of the Notice is served on the employer in question. In addition, if there is a safety representative at the relevant place of work, the Inspector must also furnish him or her with a copy of the Notice (s 66(4)). The safety representative must also be informed when the Notice is withdrawn.
3 Note that s 77(3)(k) makes it a criminal offence to contravene an Improvement Notice.

12.71 Any person aggrieved by an Improvement Notice has the right to appeal it to the District Court within 14 days from receipt of the Notice.[1]

1 Section 66(7)–(9) of the 2005 Act and s 36(4)–(7) of the 1989 Act.

Prohibition Notice[1]

12.72 A Prohibition Notice will be served in the event that an Inspector is of the view that there is, or is likely to be, a risk of serious personal injury to any person resulting from the carrying on of an activity at the place of work. The 2005 Act mandates for certain information to be contained in the Prohibition Notice.[2] In addition, the Inspector has a discretion[3] to include directions as to the measures to be taken.

1 Section 67.
2 Section 67(2).
3 Section 67(3).

12.73 The Notice will have the effect of prohibiting the activity from continuing until the matters referred to in the Notice have been remedied. This, in turn, may result in the entire business, or a portion of it, being closed down for a period to enable the problems to be resolved effectively and safely. Once the Notice is issued, the Inspector must serve it on the employer (or person in control of the place of work), and a further copy must be given to the Safety Representative (if any).

12.74 A person who feels aggrieved by the Notice may appeal it to the District Court within a period of seven days from the date of receipt.[1] The Court will decide, upon hearing the appeal, whether the Notice should be confirmed, varied or cancelled. Where the judge confirms the Notice but the appellant requests the confirmed Notice be suspended, the judge may do so in circumstances where he or she considers it appropriate.[2]

1 Whilst an appeal against the Notice will not have the effect of suspending its operation (given the risk of 'serious personal injury' believed to exist), an application can, however, be made to the Court seeking an Order to suspend the Notice until such time as the appeal is disposed of.
2 Section 67(6)–(9).

12.75 Where the matters referred to in the Prohibition Notice have been remedied, the person upon whom the Notice was served must confirm this position to the Inspector (and to the Safety Representative, if any) in writing. Within one month thereafter, and once the Inspector is happy that the issues have been addressed and remedied, he or she shall give written notice to that person, acknowledging compliance.[1]

1 Section 67(10) and (11).

12.76 The Inspector also has the power to withdraw a Prohibition Notice in circumstances where the activity to which it relates no longer involves a risk of serious personal injury, or where the Inspector is of the view that it was issued incorrectly or in error. The withdrawal of the Notice does not prevent the service of any other Prohibition Notice.[1]

1 Section 67(12)–(14).

12.77 Where an Improvement Notice and/or a Prohibition Notice is served, the person in receipt of the relevant Notice must bring it to the attention of any person whose work might be affected by it and must also prominently display a copy of the Notice at or near any place of work, article or substance affected by it.[1]

1 Note that failure to display the Notice pursuant to s 69 will be an offence under s 77(1)(a), which is one of the offences for which the penalty is liability on summary conviction to a fine not exceeding €3,000, as provided for in s 78(1).

Court Orders

12.78 Where an Inspector believes that a person is in contravention of a Prohibition Notice, the Inspector may apply, ex parte, to the High Court seeking an Order prohibiting the continuance of the activity in question. In these circumstances, the High Court may, at its discretion, make an interim or interlocutory Order as appropriate and may also include terms and conditions with regard to the payment of costs.[1]

1 See s 68.

12.79 In other circumstances, where an Inspector (or other authorised person) believes a risk exists that is so serious that the use of the place of work(or part of that place) should be restricted or prohibited until the risk is reduced to a reasonable level, that Inspector may apply, ex parte, to the High Court for the appropriate Order. At its discretion, the High Court may also, in these circumstances, make an interim or interlocutory Order. Any subsequent application to have the Order varied or revoked must be on notice, as the Inspector is entitled to be heard by the court for the purposes of such application.[1]

1 See s 71(1)–(4).

Investigations and special reports

12.80 This is the procedure for investigating and reporting on 'the causes and circumstances surrounding any accident, incident, personal injury, occurrence or situation or any other matter related to the general purposes' of the 2005 Act.[1] For the purposes of such investigation, the person carrying out the investigation has the same powers as an Inspector as provided for in the 2005 Act. Once the process is finalised, the Inspector who carries out the investigation must produce a 'special report'. The 'special report' must be presented to the Minister for Enterprise, Trade and Employment as soon as practicable, and the HSA may also make public a 'special report' in 'such manner as it considers appropriate'.[2]

1 Section 70.
2 Section 70(5).

HEALTH AND SAFETY PROSECUTIONS

Offences

12.81 The 2005 Act[1] contains an expanded list of offences (compared with those arising under the 1989 Act).[2] There are additional offences in respect of failure to discharge certain duties with regard to safety representatives,[3] and the duty to consult with and ensure the participation of employees in safety committees.[4]

1 At s 77.
2 Section 48 of the 1989 Act.
3 See s 25(4)–(6) and s 77(1)(a).
4 See s 26(1), (4)–(6) inclusive and s 77(1)(a).

12.82 Under the 1989 Act, no offence tried summarily carried a sentence of imprisonment. On indictment, only three offences carried a term of imprisonment of up to two years, as follows:

(a) failure to obey a prohibition notice;[1]
(b) unlawful disclosure of information;[2] and
(c) breaking the terms of a licence issued under the 1989 Act.[3]

1 Section 37 of the 1989 Act.
2 Section 49 of the 1989 Act.
3 Section 59 of the 1989 Act.

12.83 Given the limitations in the 1989 Act on the HSA to proceed on indictment under the Act, where either the HSA or the Director of Public Prosecutions (DPP) believed the breach was so serious that the company or individual manager or director should be proceeded against on indictment, or should potentially incur a prison sentence, the provisions of the Non-Fatal Offences Against the Person Act 1997[1] were used in addition to provisions from the 1989 Act and/or its underlying regulations. The 1997 Act created a statutory offence for anyone who 'intentionally or recklessly engages in conduct which creates a substantial risk of death or serious harm to another ...'. If found guilty, a person would be liable to a fine and/or a term of imprisonment whether on summary conviction or conviction on indictment.[2]

1 And, in particular, s 13 of the 1997 Act.
2 Not exceeding IR£1,500 and/or not exceeding 12 months on summary conviction. On conviction or indictment, s 13 provided for a fine (unlimited) or a term of imprisonment not exceeding seven years, or both.

12.84 By contrast, the 2005 Act has altered the position by increasing the number of offences punishable by a term of imprisonment. Effectively, offences can now be split into three categories:

(1) summary offences attracting a fine of up to €5,000 only;[1]
(2) summary offences for which a fine to a maximum of €5,000 and/or a term of imprisonment of up to 12 months may be imposed;[2] and
(3) indictable offences attracting a fine of up to €3 million and/or a term of imprisonment not exceeding two years.

1 These offences are provided for in s 77(1). The 2005 Act provided for a fine of €3,000 for the summary offences and a prison sentence of up to six months. However, these penalties were revised upwards by s 12 of the Chemicals (Amendment) Act 2010.
2 These 'hybrid' offences are provided for in s 77(2)–(9)(a) of the 2005 Act. In deciding on the forum for prosecution, the HSA may opt to prosecute any one of these offences summarily. In the alternative, the DPP may authorise prosecution on indictment. Clearly, the circumstances of each breach will guide the decision makers.

12.85 Either of the categories of summary offences can be brought by the HSA or a person prescribed under s 33 of the 2005 Act[1] and will be prosecuted in the District Court. Insofar as indictable offences are concerned, it will be a matter for the DPP to decide if the offences are of such a serious nature that they merit prosecution on that basis.

1 See s 82.

On-the-spot fines

12.86 In addition, the 2005 Act has created the possibility of what have become commonly known as 'on-the-spot' fines.[1] However, further regulations are required to identify the offences to be covered by this provision before such fines can be imposed. If the regulations are enacted, an Inspector who has reasonable grounds for believing that an offence is being, or has been, committed may serve a notice stating that the person to whom the notice is addressed is alleged to have committed the offence and that, if payment of a fine of up to €1,000 (as set out in the notice) is made within 21 days, no prosecution will be brought in respect of the alleged offence.

1 See s 79.

Penalties for breach

12.87 In *The People (DPP) v Roseberry Construction Ltd and McIntyre*,[1] Mr McIntyre (the second defendant) was the managing director of Roseberry Construction (the first defendant), which was constructing a development of 90 houses in County Kildare. Although the first defendant had control of the site in question, it was not the employer. The accident arose due to a failure to adequately shore up a trench, resulting in a collapse in which two employees of a sub-contractor were fatally injured.

1 [2003] 4 IR 338 (CCA).

12.88 Three individuals, namely the second defendant, the foreman for the sub-contractor and the driver of the digger, which was working close to the trench at the time of the collapse, were initially charged with manslaughter as a result of the accident. However, when the cases were dealt with in 2001, the manslaughter charges had been dropped.[1] Instead, the individuals pleaded guilty to charges under the 1989 Act and the Safety, Health and Welfare at Work (Construction) Regulations 1995. At that time, Groarke J imposed the following:

- a fine of IR£200,000 (approximately €254,000) on the company for a number of offences. The court placed particular emphasis on the company's failure to have a safety statement in place and considered that, had this document been prepared, the risk of the trench collapse would have been properly considered and the accident may not have occurred;
- a fine of IR£40,000 (approximately €50,800) on the second defendant pursuant to s 48(19) of the 1989 Act;[2] and
- an 18-month suspended sentence on the foreman on a charge of reckless endangerment pursuant to s 13 of the Non-Fatal Offences Against the Person Act 1997.[3]

1 To date, there have been no convictions in Ireland against individuals for manslaughter arising from health and safety offences.

2 Now replaced by s 80 of the 2005 Act and provides for the prosecuting individual directors and/or managers. Section 48(19) has been repeated (within some minimal rewording) in s 80(1) of the 2005 Act. Section 80(2) of the Act is new (see **12.60** to **12.61** above).

3 See **12.83** above in relation to the use of the Non-Fatal Offences Against the Person Act 1997.

12.89 The company appealed the severity of the fine to the Court of Criminal Appeal (the CCA) and, ultimately, the CCA refused the appeal. One of the arguments put forward by the company (appellant) was that the fines imposed against both it and the director involved some degree of double-counting, in the sense that the company and the director were basically the same. In refusing the appeal, the CCA commented that the director had incorporated the company and, in so doing, had drawn down the corporate veil and the purpose was to create the company as a separate entity. The CCA also referred to s 48(19) of the 1989 Act, which, in any event, provided a statutory right to proceed against the director on an individual basis in addition to proceeding against the company.

12.90 The CCA stated that the fact that the sub-contractor was in charge of the area where the fatality occurred did not mitigate the liability of the company, because ultimately the company had retained control of the site.

12.91 Having regard to the submissions as to the level of fines imposed, the CCA stated that the applicant was a medium to large company and the level of fine was not going to drive it out of business. The CCA applied the sentencing principle as established in *The People (DPP) v Redmond,*[1] *'that a fine is neither lenient, nor harsh, in itself but only in terms of the circumstances of the person who must pay it'* (emphasis added).[2]

1 [2001] 3 IR 390.

2 Having regard generally to case law in relation to fines imposed by the courts, it is a common theme that successful appeals against the severity of such fines are quite often based upon the fact that, if the company were forced to pay the fine initially imposed, it would result in undue financial hardship on the company and may also, in fact, impact on its viability.

12.92 The CCA went on to approve a list of aggravating and mitigating factors to be taken into account – regarding the level of fines to be imposed, as set out in *R v Howe & Son (Engineers) Ltd*[1] – as follows:

(a) *aggravating factors*
- death resulting in consequence of a breach;
- failure to heed warnings; and
- risks run specifically to save money.

(b) *mitigating factors*
- prompt admission of responsibility and a timely plea of guilty;
- steps to remedy the deficiencies; and
- a good safety record.

In this case, the company had no safety statement and the CCA opined that, if they had had such a document, the risk would have been considered and acted upon. The CCA went on to state that 'it was the failure of any party to take simple remedial measures that gave rise to the substantial legal and moral guilt which must be regarded as attaching in the circumstances of the case'.

1 [1992] 2 All ER 249 (English Court of Appeal decision).

12.93 In the further case of *The People (DPP) v Oran Pre-Cast Ltd,*[1] three employees of the defendant company (the crane operator, acting foreman and the employee who was killed) had successfully removed a cracked roof gutter (which weighted in excess of 2.5 tonnes) and were in the process of installing the replacement gutter. Access was restricted due to the proximity of the adjoining building. The foreman and employee were located on the platform (equipped with a guard rail) of a mobile tower. As they moved the new gutter into position, it became clear that insufficient roof sheets had been removed. It was impossible to undertake the task without climbing onto the roof, which the foreman and employee proceeded to do. The foreman had no fall protection. The employee was wearing a harness but had failed to secure it. In addition, the employee had not yet attended the statutory safety awareness ('Safe Pass') course. The employee fell from the roof and was killed.

1 Court of Criminal Appeal, 16 December 2003.

12.94 The company pleaded guilty to offences under s 6 of the 1989 Act (for failure to provide a safe system of work and failure to provide adequate training in fall protection systems) and to other charges brought under the 1995 Construction Regulations (for not taking account of the directions of the Project Supervisor Construction Stage, failing to take appropriate measures for the safety of employees, and failing to provide safety nets or harnesses).

12.95 Although the HSA was willing to prosecute summarily, the District Court refused jurisdiction and, therefore, the prosecution had to proceed on indictment. Had the matter proceeded summarily, the maximum fine would have been IR£1,500 but, on indictment, there was no limit on the fine which could be imposed. As it transpired, in the court of first instance, the company was fined IR£500,000.

12.96 The level of fine was appealed by the company, having regard to fines in previous cases involving similar circumstances[1] and having regard to the aggravating and mitigating considerations set out in *Howe*.[2] In giving its judgment on appeal, the CCA commented that:

> 'where a fine is unlimited, there must be care and restraint in the exercise of the power to fine. The actual level of fault is the principal consideration, and the financial state of the company which is to bear the fine cannot be irrelevant and cannot be ignored, unless perhaps the Court were to be told that any level of fine could be absorbed.'

1 The *Roseberry Construction* case (see **12.87** above) was specifically referred to.
2 See reference to this in UK section above.

12.97 Having decided that there were clear errors in principle in this case, the CCA, in particular, noted that it could not find any reason for the huge disparity between the fines imposed on the company in this case and those imposed on other organisations in similar circumstances, as evidenced by the case law both in this jurisdiction and in the UK.[1] There was simply no feature which could distinguish this case to explain the huge fine.

1 At the time of the appeal, the fine in this case was significantly greater than the largest fine imposed on a corporation in Ireland by that date – the fine in the *Roseberry Construction* case was the closest.

12.98 Consequently, the CCA decided that the fine was not reasonable and was lacking in a reasonable proportion to the events proved and to the previous cases, and it reduced the fine to €100,000. It included in its reasoning that the failure to obtain evidence of the means of the company was an error of principle. It also approved the aggravating and mitigating factors as established in *Howe*.[1]

1 See above.

12.99 Until recently, the largest fine imposed against a company in Ireland was in 2004, where Smurfit News Press Ltd was fined a total of €1 million after the company had pleaded guilty to six of the eight charges levelled against it under the 1989 Act and various regulations made under that Act.[1]

1 Central Criminal Court, 29 October 2004.

12.100 In this case, there were two serious accidents, two weeks apart, which arose from identical circumstances. In the first accident, an employee's leg became trapped in a printing press and he had to have his leg amputated below the knee. The second accident involved another employee whose hand became trapped in a printing press and, as a result, he suffered a de-gloving injury to his left wrist. The court was informed that both accidents were due to inadequate guarding of the printing presses and failure to prepare an adequate risk assessment of the hazards involved in working on them.

12.101 In handing down his judgment, Groarke J commented that the company in this case had placed the pursuit of profit over the safety of its employees and, had the first accident been properly dealt with, the second would not have occurred. The severity of the fine reflected the court's finding of a 'cavalier attitude' on the part of the company, particularly in the aftermath of the first accident where it was clear that lessons had not been learnt.

12.102 In a more recent decision, Bus Eireann was fined a record €2 million following a bus crash resulting in the death of five teenage girls. Bus Eireann had been charged with failure to ensure the maintenance of the bus and with failing to provide information to ensure, so far as is reasonably practicable, the safety of its employees. The accident was caused by the failure of the ABS brakes, notwithstanding that the bus had recently undergone maintenance work. In addition to the bus company, the company which carried out the maintenance work on the bus was also convicted, although to a less significant extent, as that company had not been asked by Bus Eireann to check the ABS brakes. Further, the local council, who were in the process of carrying out road works in the area where the accident occurred, was also convicted and fined. The latter entities were each fined €100,000. The judge in the case described the accident as 'entirely avoidable'.

12.103 In the case of *The People (DPP) v O'Flynn Construction Co Ltd*,[1] the defendant company pleaded guilty to breaches of the 1989 Act[2] and the Construction Regulations 1995.[3]

1 [2006] IECCA 56, Court of Criminal Appeal, 6 April 2006.
2 Section 7(1) of the 1989 Act (General duties of employers to persons other than employees).
3 Schedule 4, paragraph 18 (regarding the signposting of the surroundings and perimeter of the site).

12.104 Here, a 210-litre barrel of wood preservative delivered to the building site was leaking. There were children playing on the site, and one of the boys set fire to the barrel; the flames engulfed another boy, who died as a result. An HSA Inspector told the court that the site was adjacent to a housing estate and it had not been properly fenced to prevent children from getting in. In addition, the barrel should not have been left where it was.

12.105 The company alleged that it was not aware that children were coming into the site. However, there was evidence of a defined track leading from the housing estate where the boy lived to the building site. The court held, therefore, that the company ought to have been aware of this.

12.106 In the circumstances, a fine of €200,000 was imposed. This was then appealed by the company and, on appeal, the CCA held that, while the breaches of the legislation by the company were not the direct cause of the boy's death, they were significant contributory factors. The court noted that, although there were no reports regarding the presence of children on the site, it is foreseeable that, where a building site is adjacent to an area in which young people live, they will enter the site to play and explore, and will get up to mischief or engage in dangerous activities that a reasonable adult would not do.

12.107 In refusing the appeal, and having taken due account of the wealth and resources of the company in assessing the severity of the fine, the CCA held that the fine imposed was 'proportionate to and reflects the seriousness of the default of the company'.

12.108 A more recent decision, which indicates the court's willingness to impose significant fines where warranted,[1] followed an accident where a lorry and a trailer shed a load of coils as the lorry approached a bend on the road. This resulted in the death of two people and the injury of others. Nolan Transport pleaded guilty to failing to manage its undertaking in such a way as to ensure, so far as is reasonably practicable, that, in the course of its work, individuals at the place of work (not being its employees) were not exposed to risks to their health and safety (s 12 of the 2005 Act). The company was fined €1 million which, in proportion to the size of the company (both in terms of turnover and numbers employed), is the highest fine ever imposed by an Irish court in a health and safety case.

1 *DPP for HSA v Nolan Transport* (Wexford Circuit Criminal Court, February 2013).

12.109 One final case that is worth mentioning is the case involving the dismantling of scaffolding on Grafton Street in Dublin on 27 March 2003. During the dismantling process, one of the scaffolding bars fell and hit a pedestrian who sustained serious head and foot injuries (resulting in the amputation of some of his toes). As a result of this accident, the company (Paul Byrne t/a P Byrne Scaffolding) was prosecuted and came before the Dublin Circuit Criminal Court in March 2006. The court found the company guilty for failing to provide a safe system of work and failing to ensure that persons not employed by it were not exposed to danger, and ultimately fined the company €10,000.

12.110 A separate prosecution was undertaken against the supervisor and employee of the company, Mr Derek Daly, who pleaded not guilty and, when the matter was before the court in March 2006, the jury was unable to reach a decision. The matter was scheduled for retrial in October 2006 and, on that occasion, Mr Daly was found guilty and fined €10,000, the court having taken into account the seriousness of the offence and Mr Daly's financial circumstances.

Interestingly, Mr Daly's fine was the same as that imposed upon the company. Having regard to that fact, Mr Daly appealed, and the CCA confirmed that Mr Daly's fine should be reduced from €10,000 to €5,000. It appears that the basis of this decision is that it would be inequitable to fine an employee at the same level as the employer.[1]

1 *DPP at behest of HSA v Daly* (Dublin CCC, November 2006). Appeal to CCA (May 2007).

12.111 Two recent high-profile cases are worth mentioning. The first of these saw Wicklow County Council being prosecuted by the DPP for breaches of the 2005 Act in connection with the deaths of two fire fighters in September 2007. This case was heard at Dublin Circuit Criminal Court and concluded in June 2013.

The DPP's case was based upon a failure to comply with: the general duty to ensure the safety, health and welfare at work of employees (s 8); risk assessments (s 19); safety statements (s 20); and training (ss 8(2)(g) and 10). Initially, the council pleaded not guilty to all four counts. However, during the trial the council changed its plea to guilty and will be sentenced on 25 October 2013. The maximum penalty for a prosecution on indictment is €3 million per count.[1]

The second case resulted from an accident in June 2010 when a paramedic was killed after falling from a moving ambulance during the transfer of a patient from Cavan to Dublin. He was attempting to close the door of the ambulance fully when he was pulled out of the moving ambulance. He died as a result of serious head injuries. The Health Service Executive ('HSE') was charged with failing to provide information, instruction, training and supervision to ensure, so far as is reasonably practicable, the safety of its employee, and failing to have a written risk assessment relating to the rear-hinge side door of the ambulance. The court was told that the HSE was aware of the risk, as a similar incident occurred in Kerry in 2007 and several recommendations had been made to the HSE following that accident. The judge imposed a fine of €350,000 on the charge of not providing training and €150,000 in relation to the risk assessment.[2]

1 *DPP for HSA v Wicklow County Council* (Dublin Circuit Criminal Court, June 2013).
2 *DPP for HSA v HSE* (Dublin Circuit Criminal Court, June 2013).

12.112 From the case law referred to above, it is clear that the courts will not shy away from imposing significant fines on companies where the circumstances dictate it. The courts have also shown a willingness to fine individual directors or managers where it can be shown that their acts or omissions have resulted in the breach.

Criminal prosecutions – corporate and personal liability

12.113 The first prosecution dealing with liability of directors or corporate officers was seen in *National Authority for Occupational Safety and Health v Noel Frisby Construction Ltd and Noel Frisby*.[1]

1 District Court, 1998: see the HSA Annual Report 1998.

12.114 In this instance, the HSA Inspector had observed repeated contraventions of the 1995 Construction Regulations on one of the company's building sites. Both the company and Mr Frisby (one of its directors) were prosecuted.[1] Both pleaded guilty and were fined €2,032 and €508 respectively.[2]

1 Under s 6 of the1989 Act (employers' general duties) and regulation 8 of the 1995 Construction Regulations (duties of contractors).
2 Pursuant to s 48(19) of the 1989 Act.

12.115 The case of *The People (DPP) v Kildownet Utilities Ltd and Others* (2006) is a good example of the court's willingness to impose prison sentences on a director/manager, albeit case law will show that all such sentences have, to date, been suspended. This trend has continued with a number of convictions now secured under s 80 of the 2005 Act, including *DPP v Clare County Council and Michael Scully* (2010) (in which the senior engineer with the county council, who had overall responsibility for the road works on which a fatal accident occurred, was convicted[1] together with the county council), *DPP v Sean Doyle, Owencrest Properties Limited and Roscommon Building Company* (2009) (in which the director of the two defendant companies was convicted and fined €50,000 following a fatality on site), and *DPP v M&P Construction Limited and David Lumley* (2010) (in which the site manager was convicted as well as the company, following the collapse of a trench on site; the site manager was fined €5,000 on the basis that he was the main person with responsibility for safety on the site).

1 A 12-month sentence was imposed, suspended for two years.

12.116 Following a fatal accident at Diamond Valley Apartment Development in Bray in February 2003 where an employee of Kildownet Utilities Ltd was electrocuted, a director of that company (Mr Brian Molloy) and the site manager (Mr Joseph Byrne) for Cormac Building Contractors Ltd received suspended prison sentences – the former receiving two years and the latter receiving three years.[1] Mr Byrne was also ordered to pay compensation of €10,000 to be held in trust for the deceased's daughter. In addition, Kildownet Utilities was fined €100,000, and Cormac Building Contractors was fined €150,000 having pleaded guilty.[2] There is no doubt but that the time is coming when the courts will impose a prison sentence on an individual whom the court believes is culpable in circumstances which are serious enough to warrant a custodial sentence. In addition, and although the 2005 Act[3] provides for a prison sentence to a maximum of two years, there is nothing to prevent the DPP from continuing to rely on the Non-Fatal Offences Against the Person Act 1997 in undertaking a prosecution to enhance the potential of custodial sentences.

1 For breach of s 13 of The Non-Fatal Offences Against the Person Act 1997. Decision, 23 November 2006, Dublin Circuit Criminal Court.
2 For breach of s 6 of the 1989 Act (employer's general duties to employees) and s 7 of the 1989 Act (duties to those other than employees), respectively.
3 Refer to ss 78 and 80, and see comments earlier in this chapter.

12.117 In the Australian case of *Ken Kumar v David Aylmer Ritchie*,[1] the chief executive of a group of companies, the Owens Group, was found guilty of failing in his duty as a director and senior manager to ensure the health and safety of employees when an employee was killed in an accident involving the cleaning of a shipping tank with a highly volatile cleaning agent. Mr Ritchie, the director in question, argued that, as he was not involved in the day-to-day management of safety matters, he could not be blamed for health and safety breaches. He stated that the Owens Group owned approximately 30 companies and had staff located in approximately 80 work sites. He contended that, in his position as Chief Executive Officer of such a large group, he was required to rely upon the expertise of divisional and site managers to deal with the detail of occupational health and safety. Mr Ritchie lived and principally worked in New Zealand and only spent two to three days every two months attending to business in Australia. Essentially, Mr Ritchie's defence pointed to the unachievable level of knowledge about the company's operations that would be involved in him being informed of all health and safety matters.

1 [2006] NSWIR Comm 323.

12.118 However, in finding him guilty of failing in his duty as a director, the court found that the very nature of Mr Ritchie's role involved him having a position of influence over the company's practices, and that he could actually influence the conduct of the company in relation to the breach but elected not to do so, because he wished to concentrate on other matters.

12.119 Although this is an Australian case, the Irish courts could well adopt a similar approach. In Ireland, it certainly echoes the HSA's view.

DISQUALIFICATION OF DIRECTORS

12.120 Section 160 of the Companies Act 1990 provides for the disqualification of directors who have been convicted of an indictable offence. To date, s 160 seems to have been somewhat overlooked with regard to health and safety prosecutions. However, it is increasingly relevant, given the expanded provisions whereby directors can be convicted of indictable offences under the health and safety legislation. Section 160 specifically provides that, where a person is convicted on indictment of any indictable offence in relation to a company, for the five years following the conviction, that person cannot act as an auditor, director or other officer, or be in any way concerned or take part in the promotion, formation or management of any company. The five-year disqualification order arises automatically. However, the court also has a discretion in relation to the order made, on the application of the prosecutor, having regard to all the circumstances of the case.

12.121 An application for a disqualification order can be made under s 160(2) by the DPP or by any member, contributory, officer, employee, receiver, liquidator, examiner or creditor of any company in circumstances where the person, the subject of the application, has been or is acting or is proposing to act as an officer, auditor, receiver, liquidator or examiner or has been or is concerned or taking part, or is proposing to be concerned or take part, in the promotion, formation or management of any company.[1] The applicant must give at least ten days' notice of his intention to make an application under this subsection.

1 Section 160(4) of the Companies Act 1990.

12.122 Given the provisions of s 160, company directors need to be even more vigilant in relation to the health and safety requirements provided for in the relevant legislation. (See also the Australian case involving the prosecution of the CEO of a large corporation, *Ken Kumar v David Aylmer Ritchie* (2006)).[1]

1 See **12.117** above.

RELATIONSHIP WITH CIVIL LIABILITY

12.123 The 2005 Act has changed the dynamic between criminal and civil liability arising from health and safety offences by remaining silent on whether breach of a statutory duty could give rise to a civil claim in damages.[1] This means that it is open to individuals to initiate a civil cause of action for damages resulting from a breach of any of the statutory health and safety duties, where injury has occurred as a consequence.

1 Section 60 of the 1989 Act expressly provided that contravention of any of the duties under ss 6 to 11 (corresponding to the employer's duties under s 8 of the 2005 Act, duties owed to persons other than one's own employees under s 12 of the 2005 Act, duties of persons in control of a place of work under s 15 of the 2005 Act, duties of employees under s 13 of the 2005 Act, duties of designers, manufacturers and suppliers under s 16 of the 2005 Act, and duties of persons who design or construct places of work (substantially amended but corresponding to s 17 of the 2005 Act)) would not confer any right in civil proceedings.

CORPORATE MANSLAUGHTER

12.124 The Law Reform Commission (LRC)[1] issued a Report on Corporate Killing in October 2005,[2] in which it recommends that companies should be subject to liability for corporate killing, which would ensure an undertaking could be held responsible for a death arising from gross recklessness. To date, this has proven to be very difficult where large enterprises were concerned, given the inability to identify a single individual who might be held responsible where many minds were involved in the managerial workings of the undertaking. It is acknowledged that, where a company is held criminally liable for corporate manslaughter, this may be the result of either a level of gross negligent manslaughter by individual(s) within the company or culpability to a degree that warrants a sanction without being liable for manslaughter.

1 The Law Reform Commission (LRC) is an independent statutory body which was established pursuant to s 3 of the Law Reform Commission Act 1975. Its aim is to keep the law in Ireland under review and to make practical proposals for its reform.

2 The 2005 Report followed and built upon the LRC's work in preparing the 'Corporate Killing Consultation Paper' (LRC CP 26–2003) published in October 2003. It also took account of relevant comments made by interested parties during the consultation process and, in particular, at the LRC's seminar on corporate killing held in December 2004.

Corporate manslaughter – the proposals

12.125 The LRC recommends that a precise statutory definition be introduced and that the statutory offence be called 'corporate manslaughter'. Rather than having the offence apply to a 'company' (ie one incorporated under the Companies Acts), the LRC recommends that it should also apply to other organisations of a similar nature, and uses the term 'undertaking' which is defined as 'a person being a body corporate or an unincorporated body of persons engaged in the production, supply or distribution of goods or the provision of a service, whether carried on for profit or not'.

12.126 The offence of 'corporate manslaughter' would make an undertaking responsible for a death arising from its gross negligence. In turn, gross negligence means that the undertaking was reckless as to the danger involved in an act, or omission, when the danger would have been obvious to any reasonable person.

The LRC goes on to recommend two new statutory offences:

(1) statutory offence of corporate manslaughter for corporate entities; and
(2) secondary offence for corporate managers who play a role in the commission of the corporate offence.

12.127 Corporate manslaughter would involve gross negligence, as it is the most appropriate mental element in that it only applies to manslaughter. The leading case in this area is the case of *AG v Dunleavy*[1] which related to the prosecution of a taxi driver for causing death by 'negligent driving'. At first instance, the court directed the jury that:

> '… in order to establish criminal liability the facts must be such that in the opinion of the jury the negligence of the accused went beyond a mere matter of compensation between subjects, and showed such a disregard for the lives and safety of others as to amount to a crime against the State and conduct deserving punishment …'

The accused was convicted to 18 months' imprisonment. He applied to the CCA for leave to appeal and, during the hearing of his application, the CCA decided that the trial judge had erred in his direction to the jury. The CCA went on to state that, when considering whether an act constituted gross negligence, a court should consider whether each of the following four elements exist:

(1) ordinary negligence;
(2) negligence causing death;
(3) negligence of a very high degree; and
(4) negligence involving a high degree of risk or likelihood of substantial personal injury.

1 [1948] IR 95.

12.128 In its Report, the LRC recommends that three elements should form the basis of the test of corporate liability for manslaughter, as follows:

(1) the undertaking was negligent;
(2) the negligence was of a sufficiently high degree to be characterised as 'gross' and so warrant criminal sanction; and
(3) the negligence caused death.

12.129 The corporate offence is based around the definition of gross negligence in *Dunleavy*, and the proposed statutory offence would hold an undertaking liable where:

(a) it owed a duty of care to the deceased;
(b) it failed to meet the standard of care;
(c) the breach was of a very high degree and involved a significant risk of death or serious personal harm; and
(d) the breach of duty caused the death.

12.130 As to whether a duty of care is owed, consider the common law or statutory duties imposed on the undertaking (such as duties owed as an employer, occupier of land or producer of goods). In addition, when assessing whether a high managerial agent ought to have known of a risk, regard should be had to the high managerial agent's actual and stated responsibilities.

12.131 Gross negligence in criminal law is based on the same principles as ordinary negligence in tort law, as both require:

(a) a duty of care;
(b) breach of that duty;
(c) causation; and
(d) damage.

However, the conduct involved must be sufficiently egregious to constitute criminal wrongdoing, and the damage involved must be death.

12.132 Regarding the standard of care, the LRC recommended that, where the standard is assessed, reference should be had to the particular area of the profession that is involved (ie the standard of care required of a corporate entity can be assessed by reference to the standards used in that particular field).

12.133 Actions and decisions of individuals who have a significant influence over the operations of the undertaking must also be considered in assessing corporate liability. Therefore, the LRC recommends that provision should be made for the secondary liability of highly placed individuals who contribute to the commission of the corporate manslaughter.

12.134 The Report sets out a non-exhaustive list of factors to be taken into account by the court when considering whether the undertaking has met the standard of care, to include: the way in which the activities of the undertaking are managed or organised by its high managerial agents; the allocation of responsibility within the undertaking; its procedural decision-making rules and policies; the training and supervision of employees by the undertaking; and the response of the undertaking to previous incidents involving a risk of death or serious personal death, etc.

12.135 The Report also gives consideration to various ways in which an undertaking may be sanctioned. It includes, in the recommendations, that a pre-sanction report be prepared in advance of imposing a sanction. The sanctions recommended include:

(a) unlimited fine;
(b) remedial order;[1]
(c) community service orders;
(d) adverse publicity orders;
(e) restraining orders; or
(f) injunctions.

The hope is that some of these options would enable the tackling of the source or cause of the manslaughter, which, in turn, might reduce the risk of the same situation occurring again.

1 The LRC gives a list of suggested actions which might be considered by way of any remedial order made.

Secondary liability of high managerial agents

12.136 The LRC has proposed a secondary offence, and the proposed term assigned to this offence is 'grossly negligent management causing death'. Where an undertaking has been convicted of corporate manslaughter, its high managerial agents would also be guilty if:

(a) they knew or ought to have known of a substantial risk of death or serious personal harm;
(b) they failed to make reasonable efforts to eliminate that risk;
(c) that failure fell far below what could reasonably be expected in the circumstances; and
(d) that failure contributed to the commission of the corporate offence.

12.137 In defining who is a 'high managerial agent', the LRC considered the definition of a director, manager or other person purporting to act in that capacity, as provided for in section 80 of the 2005 Act.[1] The LRC commented in its Report that the definition in the 2005 Act excluded outside specialists not in the direct employment of the undertaking, and upon whose advice undertakings regularly rely, and it felt that this was not a satisfactory position where 'failure by the outside expert could be a substantial contributing factor to the commission of a corporate manslaughter'. It therefore recommended the following revised definition of a 'high managerial agent':

> 'A person being a director, manager or other similar officer of the undertaking, or a person who purports to act in any such capacity, *whether or not that person has a contract of employment with the undertaking*'. (emphasis added)[2]

This extension, if included in the final enactment, would have serious repercussions for all professional advisors or contractors.

1 See **12.60** to **12.61** above.
2 The first part of this definition is a repeat of the 2005 Act definition of a 'director'; the portion in italics is the extension to this definition proposed by the LRC.

12.138 The proposals go on to recommend that, should a high managerial agent be found guilty, it would render them liable to: unlimited fines; a term of imprisonment for up to 12 years; and/or disqualification from acting in a managerial capacity for such period of time as the court see fits.[1]

1 Nothing in the Report, however, should prevent the prosecution of an individual for the primary common law offence of manslaughter where that individual has sufficient culpability for that offence.

12.139 The LRC also makes certain recommendations for action to be taken where an individual is found to be in breach of a disqualification order[1] and that, where an individual knew someone was acting in breach of such order, that individual should, in turn, be subject to a disqualification order too.

1 Such as a period of imprisonment not to exceed two years, a fine not exceeding €1 million and/or a further period of disqualification of ten years.

12.140 Although the LRC Report has been around since October 2005, it was not until 4 April 2007 that the draft Bill (which was appended to the Report and which has undergone very slight amendment) was put before the Irish Government for the purpose of going through the various stages towards complete enactment. The Bill reached the committee stage,[1] overseen by delegates from 12 different governmental departments, but did not progress. Given this lack of progress, a Private Member's Corporate Manslaughter Bill was put forward by a member of the Senate in May 2013. Despite this effort to push the legislation forward, it has only reached the second stage of the legislative process and does not appear to have picked up any momentum. The political and social pressure for creating an offence of corporate manslaughter appears to have waned.

1 This is the third of five stages of the legislative process in Ireland. By the time this stage comes around, the Bill has gone through stage one (introduction of the Bill) and stage two (the debating of the proposed Bill). The Bill is yet to go through the 'report stage' and the final (enactment) stage.

CONCLUSION

12.141 As can be seen, the Irish framework for health and safety legislation, whilst developed from a similar premise, differs significantly from the UK equivalent framework. In more recent years, the rate of legislative change has slowed, although there have continued to be a number of important developments. This has happened largely through the nature and extent of prosecutions brought for health and safety offences, with a notable emphasis on securing greater personal accountability, particularly for those in managerial positions or with responsibility for safety. In the next few years, this trend is likely to continue, and we are also likely to see further refinements of the various statutory provisions regulating health and safety in Ireland. The HSA is also likely to continue its use of Codes of Practice as a means for setting standards and providing guidance to assist compliance in specific sectors or in respect of specific activities. We expect health and safety regulation in Ireland to continue to be very dynamic.

Chapter 13*

Scotland

Gilles Graham, Simpson & Marwick, Glasgow

* Chapter originally contributed by Paul Wade, Simpson & Marwick, Glasgow

INTRODUCTION

13.1 The Scottish legal system remains a mystery to many from the rest of the UK. Many lawyers in Scotland still believe that legal proceedings, like opera and church services, are at their most impressive when conducted in a language which the audience does not understand!

13.2 There are many differences of terminology, which usually have direct equivalents in other systems. For example, someone who is the subject of a prosecution is known as the 'accused', if he is being prosecuted summarily and in the first phase of a prosecution on indictment; once the indictment has been served, he is known as 'the panel'. 'Judicial expenses' are the equivalent of 'costs'.

13.3 Putting differences of terminology aside, what other differences are there in this branch of the law? Adjectival law – procedure, personnel, the court system, etc – is significantly different. The substantive law, where it is based on common law, is fundamentally different. The substantive law, where based on statute or regulation in the field of health and safety, is essentially identical. English authorities on the interpretation of UK statutes and regulations, although not binding on Scottish courts, are highly persuasive.

13.4 As any UK lawyer will be generally aware, the number of landmark Scottish judicial decisions generally is disproportionate to the population. The same is true in health and safety law. Many concepts such as that of 'reasonable practicability' were tested at an early stage, when Scotland had a high volume of the heavy industry to which the statutes and regulations containing these expressions were directed.

13.5 In 1999, Scotland acquired its own Executive and Parliament. The Scottish Parliament is entitled to legislate on any issue which is not reserved to Westminster. However, as 'health and safety' is a reserved matter, it is unlikely that, in relation to health and safety statutes and regulations, the law applicable in Scotland will diverge from that in the rest of the UK.

13.6 Curiously, one of the first areas of confusion over what was or was not reserved appears to have arisen in the context of the reform of the so-called corporate killing law. As noted above, health and safety is an issue reserved for Westminster. The general criminal law is devolved to the Scottish Parliament.

13.7 In 2005 the then Scottish Justice Minister set up a 'Panel of Experts' to consider reform of corporate killing law in Scotland. This was presumably on the basis that it was perceived that the Scottish Parliament could legislate in this area because it was part of the general criminal law. The Panel produced a report on 17 November 2005, which contained proposals for reform including many proposals that were regarded by some at least as draconian. The disadvantage of having different corporate killing regimes in force in Scotland than in the rest of the UK was met by the Executive's repeat of the mantra 'Scottish solutions for Scottish problems'.

13.8 Ultimately, however, Westminster took the view that corporate killing reform fell under the umbrella of health and safety and, as such, was a reserved matter – hence the Corporate Manslaughter and Corporate Homicide Act 2007 was introduced, which applies throughout the UK.

13.9 There was, nevertheless, one practical consequence of this debacle. The 2007 Act provides that it is no longer competent to attempt to prosecute an organisation in England and Wales for the common law crime of manslaughter. However, there is no equivalent provision in the 2007 Act for Scotland, and it remains competent to prosecute an organisation for culpable homicide (the Scottish equivalent of manslaughter) or indeed any other common law crime. Presumably, the reason for this apparent anomaly is that any provision affecting the common law in Scotland would be devolved to the Scottish Parliament.

THE COURTS AND PERSONNEL

13.10 Some matters are important to recognise immediately. In Scotland, in criminal proceedings, no awards of judicial expenses (or costs) can be made at first instance. In an appeal in summary procedure, the High Court may award expenses both in the High Court and Appeal Court as it may think fit.[1] This power is seldom exercised.

1 Criminal Procedure (Scotland) Act 1995, s 183(9).

13.11 In solemn proceedings, there are no awards of expenses, either at first instance or on appeal.[1]

1 Criminal Procedure (Scotland) Act 1995, s 128.

13.12 In health and safety prosecutions, the judge will be legally qualified. Lay magistrates are restricted to District Courts where such prosecutions are not generally raised.

13.13 All prosecutions in Scotland are presented by legally qualified professional prosecutors. HSE inspectors have no locus to prosecute or decide whether a prosecution takes place.

13.14 In HSE prosecutions, there is no effective remit available from one procedure or court to another, whether at the instance of the prosecution or defence or the court itself.

The courts

13.15 The criminal court system in Scotland is wholly different. There are three courts with criminal jurisdiction, as set out below.

District Courts

13.16 The District Court is the lowest in status of the criminal courts. It is presided over by one or more lay justices or, in a few areas, by a stipendiary magistrate. It can only deal with prosecutions by way of summary complaints. Although health and safety prosecutions are competent in the District Court, as a matter of practice, they are not initiated there. If, exceptionally, a health and safety prosecution was to be raised in the District Court, the procedure would be similar to that for summary procedure in the Sheriff Court.

Sheriff Courts

13.17 Scotland is divided into six Sheriffdoms based on local government areas. Each Sheriffdom has a Sheriff Principal who is responsible for the administration of the court business within the Sheriffdom and the hearing of civil appeals.

13.18 Each Sheriffdom has within it a number of Sheriff Court districts. In all, there are 49 districts, each of which has a Sheriff Court. The size of the various courts ranges from courts such as Glasgow, said to be the busiest court in Europe, with more than 25 full-time Sheriffs, to courts, such as Lochmaddy in the Western Isles, which might sit only one or two days each week.

13.19 The Sheriffs who preside are appointed from the ranks of experienced advocates or solicitors.

13.20 The jurisdiction of the Sheriff Court is wide. Its civil jurisdiction includes every type of financial claim without limit of value.

13.21 It also has an extensive criminal jurisdiction. In summary proceedings the Sheriff can impose penalties including a fine (up to the 'prescribed sum', currently £20,000) or imprisonment up to 12 months. The Sheriff can also deal with solemn proceedings (or proceedings on indictment) and, on conviction, impose penalties including a fine (without limit) and imprisonment of up to five years. These limits are subject to any limits imposed in the statute or regulation which creates the offence.

13.22 Because of its wide jurisdiction, it is the Sheriff Court which deals with the overwhelming majority of health and safety prosecutions. However, it should be clear that proceedings for corporate homicide can only be commenced in the High Court of Justiciary.[1]

1 Corporate Manslaughter and Corporate Homicide Act 2007, s 1(7).

13.23 In addition to a wide civil and criminal jurisdiction, the Sheriff Court has a number of additional functions. Most important of these, in the present context, is the hearing of Fatal Accident Inquiries (the nearest Scottish equivalent to an inquest).

13.24 By way of illustration, if a fatal accident was to occur, the same Sheriff Court (and potentially the same Sheriff) might:

(a) preside over a health and safety prosecution, either on summary complaint or indictment;

(b) hear a Fatal Accident Inquiry into the circumstances of the death and make determinations; and
(c) deal with and determine civil proceedings by the family of the deceased.

The High Court of Justiciary

13.25 The High Court of Justiciary is the superior criminal court and has jurisdiction over all Scotland. It consists of the Lord Justice General, the Lord Justice Clerk and a number of Lords Commissioners of Justiciary. It hears only solemn proceedings of the most serious nature. The High Court is based in Edinburgh but it also goes on circuit as required.

13.26 In a health and safety context, the only offence which requires to be prosecuted in the High Court is corporate homicide.

13.27 In recent years, although the majority of health and safety prosecutions are raised in the Sheriff Court, prosecutions in cases involving multiple fatalities or high degrees of culpability have occasionally been raised in the High Court.

13.28 The High Court also sits as an appellate court, hearing appeals from summary cases in the District and Sheriff Courts and from solemn proceedings in the Sheriff Court and the High Court.

13.29 There is no further appeal to the Supreme Court from the High Court of Justiciary, except on devolution issues.

Remits between courts

13.30 Summary proceedings, whether initiated in the District or Sheriff Court, cannot be remitted to another court or to another procedure, whether for sentence or otherwise.

13.31 In solemn proceedings in the Sheriff Court, the Sheriff has power to remit to the High Court for sentence. However, this power can only be exercised if the Sheriff considers that any competent sentence he can impose is inadequate.[1] A purported remit to the High Court, where the maximum penalty for the particular offence was within the Sheriff's powers, was held incompetent.[2] Accordingly, having regard to the Sheriff's powers and the penalties available in the Health and Safety (Offences) Act 2008, remits in health and safety prosecutions should not take place.

1 Criminal Procedure (Scotland) Act 1995, s 195(1).
2 *HMA v Stern* 1974 JC 10.

13.32 It should also be noted that the accused has no right to elect for jury trial in summary proceedings. The decision in which procedure, and in which court, to raise a health and safety prosecution is a matter exclusively for the Crown.

Prosecution personnel

13.33 In Scotland, in all cases, prosecutions are conducted by professionally qualified prosecutors. The HSE has no power to initiate prosecutions even at summary level. Private prosecutions are extremely rare.

13.34 In practice, the HSE (and, in some cases, the police) will submit a report to the Crown and it is the Crown which decides whether proceedings are to be initiated and, if so, under which procedure and in which court.

13.35 The Lord Advocate and the Solicitor General (known as the 'Law Officers') are members of the Scottish Executive and are responsible for the conduct of prosecutions in Scotland. The Lord Advocate has a number of assistants known as Advocates Depute. They are experienced advocates or solicitors with extended rights of audience. The Law Officers and the Advocates Depute are referred to as 'Crown Counsel', and they and their administration are referred to as 'Crown Office'.

13.36 A Procurator Fiscal, who also usually has a number of Deputes, is based in each Sheriff Court district. The Deputes too are legally qualified. They are known, virtually universally, as 'PFs'. PFs hold a commission from the Lord Advocate.

13.37 In a health and safety context, where an investigation takes place, the investigating agency – typically the HSE, Police, Scottish Environment Protection Agency or Environmental Services Department of a local authority – report to the PF. In serious cases, there may be an interim report and the PF may involve himself in the investigation. When the investigation is complete, a report is sent to the PF who considers it. He may at that stage decide further investigation is required and direct that it be undertaken by one or other of the investigation agencies. When he is satisfied with the investigation, he will decide on the next stage. If he decides not to prosecute, that is an end of the matter. If he decides to prosecute under summary procedure, he may initiate proceedings without reference to Crown Counsel. If he considers that a prosecution on indictment is appropriate, or if the accident involved a fatality, he will refer the matter to Crown Counsel.

13.38 One of the important functions of Crown Counsel is to 'mark cases', that is to read the report and the recommendations of the PF and decide whether to commence proceedings and, if so, under which procedure and in which court. Crown Counsel might also require further investigations to be undertaken before reaching a decision, in which event they refer the matter back to the PF.

13.39 If proceedings are to be commenced by indictment, the PF is instructed and authorised to proceed and direct whether the proceedings are to be in the Sheriff Court or High Court. If the proceedings are in the Sheriff Court, the PF appears as prosecutor. If proceedings are raised in the High Court, Crown Counsel appears as prosecutor.

13.40 The PF operates through a dedicated specialist Health and Safety Division for the investigation and prosecution of all cases reported to the PF by the HSE and other investigation agencies. The Division consists of three teams based in Aberdeen, Edinburgh and Glasgow. It deals solely with potential health and safety prosecutions, and liaises closely with the HSE and other enforcement bodies.

13.41 It will be seen that the decision whether to prosecute at all and, if so, under which procedure and in which court is exclusively a matter for the Crown. The Crown Office have issued Prosecution Policy and Guidance,[1] which gives guidance of a general nature, particularly in regard to what is or is not 'public interest'.

1 See www.crownoffice.gov.uk/publications/prosecutions-policy-and-guidance.

Defence personnel

13.42 Accused persons in Health and Safety prosecutions, like any other type of prosecution, can be represented in the Sheriff Court by an advocate or solicitor. In the High Court, they can be represented by an advocate or solicitor with extended rights of audience.

13.43 The primary area of difficulty which arises is in relation to potential conflicts of interest where there are multiple accused or where the HSE is conducting interviews under caution with a number of representatives of the one organisation. The Law Society of Scotland requires that solicitors do not act for two parties where there is a conflict of interest. However, this leaves open, in any particular case, whether a conflict of interest actually arises.

13.44 There is, for example, no blanket ban on acting for an organisation and one of its managers at the same time, provided that there is no conflict between the interests of the two parties. Each case requires to be examined on its own merits and, in case of doubt, solicitors might well wish to confer with the Law Society of Scotland's Professional Practice Department and the Faculty of Advocates with the Dean of Faculty.

The Health and Safety Executive

13.45 The Health and Safety Executive ('HSE') has no locus to appear before the criminal courts in Scotland as a party. Its role is limited to investigation and the provision of a report to the Procurator Fiscal. Frequently, the HSE appears as a witness of fact and expert witness in trials.

13.46 The powers of HSE Inspectors contained in s 20 of HSWA 1974 apply in Scotland without modification.

13.47 Although not entitled to appear in court in connection with a prosecution, an HSE Inspector is entitled to appear and call evidence at a Fatal Accident Inquiry.[1]

1 Fatal Accident and Sudden Deaths Inquiry (Scotland) Act 1976, s 4(2).

SUBSTANTIVE LAW

13.48 It might be useful for those involved in health and safety prosecutions in Scotland to classify the substantive law as falling into one of three 'blocks'.

13.49 The first block is the common law, which is different in form and content from the English common law. Common law offences are relevant to individuals and are still, potentially, relevant to organisations. Whether an accident is necessary for the offence to be committed depends on the nature of the offence (see below). Mens rea (or 'wicked intention') is a requirement of all common law offences in Scotland.

13.50 The next block is the HSWA 1974 and the 'relevant statutory provisions'.[1] They are relevant both to organisations and individuals. Generally, they are 'result based' in the sense that they require a certain result to be achieved (eg that safety of employees is ensured). They are strict liability offences, in that there is no requirement for mens rea. Where defences exist, they are usually limited (such as reasonable practicability). The offence is generally committed where the result prescribed is not achieved, whether an accident occurs or not.

1 As defined in HSWA 1974, s 53(1).

13.51 The third block is the offence of 'corporate homicide' created by the Corporate Manslaughter and Corporate Homicide Act 2007. As already noted, this applies only to certain organisations; it has no application to individuals. Although there is no requirement for mens rea, a conviction implies a systemic failure on the part of senior management. A fatal accident has to be caused by the failure.

Common law

13.52 Scots law ultimately did follow English law in holding that it is possible to convict a company of a common law offence by using the concept of the 'controlling mind'.[1]

1 *HMA v Transco plc* 2004 JC 29.

13.53 In England, it is no longer possible to prosecute an organisation for the common law crime of manslaughter.[1] There is no record in England of any attempt to prosecute an organisation for any other common law crime.

1 CMCHA 2007, s 20.

13.54 In Scotland, the position is different. CMCHA 2007 does not prevent a prosecution of an organisation for culpable homicide at common law. Moreover, there are two recorded instances in Scotland of companies being prosecuted for other common law offences.

13.55 The first was the case of *Dean v John Menzies* (*Holdings*) *Ltd*.[1] In Scotland, there exists a rather quaint common law crime of 'shameless and indecent conduct'. A limited company, operating as a newsagent, was charged with this offence for having for sale various magazines regarded by the prosecution as being too sexually explicit.

1 1981 JC 23.

13.56 The company was prosecuted under summary procedure for shameless and indecent conduct. It took a preliminary objection (a plea to the competency) which the Sheriff sustained and dismissed the complaint. The Crown appealed to the High Court, which by a majority dismissed the appeal (ie it agreed that the complaint should be dismissed without trial). Although the decision came after the House of Lords decision in *Tesco Supermarkets v Nattrass*,[1] the Scottish Appeal Court did not apply that approach to the particular offence charged. As one of the judges in the majority put it, 'many as are the attributes which have been imputed to a company, a sense of shame has never been regarded as one of them'. There was, however, a powerful dissenting judgment by Lord Cameron.

1 [1972] AC 153.

13.57 The second case was *Purcell Meats* (*Scotland*) *Ltd v McLeod*.[1] In that case, a limited company was charged, again on summary complaint, with the common law crime of fraud. The thrust of the charge was that the company had obliterated markings on beef carcasses, which were relevant to a subsidy then available. The preliminary objection by the accused company to the complaint was repelled both by the Sheriff and then on appeal by the High Court. The court's decision was that a company could commit a fraud. Whether the requisite mens rea could be attributed to the company depended on the evidence and the principles laid down in *Tesco Supermarkets* (above). The case was sent for trial. In the end, the case did not proceed to trial. There is no report of what became of the case.

1 1987 SLT 528.

13.58 The only attempt to prosecute an organisation in Scotland for a common law offence as a result of a death was *HMA v Transco plc*.[1] In that case (which was initiated under solemn procedure in the High Court of Justiciary), the charge was culpable homicide at common law resulting from an explosion caused by a leaking gas main. The judge at first instance repelled the defence objection. The matter was then appealed to the High Court who sustained the objection and dismissed as irrelevant the culpable homicide charge. It held that mens rea remains a necessary and significant element in culpable homicide. It was competent to charge an organisation with a common law offence. However, in this case the Crown had attempted to 'aggregate the separate states of mind of various individuals and various groups of individuals'. That was not legitimate and was contrary to the basic tenets of Scottish criminal law.

1 2004 JC 29.

13.59 In summary, therefore, in relation to companies and other organisations the common law is still potentially relevant. In practice, however, in any accident involving a fatality, it is likely that, if the degree of culpability is sufficiently serious, the Crown would bring a charge of 'corporate homicide' against the company or other organisation under the CMCHA 2007.

13.60 What is perhaps more likely in this context is a prosecution against a company or other organisation where a serious accident occurs, not causing a fatality, but involving a high degree of culpability, attributable to a controlling mind – in other words a 'near miss'. In that event, a charge against the company of the common law crime of 'culpable and reckless conduct' might be considered.

13.61 Culpable homicide and culpable and reckless conduct are the two common law offences considered below. They are certainly relevant in a health and safety context to individuals and, as noted above, may be relevant to companies and other organisations if the necessary state of mind can be attributed to their controlling mind.

Culpable homicide

13.62 Only in the broadest terms is the common law crime of culpable homicide equivalent to the English crime of manslaughter.

13.63 A charge of culpable homicide can arise in a number of ways. If, at the time when the accused kills, he is suffering from diminished responsibility or acting under provocation, he would be convicted of culpable homicide rather than murder.

13.64 In cases where the accused assaulted his victim but had no intention of killing him, but does, he may be convicted of culpable homicide.

13.65 There is, however, another category of culpable homicide – the so-called 'lawful act' culpable homicide. This is where the accused acts in a way which would otherwise be lawful but, in the circumstances of the particular case, is criminal and a death occurs.

13.66 When the definition of culpable homicide was last considered by the High Court in Scotland, it was said that it was 'fundamentally different' from the definition in England. It was said that the English definition involves:

> 'first, the application of the ordinary principles of the civil law of negligence, in order to ascertain whether or not the defendant had been in breach of a legal duty of care towards the deceased victim. The Scottish definition contains no counterpart to that ... Thereafter the question which arises in the application of the English definition is whether the breach of duty should be characterised as gross negligence and therefore a crime. While both definitions share the requirement that the conduct under consideration must be of a grave kind, in order to be classified as criminal ... in some degree, the English definition involves circularity. Happily the most modern formulation of the test adopted in Scotland has avoided that problem. Furthermore the Scottish formulation implies clearly ..., a certain state of mind on the part of the perpetrator that is to say, mens rea, in accordance with the basic principles of Scots criminal law. On the other hand the English approach seems to involve an objective assessment of the conduct under consideration alone.'[1]

The state of mind has to be that of someone who shows 'utter disregard of what the consequences of the act in question may be so far as the public are concerned' or, alternatively, 'recklessness so high as to involve an indifference to the consequences for the public generally'.

1 *HMA v Transco plc* 2004 JC 29.

Culpable and reckless conduct

13.67 In the context of health and safety, this can be regarded as the 'near miss' common law offence.

13.68 In *Normand v Robinson*,[1] two individuals were charged on summary complaint with culpable and reckless conduct. The substance of the offence was that they promoted and organised what was then called a 'rave' in a derelict warehouse which was awaiting demolition. It was said that there was no mains electricity, lighting was provided by candles, there was no means of alarm for fire, no fire escape, or fire-fighting equipment, the floors had holes and the place was littered with debris. It was not suggested anyone was in fact injured. The High Court agreed that this was a relevant charge and the case was sent for trial (no report exists of the result of the trial).

1 1994 SLT 558.

13.69 Another unreported example was a charge of culpable and reckless conduct brought against a butcher at the centre of the e-coli outbreak in North Lanarkshire. The thrust of the charge against him was that he supplied meat when he was aware of the possibility that the meat may contain the e-coli organism. In the result, none of those to whom the meat was supplied had died (otherwise, the charge would presumably have been culpable homicide). The legal relevancy of the charge was not challenged. The case proceeded to trial and the butcher was, on the facts, acquitted.

13.70 Although there is no reported instance of a charge of culpable and reckless conduct being brought against an organisation, presumably if the butcher had formed himself into a limited company a charge could have been brought against it in similar terms (subject to satisfying the 'controlling mind' criteria).

The HSA and the relevant statutory provisions

13.71 As already noted, HSWA 1974 and the relevant statutory provisions apply without modification within Scotland. In practice, English and Scottish authorities on the interpretation of the statute and regulations are cited interchangeably.

Corporate Manslaughter and Corporate Homicide Act 2007

13.72 This Act applies throughout the UK.

13.73 The offence which it creates is, in Scotland, known as 'corporate homicide', and in the rest of the UK as 'corporate manslaughter'. There is no other difference – the constituent elements of the offence are precisely the same.

13.74 It is to be noted that the offence is closely based on the English common law of manslaughter. This will involve the Scottish High Court applying concepts quite new to Scottish criminal law (the existence of a duty of care, gross negligence, etc).

13.75 In Scotland, corporate homicide can only be indicted under solemn procedure in the High Court of Justiciary.

INVESTIGATION

13.76 The police are usually first on the scene of an accident and carry out initial investigations. They usually cede control of the investigation at an early stage to the HSE or to the Environmental Health Department once the immediate circumstances are established and the police are satisfied that any criminal offence which has been committed falls under health and safety legislation.

13.77 For the investigation of work related deaths, there is a protocol for investigation between the police, HSE and others, including some local authorities.[1]

1 *Work-related Deaths: A Protocol for Liaison* (HSE).

13.78 Both the police and the HSE report to the Procurator Fiscal. In the case of a serious or fatal accident, the Procurator Fiscal will be informed immediately. He may visit the locus and personally give some direction to the investigation. The role of the HSE and police in relation to the Procurator Fiscal is similar: they investigate and report to him.

13.79 Usually, the more detailed investigation is undertaken by the HSE. That is essentially for two reasons. First, it is assumed that the HSE has the expertise and resources to undertake what may be a technical investigations. Secondly, it has statutory powers to demand information from witnesses.

HSE INTERVIEWS

13.80 An interview of an employee or director by an HSE Inspector can take one of three forms:

(1) First, the interview can be informal. The witness may answer questions in which event his answers are usually noted and he may be asked to sign an account of the interview.
(2) Secondly, the interview can be conducted under s 20 of the HSWA 1974. As has already been noted elsewhere, the witness is obliged to answer the questions put and is obliged to sign a statement certifying the truth of his answers.
(3) Finally, the interview may be under caution.

13.81 The terms of the caution are not statutorily regulated but are to the effect that the suspect is not obliged to say anything but anything he does say will be noted and may be used in evidence.

13.82 The question of whether a statement made by a suspect is admissible in subsequent proceedings against him depends on whether it was fairly obtained and the caution is only one of a number of factors to be considered. Following the decision of the Supreme Court in the *Cadder* case,[1] it seems unlikely that a statement made by a suspect will be admissible unless the suspect was given the opportunity to take advice from a solicitor before being interviewed. The Police and Criminal Evidence Act 1984 does not apply in Scotland.

1 *Cadder v HM Advocate* [2010] UKSC 43.

13.83 The product of each of these interviews – informal, s 20 and under caution – is the production of a 'statement'.

13.84 Scots law draws a distinction between a 'statement' and a 'precognition'. The distinction is not always an easy one to maintain. The theory of the distinction is that a statement contains the words of the witness, without explanation or embellishment, usually taken at an early stage by someone who is investigating the case and before any charge has been brought. A precognition, on the other hand, is the evidence of the witness generally taken at a later stage when a case is being constructed against an accused and usually taken by someone who is trying to build or defend a charge.

13.85 Generally, the police and HSE, when they interview a witness, take a statement.

13.86 If the Procurator Fiscal or the defence solicitor chooses to interview a witness to obtain further information, that would normally be a precognition.

13.87 A witness may have a statement put to him if its content is different from the evidence given by him in court. If his response is that the evidence given by him in court is the truth and that what is contained in the statement is, for whatever reason, not correct, then the terms of the statement do not add to the case against the accused.[1] However, such an exchange is likely, and often calculated, to damage the credibility or reliability of a witness upon whom the defence may be relying. The usual purpose of putting a prior statement to a witness is to get the witness to adopt what is said in it. This is usually achieved by putting the matter to the witness in instalments ('Were you interviewed by the HSE? Did you tell them the truth? Did they write down what you told them? Did you sign the statement and certify that it accurately recorded what you told them? Is this the statement?' etc). Most witnesses accept the accuracy of their statements, in some cases perhaps for fear of the perceived consequences of maintaining that it is not accurate. This underlines the need to stress to witnesses that they must be very careful in what they say and sign when interviewed by the police or HSE.

1 See generally Criminal Procedure (Scotland) Act 1995, s 263.

13.88 The circumstances where a statement can be introduced as evidence against an accused are very limited and mainly restricted to situations where a witness cannot be called to give evidence.[1]

1 See generally Criminal Procedure (Scotland) Act 1995, s 259.

13.89 Information given to the defence is usually in the form of precognition and therefore cannot normally be put to witnesses if the evidence of the witness in court varies. By the time that defence solicitors are instructed, a charge has usually been decided upon or brought. Information given to defence lawyers by a witness at interview will usually be regarded as a precognition.

13.90 If, in the course of the defence investigation, it is considered that a witness is likely to change his evidence, it is worthwhile asking him to read over his precognition, correct it and sign it. Although the matter is not without doubt, that might convert it from a precognition to a statement and enable it to be put to the witness, should his evidence in court diverge from what he told the defence.

13.91 The other significance of the distinction between statements and precognitions is in relation to disclosure. The Crown are now obliged to disclose to the defence the statements which they have obtained insofar as they are likely to be of material assistance to the defence. As a matter of practice, usually all statements are disclosed. There is no duty to disclose precognitions although, as part of the Crown's general duty of disclosure, they should advise the defence of any relevant information even though it only appears in a precognition.

13.92 Quite complex issues can arise when a solicitor is instructed at the initial stage of an HSE investigation into a serious accident.

13.93 The HSE has no obligation to allow a legal representative to be present at an informal interview or an interview under s 20. However, some HSE Inspectors are prepared to permit a solicitor to be present if the witness is agreeable to that. This is a sensible attitude in most cases, because it gives the solicitor an idea of the nature of the investigation being undertaken by the HSE and prevents the witness from having to repeat his evidence at a time where he may be extremely upset by the recent accident circumstances.

13.94 A solicitor permitted to be present at such an interview should, of course, bear in mind that he is not there of right and that he has no locus to intervene.

13.95 He should also consider carefully any issue of conflict of interest. Some conflicts are patent, as where an individual seeks to shift responsibility to a fellow employee or to his employer. Some, however, may be latent. A manager may be put forward as, in effect, the spokesman for his employer. He may have reassured his colleagues following the incident being investigated that the organisation's procedures are in order. Faced with contrary evidence at interview, his position may change. He may blame someone else or his employers. This will raise a potential conflict between the interests of the employers on the one hand and the employee on the other. The interview may require to be terminated and separate representation sought for the employee.

13.96 A solicitor is typically instructed by the employer and/or the employer's insurers. At present, the Law Society of Scotland does not preclude the same solicitor acting for a company and its employees, provided no conflict of interest exists. This matter is currently under consideration by the Law

Society of Scotland, as prosecuting authorities are seeking the same situation as prevails south of the border. If an employee is or should be concerned about the possibility of an individual prosecution, the solicitor should proceed with the utmost care. He should explain to the company and to the individual what he is there to do and identify whether either of them has any objection or concern about him representing both. He should also exercise his own judgement as to whether there is a conflict.

13.97 There is also a place for pragmatism in this area. An HSE investigation may involve the interviewing of a substantial number of employees, any of whom may ultimately be proceeded against individually. Especially where there are technical issues, the involvement of a separate solicitor for each individual, who is properly briefed, may be simply impracticable.

13.98 In some cases, the appointment of a second firm of solicitors, whom witnesses are invited to consult on a wholly confidential basis, may be worth considering.

13.99 A solicitor funded by the employer's insurers and instructed by the employer may feel that he is under pressure, for reasons of economy or otherwise, to deal with the whole matter himself, without the expense of involving another firm. That pressure needs to be resisted if there is any possibility of a conflict of the interests of the parties to be represented. On the other hand, from the standpoint of fairness to the employee who may not understand the importance of a s 20 interview under caution, a lack of representation could prove important, and the solicitor may require carefully to consider what is best for both the employer and the employee. Where there is doubt, the solicitor should arrange for the employee to secure separate representation.

13.100 Typically the solicitor will be asked to speak to witnesses who are concerned about their own individual positions. The witnesses will seek advice about the types of interview, which they may encounter.

13.101 The solicitor will no doubt wish to advise witnesses to be interviewed under s 20 that nothing that they say in the course of that interview can be used in evidence against them. However, if a witness volunteers information which incriminates him, that can nevertheless be to the witness's disadvantage. The HSE, if it had not reviewed it before, will certainly consider the possibility of prosecution of the witnesses. It may be that it will be able to gather the factual basis for the prosecution from other witnesses. The HSE will also have the confidence of knowing that the witnesses have no answer to the charge.

13.102 Interviews under caution can be particularly anxious. A 'no comment' response is not always the right approach. If the HSE can readily obtain the answers to its questions from other sources, a refusal to answer may antagonise it and create the impression that the individual concerned either does not recognise, or seeks to conceal, his own part in the events leading to the accident. On the other hand, some witnesses are too keen to tell their stories and do not always understand the nature of the charge likely to be brought against them.

13.103 When the HSE has completed its investigation, it will produce a report with statements and productions attached and despatch it to the Procurator Fiscal. The report will generally make a recommendation. If the report recommends proceedings, usually the potential accused are advised of that recommendation by the HSE.

13.104 In cases involving serious accidents or death, the PF is obliged to report the matter to Crown Counsel who take the decision about a prosecution. In less serious cases, if the PF decides to initiate summary proceedings, he can do so himself without involving Crown Counsel. In either case, further enquiries may be made. Usually they are directed to the HSE. In some cases, they are directed to the police.

13.105 There is no time limit for this part of the process. It is not at all unusual, at least in the central belt of Scotland, for three years to elapse between the time of a fatal accident and the decision to initiate proceedings.

13.106 The time limit of six months for summary proceedings is disapplied unless the offence is 'triable only summarily'.[1]

1 Criminal Procedure (Scotland) Act 1995, s 136(2).

13.107 Ultimately, a decision will be taken on whether there is to be a prosecution, what the charges will be, whether the procedure will be solemn or summary, and in which court.

13.108 In the case of fatal accidents, the Crown can decide to hold a Fatal Accident Inquiry. Usually, if there is to be a prosecution, that takes place before any Fatal Accident Inquiry. However, that is not always the case. Although it is unusual, there have been instances where the Crown has initiated a prosecution after a Fatal Accident Inquiry has been held.

CHOICE OF PROCEDURE IN COURT

13.109 It is not intended here to attempt any comprehensive review of Scottish criminal procedure.[1] What is attempted is to outline the main areas of Scottish criminal procedure which are relevant to health and safety prosecutions.

1 For detailed information, it is suggested that the reader consults *Renton & Brown's Criminal Procedure* (Sweet & Maxwell, 6th edn, looseleaf).

13.110 Except in the case of a prosecution for corporate homicide under the CMHCA 2007 (where proceedings must be raised on indictment in the High Court of Justiciary), the prosecution has a choice:

(a) it can raise summary proceedings in the Sheriff Court; or
(b) it can raise proceedings on indictment either in the Sheriff Court or in the High Court.

As already noted, once proceedings are raised, the court cannot remit or transfer the case from one procedure or court to another.

13.111 For the accused, the practical consequences are obvious:

(a) in summary procedure, if the case proceeds to trial, it does so before a Sheriff alone; whereas
(b) trial on indictment is always a jury trial, whether in the Sheriff Court or in the High Court.

13.112 In the event that the accused pleads or is found guilty, the maximum penalty available to the court under most health and safety offences is limited, on summary prosecution, to a maximum fine of £20,000 (see above).

13.113 On indictment, most fines are not subject to any limitation.

13.114 Imprisonment for individuals convicted is now a competent penalty for most offences.[1]

1 Health and Safety Offences Act 2008.

13.115 There is no requirement for consultation or discussion in advance of the initiation of the prosecution, either about the type of procedure or the court in which proceedings are to be raised.

13.116 Unless an accused is notified by the HSE or takes proactive steps to discuss the position with the Procurator Fiscal, the first he will know about a prosecution is when he is served with the summary complaint or indictment.

Summary procedure

13.117 The majority of health and safety prosecutions are under the summary procedure in the Sheriff Court.

13.118 The prosecution is initiated by a complaint lodged by the Procurator Fiscal with the court and served on the accused. The complaint is at the instance of the Procurator Fiscal himself and contains the factual background which the Procurator Fiscal seeks to prove and a reference to the statutory provisions said to have been infringed.

13.119 When the complaint is served on the accused, he will also be advised of the pleading diet, which is the first time the case will call in court. The pleading diet takes place along with many other cases of every kind. The court in which these cases call in usually called the 'Diet Court'. The procedure is intended to be fairly straightforward. If the accused wants to plead guilty, he can do so at the pleading diet. In that event, the court can proceed to sentence there and then.

13.120 If the accused needs further time, he can ask the court to continue the case without plea, usually for three or four weeks, to the next Diet Court. Sometimes a number of continuations without plea take place. Each is at the discretion of the Sheriff.

13.121 At the pleading diet or any continuation thereof, the accused can plead not guilty. In that event, his plea is noted and a trial is fixed, usually two or three months ahead, depending on the individual court.

13.122 If the accused is charged in a 'special capacity' and he denies that it applies to him, he must take exception to that special capacity before he pleads not guilty to the charge.[1] Health and safety legislation is peppered with various requirements about special capacity and defence lawyers should be alert to this. Sometimes it is not easy to distinguish an allegation of a special capacity, as against the substance of the charge. For example, the Supply of Machinery (Safety Regulations) 2008 place obligations on a 'responsible person' in relation to machinery. That phrase is defined in reg 2 as being the manufacturer or the manufacturer's authorised representative. If an accused was charged with breach of reg 7 on the basis that he was a 'responsible person' and if his defence was, say, that he was not the authorised representative of the manufacturer, his legal representatives would be well advised to intimate that to the court. If he fails to intimate it, he runs the risk that the court takes the view that he is charged in a special capacity and, having failed to intimate objection, is held to have admitted that.[2] It is to be noted that this provision also applies to proceedings on indictment.

1 An example of this is found in HSWA 1974, s 3. In order to commit the offence, the accused must be 'an employer' – that is a special capacity. If the accused's defence is that he does not employ anyone, he must take exception to that special capacity at or before the time he pleads not guilty.

2 See Criminal Procedure (Scotland) Act 1995, s 255.

13.123 If the accused considers that the charge against him is either legally irrelevant or lacking in specification or, unusually, incompetent, he needs to intimate (verbally) a plea to that effect before he pleads to the charge. In that event, the court would fix a 'Debate', ie a legal hearing on the relevancy, specification or competence of the complaint. Examples of preliminary objections are often to be found in health and safety prosecutions. The two cases referred to above[1] came before the Appeal Court by way of such preliminary objections. In the case of *Dean*, the complaint was dismissed without the need to hear any evidence.

1 *Dean v John Menzies (Holdings) Ltd* 1981 JC 23 and *Purcell Meats (Scotland) Ltd v McLeod* 1987 SLT 528.

13.124 A similar procedure is available in cases on indictment. Transco plc were charged with culpable homicide (manslaughter) and took a preliminary objection. That preliminary objection was sustained by the Appeal Court,[1] which dismissed the charge of culpable homicide without the requirement to hear any evidence.

1 *HMA v Transco plc* 2004 JC 29.

13.125 The procedure for summary health and safety prosecutions is identical to the procedure for all summary prosecutions.

13.126 If the matter is ultimately resolved by a plea of guilty, whether at the pleading diet, intermediate diet or indeed the trial diet, the Procurator Fiscal addresses the court. Usually, if the plea of guilty has been adjusted as a result of discussion between the defence solicitor and the Procurator Fiscal, a 'narrative' will have been agreed at least in outline. This should avoid any unpleasant surprise for the defence at this stage. The basis upon which the plea is tendered will have been agreed.

13.127 The defence is then entitled to address the court in mitigation.

13.128 Usually about two weeks or so before the trial, an intermediate diet is fixed. The purpose of the intermediate diet is to check on both parties' state of readiness for the trial itself. If, for example, some essential witness is missing, or some report is delayed, an application can be made at the intermediate diet for the trial itself to be adjourned. If the accused wishes to change his plea to one of guilty, he can do so at the intermediate diet and, again, the court can proceed to sentence. The earlier that a plea of guilty is made, the more likely it is that the court will be prepared to discount the sentence.

13.129 Pleas in mitigation in health and safety prosecutions can present their own challenges.

13.130 In the context of a plea in mitigation, the court must accept what is said on behalf of the accused, provided that it is consistent with a plea of guilty. (This is subject to the exception regarding a proof in mitigation – see below.) Accordingly, the representative must know precisely what does and does not constitute a defence to the charge. Sometimes it is worth reminding the court about the law in relation to the charge.

13.131 The court can, however, even in the context of a plea of guilty, demand evidence to substantiate that plea. Such a course is competent but unusual. It is unusual because most pleas of guilty are preceded by discussion and agreement between prosecution and defence, often resulting in an agreed Crown narrative which is put before the court. Such a narrative would usually include an agreed account of the circumstances of the offence and what led up to it.

13.132 Occasionally, however, although the defence accepts that an offence has been committed, an agreement cannot be reached on the circumstances that gave rise to it. For example, if an employee was injured because of the absence of fixed guards on a machine, his employer might thereby be guilty of a breach of the Provision and Use of Work Equipment Regulations 1998, reg 11. Regulation 11 is absolute in its terms. The employer's position may be that it was the employee who unwarrantably removed the guard just before the accident. The Crown position may be that the guard had never been on the machine, and the occasion of the accident was the first time it was used. In the absence of an agreement, the court would be faced with materially different accounts of events, the choice of one or the other having a significant effect on the accused's culpability. In those circumstances, if the matter could not otherwise be resolved, the court might consider requiring a proof in mitigation.

Trial

13.133 The trial procedure in Scotland is essentially an oral procedure. Statements and precognitions are not routinely produced. If agreement on certain facts is reached, the parties enter into a Joint Minute admitting the agreed facts.

13.134 At the commencement of the trial, the Sheriff will have available to him only the complaint. He will know the previous procedural history of the case from the minutes on the complaint itself.

13.135 It is not necessary to lodge productions in advance.

13.136 Preparation for the trial will certainly involve a request to the Procurator Fiscal for production of the statements and other documents which have been acquired in the course of investigation.

13.137 There will also be a request for a list of witnesses. The Crown responds to that request with a copy of the list of witnesses, given on the condition that a similar list of any defence witnesses is made available 72 hours before the commencement of the trial.

13.138 The provisions in the Criminal Justice and Licensing (Scotland) Act 2010, if enacted fully, may go some way to focus the issue in health and safety trials.

13.139 In criminal procedure in Scotland, there is a requirement that all material facts necessary to establish the guilty of the accused are corroborated. A second source of evidence must be found. However, proposals have been drawn up by the Scottish Government to introduce legislation to the Scottish Parliament to withdraw the requirement for corroboration in all criminal cases. The Parliament commissioned a report from Lord Carloway,[1] which was published on 17 November 2011 and which contains 76 recommendations, which seek to incorporate a rights-based approach in line with the European Convention on Human Rights, in particular Article 5 (right to liberty and security of the person) and Article 6 (right to a fair trial). Its recommendations include removing existing rules in relation to the admission of different types of statement which were made by suspects.

1 Available at www.scotland.uk.gov.uk/About/Review/CarlowayReview.

13.140 As most prosecutions are under the flagship provisions contained in HSWA 1974, ss 2 and 3, where the Crown is required to prove very little to shift the onus onto the defence, the requirement for corroboration is not a major concern. Most of such prosecutions follow an incident or, even more usually, an accident in which injuries are sustained. In the case of an employee who has sustained injuries, in the absence of any intimation that the accused is not an employer, that will be held to be admitted. Evidence may still be required that the injured party was an employee of the accused. There will usually be evidence that an accident occurred in which the employee sustained injury. In most cases, to obtain a sufficiency of evidence, the Crown need not go beyond that. It is then for the accused to establish their defence. Of course, generally, the prosecution case will include other evidence, particularly the evidence of the HSE. That can often be seen in the context of the Crown attempting to contradict the anticipated line of defence. However, it is unlikely that the requirement to corroborate the evidence of the essential facts necessary to transfer the onus to the defence will represent a major challenge to the prosecution.

13.141 The requirement for corroboration does not apply to the defence. If the defence is, say, that all reasonably practicable steps were taken, it is not necessary to corroborate the evidence on which the defence is based.

13.142 When the trial commences, the Sheriff has before him only the complaint. If any documents have been lodged by either party, strictly he should pay no attention to those until they are spoken to by a witness. There is no opening speech by the parties. The trial commences with the Procurator Fiscal calling his first witness.

13.143 The procedure for examination, cross-examination and re-examination of witnesses is standard.

13.144 At the conclusion of the Crown case, the defence is entitled to make a submission of 'no case to answer'. If that submission is not made or is made and rejected, the defence then calls any evidence available to it.

13.145 At the conclusion of the trial, both parties then make submissions to the court and the Sheriff decides.

13.146 He can consider the matter by continuing the trial to another date. His decision is expressed orally, although in some health and safety cases he might be tempted to write a note explaining the basis for his decision. He can find the accused not guilty, he can find the charge not proven or he can convict the accused. The first two verdicts are verdicts of acquittal and there is no practical difference between them. Curiously, Scottish authorities are not very precise about the distinction between 'not guilty' and 'not proven'. In cases before a jury, the Appeal Court has held that the jury should not be told the meaning of the 'not proven' verdict.[1]

1 See eg *McDonald v HM Advocate* 1989 SLT 298, where the Appeal Court quashed the conviction of an accused person because, in the course of charging the jury on the verdicts which were available to them, the Sheriff directed that, if the verdict of 'not proven' were not available, their verdict would almost certainly be 'guilty'. The Appeal Court observed that it was highly dangerous to explain to a jury what the 'not proven' verdict was in relation to a 'guilty' verdict).

13.147 If the Sheriff convicts the accused, he proceeds to sentence.

13.148 Either party can appeal the decision. Appeal is to the High Court of Justiciary by means of a stated case. For practical purposes, the appeal is limited to questions of law. The only record of the evidence which is kept is by the Sheriff (there is no tape recording or stenographer present). Parties, therefore, have to content themselves with the Sheriff's account of what took place. In a summary prosecution, proceedings 'shall be conducted summarily viva voce and … no record need be kept of the proceedings other than the complaint … a note of any documentary evidence produced and the conviction and sentence or other finding of the court'.[1] There is no provision under which an accused might privately seek to have the evidence recorded. This would appear to be a matter for the discretion of the court, but a private recording would have no status.

1 Criminal Procedure (Scotland) Act 1995, s 157.

Solemn procedure

13.149 As in summary procedure, the rules of solemn procedure are not adapted particularly for health and safety prosecutions.

13.150 Most serious solemn procedure cases are generally initiated by the service of a petition. It is only later that an indictment is served. The petition procedure is primarily intended for serious common law crime. Its main purpose is to create a procedure in which various warrants (eg search) are granted. Petition procedure also involves a means of determining whether the accused is to be given bail and whether the accused is to be judicially examined. In the context of an HSE prosecution, these considerations seldom arise and, for that reason, the service of a petition to initiate the proceedings is unnecessary (although sometimes it still happens).

13.151 For the accused, there is one potential advantage to petition procedure. The service of a petition also marks the date where the proceedings become 'active' for the purposes of the Contempt of Court Act 1981. In the case of *Transco*, the Lord Advocate made a press statement to the effect that proceedings for culpable homicide (at common law) were going to be taken against Transco. However, no petition or indictment was served at that time. In the immediate aftermath of that statement, a significant number of potentially prejudicial press reports appeared. Transco's legal representatives requested the Crown to serve a petition purely for the purpose of preventing such reports. The Crown agreed, and a petition was served on 6 March 2003.

13.152 One of the advantages of a petition is that it does impose a time limit of sorts. The Crown is required to serve an indictment with a trial fixed within one year of the date of service of the petition.

13.153 Because the service of a petition is relatively unusual in an HSE prosecution, it is not intended to deal in detail with petition procedure. Suffice to say that, if a petition is served, there is a procedure by which the accused company can, if it wishes, intimate to the Crown that it would intend to plead guilty to the charge under s 76 of the Criminal Procedure (Scotland) Act 1995 (CPSA). This could be of significance for health and safety prosecutions where the accused wishes to plead guilty at the earliest opportunity and thereby secure the greatest discount. Section 196(1) of CPSA allows the court to take into account the stage in the proceedings for the offence at which the offender indicated to plead guilty.

13.154 Accordingly, in the majority of cases, solemn procedure is instigated by the service of an indictment on the accused. The indictment runs in the name of the Lord Advocate and contains a list of witnesses and productions.

13.155 The disclosure requirements are the same as in summary procedure.

13.156 The defence is required to lodge lists of witnesses and productions at or before a preliminary hearing. In terms of s 70A of CPSA, a requirement now exists for the accused to lodge a defence statement. The statement should cover broadly the legal and factual basis for the defence, along with any indication of factual matters in dispute and what further disclosure is required from the Crown. The statement requires to be lodged at least 14 days in advance of the first diet, and then updated as necessary prior to subsequent callings.

13.157 If the case proceeds to trial, it does so before a jury presided over by the Sheriff or by a judge in the High Court.

13.158 In the event of conviction, the Crown are obliged to move for sentence.

13.159 Appeal is to the High Court by way of Note of Appeal.

REFORM OF SCOTTISH CRIMINAL PROCEDURE

13.160 The Criminal Justice and Licensing (Scotland) Act 2010 received Royal Assent on 6 August 2010, containing a miscellany of provisions, some of which are relevant to health and safety prosecutions. The first part of the Act has not yet received a commencement order, but deals with sentencing and the establishment of the 'Scottish Sentencing Council', who will be charged with the duty of promoting consistency in sentencing and assisting the development of policy in relation to sentencing. They will be able to provide sentencing guidelines and the court should take them into account.

13.161 A partner of a partnership (other than a limited liability partnership) is guilty of a corporate offence where the partnership is guilty of that offence and it is proved that the offence was committed with the consent or connivance of the partner or is attributable to his neglect.[1]

1 Criminal Justice and Licensing (Scotland) Act 2010, s 53.

13.162 It is proposed that limited use will be allowed to be made of witness statements by permitting the witness, in certain circumstances, to be provided with his statement prior to giving evidence.[1]

1 Criminal Justice and Licensing (Scotland) Act 2010, s 54.

SENTENCING

13.163 The statutory provisions regarding the maximum sentence in health and safety offences apply without modification in Scotland.

13.164 In determining sentence, the court is entitled to discount, having regard to any plea tendered and the timing of such a plea.[1] Detailed guidance has been provided more recently by the Appeal Court in *Gemmell v HMA*.[2] The Appeal Court found that an accused had no entitlement to any particular discount in return for a plea of guilty, but it was desirable that the discretion of the court should be exercised in accordance with broad, general principles including (but not restricted to): savings in jury costs, time and inconvenience;

the sparing of witnesses from giving evidence; and other administrative benefits, including the preparation of the case by the Crown.

1 *Du Plooy v HMA* 2005 JC 1; *Spence v HMA* 2008 JC 174.
2 2012 SLT 484.

13.165 In general terms, a plea at the earliest stage might attract a discount of one-third. It has been suggested that even a last-minute plea should entitle the accused to some discount.

13.166 The importance, in negotiation with the Crown, not only of adjusting a plea but also seeking confirmation that the Crown will agree that, in the circumstances, the plea was tendered at the earliest date practicable, is important, particularly with proceedings on indictment.

13.167 The court is obliged to state the amount of any discount and, if it refuses to discount, to state the reasons why.

13.168 In Scotland, sentencing principles in health and safety cases had not come before the Court of Appeal until January 2009 when the court considered two appeals, one at the instance of the accused and the other at the instance of the Crown.

13.169 In *HMA v LH Access Technology Ltd and Border Rail and Plant Ltd*,[1] two companies pled guilty to contraventions of HSWA 1974, ss 2 and 3 that resulted in an employee of one of the companies being fatally injured. The proceedings were taken on indictment in the Sheriff Court. The Sheriff imposed fines on each company of £240,000. In each case, the fine represented a penalty of £300,000 discounted by 20 per cent to reflect the plea of guilty. Both appeals were refused. The court took the view that there was a high level of culpability in both cases. In one case, the fine was greater than the company's net current assets. In the other case it represented considerably more than the company's net profits for the previous three years.

1 (29 January 2009) [2008] HCJAC 11.

13.170 In *HMA v Munro & Son (Highland) Ltd*,[1] proceedings were raised in the High Court on indictment against an employer alleging a breach of s 3(1) of HSWA 1974. The accused pled guilty at an early stage. An accident had occurred when a large load broke free from a transporter, injuring a motorist and killing her passenger. Throughout the proceedings, it was accepted that culpability was great but the accused pled guilty at the earliest date. The judge at first instance imposed a fine of £3,750 discounted from £5,000.[2] The Crown appealed on the basis that the fine was too lenient.

1 [2009] HCJAC 10.
2 It is to be noted that such a fine could have been imposed in summary proceedings in the Sheriff Court.

13.171 More recently, in *HMA v Discovery Homes (Scotland) Ltd*,[1] a company and its director pled guilty to a contravention of s 2(1) of HSWA 1974 where an employee of the company had been fatally injured following a fall in a smoke extraction shaft in a dwelling house which was under construction. The proceedings were taken on indictment in the Sheriff Court. The Sheriff imposed fines on the company of £5,000 and, in respect of the individual, £4,000. In each case, the fine had been discounted by approximately one-third to reflect the plea of guilty. The Crown appealed on the basis that the fine was too lenient.

1 [2010] HCJAC 47.

13.172 The Appeal Court upheld the Crown appeal. It observed that, as at June 2009, the level of fines for health and safety contraventions resulting in deaths in Scotland had materially increased from those ordinarily imposed not many years previously. It substituted a fine of £40,000 in respect of the company, discounted from £60,000. The court stated that, where a company intends to place financial material before a sentencing judge, it should do so in a way that allows that material adequately to be tested and explored before the sentencing court, both as to its completeness and as to its implications. It provided that, in these circumstances, a company should intimate copies of any material on which it sought to rely to the Crown at least 14 days prior to the sentencing diet, together with an indication of whether it intended to lead evidence from an accountant or any other person at that diet.

13.173 In *Dundee Cold Store Ltd v HMA*,[1] three companies pled guilty to breaches of HSAW relating to an incident whereby a worker fell through a roof light to his severe injury. The trial Sheriff imposed fines of £135,000 (modified from £200,000 following an early plea of guilty) in respect of each of two of the companies, and £66,000 (modified from £100,000) in respect of the third company. The companies appealed, the fines were quashed, and £50,000 substituted in respect of each of the first two companies and £44,000 in respect of the third. In arriving at its determination, the Appeal Court found that the Sheriff had been unduly distracted by the serious injuries sustained by the worker, as well as having been driven by her perception of a continuing failure by courts to punish those who breached the relevant statutory provisions without apparently taking into account the fact that fines had considerably increased in the recent past.[2] The Appeal Court found that the exercise to be carried out by a sentencing judge involved: assessing, in the first place, the seriousness of the breach and the incident, which would give the court a provisional view of the potentially appropriate level of penalty; then proceeding to consider the relative aggravating and mitigating factors which might require to increase or decrease that level of penalty; and, finally, requiring to have as full information as possible as to the accused's financial position at the time of sentencing.

1 [2012] HCJAC 102.

2 See eg *HMA v Munro & Son (Highland) Limited*, above.

13.174 The Definitive Guideline issued by the Sentencing Guidelines Council does not have statutory effect in Scotland, but it may be considered by a Scottish court for the purposes of sentencing similar cases in Scotland.[1]

1 *Scottish Sea Farms Ltd & Logan and Inglis Ltd v HMA* [2012] HCJAC 11; *Dundee Cold Stores & Others v HMA* [2012] HCJAC 102; and *HMA v Discovery Homes (Scotland) Ltd* [2010] HCJAC 47.

13.175 As far as the financial position of the accused was concerned, the court emphasised that it was for the accused to place before the court sufficiently detailed information to enable the court to see the complete picture without resorting to speculation. In some cases, it might be appropriate to lead the evidence of an accountant.

13.176 In summary, therefore, it would appear that the approach to sentencing in Scotland is the same as in England and Wales.

HUMAN RIGHTS ISSUES

13.177 It is worth noting that, in Scotland, there has already been a challenge to the procedure which is followed in health and safety cases based on an apprehended violation of Art 6 of the European Convention on Human Rights which, of course, provides a guarantee of a 'fair trial'.

13.178 In the case of *Transco plc v HMA* (*No 2*),[1] the accused was originally charged with the common law crime of culpable homicide and, as an alternative, with a breach of HSWA 1974, s 3(1). The circumstances of the alleged offence arose from the escape of gas from a main, causing a massive explosion which destroyed a dwelling-house, killing all four occupants.

1 2005 1 JC 44.

13.179 The accused company raised various objections, including an argument that the culpable homicide charge was irrelevant. The Appeal Court sustained that argument, leaving as the only charge the alleged breach of HSWA 1974, s 3(1).

13.180 The accused argued a Minute under the Scotland Act 1998 that a trial before a jury would breach their right to a fair trial under Art 6(1). They pointed to the fact that: the evidence in the case was likely to be long and last a number of months; the legal issues in the case were complex; there was a split onus of proof; there were no procedural safeguards to achieve fairness; there was likely to be voluminous and complicated evidence about financial resources; because of the absence of any opening speeches in the Scottish procedure, the jury might not realise the significance of evidence until the closing speeches or the Judge's Charge, which would take place several months after the evidence was heard; and the case would present a series of issues which would be complex even for a judge sitting alone to determine. The jury would return with a verdict consisting of either one or two words. Whether the verdict was one of acquittal or conviction, the accused argued that they would not be seen to receive a fair trial. In the absence of reasons for the verdict, no one could be satisfied that the jury had properly addressed their minds to the complex issues involved.

13.181 The Appeal Court decided, unanimously, to repel the objection and allow the trial to proceed.

13.182 The decision of the Appeal Court is interesting in a number of respects. They decided, in order to find in favour of the accused before the trial took place, there had to be 'practical certainty' or 'inevitability' that there would be a breach of Art 6(1). They felt the difficulties that the jury would face might not be as great as the accused suggested. Some of these difficulties might be overcome, the Appeal Court said, by the trial judge apprising them of the main issues at the outset of the trial. They also indicated that the court might consider making available modern techniques, such as 'livenote', whereby an agreed transcript of the evidence could be available. The absence of reasons for the jury's verdict might well be less important because it was likely, in the event of there being a verdict of guilty, that various parts of the long indictment would be deleted and that would give some indication of the jury's reasoning.

13.183 In the result, the trial judge did not give any indications to the jury in advance of the commencement of the case of the likely issues. He rejected an application by the defence for the livenote system to be made available to the court. The jury convicted the accused and made no deletions from the indictment.

13.184 The issue of whether Scottish procedure does provide a fair trial to a corporate accused in a complicated health and safety prosecution has perhaps not yet been fully explored. In the *Transco* case, there was no subsequent appeal either against conviction or sentence.

FATAL ACCIDENT INQUIRIES

13.185 It is not proposed to deal in detail here with the procedure in Fatal Accident Inquiries (FAIs).[1]

1 Reference can be made to Ian H B Carmichael, *Sudden Deaths and Fatal Accident Inquiries* (3rd edn).

13.186 The procedure for FAIs has been the subject of review by Lord Cullen, who reported to the Scottish Government in November 2009. Thus far, his recommendations have not been taken up by the Scottish Parliament, but it is believed this is due to pressure of business rather than any objections to the conclusions of the report.

Under the Coroners and Justice Act 2009, a new power is given to the Chief Coroner (England and Wales) from September 2012 to recommend that the death of someone in the Armed Forces might be investigated at a Fatal Accident Inquiry. Hitherto, it is understood that Coroner's Inquests were undertaken within the place where the remains of the deceased were received into the UK.

13.187 FAIs are regulated by the Fatal Accidents and Sudden Deaths Inquiry (Scotland) Act 1976 ('FASDIA 1976'). They are mandatory in the case of deaths in the course of employment or deaths in custody.[1] However, even in cases of deaths in the course of employment, if criminal proceedings have been concluded in respect of the death or any accident from which the death resulted, and the Lord Advocate is satisfied that the circumstances of the death have been sufficiently established in the course of those proceedings, he may direct there is no FAI.[2] This qualification applies whether the proceedings result in conviction or acquittal and, if conviction, whether or not the case proceeds to trial. Accordingly, for a work related death, there will always be an FAI or a prosecution, and sometimes both.

1 FASDIA 1976, s 1(1)(a).
2 FASDIA 1976, s 1(2).

13.188 Deaths which do not fall within s 1(1)(a) which are sudden, suspicious or unexplained, or which occur in circumstances giving rise to serious public concern, may be the subject of an FAI. The decision is that of the Lord Advocate on receiving a report from the PF.

13.189 If an FAI is mandatory, or if the Lord Advocate directs that one should be held, an application is made to the Sheriff by the PF to hold the FAI. The Sheriff must grant the application and fix a date, time and place for the hearing. The arrangements are advertised and, in addition, notified to the deceased's family, the deceased's employer and the Health and Safety Commission. The PF will also intimate the arrangements to any party having an interest whose actions may come under scrutiny at the FAI. This frequently includes other contractors and medical staff.

13.190 At the FAI the Procurator Fiscal leads the evidence on behalf of the Crown. The other parties who have a right to appear are the family and the employer of the deceased, and an HSE Inspector. All other parties who wish to appear must seek leave under s 4(2) of FASDIA 1976 by showing they have an interest. Parties appearing at the FAI may appear personally or be represented by an advocate or solicitor or, with leave of the court, by any other party.

13.191 A party may cross-examine witnesses who have been led by the Procurator Fiscal. He may also lead evidence in the form of any new witnesses he considers relevant to the Sheriff's task. Proceedings at the FAI are recorded, either by a shorthand writer or by tape recording.

13.192 No witness can be compelled to answer any question tending to show that he is guilty of any crime or offence.[1] Some Sheriffs are inclined to issue 'warnings' to witnesses at the commencement of their evidence that they need not answer questions tending to show that they are guilty of any offence.

1 FASDIA 1976, s 5(2).

13.193 At the conclusion of the FAI and after hearing submissions, the Sheriff makes a determination. The provisions for the determination are as follows:[1]

(a) where and when the death and any accident resulting in the death took place;
(b) the cause or causes of such death and any accident resulting in the death;
(c) the reasonable precautions, if any, whereby the death and any accident resulting in the death might have been avoided;
(d) the defects, if any, in any system of working which contributed to the death or any accident resulting in the death; and
(e) any other facts which are relevant to the circumstances of the death.

It could be seen that paras (c), (d) and (e) provide ample scope for criticisms of individuals and organisations, if that is justified on the evidence.

1 FASDIA 1976, s 6.

13.194 The whole proceedings are public and often attract publicity.

13.195 The Determination of the Sheriff will not be admitted in evidence or founded upon in any judicial proceedings arising out of the death or out of any accident from which the death resulted.[1] A party who obtains a favourable Determination following an FAI is precluded from founding upon that in any subsequent proceedings for damages or, indeed, in any subsequent prosecution.

1 FASDIA 1976, s 6(3).

13.196 The procedure for FAIs in Scotland is a significant part of the process following a fatal accident. Generally, most organisations which might conceivably have any part to play in the occurrence of the accident or the death choose to be represented. Usually the family are also represented. Accordingly, employees of organisations which are involved in FAIs are subject to questioning not only by the Crown but also by the family. Questions are unlikely to be significantly restricted, unless the line of questioning can be seen to be wholly irrelevant to the Sheriff's task. The Productions lodged at the Inquiry typically contain the whole investigation file from the Health and Safety Executive, including copies of statements that have been given and documents recovered from the organisations in question. There is therefore a probability that the processes and procedures in place in organisations will be held up to close and detailed public scrutiny.

13.197 The scope of evidence in an FAI is potentially much wider than the evidence in a health and safety prosecution. For that reason, some organisations in some circumstances would prefer a discrete prosecution to an open-ended FAI, which might have a more profound reputational effect.

13.198 Individuals who seek representation at an FAI can apply for civil legal aid. The application should focus on why the individual seeks separate representation at the FAI. Guidance is offered by the Scottish Legal Aid Board in their Civil Legal Assistance Handbook.[1]

1 Available at www.slab.org.uk/providers/handbooks/Civil-Handbook.

CONCLUSION

13.199 Although prosecutions in respect of breaches of health and safety legislation in Scotland have often operated at a snail's pace in recent years, it is hoped that, as the dedicated Health and Safety Division continues to develop its expertise, and guidance on sentencing has been issued by the Appeal Court,[1] the pace of investigation by the Crown and prosecution, where appropriate, will increase significantly, which has to be of benefit to both the Crown and potential accused alike.

1 See *Dundee Cold Stores Ltd v HMA*, above.

Chapter 14

Corporate manslaughter: an international perspective*

Anne Davies, Senior Counsel, Withers LLP

Bruce Hodgkinson SC, Denman Chambers, Sydney, Australia

Michael Tooma, Partner, Norton Rose Fulbright, Australia

Maria Grazia Saraceni, Barrister, Francis Burt Chambers, Perth, Western Australia

* Second edition (2010) authors:
Victoria Howes, Senior Lecturer in Law, Salford Law School, University of Salford;
Dr Tim Marangon, Lecturer in Law, Salford Law School, University of Salford

INTRODUCTION

14.1 The issue of corporate criminal liability has been a concern not only in the UK but also in other jurisdictions worldwide. Deaths resulting both from accidents at work and from the failure of corporations to ensure safe working conditions and practices are increasingly the subject of scrutiny by legislators.

14.2 Globally, it is estimated that there are 2.3 million deaths per year as a result of occupational accidents or work related diseases and 317 million accidents at work. These figures have increased since the last edition of this book was published, and the economic burden of poor occupational health and safety practice is now estimated at 4 per cent of global gross domestic product each year.[1]

The world's worst industrial accidents have led to the imposition of more stringent controls over workplace activities. These include the Seveso disaster, when, in 1976, an explosion occurred in a TCP reactor of the ICMESA chemical plant on the outskirts of Meda, a small town about 20 km north of Milan, Italy. A toxic cloud containing TCDD,[2] then widely believed to be one of the most toxic man-made chemicals, was accidentally released into the atmosphere. Toxic materials were released over an area of 2.8 square km. Four per cent of the domestic animals living in the contaminated zones died spontaneously, and the remaining 77,716 animals were slaughtered as a protective measure to protect the food chain. Some 736 people were exposed to relatively high doses of TCDD.[3]

1 Figures as reported by the International Labor Organisation (ILO) at www.ilo.org/global/topics/safety-and-health-at-work/lang–en/index.htm
2 2,3,7,8-tetrachlorodibenzo-*p*-dioxin.
3 L Conti, *Visto da Seveso* (1977); F Pocchiari, V Silano and G Zapponi, 'The Seveso accident and its aftermath' in P Kleindorfer and H Kunreuther (eds), *Insuring and Managing Hazardous Risks: From Seveso to Bhopal and Beyond* (1987) pp 60–78.

14.3 A much worse accident took place at the Union Carbide chemical plant at Bhopal, India in December 1984. In the early hours of 3 December 1984, gas leaked from a tank of methyl isocyanate (MIC) at a plant in Bhopal, India, owned and operated by Union Carbide India Ltd (UCIL). The state government of Madhya Pradesh reported that approximately 3,800 persons died, 40 persons experienced permanent total disability and 2,680 persons experienced permanent partial disability. The US-based corporation was able to transfer proceedings from the US courts to India, where a final settlement was reached with the Indian government for $470 million.[1] In June 2010, seven former members of Union Carbide's senior management in India were convicted of criminal negligence and sentenced to two years' imprisonment and fined RPs 100,000 (£1,000) each. An eighth manager was also convicted but died before the judgment was passed.[2]

1 P Shrivastava, *Bhopal: Anatomy of a Crisis* (Chapman Pub, 1992); T Jones, *Corporate Killing: Bhopals Will Happen* (London: Free Association Books, 1988); K Fortun, *Advocacy after Bhopal: Environmentalism, Disaster, New Global Orders* (2001); and *Re Union Carbide Corpn Gas Plant Disaster at Bhopal, India in December 1984* 809 F 2d 195 (2nd cir, 1987).
2 J Lamont, 'Eight Top Managers Convicted over Bhopal Gas Leak Disaster', *Financial Times* (8 June 2010), p 1.

14.4 The world's most serious nuclear power accident took place in Chernobyl, Ukraine, on 26 April 1986. Thirty people were killed immediately and, as a result of the high radiation levels in the surrounding 20-mile radius, 135,000 people had to be evacuated. There was a prolonged release to the atmosphere of large quantities of radioactive substances. Activity transported by the multiple plumes from Chernobyl was measured not only in Northern and Southern Europe, but also in Canada, Japan and the US. Only the Southern hemisphere remained free of contamination.[1]

More recently, the Fukushima 1 nuclear power plant was severely damaged by the tsunami produced by the Tōhoku earthquake in March 2011, in what is considered to be the worst nuclear disaster since Chernobyl. Although there were no direct deaths from the accident, the full consequences are yet to be determined. An independent investigation published in July 2012 by the National Diet concluded that the accident was man-made and could have been foreseen and prevented.[2] In September 2013, it was announced that a prosecution of criminal negligence against the government and factory management would not be pursued, due to a lack of clear evidence.[3]

1 Chernobyl Nuclear Disaster, www.chernobyl.co.uk.
2 NAIIC report: http://warp.da.ndl.go.jp/info:ndljp/pid/3856371/naiic.go.jp/en/.
3 http://the-japan-news.com/news/article/0000605540.

14.5 The 21st century has seen its fair share of large-scale industrial disasters. In March 2005 an explosion at BP's Texas City Refinery in Texas, USA left 180 injured and 15 dead.[1] BP PLC were prosecuted for breaches of the Clean Air Act.[2] The company admitted guilt as part of a plea bargain agreement[3] which saw them fined $50 million (at the time, both the largest fine ever imposed under the Clean Air Act and the largest criminal fine ever levied by the US Department of Justice) and sentenced to three years' probation.[4]

BP faced further fines following the explosion of the Deepwater Horizon oil rig on 20 April 2010, which killed 11 people and caused the equivalent of millions of barrels of oil to be discharged into the Gulf of Mexico. In November 2012, BP admitted negligence, reaching a new record $4.5 billion settlement with the US Department of Justice in relation to the criminal charges from the spill. A trial under the Clean Water Act commenced in February 2013 and is expected to finish in 2014, when the penalties will be decided.

In April 2013, Rana Plaza, an eight-storey commercial building just outside Dhaka, Bangladesh, collapsed, killing 1,129 people and injuring more than 2,500. It is considered to be the deadliest garment-factory accident in history, as well as the deadliest structural failure. The accident was just months after a factory fire, also on the outskirts of Dhaka, which killed at least 117 people. Both factories were involved in the production of clothing for a number of UK and US retailers, and the disasters sparked global interest in working conditions in Bangladeshi factories. As a result, an Accord on Factory and Building Safety in Bangladesh was signed in May 2013 by a number of European retailers, with US companies forming an Alliance for Bangladesh Worker Safety.

1 Figures as reported by the US Chemical Safety Board (CSB) in their final report into the incident, Contributory factors identified in BP's investigation into the incident included a failure by management to set and consistently reinforce process safety, operations performance and systematic risk reduction priorities as well as confusion amongst the workforce as to roles and responsibilities caused by numerous changes within the organisation which led to a lack of clear accountabilities and poor. Rather damningly the CSB also noted in their report (at p 210) that managers at the Texas City Refinery failed to 'create an effective reporting and learning culture'. It goes on to note that 'reporting bad news was not encouraged' and that 'incidents were often ineffectively investigated and appropriate corrective actions [were] not taken'.
Additionally, the final report of the BP US Refineries Independent Safety Review Panel into the Texas City Refinery Explosion (commissioned by BP on the recommendation of the CSB) is accessible at www.bp.com/liveassets/bp_internet/globalbp/globalbp_uk_english/SP/STAGING/local_assets/assets/pdfs/Baker_panel_report.pdf.

2 As an interesting aside, the offence that BP was convicted of was enacted specifically in light of the horrors that unfolded in Bhopal, India following the gas leak at the Union Carbide Plant (discussed above).

3 The plea bargain saw BP admit liability for the Texas City Refinery explosion as well as liability for a pipeline leak in Alaska and for entering into energy price-fixing deals, as reported by the US Department of Justice in a press release distributed on 25 October 2007.

4 As reported by the US Department of Justice in a press release distributed on 25 October 2007. Both the families of the deceased and surviving victims of the explosion appealed against the fine initially set by District Judge Lee Rosenthal on the grounds that the punishment was too lenient and that federal prosecutors had violated the rights of victims to be consulted before any plea bargain was agreed, contrary to the provisions of the Crime Victims Act 2004. A 5th US Circuit Court of Appeals panel agreed that there had been a violation of the victims' rights under the 2004 Act but did not go so far as to require District Judge Rosenthal to reject the plea bargain. In March 2009, District Judge Rosenthal finally accepted the deal.

14.6 The potential for causing large-scale casualties amongst the general public is not limited to the primary and secondary sectors of industry. The tertiary sector has also claimed a substantial number of victims. We have, for example, witnessed several large-scale air disasters, such as: the loss of ValuJet Flight 592, which came down in the Florida Everglades in May 1996 after the plane was engulfed in flames shortly after take-off;[1] Trans World Airlines Flight 800, which exploded minutes into its flight from New York to Paris and crashed into the Atlantic Ocean in July 1996;[2] the massive loss of life caused by the events which unfolded in the USA on 11 September 2001;[3] and the crash of American Airlines Flight 587 in November 2001.[4]

1 The fire originated in the plane's cargo hold. It was caused by the incorrect packing of several oxygen tanks which subsequently went off. These tanks had been stored on board in spite of Federal Aviation Administration (FAA) regulations which prohibited the carrying of hazardous

materials in the plane's cargo hold. The US National Transport Safety Board (NTSB) placed the blame for the incident squarely at the feet of three parties: (1) SabreTech (ValuJet's maintenance contractor) for failing to properly repair and package the oxygen tanks for flight; (2) ValuJet for failing to properly supervise the activities of its maintenance contractor to ensure that SabreTech complied with requirements and accepted practices in relation to maintenance, maintenance training, and the handling/transportation of hazardous materials; and (3) the FAA for failing to require smoke detection and fire suppression systems in cargo holds such as the one present on Flight 592. All 105 passengers onboard died. The subsequent prosecution of SabreTech is discussed in more detail at **14.86** onwards below.

2 The explosion resulted in the loss of 230 lives. The US National Transport Safety Board (NTSB) concluded that the explosion resulted from the ignition of the volatile fuel/air mixture contained in the Central Wing Tank. The cause of the ignition could not, however, be determined conclusively. The NTSB's report final report into this incident can be accessed at www.ntsb.gov/doclib/reports/2000/AAR0003.pdf.

3 On 11 September 2001, terrorists hijacked two American Airlines planes (Flights 11 and 77) and two United Airlines planes (Flights 93 and 175). These were subsequently used in a series of coordinated terrorist attacks. At 8.45am, American Airlines Flight 11 was deliberately crashed into the North Tower of the World Trade Center. At 9.05am, United Airlines Flight 175 was crashed into the South Tower of the World Trade Center. American Airlines Flight 77 was piloted into the Pentagon soon after. Finally, United Airlines Flight 93 came down in Shanksville, Pennsylvania at 10.03am. Famously, Flight 93 reportedly saw an attempt made by the plane's crew and passengers to regain control of the flight from the hijackers, leading to the hijackers deciding to deliberately crash the plane short of its intended target, thought to have been either the United States Capitol building or the White House. For a detailed breakdown of events as they unfolded and the aftermath of the 2001 attacks, you are advised to read 'The 9/11 Commission Report: Final Report of the National Commission on Terrorist Attacks Upon the United States', accessible online at http://govinfo.library.unt.edu/911/report/911Report.pdf.

In total, over 2,600 civilians lost their lives at the World Trade Center, 125 people lost their lives at the Pentagon, and 256 passengers died on the four planes (figures taken from 'The 9/11 Commission Report: Final Report of the National Commission on Terrorist Attacks Upon the United States').

On 9 September 2003, US district judge Alvin Hellerstein gave the go-ahead to families of those who perished in the 9/11 attacks to sue American Airlines, United Airlines, Boeing (who manufactured all four of the hijacked planes) as well as the New York Port Authority, the owners of the World Trade Center, for negligence. It is alleged that the airlines were negligent in failing to carry out proper safety checks, which might have prevented the hijacking. It is further claimed that Boeing had been negligent in failing to design a cockpit door that hijackers could not break in through. Finally, it was alleged that the New York Port Authority were negligent in failing to design buildings with adequate escape routes.

In his ruling, Judge Hellerstein announced that he was of the view that the hijacking of commercial jets was a foreseeable risk that the airline industry should have guarded against. According to the judge: 'The intrusion by terrorists into the cockpit, coupled with the volatility of a hijacking situation, creates a foreseeable risk that hijacked airplanes might crash, jeopardising innocent lives on the ground as well as in the airplane.' – see D Teather, 'Families of 9/11 Victims Win Right to Sue Airlines,' *Guardian*, 9 September 2003.

It had been reported that 95 claims were initially filed on behalf of 96 victims of the events which unfolded on 11 September 2001 – A Hartocollis, 'Little Noticed 9/11 Lawsuits Will Go to Trial', *New York Times*, 4 September 2007. Whilst the claims for 'personal injury and wrongful death' filed at the US District Court for the Southern District of New York might come as no surprise, further claims were made by workers involved in the post 9/11 cleanup who alleged that they had developed respiratory diseases as a consequence. Key judicial opinions related to the 11 September litigation can be accessed through the website for the US District Court for the Southern District of New York at www.nysd.uscourts.gov/sept11. Many of these claims have since been settled, but a trial date has not been set – as reported by the Reuters news agency (see C Kearney, 'September 11 Families Argue for Security Papers Release', 25 March 2009.Interestingly, Larry Silverstein, a New York Real Property Developer, subsequently filed a claim against United Airlines and American Airlines. Larry Silverstein had purchased 99-year leaseholds for World Trade Center towers 1, 2, 4 and 5 two months before the September 2001 attacks for the sum of $2.805 billion. The hijacked planes struck towers 1 and 2, whilst towers 4 and 5 (along with other surrounding properties owned by Mr Silverstein's company – World Trade Center Properties LLC) were destroyed by the subsequent collapse of towers 1 and 2. Once again, it was alleged that the defendants had been negligent in failing to take sufficient steps to prevent the hijackers taking control of the four doomed flights.

In an opinion delivered on 10 October 2008, Judge Hellerstein declared that, if the defendant airlines were found liable, the sum of their liability would be limited to the market values of the four 99 leaseholds, as they stood on 11 September 2001 (see the official copy of Judge Hellerstein's opinion at www.nysd.uscourts.gov/docs/rulings/21MC101%20DEC%2011%202008%201215%20TS.pdf). This decision was challenged by the airlines, but their motion to get the decision reconsidered was rejected in June 2009 (see an official copy of Judge Hellerstein's opinion at www.nysd.uscourts.gov/docs/rulings/21mc101_order_042909.pdf).

4 American Airlines Flight 587 crashed into Belle Harbor, a residential area of New York City, USA, minutes into a flight from New York to the Dominican Republic. All 260 passengers onboard the plane were killed, as were five civilians on the ground. The cause of the crash was eventually found to be pilot error.

It is reported that American Airlines Flight 587 took off soon after a Japan Airlines flight and was immediately caught up in the earlier plane's wake turbulence. In an attempt to keep control of the plane, the First Officer made a number of 'unnecessary and excessive rudder inputs'. The strength of the air flowing over the rudder placed a greater stress on the plane's vertical stabiliser than it was designed to bear. This eventually resulted in the plane's rudder shearing off. The National Transport Safety Board's report into the accident can be found online at www.ntsb.gov/doclib/reports/2004/AAR0404.pdf.

14.7 Closer to home, in July 2000, Air France Flight 4590 (Concorde) caught fire soon after taking off from Charles de Gaulle Airport, Paris, France and crashed into a hotel in Gonesse, France killing 100 passengers, 9 flight crew and 4 civilians on the ground.[1] Finally, Air France Flight 447 from Rio de Janeiro to Paris disappeared into the Atlantic Ocean in June 2009 with the loss of 228 lives.[2]

1 An authorised English translation of the official report into the crash was published by the Bureau d'Enquêtes et d'Analyses pour la Sécurité de l'Aviation, France's air accident investigation bureau.

The report concludes that the probable cause of the accident was a tyre blowout caused when one of the Concorde's tyres passed over a strip of metal that had fallen from a DC-10 aircraft, owned by Continental Airlines, during take-off five minutes earlier. Rubber from the tyre flew up and hit one of the plane's under-wing fuel tanks. The impact was sufficiently severe to rupture the fuel tank and trigger a leak of aviation fuel. The leaking fuel was ignited and the fire eventually led to a loss of power in two of the engines. The destabilisation that this caused was exacerbated by the fact it had become impossible to retract the plane's landing gear. The pilots eventually lost control of the plane and it came down.

In July 2008 a French judge ordered Continental Airlines and five individual defendants to stand trial over the crash, on the grounds that it had been negligent in the maintenance of its fleet of DC-10 aircraft. The five individuals included Henri Perrier, former head of testing at Concorde, and another former Concorde engineer, two Continental Airlines mechanics who were accused of negligently failing to follow normal procedures when repairing the DC-10, and Claude Frentzen, an employee of France's civil aviation authority at the time of the crash. It is alleged that France's civil aviation authority were negligent in failing to enforce design safety requirements on Concorde who did not fit extra protection to its under-wing fuel tanks until after the 2000 crash. See C Balmer, 'French Court to Try US Airline Over Concorde Crash', 3 July 2008, accessed online at www.reuters.com/article/latestCrisis/idUSL03332623.

In December 2010, Continental Airlines was found criminally responsible for the crash and fined €200,000. One of the Continental mechanics was also given a 15-month suspended sentence, but the other accused parties were found not guilty. The appeal court overturned the first instance manslaughter decision in November 2012, but criticised Continental and the French aviation authorities for the part they had played in the accident.

2 Investigation into the accident by the Bureau d'Enquêtes et d'Analyses pour la Sécurité de l'Aviation (BEA) found that the captain of the plane had been asleep when the plane hit turbulence during a tropical thunderstorm and had been too tired to be able to assist his co-pilots in bringing the plane under control. Read the BEA report here: www.bea.aero/en/enquetes/flight.af.447/rapport.final.en.php. A criminal inquiry for alleged manslaughter has been launched against Air France and Airbus, and is still ongoing in September 2013.

14.8 Breaches of health and safety regulations usually result in administrative or regulatory sanctions imposed by the health and safety authorities of the country concerned. They may also be relied on in civil claims as evidence of negligence. When regulatory offences result in fatal accidents, however, the question of criminal responsibility falls to be addressed.

14.9 Discussion in this chapter shows the wide variety of approaches that have been adopted by various countries when deciding whether to criminalise corporate behaviour for death at work, perhaps by imposing criminal liability on corporations for the offence of manslaughter. In general, a corporation is in the same position in relation to criminal liability as a natural person and may be convicted of criminal offences, *including* those requiring proof of mens rea. There are, however, some crimes which a corporation is incapable of committing, for example, reckless driving. Nor can a corporation be found guilty of a crime for which imprisonment is the only punishment, for example, murder. As will be seen, the question of whether a corporation can be guilty of manslaughter is a complex one.

14.10 Generally, a natural person will be guilty of gross negligent manslaughter if his or her recklessness or negligence was the substantial cause of death. The liability of a natural person has been well established in the criminal laws of all the jurisdictions we will consider and is thus not an issue. Since corporations have no physical existence, however, prosecuting them for homicide presents legislators with a number of challenges.

14.11 Criminal offences consist of two main elements: a physical element and a mental element. Since a corporation has no body, it cannot carry out the physical element; and since it has no brain, it cannot form a state of mind. The challenge becomes, therefore, how to attribute the actions and state of mind of natural persons to the corporation. This dilemma has been a recurrent theme throughout this book and remains a common theme for discussion in this chapter.

14.12 We will be discussing the development of criminal laws dealing with corporate liability for the crime of manslaughter. We will start by considering the position in common law countries, such as the US, Canada and Australia. As one of the world's largest industrialised nations, it makes eminent sense to discuss how courts and legislators in the US have attempted to deal with the issue of corporate criminality.[1] There is a new section on the Far East, which reflects the growing global importance of this region; its safety legislation is developing, as is the focus on increased enforcement against corporate entities. Canada and Australia have been chosen for discussion for two reasons. First, as Commonwealth countries, the development of the law in both countries has been strongly driven by a common law system, which means both countries can be (relatively) easily compared to the legal position in England and Wales. Secondly, as countries which have both federal and state laws, Canada and Australia also provide useful comparators for the US.

1 The topic of corporate (or 'white collar') crime has been receiving increased attention in the US in what has been labelled the 'post-Enron era' – a label which symbolises 'the image of major corporations brought low by the [criminal] practices of its executives': J Hasnas, 'Symposium: Corporate Criminality: Legal, Ethical and Managerial Implications' (2007) 44 American Crim L Rev 1269 at p 1269.

14.13 Once we have concluded our discussion of the legal position in the US, the Far East, Australia and Canada, we look at some European countries, such as France, Italy, The Netherlands, Germany, Finland, Sweden, Norway and Denmark. Ultimately, our aim is to analyse the position of the development of the law of corporate manslaughter from both an international and a European perspective.

14.14 These particular countries have been chosen for a number of reasons. First, the countries are located in different geographical areas and have developed from quite different political and legal backgrounds. Secondly, they are members of similar international and supranational bodies, such as the Council of Europe and European Union, albeit joining them at different times. Such memberships lead to harmonisation of laws and, even in areas where bodies have no jurisdiction to legislate soft laws, create certain pressures towards changes. Finally, these countries represent a variety of different approaches to law relating to corporate criminal liability, which provides a good scope for analysis and discussion.

THE US

Introduction

14.15 A common feature of all the first three jurisdictions we consider in this chapter is that the US has a federal system of government in place. Laws, including criminal laws, can be introduced at both federal (ie national) and state level. Most prosecutions, however, are brought under state criminal laws.[1]

1 Arthur Allens Robinson, 'Corporate Culture as a Basis for the Criminal Liability of Corporations' (2008) at p 29, accessible online at 198.170.85.29/Allens-Arthur-Robinson-Corporate-Culture-paper-for-Ruggie-Feb-2008.pdf (accessed 12 December 2013).

14.16 It is no surprise that one of the world's greatest industrial nations should find its organisations prosecuted for deaths caused by their actions.[1] Who can forget, for example, the prosecution of Ford for injuries and deaths caused by its defective model, the Pinto. Although the prosecution was unsuccessful, it played a huge role in changing the attitude of those parts of American society which still believed that a corporation was not capable of intentionally committing homicide.[2]

1 In August 2012 the US Bureau of Labor Statistics (BLS) reported that 4,383 fatal work injuries were recorded in the US in 2012. This figure represented a drop of seven per cent from the total of 4,693 fatal work injuries recorded in 2011 and the second lowest total since figures were first collated in 1992The figures are reported in the BLS 'Census of Fatal Occupational Injuries 2012'

2 D Miester, 'Criminal Liability for Corporations that Kill' (1990) 64 Tulane L Rev 919.

14.17 As will be seen in this section of the chapter, crimes under US law, in common with the position under English law, consist of two key elements: a physical element; and a mental element (the mens rea). Corporations are capable of being convicted of criminal offences, but the age-old problem remains of how to attribute the actions and state of mind of natural persons to the defendant company. As we will see, the mechanism used for attribution varies according to whether the corporation is charged with a federal criminal offence or a state-level criminal offence.

Criminal liability of corporations at federal level

14.18 Perhaps the first case in which it was established at federal level that a corporation could be convicted of a crime is the oft-cited case of *United States v Van Schaik*.[1] That case concerned the prosecution of a company following the death of 900 passengers when a steamship owned by the company caught fire and sank. The company had failed to provide sufficient life preservers, contrary to statutory obligations. Managers of the company were also prosecuted for aiding and abetting the company in committing the offence. The managers appealed, on the grounds that they could not be indicted for aiding and abetting an offence which a corporation was incapable of committing. The basic problem was that the prescribed punishment for the offence in question was imprisonment and, since a corporation has no physical existence, imprisonment was a logistical impossibility. The argument advanced by the managers, therefore, is that the failure of Congress to include a punishment suitable for a corporate offender was indicative of the fact that Congress never intended the offence to be applied to corporations. This argument was rejected, however, on the grounds that the social utility of allowing the prosecution of corporations for criminal offences far outweighed the 'inadvertent oversight' of Congress to legislate for such an eventuality.

1 (1904, CCNY) 134F 592.

14.19 Where the corporation is being prosecuted for a federal offence, the mechanism of attribution used is vicarious liability. Consequently, a corporation can be held liable for the crimes of its officers, employees or agents where it can be established that:

> 'the individual's actions were within the scope of their duties and the individual's actions were intended, at least in part to benefit the corporation'.[1]

A very early example of this in action was the case of *New York Central and Hudson River Railroad Company v United States*,[2] which was heard soon after the decision in *Van Schaik*.

1 Arthur Allens Robinson, 'Corporate Culture as a Basis for the Criminal Liability of Corporations' (2008) at p 29, accessible online at: 198.170.85.29/Allens-Arthur-Robinson-Corporate-Culture-paper-for-Ruggie-Feb-2008.pdf (accessed 12 December 2013).

It is clear that, if the corporation is to be held liable, the crime must not be a personal aberration of the employee acting on his or her own. It must in some way fall within the employee's scope of employment. Indeed, in the case of *United States v Inv Enter Inc* 10 F 3d 263, 266 (5th Cir 1998) the court stated '[A] corporation is criminally liable for the unlawful acts of its agents, provided that the conduct is within the scope of the agent's authority, whether actual or apparent'. The law of agency clearly plays a significant role in this context (see *United States v American Radiator and Standard Sanitary Corpn* 433 F 2d 174 (3rd Cir, 1970)). It is for the government to prove the existence of any agency relationship (see *United States v Bainbridge Mgmnt*, Nos 1 CR 469–1, 01 CR 469–6, 2002 WL 31006135). It is important to note that the federal courts have proved willing to find that an employee's actions can fall within the scope of their employment even where the corporation has implemented policies explicitly prohibiting that behaviour. The 'US Department of Justice Memorandum: Principles of Federal Prosecution of Business Organisations' notes, at p 15, 'The existence of a corporate compliance programme, even one that specifically prohibited the conduct in question, does not absolve the corporation from criminal liability …'. The guidance also cites the case of *United States v Potter*, 463 F 3d 9 (1st Cir 2006) in which the court clearly stated that a corporation could not 'avoid liability by adopting abstract rules' that forbid its agents from engaging in illegal acts because 'even a specific directive to an agent or employee or honest efforts to police such rules do not automatically free the company from the wrongful acts of its agents'.

Furthermore, whilst the act must in some way have been in order to benefit the company, it need not be exclusively for its benefit. If the agent is acting in violation of the fiduciary duty owed to the corporation (in order to gain personal profit, for example), the corporation will not be liable (see *Standard Oil Co of Texas v US* (1962)). The court has gone so far as to say that vicarious liability existed even where the corporation did not, in the event, benefit, where the employee had the intention of benefiting the company (see *United States v Carter* 311 F 2d 934 (6th Cir); *Old Monastery Co v United States* 147 F 2d 905 (4th Cir)). The 'US Department of Justice Memorandum: Principles of Federal Prosecution of Business Organisations' refers to the case of *Automated Medical Laboratories* 770 F 2d at 407 as authority for this point, in which the court stated:

> 'Benefit is not a "touchstone of criminal corporate liability; benefit at best is an evidential, not an operative, fact." Thus, whether the agent's actions ultimately redounded to the benefit of the corporation is far less significant than whether the agent acted with the intent to benefit the corporation. The basic purpose of requiring that an agent have acted with the

intent to benefit the corporation, however, is to insulate the corporation from criminal liability for actions of its agents which be inimical to the interests of the corporation or which may have been undertaken solely to advance the interests of that agent or of a party other than the corporation.'

2 (1909) 212 US 481.

14.20 The defendant company had been charged for breaching the provisions of the Elkins Act 1903. Section 1 of that Act stipulated that the acts of a carrier's officers, agents and employees could be attributable to the carrier. The Supreme Court realised that many offences would go unpunished if only individuals were capable of being subject to the provisions of the criminal law. Consequently, the court upheld the natural reading of the section, rejecting arguments that a company's liability would be unconstitutional.

14.21 In *United States v Time-DC Inc*, the corporation was successfully prosecuted under an International Chamber of Commerce regulation, which forbade lorry drivers from driving when ill. One employee knew that the driver concerned in the case had telephoned to say that he could not work but had then, when learning of the company's absentee policy, which required a doctor's note, changed his mind. It was held that the corporation could be found to have 'knowingly and willingly violated the regulation'. The court summarised the principle as follows:

> '... knowledge acquired by employees within the scope of their employment is imputed to the corporation. In consequence, a corporation cannot plead innocence by asserting that the information obtained by several employees was not acquired by any one individual employee who then should have comprehended its full report. Rather, the corporation is considered to have acquired the collective knowledge of its employees and is held responsible for their failure to act accordingly.'

1 *US v American Stevedores Inc* (1962).

14.22 A more recent case discussing the issue of corporate vicarious liability for criminal offences governed by federal law is *US v Potter*.[1] A general manager of the defendant corporation had paid a bribe to the Speaker of the Rhode Island House of Representatives. It transpired that the company's president had considered the manager's proposed action but had expressly ordered him not to proceed. The court reiterated the vicarious liability test and observed that the president's prohibition was not enough to absolve the corporation of liability,[2] stating at pp 42–43:

> 'The principal is held liable for acts done on his account by a general agent which are incidental to or customarily a part of a transaction which the agent has been authorised to perform. And this is the case, even though it is established fact that the act was forbidden by the principal ... despite the instructions [the general manager] remained the high-ranking official centrally responsible for lobbying efforts and his misdeeds in that effort made the corporation liable even if he overstepped those instructions.'

1 463 F 3d 9 (1st Cir, 2006), cited in Arthur Allens Robinson 'Corporate Culture as a Basis for the Criminal Liability of Corporations' (2008) at p 29.

2 Further cases considering this very point include *United States v Twentieth Century Fox Film Corpn* 882 F 2d 656 (2nd Cir, 1989) and *United States v Hilton Hotels Corpn* 467 F 2d 1000 (9th Cir, 1972).

14.23 It has been clear from a very early stage, therefore, that corporations are capable of committing (and being convicted of) criminal offences governed by federal law. This does not, however, get us any further in our discussion of laws governing corporate liability for deaths caused by its activities, because the last federal statute dealing specifically with homicide was repealed by Congress quite some time ago.

14.24 The only remaining federal statute in the context of manslaughter is the Occupational Safety and Health Act 1970.[1] Section 5(a) of that Act sets out the general duty which requires an employer to 'furnish to each of his employees employment and a place of employment which are free from recognized hazards that are causing or are likely to cause death or serious physical harm to his employees' and to 'comply with occupational safety and health standards promulgated under [the Occupational Safety and Health Act]'.[2]

1 A copy of this legislation is accessible online. Visit the website for the Occupational Safety and Health administration (a division of the United States Department of Labor). Section 3 of the Act defines 'person' as 'one or more individuals, partnerships, associations corporations, business trusts, legal representatives, or any organised group of persons'.

2 Section 5(b) places a duty on employees to 'comply with occupational safety and health standards and all rules, regulations, and orders issued pursuant to this Act which are applicable to his own actions and conduct'.

14.25 Section 17 of the Occupational Safety and Health Act goes on to set out the penalties that can be imposed on an employer found to be in breach of obligations placed upon them, either directly under the Act or by regulations passed in pursuance of the objectives of the Act. In the event of an employer receiving one or more citations for breaches which are not deemed to be of a 'serious nature', the employer will face a fine of up to $7,000 for each citation.[1] Where the employer receives one or more citations for breaches which are deemed to be a 'serious violation', it also faces a fine of up to $7,000 for each citation.[2] Where, however, the employer 'wilfully or repeatedly' breaches the obligations placed upon them by the Act or associated regulations, the employer faces a fine of between $5,000 and $70,000 for each wilful violation.[3]

1 Section 17(c) of the Occupational Safety and Health Act.

2 Section 17(b) of the Occupational Safety and Health Act. Note that, in accordance with the provisions of s 17(k), a violation will be deemed to be a 'serious violation' where 'there is a substantial probability that death or serious physical harm could result from a condition which exists, or from one or more practices, means, methods, operations, or processes which have been adopted or are in use, in such place of employment unless the employer did not, and could not with the exercise of reasonable diligence, know of the presence of the violation'.

3 Section 17(a) of the Occupational Safety and Health Act.

14.26 The US Department of Justice Guidance ('US Department of Justice Memorandum: Principles of Federal Prosecution of Business Organisations', 28 August 2008) cites the only case addressing the issue of 'wilfulness' for the purposes of finding a criminal violation under the Act, *United States v Dye Construction*,[1] as authority for the proposition that the government need not prove that the employer entertained a specific intent to harm the employee or that the employer's actions involved moral turpitude; rather, the court approved the following jury direction:

> 'The failure to comply with a safety standard under the Occupational Safety Health Act is wilful if done knowingly and purposely by an employee who, having free will or choice', either intentionally disregards the standard or is plainly indifferent to its requirement. An omission or failure to act is wilful if done voluntarily and intentionally.'

1 510 F 2d 78, available at https://bulk.resource.org/courts.gov/c/F2/510/510.F2d.78.74–1176.html. Indifference can also be sufficient to show 'wilfulness' for the purposes of the statutory offence – see *Georgia Electric Co v Marshall* 576 F 2d and *FX Messina Construction Co v OSHRC* 522 F 2d. For a discussion of these cases, see Criminal Resource

Manual 2012 OSHA 'Wilful Violation of a Safety Standard which Causes Death to an Employee' at www.justice.gov/usao/eousa/foia_reading_room/usam/title9/crm02012.htm.

14.27 Despite pledges from the federal Occupational Safety and Health Administration to 'crack down' on those responsible for workplace fatalities and to pursue criminal charges under the Occupational Safety and Health Act 1970, the reality seemed to be much different. Figures obtained by the *New York Times* in 2003 noted:

> 'Over a period of two decades, from 1982 to 2002, OSHA investigated 1,242 [cases] … in which the agency itself concluded that workers had died because of their employer's "wilful" safety violations. Yet in 93 percent of those cases, OSHA declined to seek prosecution … OSHA's reluctance to seek prosecution … persisted even when employers had been cited before for the very same safety violation. It persisted even where the violation caused multiple deaths, or when the victims were teenagers. And it persisted even where reviews by administrative judges found abundant proof of wilful wrongdoing.'[1]

1 D Barstow, 'U.S. Rarely Seeks Charges for Deaths in Workplace', (2003) *New York Times*, 22 December. In the same article, Barstow goes on to cite a Congressional report in 1988 which commented: 'A company official who wilfully and recklessly violates federal OSHA laws stands a greater chance of winning a state lottery than being criminally charged'.

14.28 The *New York Times*' investigation[1] went on to suggest that the lack of prosecutions under criminal provisions in such cases could be attributed to both a 'simple lack of guts and political will' as well as evidential problems and a high threshold of proof.[2] The *New York Times*' investigation, however, was completed ten years ago, and there has since been a major change in the OSHA's approach to prosecuting workplace deaths through the criminal law.

1 The newspaper proudly reported that their investigation – 'based on a computer analysis of two decades worth of OSHA data, as well as hundreds of interviews and thousands of government records' – was 'the first systematic accounting of how [the US] confronts employers who kill workers by deliberately violating workplace safety laws'. Furthermore, the article claims that: 'Until presented with the results of the *Times* examination, the [OSHA] had never done a comprehensive study of how often workers were killed by wilful safety violations' – see D Barstow, 'U.S. Rarely Seeks Charges for Deaths in Workplace', (2003) *New York Times*, 22 December.

2 A former OSHA inspector, Jeff Brooks, is reported as saying that '[the violation] can't just be willful it has to be obscenely willful' – see D Barstow, 'U.S. Rarely Seeks Charges for Deaths in Workplace', (2003) *New York Times*, 22 December.

14.29 A review of recent press releases on the OSHA website[1] seems to show that there has been a greater willingness on behalf of federal prosecutors to charge employers for wilful violations of health and safety law. In 2013, fines have been issued by the OSHA website on a regular basis, with amounts varying from tens of thousands of dollars through to millions for more severe cases. One such case was the $1.1 million fine given to Republic Steel in August 2013 for 15 wilful violations, eight serious violations and one repeat violation.

1 Accessible online at www.osha.gov/newsrelease.html.

14.30 Perhaps the highest-profile federal prosecution of an organisation for breach of its health and safety responsibilities, however, would be the series of multi-state actions brought against McWane Inc, a major manufacturer of cast iron water pipes, which began in 2003. The prosecution of McWane Inc was novel in the sense that it stemmed from a new partnership between the Occupational Health and Safety Administration, the Environmental Protection Agency and a select band of Justice Department prosecutors. In April 2007 a federal jury found Atlantic States Cast Iron Pipe Co, a pipe foundry owned by

McWane Inc, and four of its managers guilty of 'conspiring to evade workplace safety and environmental laws by lying to regulators, tampering with evidence and bullying employees into silence about dangerous working conditions'.[1]

1 D Barstow, D, 'Guilty Verdicts in New Jersey Worker-Safety Trial', (2006) *New York Times*, 27 April.

14.31 By this stage, McWane had already been ordered to pay $19 million in fines and restitution, and had also seen several current or former managers fined or sentenced to probation, and McWane Inc's working practices had been repeatedly criticised by the prosecution.[1] In April 2009, sentence was finally passed. McWane Inc was ordered to pay $8 million in fines and was additionally made the subject of a four-year probation order. Crucially, however, the four managers facing sentence alongside the organisation were reportedly sentenced to prison sentences ranging from 6 months to 70 months.[2] The case was reheard in 2012 by the Court of Appeal, who rejected all claims and upheld the fines and prison terms.

1 D Barstow, 'Guilty Verdicts in New Jersey Worker-Safety Trial', (2006) *New York Times*, 27 April. In the same report, Norv McAndrew, an assistant US attorney, is quoted as describing 'the McWane way' as one where 'production is priority number one – everything else is incidental'.
2 D Barstow, 'Iron Pipe Maker is Fined $8 million for Violations', (2006) *New York Times*, 25 April.

14.32 There has also been some further progress when it comes to prosecuting corporations for wilful breaches of health and safety law at state level. Whilst the Occupational Safety and Health Act 1970 generally pre-empts state workplace health and safety laws, s 18 of the OSHA 1970 encourages individual states to develop and operate their own safety and health standards.[1] To receive federal approval, the proposed state plan has to be deemed to be as effective as the federal program. At present, there are 27 states and territories which operate state plans covering both private sector and public sector workers, and four states and territories with plans which cover public sector workers only. Together, officials from the 27 states with OHSA-approved state plans form the Occupational Safety and Health State Plan Association (OSHSPA).[2] The OSHSPA publishes an annual report highlighting how each OSHSPA state and territory has taken steps to enhance the health and safety of America's workforce.[3]

1 This includes a delegated power to prosecute breaches of occupational safety and health legislation.
2 A body which holds regular meetings with the Federal Occupational Safety and Health Agency, gets together to share good practice amongst members, and offers advice to other states thinking of applying for State Plan status.
3 Of particular relevance to the current discussion is the section of the report highlighting the enforcement activities of the OSHPA states in the previous year and noting key prosecutions brought during that period. A copy of the latest OSHSPA report, 'Grassroots Worker Protection 2011 OSHSPA Report', is accessible online at www.oshspa.org/Files/grass2011.pdf.

14.33 These developments seem to be indicative of a greater willingness on behalf of prosecutors, at both federal and state level, to pursue corporations through the courts where their activities have resulted in serious injury or even death. However, our main area of interest in this chapter is the prosecution of corporations under the criminal law for the offence of manslaughter. Since corporate manslaughter is not a specific criminal offence at federal level, we will need to move on and consider whether the position is any different at state level.

Criminal liability of corporations at state level

14.34 Early attempts to prosecute corporations for manslaughter under state laws met with mixed results. In the case of *State v Lehigh Valley Railroad Company*[1] it was held that a corporation *could* be prosecuted for the offence of homicide. It was said in that case:

> 'A corporation may be held liable for criminal acts of misfeasance or non-feasance unless there is something in the nature of the crime, the character of the punishment prescribed therefore, or the essential ingredients of the crime which make it impossible for a corporation to be held liable. Involuntary manslaughter does not come within any of these exceptions.'

1 (1917) 90 NJL 372, 103 A 685.

14.35 In the case of *People v Rochester Railway & Light Co*,[1] on the other hand, the court rejected the idea that a corporation could be held liable for the offence of homicide, based on a restrictive interpretation of the statutory provision governing the offence of homicide in the State of New York. The offence was defined in terms which required the killing of one human being by another human being. This clearly precluded corporations from the ambit of the offence.

1 (1908) 195 NY 102, 88 NE 22.

14.36 The court in *People v Rochester Railway & Light Co* did, however, adopt an approach similar to that taken later on by the court in *Commonwealth v Illinois Central Railway Co*,[1] a case heard in the State of Kentucky. In this case the court declared that there was no good reason why the legislature could not enact a piece of legislation defined in such a manner as to allow corporations to be found guilty of homicide.

1 (1913) 152 Ky 320, 153 SW 459.

14.37 What the decisions in *State v Lehigh Valley Railroad Co* and *People v Rochester Railway & Light Co* show us is that a court could vary dramatically from state to state in its attitudes towards prosecuting corporations for manslaughter. This changed somewhat, however, with the introduction of the Modern Penal Code in the 1960s.

14.38 The code was devised by the American Law Institute over a period of several years and was eventually adopted by that organisation in 1962. It was the intention of the Institute that their model code should be considered by state legislators when reviewing or reforming their criminal laws.

14.39 Corporate liability for criminal offences was justified on the grounds of 'deterrence efficiency'. The code also purported to put forward a more systematic approach to imposing criminal liability on corporations for offences such as manslaughter. The 'alter ego' doctrine, as adopted by the English courts, had a significant influence upon the approach taken by the code in this regard, limiting prosecutions solely to those instances where direct 'high-level' involvement could be shown. Section 2.07 states:

> 'A corporation may be convicted of an offence if …
>
> (c) The commission of the offence was authorised, requested, commanded, performed or recklessly tolerated by the board of directors or by a high managerial agent acting on behalf of the corporation within the scope of his office or employment.'

14.40 The Model Penal Code was adopted by most state legislatures. An examination of some of the cases that have been considered shows that the state courts, when dealing with the question of corporate homicide, have faced three major problems, namely: (1) the definition of the offence; (2) establishing that the corporation possessed the necessary intention in order for them to be held liable for the offence; and (3) the lack of an appropriate punishment for the offence.

14.41 In relation to the first problem, we will see that, in some states, the offence of homicide is defined in terms of the 'killing of one human being by another human being'. As we have seen, this is problematic, because it clearly excludes corporations from the ambit of the offence. This problem is remedied in other states by defining the offence in terms of the killing of one *person* by another person. In these states, the term 'person' is defined as including both natural and legal persons, ie humans and corporations.

14.42 What about the difficulties involved in punishing a corporation? Corporations present a problem if the only prescribed form of punishment is death or imprisonment, or any other form of punishment which is inapplicable to a corporation. This has proved problematic in some states, as the courts have perceived the failure to provide an 'appropriate' form of punishment as a lack of legislative intent to hold corporations liable for criminal offences. In some states, this problem has been dealt with by setting out the penalties available in terms of both fines and prison sentences. In other states, the legislature has enacted specific penalty provisions applicable to corporations.

14.43 In the next section, we consider the law as it stands in the States of New York, California, Kentucky, Pennsylvania, Florida, Texas and Michigan. These states have been chosen on the grounds that they represent a broad spectrum of demography, geography, political climate and industry.

Corporate criminal liability in the State of New York

14.44 A recent report published by the New York Committee for Occupational Safety and Health revealed that 219 workers were killed as a result of occupational injuries in New York State in 2007. Of those deaths, 81 were in New York City alone.[1] Even prior to the introduction of the current version of the New York Penal Code, it was clear that the attitude of the court towards the possibility of a corporation being held guilty of manslaughter had changed from that evident in *People v Rochester Railway & Light Co*, as outlined earlier.

1 New York Committee for Occupational Safety and Health, 'Dying for Work in New York' (June 2009). The report notes that the fatality figures for 2007 were an improvement on the 2006 figures, which recorded the deaths of 234 workers in New York State, of which 99 were in New York City. The report also notes that immigrant, minority and non-union workers continue to be particularly at risk of suffering a work related fatality.

14.45 In the case of *People of the State of New York v Ebasco Services Incorporated*[1] the defendant had been charged with negligent homicide following the death of two construction workers who died when a cofferdam they were working in collapsed and flooded. The defendant company initiated a motion to get the indictment against them dismissed.

1 (1974) 77 Misc 2d 784, 354 NYS 2d 807.

14.46 On appeal, the court held that a corporation could be indicted for the offence of corporate homicide under s 125.10 of the New York State Penal Code. That section stated that:

> 'A person is guilty of criminally negligent homicide when, with criminal negligence, he causes the death of another person.'

The definition of the offence created a problem for the courts, however, as s 125.05 (1), the homicide section, defined the term 'person' as 'a human being who was born alive and well'. Unsurprisingly, the defendants contended that, since the definition of person made reference to human being, this meant that a corporation could not commit homicide. This contention was rejected by Wallace J, who claimed that the defendant's approach 'flies in the face of the statute which equates "persons" with human beings only in regard to the victim of the homicide'. Wallace J went on to argue that there was no requirement in the statute that the person committing the offence had to be a human being and, as such, the provision's reference to the term 'human being' was of limited application.

14.47 Since no other definition of 'person' was available elsewhere in the provisions dealing with the offence of homicide, the court moved on to consider the overall definitional provision contained in s 10.00 (7), which provided that the term 'person' could include corporations except in those circumstances where it would be deemed inappropriate. Since it had already been stated in the earlier decision of *People v Rochester Railway and Light Co* that it was entirely appropriate to indict a corporation for manslaughter, the defendant company was clearly within the ambit of the homicide offence as defined in the Penal Code. The case was, however, dismissed on a technicality, namely that the indictment was defective in that it was not particular enough in the alleged facts.

14.48 There have been some amendments to the New York Penal Code. Section 125 is still the section that deals with homicide. Section 125.00 defines homicide as:

> 'conduct which causes the death of a person or an unborn child with which a female has been pregnant for more than twenty-four weeks under circumstances constituting murder, manslaughter in the first degree, manslaughter in the second degree, criminally negligent homicide, abortion in the first degree or self-abortion in the first degree'.[1]

Furthermore, s 125.05 states that the term person 'when referring to the victim of a homicide, means a human being who has been born and is alive'.[2]

1 Notice that there is no mention of a requirement for the killing to be carried out by a 'person'.
2 Notice how the definition only requires the *victim* of the homicide to be a human being.

14.49 Looking at more specific homicide offences, a 'person' will be convicted of criminally negligent homicide when 'with criminal negligence, he causes the death of another person'.[1] Similarly, a 'person' will be guilty of manslaughter in the second degree where they 'recklessly cause the death of another person'[2] or manslaughter in the first degree where 'with intent to cause serious physical injury to another person, he causes the death of such person or of a third person'.[3]

Theoretically speaking, a corporation can be convicted of all of these offences, although first degree manslaughter in the case of a corporation would probably be extremely rare.

1 Section 125.10 of the New York Penal Code. The general definition of the term 'person' is still contained in s 10.00. The term 'person' means 'a human being, and where appropriate, a public or private corporation, an unincorporated association, a partnership, a government or a governmental instrumentality'. Terms such as 'criminal negligence', 'intentionally' and 'recklessly' are defined in s 15.05 of the Code.
2 Section 125.15 of the New York Penal Code.
3 Section 125.20 of the New York Penal Code.

14.50 There have been other significant prosecutions for homicide offences under New York State Law. On 21 November 1976 a huge explosion ripped through four floors of the American Chicle gum factory in Queens, New York, USA. A total of six workers died and dozens more were injured. The owner of the factory, Warner-Lambert Company, and four other employees were indicted on charges of reckless manslaughter and criminally negligent homicide. The explosion was caused when magnesium stearate (MS) dust, which was used to dust the chewing gum manufactured at the plant, ignited.

14.51 A subsequent investigation revealed that Warner-Lambert had been advised by their insurers that the presence of MS dust in the factory created an explosion risk. Rather than take the steps recommended by their insurers, Warner-Lambert opted to try and eliminate the presence of MS dust in the atmosphere by modifying their work equipment. They were in the process of carrying out these modifications when the explosion occurred.

14.52 Although it was clear from this evidence that the defendants were aware of a broad risk of explosion from MS dust in the plant, the indictments against the company and the four employees were dismissed after the court held that, whilst the explosion of MS dust might have been foreseeable, the actual, immediate triggering cause of the explosion was not.[1]

1 *People v Warner-Lambert Co*, 51 NY 2d 295, 414 NE 2d 660, 434 NY S 2d 159 (1980).

14.53 The outcome in *People v Warner-Lambert* can be compared with that in *People v Roth*.[1] In that case an employee of the corporate defendant died when petroleum vapours exploded as he was cleaning a tanker trailer. The trigger for the explosion was an electric spark from a non-explosion proof 'troublelight'. The evidence suggested that the spark was caused when it was struck by a jet of water from the high pressure washer operated by the victim.

1 *People v Roth*, 80 NY 2d 239, 604 NE 2d 92, 590 NY S 2d 30 (1992).

14.54 The defendant corporation and the defendants, Roth and Wilson, the district manager and operations manager of the facility, were indicted for second degree manslaughter, criminally negligent homicide, reckless endangerment, endangering public health, safety or environment in the second degree, and a variety of other charges relating to their handling and documentation of hazardous waste. However, the homicide charges against the defendants were dismissed.

14.55 In line with the decision in *People v Warner-Lambert*, the court held that:

> 'it was not enough to show that, given the variety of dangerous conditions existing at the site, an explosion was foreseeable; instead the People were required to show that it was foreseeable that the explosion would occur in the manner that it did.'

The court was forced to conclude, on the basis of the case as presented to it, that:

> 'the evidence before the grand jury is insufficient to support the conclusion that the defendants should have foreseen that their employee would place the

unprotected trouble light in the path of the high pressure washer during the tank cleaning operation and that an explosion-causing spark would result from this combination.'

Once again, therefore, a corporation avoided liability for homicide owing to a lack of a causal link between the corporation's actions and the victim's death.

14.56 In the late 2000s, New York State witnessed a flurry of indictments against corporate defendants for homicide offences. In August 2007, fire-fighters were called to attend a fire at the Deutsche Bank skyscraper in Manhattan, New York, USA.[1] The immediate cause of the fire was subsequently identified as a cigarette which had been carelessly disposed of. The building had been empty since the terrorist attack on the World Trade Center in 2001. Following the collapse of the 'Twin Towers' the building had been left severely damaged and filled with toxic debris. Owing to this contamination the building was identified for demolition, and plans were put in place to decontaminate and demolish the building at the same time, something which had never been attempted before in the City of New York. Two fire-fighters lost their lives in the fire and 105 fire-fighters were injured.

1 Facts taken from R Rivera and F Santos, '2 Fire-fighters are Killed in Blaze at Ground Zero', (2007) *New York Times*, 19 August, and Robert Morgenthau, the Manhattan District Attorney's report into the fire – 'Statement by the District Attorney: The Deutsche Bank Fire', published on 22 December 2009.

14.57 The Manhattan District Attorney's Office released a statement in December 2008 identifying a number of factors which contributed to the magnitude of the blaze and made it harder to combat effectively. As part of the demolition process the building's sprinkler system had been permanently disabled. This meant that the only way to get water into the building to fight any fire was through the building's standpipe system.[1] However, a 42-foot section of the building's standpipe had been removed[2] which, in turn, meant that fire-fighters' attempts to get water through the building's riser system to their colleagues on the upper floors who were trying to tackle the fire were fruitless. Instead, they were forced to use a construction hoist and physically haul hose line up to the 14th floor outside the building. It took 61 minutes before those fire-fighters who first entered the building got any water.

1 The standpipe rises up through the building, allowing fire-fighters to access water on each floor.

2 This section of the pipe had become damaged during the decontamination process. The bosses at Galt, the company contracted to carry out the demolition and decontamination of the Deutsche Bank skyscraper, had consistently told their workers that they needed to speed up the process of cleaning the pipes.

14.58 The fire-fighters had entered the building on the 15th floor, two floors below where the fire had started, in accordance with Fire Department procedures, and set up a staging area there. When they attempted to move up to the 16th floor, however, their progress was hampered by containment barriers which blocked the stairwells.[1] By the time water finally arrived, however, smoke and fire began to move rapidly downwards, the thick smoke reducing visibility to near zero. Fire-fighters moved down to the 14th floor, to try and escape the smoke and choking conditions, but the 14th floor also quickly became smoke filled. At this point the fire-fighters tried to move to the 13th floor but were confronted by another barrier blocking their route down the stairwell. At this point, chaos ensued.

1 These had been installed as part of procedures introduced by Galt, which required those floors that were being decontaminated to be sealed off from the other floors. It reportedly took the fire-fighters 20 minutes to get through the barrier between the 15th and 16th floors.

14.59 On 22 December 2008 the Manhattan District Attorney's Office announced the indictment of three individuals and one corporate defendant on two counts of manslaughter, criminally negligent homicide and reckless endangerment. The findings of the District Attorney's Office investigation into the causes of the deaths of the two fire-fighters makes for grim reading. The report identifies the damage to the standpipe as the ultimate cause of the fire-fighters' deaths and highlights failings on the part of: Bovis Lend Lease LMB, Inc (Bovis), the company selected by the building's owner (the Lower Manhattan Development Corporation) to be the construction manager for the demolition and decontamination project; the John Galt Corporation (Galt),[1] the company sub-contracted by Bovis to actually carry out the decontamination and deconstruction work; and the City of New York, for the failure of its agencies, the Fire Department of New York (FDNY)[2] and the Department of Buildings (DOB).[3] As the District Attorney, Robert M. Morgenthau commented at the end of his December press release, 'In summary, everyone failed at the Deutsche Bank Building'.

1 The report revealed that the standpipe was damaged when it was cleaned as part of the decontamination process. This, in turn, was the result of orders from above which made it clear to the workers involved that they needed to work faster. Once the pipe was damaged, however, this was not reported, despite the fact that those involved in directing the clean-up should have been aware of the importance of the standpipe.

This problem was exacerbated by the fact that one of the individual defendants, Jeff Melofchik (the site safety manager for Bovis on this project), had not been completing his daily safety checklist properly. It appears that the boxes were just being ticked without an inspection actually being carried out. This meant that the paperwork suggested that the building's standpipe was in working order around the time of the fire when, in fact, the pipe had been damaged since autumn of the year before.

A further consequence of Jeff Melofchik's sub-standard approach to completing his duties was that the safety checklist indicated that 'the stairways were clean and clear' and that 'paths of emergency egress' were clear, in spite of the fact that parts of the stairwell were blocked off, creating what was subsequently described as 'maze like conditions' (R Rivera, and F Santos, '2 Fire-fighters are Killed in Blaze at Ground Zero', (2007) *New York Times*, 19 August).

The report also noted that it was common knowledge amongst the Bovis site superintendents that people other than Jeff Melofchik signed off the checklist (using Melofchik's initials). All this points to a poor safety culture and, crucially, a poor safety culture that was tolerated by more senior members of Bovis' management.

2 The failings of the FDNY were, in many ways, just as severe as those of Bovis and Galt. The report revealed that the '15-day rule', which required fire-fighters to inspect buildings undergoing construction and demolition 'at least every 15 days, but more when conditions dictate', was being regularly ignored by fire companies all across the City of New York. Furthermore, high-ranking FDNY officers knew this to be the case. Secondly, despite receiving a number of recommendations to do so, the FDNY had failed to develop a special fire-fighting plan for the Deutsche Bank building.

The consequences of FDNY's failings were twofold. First, FDNY missed an opportunity to discover the damage to the Deutsche Bank building's standpipe, which would have been checked as part of any inspection under the 15-day rule. Secondly, since no inspection was carried out on the Deutsche Bank building since the time of the attacks on the World Trade Center, and since there was no fire-fighting operations plan in place for that building, fire-fighters were sent into the building to combat a fire with no knowledge of what they would face.

3 The Department of Buildings was found to be guilty of two key failings. It was discovered that the buildings inspectors missed an opportunity to detect the damage to the standpipe because they never once ventured into the section of the basement where the damaged pipe was located. The report notes that, if the investigators had just once traced the standpipe connection in that section of the basement, the damage would have been discovered and repaired.

The second key failing stems from the fact that many of the investigators assigned to monitor this project were inexperienced. As such, although the inspectors were aware that sections of the stairwell were blocked off, they may not have been aware that such practices were in breach of the City of New York's building code.

14.60 Although the case against Galt is yet to be resolved, the Manhattan District Attorney's Office entered into non-prosecution agreements with both the City of New York and Bovis.[1] The decision to enter into a plea agreement with the City of New York was a pragmatic one.[2] Because of the principle of sovereign immunity the City was protected from indictment unless it waived that immunity. Whilst the District Attorney had the option of litigating the issue through the courts, the decision was made to focus on getting the City to implement a wide variety of reforms designed to address the failings that lead to the fire-fighters' deaths.[3]

1 The concept of plea agreements is discussed below (with specific focus on Deferred Prosecution Agreements (DPAs)). In a Non-Prosecution Agreement the District Attorney will agree not to file criminal charges against the accused on condition that they fulfil certain obligations as set out in the terms of the agreement (P Spivack and S Raman, 'Essay – Regulating the "New Regulators": Current Trends in Deferred Prosecution Agreements' (2008) 45 American Crim L Rev 159 at p 160).

2 A copy of the non-prosecution agreement between the District Attorney's Office and the City of New York can be found online at www.thebravest.com/FDNYNewsArchiveDown/09/0709/DeutscheBank/Bovis%2520agreement%2520with%2520Fire%2520Safety%2520Initiatives.pdf. As part of the terms of the agreement, the City of New York publicly admitted responsibility for the failings which led to the deaths of the two fire-fighters. This move surprised many, with some lawyers reportedly suggesting that the City's statement might amount to an admission of liability or culpability in any future civil claims that might be brought by the families of those killed or injured in the fire (see WK Rashbaum and CV Bagli, '3 Men Indicted in Tower Blaze, but Not the City' (2008) *New York Times*, 23 December).

The agreement also required the City to set up a specialist inspectorate whose only job would be to conduct inspections at the site of buildings undergoing construction or demolition (to be known as the CDA). The team would pay particular attention to the condition of the buildings' standpipes. Additionally, local fire companies would be expected to carry out familiarisation drills at the site of all buildings under construction or demolition in their area, although this might be changed to a drill every 30 days in the case of particularly risky projects. The fire companies would be assisted by the CDA teams in carrying out follow-up investigations. Finally, the FDNY was also required to add two new civilian members to its Fire Prevention Suppression Unit to carry out random audits of all inspections of buildings under construction or demolition, as well as auditing standpipe systems in ten per cent of those buildings under construction or demolition in the city.

Six months after entering into this agreement, Francis X Gribbon, chief spokesman for the Fire Department, announced that there was now a Fire Battalion Chief assigned full time to oversee inspections at the Deutsche Bank building. Furthermore, he announced that compliance with the '15-day rule' was now 'virtually at 100%' – as reported in the *New York Times* (A Baker, 'City Agencies Faulted in Deutsche Bank Fire', (2009) *New York Times*, 20 June).

Similarly, Tony Sclafani, chief spokesman for the Department of Buildings, announced that buildings inspectors had received more joint training with fire-fighters and inspectors form the Department of Environmental Protection than ever before. He also asserted that the Department of Buildings had strived to implement tough new safety standards to ensure greater safety at job sites (A Baker, 'City Agencies Faulted in Deutsche Bank Fire', (2009) *New York Times*, 20 June).

Unlike the City of New York, Bovis did not make any admission of civil or criminal liability in their public statement released as part of the plea agreement. Instead, Bovis simply said that they did not challenge 'the factual conclusions of the investigation and acknowledged that some of its supervisors had not conducted inspections that would have revealed that the standpipe was not in working order'.

This point was confirmed by the New York courts in January 2013, when it was found that the non-prosecution agreement could not be deemed admissible as proof of liability. As such, a partial summary judgment against the company was not allowed.

The non-prosecution agreement imposed a number of requirements on Bovis. First, they were required to develop a comprehensive standpipe, smoking prevention and first responder safety programme at the Deutsche Bank building site and at all other Bovis sites across New York.

Secondly, Bovis were required to agree to a number of changes to their management and staff in order to demonstrate their commitment to fire safety. The required changes included hiring a Senior Fire Safety Manager approved by the District Attorney, hiring a new Regional Safety Director, assigning executive responsibility for direct supervision of New York operations, including safety, to the Chief Operating Officer of Bovis, and firing all of those responsible for Bovis' shortcomings at the Deutsche Bank site.

Thirdly, Bovis agreed to appoint an independent monitor, approved by the District Attorney and paid for by Bovis, to oversee the implementation and effectiveness of Bovis' safety initiatives over the next five years. The monitor would be required to report back his findings to the District Attorney once every six months. Fourthly, Bovis were required to pay out $2 million to fund the setting-up of a Fire Safety Academy for the training of New York City construction workers. The Academy would also carry out research into the development of fire safety initiatives.

Finally, Bovis set up a memorial fund containing $10 million ($5 million for the family of each of the deceased fire-fighters). Bovis agreed that this figure would not be used as a set-off against any additional damages they might be required to pay in the event of a civil suit being brought against them by the families. Furthermore, Bovis had to agree that they would not seek to claim the payment of these monies as a tax deduction at any time, nor would they seek to have this sum paid by their insurers.

3 As reported by W K Rashbaum and C V Bagli, '3 Men Indicted in Tower Blaze, but Not the City', (2008) *New York Times*, 23 December.

14.61 The criminal prosecution of the three individuals and Galt came to an end in July 2011, when all parties were cleared of charges of manslaughter and criminally negligent homicide. If convicted on the counts of manslaughter in the second degree, each individual defendant would have faced between 5 and 15 years in jail for each count. If convicted on the two counts of criminally negligent homicide, each individual defendant faced one year and four months to four years in jail for each count. The family of one fire-fighter still has a pending civil suit filed against Galt.

1 J Riley, 'Union seeks new Deutsche probe', (2011) *Newsday*, 8 July.

14.62 Recent developments have seen both families of the fire-fighters accepting their share of the memorial fund set up by Bovis as part of the non-prosecution agreement they entered into with the Manhattan District Attorney. The total amount of compensation paid to the two families by Bovis and New York City was $16 million. One family still has a civil lawsuit pending against Galt.[1] The misery continued for the John Galt Corporation after their former purchasing agent was arrested and charged with stealing more than $1 million by filing false invoices.[2] In April 2009 it was announced that the prosecution had released evidence to the defence in the form of statements from electricians who had worked on the Deutsche Bank building. The electricians claim that the standpipe was still intact at the time the John Galt Corporation completed working in that part of the building in 2006.[3] Finally, seven Fire Department Officers received a censure on Wednesday 24 June 2009 for failing to ensure timely inspections at the Deutsche Bank building in the period between the attacks on the World Trade Center and the fatal fire.[4]

1 G B Smith, '$10m for her Brave Hubby: Deutsche Widow Settles with Firm and City', (2012) *Daily News*, 20 May.

2 See J Eligon, '$1 Million theft Charges at Troubled Demolition Project', (2009) *New York Times*, 7 January.

3 J Eligon, 'Prosecution Gives Defense New Evidence in Fatal Fire at Deutsche Bank', (2009) *New York Times*, 24 April. In spite of this fresh development, John Eligon reports that the Manhattan District Attorney's Office remain confident that they will succeed in their case.

4 A Baker, '7 Fire Dept. Officers Censured in Bank Blaze Inquiry', (2009) *New York Times*, 25 June.

14.63 A second unresolved indictment of a corporation for criminally negligent homicide followed the collapse of a 22-storey crane on 15 March 2008 at a construction site at 303 East 51st Street, Manhattan, New York, USA. The episode left seven people dead, namely six construction site workers (including the crane operator) and a tourist. A report from the New York City Buildings Department[1] concluded that poor rigging practices had caused the incident.

1 New York City Department of Buildings '51st Street Crane Investigation Report' (March 2009).

14.64 As part of a process known as 'jumping the crane', in which the crane rigging crew install several additional sections to a crane in order to increase its height, the crane's rigging crew had to install a steel collar around the mast of the crane. The collars are then attached to an adjacent building with steel ties in order to provide the crane with additional lateral support. Whilst the collar is being connected to these steel ties, it needs to be temporarily held in place with slings, and slings were used in this case. When these slings snapped, however, the now unsupported collar, which weighed approximately 12,000 pounds, came sliding down the crane's mast and crashed into a second collar attached lower down. The first loose collar crashed into the second collar with enough force to dislodge it. Both collars then slid down the crane's mast and crashed into a third collar. This left the crane totally unsupported and the crane, which by this stage was unstable, rocked by the force of the collars colliding and top-heavy, toppled over.[1]

1 Facts taken from New York City Department of Buildings '51st Street Crane Investigation Report' (March 2009), see pp 238–240.

14.65 On 5 January, 2009, the Manhattan District Attorney's Office announced that William Rapetti and a company, Rapetti Rigging Services Inc, had been indicted on multiple counts of criminally negligent homicide, assault, reckless endangerment and failing to file tax returns. The prosecution points to several key failings which led to the incident, namely: (1) the decision to use only four polyester slings, even though the crane's manufacturer had recommended the use of eight chain blocks during this procedure; (2) of the four slings used, one was badly worn showing evidence of cuts and discoloration, issues that should have been spotted by Mr Rapetti if he had inspected the slings properly; (3) one of the slings had been tied around sharp metal edges of the crane without using protective padding; and (4) the knot used by Mr Rapetti is the weakest of the three varieties of knots typically used during the 'jumping' process.

14.66 Mr Rapetti and his company, Rapetti Rigging Services Inc, were both found not guilty in July 2010 of charges of second-degree manslaughter (which would have resulted in up to 15 years in jail if found guilty), criminally negligent homicide, second-degree assault, reckless endangerment and failure to submit a tax return. Civil charges were also brought against Rapetti Rigging Services, although these were settled out of court for undisclosed sums.[1]

1 J Eligon, 'Rigging Contractor is Acquitted in the Collapse of a Crane', (2010) *New York Times*, 23 July. The investigations which followed the collapse of the crane on 51st Street, and the collapse of a second crane just two months later in Manhattan's 91st Street (see M M Grynbaum, 'Crane Collapse Kills Two and Unsettles New York', (2008) *New York Times*, 31 May), revealed widespread mismanagement, corruption and poor practices within the City of New York's Buildings Department – which was, of course, subject to criticism following the Deutsche Bank building fire, as discussed earlier – particularly within the Department's Cranes and Derricks Unit. This subsequently resulted in arrest and indictment of a buildings inspector and of the City of New York's chief crane inspector.

Edward J Marquette, a buildings inspector, was indicted on charges of tampering with public records, filing a false instrument, falsifying business records and official misconduct. Marquette had originally told investigators that he had inspected the crane at 51st Street on 4 March 2008 (11 days before it collapsed) and found nothing wrong. He eventually confessed to lying about the investigation and to filing false documentation to support his claims. At the time, the buildings commissioner, Patricia J Lancaster, announced that it was 'highly unlikely' that a properly conducted inspection would have prevented the tragedy – see W Neuman and A Hartocollis, 'Inspector is Charged with Filing False Report Before Crane Collapse', (2008) *New York Times*, 21 March.

Marquette was later linked to two further false inspection reports for two other cranes (see J Eligon, 'Crane Inspector Is Tied to More False Reports', (2008) *New York Times*, 4 July). The suggestion was subsequently raised that Mr Marquette's 'corner cutting' was linked to a combination of under-resourcing, a lack of experienced inspectors and an unprecedented building boom (see W K Rashbaum, 'Officials Broaden Focus of Inquiry into Corruption in Crane Inspections', 12 June 2008). These were, of course, factors which, it was suggested, led to inexperienced inspectors monitoring proceedings at the Deutsche Bank building.

The City of New York's chief crane inspector fared little better. On 7 October 2008 the Manhattan District Attorney's Office announced the indictment of James Delayo (along with other defendants) on charges relating to a bribery scheme. Mr Delayo stood accused of taking thousands of dollars in bribes, dating back to 1996, for allowing cranes to pass inspections and for helping the employees of one company to ensure that they passed a licensing test to operate smaller cranes. It transpires that Mr Delayo had sold a copy of the test and its answers to the crane company in question (see W K Rashbaum, 'Officials Broaden Focus of Inquiry into Corruption in Crane Inspections', 12 June 2008).

14.67 A successful 2009 verdict against a company for criminally negligent homicide and reckless endangerment was overturned by the State Supreme Court in 2010, in what was seen as an unusual step. The conviction followed the death of two fire-fighters in a building fire on the morning of 23 January 2005.[1] The two fire-fighters died after throwing themselves from a fourth floor window to avoid being burnt. The fire-fighters had become trapped in the building by partitions which had been installed by two of the building's tenants in order to subdivide their apartments into further rooms.[2] These rooms were then illegally sub-let by the tenants. As a consequence, the fire-fighters were prevented from gaining access to a fire escape.

1 See the press release from the Bronx District Attorney's Office online at http://bronxda.nyc.gov/information/2009/case5.htm.

2 This fact led to a separate prosecution of the tenants, Rafael Castillo and Caridad Coste, who were accused of contributing to the deaths by illegally erecting partition walls within their apartments. Both defendants were charged with manslaughter, criminally negligent homicide and reckless endangerment, but the jury acquitted them on all counts. It was later revealed that the jury took the view that the blame lay with the Fire Department rather than the tenants (J Eligon, '2 Bronx Tenants Acquitted in Fire-fighter Deaths', (2009) *New York Times*, 14 February). An earlier internal investigation carried out by the FDNY into events surrounding the deaths of the fire-fighters had revealed shortcomings in its equipment and procedures. It should be pointed out that the case against the tenants was heard by a separate jury from that hearing the case against Cesar Rios and 234 East 178th Street LLC. This allowed some evidence to be heard by both panels, but other evidence to be heard by only one of the panels, on the ground that some of the testimony might have been prejudicial to a co-defendant (as reported by John Eligon, '2 Bronx Tenants Acquitted in Fire-fighter Deaths', (2009) *New York Times*, 14 February).

14.68 The prosecution submitted that 234 East 178th Street Limited Liability Company and Cesar Rios, the building manager who was deemed by the prosecution to be an agent of the defendant corporation, had been aware of the illegal partitions yet had done nothing about them. According to the prosecution, this amounted to a 'reckless toleration' of dangerous conditions.[1] The defence argued, however, that the defendants were not liable. They suggested that the defendants were not aware of the presence of the illegal partitions. In the case of Mr Rios, his lawyer claimed that his client had done everything within his powers, short of '[taking] it down with his bare hands'.[2] The lawyer for 234 East 178th Street LLC also argued that the blame being put on his client was misplaced. Mr Comer contended that, whilst the company owned the building, 'They don't know what's going on in each apartment'.[3]

1 As quoted in the press release from the Bronx District Attorney's Office online at http://bronxda.nyc.gov/information/2009/case5.htm.

2 As quoted in R Blumenthal, 'At Trial, Prosecutor Faults Tenants and Landlord in '05 Firefighter Deaths', (2009) *New York Times*, 6 January (see also J Eligon and M R Warren, 'Bronx Landlords Guilty in 2 Firefighters' Deaths', (2009) *New York Times*, 19 February). Evidence had been presented during the trial that Mr Rios had interrogated the daughter-in-law of one of the defendant tenants, Rafael Costillo, a mere four days before the fire. Mr Rios had, reportedly, demanded to know whether the illegal partition he had ordered Mr Costillo to tear down a year before had been rebuilt after he had heard rumours that this was indeed the case. The daughter-in-law testified that she told Mr Rios she did not know whether the wall had been rebuilt because she was afraid of getting Mr Costillo into trouble and evicted. See M Grace, 'Black Sunday Super Didn't Know About Illegal Wall, Tenant Says', (2009) *New York Daily News*, 22 January.

3 As quoted in R Blumenthal, 'At Trial, Prosecutor Faults Tenants and Landlord in '05 Firefighter Deaths', (2009) *New York Times*, 6 January (see also J Eligon and M R Warren, 'Bronx Landlords Guilty in 2 Firefighters' Deaths', (2009) *New York Times*, 19 February).

14.69 Following the conviction for criminally negligent homicide, lawyers for Mr Rios and the defendant company announced they intended to seek a motion to get the convictions overturned. It was reported that the lawyer for the corporation, Mr Neal Cromer, had announced his surprise at the verdict in light of the acquittal of the two tenants who had installed the partition walls which created the immediate problem. Mr Cromer's surprise was mirrored by the lawyer representing the acquitted tenant, Caridad Coste. Francisco Knipping-Diaz is reported to have described the different verdicts as akin to this: 'I lend my car to you, you have an accident, and we have a trial, and they find you not guilty of the accident and find me guilty'.

The case was overturned in February 2010 when the judge found that the prosecution had failed to prove that the company and Cesar Rios knew about the partitions: 'An individual or entity cannot be convicted of a crime without evidence of actual knowledge'.

Corporate criminal liability in the State of California

14.70 In the *Census of Fatal Occupational Injuries 2007* released by the US Bureau of Labor Statistics, the State of California had the second highest rate of workplace fatalities in 2007 of all the American States.[1] The starting point for our discussion of corporate liability for homicide in the State of California is s 17 of California's Penal Code. This section simply states that 'person' 'includes a corporation as well as a natural person'. In the case of corporations we might normally expect to be dealing with the offence of involuntary manslaughter, which is defined as 'the unlawful killing of a human being without malice … in the commission of an unlawful act, not amounting to a felony; or in the commission of a lawful act, which might produce death, in an unlawful manner, or without due caution and circumspection'.[2] In theory, therefore, it should be possible to prosecute a corporation for homicide. A potential difficulty arises, however, in the form of s 193 of the Californian Criminal Code.

1 The census shows that California had 339 workplace fatalities in 2012 compared to the 390 workplace fatalities recorded in 2011. California came second only to Texas. The census figures can be viewed online at www.bls.gov/news.release/cfoi.t05.htm.

2 Section 192 of the California Penal Code.

14.71 Section 193 of the Code sets out the available sanctions for the main categories of manslaughter. The only punishment available for a defendant convicted of involuntary manslaughter is imprisonment in the state prison for two, three or four years. A corporation, having no physical existence, cannot be incarcerated. This issue came up for consideration in the case of *Granite Construction Co v The Superior Court of Fresno County.*[1]

1 (1983) 149 Cal App 3d 465.

14.72 The corporation in that case had been prosecuted following the death of seven workers during the construction of a power plant. The court, however, was presented with a problem. Section 7 of the Californian Penal Code, as we have seen, gives a definition of the term 'person' which includes corporations. Furthermore, the offence of involuntary manslaughter we have already considered is not limited in its definition to human offenders. The lack of provision in s 193 for any penalty apart from imprisonment, however, could have been taken as an indication that a corporation was never meant to be prosecuted for the offence.

14.73 The way the court got around this problem in *Granite Construction* was by making reference to s 672 of the Californian Penal Code, which provided an alternative mechanism for corporate criminal offenders, namely a catch-all fine system enacted in 1872. Since there was a mechanism available to punish a corporation, even where imprisonment is the only prescribed penalty, there was nothing to prevent the court from convicting the defendant. Section 672 of the Californian Code, which is still in force today, states:

> 'Upon a conviction for any crime punishable by imprisonment in any jail or prison, in relation to which no fine is herein prescribed, the court may impose a fine on the offender not exceeding one thousand dollars ($1,000) in cases of misdemeanors or ten thousand dollars ($10,000) in cases of felonies, in addition to the imprisonment prescribed.'

The most recent example available of a successful prosecution for manslaughter following a workplace fatality in the State of California dates back to 2002, when Christie Binn Chung, a Californian roofing contractor, was charged and pleaded guilty to 52 felony violations, including a charge of involuntary manslaughter[1] after a worker fell through an unsafe roof and plummeted 40 feet to his death. The deceased's 'fall protection', such as it was, reportedly consisted of a rope tied to his waist and leashed to a fixed point. The rope was untied at the time of the fall, as the deceased moved from one section of the roof to another.

1 It should be noted that the contractor's firm, 101 Roofing, was not charged with involuntary manslaughter, although it was charged with violations of the State Occupational Safety and Health Act.

Corporate criminal liability in the State of Kentucky

14.74 The position in Kentucky was originally set out in the case of *Commonwealth v Illinois Central Railway Co.*[1] The court refused to allow the prosecution of a corporation for homicide in this case because the offence was defined in terms which required evidence of the killing of one human being by another human being. Corporations clearly did not fall within this definition.

1 (1913) 152 Ky 320, 153 SW 459.

14.75 An attempt was made by the defendant corporation in the case of *Kentucky v Fortner LP Gas Company Inc*[1] to rely on the judgment in *Illinois Central Railway Co*. The defendants had been charged with manslaughter after two children were run over by a truck, owned by the defendant company as they were, getting off a school bus. The truck driver had slowed down and applied the brakes but the truck failed to stop. A subsequent investigation revealed that the truck's brakes were defective.

1 610 SW 2d 941.

14.76 The defendant corporation claimed that the indictment against them should be dismissed following the decision in *Illinois Central Railway Co*, where the court stated that 'corporations cannot be indicted for offences which derive their criminality from evil intention or which consist in a violation of those social duties'.

14.77 Gant J dismissed this contention on the grounds that the case was heard a long time ago and that there had been significant statutory developments since that date. Gant J also observed that the Kentucky Penal Code provided that a corporation could be held liable for criminal offences, as well as stating that the term 'person' could include a corporation within its definition.

14.78 Bearing all this information in mind, Gant J was of the opinion that this was indicative of a strong legislative intent to hold corporations liable for criminal offences. His beliefs were given added weight by two key provisions within the Penal Code, the first of which was s 502.050. This sets out the general provision dealing with corporate liability, and states that a corporation is guilty of an offence where the conduct constituting the offence:

'(a) … consists of a failure to discharge a specific duty imposed on corporations by law; or
(b) … is engaged in, authorized, commanded or wantonly tolerated by the board of directors or by a high managerial agent acting within the scope of his employment in behalf of the corporation; or
(c) … is engaged in by an agent of the corporation acting within the scope of his employment and in behalf of the corporation and:
1. The offense [sic] is a misdemeanour or a violation; or
2. The offense [sic] is one defined by a statute which clearly indicates a legislative intent to impose such criminal liability on a company.'[1]

1 For the purposes of this section, the term 'agent' means 'any officer, director, servant or employee of the corporation or any other person authorized to act on behalf of the corporation' (see s 502.050 (2)(a)). Similarly the term 'high managerial agent' means 'an officer of a corporation or any other agent of a corporation who has duties of such responsibility that his conduct reasonably may be assumed to represent the policy of the corporation'.

14.79 Section 502.050 leaves little doubt that corporate offenders were intended to fall within the ambit of the criminal law.[1] Gant J also referred, however, to s 534.050 which sets out fines specifically designed for application in the case of corporate offenders. Considered together, Gant J would have found it difficult to conclude anything other than the fact that the state legislature intended to allow the prosecution of corporations for criminal offences including homicide.

1 The influence of the 'alter ego' doctrine, as used for a long time by the courts in England and Wales in relation to those corporations charged with manslaughter, is clearly evident here.

14.80 The different categories of homicide offences under Kentucky State law are found in s 507. The offence of first degree manslaughter will be committed where a person 'with the intent to cause serious injury to another person … causes the death of such a person or of a third person'.[1] Second degree homicide (much more likely than first degree homicide in the case of corporate offenders) will be committed where 'a person … wantonly causes the death of another person.'[2] Finally, reckless homicide will be committed where a person causes the death of another person with recklessness.[3]

1 Section 507.030 (1)(a) of the Kentucky Penal Code. According to s 507.030 of the Code, first degree manslaughter is a class B felony. A reading of s 534.50 tells us that a corporation convicted of such an offence faces a fine of up to $5,000 or double the amount of the defendant's gain from the commission of the offence, whichever is the greater (see s 534.050 (1)(c) and (e)). The term 'intentionally' is defined in s 501.020 (1) of the Kentucky Penal Code.
2 Section 507.040 (1) of the Kentucky Penal Code. According to s 507.040 (2) of the Code, second degree manslaughter is a class C felony. A reading of s 502.50 tells us that a corporation convicted of such an offence faces a fine of up to $20,000 or double the amount of the defendant's gain from the commission of the offence, whichever is the greater (see s 502.050 (1)(a) and (e)). The term 'wantonly' is defined in s 501.020 (3) of the Kentucky Penal Code.
3 Section 507.050 (1) of the Kentucky Penal Code. According to s 507.050 (2) of the Code, reckless manslaughter is a class D felony. A reading of s 502.50 tells us that a corporation convicted of such an offence faces a fine of up to $20,000 or double the amount of the defendant's gain from the commission of the offence, whichever is the greater (see s 502.050 (1)(a) and (e)). The term 'recklessly' is defined in s 501.020 (4) of the Kentucky Penal Code.

Corporate criminal liability in the State of Pennsylvania

14.81 The Pennsylvanian Criminal Code is a little different from any of the codes we have considered so far. A key difference is the simple fact that the code contains no definition of the term 'person'; however, s 307 of the Code is dedicated to the liability of organisations and 'related persons'. It states explicitly that:

'(a) Corporations generally – A corporation may be convicted of the commission of an offence if

1. the offence is a summary offence or the offence is defined by a statute other than this in which a legislative purpose to impose liability on corporations plainly appears and the conduct is performed by an agent of the corporation acting on behalf of the corporation within the scope of his office or employment, except that if the law defining the offence designates the agents for whose conduct the corporation is accountable or the circumstance under which it is accountable, such provisions shall apply;
2. the offence consists of an omission to discharge a specific duty of affirmative performance imposed on corporations by law; or
3. the commission of the offence was authorized, requested, commanded, performed or recklessly tolerated by the board of directors or by a high managerial agent acting on behalf of the corporation within the scope of his office or employment.'

14.82 There has certainly been no shortage of cases in the state dealing with the criminal liability of corporations for homicide. In the case of *Commonwealth v McIllwain School Bus Lines Inc*,[1] for example, the defendant company was found guilty of the offence of vehicular manslaughter, despite the fact that the legislature had 'inadvertently' failed to provide a suitable punishment for a corporate offender.

1 (1980) 283 Pa Super 350, 443 A 2d 1157.

14.83 Five years later, a corporation was convicted, inter alia, of involuntary manslaughter following the death of an underage drinker in a road accident.[1] It was alleged that the president of the defendant corporation had knowingly served drink to an underage patron at a college students' dinner held at his resort. Furthermore, the president was accused of failing to require proof of the patron's age and of serving him alcohol whilst the patron was visibly intoxicated. According to the court, this all pointed to the fact that the president, and therefore the corporation, had acted in a 'reckless or grossly negligent manner'.

1 *Commonwealth of Pennsylvania v Penn Valley Resorts Inc* 343 Pa Super 387, 494 A 2d 1139.

14.84 In its defence the corporation claimed that it was not a 'person' and, as such, it could not be convicted of the offence of involuntary manslaughter or reckless manslaughter. Furthermore the defendant company claimed that it could not be liable because the president's actions were not 'condoned, sanctioned or recklessly disregarded by the Board of Directors'. This would, of course, have fallen foul of s 307(a)(3) as outlined above.

14.85 It had been declared in the earlier case of *Commonwealth v Schomaker*[1] that the provisions of s 307 could be applied to all offences in the criminal code, meaning that a corporation could be convicted of homicide. In this case, however, the court made no reference to the principles contained in that section, and held instead that the question was simply one of agency and, as such, liability was not contingent upon the Board of Directors condoning the president's actions. The corporation was deemed capable of being a 'person' within the meaning of the definition of the homicide offence and, as such, was convicted.

1 (1981) 293 Pa Super 78, 437 A 2d 999.

Corporate criminal liability in the State of Florida

14.86 On 11 May 1996, ValuJet Flight 592 caught fire and came down in the Florida Everglades. The subsequent Air Accident Report[1] (AAR) published by the National Transportation Safety Board (NTSB) split responsibility for the accident three ways between ValuJet, SabreTech (a maintenance company contracted by ValuJet) and the US Federal Aviation Administration. The NTSB concluded:

> '... the probable causes of the accident, which resulted from a fire in the planes class D cargo compartment that was initiated by the actuation of one or more oxygen generators being improperly carried as cargo, were (1) the failure of SabreTech to properly prepare, package and identify, unexpended chemical oxygen generators before presenting them to ValuJet for Carriage;[2] (2) the failure of ValuJet to properly oversee its contract maintenance programme to ensure compliance with maintenance, maintenance training, and hazardous materials requirements and practices; and (3) the failure of the Federal Aviation Administration (FAA) to require smoke detection and fire suppression systems in class D cargo compartments.'[3]

1 The Air Accident Report can be accessed online at www.ntsb.gov/doclib/reports/1997/aar9706.pdf.

2 Damningly, it was subsequently reported that it would have cost SabreTech a mere $9.16 to fit all the chemical oxygen generators with yellow plastic caps which would have sealed the oxygen tanks correctly. This may well have prevented the fire and the subsequent loss of life on ValuJet Flight 592; see National Transportation Safety Board, 'Aircraft Accident Report: In-Flight Fire and Impact with Terrain, ValuJet Airlines Flight 592, DC-9–32, N904VJ, Everglades, Near Miami, Florida, May 11 1996' at p 115, accessed online at www.ntsb.gov/doclib/reports/1997/aar9706.pdf.

3 National Transportation Safety Board 'Aircraft Accident Report: In-Flight Fire and Impact with Terrain, ValuJet Airlines Flight 592, DC-9–32, N904VJ, Everglades, Near Miami, Florida, May 11 1996' at p x, accessed online at www.ntsb.gov/doclib/reports/1997/aar9706.pdf.

14.87 The NTSB's report also highlighted a number of contributory factors, namely:

> '... the failure of the FAA to adequately monitor ValuJet's heavy maintenance programmes and responsibilities, including ValuJet's oversight of its contractors and SabreTech's repair station certificate; the failure of the FAA to adequately respond to prior chemical oxygen generator fires with programmes to address the potential hazards; and ValuJet's failure to ensure that both ValuJet and contract maintenance facility employees were aware of the carrier's 'no carry' hazardous materials policy and had received appropriate hazardous materials training.'[1]

1 National Transportation Safety Board 'Aircraft Accident Report: In-Flight Fire and Impact with Terrain, ValuJet Airlines Flight 592, DC-9–32, N904VJ, Everglades, Near Miami, Florida, May 11 1996' at p x, accessed online at www.ntsb.gov/doclib/reports/1997/aar9706.pdf.

14.88 In July 1999 the Miami-Dade Office of the State Attorney announced the return of 221 criminal indictments against SabreTech, consisting of one count of carrying hazardous waste, a third degree felony, on the grounds that SabreTech and its employees violated Florida State criminal law in the transportation and disposal of hazardous waste (the chemical oxygen generators). Secondly, Sabretech was indicted with 110 counts of third degree felony murder (one for each deceased passenger), on the basis that the 110 deaths were a direct consequence of the illegal transportation and disposal of the oxygen generators. Finally, SabreTech was charged with 110 counts of manslaughter as a result of the criminal and negligent acts of SabreTech and its employees in loading the oxygen generators onto the ValuJet flight.[1] ValuJet, however, was not criminally charged. The prosecution of SabreTech is of historical importance as it represents the first criminal prosecution to arise from an air disaster in the US.

1 See www.miamisao.com/press/1999.htm#071399 – accessed 7 September 2009. The full 221-count criminal information laid against SabreTech can be read online at www.miamisao.com/pdfs/sabretech.pdf.

14.89 The State prosecution drew to a close in 1999 when the State dropped the 220 charges of third degree felony murder in return for a plea of no contest to the charge of carrying hazardous waste. However, SabreTech and three of its mechanics had also been made the subject of federal charges relating to false statements on maintenance records, transporting hazardous materials on a plane and failing to train its employees properly in the handling of such materials. Whilst the charges against the three mechanics were eventually dropped, SabreTech was found guilty of 9 of the 23 charges it faced and was sentenced to pay $2 million in fines and over $9 million in restitution. Eight of these convictions were, however, subsequently overturned on appeal, and the case was sent back for resentencing on the sole remaining charge of failing to properly train employees in the handling of hazardous materials. As a result, the company was sentenced to pay the maximum fine for the offence and made subject to a three-year probation order.

Corporate criminal liability in the State of Texas

14.90 The penultimate State for us to consider in our discussion of corporate liability for homicide offences under state criminal law is the State of Texas. Our starting point is s 1.07 of the Texas Penal Code which defines a 'person' to mean an 'individual, corporation, or association'. This is clearly helpful as it suggests that Texas criminal law is also applicable to corporations. A consideration of the definitions of homicide offences as outlined in the Texas Criminal Code also indicates that a corporation is capable of being convicted of homicide.

14.91 The first key offence is the offence of manslaughter. Section 19.04 provides that 'A person commits an offense if he recklessly causes the death of an individual'. The term 'recklessly' is defined in s 6.03 (c). Alternatively, s 19.05 states that a person commits the offence of criminally negligent homicide where 'he causes the death of an individual by criminal negligence'. The term 'criminal negligence' is defined in s 6.03 (d).

14.92 The issue of sentencing corporate offenders is dealt with in s 12.51. The offence of manslaughter set out in s 19.04 is categorised as a felony of the second degree.[1] Accordingly, the prescribed punishment in the case of an individual offender would be facing imprisonment in the institutional division for a period of between two and 20 years as well as a possible fine for up to $10,000.[2] Section 12.51 tells us that, in the case of a corporate offender convicted of manslaughter, the prescribed penalty would be a fine of up to $20,000[3] or, if the corporation gained money or property as a result of the conduct which resulted in death, the court could set a fine at a level not exceeding double the amount gained.[4] The court also has the power to order the corporation to advertise its conviction in addition to any fine imposed.[5]

1 Section 19.04 (b) of the Texas Penal Code.
2 Section 12.33 (a)-(b) of the Texas Penal Code.
3 Section 12.51 (b)(1) of the Texas Penal Code.
4 Section 12.51 (c) of the Texas Penal Code.
5 Section 12.51 (d) of the Texas Penal Code. This is very similar to a sentencing option made available by the Corporate Manslaughter and Corporate Homicide Act 2007 (CMCHA 2007). When considering what sentence to impose on a corporate defendant who has been convicted of the offence of corporate manslaughter, the court has an option to impose a publicity order under s 10 of the CMCHA 2007. Any publicity ordered by the court can require the convicted corporation to publicise:
- the fact it has been convicted of an offence;
- specified particulars of the offence;
- the amount of any fine imposed; and
- the terms of any remedial order made.

When determining the terms of any proposed publicity order, s 10(2) of the CMCHA 2007 requires the court to take into account:
- the views of such enforcement authority (or authorities) as it sees fit; and
- any representations made by the prosecution or by the defence on behalf of the convicted organisation.

The court can specify a deadline by which the terms of the publicity order must be complied with. Section 10(4) of the CMCHA 2007 notes that a failure to comply with the deadline is a criminal offence punishable by way of a fine.

14.93 Should a corporation be convicted of criminally negligent homicide, the same sanctions apply. The offence is categorised as a state jail felony.[1] An individual convicted of criminally negligent homicide would be facing 180 days to two years in a State jail as well as a possible fine of up to $10,000.[2] Once again, s 12.51 tells us that, in the case of a corporate offender convicted of manslaughter, the prescribed penalty would be a fine of up to $20,000[3] or, if the corporation gained money or property as a result of the conduct which resulted in death, the court could set a fine at a level not exceeding double the amount gained.[4] As was the case following a conviction for manslaughter, the court also has the power to order the corporation to advertise its conviction in addition to any fine imposed.[5]

1 Section 19.05 (b) of the Texas Penal Code.
2 Section 12.35 (a)-(b) of the Texas Penal Code.
3 Section 12.51 (b)(1) of the Texas Penal Code.

4 Section 12.51 (c) of the Texas Penal Code.
5 Section 12.51 (d) of the Texas Penal Code.

14.94 We can see the provisions of the Texas Penal Code in action in the case of *Vaughan and Sons Inc v The State of Texas.*[1] It was alleged that the defendant corporation had caused the death of two individuals in a car crash through the actions of two of its agents. The corporation argued that the court had erred in finding that a corporation could be guilty of criminal homicide under the provisions of the Texas Penal Code. The Court of Appeal reversed the finding of the lower court.

1 (1987) 737 S W 2d 805.

14.95 The case eventually reached the Court of Criminal Appeals on the basis of the state's contention that the Court of Appeal was wrong in holding that the legislature did not intend for the homicide provisions set out in s 19.07 (as discussed above) to apply to corporations.[1]

1 Section 19.07 was quoted in the case report for *Vaughan* (1987) but is now renumbered as 19.05(b) in the latest version of the Texas Penal Code.

14.96 Before the enactment of the Texas Penal Code in 1974 and its subsequent amendments, the State of Texas only recognised corporate criminal liability in very limited circumstances. Under s 1.07 (a) (as we have seen) of the Penal Code, however, the term 'person' was defined in a way which included corporations within its meaning. Furthermore, s 7.22 makes it clear that a corporation could be held criminally liable for any of the offences contained in the Code.[1] Finally, s 12.51 deals with the punishment of corporations, and permits the use of fines in those cases where the only prescribed punishment for the offence is imprisonment.

1 Section 7.22 deals generally with the criminal liability of corporations. It allows for the actions of certain members to be attributed to the corporation (s 7.22 (a)); however, liability will only be imposed on the corporation where those actions were 'requested, commanded, performed, or recklessly tolerated by … a majority of the governing board acting in behalf of the corporation …; or a high managerial agent acting in behalf of the corporation … and within the scope of his office or employment' (7.22 (b)).

14.97 With all these statutory provisions, it is not surprising that the Court of Criminal Appeals took this as indicative of strong legislative intent that corporations should be held liable for criminal offences in the State of Texas. The court concluded that, since the statutory definition of 'person' included corporations, and since the offence of criminally negligent homicide could be committed by a person, logic dictated that a corporation could commit the offence of homicide. The corporation was convicted accordingly.

Corporate criminal liability in the State of Michigan

14.98 The final state that falls for our consideration is the State of Michigan. Unlike any of the states we have considered up until this point, the definition of the offence of manslaughter in the State of Michigan has been left exclusively to the common law. Consequently, there are two 'species' of manslaughter: voluntary manslaughter and involuntary manslaughter.[1] A corporation cannot be convicted of voluntary manslaughter, but it can be convicted of involuntary manslaughter.

1 The law has, in this sense, developed in much the same way as the offence of manslaughter in the English common law, where the offence of voluntary manslaughter is committed where the defendant has the requisite mens rea and has committed the requisite actus reus for the offence of murder, but they have a partial defence which, whilst it does not warrant a pardon, warrants a

reduction in the offence charged. Involuntary manslaughter, on the other hand, arises where the defendant did not intend to kill the deceased – indeed, they may have actively taken steps to prevent it – but nevertheless death occurred as a result of their actions. For a more detailed discussion of the law governing manslaughter in England and Wales, you are advised to refer to a criminal law text such as D Ormerod (ed), *Smith & Hogan Criminal Law: Text and Materials* (11th edn, OUP, 2011). Similar definitions of the offences of voluntary and involuntary manslaughter in Michigan State Law can be found in the *Michigan Criminal Jury Instructions* (2nd edn).

14.99 According to the decision in the case of *People v Zak*, an individual will be found guilty of involuntary manslaughter only where it can be shown that the victim's death was caused by the defendant's 'gross negligence'.[1] The term 'gross negligence' is further defined in guideline 16.18 of the *Michigan Criminal Jury Instructions* (2nd edn), as meaning 'more than carelessness'. Rather, it means 'wilfully disregarding the results to others that might follow from an act or failure to act'. Guideline 16.18 goes on to explain that an individual will be deemed to be 'grossly negligent' where:

(a) the defendant knew of the danger to another, that is, he knew there was a situation that required him to take ordinary care to avoid injuring another;
(b) the defendant could have avoided injuring another by using ordinary care; or
(c) the defendant failed to use ordinary care to prevent injuring another when, to a reasonable person, it must have been apparent that the result was likely to be serious injury.[2]

1 *People v Zak* 184 Mich App 1, 457 N.W.2d 59, 62 (1990). This much is also made clear in the *Michigan Criminal Jury Instructions* (2nd edn), guideline 16.10 (3).

2 This definition mirrors that developed through the case law in Michigan courts, such as the decisions in *People v Zak*, 184 Mich App 1, 457 N W 2d 59, 62 (1990) and in *People v Moye*, 194 Mich App 373, 487 N W 2d 777, 778–779 (1992). Again, the approach taken to defining 'involuntary manslaughter' in the State of Michigan bears some similarities to that taken in English law, where proof of 'gross negligence' (as defined in the case of *R v Adomako* [1995] 1 AC 171) is required to convict a defendant of involuntary manslaughter.

14.100 What is not immediately clear from this discussion, however, is whether an organisation can be convicted of involuntary manslaughter. It is suggested that the approach taken by the courts in tackling this question would be very much like that taken in the State of Texas.

14.101 Section 321 of the Michigan Penal Code states:

> 'Any person who shall commit the crime of manslaughter shall be guilty of a felony punishable by imprisonment in the state prison, not more than 15 years or by fine of not more than 7,500 dollars, or both, at the discretion of the court'.

Crucially, s 10 of the Code states that:

> 'The words "person", "accused", and similar words include, unless a contrary intention appears, public and private corporations, co-partnerships, and unincorporated or voluntary associations'.

When the definition of 'person' is read in conjunction with the provision of a financial penalty for the offence of manslaughter, there appears to be the same sort of legislative intent identified by the Texan Court of Criminal Appeals (as outlined above) to hold corporations liable for the offence of manslaughter.

14.102 This supposition has been borne out by the successful prosecution in (relatively) recent years of two organisations for manslaughter following workplace fatalities. In a landmark case brought by the Consumer and Industry

Services Bureau of Safety and Regulation in November 2000, Midland Environmental Services Inc and the company's owner, Edmond Woods, entered guilty pleas to several citations arising out of a workplace fatality. The importance of this case lies in the fact that it represented the first criminal case in the history of prosecutions for breaches of health and safety regulations in the State of Michigan where both an owner and a corporation had been convicted of criminal offences following a workplace fatality.

1 The MIOSHA Michigan Occupational Safety & Health Administration (part of the Department of Licensing and Regulatory Affairs) is the body responsible for administering the Michigan Occupational Safety and Health Act.

14.103 Both the organisation and the owner pleaded guilty to the (curious) charge of attempted involuntary manslaughter[1] as well as to citations following wilful criminal breaches of the Michigan Occupational Safety and Health Act. Midland Environmental Services Inc was subject to a fine of $17,500. The owner, Edmond Woods, was ordered to pay a fine of $17,500 and was additionally sentenced to 200 hours of community service and five years' probation.[2]

1 Section 92 of the Michigan Penal Code deals with penalties to be imposed where the defendant has been convicted of attempting to commit an offence but there is no provision dealing specifically with punishment for such an attempt. An attempt would arise where the defendant 'does any act towards the commission of the [full offence], but … [fails] in the perpetration, or [is] intercepted or prevented in the execution of the same'.

2 Michigan Department of Energy, Labor and Economic Growth, 'CIS Announces Owner and Firm Sentenced in Worker's Death; Owner Edmond Woods and Midland Environmental Services, Inc. Sentenced for attempted Involuntary Manslaughter in Fatal Explosion', 19 December 2000.

14.104 In October 2002, Michigan's CIS Bureau of Safety and Regulation announced a second successful prosecution of a corporation for involuntary manslaughter. The deceased, a truck driver, had been directed by the site foreman to deliver crushed gravel into an area located under a 7,600-volt live power line. This instruction was delivered in spite of a clear warning given to the company that it was not to carry out work under the power line. The driver was electrocuted when he raised the truck bed, again following orders from the site foreman, which made contact with the live overhead wire.

14.105 J A Morrin Concrete Construction Company entered a plea of no contest to the involuntary manslaughter charge and to the charge relating to a wilful violation of the Michigan Occupational Safety and Health Act. The site foreman, who was also charged following the fatality, pleaded guilty to the wilful violation charge. For the offence of involuntary manslaughter the corporation was sentenced to pay a fine of $156,903 and was sentenced to five years' probation.[1] For the wilful violation the company was further subjected to a fine of $10,350 and two years' probation. The foreman was sentenced to a fine of $1,000, 360 days in jail and three years' probation.[2]

1 The corporation was also ordered to pay a monthly supervision fee to the Michigan Department of Corrections of $135, totalling $8,100 over the course of five years.

2 As with the corporation, the defendant, Jim Morrin Jr, was additionally required to pay a monthly supervision fee of $100, totalling $3,600 over the course of his probationary period: Michigan Department of Energy, Labor and Economic Growth 'CIS Announces Foreman Sentenced to 360 Days in Jail for Worker Fatality', 10 October 2002.

Prosecution and sentencing

14.106 The criminal justice system in the US pays little attention to corporate manslaughter.[1] Investigation is carried out at the state level by health and safety officers under the Department of Labor (for example, in New York State) or the equivalent state department. Undoubtedly the police will be called, but there are not specialist sections of the police force designated to deal with corporate manslaughter.

1 Eg D Neubauer, *America's Courts and the Criminal Justice System* (1988), Ch 2.

14.107 The federal body responsible is the Occupational Safety and Health Authority (OSHA). State departments of OSHA have established local Fatality Assessment and Control Evaluation (FACE) schemes. These report on fatalities, inspecting the scene of the incident and preparing a report. They do not address the issue of liability. The reports set out means of avoiding future incidents. However, a corporation's cooperation may affect sentencing and possibly liability in relation to a prosecution. Departments of Health in conjunction with Departments of Labor also produce statistics in relation to fatalities.[1] This work is picked up at the state level by bodies such as the Center for Disease Control and Prevention (CDC) and the National Institute of Occupational Safety and Health (NIOSH) through its National Occupational Research Agenda (NORA).[2] These are of some assistance in evaluating the impact of corporate criminal litigation.

1 See eg www.health.ny.gov/environmental/investigations/face/.
2 See www.cdc.gov/niosh/nora/about.html.

14.108 Public interest groups also play an important role in investigating and reporting workplace fatalities.[1]

1 Eg *Corporate Crime Reporter*, available at www.corporatecrimereporter.com.

14.109 Whilst civil prosecutions may be brought by individuals or by local regulatory bodies, criminal prosecutions are only brought by District Attorneys. They are assisted by the *US Department of Justice Memorandum: Principles of Federal Prosecution of Business Organisations* (2008). This sets out the general principle that:

> 'Corporations should not be treated leniently because of their artificial nature nor should they be subject to harsher treatment. Vigorous enforcement of the criminal laws against corporate wrongdoers, where appropriate results in great benefits for law enforcement and the public … Indicting corporations for wrongdoing enables the government to address and be a force for positive change of corporate culture and a force to prevent, discover and punish serious crimes.'[1]

1 *US Department of Justice Memorandum: Principles of Federal Prosecution of Business Organisations* (2008) at p 2. Accessible at www.justice.gov/opa/documents/corp-charging-guidelines.pdf.

14.110 The Guidance lists nine factors to be considered in deciding whether to prosecute. These focus on the seriousness and pervasiveness of the act, any history of similar conduct, voluntary disclosure, cooperation, collateral consequences of the act, remedial action and the adequacy of alternatives such as prosecution of individual employees, civil or regulatory enforcement actions.[1] The Guidance goes into some depth on each of these. Indictment of the corporation is not, however, the only charging option available to prosecutors. Negotiated Plea Agreements, Non-Prosecution Agreements and Deferred Plea Agreements are also to be found amongst the prosecutor's options. It is this last alternative which warrants further examination.[2]

1 *US Department of Justice Memorandum: Principles of Federal Prosecution of Business Organisations* (2008) at pp 3–4. Accessible at www.justice.gov/opa/documents/corp-charging-guidelines.pdf.

2 In the section that follows, reference is made to the following excellent articles: P Spivack and S Raman, 'Essay – Regulating the "New Regulators": Current Trends in Deferred Prosecution Agreements' (2008) 45 American Crim L Rev 159; P H Bucy, 'Why Punish? Trends in Corporate Criminal Prosecutions' (2007) 44 American Crim L Rev 1287; BA Bohrer and B L Trencher, 'Prosecution Deferred: Exploring the Unintended Consequences and Future of Corporate Cooperation' (2007) 44 American Crim L Rev 1481; and C J Christie and R Hanna, 'A Push Down the Road of Good Corporate Citizenship: The Deferred Prosecution Agreement Between the US Attorney for the District of New Jersey and Bristol-Meyers Squibb Co.' (2006) 43 American Crim L Rev 1043.

14.111 At the start of the 21st century, America (and the rest of the world) witnessed a large number of corporate fraud scandals, including the collapse of Enron[1] and World Com.[2] Against a background which saw an increasing willingness to pursue such corporations for their wrongdoing,[3] the US government went after the accounting giant Arthur Andersen for its role in the collapse of Enron. What followed has been described as 'an unmitigated disaster'.[4]

1 At its peak, Enron was America's seventh largest company. Enron filed for bankruptcy in December 2001 after it transpired that Enron had systematically lied about its profits and hidden its debts. At the time it was the largest corporate bankruptcy on record; a record which was subsequently eclipsed by the collapse of WorldCom. For a thorough discussion of Enron's legacy, in terms of the impact it had on both the regulation of financial institutions and on the enforcement environment surrounding such organisations, and for an insight into some other large-scale corporate fraud scandals, you are advised to read K F Brickey, 'Enron's Legacy' (2004) 8 Buffalo L Rev 221.

2 In June 2002, WorldCom revealed a colossal corporate fraud involving billions of dollars. As well as accounting irregularities, which saw profits reported when the company should have announced losses, the company had also improperly reported much of its earnings. At the time it was recognised, WorldCom's fraud was identified as the largest corporate fraud in American history. In July 2002 the company was responsible for the largest bankruptcy filing in US history, beating the record set just a year previously by Enron – see K F Brickey, 'Enron's Legacy' (2004) 8 Buffalo L Rev 221 at p 226.

3 2003, for example, saw the publication of what became known as 'the Thompson memo' (*'Memorandum from Larry D Thompson, Deputy Attorney General, to Heads of Department Components, United States Attorneys'* (23 January 2003), in which the Deputy Attorney General talks about the benefits that flow from charging a corporation with a criminal offence as follows:

> 'First and foremost, prosecutors should be aware of the important public benefits that may flow from indicting a corporation in appropriate cases. For instance, corporations are likely to take immediate remedial steps when one is indicted for criminal conduct that is pervasive throughout a particular industry, and thus an indictment often provides a unique opportunity for deterrence on a massive scale. In addition, a corporate indictment may result in specific deterrence by changing the culture of the indicted corporation and the behavior of its employees. Finally, certain crimes that carry with them a substantial risk of great public harm, e.g., environmental crimes or financial frauds, are by their nature most likely to be committed by businesses, and there may, therefore, be a substantial federal interest in indicting the corporation.'

4 P Spivack and S Raman, 'Essay – Regulating the "New Regulators": Current Trends in Deferred Prosecution Agreements' (2008) 45 American Crim L Rev 161 at p 165.

14.112 As Bucy notes:

> 'Arthur Andersen … was convicted for obstruction of justice, went out of business as a result of its prosecution, and then saw its conviction overturned by the United States Supreme Court.'[1]

1 P H Bucy, 'Why Punish? Trends in Corporate Criminal Prosecutions' (2007) 44 American Crim L Rev 1287 at p 1287.

14.113 The collapse of Arthur Andersen brutally brought home the reality that the US Department of Justice is faced with a huge problem when investigating and prosecuting organisations involved in large-scale financial wrongdoing (although these considerations apply when investigating/prosecuting any organisation for any crime); how to strike a balance between trying to investigate/prosecute/deter corporate crime whilst, at the same time, taking care to avoid the potentially severe consequences of indicting a corporation.[1] Indeed, in light of the potential economic ramifications of a corporation collapsing as a result of a criminal investigation/prosecution, questions must surely be raised about whether such activities are counter-productive? It is for this very reason that prosecutors are reportedly making increased use of Deferred Prosecution Agreements (DPAs).[2]

1 The potential ramifications of a criminal prosecution of a corporation are ably set out by Bucy, who notes:

> '… there is no question that criminal prosecution of a corporation has a tremendous impact on the corporation and its community, employees, customers and lenders. For starters, the tangible and intangible costs of responding to any corporate criminal investigation are significant. Company employees must gather thousands of documents in response to subpoenas. Prior to supplying subpoenaed documents, legal counsel must review each document to verify compliance and to ensure that privileged information is not being released. This process is time consuming and expensive. In addition, any company under investigation should undertake its own internal investigation. If outside counsel is hired to do this investigation, the legal fees are large. If in-house counsel undertakes the investigation, counsel is diverted from other corporate projects and tasks and this diversion hurts the company in small and large ways. Also, once the existence of an investigation becomes public, stock prices of public traded companies also drop …
>
> Additionally, lenders may raise short term interest rates, terminate lines of credit or call in loans. Moreover, business is also disrupted by an investigation. Deals and plans are put on hold because of the uncertainty surrounding the targeted company. Employee morale plummets. Competing businesses swoop in and lure away star employees who are reluctant to remain with a business under investigation. Customers leave for competitors.
>
> In short, a company may not survive the distractions, costs and damage to its reputation and business that a corporate criminal investigation brings.'

See P H Bucy, 'Why Punish? Trends in Corporate Criminal Prosecutions' (2007) 44 American Crim L Rev 1287 at pp 1288–1289.

2 See P Spivack and S Raman, 'Essay – Regulating the "New Regulators": Current Trends in Deferred Prosecution Agreements' (2008) 45 American Crim L Rev 159, and B A Bohrer and B L Trencher, 'Prosecution Deferred: Exploring the Unintended Consequences and Future of Corporate Cooperation' (2007) 44 American Crim L Rev 1481. Spivack & Raman in particular describe the marked increase in the use of Deferred Prosecution Agreements and Non-Prosecution Agreements as 'the most profound development in corporate white collar criminal practice over the past five years' (at p 159).

14.114 In basic terms, a DPA will see a prosecutor filing charges against the defendant organisation, but then enter into an agreement with the organisation not to proceed with the prosecution if they comply with the terms of the deferral agreement for a set period of time. The use of DPAs as an alternative to indicting a corporation was first explicitly identified as a valid option for prosecutors in the Thompson memo. The Deputy Attorney General noted that: 'In some circumstances … granting a corporation immunity or amnesty or *pre-trial diversion*[1] may be considered in the course of the government's investigation' (author's emphasis). Such an option was to be used, effectively speaking, as a 'reward' where the organisation under investigation voluntarily disclosed its wrongdoing in a timely fashion and/or cooperated fully with any investigation.[2] In effect, therefore, the option of entering into a DPA or an NPA provides the prosecutors with an alternative 'somewhere in between the "all or nothing" choice between indicting (and destroying) a company and giving it a complete "pass" '.[3]

1 That is, a DPA or an NPA.

2 This approach ties in with this key message espoused by the US government:

> 'The message we're sending to corporate America on this point is two-fold: Number one, you'll get a lot of credit if you cooperate, and that credit will sometimes make the difference between life and death for a corporation. Number two, if you want to ensure that credit, your

cooperation needs to be authentic: you have to get all the way on board and do your best to assist the government.'

Per Christopher A Wray, Assistant Attorney General, US Department of Justice, Remarks to the Association of Certified Fraud Examiners Mid-South Chapter (2 September 2004), as cited and quoted in B A Bohrer and B L Trencher, 'Prosecution Deferred: Exploring the Unintended Consequences and Future of Corporate Cooperation' (2007) 44 American Crim L Rev 1481 at p 1487.

3 P Spivack and S Raman, 'Essay – Regulating the "New Regulators": Current Trends in Deferred Prosecution Agreements' (2008) 45 American Crim L Rev 159 at p 166, quoting C A Wray and R K Kur, 'Corporate Criminal Prosecution in a Post-Enron World: The Thompson Memo in Theory and in Practice' (2006) 43 American Crim L Rev 1095 at p 1103.

14.115 There are clearly a number of advantages associated with the use of DPAs. For a start:

> 'the government is able to address corporate wrongdoing without the seeking an indictment and the resulting severe consequences to both the company and its employees who were uninvolved in the allegedly wrongful conduct.'[1]

Secondly, the option of making use of pre-trial diversions broadens the role of prosecutions in this area; rather than having a purely punitive role (ie a prosecution is a form of 'punishment' flowing from the corporation's wrongdoing), it also encourages greater transparency and openness in that such diversions encourage corporations to self-police and readily disclose when they are guilty of wrongdoing. Thirdly:

> 'In contrast to the far more rigid criminal sentencing process, deferred prosecution agreements allow prosecutors and companies to work together in creative and flexible ways to remedy past problems and set the corporation on the road of good corporate citizenship'.[2]

Finally, it is possible to require the payment of a fine in a DPA much larger than that which could be prescribed by the court under the terms of the United States Sentencing Guidelines.[3] Despite these potential benefits of making use of DPAs, there have also been a number of key criticisms levelled at their use. A brief consideration of these criticisms concludes this section of the chapter.

1 B A Bohrer and B L Trencher, 'Prosecution Deferred: Exploring the Unintended Consequences and Future of Corporate Cooperation' (2007) 44 American Crim L Rev 1481 at p 481.

2 C J Christie and R Hanna, 'A Push Down the Road of Good Corporate Citizenship: The Deferred Prosecution Agreement Between the US Attorney for the District of New Jersey and Bristol-Meyers Squibb Co.' (2006) 43 American Crim L Rev 1043 at p 1043. The authors go on to comment (at p1043) that Deferred Prosecution Agreements 'permit us to achieve more than we could through court imposed fines or restitution alone. These agreements, with the broad range of reform tools, permit remedies beyond the scope of what a court could achieve after a criminal conviction'. This article provides the reader with a clear discussion of the thought processes which inform prosecutors' decisions about whether to enter into a DPA and what that agreement should contain.

3 P Spivack and S Raman, 'Essay – Regulating the "New Regulators": Current Trends in Deferred Prosecution Agreements' (2008) 45 American Crim L Rev 159 at p 182. The authors do, however, go on to note at p 183 that 'prosecutors seem not to be abusing the tool of diversion to extract more out of a company than the entity would owe after conviction and sentencing'.

14.116 Whilst the Thompson memo advocated the use of pre-trial diversions as an option when deciding how to deal with corporate offenders, it did not provide any guidance on when it would be appropriate to make use of such diversions, nor did it provide any guidance on the content of such agreements.[1] This creates a huge potential problem, namely that the US Attorney's Offices have been left to make use of DPAs with no central guidance from the US Department of Justice, meaning that each jurisdiction has discretion (to be exercised within the confines of the general principles contained in the *Principles of Federal Prosecution of Business Organisations*) about when to enter into a DPA with a corporate defendant and about what obligations might be imposed upon the corporation.[2] A key consequence of this lack of central guidance, therefore, is that there are inconsistencies in the use of DPAs from jurisdiction to jurisdiction.[3]

1 This lack of guidance was similarly a feature of both: (1) the subsequent 'McNulty memo' (*Memorandum from Paul J McNulty, Deputy Attorney General, to Heads of Department Components, United States Attorneys* (12 December 2006), accessible online at www.justice.gov/dag/speeches/2006/mcnulty_memo.pdf; and (2) the most recent edition of the *Principles of Federal Prosecution of Business Organisations*, as set out in the 'Filip memo' (*Memorandum from Mark R. Filip, Deputy Attorney General, to Heads of Department Components, United States Attorneys* (28 August 2008).

2 This criticism has been addressed in part by the release on 7 March 2008 of a memorandum from the Acting Deputy Attorney General, Craig S Morford, which provides guidelines which must be followed where the DPA requires the appointment of an independent corporate monitor. These guidelines set out the factors that must be taken into account when deciding who to appoint to that role. The 'Morford memo' (*Memorandum from Craig S. Morford, Acting Deputy Attorney General, to Heads of Department Components, United States Attorneys* (7 March 2008) can be accessed online at www.justice.gov/dag/morford-useofmonitorsmemo-03072008.pdf.

3 This very point is made in P Spivack and S Raman, 'Essay – Regulating the "New Regulators": Current Trends in Deferred Prosecution Agreements' (2008) 45 American Crim L Rev 159 at pp 171–173.

14.117 A second problem that has been highlighted in the literature is that a feature of some the DPAs that have been entered into is that they have included terms which have 'little to do with the underlying misconduct'.[1] This issue leads neatly into the final key criticisms of the use of DPAs.

1 P Spivack and S Raman, 'Essay – Regulating the "New Regulators": Current Trends in Deferred Prosecution Agreements' (2008) 45 American Crim L Rev 159 at p 174. Spivack & Raman give a number of examples of terms which require the organisation to take remedial actions which are unconnected to the original offence such as 'the provision in the June 2005 DPA between the [United States Attorney's Office] of the District of New Jersey and Bristol-Meyers Squibb that required the company to endow a chair in business ethics at the United States Attorney's law school alma mater, Seton Hall' (at p 174). This may well be a consequence resulting from the lack of central guidance on the use of DPAs.

Spivack & Raman do note, however, that, as their 2007 data set reveals, 'recent diversion agreements tend not to contain unrelated terms'. They describe this as 'a positive trend perhaps attributable to the hostile reaction to some of the earlier agreements elicited'. See P Spivack and S Raman, 'Essay – Regulating the "New Regulators": Current Trends in Deferred Prosecution Agreements' (2008) 45 American Crim L Rev 159 at p 174.

14.118 The increasing use of DPAs by prosecutors is seen by Spivack & Raman to be a by-product of a shift in the US Department of Justice's policy in line with their 'evolving view of the purpose and function of the criminal law in the corporate context'.[1] As the authors note:

> 'DOJ officials seem to believe that the principal role of corporate criminal enforcement is to reform corrupt corporate cultures – that is, to effect widespread cultural reform – rather than to indict, to prosecute, and to punish'.[2]

This raises major questions, however, about the scope of the prosecutor's role. One of the requirements that seems to be a common feature of DPAs is that organisations are obliged to improve their compliance efforts.[3] This may involve requiring the firm to fire those individuals who were involved in the criminal activity, to reform the organisations management structure, or to implement a more rigorous compliance programme.[4] Additionally, many DPAs require the organisation 'to hire, at its own expense, a federal monitor to help ensure compliance and reduce the number of future violations'.[5] The problem that this creates is all too evident. As Spivack & Raman note, 'prosecutors these days are "fashioning themselves as the new corporate governance experts" – a position for which they are singularly unqualified'.[6]

1 P Spivack and S Raman, 'Essay – Regulating the "New Regulators": Current Trends in Deferred Prosecution Agreements' (2008) 45 American Crim L Rev 159 at p 161.

2 P Spivack and S Raman, 'Essay – Regulating the "New Regulators": Current Trends in Deferred Prosecution Agreements' (2008) 45 American Crim L Rev 159 at p 161. Spivack & Raman cite the work of Henning to support this contention (P J Henning, 'The Organisational Guidelines: RIP?' (2007) 116 Yale Law Journal Pocket Part 312 – accessible at http://yalelawjournal.org/images/pdfs/528.pdf) who comments at p 315, 'The purpose of corporate prosecutions is not to punish but instead to change corporate cultures through agreements that deal directly with internal governance … The focus on how businesses will operate in the future is now a central feature of corporate criminal investigations'.

3 A similar approach is taken in the context of prosecutions for breaches of health and safety regulations by the Occupational Safety and Health Agency States (as discussed earlier in the context of the prosecution of breaches of health and safety laws at state level), except in these cases the agreements are referred to as settlement agreements. In the case of the prosecution of J A Concrete Construction Company in the State of Michigan (also discussed earlier), for example, the Michigan Occupational Safety and Health Agency (MIOSHA) entered into a settlement agreement with the defendant corporation in which the corporation agreed as follows:

- To provide notification to MIOSHA not less than ten calendar days (but under no circumstances less than two hours) prior to commencement of any construction work activity within the State of Michigan for a period of three years.
- To establish and implement an ongoing safety and health programme, provide documentation of training received by employees, and maintain records of training and refresher training.
- To pay a civil penalty to MIOSHA of $50,000.
- To establish and implement a structured form to serve as a 'checklist' for inspecting all future worksites prior to the commencement of work activity.
- To conduct safety and health awareness training for all its employees at quarterly intervals for a period five years.
- To donate $5,000 to the Safety Council of Northwest Ohio in memory of Robert Sorge, which is to be used for training purposes focusing on the hazards of home building. A plaque will be placed with the Safety Council of Northwest Ohio memorialising the donation, and will state that Robert Sorge lost his life through no fault of his own.

This agreement provided the MIOSHA with the ability to monitor the future actions of the company in greater detail to prevent future violations. Details of the settlement agreement in this case are as reported in the following press release: Michigan Department of Energy, Labor and Economic Growth, 'CIS Announces Foreman Sentenced to 360 Days in Jail for Worker Fatality', 10 October 2002.

4 P Spivack and S Raman, 'Essay – Regulating the "New Regulators": Current Trends in Deferred Prosecution Agreements' (2008) 45 American Crim L Rev 159 at p 184. The DPA entered into between the District of New Jersey and Bristol-Meyers Squibb Co provides a good example of an agreement which imposed structural reforms on the organisation as a term of the agreement. Christie and Hanna (C J Christie and R Hanna, 'A Push Down the Road of Good Corporate Citizenship: The Deferred Prosecution Agreement Between the US Attorney for the District of New Jersey and Bristol-Meyers Squibb Co.' (2006) 43 American Crim L Rev 1043) report that one of the potentially problematic issues present in Bristol-Meyers' corporate structure was the fact that, like many US companies, the organisation's top leader held both the roles of chairman of the board and of chief executive officer (CEO). This particular approach meant that there were fewer checks and balances in place than might have been ideal and that there was a potential bottleneck of information between the corporate officers and the board of directors (see pp 1051–1052). As part of the DPA, therefore, Bristol-Meyers were required to split the roles of board chair and chief executive. Furthermore, Bristol Meyers were also required to appoint an additional non-executive Director acceptable to the US Attorney's Office.

5 P Spivack and S Raman, 'Essay – Regulating the "New Regulators": Current Trends in Deferred Prosecution Agreements' (2008) 45 American Crim L Rev 159 at p 184.

6 P Spivack and S Raman, 'Essay – Regulating the "New Regulators": Current Trends in Deferred Prosecution Agreements' (2008) 45 American Crim L Rev 159 at p 184, quoting Former US Attorney Mary Jo White (M J White, 'Corporate Criminal Liability: What has Gone Wrong?' (2005) 2 37th Annual Institute on Securities Regulation 815 (*PLI Corporate Law &*

Practice, Course Handbook Series No B-1517) at p 818. This same point is made by Bohrer & Trencher (B A Bohrer and B L Trencher, 'Prosecution Deferred: Exploring the Unintended Consequences and Future of Corporate Cooperation' (2007) 44 American Crim L Rev 1481) who ask, at p 1501, whether 'prosecutors have the institutional competence to oversee personnel and business decisions of corporations'.

This particular issue is further exacerbated by the fact that the typical monitor appointed is a former judge or prosecutor, someone who is not particularly qualified to make business decisions or decisions about corporate governance (P Spivack and S Raman, 'Essay – Regulating the "New Regulators": Current Trends in Deferred Prosecution Agreements' (2008) 45 American Crim L Rev 159 at pp 185–186).

14.119 This issue has been mitigated, to a degree, by the fact that experts from the regulatory agencies are becoming more involved in the investigation and resolution of corporate criminal activities, as such individuals are more knowledgeable about corporate governance issues than prosecutors.[1] But even this development is far from uncontroversial as questions are being raised about whether 'prosecutors and regulators might be manipulating the criminal justice system to achieve, illegitimately, the ends of sweeping civil structural reform'.[2] Nevertheless, it seems that DPAs are fast becoming a staple of the prosecutor's arsenal and, whilst DPAs have primarily been used to date in the context of accounting and securities fraud, there is an increasing use of DPAs outside this area,[3] so it is well worth the reader acquainting themselves with the key benefits and pitfalls surrounding their use.

1 P Spivack and S Raman, 'Essay – Regulating the "New Regulators": Current Trends in Deferred Prosecution Agreements' (2008) 45 American Crim L Rev 159 at p 183. Spivack and Raman report at p 184 (fn 145) that their data shows that at least 28 of the DPAs entered into in the previous year reflected the combined efforts of prosecutors and regulators.

2 P Spivack and S Raman, 'Essay – Regulating the "New Regulators": Current Trends in Deferred Prosecution Agreements' (2008) 45 American Crim L Rev 159 at pp 183–184. For a further, and more detailed, commentary on the issues raised by this extension of the prosecutors' role, read B A Bohrer and B L Trencher, 'Prosecution Deferred: Exploring the Unintended Consequences and Future of Corporate Cooperation' (2007) 44 American Crim L Rev 1481.

3 P Spivack and S Raman, 'Essay – Regulating the "New Regulators": Current Trends in Deferred Prosecution Agreements' (2008) 45 American Crim L Rev 159 at p 176.

14.120 Should a prosecution go ahead and should the corporation be convicted of an offence, the corporation will need to be sentenced. Some consideration needs to be given, therefore, to the provisions of the *Federal Sentencing Guidelines Manual* relating to the sentencing of defendants who are organisations.[1] Within the context of our discussion of the criminal liability of corporations that kill, the two key sections are s 8C2.5 and s 8C4.2.

1 The *United States Sentencing Guidelines Manual* is accessible on the website of the United States Sentencing Commission at www.ussc.gov/Guidelines/2012_Guidelines/Manual_PDF/index.cfm. The latest version of the guidelines is applicable from 1 November 2012. Factors impacting upon the sentencing of corporate offenders are dealt with in Chapter 8 of the guidelines. For a more detailed, yet relatively straightforward, account of how the sentencing guidelines are applied, read A H Lipman, 'Corporate Criminal Liability' (Twenty-Fourth Annual Survey of White Collar Crime) (2009) 46 American Crim L Rev 359. Lipman's article has had a strong influence on this section of the chapter.

The term 'organisation' is defined in s 8A1.1 as meaning 'a person other than an individual'. This definition is said to include 'corporations, partnerships, associations, joint-stock companies, unions, trusts, pension funds, unincorporated organizations, governments and political subdivisions thereof, and non-profit organizations'.

14.121 Where the punishment that the court is planning to impose is a fine,[1] the court's first step will be to determine the 'base offence level'. This is done 'by locating the count upon which the conviction was obtained in the applicable offence guideline'.[2]

1 The US has a very varied sentencing regime available in relation to corporate offenders including corporate probation. Regardless of what punishment the court eventually decides to settle on, their final decision must always take into account the following key principles underpinning (and set out in) the sentencing guidelines:
  (a) The court must, whenever practicable, order the firm (organization) to remedy any harm caused by the offence. The resources expended to remedy the harm should not be viewed as punishment, but rather as a means of making victims whole for the harm caused.
  (b) If the organisation acted primarily for a criminal purpose or primarily by criminal means, the fine should be set sufficiently high to divest the organisation of all its assets.
  (c) The fine range for any other organisation should be based on the seriousness of the offence and the culpability of the organisation. The seriousness of the offence generally will be reflected by the greatest of the pecuniary gain, the pecuniary loss, or the amount in a guideline offence level fine table. Culpability generally will be determined by six factors that the sentencing court must consider. The four factors that increase the ultimate punishment of an organisation are: (i) the involvement in or tolerance of criminal activity; (ii) the prior history of the organisation; (iii) the violation of an order; and (iv) the obstruction of justice. The two factors that mitigate the ultimate punishment of an organisation are: (i) the existence of an effective compliance and ethics programme; and (ii) self-reporting, cooperation, or acceptance of responsibility.
  (d) Probation is an appropriate sentence for an organisational defendant when needed to ensure that another sanction will be fully implemented, or to ensure that steps will be taken within the organisation to reduce the likelihood of future criminal conduct.

  As an aside, where the court is considering imposing a corporate probation order, it will need to consider the factors set out in s 8D1.1.

2 A H Lipman, 'Corporate Criminal Liability' (Twenty-Fourth Annual Survey of White Collar Crime) (2009) 46 American Crim L Rev 359. The offence guidelines in question can be found in Chapter 2 of the *United States Sentencing Guidelines* (see s 8C2.3 (a)).

14.122 The next step when calculating the fine is to determine the 'base fine'.[1] The base fine is intended to reflect both the type and severity of the offence and is intended to be set at a level which will 'deter organisational criminal conduct and … provide incentives for organisations to maintain internal mechanisms for preventing, detecting and reporting criminal conduct'.[2] The level at which the base fine is set is determined by:[3] (1) the amount prescribed by the table found at s 8C2.4 (d),[4] (2) the pecuniary gain to the organisation from the offence; or (3) the pecuniary loss caused by the offence 'to the extent that such loss was caused intentionally, knowingly or recklessly',[5] whichever sum is the greatest. However, s 8C2.4 (c) provides that, where the process of calculating the pecuniary gain or pecuniary loss resulting from the organisation's criminal activities would 'unduly complicate or prolong the sentencing process', that amount (ie the gain or loss caused, as appropriate) will not be used for determining the base fine.

1 The determination of the base fine is governed by s 8C2.4.
2 See the commentary which accompanies s 8C2.4 in the *United States Sentencing Guidelines*.
3 As set out in s 8C2.4 (a)(1)–(3).
4 Before using this table, the court will first need to determine the base offence level, as discussed above. Once the court has determined the offence level (determined under 8C2.3 (offense level)), it can look to the table contained in s 8C2.4 (d) and find the prescribed base fine for offences at that offence level.
5 See the commentary which accompanies s 8C2.4 in the *United States Sentencing Guidelines*.

14.123 Once the base fine has been determined, the third step is to determine the organisation's 'culpability score'. The culpability score provides the courts with a means of quantifying any aggravating or mitigating features of the organisation's offence. The organisation will be assigned an initial culpability score of five points.[1] The court will then add or subtract points, depending on whether there are particular aggravating or mitigating factors present in the offence.[2]

1 Section 8C2.5 (a).

2 To do this, the courts will have to work their way through s 8C2.5 (b)–(g). Each aggravating and mitigating factor has a designated value to be added to the starting culpability score or subtracted from it as required; for example, if the organisation had 50 or more employees, and an individual within substantial authority personnel participated in, condoned or was wilfully ignorant of the organisation's offence, s 8C2.5 (b)(4) prescribes that two points should be added to the defendant's culpability score. If, on the other hand, the offence occurred in spite of the fact that the defendant organisation had an effective compliance and ethics programme in place at the time of the offence, s 8C2.5 (f) (1) states that the court should deduct three points from the defendant organisation's culpability score.

14.124 Section 8C2.5 outlines a list of aggravating factors which are deemed to increase the organisation's 'culpability score'. Aggravating factors listed in s 8C2.5 (b)–(e) include the high-level personnel's involvement in or tolerance of criminal activity,[1] a recent history of similar conduct,[2] violation of a condition of probation,[3] and obstruction of justice.[4]

1 See s 8C2.5 (b). Involvement can take the form of participating, condoning, or wilfully ignoring the commission of the offence. Note that the organisation's level of culpability (and thus the prescribed culpability score modifier) increases in proportion to the size of the defendant organisation; thus, to return to the earlier example, if the organisation had 50 or more employees, and an individual within substantial authority personnel participated in, condoned or was wilfully ignorant of the organisation's offence, s 8C2.5 (b)(4) prescribes that two points should be added to the defendant's culpability score. If, on the other hand, the organisation had 1,000 or more employees, and an individual within high-level personnel of the organisation participated in, condoned or was wilfully ignorant of the offence, s 8C2.5 (b)(2)(A)(i) states that the court should add four points to the defendant organisation's culpability score.

2 See s 8C2.5 (c). If the organisation has committed a similar offence within the last 10 years, s 8C2.5 (c)(1) requires the court to add one point to the defendant organisation's culpability score. That rises to two points if the organisation has committed a similar offence within the past five years (s 8C2.5 (c)(2)).

3 See s 8C2.5 (d). If the organisation has breached the terms of its probation, the court will ordinarily add one point to its culpability score (s 8C2.5 (d)(2)). Where, however, the organisation's breach of the terms of its probation stems from the very type of conduct for which it was put on probation to begin with, in those circumstances the court will add two points to the defendant organisation's culpability score (s 8C2.5 (d)(1)).

4 See s 8C2.5 (e). This section states that, where the organisation wilfully obstructs or attempts to obstruct the course of justice during the conduct of any investigation, prosecution or sentencing of an offence, the court will add three points to the defendant organisation's culpability score.

14.125 Section 8C2.5 also contains a number of mitigating factors which will have a positive impact on the organisations 'culpability score'. The mitigating factors can be found in s 8C2.5 (f)–(g). The severity of the offence (and, in turn, the size of any fine imposed) will be mitigated where the defendant organisation had in place effective corporate compliance programmes at the time of the offence which were designed to prevent and detect any breaches of the law.[1] The organisation's culpability will also be mitigated where the organisation promptly self-reports the fact it has committed an offence, cooperates in any subsequent investigation into the offence and accepts responsibility for the offence.[2]

1 See s 8C2.5 (f). The presence of an effective compliance programme can result in a reduction of the organisation's culpability score of three points (s 8C2.5 (f)(1)). Furthermore, an organisation can still benefit from this reduction even where its compliance programme failed to prevent the commission of a criminal offence (s 8C2.5 (f)(1)). This reflects a recognition in the *Principles of Federal Prosecution of Business Organisations* that 'no compliance program can ever prevent all criminal activity by a corporation's employees' (www.justice.gov/opa/documents/corp-charging-guidelines.pdf at p 15). The factors for the court to consider, when determining whether or not the defendant organisation has put in place an 'effective compliance and ethics program', are set out in s 8B2.1. The term 'compliance and ethics program' is further defined in the commentary accompanying this section as meaning 'a program designed to prevent and detect criminal conduct'.

In order to satisfy this requirement, the organisation must 'exercise due diligence to prevent and detect criminal conduct' and 'otherwise promote an organisational culture that encourages ethical conduct and a commitment to compliance with the law' (ss 8B2.1 (a)(1) and 8B2.1

(a)(2) respectively). Section 8B2.1 (b) goes further and sets out a number of minimum requirements which any corporate compliance and ethics programme must satisfy if it is to be deemed to demonstrate that the organisation has exercised due diligence to prevent and detect criminal conduct, and otherwise promote an organisational culture that encourages ethical conduct and a commitment to compliance with the law (as required by s 8B2.1 (a)), namely:

(a) The organisation must establish procedures to detect and prevent criminal conduct (s 8B2.1 (b)(1)).

(b) The organisation must assign high-level personnel to oversee and monitor the implementation and effectiveness of any compliance and ethics programme (s 8B2.1 (b)(2)). The commentary contained in the guidelines which relates to this section notes that 'High-level personnel and substantial authority personnel of the organisation shall be knowledgeable about the content and operation of the compliance and ethics program, shall perform their assigned duties consistent with the exercise of due diligence, and shall promote an organisational culture that encourages ethical conduct and a commitment to compliance with the law'.

It is important to note, however, that s 8C2.5 (f) will not apply to reduce a defendant organisation's culpability score where a high-level manager of the organisation or a person responsible for the administration or enforcement of the organisation's compliance and ethics programme 'participated in, condoned, or was wilfully ignorant of the offence' (s 8C2.5 (f)(3)). Where high-level personnel are involved in the offence (either because they participated in it, condoned it, or wilfully ignored it), this gives rise to a rebuttable presumption that the organisation's compliance and ethics programme was, in fact, ineffective (s 8C2.5 (f)(3)(B)).

(c) The organisation must make 'reasonable efforts not to include within the substantial authority personnel of the organisation any individual whom the organisation knew, or should have known through the exercise of due diligence, has engaged in illegal conduct or other conduct inconsistent with an effective compliance or ethics program' (s 8B2.1 (b)(3)).

The commentary accompanying s 8B2.1 (b)(3) points out that nothing within that section 'is intended to require conduct inconsistent with any Federal, State, or local law governing employment or hiring practices'. As such, an organisation will not be expected to make criminal background inquiries where such inquiries would be in breach of any relevant employability/equality legislation.

The commentary accompanying this section further points out that, when deciding the suitability of an individual who may have previously been guilty of committing illegal activities or other misconduct for a post within the high-level personnel (or substantial authority personnel) of the organisation, the organisation will need to take into account that individual's ability to perform their assigned duties in a manner which is consistent with due diligence and with the promotion of an organisational culture that encourages ethical conduct and a commitment to compliance with the law as required by s 8B2.1 (a). According to the commentary, factors that may affect that individual's ability to fulfil those obligations include: how related their previous illegal activities/misconduct was to specific responsibilities they are likely to be given in their new role; how recently the individual in question committed the illegal acts/was involved in misconduct; and whether that individual has engaged in other such illegal activities or misconduct.

(d) The organisation must 'take reasonable steps to communicate periodically and in a practical manner its standards and procedures, and other aspects of its compliance and ethics program' to its employees and agents. In order to achieve this goal, the organisation will be expected to conduct training programmes and otherwise disseminate information to its employees and agents about their respective roles and responsibilities (s 8B2.1 (b)(4)). Note that the term 'standards and procedures' is further defined in the commentary accompanying this section as 'standards of conduct and internal controls that are reasonably capable of reducing the likelihood of criminal conduct'.

(e) The organisation is required to take reasonable steps to ensure the effectiveness of its corporate compliance and ethics programme by monitoring compliance with the scheme, evaluating its efficacy (and presumably taking steps to address any shortcomings the organisation discovers) and by implementing a system which allows 'whistleblowers' to notify any breaches/potential breaches of the scheme/the law (possibly anonymously) without fear of repercussions.

(f) The organisation should actively promote compliance with its compliance and ethics programme and continuously emphasise the importance of fulfilling the requirements imposed on its employees/agents by such a programme. The importance and value of the system can be emphasised in two key ways. First, by adopting an (appropriate) incentive scheme to encourage compliance with the programme's requirements. Secondly, by taking appropriate disciplinary action against those who engage in criminal conduct or who fail to take reasonable steps to prevent or detect criminal conduct (s 8B2.1 (b)(6)). The commentary relating to this section notes that what will be deemed 'appropriate' disciplinary action will be case specific.

As the commentary points out, this particular element is requiring the organisation to take a proactive approach to detecting and preventing crime. The organisation will be expected to periodically: (1) assess the risk that it may become involved in criminal activity; (2) revise and prioritise any actions it can take to demonstrate that it has an effective compliance and ethics programme in place; and (3) take any actions deemed necessary to reduce the risk that the organisation might become involved in criminal activity.

It should be noted that there is a further benefit attached to an organisation having in place an effective compliance and ethics scheme (in addition to a potential reduction in its culpability score and, thus, subsequent fine). The existence of such a scheme may impact upon a prosecutor's decision not to prosecute. This point is further discussed in the section dealing with deferred prosecution agreements and non-prosecution agreements (see **14.113** onwards).

2 An organisation can have its culpability score reduced by up to five points if it promptly reports the commission of an offence (of its own free will), if it cooperates in any subsequent investigation into its activities, and if it admits responsibility for any criminal offence that it might have committed.

To qualify for any such reduction in its culpability score, the organisation must report the offence to the 'appropriate governmental authorities'. According to the commentary accompanying this section, 'appropriate government authorities' are 'federal or state law enforcement, regulatory or program officials having jurisdiction over such matter'. The commentary also points out that the organisation will only qualify for such a reduction in their culpability score if their revelation is 'timely and thorough'. As such, the disclosure must be made to the appropriate authorities pretty much as soon as the organisation itself becomes aware that a potential crime is under investigation, and the information provided must be 'sufficient for law enforcement personnel to identify the nature and extent of the offence and the individual(s) responsible for the criminal conduct'.

According to the commentary accompanying this section, the court should measure the organisation's level of cooperation rather than the level of cooperation of individuals within the organisation. As the commentary points out, this effectively means that an organisation as a whole may be deemed to have fulfilled the requirement of cooperation in an investigation, even though some individual(s) fail to cooperate.

Finally, it is noted in the commentary accompanying this section that an organisation will be deemed to have accepted responsibility for its criminal actions where: (1) it has pleaded guilty to the offence before the trial began (it will be insufficient for the organisation to admit guilt and express remorse for its actions after it has been found guilty by a court); and (2) it admits being involved in the offence and any related conduct.

It must always be remembered that an organisation will only qualify for the five-point subtraction from its culpability score where it promptly reports the commission of an offence, and cooperates in any subsequent investigation into its activities, and admits responsibility for any criminal offence it might have committed. If the organisation fails to promptly report the commission of an offence to the appropriate authorities, but cooperates fully with any subsequent investigation and accepts responsibility for its criminal conduct, it will only qualify for a two-point reduction in its culpability score (s 8C2.5 (g)(2)). Similarly, if the organisation fails to promptly report the commission of an offence to the appropriate authorities and fails to cooperate fully with any subsequent investigation, but accepts responsibility for its criminal conduct, it will only qualify for a one-point reduction in its culpability score (s 8C2.5 (g)(3)).

14.126 Once the court has taken into account any aggravating/mitigating factors of the offence and determined the organisation's culpability score, it will then need to make reference to the chart contained in s 8C2.6 and find the applicable minimum and maximum fine multipliers for that culpability score. The court will then determine the minimum and maximum fine range for the offence by multiplying the base fine by the minimum and maximum fine multipliers.[1] It is important to point out that, if the organisation has been convicted of being involved in bid-rigging, price-fixing or market-allocation agreements among competitors, special rules apply which set a floor for minimum and maximum multipliers to be applied in such cases.[2]

1 See s 8C2.6. It may be helpful at this stage to provide an example to illustrate the process that the court is engaged in when determining the level at which to set any fine. In this example the organisation has been found guilty of manufacturing an eavesdropping device for pecuniary

gain. A quick reference to Chapter 2 of the *United States Sentencing Guidelines* (specifically s 2H3.2) reveals that the base offence level is 9.

The court then makes reference to s 8C2.4 in order to determine the base fine. In this case the table at s 8C2.4 (d) sets the base fine for a level 9 offence at $15,000 (for argument's sake, this sum is greater than both the pecuniary gain and/or any pecuniary loss caused).

The court's next step is to determine the organisation's culpability score, which starts at 5. The court works its way through the list of aggravating factors and decides that it is dealing with an organisation that had 10 or more employees, and an individual within substantial authority personnel had participated in and condoned the commission of the offence. The court adds one point to the organisation's culpability score (s 8C2.5 (b)(5)). The court then looks to see whether there are any mitigating factors. The court notes that the organisation failed to self-report that an offence had been committed and failed to cooperate with an investigation into its actions, but did ultimately demonstrate recognition and affirmative acceptance of responsibility for its criminal conduct. The court deducts one point from the organisation's culpability score (s 8C2.5 (g)). At the end of this process, the organisation's culpability score is 5 (ie (5 + 1) – 1 = 5).

Now that the court knows the base fine ($15,000) and the organisation's culpability score (5), the next step is to determine the fine range for the offence. To do this, it will need to determine the applicable minimum and maximum fine multipliers. The court refers to s 8C2.6 and discovers that the minimum fine multiplier for a culpability score of 5 is 1.00 and the maximum fine multiplier at that level is 2.00. This means that the defendant organisation may be sentenced to a fine of no less that £15,000 (ie 15,000 × 1) and no more than $30,000 (15,000 × 2).

2 That is to say that there is a level below which the minimum and maximum fine multipliers for that offence cannot drop, regardless of what the recommended multipliers are for organisations with the defendant's culpability score according to s 8C2.6. This much is evident from s 2R1.1.

14.127 The court's final job is to determine where, within that fine range, to set the penalty for the offence. The court is aided in this task by s 8C2.8. This sets out a list of factors which the court should consider when making this decision including:

> 'the need for the sentence to reflect the seriousness of the offence, promote respect for the law, provide just punishment, afford adequate deterrence, and protect the public from further crimes of the organisation'.[1]

The list contains factors which might encourage the courts to set the fine closer to the maximum level, such as the failure of the organisation to have in place an effective compliance an ethics programme,[2] just as it contains factors which might encourage the court to set the fine closer to the minimum level, such as the organisation having a culpability score of less than 0 (which is possible).[3] The court may, additionally, take into account:

> 'the relative importance of any factor used to determine the range, including the pecuniary loss caused by the offence, the pecuniary gain from the offence, any specific offence characteristic used to determine the offence level, and any aggravating or mitigating factor used to determine the culpability score.'[4]

1 See s 8C2.8 (a)(1).

2 See s 8C2.8 (a)(11). This section also points out that the question of whether the organisation has in place an *effective* compliance and ethics programme or not is to be determined by reference to the factors set out in s 8B2.1 (as discussed earlier).

3 See s 8C2.8 (a)(8).

4 See s 8C2.8 (b). The commentary accompanying this section explains its significance as follows: 'This [section] allows for courts to differentiate between cases that have the same offence level but differ in seriousness … Similarly, this [section] allows for courts to differentiate between two cases that have the same aggravating factors, but in which those factors vary in their intensity …'.

14.128 Even this is not quite the end of the matter. In some circumstances, the courts are required to increase the amount of any fine, once they have applied the minimum and maximum multipliers and determined a fine within the prescribed range, by adding 'any gain to the organisation from the offence that has not and will not be paid as restitution or by way of other remedial measures'.[1]

1 The commentary accompanying this section notes that this section is designed to 'ensure that any amount of gain that has not and will not be taken from the organisation for remedial purposes will be added on to the fine'. The commentary goes on to explain that this section will generally apply where 'the organisation has received gain from an offence but restitution or remedial efforts will not be required because the offence did not result in harm to identifiable victims, e.g. money laundering, obscenity and regulatory reporting offences'.

14.129 The final key section that needs to be highlighted in this discussion of the US sentencing regime is s 8C4.2. This section comes into play where the organisation's conduct has resulted in a death. In situations where the offence 'resulted in death or bodily injury, or involved a foreseeable risk of death or bodily injury', the court is permitted to consider an 'upward departure' from the guideline fine range for the offence, that is to say that the court can consider going above the maximum prescribed fine following the application of the maximum fine multiplier. Factors that should be considered by the court, in deciding the extent to which it should increase the maximum prescribed fine, include 'the nature of the harm and the extent to which the harm was intended or knowingly risked' as well as the extent to which such harm or risk is normally taken into account when setting a fine within the general applicable guideline figure range. This is clearly relevant in the context of deaths resulting from breaches of the obligations contained in the Occupational Safety and Health Act that we have just considered.

14.130 The courts have been creative in their use of sentencing powers. In *United States v Missouri Valley Construction Co*,[1] for example, the court ordered the convicted corporation to endow a chair of ethics at the state university. This touches on the border with the growing field of social responsibility law and practice.

1 741 F 2d 1542 (8th Cir).

14.131 Emphasis in both the Sentencing and Prosecution Guidelines is placed on corporate compliance programmes. Consultancy can provide this commercially, and large corporations are unlikely to overlook the advantages of having a programme in place. The court's general approach has been to attempt to prevent incidents occurring and to achieve remediation first and foremost, reducing fines where this has been assured.[1] However, this has of course not manifested itself to the same degree in relation to corporate manslaughter. Having in place a corporate compliance programme may support a defence that the corporation was not liable to begin with. This applies particularly in relation to strict liability, regulatory offences with criminal sanctions. Here there is often a due diligence defence which a company with some foresight should be able to take advantage of.

1 Eg *Logan County Farm Enterprises Inc* OSHRC Docket no 78–4535.

14.132 The concept of vicarious liability and corporate manslaughter under the relevant state statutes does not exclude liability of individual employees, directors and officers. The guidelines on prosecution and sentencing explicitly recognise this. Under the 'responsible corporate officer' doctrine, a corporate official who holds a position of 'responsibility and authority' within a company can be held criminally liable for not exercising the 'highest standard of foresight and vigilance' to prevent a crime from occurring in the sphere of the company's operations for which the officer had responsibility.[1]

1 *United States v Park* 421 US 658 (1975), following *United States v Dotterweich* 320 US 277 (1943).

14.133 Under the Occupational Health and Safety Act 1970, for the purposes of criminal enforcement only, an individual who is a corporate officer or director may be an 'employer' within the meaning of the Act and consequently found personally guilty of the offences set out in the Act.[1] This is particularly the case where the officer's role in running the company is so pervasive that it is, in effect, a sole proprietorship.[2]

1 *United States v Doig* 950 F 2d 411.
2 *United States v Cusack* 806 F Supp 47 (DNJ, 1992).

14.134 Whilst directors and officers cannot be charged as aiders and abettors under the Occupational Health and Safety Act 1970, they regularly, if infrequently, are in relation to other offences, and corporate manslaughter is very much the type of offence that individual directors and officers are likely to be investigated at least in relation to.

14.135 In *Sabine Consol Inc v State*[1] a reporting supervisor was successfully charged with the offence of criminally negligent homicide when trench walls collapsed.[2]

1 (1974) 806 SW 2d 553 (Tex Crim App, 1991).
2 See also *People v Pymm* 563 NE 2d 1,3 (NY, 1990); *Chicago Magnet Wire Corporation* 534 NE 2d 963, *People v Hegedus* 443 NW 2d 127, 128 (Mich 1989); *Cornellier* 425 NW 2d at 22.

Conclusion

14.136 The danger of corporate compliance programmes and individual programmes, as well as corporate crime in general, is that this can induce a minimal standard from corporations and undermine their willingness to increase their investment in safety if they can protect themselves through delegating responsibility to officers or to consultants providing compliance programmes.[1]

1 W Laufer 'Corporate Liability, Risk Shifting and the Paradox of Compliance' (1999) Vand LR 1344.

14.137 The Census of Fatal Occupational Injuries has suggested, year on year, a decrease in the level of workplace fatalities,[1] not including victims of 11 September 2001. There is very wide variation between states. The media regularly reports incidents of corporate killing and the like. Non-governmental groups monitor workplace fatalities and provide information about the worst 100 corporations and reports of fatalities. In the words of Robert Weissman, there is a relative academic Siberia in the field of statistics regarding corporate manslaughter.[2]

1 Bureau of Labour Statistics Census of Fatal Occupational Injuries.
2 R Weissman, editor of Multinational Monitor, letter to Corporate Crime Reporter, December 1998.

14.138 The 1990s saw global actors concerned about human rights shift their attention from abuses committed by governments to close scrutiny of the activities of business enterprises, particularly multinational corporations.[1] In the light of Enron, WorldCom and the relative weakness of regulatory bodies in the field, corporate criminal prosecutions are unlikely to reduce. There remains much work to be done in assessing their effectiveness. None of the available statistics show a breakdown in the numbers of civil or criminal prosecutions. It is impossible without an in-depth study to assess the impact on the level of workplace fatalities. The cost of criminal sanctions, whether financial or otherwise, can be enormous and cannot be justified on the basis of unconsidered assumptions which seem to underpin the present law.[2]

1 (2001) 111 Yale LJ 443.
2 V Khanna, *A Political Theory of Corporate Crime Legislation* (2003).

THE FAR EAST

China

Introduction

14.139 China's fast economic growth has led to widespread concerns about workplace safety – in July 2010, the government reported that an average of 187 people die per day in industrial accidents in China.[1] Although legislation has been implemented to address some of the prevalent issues in workplace safety, such as collusion between local officials and factory owners, and the failure to properly report and investigate accidents, it is not clear how successfully these are enforced.

1 Wall Street Journal.

Health and safety legislation

14.140 Health and safety law in China is governed by the Law on Work Safety (No 70 of 2002). Separately to this, the Chinese Criminal Code sets out a number of health and safety offences which carry criminal liability for individuals and corporations.

14.141 The Law on Work Safety imposes a wide range of requirements on local inspectors, owners of and investors in businesses, administrative and managerial staff, production supervisors, and individual workers. For offences that are not sufficiently serious to carry criminal liability, administrative penalties will be imposed. These include warnings, fines, dismissal, confiscation of illegal gains, cessation of production, and factory closure.

Criminal liability

14.142 Criminal liability is attached to certain offences under the Chinese Criminal Code. Consequently, liability is for the specific offence created by the Criminal Code, rather than health and safety breaches leading to liability for existing offences such as manslaughter.

14.143 The Criminal Code specifically provides that a company can bear criminal responsibility.[1] The Code also provides that this will result in a fine for the company and criminal punishment for the persons directly in charge and also the other persons directly responsible for the crime.

1 Art 30, Chinese Criminal Code.

14.144 However, the particular offences in the Criminal Code do not specifically provide for punishment of the company itself, but rather the personnel directly responsible. There is debate in Chinese legal circles as to whether the personnel are responsible by virtue of the corporation being found guilty of an offence, therefore leading to the possibility of high-ranking management potentially being responsible for accidents.

14.145 Unlike other jurisdictions where liability can be ascribed to a corporation through its 'controlling mind', it is thought that companies cannot be responsible for 'intentional' offences under the Criminal Code, ie offences requiring a *mens rea.* The Criminal Codes wording of 'directly responsible', however, appears to preclude prosecution of, for example, management who might otherwise be liable for failing to implement an adequate system to prevent accidents, or where the corporate governance and guidance has given rise to an omission which caused the accident.

Sanctions and sentencing

14.146 The Criminal Code provides for a number of health and safety offences.

14.147 If an employee disobeys management or violates rules and regulations, or forces another employee to work under hazardous conditions, and causes an accident with heavy casualties or other serious consequences, he can be sentenced to up to three years in prison, or up to seven years for particularly severe offences. There are no guidelines in the legislation as to what constitutes 'heavy casualties' or other 'severe consequences', and hence when criminal proceedings will be appropriate.

14.148 If a company is warned that its working conditions are unsafe, and takes no steps to remedy this, then if an accident occurs involving heavy casualties or other serious consequences, the person directly responsible for the accident can be imprisoned for the same terms as above.

14.149 The same sentence also applies to offences involving the control of hazardous materials, where regulations are not followed and a serious accident is caused during the production, storage, transportation or use of those materials.

14.150 If a company lowers the quality standard of a product and causes a serious accident, the person directly responsible for the accident can be sentenced to up to five years in prison (or up to 10 years for especially serious offences) and can also be fined.

Indonesia

Introduction

14.151 Indonesia's employment legislation enshrines the right of every worker to receive occupational safety and health protection.[1] Despite this, Indonesia has a poor work safety record compared to other countries in the region. In 2010, the National Safety and Health Council of Indonesia released statistics showing that about 23 out of every 100,000 workers die in work-related accidents each year, compared to six casualties per 100,000 workers in Malaysia and Thailand.[2]

1 Article 86(1) of Law No 13 of 2003 concerning Manpower.

2 'Ministry Moves to Improve Indonesia's Poor Work-Safety Record' (2010) Jakarta Globe, 16 January.

Health and safety legislation

14.152 Indonesia has a variety of laws relating to health and safety in the workplace, including Law No 1 of 1970 regarding Occupational Safety, which sets out the general principles, and further regulations and decrees.

14.153 An employer is subject to the requirements of health and safety legislation if they employ 100 or more individuals, or require its employees to carry out high-risk work/activities. The Act imposes requirements on managers, including the requirement to provide guidance on working safely. These specific requirements include displaying information about health and safety (including pictures) in easily seen places, and providing workers with free safety equipment and guidance for using such equipment.

14.154 Employees are also required to obey all safety instructions and to use personal protective equipment, and have the right under law to remove themselves from danger and refuse to carry out or continue work in conditions that present an imminent and serious threat to their life or health.

14.155 The Department of Manpower and Transmigration is responsible for the supervision of companies' health and safety systems, which includes regular assessments.

Criminal liability

14.156 Individual managers can be subject to up to three months' imprisonment for breaches of health and safety legislation, as well as fines. However, there appears to be no mechanism in Indonesian law for imposing such sanctions or criminal liability on the 'controlling mind' of a company or those higher up in the management hierarchy.

14.157 Under the Indonesian Penal Code, corporations are not subject to criminal liability. A corporate body can be subject to criminal liability only under the specific statute. The two statutes that currently provide for corporate criminal liability include Law No 32 of 2009 regarding Environmental Protection and Management. (The other statute, Law No 8 of 2010 regarding Prevention and Eradication of Money Laundering Crime, does not relate to occupational safety.) The Law on Environmental Protection and Management includes provisions relating to the handling of hazardous and toxic materials and also toxic waste. As indicated by the title, the purpose of the law is to protect the environment rather than individual workers. This leads to an inconsistency in the law, whereby corporate criminal liability could be possible for an accident caused by the improper handling of toxic waste, but not for an accident caused by unsafe procedures or faulty equipment used in the handling of such material.

Sanctions and sentencing

14.158 Sanctions under the Occupational Safety Act and related regulations include a prison term of up to three months and a fine of up to 100,000 rupiah.

14.159 Criminal sanctions under the Law on Environmental Protection and Management include seizure of profits and business closure. In addition, personnel in charge of businesses in breach of the law can be subject to up to a year's imprisonment and a maximum fine of 1 billion rupiah.

Malaysia

Introduction

14.160 While Malaysia has a better workplace safety record than some of its neighbouring countries, the issue of corporate liability has become relevant not only for occupational safety and worker protection but also for public safety and accident prevention. There are a high number of accidents involving public transport such as buses, trains and ferries, some with heavy public casualties. These accidents have prompted some legal commentators in Malaysia to look to the comparable legislation in both the UK and in Australia that can impose criminal liability on the corporations responsible for such accidents.

Health and safety legislation

14.161 The main Act dealing with health and safety is the Occupational Safety and Health Act 1994 and its associated regulations. This Act is designed to safeguard both employees and visitors to workplaces. It imposes obligations on employers, which includes managers, agents and occupiers of a place of work, and also government departments.

14.162 Employers are required to provide safe equipment, workplaces and systems of work, and provide such information and training as is necessary to ensure the safety and health of employees. The requirements under the Act are subject to the qualification 'so far as is practicable', which the Act states must be considered having regard to the severity of the risk, the knowledge of the risk, and the availability and cost of ways to remove or mitigate the risk.

14.163 Employers employing 40 or more persons must establish a health and safety committee to keep such measures under review and investigate any accidents. Some employers are required to employ a competent person to act as a safety and health officer for the exclusive purpose of ensuring that the regulations are complied with. This includes large construction projects and other heavy industrial workplaces employing more than 100 employees.

Criminal liability

14.164 A company is recognised as a separate legal entity and can be prosecuted. Sanctions include fines and compounding. It is possible to charge and convict a company without also charging its officers. There are also a number of statutes, such as the Factories and Machineries Act 1967, that have deeming provisions, whereby an offence committed by an individual is deemed to have been have been committed by a company.

14.165 There are also a number of offences in Malaysia's Penal Code relating to public health and safety, which carry criminal liability and sanctions. These include: making the atmosphere noxious to health (which carries a fine); rash navigation of a vessel (imprisonment of up to six months or a fine); and conveying persons by water for hire in a vessel overloaded or unsafe (up to six months or a fine). However, it is thought that this will result in individual, rather than corporate, liability.

14.166 There are very few reported cases dealing with the concept of corporate criminal liability, and so it is not yet clear how the doctrine is developing. However, the majority of cases impute intent to the company by the 'directing minds' theory, that is by identifying the personnel who can be said to be the embodiment of the company itself. This is the approach taken in *Public Prosecutor v Kedah & Perlis Ferry Service Sdn Bhd*,[1] in which Syed Agil Barakbah J said:

> 'a limited company … could not be found guilty of the offence without proof of mens rea of its agents or officers. The persons whose knowledge would be imputed to the company would be those who were entrusted with the exercise of the powers of the company.'

1 [1978] 2 MLJ 221.

14.167 There have also been cases imputing vicarious liability to a corporation for the acts of its employees,[1] and it is thought that this approach is better suited to strict liability offences.

1 See eg *Public Prosecutor v Teck Guan Co Ltd* [1970] 2 MLJ 141.

14.168 A company can offer a defence to such charges if it has taken reasonable care to exert control on its directors or employees and those staff have acted ultra vires.

14.169 Liability will not extend to parent or other group companies, as these are regarded as separate entities, unless there is evidence of collusion.

Sanctions and sentencing

14.170 Criminal sanctions for companies include fines and compounding, and the scope of such fines is determined by the individual legislation under which the charges are brought. There is no mechanism by which to imprison or fine an individual director for a conviction against the company. However, individual directors can face liability for offences under legislation such as the Occupational Safety and Health Act, and offences under that statute can carry prison sentences of up to two years.

14.171 The legislation also imposes liability on employees for failing to follow instructions and take reasonable care for the safety of himself and others, with possible sanctions includes fines and also imprisonment for up to three months.

Thailand

Introduction

14.172 Thailand has recently (2011) revised its laws relating to health and safety at work, following some high-profile accidents. These included a fire at the Kader toy factory in Bangkok in 1993, resulting in the deaths of 188 people, and a fire at the Santika nightclub in Bangkok in 2009, resulting in 66 deaths.

Health and safety legislation

14.173 The main Act is the Safety, Health and Workplace Act 2011, which imposes duties upon employers for health and safety in the workplace. These include notifying employees of possible dangers and distributing operational manuals, arranging necessary training, displaying warning signs and safety directives, and providing protective equipment.

14.174 Employers wishing to operate a factory in Thailand must obtain a factory licence, which will stipulate the size and operation of the factory, including safety procedures. Licences must be periodically renewed.

Criminal liability

14.175 The Civil and Commercial Code of Thailand recognises that a corporation is a separate legal entity, and that its rights and duties are distinguished from those of its shareholders. Specific statutes impose criminal liability on corporations and specific individuals for breach of their requirements. For example, the Anti-Money Laundering Act[1] imposes criminal liability on a corporation. The Factory Act[2] imposes criminal liability on 'any person who has been granted a licence to operate a factory'. The Act on the Transportation by Land[3] imposes criminal liability on 'any person who has been granted a licence to operate transportation'.

1 BE 2542 (1999).
2 BE 2512 (1969).
3 BE 2522 (1979).

14.176 However, the Penal Code, which is the main piece of criminal legislation in the country, is silent on whether a corporation can be criminally liable for the offences set out in the Code. Thailand's Supreme Court has considered this and has held[1] that, where the representative of a corporation acted in the scope of his authority and in accordance with the objectives of the corporation in a manner such that the corporation benefited from the act, the corporation was accountable and held criminally liable for the act. Consequently, a corporation can be liable for offences under the Penal Code.

1 In decision nos 787–788/2506.

14.177 Although it is not necessary to identify an individual offender in order to prosecute a company, criminal liability of a corporation is usually linked to criminal liability of its executives. For offences under certain Acts, liability will attach to the named individual on the operating licence.

14.178 Consistent with the approach that a corporation is a separate legal entity, a parent company cannot be prosecuted for offences committed by a subsidiary or group company.

Sanctions and sentencing

14.179 The Safety, Health and Workplace Act 2011 imposes fines and, in some cases, daily fines and imprisonment for breaches of the Act.

Singapore

Introduction

14.180 Singapore has set itself ambitious goals in relation to health and safety: the National Strategy for Workplace Safety and Health for Singapore (WSH 2018) aims to bring down the national fatality rate to less than 1.8 per 100,000 workers by 2018.[1]

1 www.mom.gov.sg/workplace-safety-health/Pages/default.aspx.

Health and safety legislation

14.181 The Workplace Safety and Health Act sets out obligations and responsibilities for different stakeholders: employers, principals, occupiers, manufacturers and suppliers, installers or erectors of machinery, employees, and the self-employed. All stakeholders have an obligation to ensure, as far as is reasonably practicable, the safety and health of employees.

14.182 Enforcement is monitored by the Commissioner for Workplace Safety and Health, assisted by Deputy Commissioners and inspectors. Inspectors are empowered to enter and inspect any workplace at any time, and to take copies or samples from the workplace. An inspector may also take any article into custody for the purpose of an investigation or an inquiry. The Commissioner may issue a Remedial Order in respect of any workplace, under which the occupier must take required measures to remedy any danger in the workplace. The Commissioner may also issue a Stop Work Order, under which the specified work must cease until measures have been taken to ensure that work can be carried out safely.

14.183 Certain people and businesses are required to register with the Ministry of Manpower (MOM) before commencing operations. Factories carrying out lower-risk activities need only make a one-off notification before commencing operations, including a Risk Management declaration. Higher-risk factories must register with the Ministry and may be subject to regular renewals. Registration is also required for equipment operators such as crane and scaffolding contractors. Safety professionals, such as examiners of lifting equipment, examiners of pressure vessels, workplace safety and health professionals, training providers and risk consultants, are also licensed by the Ministry of Manpower.

Employers now come under new legislation and have to report to the MOM if their workers cannot work for more than three days due to the same injury, even if these were not consecutive days. This has been put in place to discourage some employers from breaking up the medical leave of injured workers to avoid reporting incidents.

The rule change was first announced in July 2013, and public consultations were held on the proposed amendments in September 2013, and MOM said that there was 'a consensus supporting the amendments'. Before this, employers only had to report work-related accidents that resulted in medical leave of more than three consecutive days. However, MOM received industry feedback that employers were avoiding this requirement by breaking up medical leave.

MOM said that such a practice could 'affect employees' recovery process' and 'obscure the actual severity of the accidents that have occurred'. Accidents covered by this rule must be reported by employers within 10 calendar days from the fourth day of medical leave. Employers also have to report all work-related traffic accidents involving their workers to the Commissioner for Workplace Safety and Health.

The maximum penalty for failing to make an incident report, as required by law, is a fine of $5,000 for a first-time offence and a $10,000 fine and a six-month jail term for subsequent offences.

Criminal liability

14.184 Criminal liability can be imputed to a company through the acts of those directors and officers that constitute its 'directing mind and will', on the grounds that their acts are regarded as the company's acts.

14.185 The Workplace Safety and Health Act also provides for sentences on both the individual person and the corporate body.

Sanctions and sentencing

14.186 The Commission for Workplace Safety and Health may order composition fines for breaches of the Workplace Safety and Health Act.

14.187 Companies that fail to comply with a Remedial Order can be subject to a maximum fine of S$50,000, plus a daily fine of S$5,000 if the offence continues. In addition, the offence carries a prison sentence of up to 12 months.

14.188 Companies that fail to comply with a Stop Work Order can be subject to a maximum fine of S$500,000, plus a daily fine of S$20,000. The maximum prison sentence is 12 months.

14.189 In the case of other offences under the Workplace Safety and Health Act, general penalties will apply. For individuals, this can be a fine of up to S$200,000 (for a first conviction) or S$400,000 (for repeat offenders) and/or a prison sentence of up to two years. A corporate body can be fined up to S$500,000 for a first offence or S$1 million for repeat offenders.

The Philippines

Introduction

14.190 Health and safety is enshrined in the Philippine Constitution of 1987 as 'just and humane terms and conditions of work', and the government has instigated the Zero Accident Programme as a long-term strategy to promote the health and safety of workers.

Health and safety legislation

14.191 Book IV of the Philippine Labor Code covers prevention and compensation of work related injuries and illnesses. This obliges all employers, in both the public and private sectors, to keep workplaces free from hazards that are likely to cause physical harm to employees or damage to property. Further, the Occupational Safety and Health Standards (OSHS) were developed in 1978. This sets out rules relating to, among other elements, health and safety training, record keeping, materials handling and storage, fire protection and control, personal protective equipment and devices, and occupation health services.

14.192 The lead agency in implementing and enforcing occupational health and safety policies is the Department of Labor and Employment (DOLE).

14.193 The state maintains the State Insurance Fund, a tax-exempt employees' compensation programme to provide benefits in the case of disability or death arising from work-connected accidents. This is funded by contributions from both private- and public-sector employers.

14.194 Construction firms must be licensed with the Philippine Contractors Accreditation Board (PCAB).

Criminal liability

14.195 Unlike other jurisdictions in the region, there are no specific health and safety offences listed in the Penal Code of the Philippines. However, there is scope in the law for a corporation to be criminally liable because of the activities of its officers. The Supreme Court, in the case of *Philippine National Bank v Court of Appeals*, commented that 'the rules governing the liability of a principal or master for a tort committed by an agent or servant are the same whether the principal or master be a natural person or a corporation, and whether the servant or agent be a natural person or artificial person'.[1]

1 (1978) 83 Supreme Court Reports Annotated 237.

14.196 In another case, the Supreme Court held that, when a criminal statute forbids a corporation from an act, this prohibition (and potential liability) will extend to each director of the company. Directors will have a defence, however, if they act within their authority and act in good faith, and do not assent to unlawful acts of the company or act negligently. Consequently, there is scope in the country's legislation to introduce corporate criminal liability, extending to a company's directors, for breaches of health and safety legislation in future.

Sanctions and sentencing

14.197 Any employers in breach of labour standards and safety regulations can be reported to the PCAB. The DOLE regional director can also issue a stop order in the case of imminent danger, forcing employers to cease operation.

Hong Kong

Introduction

14.198 Hong Kong faces a number of occupational health and safety challenges, due in part to its sizeable construction industry and the number of large projects that are currently reaching their peak. The Hong Kong branch of the Institute of Safety and Health reported that, in 2011, there were over 3,000 industrial accidents in the construction industry in Hong Kong, with a sharp rise in the number of electric accidents.[1]

1 www.iosh.co.uk/news/latest_news_releases/55_hong_kong_safety_warning.aspx.

Health and safety legislation

14.199 There are two main pieces of health and safety legislation in Hong Kong: the Occupational Safety and Health Ordinance (Cap 509) and the Factories and Industrial Undertakings Ordinance (Cap 59), both of which have subsidiary regulations.

14.200 The former piece of legislation established the Occupational Safety and Health Council in 1988, which is the statutory body for promoting health and safety at work.

14.201 The Occupational Safety and Health Ordinance imposes duties on employers, occupiers and employees to ensure a safe and healthy workplace.

There are a number of exceptions to the law, including an aircraft or vessel in a public place, and the driver (but not other staff) of a land transport vehicle when in a public place. Domestic premises employing only domestic servants and self-employed people are further exceptions.

14.202 Duties include providing all necessary information and safety equipment, maintaining plant and work systems that do not endanger safety and health, and ensuring the safety of work premises and access ways.

14.203 The Commissioner for Labour can issue improvement notices and suspension notices where workplace conditions create an immediate hazard to employees.

14.204 The Factories and Industrial Undertakings Ordinance provides for the safety and health of workers in the industrial sector. Employers must register with the Commissioner for Labour before commencing any industrial process, and notify it of any change in the location, nature of activities, and/or person having control of the workplace. If the Commissioner considers that the workplace is not suitable, he issues a prohibition notice.

Criminal liability

14.205 Both the Occupational Safety and Health Ordinance and the Factories and Industrial Undertakings Ordinance specifically provide that companies can be convicted of offences under the legislation. Furthermore, where it is proved that the offence was committed with the consent of, or is attributable to neglect on the part of, any officer of the company, that officer will be guilty of the same offence.

14.206 The Factories and Industrial Undertakings Ordinance provides that the proprietor of every industrial undertaking in (or in respect of) which an offence has been committed will be guilty of a like offence, and liable to the penalty prescribed. It is not a defence that the offence was committed with the proprietor's knowledge.

Sanctions and sentencing

14.207 Under the Occupational Safety and Health Ordinance, failure to comply with an improvement notice carries a fine of up to HK$200,000 and imprisonment of up to 12 months. Failure to comply with a suspension notice is punishable by a fine of up to HK$500,000 and imprisonment of up to 12 months.

14.208 Prosecutions for offences under the Occupational Safety and Health Ordinance may be brought by an occupational safety officer in the name of the Commissioner for Labour. These further offences include interfering with or misusing an article provided at the workplace in the interests of health and safety, which is punishable by a fine, and attempting to charge employees for measures taken under the Ordinance, also punishable by a fine. It also an offence to prevent aid being given to an employee at a workplace, punishable by a fine and imprisonment for up to six months.

14.209 Failure to notify the Commissioner under the Factories and Industrial Undertakings Ordinance carries a fine of up to HK$10,000. Failure to comply with a prohibition notice carries a fine of up to HK$200,000 and imprisonment for up to six months. There is also a daily fine of HK$5,000 for continuing offences.

Myanmar

Introduction

14.210 The UN General Assembly has noted its concern at Myanmar's failure to protect workers, along with other human rights breaches. Myanmar is currently in the process of drafting new health and safety legislation, and it remains to be seen how robust this legislation and its enforcement will be in comparison with other countries in the region.

Health and safety legislation

14.211 Under Myanmar's constitution, the government is required to safeguard the interests of the working people. Myanmar has some legislation in this area, including the Factories Act 1951, which imposes requirements for safe working conditions, the Mines Law 1994, relating to the mining industry, and the Workmen's Compensation Act 1923, which requires employers to pay compensation for employees' work related injuries. However external observers, such as the US Bureau of Democracy, Human Rights, and Labor, observed that, in practice, the government did not adequately enforce these regulations.[1]

1 'Country Reports on Human Rights Practices – 2005', released by the Bureau of Democracy, Human Rights, and Labor, 8 March 2006.

14.212 In December 2012, the Ministry of Labour announced that a new law on safety and health in workplaces was being drafted and would be promulgated in 2013. The law will aim to prevent air and water pollution and improve safety at worksites.[1]

1 'Myanmar to draft first labour safety law'(2012) The Myanmar Times, 21 December.

Corporate criminal liability

14.213 The Myanmar Penal Code expressly states, at Article 11, that 'the word "person" includes any company or association, or body of persons, whether incorporated or not'. Consequently, it appears that the offences listed in the Penal Code can apply to companies.

14.214 Although Myanmar has yet to implement civil health and safety legislation, there is at least a theoretical possibility that prosecution for certain offences could be brought under the Penal Code. For example, it is an offence to knowingly or negligently convey for hire a person by water in any vessel that is so loaded as to endanger the life of that person, punishable by up to six months in prison and/or a fine. It is also an offence to act in a 'rash or negligent' manner to endanger life, or be likely to cause hurt or injury, when dealing with poisonous substances, fire or combustible matter, explosive matter, or machinery. These offences are also punishable by a fine or imprisonment. Although prosecutions for health and safety offences are theoretically possible, it is not apparent whether any prosecutions have in fact been brought, and the law as it stands under the Penal Code could lead to inconsistencies. For example, it is an offence to knowingly or negligently fail to take 'such an order as sufficient to guard against any probable danger to human life' when pulling down or repairing a building – but not, seemingly, when constructing one in the first place.

14.215 Corporations may also have criminal liability outside the Penal Code, and this liability can extend to the officers of the company. For example, the Foreign Exchange Regulation Act states that, where a company commits an offence under that Act, every director, manager and secretary will be punishable as if they committed the offence, unless they can prove that they had no knowledge of the offence. Consequently, there is a well-established legal framework, at least in theory, for Myanmar's health and safety legislation to impose criminal liability on companies for work related deaths.

Japan

Introduction

14.216 Among the many challenges facing the Japanese economy, one of the most pressing has been the fallout and recovery from the Great East Japan Earthquake, which hit Japan on 11 March 2011. Recovery workers and volunteers are frequently subject to accidents and injuries, including ongoing exposure to asbestos in destroyed buildings and debris, and radiation from the Fukushima Daiichi Nuclear Plant. Another ongoing risk, in less obviously dangerous environments, is the risk of heart attacks, strokes and suicides due to stress – while Japan is not the only country to experience this, it is perhaps the only country to report the statistics in a separate category and to specifically label it (Karoshi, or death from overwork).

Health and safety legislation

14.217 Occupational health and safety is governed by a range of legislation, most importantly the Labour Standards Law and the Industrial Safety and Health Law.[1] The legislation imposes a requirement on employers to 'not only comply with the minimum standards for preventing industrial accidents' but also to 'endeavour to ensure the safety and health of workers in workplaces through creating a comfortable working environment'. This includes requirements to actively monitor employees' health and to take necessary measures to promote health by making provision for physical exercise and recreation.

1 No 57 of 1972.

14.218 Workplaces are required to appoint a general safety and health manager to oversee safety measures, training of workers, medical examination workers, and investigation of accidents.

Corporate criminal liability

14.219 Under the modern Japanese legal system, which was based on French and German civil law systems, it was originally thought that only natural persons could commit crimes. However, in 1932, the Act Preventing Escape of Capital to Foreign Countries was passed and introduced the concept of 'Ryobatsu-Kitei', or 'double punishment', into Japanese law. Ryobatsu-Kitei are legislative provisions declaring that, in the event of a natural person committing an offence, an associated legal person may also be punished. Examples include certain provisions of the Labour Standards Law and the Industrial Safety and Health Law. There is also a rare provision known as 'Sanbatsu-Kitei', or 'triple punishment', which provides for punishment of the natural person, the associated legal person, and the representative(s) of the legal person. Examples are found in the Vessel Safety Act and Food Sanitation Act.

14.220 In the case of an offence committed by an employee, it must be shown that there was negligence on the part of the legal person in appointing or supervising the employee. There is, however, an assumption of negligence unless rebutted, which has led to some commentators suggesting that corporate criminal liability in Japan is a form of strict liability.[1]

1 'Survey Response, Laws of Japan (Human Rights Now), "Commerce, Crime and Conflict: A Survey of Sixteen Jurisdictions" (2006)', Fafo AIS, available at www.fafo.no/liabilities/CCCSurveyJapan06Sep2006.pdf.

14.221 There are very few precedents in which a presumption of negligence is overridden. It is clear that a business owner must make efforts to give active and specific instructions. For example, there is a reported case in which an ironwork company was prosecuted for the death and injury of employees, violating article 42 of the Labour Standards Law (measure of prevention of danger). Although the court found that, in that case, the employer had breached the prevention requirement, it commented that 'general and abstract warnings and cautions against violations by a business owner are not sufficient for the discharge on the ground of absence of negligence, and the business owner is required to take appropriate and specific measures enough to prevent effectively illegal conduct'.[1]

1 Takamatsu High Court, judgment of 9 November 1971, 275 HANREI TAIMUZU 291.

14.222 Although it is possible to attribute an individual's act to a business entity, there is no provision whereby an individual can be liable for an offence primarily committed by a corporation. Consequently, individual liability and any punishment will lie with the individual manager or owner who was in charge of the business activity in question.

Sanctions and sentencing

14.223 Offences under the health and safety legislation are punishable by fines and prison sentences.

14.224 Criminal sanctions for corporations are limited to fines and confiscation of property.

AUSTRALIA

Introduction

14.225 Since the publication of the 2nd Edition of *Corporate Liability: Work Related Deaths and Criminal Prosecutions,* the Australian Occupational Health and Safety (OHS) legislative landscape has changed dramatically. The introduction of harmonised work health and safety (WHS) laws in Australia represents a significant milestone in the development of Australian OHS regulation. For the first time, the States and Territories of Australia will be operating under laws that impose the same rights, duties and obligations in the occupational health and safety field.

14.226 The model laws are a culmination of a long process, which gained renewed impetus with a National Review commissioned by the Federal Government on behalf of all Australian governments in April 2008.[1] The harmonisation of WHS laws has been part of the Council of Australian Governments' National Reform Agenda, seeking to achieve equity between workers in all Australian jurisdictions, and to support Australian business by decreasing costs and increasing ease of compliance associated with having one national OHS legal regime.

1 Richard Johnstone & Michael Tooma, *Work Health & Safety Regulation in Australia – The Model Act* (2012), at 1.

14.227 The harmonised laws are enactments by each Australian jurisdiction of the *Model Work Health and Safety Bill* (WHS Act). Currently, all Australian jurisdictions[1] are in the process of harmonising their occupational health and safety laws by introducing the WHS Act and, in certain jurisdictions, the Acts commenced operation on 1 January 2012 – namely the Commonwealth,[2] New South Wales,[3] Queensland[4] and the two Territories (being the Australian Capital Territory[5] and the Northern Territory).[6] In South Australia[7] and Tasmania[8] the WHS Act commenced on 1 January 2013. The balance of Australian jurisdictions – Victoria and Western Australia – have not implemented the Model WHS Act.[9]

1 New South Wales, Victoria, Queensland, South Australia, Western Australia, Tasmania, Australian Capital Territory and Northern Territory.
2 *Work Health and Safety Act 2011* (Cth).
3 *Work Health and Safety Act 2011* (NSW).
4 *Work Health and Safety Act 2011* (Qld).
5 *Work Health and Safety Act 2011* (ACT).
6 *Work Health and Safety (National Uniform Legislation) Act 2011* (NT).
7 *Work Health and Safety Act 2012* (SA).
8 *Work Health and Safety Act 2012* (Tas).
9 The Victorian Government has not yet introduced a Bill into Parliament and has further announced that it would delay the implementation of the harmonisation process. The Western Australian Government is currently seeking public comment.

Overview of the Model WHS Act

14.228 The recommendations in the First and Second Reports of the National Review, as modified by the responses of the Workplace Relations Ministers' Council (WRMC), became the basis of the Safe Work Australia's drafting instructions to the Australasian Parliamentary Counsel's Committee (PCC). After a first draft of the model legislation was released for public comment in September 2009, a draft Bill was endorsed in December 2009 by the WRMC.

14.229 The Model WHS Act largely follows the architecture of post-Robens Australian work health and safety statutes. It imposes broadly based health and safety duties without defining how the duty holder will discharge the duty. It is structured in 14 Parts: Part 1 deals with objects and definitions; Part 2 contains general duties; Part 3 deals with incident notification; Part 4 deals with authorisations to perform certain work; Parts 5–7 contain provisions governing consultation and worker representation and participation; and the last parts, Parts 8–14, deal with inspection and enforcement.

14.230 While the Model WHS Act adopts many of its predecessor's features, certain aspects of the Model laws are unique. The main distinguishing feature of the Model WHS Act is the breadth of its scope and application.[1]

1 Richard Johnstone & Michael Tooma, *Work Health & Safety Regulation in Australia – The Model Act* (2012), at 1.

PCBU

14.231 The principal duty holder under the Model WHS Act is a 'person conducting a business or undertaking' (PCBU). The definition of PCBU is extremely broad. The term 'person' is not defined but will take its meaning from the *Acts Interpretation* statutes in each jurisdiction; that is, it will include individuals and bodies corporate. The phrase 'business or undertaking' is intended to be read broadly, and covers businesses or undertakings conducted by persons including employers, principal contractors, head contractors, franchisors and the Crown.

14.232 A person may be a PCBU, whether the person conducts a business or undertaking alone or with others, and whether the business or undertaking is conducted for profit or gain or not.[1]

1 *Work Health and Safety Act 2011* (Cth), s 5(1).

Primary duty of care

14.233 A PCBU's primary duty of care is for the health and safety of:

(1) workers at work; and
(2) other persons who might be put at risk from work carried out by the PCBU.[1]

1 *Work Health and Safety Act 2011* (Cth), s 19.

14.234 The term 'workers' is defined very broadly, and includes employees, contractors or subcontractors and their employees, labour hire employees, outworkers, students gaining work experience and volunteers.

14.235 The duty to ensure, so far as is reasonably practicable, the health and safety of workers and other persons will be referred to as the 'primary duty'.

14.236 It is clear from the wording of s 19(1) that the relationship between the PCBU's primary duty and protected workers does not have to be a direct contractual relationship – at a minimum, the worker must have been 'engaged' or 'caused to be engaged' by the PCBU, or the worker's work activities must be 'influenced or directed' by the PCBU.[1] It is likely that the courts will interpret 'engaged' broadly, to include not only contractors engaged by the person, but also sub-contractors, sub-sub-contractors and so on.[2]

1 See *Explanatory Memorandum – Model Work Health and Safety Bill,* Safe Work Australia, Canberra, 2 December 2010, at [77].
2 See *R v ACR Roofing Pty Ltd* (2004) 11 VR 187; and *Moore v Fielders Roofing Pty Ltd* [2003] SAIRC 75.

14.237 The primary duty is phrased to ensure that the persons specified in s 19(1) are protected whilst carrying out work for the PCBU. The duty applies wherever the work is being carried out; it is not confined to workplaces owned or controlled by the PCBU for whom the work is being done. Therefore, a PCBU will owe a duty to those workers who are working off site or at workplaces controlled by others. The PCBU will also owe a duty to other persons affected by the work carried out by its workers in those other locations.[1]

1 The *Explanatory Memorandum – Model Work Health and Safety Bill,* Safe Work Australia, Canberra, 2 December 2010, at [79], makes it clear that the primary duty 'is tied to work activities wherever they occur and is not limited to the confines of the physical workplace'.

14.238 The duty is to 'ensure', so far as is reasonably practicable, the health and safety of workers and other persons. There is little to suggest that 'ensure' should not be given its 'ordinary meaning of guaranteeing, securing, making certain'.[1] Section 17 provides that a duty imposed on a person to ensure health and safety requires that person:

(1) to eliminate risks to health and safety, so far as is reasonably practicable; and
(2) if it is not reasonably practicable to eliminate risks to health and safety, to minimise those risks so far as is reasonably practicable.[2]

1 *Carrington Slipways Pty Ltd v Callaghan* (1985) 11 IR 467 at 469–470.
2 *Work Health and Safety Act 2011* (Cth), s 17. See also *How to Manage Work Health and Safety Risks,* Code of Practice, Safe Work Australia, Canberra, 2011.

14.239 Section 19(3) of the WHS Act also sets out further duties for PCBUs that carry out certain tasks or who have a greater degree of control over or responsibility for particular stages of work, including:

(1) the management or control of workplaces;
(2) the management or control of fixtures, fittings or plant at a workplace;
(3) the design, manufacture, importation or supply of plant, substances or structures for use at or as a workplace; and
(4) the installation, construction, commissioning, decommissioning or dismantling of plant or structures at a workplace.

14.240 These duties are incorporated within the primary duty; however, by expressly setting out these further duties under separate sections of the WHS Act, the WHS Act aims to ensure that PCBUs know that, in order to comply with their primary duty, they will need to take certain steps or actions if they have a greater degree of control or responsibility for certain aspects of workplaces, tasks or processes, including the design process.

Who are 'other persons'?

14.241 As outlined above, in addition to protecting 'workers', a PCBU's primary duty is also to ensure that the health and safety of 'other persons' is not put at risk by work carried out as part of the conduct of the PCBU.[1] An 'other person' may include, for example, a visitor to a worksite or a member of the public.

1 *Work Health and Safety Act 2011* (Cth), s 19(2).

14.242 This is a substantially broader duty than that found in occupational health and safety laws that previously applied in the harmonised States and Territories. As a consequence, acts or omissions, which in the past may have only been actionable under public liability laws or negligence laws, may now also constitute a breach of the WHS legislation.

What is 'reasonably practicable'?

14.243 A PCBU's primary duty is qualified by the concept of reasonable practicability.

14.244 A PCBU is required to consider and weigh up 'all relevant matters' when determining what is reasonably practicable in relation to a particular hazard or risk. 'Relevant matters' will include (but are not limited to):

(1) the likelihood of the hazard or risk concerned occurring;
(2) the degree of harm that might result from the hazard or the risk;

(3) what the person concerned knows, or ought reasonably to know, about:
(a) the hazard or risk;
(b) ways of eliminating or minimising the risk; and
(c) the availability and suitability of ways to eliminate or minimise the risk; and

(4) after assessing the extent of the risk and the available ways of eliminating or minimising the risk, the cost associated with the available ways of eliminating or minimising the risk. This cost assessment should take into account whether the cost is grossly disproportionate to the risk.[1]

1 *Work Health and Safety Act 2011* (Cth), s 18.

14.245 Cost must always be the last consideration in determining what is reasonably practicable and must be weighted accordingly. This step may only be taken *after* considerations (1) to (3) above have been taken into account. The more likely or greater potential harm of the risk or hazard, the less costs considerations should be given weight in determining whether or not that particular action is reasonably practicable. The question should be asked whether the costs of eliminating the risk is grossly disproportionate to the risk and, in this manner, there is a clear presumption of safety ahead of costs. The conclusion that a control measure is 'grossly disproportionate' does not mean that nothing may be done regarding that risk or hazard; rather, a less expensive control should be used instead.[1]

1 Michael Tooma, *Tooma's Annotated Work Health and Safety Act 2011,* (2012) Lawbook Co Thomson Reuters (Professional) Australia Limited, Sydney, at 37.

Officers

14.246 The issue of personal liability of *management* for health and safety offences is a vexed one in WHS regulation. While there is a broad consensus about the role of *senior management* in delivering health and safety outcomes, the regulatory culture in Australia is yet to reach the stage where *senior managers'* accountability for health and safety concerns is universally accepted.[1] As part of the harmonisation process, there has been a concerted attempt to move this debate forward by imposing a separate positive duty to exercise due diligence to ensure that the PCBU complies with their obligations under the WHS Act on all 'officers' of a corporate PCBU. This increases the accountability of *senior management* for health and safety outcomes.[2]

1 Richard Johnstone & Michael Tooma, *Work Health & Safety Regulation in Australia – The Model Act* (2012), at 98.
2 Research shows that senior management leadership of the work health and safety agenda is critical to positive health and safety outcomes; see C Gallagher, *Health and Safety Management Systems: An Analysis of System Types and Effectiveness,* Report to NOHSC, NOHSC, Sydney, 1997; and Neil Gunningham, *CEO and Supervisor Drivers: Review of Literature and Current Practice*, Report to NOHSC, NOHSC, Sydney, 1999, at 11.

14.247 The current position is in stark contrast to that under previous occupational health and safety legislation,[1] where directors and officers of companies did not have a positive, proactive duty with respect to health and safety, but rather had limited liability attributed to them for the conduct of their company in certain circumstances.[2]

1 Richard Johnstone & Michael Tooma, *Work Health & Safety Regulation in Australia – The Model Act* (2012), at 102.
2 See eg *Occupational Health and Safety Act 1984* (WA), s 55; *Occupational Health and Safety Act 1985* (Vic), s 52; *Occupational Health, Safety and Welfare Act 1986* (SA), s 59C;

Workplace Health and Safety Act 2007 (NT), s 86; *Work Safety Act 2008* (ACT), s 219; *Occupational Health and Safety Act 2000* (NSW), s 26; Workplace *Health and Safety Act 1995* (Qld), s 167; and *Workplace Health and Safety Act 1995* (Tas), s 52.

Who is an 'officer'?

14.248 Under the harmonised WHS legislation, the duty of due diligence is a proactive duty imposed on 'officers', the definition of which has been adapted from the *Corporations Act 2001* (Cth).[1] For the private sector, s 4 of the WHS Act adopts the definition of 'officer' under s 9 of the *Corporations Act 2001* (Cth), namely:

(1) a director or secretary of a corporation;
(2) a person who makes or participates in making, decisions that affect the *whole*, or a *substantial part*, of the business;
(3) a person who has the *capacity* to affect significantly the corporation's financial standing;
(4) a person in accordance with whose instructions or wishes the directors of the corporation are accustomed to act (excluding advice given by the person in the proper performance of functions attaching to the person's professional capacity or their business relationship with the directors or the corporation);
(5) a receiver, or receiver manager of the property of the corporation;
(6) an administrator of the corporation;
(7) an administrator of the a deed of company arrangement executed by the corporation;
(8) a liquidator of the corporation; or
(9) a trustee or other person administering a compromise or arrangement made between the corporation and someone else (emphasis added).[2]

1 Richard Johnstone & Michael Tooma, *Work Health & Safety Regulation in Australia – The Model Act* (2012), at 127–8.
2 *Corporations Act 2001* (Cth), s 9. Additionally, it should be noted that the s 9 definition of 'officer' excludes a partner in a partnership as an officer.

14.249 It is clear from the above that the upper echelons of corporate management will be caught by this definition of 'officer'. Such positions include directors, senior executives and potentially even middle managers (should they make, or *participate in making*, decisions that affect the whole, or a *substantial part*, of the business). In other words, the WHS legislation is brought into conformity with contemporary Australian corporate law by imposing responsibility for the actions of a corporation on a wide group of personnel involved in the decision-making process of the corporation.

14.250 There is no difference in the definition of an 'officer' between the different state and territory jurisdictions. However, it should be noted that full harmonisation across the state jurisdictions has not yet been completed. To date, the Commonwealth, New South Wales, Queensland, Northern Territory, South Australia, Tasmania and the Australian Capital Territory have all adopted the harmonised legislation and, as such, have the same definition of 'officer'.

14.251 From what little case law does currently exist on the definition of an 'officer', the Australian courts appear to endorse an expansive interpretation of it as it appears in the *Corporations Act 2001* (Cth) and are likely to do the same in relation to the *WHS Act 2011* (Cth). The NSW Court of Appeal in *Morley v Australian Securities and Investments Commission* observed that:

'It is a reality of corporate life that board and other important decisions involve many persons other than the ultimate decision-makers. Just as s 9(b) (ii) of the law recognised the reality that a person may have "the capacity to affect significantly the corporation's financial standing", that being sufficient for the status of an officer as defined, so s 9(b) (i) recognised the reality of participation in decision-making. But it required participation in making decisions affecting the whole or a substantial part of company's business.'[1]

1 [2010] NSWCA 331 at [897] per Spigelman CJ, Beazley and Giles JJA.

14.252 On appeal, the High Court of Australia confirmed this approach to the definition of participating in making decisions, affecting the whole or a substantial part of the business of the corporation, stating that:

'participating in making decisions should not be understood as intended primarily, let alone exclusively, to deal with cases where there are joint decision makers … The idea of "participation" directs attention to the role that a person has in the ultimate act of making a decision, even if that final act is undertaken by some other person or persons. The notion of participation in making decisions presents a question of fact and degree in which the significance to be given to the role played by the person in question must be assessed.'[1]

1 *Shafron v Australian Securities and Investments Commission* [2012] HCA 18 per French CJ, Gummow, Hayne, Crennan, Kiefel and Bell JJ at [23].

14.253 Furthermore, the definition of 'officer' is not restricted to decisions that affect the whole of the business of the corporation, but rather may be triggered by decisions that affect a 'substantial' part of the business of the corporation. 'Substantial' is not defined in either Act and therefore takes on its ordinary meaning of 'considerable or significant'. The *Macquarie Dictionary* (5th ed) defines the word 'substantial' to be 'of ample or considerable amount, quantity, size'.[1]

1 Richard Johnstone & Michael Tooma, *Work Health & Safety Regulation in Australia – The Model Act* (2012), at 129.

14.254 When one considers the expansive definition of 'officer' in the context of a typical corporate structure, it is not difficult to imagine the positions in addition to that of director that would fall within the scope of the definition. As indicated above, it is clear that the chief executive officer (CEO), chief financial officer (CFO), chief operation officer (COO) and company secretary are caught by the definition of 'officer'. A general counsel will typically participate through advice in making decisions affecting the whole of the corporation. Furthermore, it is likely that human resources managers, occupational health and safety managers, any other managers of a significant division within the business, and any legal practitioner working in the role of receivers, receiver managers, liquidators, administrators or trustees would fall within the definition of an 'officer', and thus be liable under s 27 for a personal and proactive duty to take reasonable steps to exercise due diligence in ensuring that the PCBU fulfils his or her obligations under the WHS Act.

Differences in the public sector

14.255 An officer in the public sector is defined as a person 'who makes, or participates in making, decisions that affect the whole, or a substantial part, of a business or undertaking of the Commonwealth.[1] Subsection 247(2) expressly excludes ministers of a State or Commonwealth from the definition of officer. This exclusion is out of step with the inclusive nature of the definition of 'officer', given the role that ministers play in resource allocation within the public sector and the inclusion in the definition of due diligence of a component concerned with resource allocation.[2]

1 *Work Health and Safety Act 2011* (Cth), s 247.
2 Richard Johnstone & Michael Tooma, *Work Health & Safety Regulation in Australia – The Model Act* (2012), at 132.

Due diligence

14.256 Under the newly introduced WHS Act, a proactive duty of due diligence is imposed on officers of PCBUs. This proactive duty, which is criminal in nature, has attracted a good deal of interest from practitioners and academics alike.[1]

1 See eg Michael Tooma, *Due Diligence: Duty of Officers* (2012).

14.257 It has been remarked that 'few areas instil fear amongst executives as much as duties and liabilities for work health and safety'.[1] While the concept of personal liability for work health and safety offences is not in and of itself a novel concept, the nature and scope of the due diligence duty under the new legislation is unprecedented.

1 Michael Tooma, *Due Diligence: Duty of Officers* (2012).

The nature of the duty

14.258 The officer's duty is enshrined in s 27 of the WHS Act. Liability under s 27 arises where an officer fails to exercise due diligence. The test is concerned with the officer's individual conduct. The officer's failure is not assessed with the officer's individual conduct of the PCBU. If an officer is shown to have exercised due diligence, he or she will not be liable regardless of the PCBU's conduct.

14.259 Unlike previous duties imposed on corporate officers, under the new legislation, liability is not simply attributed to officers for a corporation's unlawful behaviour. Rather, officers have a duty of their own. The new duty is consistent with the general duty of officers under the *Corporations Act 2001* (Cth) to exercise care and diligence with respect to corporate issues.

14.260 The obligation is both positive and proactive in nature and, in summary, the duty set out in s 27[1] requires officers of PCBUs to apply due diligence to ensure that the PCBU acts in accordance with its obligations under the WHS legislation.

1 *Work Health and Safety Act 2011* (Cth), s 27. Pursuant to s 27, an officer is required to exercise due diligence to ensure his or her PCBU is compliant with its obligations under the WHS Act.

Definition of 'due diligence'

14.261 Under the WHS Act, the officer's duty is comprised of the following six elements:

(1) knowledge of work health and safety matters;[1]
(2) understanding of the nature of the operations of the business and the hazards associated with those operations;[2]

(3) resources and processes;[3]
(4) information regarding incidents, hazards and risks and responding in a timely way to that information;[4]
(5) legal compliance;[5] and
(6) verification of the provision and use of the resources and processes.[6]

1 *Work Health and Safety Act 2011* (Cth), s 27(5)(a).
2 *Work Health and Safety Act 2011* (Cth), s 27(5)(b).
3 *Work Health and Safety Act 2011* (Cth), s 27(5)(c).
4 *Work Health and Safety Act 2011* (Cth), s 27(5)(d).
5 *Work Health and Safety Act 2011* (Cth), s 27(5)(e).
6 *Work Health and Safety Act 2011* (Cth), s 27(5)(f).

14.262 Each and every officer must take 'reasonable steps' in relation to all six elements. As such, an officer is not absolved of their duty by referring to the due diligence exercised by their fellow officers.

Element one

14.263 Element one imposes a requirement on officers to take reasonable steps to acquire and keep up-to-date knowledge of work health and safety matters. Significantly, knowledge requires something more than simply obtaining 'information'. It means approaching WHS matters with an 'enquiring mind'. Whilst there is no strict definition, it means that officers should actively seek out work health and safety information in addition to relying on passive receipt of work health and safety data. It also means ensuring that officers understand – and, where necessary, question – the work health and safety information provided to them.

14.264 In *Statewide Tobacco Services v Morley*,[1] Ormiston J said:

> 'In light of the various duties now imposed upon directors, it would not appear unreasonable that they should apply their minds to the overall position of the company … Directors are entitled to delegate to others the preparation of books and accounts and the carrying on of the day-to-day affairs of the company. What each director is expected to do is to take a diligent and intelligent interest in the information either available to him or which he might with fairness demand from the executives or other employees and agents of the company.'

1 (1990) 2 ASCR 405, at 431.

Element two

14.265 Element two requires officers to take reasonable steps to gain an understanding of the nature of the PCBU's operations, and the hazards and risks associated with those operations. In *AWA Ltd v Daniels*,[1] Rogers CJ said:

> 'A director is obliged to obtain at least a general understanding of the business of the company and the effect that a changing economy may have on that business. Directors should bring an informed and independent judgment to bear on the various matters that come to the Board for decision.'

1 (1992) 7 ASCR 759 at 864.

14.266 The obligation to acquire an understanding builds on the requirements imposed by element one. Having acquired knowledge relating to health and safety, officers must then contextualise that knowledge by relating it to their PCBU's operations.

14.267 On the basis of this understanding, officers are required to identify and assess the hazards and risks arising from their PCBU's work and then take those hazards and risks into account in all decision-making processes. Officers are required to develop an understanding of not only the business of the PCBU but also the nature of the operations conducted as part of the business. Further, they must identify and consider the 'hazards associated with those operations'. This will include each individual officer bringing an informed mind to the way in which the business is conducted, so as to understand that the hazards arising from the operation constituting the business are being dealt with in health and safety terms.

This should include a detailed understanding of how the business is identifying risks and hazards and, further, the methodologies being utilised to eliminate – or, if not reasonably practicable, minimise – those risks and hazards. This should also include a detailed understanding of the circumstances surrounding past safety issues (including how those issues have been managed) and WHS performance measures and statistics.

Element three

14.268 To satisfy element three, an officer must take reasonable steps to ensure that his or her PCBU has available for use appropriate resources and processes and uses those resources and processes to eliminate and minimise risks to health and safety from work carried out by the PCBU.[1]

1 Michael Tooma, *Due Diligence: Duty of Officers* (2012), at 63–77.

14.269 In this context, 'resources' are taken to refer to personnel, and physical and financial resources. Put simply, officers are required to ensure that the right people, with the right experience and skills, are in the right places, with adequate resources, to enable health and safety risks to be identified and eliminated or minimised.[1]

1 Michael Tooma, *Due Diligence: Duty of Officers* (2012), at 51–63.

14.270 Officers will be required to keep abreast of developments of all kinds, that might enhance or alter the resources and processes utilised in the business, so as to satisfy themselves that they remain 'appropriate'.

Element four

14.271 Officers are required, pursuant to the fourth element, to ensure that a PCBU has the appropriate processes for receiving, considering and responding in a timely way to information concerning both incidents and also hazards and risks.

14.272 The fourth element, by referring to 'incidents, hazards and risks', addresses the importance of promoting a culture of learning and reflects the fact that officers cannot merely rely on staff to draw matters to their attention. They must monitor the performance of the business and make appropriate inquiries to satisfy themselves of the decisions being made.[1] By referring to 'risks', it imposes on an officer an obligation to deal with the adverse possibilities created by the existence of risk associated with the operations of the business. Armed with this type of information, an officer is required to ensure that steps are taken to eliminate – or, if not reasonably practicable, minimise – those risks before they create a health and safety problem.

1 See eg *ASIC v Healey* [2011] FCA 717, at [576].

Element five

14.273 The fifth element requires officers to take reasonable steps to ensure that the PCBU has and implements processes for complying with any duty or obligation of the PCBU under the WHS Act. The fifth element therefore requires a process for conducting regular legal compliance audits, where the standard against which the system is measured is the law itself.[1]

1 Michael Tooma, *Due Diligence: Duty of Officers* (2012), at 71–77.

14.274 Legal compliance includes complying with incident reporting requirements, consulting with workers and other PCBUs, preventing victimisation of people who raise health and safety concerns, and cooperating with regulatory investigations.

Element six

14.275 Due diligence means taking reasonable steps to verify the provision and use of the resources and processes referred to in the other elements. The assessment and performance of a safety management system is a crucial part of any effective management system and is the gateway to the continuous improvement process. It is the essence of the sixth element.[1]

1 Michael Tooma, *Due Diligence: Duty of Officers* (2012), at 79–87.

Penalties

14.276 Offences under the WHS Act are criminal in nature, involving monetary penalties and, in category 1 cases, custodial sentences. Division 5 of Part 2 of the WHS Act provides for a tiered penalty regime for contraventions of health and safety duty provisions, with the level of maximum financial penalty determined by the category of the offence. These tiered penalties are summarised below:

Offence[1]	**Penalty**
Category 1 offence – Reckless conduct A person commits a Category 1 offence if: (a) the person has a health and safety duty, and (b) the person, without reasonable excuse, engages in conduct that exposes an individual to whom that duty is owed to a risk of death or serious injury or illness, and (c) the person is reckless as to the risk to an individual of death or serious injury or illness.	**Maximum penalty:** (a) in the case of an offence committed by an individual (other than as a PCBU or as an officer of a PCBU)—$300,000 or five years' imprisonment or both, or (b) in the case of an offence committed by an individual as a PCBU or as an officer of a PCBU —$600,000 or five years' imprisonment or both, or (c) in the case of an offence committed by a body corporate—$3,000,000.

Category 2 offence – Failure to comply with health and safety duty A person commits a Category 2 offence if: (a) the person has a health and safety duty, and (b) the person fails to comply with that duty, and (c) the failure exposes an individual to a risk of death or serious injury or illness.	**Maximum penalty:** (a) in the case of an offence committed by an individual (other than as a PCBU or as an officer of a person conducting a business or undertaking)—$150,000, or (b) in the case of an offence committed by an individual as a PCBU or as an officer of a PCBU—$300,000, or (c) in the case of an offence committed by a body corporate—$1,500,000.
Category 3 offence – Failure to comply with health and safety duty A person commits a Category 3 offence if: (a) the person has a health and safety duty, and (b) the person fails to comply with that duty.	**Maximum penalty:** (a) in the case of an offence committed by an individual (other than as a PCBU or as an officer of a PCBU)—$50,000, or (b) in the case of an offence committed by an individual as a PCBU or as an officer of a PCBU—$100,000, or (c) in the case of an offence committed by a body corporate—$500,000.

1 *Work Health and Safety Act 2011* (Cth), ss 31–33.

14.277 The provisions of the WHS Act do not impose higher penalties if a contravention results in death or injury; rather, the WHS Act takes the more principled position that, in preventive legislation, maximum penalties should be increased where there is recklessness and the contravention results in a person being exposed to the risk of death or serious injury or disease.[1] It should be noted that custodial sentences are available for officers and workers committing category 1 offences (the most serious offences, involving recklessness and a risk of serious harm).

1 Richard Johnstone & Michael Tooma, *Work Health & Safety Regulation in Australia – The Model Act* (2012), at 238.

Other offences

14.278 For offences against provisions other than the health and safety duties in Part 2 of the WHS Act, the maximum penalties for individuals and corporations are set out in the provisions creating the offence.

Non-financial penalties

14.279 In addition to financial penalties or imprisonment that may be imposed in relation to an offence, where a court finds a person guilty of an offence against the WHS Act, the court can also make one or more orders under Division 2 of Part 13 for non-financial penalties.[1] Such non-financial penalties can include:

(1) adverse publicity order (s 236);

(2) restoration orders (s 237);
(3) WHS project orders (s 238);
(4) WHS undertakings (s 239);
(5) injunctions (s 240); and
(6) training orders (s 241).

1 See *Work Health and Safety Act 2011* (Cth), ss 234 and 235.

Prosecutions

14.280 The WHS Act authorises nominated representatives of the regulators to initiate prosecutions for contravention of the WHS Act and the regulations. Under s 230(1), proceedings for offences under the WHS Act may only be brought by the regulator or by an inspector with either the general or particular written authorisation of the regulator. However, nothing in the WHS Act prevents the Director of Public Prosecutions (DPP) from bringing proceedings for an offence against the Act (s 230(4)).

14.281 The WHS Act is not prescriptive, in that it does not define what must be done by a particular duty holder in a particular situation. Rather, by providing a broadly described set of duties, it imposes obligations on duty holders, which operate in all circumstances facing the particular business or undertaking. In framing a charge, the prosecutors are required to identify a particular measure which they allege would have addressed the risk identified in the charge. There is, however, no limitation as to what the prosecutor might describe as an appropriate measure and, as a consequence, charges are often very broadly based.

Limitation periods

14.282 Proceedings for an offence against the Act may be brought within the latest of the following periods to occur:

(1) within two years after the offence first comes to the notice of the regulator;
(2) within one year after a coronial report was made or a coronial inquiry or inquest ended, or an official inquiry ended if it appeared from the report or the proceedings at the inquiry or inquest that an offence had been committed against this Act (it should be noted that, under Australian law, a coronial inquiry or inquest can be brought at any time);
(3) if a WHS undertaking has been given in relation to the offence, within six months after:
    (a) the WHS undertaking is contravened; or
    (b) it comes to the notice of the regulator that the WHS undertaking has been contravened; or
    (c) the regulator has agreed under s 221 to the withdrawal of the WHS undertaking.[1]

1 *Work Health and Safety Act 2011* (Cth), s 232(1).

14.283 However, it should be noted that a proceeding for a Category 1 offence can be brought after the end of the applicable limitation period if fresh evidence relevant to the offence is discovered and the court in which the proceedings are brought is satisfied that the evidence could not reasonably have been discovered within the relevant limitation period.[1]

1 *Work Health and Safety Act 2011* (Cth), s 232(2).

Officer prosecution trends

14.284 While prosecutions under the personal liability provisions were used sparingly in the early stages of the modern Australian occupational health and safety law, prosecutions of directors and senior managers have become more prevalent in recent years.[1]

1 Richard Johnstone & Michael Tooma, *Work Health & Safety Regulation in Australia – The Model Act* (2012), at 105.

14.285 Early targets of prosecutions were directors of smaller companies. These prosecutions often focused on serious incidents where the company was small and its directors were actively involved in work activities. *Inspector Yeung v Herring*[1] is a typical example of such prosecutions. The director of Forsyth Pty Ltd was convicted and fined $9,750 after a fatal incident where an employee of a tree lopping sub-contractor was run over by an excavator while removing a tree as part of the construction of an industrial site.[2] In *Inspector Casto v Gremmo*[3] a director was prosecuted after an incident where one of his company's employees tripped over a piece of wood and fell 2.4 metres down a stairwell while carrying a spa bath that was to be installed in the premises. In *Inspector Maddaford v Elomar*[4] a director was convicted after an incident where an employee of the company was struck by a three-metre-long metal beam that he was moving with an overhead crane.

1 [2005] NSWIRComm 266.
2 See also *Inspector Drain v Keledijian* [2004] NSWCIMC 93, where the director of Manly Valley Pty Ltd (trading as Randwick Car Centre) was convicted in relation to an accident that occurred at the company's motor repair workshop in which an employee fell 3.8 metres off a ladder while painting the workshop.
3 [2004] NSWCIMC 51.
4 [2006] NSWCIMC 9.

14.286 More recently, senior managers of larger companies have also been prosecuted and convicted for offences under occupational health and safety legislation. The prosecutions against a director of Blue Haven Pools, Mr Awadallah,[1] the prosecution against the managing director of Owens Container Services, Mr Ritchie,[2] and the unsuccessful prosecution against a senior manager of the Alesco Group, Mr Ryan,[3] are three recent examples of this trend.[4] While most of the early prosecutions were responses to serious, often fatal, injuries, many recent prosecutions have not involved significant injuries. For example, a director of a company was recently convicted in relation to housekeeping and trip hazards that existed at the company's premises.[5] In another case, a director of a small construction company was convicted after an incident in which the brick boundary of his company's building site collapsed on an adjacent commercial property.[6] Two directors of a small furniture manufacturer were convicted of three offences arising from their company's failure to comply with an improvement notice and two prohibition notices.[7] A director of another construction company was convicted for health and safety breaches identified during site inspections by a WorkCover inspector.[8] Similarly, a director of JB Roofing Pty Ltd was convicted and fined $5,000 after an incident in which four employees were observed to be working on a roof without a harness.[9]

1 In *WorkCover Authority (NSW) (Inspector Cooper) v Awadallah* [2006] NSWCIMC 41, a director of Blue Haven Swimming Pools and Spas Pty Ltd was convicted and fined $3,375 after an incident in which a six-year-old boy suffered severe abrasions to his face when he fell from a push bike into a cement shell of a residential swimming pool that was under construction at Long Point in New South Wales.

2 In *Inspector Kumar v Ritchie* [2006] NSWIRComm 384, the managing director of Owens Container Services Australia Pty Ltd was convicted and fined $22,500 in relation to a fatal accident involving an employee who was cleaning a container.

3 *Inspector James v Ryan* [2009] NSWIRComm 215, which was upheld on appeal in *Inspector James v Ryan (No 3)* [2010] NSWIRComm 127. In that case, the deputy CEO of Alesco Corporation Ltd was found not to have been appointed as a director of a subsidiary, Dekorform, and was therefore acquitted of charges against him in relation to a fatal accident involving an employee of that subsidiary.

4 Richard Johnstone & Michael Tooma, *Work Health & Safety Regulation in Australia – The Model Act* (2012), at 106.

5 *Inspector de Silva v Mudaliar* [2006] NSWCIMC 70. The director was fined $1,575.

6 *Inspector Moulder v Nicolas* [2007] NSWIRComm 195. The directors received a good behaviour bond under s 10 of the *Crimes (Sentencing Procedure) Act 1999* (NSW).

7 *Inspector James v Huang* [2004] NSWCIMC 32. The two directors were fined $9,499 and $8,312 respectively.

8 *Inspector Kent v Reitsma* [2006] NSWCIMC 37.

9 *Inspector Jones v Denson* [2006] NSWIRComm 234.

14.287 While, for the most part, fines imposed on directors in early cases have been small, reflecting the fact that the burden of the corporate fine would, for practical purposes, also fall on the individual director, more significant fines have been imposed in some cases.[1] In *WorkSafe Victoria v Smith*,[2] the managing director of Orbit Drilling was convicted and fined $120,000 after a fatal incident involving one of Orbit Drilling's employees. The employee was killed when the truck that he was driving went out of control on a steep slope and overturned, crushing him as a result of the truck's defective brakes. In addition, the employee was inexperienced in operating such a vehicle on a steep slope and was not instructed, trained or supervised to operate the vehicle safely.

1 Richard Johnstone & Michael Tooma, *Work Health & Safety Regulation in Australia – The Model Act* (2012), at 106.

2 *WorkSafe Victoria v Smith* (Unreported), Magistrate's Court, Magistrate Rozencwajg, 9 November 2009.

14.288 In *WorkCover Authority (NSW) v Hitchcock*[1] the managing director of a transport company was prosecuted after a fatal head-on collision between two lorries on the Pacific Highway that involved one of the company's drivers. Fatigue was alleged to have contributed to the incident. For eight weeks before the incident, the deceased driver was employed by haulage company Sayogi Pty Ltd, trading as 'Jim Hitchcock Haulage'. The defendant and his wife were the sole directors of that company. In the week preceding the collision, there were two occasions on which it was impossible for the deceased driver to have had the requisite six-hour continuous break in every 24-hour period. The court found that the deceased worker was exposed to risk of driving while fatigued and, indeed, was fatigued at the time of the collision. The defendant was found guilty of two breaches of the *Occupational Health and Safety Act 2000* (NSW) and fined a total of $42,000.[2] Similarly, in *WorkCover Authority (NSW) (Inspector Wilson) v Ghafoor*,[3] a substantial fine of $21,000 was imposed against the director of All Time Security Services after the fatal shooting of a security guard.

1 [2004] NSWIRComm 87.

2 At the time, the maximum fine in New South Wales for a first offence by a director was $55,000.

3 [2005] NSWIRComm 430.

14.289 Section 38 of the WHS Act imposes a duty on the PCBU to ensure that the regulator is notified of a 'notifiable incident', which includes a death, serious injury or illness of a person, or dangerous incident. Practically speaking, the PCBU will notify both the police and the regulator of an incident.

If there is a work fatality, the State police are generally the first investigative body to attend to that incident and will inform the Coroner that a death has occurred.

14.290 The split between State police and the WHS regulator is a concurrent one. The police handle the pre-prosecution investigative steps, including preparing for any coronial inquest. Whilst the police consider any matters of criminality, inspectors from the WHS regulator at the same time investigate potential breaches of the WHS Act which – similar to the police role – would involve inspection of the incident site, interviewing witnesses, officers, and the PCBU. If an inquest is held into a workplace death, the Coroner will present his or her findings and make recommendations, which may include a recommendation for the regulator to prosecute. A prosecution may then be brought by the regulator on that recommendation, or alternatively simply through its own investigation. A prosecution under the WHS Act may bring charges against the PCBU, officers or any other person owing a duty under the Act which the regulator reasonably believes has been breached.

14.291 In Australia, the growing trend toward personal prosecutions of directors and managers has been fuelled by community expectations that directors and managers be held accountable for serious incidents.[1] Indeed, recent global disasters such as Pike River, the West Virginia mine disasters, and Deepwater Horizon have served to entrench community views on the need for greater management accountability for work health and safety.

1 Richard Johnstone & Michael Tooma, *Work Health & Safety Regulation in Australia – The Model Act* (2012), at 107. See also: ABC, 'WA Unions Call for Tougher Workplace Safety Laws', ABC online, 14 January 2012.

14.292 The consideration by the courts of the WHS laws means that considerable legal refinement of the duties and obligations, as well as the way in which breaches are found to have occurred, will take place over the next few years. The likelihood is that the breadth of language used to express the obligations in the WHS laws will be given full force, resulting in broad application of the provisions to all operations of all businesses.

The Western Australian perspective on health and safety at work

14.293 In a country of just over 22 million inhabitants (approximately the size of Istanbul's population), it makes very good sense that legislation regulating safety and health at work of those workers: is consistent across borders, whilst meeting international obligations and standards; has minimal red tape for businesses (particularly small business), and gives certainty as to what is required from them; and, most importantly, improves the health and safety outcomes for the workers.

14.294 The West Australian Liberal/National Party Coalition government does not argue against this inarguable position, but nor has it moved to legislate in accordance with the harmonised model. Its continued resistance to the harmonised safety laws must be seen in the context of one of the least populated of the Australian states – being one of the most geographically isolated states, yet one of the most prolific contributors to the economic well-being of Australia (via its resource industries) – not wanting to succumb to pressure from the Commonwealth Government (principally from the previous Labor government), without first satisfying itself that the proposed harmonised safety laws are in the best interests of businesses operating in Western Australia and their workers.

This Western Australian stance is reminiscent of its position more than 110 years ago when the State was asked to join in the federation of all other colonies. To this day, in the preamble to the federal Australian Constitution, the State of Western Australia is the only one which is not mentioned. As at the date that the others had agreed, Western Australia had not agreed to unite in one indissoluble Federal Commonwealth under the Crown of the United Kingdom of Great Britain and Ireland and under the newly established constitution. In fact, Western Australia had requested amendments to the proposed Constitution to which it had not received any substantive positive response as at the time that the Constitution Bill was introduced into the United Kingdom House of Commons. Eventually, the Queen signed a proclamation on 17 September 1900 that the Commonwealth of Australia, including Western Australia, would commence on 1 January 1901.

14.295 Not dissimilar to this is Western Australia's position on taking on board the harmonised health and safety laws. On 1 February 2008, the Workplace Relations Ministerial Council agreed to use the model legislation to harmonise the health and safety laws across the country. In July 2008 the various States and Territories of Australia signed an intergovernmental agreement for regulatory and operational reform in health and safety. Western Australia signed this agreement. Thereafter, the Federal Government appointed a three-person panel to undertake a national review. The panel issued two reports containing 232 recommendations with a view to harmonising the health and safety laws.

14.296 In May 2009 the Workplace Relations Ministerial Council made decisions on the panel's recommendations. The philosophy behind the recommendations kept alive Roben's outcome/goal-based safety outcomes rather than prescriptive legislation. Most of the recommendations were accepted. A timetable was developed, and the date for the introduction of the harmonised laws was set as 1 January 2012.

Work started in earnest to draft the Model Work Health and Safety Act and the accompanying harmonised Regulations and Codes of Practice. Participants from Western Australia were involved in the discussion and design phases of preparing the Model Act, Regulations and Codes.

14.297 In September 2009, Safe Work Australia (previously named the National Safety Commission) published the Model Work Health and Safety Act for public comment. Shortly after in December 2009, the Model Act was endorsed and then published on its website.[1] Thereafter, work continued on developing the Model Regulations and Model Codes of Practice.

1 See www.safeworkaustralia.gov.au/sites/SWA.

14.298 Running almost parallel to the work in developing the Model Act, Regulations and Codes, was work being done by other government agencies on the National Mine Safety Framework. In fact, work on the framework had started prior to conceptualising the harmonised health and safety legislation generally. That work looked at updating, improving and standardising health and safety in the mining and resources sector in Australia. Legislation covering this industry sector had, over time, become disjointed. As such, the so-called mining states of Western Australia, Queensland and New South Wales agreed to work together on revising the legislation.

14.299 What was (and, as at the date of writing, remains) the position in relation to regulating health and safety in the mining and resources sector in Western Australia? Currently, the Mines Safety and Inspection Act 1994 (MSI Act) and its accompanying Regulations and Codes of Practice regulate health and safety for 'mining operations', and the extended definitions under the Act include or exclude those that might otherwise consider themselves falling within the greater mining and resources sector. The MSI Act is modelled on the Occupational Safety and Health Act 1984 (OSH Act), which regulates health and safety for most other industry sectors. As such, the MSI Act is also modelled on the Robens goal/outcome-based safety regulation for workplaces.

14.300 The obvious question that comes to mind is why there is separate health and safety legislation for one industry sector whereas, before the creation of the MSI Act, that sector came under the umbrella of the OSH Act. Obviously, there are some details which, by virtue of the manner of conducting 'mining operations', need to be tailored to that sector, but surely these could adequately be dealt with by way of regulations. Or is there some other reason? Could it be related to the fact that the administration of one of the Acts comes under the responsibility of one state government minister and his/her department, and the other Act comes under a different minister and department? Or because each has a separate budget? Which came first? Will the answer to this question resolve the dilemma? Or will it further complicate the inarguable need for harmonisation of the health and safety laws in Western Australia?

Alternatively, has the proposed harmonisation of health and safety laws gone far enough? There are numerous other Acts of the Western Australian government regulating health and safety in different industry sectors, including dangerous goods, rail, on-shore petroleum, energy, radiation, transport, roads, food, and fire and emergency.

14.301 Maybe now is the time to look at what is happening on the ground in work health and safety in Western Australia. In 2012, one of the little-known provisions of the OSH Act was tested for the first time when the Department of Corrective Services and its private contractor company, G4S (together with two employees of G4S), were successfully prosecuted for health and safety breaches which resulted in the death of an aboriginal elder who, at that time, was a remand prisoner being transported by the contractor at the request of the department. The provision relied on was concerned with the risks brought about by the way the work was being undertaken. Both pleaded guilty and each entity received a penalty of A$285,000.

14.302 The concept of how a corporation operates and functions and, as a result, creates risks for persons (not just its direct and indirect employees) appears to be alive and well in the current OSH Act. However, this concept has been touted as revolutionary under the Model Act, where the primary duty of care will no longer be rooted in the employment relationship but rather on how the operations of a 'person (corporate and natural) conducting a business or undertaking' (PCBU) is being undertaken and the associated risks for workers (including direct and indirect employees, apprentices, trainees, vocational placements, volunteers and others). Surely there are some similarities, or by this comment is the writer showing her allegiance to the Western Australian legislation?

14.303 The Model Act definition of 'worker' is more representative of the current employment-type arrangements that exist in Australia than the obligations imposed on those who fall within the definition of 'employees' under the OSH Act. Surely that is a good thing. Or is it? Is the use of the term 'worker' more correct, as it reflects the terminology used in the ILO Occupational Safety and Health Convention 1981? Or is it to better align the occupational safety and health regime with the workers' compensation regime which extends insurance coverage to persons coming within the definition of 'worker' who sustain an injury in the course of employment under the provisions of the state-based Workers' Compensation and Injury Management Act 1981?

Given both the writer's view that health and safety and workers' compensation are two sides of the same coin – the former is where, if risks are managed effectively, there is no need for the latter, and the latter is the after-effects of failed or inadequate risk measures – and also the writer's suspicion that the next legislative changes to reduce red tape for businesses in Australia will be in the workers' compensation regime, is the current Federal government's rush to harmonised health and safety legislation part and parcel of an overhaul and harmonisation of the workers' compensation regime?

14.304 Leaving aside these questions, what other changes are proposed in the model Act which would positively impact on safety outcomes for businesses and staff in Western Australia? The positive duty of care proposed to be placed on persons who come within the definition of 'director' (as defined in the *Corporations Act 2001* (Cth)), such that they must exercise 'due diligence' to ensure that their PCBU meets its obligations for health and safety, is a first for Western Australia. It recognises that a corporate entity cannot of itself ensure, in so far as is reasonably practicable, the health and safety of staff, because the entity – not being a natural person – has 'no brains, no arms and no legs' and, as such, the only way in which the entity can meet its current statutory obligations is if its directors and senior managers do what is needed for the entity to meet its obligations. However, the corporate liability currently does not recognise this distinction. Under the current regime in Western Australia, a so-called 'director' can only be held liable for breach of the health and safety laws if, first, the corporate entity has been found guilty, and then there is proof beyond reasonable doubt that the 'director' has consented to, connived in or been neglectful in their duties. Unsurprisingly, the number of directors successfully prosecuted under this provision can be counted on one hand.

Given the economic powerhouse that Western Australia is, this recognition by the corporate world is a clear indication that, if personally held liable, 'directors' would play a more proactive role and provide the safety leadership and engender a safety culture which is sorely needed.

14.305 The results of the last state election have returned the conservative government to power, and it is unlikely that the Model Act, Model Regulations and Model Codes will become law in Western Australia. However, some changes may be made to the current statutory regime for health and safety in Western Australia, such that some elements of the Model Act are reflected in the Western Australia Act.

By so doing, not only will Western Australia come closer to the regulatory regime in the eastern states of Australia, but it will also return safety at work to a human right which a worker is entitled to, not just morally but by virtue of international law.

CANADA

Introduction

14.306 In common with Australia and the US, Canada has a federal system of government. As such, the jurisdiction for making laws is shared between the Canadian Parliament and the legislatures of the individual Provinces and Territories. Unlike Australia and the US, on the other hand, responsibility for legislating in the arena of the criminal law rests solely with the Canadian Parliament,[1] which makes the job of tracking developments in the law governing corporate killings that little bit easier.

1 Arthur Allens Robinson, *Corporate Culture as a Basis for the Criminal Liability of Corporations* (2008) at p 24, accessible online at http://198.170.85.29/Allens-Arthur-Robinson-Corporate-Culture-paper-for-Ruggie-Feb-2008.pdf (accessed 12 December 2013).

14.307 Canadian law classifies criminal offences into one of three categories, depending on the mental element of the offence. First, there are offences which require proof of mens rea; thus, the prosecution will have to show that the defendant possessed the requisite state of mind. Secondly, there are strict liability offences where the defendant's state of mind is irrelevant as such. The defendant will be held liable unless they can prove that they demonstrated due diligence in trying to prevent/avoid the commission of the offence. Finally, there are absolute liability offences. Again, the defendant's state of mind is irrelevant and no defence is available to them. In such cases, the mere fact that the physical element has been committed by the defendant is enough to warrant conviction.[1] We now need to consider how these principles impact upon attempts to impose liability on corporations for deaths caused by their actions.

1 Arthur Allens Robinson, *Corporate Culture as a Basis for the Criminal Liability of Corporations* (2008) at p 24, accessible online at http://198.170.85.29/Allens-Arthur-Robinson-Corporate-Culture-paper-for-Ruggie-Feb-2008.pdf (accessed 12 December 2013).

The application of criminal law to corporations in Canada: common law principles

14.308 The starting point for our discussion in this section is the observation that corporations can be held liable for criminal offences, such as manslaughter, which might typically be viewed as offences capable of being committed solely by individuals. This is largely down to s 2 of the Canadian Criminal Code, which defines 'person' as including 'public bodies, bodies corporate, societies, [and] companies'.[1]

1 Many of the offences under the Criminal Code, however, are not capable of being committed by a corporation. Those mens rea crimes under the Code which are, at least in theory, capable of corporate commission, including corporate manslaughter, have not traditionally been brought against corporations.

14.309 The Criminal Code contains a specific offence of criminal negligence causing death.[1] Theoretically, this is an offence which a corporation is capable of committing, although no corporation has actually been convicted of this offence to date. If we think back to our earlier discussion of the classification of criminal offences in Canada, the need to demonstrate criminal negligence marks this offence out as one which requires proof of mens rea. As usual, however, we remain dogged by the fact that a corporation has no physical existence; thus, it cannot carry out a physical act, nor can it form a state of mind, criminal or otherwise. The question becomes, therefore, how have the Canadian courts attempted to get around this issue?

1 Section 220 of the Canadian Criminal Code. Section 221 of the Code contains the offence of causing bodily harm by criminal negligence. According to s 219 of the Code, a person will be criminally negligent where 'in doing anything or in omitting to do anything that it is his duty to do[they show] wanton or reckless disregard for the lives or safety of other persons'. Notice how there is no need for the defendant to show wanton and reckless disregard specifically for the life of the actual victim(s).

14.310 It should not surprise us to learn that Canada's approach to this issue, as a Commonwealth country, has been heavily influenced by the approach taken by the English courts. As such use is made of the doctrine of identification, as espoused in *Tesco v Nattrass*,[1] this much is made clear by the decision of the Supreme Court of Canada in the case of *Canadian Dredge & Dock Co v The Queen*.[2]

1 [1972] AC 153.
2 [1985] 1 SCR 662.

14.311 The *Canadian Dredge & Dock Co* case involved four corporate defendants who had been convicted of attempting to defraud after entering into an illegal price-fixing agreement to rig the bidding process for a government contract. The issue of what mechanism should be used to attribute a state of mind to the defendant organisations was a key issue under consideration. As with the English courts, the Supreme Court rejected suggestions that vicarious liability was an appropriate means for achieving this goal on the grounds that, under the theory of vicarious liability, the 'net is flung too widely' resulting in the prospect of corporations being punished 'in instances where there is neither moral turpitude nor negligence'.[1] The chosen method for attributing liability to a corporation, therefore, was the doctrine of identification, albeit in a slightly different guise from that adopted by the English courts.

1 [1993] 1 SCR 497 at para 42.

14.312 In contemplating its final verdict, the Supreme Court made reference to the Canadian transport industry. The court recognised that the organisation of corporations involved in such activities can be geographically diffuse, thus necessitating the delegation of responsibility for management decisions. The court indicated, therefore, that a narrow interpretation of the identification doctrine did not suit the realities of life in Canada. Taking this observation into consideration, the Supreme Court concluded that, in any application of the doctrine of identification, it must be borne in mind that there can be more than one individual who may be viewed as representing the 'directing mind and will' of the defendant organisation. Consequently:

> 'The identity doctrine merges the board of directors, the managing director, the superintendent, the manager or anyone else delegated by the board of directors to whom is delegated the governing executive authority of the corporation, and the conduct of any of the merged entities is thereby attributed to the corporation'.[1]

1 [1985] 1 SCR 662 at para 50.

14.313 The doctrine of identification as set out in the case of *Canadian Dredge & Dock Co* is, therefore, broader than that espoused by the courts in England and Wales, providing that a seemingly low-level individual can be classed as representing the directing mind, and will be so classed if they have been delegated power to act on the matter in question. The Supreme Court did go on to state, however, that:

> 'the identification doctrine only operates where the Crown demonstrates that the action taken by the directing mind was (a) within the field of operation

assigned to him; (b) was not totally in fraud of the corporation; and (c) was by design or result partly for the benefit of the company'.

This is an important limitation in line with what we have already seen in our discussion of corporate liability in Australia for deaths caused by their undertakings. Thus, despite the apparent outright rejection of the doctrine of vicarious liability by the Supreme Court in *Canadian Dredge & Dock Co*, this doctrine has still influenced the courts' application of the doctrine of identification.[1]

1 *Canadian Dredge & Dock Co v The Queen* [1993] 1 SCR 497.

14.314 The Supreme Court in *Canadian Dredge & Dock Co* also discussed some of the defences that might be relied upon by a corporate defendant in order to avoid criminal liability under the doctrine of identification. As is the case under English law (see the case of *R v Supply of Ready Mixed Concrete* (*No 2*)),[1] it is no defence under Canadian law for a corporation to claim that the prohibited act in question had been carried out in direct contravention of corporate directions aimed at preventing such acts from taking place. The only defence available to Canadian corporations, facing a charge for an offence requiring proof of mens rea, is to show that the individual who carries out the prohibited act does so wholly in fraud of the company without the company gaining any benefit from it. This allows the application of the Canadian approach to the doctrine of identification (based as it is on delegation of authority) to be kept within certain confines. As the court said in *Canadian Dredge and Dock Co*:

> 'The outer limit of the delegation doctrine is … reached and exceeded when the directing mind ceases completely to act, in fact or in substance, in the interests of the corporation. The identification theory ceases to operate when the directing mind intentionally defrauds the corporation and when his wrongful actions form the substantial part of the regular activities of his office. In such a case, where his entire energies are directed to the destruction of the undertaking of the corporation, the manager cannot realistically be considered to be the directing mind of the corporation. The same reasoning can be applied to the concept of benefits … Where the criminal act is totally in fraud of the corporate employer and where the act is intended to and does result in benefit exclusively to the employee-manager, the employee-directing mind, from the outset of the design and execution of the criminal plan, ceases to be the directing mind of the corporation and consequently his acts cannot be attributed to the corporation under the identification doctrine.'[2]

1 *R v Supply of Ready Mixed Concrete* (*No 2*) [1995] 1 All ER 135.
2 *Canadian Dredge & Dock Co v The Queen* [1993] 1 SCR 497.

14.315 The broader doctrine of identification set out in *Canadian Dredge and Dock Co* was further curtailed by a later decision in the case of *The Rhone v Peter AB Widener* (*The*).[1] In that case the lead of four tugs pulling a barge ('The Widener') was travelling too fast and made a crucial turn too quickly, causing the barge to go out of control and collide with a nearby ship ('The Rhone'). The owners of The Rhone attempted to sue the owners of The Widener and the owners of the four tugs for damages suffered. The Rhone's owners also tried to sue Great Lakes Towing Company for breach of its towing contract. Great Lakes Towing Company owned two of the four tugs including the lead tug, The Ohio, which ultimately caused the collision.

1 [1993] 1 SCR 497.

14.316 One of the key issues for the court to consider, when trying to determine the liability of Great Lakes Towing Company, was whether Captain Kelch could be viewed as part of the 'directing and mind and will'. Captain Kelch was in charge of The Ohio and was acting as the de facto master of the flotilla. It was ultimately a combination of his navigational errors, along with The Ohio's defective towing equipment, which led to the collision. Interestingly, the lower court (the Federal Court of Appeal) took the view that Captain Kelch could be deemed to be part of the directing mind and will, at least for the purposes of carrying out Great Lakes Towing Company's obligations in relation to the towing of The Widener. Consequently, the court found Great Lakes Towing Company to be liable accordingly.

14.317 This approach taken by the Supreme Court, however, was somewhat different. Iacobucci J considered the decision of the Supreme Court in *Canadian Dredge & Dock Co* and noted:

> 'As Estey J's reasons demonstrate, the focus of the inquiry must be whether the impugned individual has been delegated the "governing executive authority" of the company within the scope of his or her authority. I interpret this to mean that one must determine whether the discretion conferred on an employee amounts to an express or implied delegation of executive authority to design and supervise the implementation of corporate policy rather than simply to carry out such policy. In other words, the court must consider who has been left with the decision-making power in a relevant sphere of corporate activity.'[1]

1 *The Rhone v Peter AB Widener (The)* [1993] 1 SCR 497.

14.318 In effect, what Iacobucci J appears to be saying is that, although low-level employees may be viewed as part of the directing mind and will in some circumstances, the key question for the court to consider will seemingly be whether the individual whose state of mind the prosecution wants attributed to the defendant corporation had been delegated power to dictate how the activity in question was to be carried out. If the prohibited consequence merely arose out of a failure of a low-level employee to implement policy set by those in the upper echelons of management, this in itself would not be enough. Consequently, having considered the approach of Denault J in the lower court, Iacobucci J concluded that Captain Kelch was not part of the directing mind and will of the company, stating:

> 'The key factor which distinguishes directing minds from normal employees is the capacity to exercise decision-making authority on matters of corporate policy, rather than merely to give effect such policy on an operational basis, whether at head office or across the sea. While Captain Kelch no doubt had certain decision-making authority on navigational matters as an incident of his role as master of the tug Ohio and was given important operational duties, he did not have governing authority over the management and operation of Great Lakes' tugs.'[1]

1 *The Rhone v Peter AB Widener (The)* [1993] 1 SCR 497.

14.319 A further crucial decision in the context of our discussion of Canadian case law dealing with issues of corporate liability for offences requiring proof of mens rea is the decision of the Ontario Court of Appeal in the criminal case of *R v Safety Kleen Canada Inc*.[1] The defendant corporation had been charged with the offence of filing a false shipping manifest involving hazardous waste. The false manifest was provided by a truck driver who was the defendant company's sole representative in this particular, very large, geographical area.

Furthermore, the driver was the only person responsible for the collection of waste materials, as well as being responsible for the company's bookkeeping in the area and customer relations in the area. Perhaps most interestingly of all, however, was the fact that, when this truck driver left the employment of the defendant company, they ceased their activities in that geographical area.

1 (1998) 16 CR (5th) 90.

14.320 It was apparent that the truck driver had many responsibilities. It was also clear that he had wide discretion in the exercise of said activities. It was also said that those customers who had dealings with this particular truck driver clearly viewed him as the embodiment of the defendant corporation. This was not enough in the court's eyes, however, to warrant attributing the truck driver's actions to the company.

14.321 Whilst the driver had obviously been delegated sufficient authority to allow him to carry out the many tasks that he was expected to perform and discharge the duties that had been imposed upon him by his employer, his position was not such as to allow him to formulate company policies. As such, following the decision in *Canadian Dredge and Dock Co*, as altered by the decision in *The Rhone v Peter AB Widener (The)*, the truck driver could not be identified with the corporation, for he was not part of the 'directing mind and will'.

14.322 The approach of the Canadian courts to the issue of identification bears a great similarity to that taken by the Privy Council in the case of *Meridian Global Funds Management Asia Ltd v Securities Commission*.[1] Lord Hoffmann decided in that case that the question of whose knowledge and actions could be attributed to the company was to be determined by the courts looking at the meaning of the words used in defining the offence in the Act in question and the policy underpinning it. For Lord Hoffmann, the question was one of 'construction rather than metaphysics'.

1 [1995] 2 AC 500.

14.323 The basic principles underpinning the decisions in *Meridian Global Funds Management Asia Ltd* and *Canadian Dredge & Dock Co* are, essentially, the same. The courts in both cases held that the person whom the courts should be looking for, when trying to determine who forms part of an organisation's directing mind and will, is the person who had direct control of the area of corporate activity in which the offending act was committed. Such an approach can be beneficial in any attempt to impose liability on a corporate offender, as it recognises the fact that large corporations can have very diffuse power structures. Furthermore, it makes it harder for organisations to avoid liability by delegating responsibility for potentially criminal activities to low-level workers who, ordinarily, would not be seen as sufficiently senior within the corporation to form part of its directing mind and will.

The application of criminal law to corporations in Canada: legislative changes

14.324 Up until this point, we have been considering the basics of attempts to impose liability on corporations for offences requiring proof of mens rea, and we have seen that, in theory at least, it is plausible. In this next section, we will move on to consider legislative attempts at State level to impose liability on a corporation for criminal offences.

14.325 One of the common threads that we have noted, during our consideration of attempts to legislate in the field of corporate manslaughter in Australia, is the move towards using 'corporate culture' as a mechanism for imposing liability. The same issues have been under consideration by the Canadian legislature.

14.326 A key trigger for moves in this direction was the Westray mine incident of May 1992. This was the site of a methane explosion that left 26 miners dead. Two of the mine's managers were charged with manslaughter, but the trial did not proceed after the prosecution determined there was not enough evidence to get anyone convicted. The subsequent report of the Royal Commission of Inquiry[1] into the explosion resulted in a number of recommendations, including the recommendation that the Canadian government should look into the systems governing the accountability of corporate executives for the wrongful acts of corporations and consider introducing any legislative amendments necessary to ensure that executives are held accountable for workplace safety.[2]

1 Royal Commission, *The Westray Story: A Predictable Path to Disaster* (Report of the Westray Mine Public Inquiry, 1997).

2 Arthur Allens Robinson *Corporate Culture as a Basis for the Criminal Liability of Corporations* (2008) at p 25, accessible online at http://198.170.85.29/Allens-Arthur-Robinson-Corporate-Culture-paper-for-Ruggie-Feb-2008.pdf (accessed 12 December 2013).

14.327 In the aftermath of the publication of the Royal Commission's report, an attempt was made to introduce a Private Member's Bill into the Canadian Government which would have imposed criminal liability on a corporation in situations where the 'act or omission was tolerated, condoned by the [organisation's] policies or practices' and that corporate culture of tolerance had developed as a result of the attitude of the corporation's management to its legal obligations.[1]

1 Arthur Allens Robinson, *Corporate Culture as a Basis for the Criminal Liability of Corporations* (2008) at p 26, accessible online at http://198.170.85.29/Allens-Arthur-Robinson-Corporate-Culture-paper-for-Ruggie-Feb-2008.pdf (accessed 12 December 2013). See also S Bittle and L Snider, 'From Manslaughter to Preventable Accident: Shaping Corporate Criminal Liability' (2006) 28 Law & Policy 470.

14.328 The Bill was referred to the House of Commons Standing Committee on Justice and Human Rights. In its report delivered in March 2002, the Committee concluded that legislative change was required to make it a crime for corporate managers and directors knowingly to put their employees in danger.[1] The Report presented by the Standing Committee to the House of Commons recommended 'that the Government table in the House legislation to deal with the criminal liability of corporations, directors and officers'.

1 Department of Justice, *Corporate Criminal Liability; Discussion Paper* (March 2002).

2 Arthur Allens Robinson. *Corporate Culture as a Basis for the Criminal Liability of Corporations* (2008) at p 26.

14.329 In its response to this Report the House of Commons noted that the Westray incident had made obvious that 'regulations, no matter how effective on paper, are worthless when they are ignored or trivialized by management'. The government also noted that the 'first line of defence against death and injury in the workplace is workplace safety and health regulation'. It acknowledged that criminal law can provide an additional level of deterrence but that the criminal law should be reserved for the most serious offences. However, there was also an underlying desire not to interfere with the provincial regulatory role in relation to health and safety.

14.330 The government also acknowledged the deficiencies that existed in the present system which prevented the successful operation of the criminal law in this area: the 'requirement that one individual be liable before liability can be ascribed to a corporation was seen as providing a measure of impunity to large corporations while exposing smaller corporations to conviction'. Although it agreed that there was a need for legislative change in this field, the government expressed its preference for changes to the general rules regarding corporate responsibility, rather than the creation of specific offences like the British approach to imposing liability on a corporation for manslaughter. The Australian corporate culture model was felt to require too radical a departure from established general principles.

14.331 The result of all this discussion and debate was Bill C-45 An Act to Amend the Criminal Code (Criminal Liability of Organizations).[1] This amended the provisions of the Canadian Criminal Code with effect from 31 March 2004. The Bill introduced 'new rules for attributing criminal liability to organizations for acts or omissions by their representatives', and introduced a legal duty for 'all persons directing work to take reasonable steps to ensure the safety of workers and the public'. The Bill also effectively codified the liability of corporations for offences with a fault element.

1 Often referred to colloquially as 'the Westray Bill' in reference to the incident which ultimately led to its creation.

14.332 A key amendment introduced by Bill C-45 was s 217.1. This section states that:

> 'Everyone who undertakes, or has the authority, to direct how another person does work or performs a task is under a legal duty to take reasonable steps to prevent bodily harm to that person, or any other person, arising from that work or task'.

Bill C-45 also amended the definition of 'everyone' and 'person', contained in s 2 of the Canadian Criminal Code, to include 'an organisation'.[1] Why is this significant?

1 The term 'organisation' is further defined in s 2 of the Canadian Criminal Code to mean either '(a) a public body, body corporate, society, company, firm, partnership, trade union or municipality', or '(b) an association of persons that (i) is created for a common purpose, (ii) has an operational structure, and (iii) holds itself out to the public as an association of persons'. This definition was also introduced by Bill C-45. A copy of the Canadian Criminal Code can be accessed on the Department of Justice Canada website at http://laws.justice.gc.ca/en/C-46.

14.333 Section 217.1 clearly imposes a duty on an organisation to fulfil certain obligations. Section 219.1 adds to this and states: 'Everyone is criminally negligent who (a) in doing anything, or (b) in omitting to do anything that is his duty[1] to do, shows wanton or reckless disregard for the lives or safety of other persons'. It seems, therefore, that Bill C-45 creates the potential for organisations to be held liable where there negligence has resulted in either the death of an individual,[2] or where there negligence has caused bodily harm.[3] How would the prosecution secure a conviction of a corporation for criminal negligence?

1 Section 219.2 notes that, for the purposes of s 219.1, the term 'duty' includes 'a duty imposed by law'. This clearly includes the duty contained in s 217.1.
2 Under s 220 of the Canadian Criminal Code.
3 Under s 221 of the Canadian Criminal Code. The term 'bodily harm' is defined in s 2 of the Canadian Criminal Code as meaning 'any hurt or injury to a person that interferes with the health or comfort of the person and is more than merely transient or trifling in nature'.

14.334 Where an organisation is facing a charge of criminal negligence, the prosecution has to satisfy two steps. First, they have to establish that a representative of the company had breached the s 217 duty in a 'wanton or reckless' manner (as required by s 219.1). Secondly, they can refer to s 21 or s 22[1] of the Canadian Criminal Code to help them establish that the organisation took part in the criminal act.

1 Section 22 will be used where the offence identifies criminal negligence as the requisite state of mind. Section 21 will be used where a mental state other than criminal negligence is part of the offence.

14.335 Where an individual dies in circumstances that are indicative of criminal negligence,[1] s 22.1 of the Canadian Criminal Code creates the potential for a corporation to be held liable as 'a party to [the] offence'.[2] Section 22.1 of the Criminal Code states that where proof of criminal negligence is required a corporation can be treated as 'a party to [an] offence' in one of two circumstances. The first circumstance will be where 'one of its representatives[3] is a party to the offence' and 'the senior officer who is responsible for the aspect of the organization's activities that is relevant to the offence departs – or the senior officers, collectively, depart – markedly from the standard of care that, in the circumstances, could reasonably be expected to prevent a representative of the organization from being a party to the offence'.[4]

1 The Canadian Criminal Code also includes a mechanism for allowing a corporation to be held liable as a party to an offence where the requisite state of mind is anything other than criminal negligence. This is contained in s 22.2.
2 Although this may sound redundant, the very fact that a corporation can be convicted as a party to the offence means that the individual who commits the offence can also be convicted. The individual offence is set out in ss 220 and 221.
3 The term 'representative' is defined in s 2 of the Canadian Criminal Code as including directors, partners, employees, members, agents and contractors.
4 Section 22.1(a)(i) and (b) of the Canadian Criminal Code.

14.336 Alternatively, in a slight variation, a corporation can be treated as a party to an offence where 'two or more of its representatives engage in conduct, whether by act or omission, such that, if it had been the conduct of only one representative, that representative would have been a party to the offence' and 'the senior officer who is responsible for the aspect of the organization's activities that is relevant to the offence departs – or the senior officers, collectively, depart – markedly from the standard of care that, in the circumstances, could reasonably be expected to prevent a representative of the organization from being a party to the offence'.[1]

1 Section 22.1(a)(ii) and (b) of the Canadian Criminal Code.

14.337 Whilst the mechanism for holding corporations liable as a party to an offence varies slightly, depending on whether s 22(a)(i) or (ii) is used, the common feature is that the corporation can only be held liable in a situation where a senior officer should have taken reasonable steps to prevent the company's representative(s) from committing an offence. In trying to determine whether the senior officer has failed to take the requisite steps, the key question is whether they had 'departed markedly' from standards that could be reasonably expected to be followed to prevent the employee from committing the offence.[1]

1 S Bittle and L Snider, 'From Manslaughter to Preventable Accident: Shaping Corporate Criminal Liability' (2006) 28 Law & Policy 470 at p 477.

14.338 What if the corporation is convicted of acting as a party to the offence of causing death by criminal negligence? Section 735 of the Code provides that a corporation can be punished with a fine for offences that would carry imprisonment as a sentence in the case of an individual offender. Section 220 makes it clear that criminal negligence leading to death is an indictable offence. Similarly, s 221 states that criminal negligence leading to bodily harm is also an indictable offence. The importance of this is made clear by s 735.1.

14.339 Section 735.1 states that, where an organisation is convicted of an indictable offence, the court is given complete discretion to set the level of any fine imposed. When exercising this discretion, the court is required to take into account s 718.21 of the Canadian Criminal Code, which details aggravating and mitigating factors that the court needs to take into account when setting the level of any fine, including any advantage gained by the organisation as a result of the offence, how much it cost the public authorities to investigate and prosecute the offence, and what steps had been taken by the organisation to try and prevent the commission of such an offence in future.

14.340 The really unique feature of Bill C-45, however, was the introduction of corporate probation as a form of punishment. The key section is s 732.1 of the Code. This gives the court specific powers in relation to corporate offenders. Consequently, the court can order such an offender, amongst other things, 'to establish policies, standards and procedures to reduce the likelihood of the organization committing a subsequent offence',[1] to 'report to the court on the implementation of those policies, standards and procedures'[2] or to publicise the fact the organisation had committed the offence and been convicted of it. To date, there have been three cases where charges have been laid under s 217.1 of the Criminal Code, and two successful convictions.

1 Section 732.1(3.1) of the Canadian Criminal Code.
2 Ibid.

14.341 The first case in which charges were laid under the provisions of s 217.1 occurred soon after the Bill C-45 amendments made it onto the statute books. In April 2004 a worker was found dead in a collapsed trench. Section 224 of Ontario Regulation 213/91 placed an obligation on those engaged in construction projects to comply with certain key measures and procedures. Given the nature of the soil the workmen were working in, Regulation 213/91 prohibited anyone from entering the trench unless the sides of the trench were sloped at a 45-degree angle or unless they were properly shored up. Furthermore, ss 22 and 23 of Ontario Regulation 213/91 also prohibited anyone from entering the trench unless they were wearing appropriate protective headgear (s 22) and appropriate protective footwear (s 23). It was alleged by the prosecution that Mr Domenico Fantini, the employer of the deceased, had failed to comply with these requirements, each failure putting him in breach of s 25(1)(c) of Ontario's Occupational Health and Safety Act, RSO, 1990, c.0.1. Mr Fantini was also charged with the offence of criminal negligence causing death.[1]

1 Facts as taken from the transcript of Mr Fantini's sentencing hearing.

14.342 Mr Fantini pleaded guilty to each of the three counts laid against him for breaches of Ontario's health and safety laws. Mr Fantini was fined $25,000 for the improperly constructed trench, $20,000 for failing to ensure the deceased was wearing appropriate headwear, and $5,000 for failing to ensure the deceased was wearing appropriate footwear. Finally, Mr Fantini was ordered to pay a victim surcharge of $10,000.[1] However, the criminal charge laid against Mr Fantini was dropped.

1 The victim surcharge is a levy that is imposed on top of a fine handed out by the court to those convicted of committing a criminal offence. The 'proceeds' from this levy are then put into a

special fund designed to provide programmes, services and assistance to the victims of crime within the court's province or territory.

14.343 The case of *R v Fantini* understandably created a great deal of debate amongst the legal profession in Canada. It was noted, for example, that the newly introduced criminal charge was laid against Mr Fantini in spite of the fact that he was a simply a supervisor in a small business carrying out construction work and not an organisation or even a corporate executive. This raised questions about the breadth of the wording of s 217.1.[1] Secondly, concerns were raised about the apparent link between the prosecution's decision to withdraw the criminal charges laid against Mr Fantini and his decision to plead guilty to the charges laid against him for breaches of health and safety legislation. The concern seemed to be that the inclusion of a criminal charge might be used in future in plea bargaining as leverage to ensure a guilty plea to health and safety breaches, ie a guilty plea from the defendant to charges that health and safety regulations had been breached would secure the withdrawal of the criminal charge.

1 Keith Norm, 'First Criminal Charges under Bill C-45 Mean Workplace Prosecution may be Rare' (2005) *The Lawyers Weekly*, 3 June. Certainly a review of the duty, as set out in s 217.1, is broad enough to create both individual and organisational liability, even though Bill C-45 was, theoretically, enacted in order to make it easier to prosecute corporate defendants following breaches of safety rules which led to workplace fatalities.

2 Keith Norm, 'First Criminal Charges under Bill C-45 Mean Workplace Prosecution may be Rare' (2005) *The Lawyers Weekly*, 3 June; and Anon, 'Focus on Workplace Accidents can Provoke Parallel Prosecutions'.

14.344 There were some suggestions amongst the legal profession that the prosecution of Mr Fantini would lead to a flood of further prosecutions under s 217.1. Indeed, John Mastoras, a partner at the Canadian law firm Ogilvy Renault LLP,[1] was reported soon after the *Fantini* decision as saying 'More such charges will like be laid as the Crown tests the new health and safety provisions in the Criminal Code'. This prediction proved to be, more or less, accurate and it was less than a year before a second firm was charged with the offence of criminal negligence causing death.

1 Ogilvy Renault LLP specialises, amongst other things, in business law, employment matters and health and safety law.

14.345 In July 2004, Ontario Provincial Police charged Ontario Power Generation (OPG), a dam manager and one of the dam's operators with criminal negligence following the death of a mother and her son in 2002.[1] Cindy Cadieux and her son, Aaron, were both swept to their deaths when sluice gates at a power generating station were opened without warning unleashing a torrent of water which sped down the Madawaska River. Mrs Cadieux and her son had been amongst approximately 20 people sunbathing further down the river.

1 Anon, 'Dam Operators Charged with Criminal Negligence', CBC News, 9 July 2004. It is important to note, however, that OPG and the employees were charged and tried based on the law as it stood at 2002, ie at the time of the deaths. As such the amendments subsequently introduced by Bill C-45 had not yet appeared on the statute books, and the prosecution had to prove that the dam manager was part of the 'directing mind and will of the company'.

14.346 The Crown had argued that OPG and its employees had been negligent in failing to warn the public that they planned to release the water, even though the area was known to be a popular spot for enjoying the river. Criminal charges against OPG were dropped in November 2006, however, when the court ruled that there was insufficient evidence to convict the company.[1] The two employees were subsequently found not guilty of the criminal charges in

December 2006. It is reported that the judge presiding over the case, Justice Paul Belanger, had declared that 'while there was negligence that led to the spill of water … and the deaths of [Mrs Cadieux and Aaron], it [was] a matter for the civil courts and did not meet the criminal standard'.[2]

1 Anon, 'Charges Dismissed Against Ontario Power Generation in Calabogie Deaths', CBC News, 14 November 2006. In essence, the prosecution were unable to prove that the dam manager was part of the 'directing mind and will of the organisation'. As such, the dam manager's mental state could not be attributed to the defendant organisation.

2 S McKibbon, 'Dam Tragedy Ruling: Workers Found Not Criminally Negligent', (2006) *Ottawa Sun*, 19 December.

14.347 The next prosecution of a corporation for criminal negligence causing death under the Bill C-45 offence did not arise, however, until 2006. Transpavé Inc. was scheduled to appear on 15 November 2006 to answer charges laid against it under s 217.1 of the Canadian Criminal Code.[1] The company was charged following the death of an employee who was crushed to death by a piece of machinery. A key safety device on the machinery had been disabled, and the employee in question had not been given adequate training.[2]

1 C A Edwards and R J Conlin, 'First Corporation Charged with Workplace Safety Crime Post Bill C-45' OH&S Due Diligence Update, 29 September 2006.

2 A report of the findings of the Québec Commission de la Santé et de la Sécurité Travail (Québec's Health and Safety Board) can be accessed online at http://centredoc.csst.qc.ca/pdf/ed003606.pdf (it should be noted that the report is entirely in French).

14.348 The company pleaded guilty to the criminal charge and in March 2008 was sentenced to pay a fine of $100,000 plus an additional $10,000 victim surcharge.[1] To some extent the outcome of the Transpavé prosecution is disappointing. The company's admission of guilt meant that the practicalities of applying the reformed law were not considered. On the other hand, the reasoning behind the judge's decision to accept the proposed fine of $100,000 does provide us with some additional insight into mitigating and aggravating factors which will influence the setting of any fine. Aggravating factors included the fact that a person had died as a result of the organisation's negligence. Mitigating factors, on the other hand, included the organisation's admission of guilt, the fact that Transpavé had no prior convictions for statutory or criminal offences, it had not tried to conceal any of its assets in anticipation of a possible prosecution and fine, the fact that it had not profited in any way from the offence, that it had already invested $750,000 in improving health and safety (bringing it up to a much higher standard than that generally expected of organisations in North America), and that Transpavé had shown sensitivity to the human tragedy caused by the offence by paying for psychologists to come and provide support to the organisation's employees. Crucially, it was also acknowledged by the judge that Transpavé was a family business employing 100 workers, so any fine imposed should not endanger the economic viability of the firm, so as not to put the jobs of its employees at risk.[2]

1 The figure of $100,000 had been agreed between counsel for the prosecution and counsel for the defence. The recommendation was accepted by the judge, Justice Paul Chevalier. The level of the victim surcharge is a matter entirely for the court's discretion.

2 Justice Chevalier is quoted as saying: 'The fine imposed by the Court should not put into play the viability of the company and cause the loss of jobs of about 100 employees who earn good salaries.' As reported by Luis Millán, 'Quebec Company First to be Convicted of Criminal Negligence Causing the Death of an Employee', accessed at http://quebecrulings.wordpress.com/2009/03/26/quebec-company-first-to-be-convicted-of-criminal-negligence-causing-the-death-of-an-employee/.

14.349 Thus, we had the first company convicted of the offence of criminal negligence causing death; however, observers were still no clearer about what conduct would result in a criminal rather than a regulatory prosecution of an organisation following a workplace fatality. Similarly, Transpavé's decision to plead guilty meant that the courts were robbed of an opportunity to get to grips with the practicalities of prosecutions under the amendments introduced by Bill C-45.

The largest fine to date under the Bill was given in July 2012 when Metron Construction Corporation were found guilty of criminal negligence causing death and given a $200,000 fine. This was after four Metron Construction employees died when the suspended swing that they were working on collapsed. The president of Metron was fined $90,000 after pleading guilty to four charges under the Ontario Occupational Health and Safety Act.

A key difference to the Transpavé case was that Metron and the Crown did not present a joint submission regarding fines. The Crown requested a penalty of $1 million whilst Metron argued for $100,000. The total fines paid by Metron and its president reached $342,500, which was three times Metron's earning in its last profitable year, which the court felt was sufficient to send a clear message to other businesses regarding the importance of worker safety.

The law governing corporate liability for health and safety offences

14.350 We have already considered the issue of corporate criminal liability for deaths they have contributed to; in this section, we consider corporate liability for breaches of health and safety legislation. After all, there is a good chance that the actions of a corporation which result in a death may well also be a breach of its health and safety obligations.

14.351 Currently, responsibility for health and safety of employees falls under federal, provincial and territorial labour laws rather than under the provisions of the Criminal Code. Each of the ten provinces, three territories and the federal government has its own occupational health and safety legislation. The details of the legislation and how they are enforced varies from one jurisdiction to the next.

14.352 The Canada Labour Code and provincial labour statutes impose a general duty on employers to ensure the safety and health of their employees, and they provide detail on safety regimes and inspection procedures. The Canada Labour Code, however, only applies to employees who work under federal jurisdiction, which accounts for approximately ten per cent of the Canadian workforce.[1] In most statutes, breach of these regimes is usually a punishable offence. For example, the Canada Labour Code provides that the wilful breach of health and safety standards in knowledge that serious injury or death is likely is a criminal offence.[2] Actual proof of injury or harm is not necessary. Maximum financial penalties range from $100,000 to $1,000,000. However, unlike mens rea offences, the penalties actually imposed are much lighter.

1 The Constitution of Canada did not specify which jurisdiction had responsibility for health and safety matters. This was resolved in 1925 when the Judicial Committee of the Privy Council ruled that health and safety matters are primarily a responsibility of the provinces.

2 Canada Labour Code, RS 1985 c L-2, s 148.

14.353 Similarly, the Occupational Health and Safety Act 1990 of Ontario provides that every person who contravenes or fails to comply with the Act is guilty of an offence, and further that corporations convicted of such an offence may be fined up to a maximum of $500,000.[1] The basic elements of the provincial health and safety laws are similar across all 13 provinces. However, there are variances in the detailed provisions and in the way the laws are enforced.

1 RSO 1990, s 66(2).

14.354 Several Canadian cases offer general guidance on sentencing of corporations.[1] Recent guidance on the issue of sentencing corporations for health and safety offences is provided in the case of *R v General Scrap Iron & Metals Ltd.*[2] The employer appealed against a fine of $100,000 plus 15 per cent victim surcharge out of a total maximum allowable fine of $150,000 for conviction of failure to ensure the health and safety of an employee after an employee in his scrap metal plant was killed whilst sorting metal. Watson J refused to overturn the sentence on the basis that the employer had been found grossly negligent, the accident had been preventable and the employer had previous knowledge of the risk but chose to ignore it. He held that the trial judge had reached a reasonable conclusion as to sentence, notwithstanding that the fine was close to the maximum for first offenders and upheld the paramount importance of deterrence in such cases.

1 See eg *R v United Keno Hill Mines Ltd* (1980) 10 CELR 43; *R v Hoffman La-Roche Ltd* (*No 2*) (1980) 56 CCC (2d) 563; *R v New Brunswick Electric Power Commission* 10 CELR (NS) 184; *R v Ford Motor Co of Canada Ltd* (1979) 49 CCC (2d) 1; *R v K Mart Canada Ltd* (1982) 66 CCC (2d) 329.
2 [2003] 5 WWR 99.

14.355 Watson J did consider, however, that there are some notable distinctions with regards the sentencing of individuals compared with the sentencing of corporations. In that light he suggested that three issues be taken into account when sentencing corporate offender: first, the conduct, circumstances and consequences of the offence; secondly, the objectives of the relevant regulation in the context of legitimate corporate functioning in the relevant areas; and, thirdly, the participation, attitude and character of the offender in the larger context of corporations engaged in the relevant industry in a process which identifies the aggravating and mitigating factors so as to allow rational comparison between cases,[1] in a similar manner to that set out by the English Court of Appeal in the case of *R v F Howe & Son* (*Engineers*) *Ltd.*[2] It is intended that such a principled approach to sentencing of corporate offenders would offer clearer guidance to corporate behaviour and refute allegations about the arbitrariness of sanctions for such offences.[3]

1 [2003] 5 WWR 99 at [35].
2 [1999] 2 Cr App R (S) 37.
3 [1999] 2 All ER 249.

Conclusion

14.356 The issue of corporate criminal liability and responsibility for health and safety have been exhaustively reviewed in Canada over a substantial period of time. Careful attention has been paid to the experiences of other countries in this developing area of the law. It seems generally to have been accepted that the restrictive interpretation of the identification doctrine, as expressed in the cases we have considered in our discussion of Canadian law in this field, is unacceptable as the sole basis of corporate liability.

14.357 We also considered the latest reforms introduced by Bill C-45. We saw that the Canadian government appears to have steered clear of temptation to introduce new specific offences relating to corporate liability for work related deaths and has shied away from introducing more general reforms under which the circle of liability would have been extended to those who exercise delegated authority.

14.358 Finally, we have noted reforms in the area of the sentencing of corporate offenders. In this context, the legislative amendments considered in this section show some originality in the possible use of probation as a sanction for convicted corporate offenders.

EUROPE

The European approach in general

14.359 In comparison to common law jurisdictions, such as the United States, for example, corporate criminal liability has been introduced in European countries relatively recently. Since many European countries now belong to international and supranational bodies, such as the Council of Europe and European Union, there is a certain pressure to harmonise laws relating to corporate responsibilities.

14.360 Back in 1988 the Council of Europe made a Recommendation[1] to Member States concerning the liability of enterprises, with legal personality, for offences caused in the exercise of their activities. Although it was not competent to legislate in the area of criminal law, the Council suggested the development and introduction of a system of administrative sanctions and monetary penalties for legal entities. It recommended that Member States should adopt legislation which would hold corporations criminally liable.

1 Committee of Ministers, Council of Europe, Recommendation on Liability of Enterprises Having Legal Personality for Offences Committed in the Exercise of their Activities, Recommendation no R(88) 18 (October 1988).

14.361 In particular, the Council recommended the following:

> '1. Enterprises should be able to be made liable for offences committed in the exercise of their activities, even where the offence is alien to the purposes of the enterprise.
>
> 2. The enterprise should be so liable, whether a natural person who committed the acts of omission constituting the offence can be identified or not.'

The Council also clearly expressed the difficulties that could occur in arriving at a proper solution:

> 'The difficulty, due to the often complex management structure in an enterprise, of identifying the individuals responsible for the commission of an offence …
>
> The difficulty, rooted in the legal traditions of many European states, of rendering enterprises which are corporate bodies criminally liable …
>
> The difficulty to capture the reality of the decision-making in large modern corporations companies' and to determine 'what the company has done wrong to merit being subjected to criminal prosecution'.[1]

These difficulties find their roots in the legal traditions of the [Council of Europe] Member States. Most of their criminal laws are based on two major principles: 'a person is responsible for his own acts', and 'corporations cannot make any wrong'.

1 Council Recommendation R(88) 18: see note 1 to **14.360** above.

14.362 It is quite normal for European countries to have laws which hold corporations/enterprises liable for committing environmental crimes and financial crimes. However, it is less usual and more novel to impose criminal liability on corporations for crimes, such as manslaughter. The countries which have adopted laws on corporate manslaughter have invariably used different approaches. It is not surprising, as the crime of manslaughter is usually identified with a natural person. To transfer the ingredients of the offence to a corporation with often complex structures requires certain changes in perception and, most importantly, the ability to ensure that the law can be applied in practice and can achieve its aims.

In spite of these problems, the willingness to impose corporate liability for the specific offence of manslaughter brought about changes in the law in some Member States.

France

Introduction

14.363 France has a centralised government, although its 22 regions, 95 departments, and 36,000 municipalities enjoy a certain amount of autonomy. The French legal system is governed by the principal of unity of the civil and criminal justice system, which means that the same courts (courts belonging to *l'ordre judicaire*) decide both criminal and civil cases, unless the state, a state employee or a corporation is involved. Such cases proceed to *l'ordre administratif*, or administrative courts, which are part of the executive branch of government dealing with administrative disputes. *L'ordre judicaire* courts are broken down into the *tribunal d'instance* and the *tribunal de grande instance* which both handle civil cases; and courts which have jurisdiction over criminal cases, including the *tribunal de police,* for small offences such as parking violations, *tribunal correctionnel* for more serious cases, and the *cours d'assises* (assize courts) which deals with the most serious cases.

14.364 Under both the Penal Law and the Penal Procedure, offences may be classified according to their respective seriousness: crimes, misdemeanours, and violations.[1] There are distinctions between completed and attempted acts for crimes and misdemeanours, but not for violations. The punishment for crimes and misdemeanours can result in up to 30 years' imprisonment. Crimes are also classified into crimes against the person, crimes against property and crimes against public security.[2] The most serious crimes, for which the penalties may range up to life imprisonment, are tried in assize courts (*cours d'assises*). They do not sit regularly but are called into session when necessary.[3]

1 New Penal Code, art 111–1.
2 New Penal Code books 2–4.
3 See also www.justice.gouv.fr, accessed on 19 September 2013.

14.365 'Corporate liability' is a relatively new provision in French criminal law. The former Criminal Code stated that only individuals could be held liable for criminal offences such as manslaughter if they caused death. There was no criminal legislation which governed liability for legal entities. The Penal Code was entirely reformed in 1992. Since then,[1] the French New Penal Code provides that all legal personalities may be held criminally liable. Legal persons can be liable for most of the offences provided by the Penal Code.

1 The New Penal Code came into force in 1994.

Law on corporate criminal liability

14.366 The Penal Code, art 121–1 provides that no person may be held criminally liable except for his or her own conduct. The principle of criminal liability of legal persons has been introduced in the Penal Code by virtue of art 121–2,[1] which states that juridical persons, with the exception of the state, are criminally liable for the offences committed on their account by their organs or representatives, according to the distinctions set out in arts 121–4 and 121–7 and in the cases provided for by statute or regulations. Article 121–4 provides that the perpetrator of an offence is the person who commits the criminally prohibited act; and attempts to commit a felony or, in cases provided for by statute, a misdemeanour. Article 121–7 provides that the accomplice to a felony or a misdemeanour is the person who knowingly, by aiding and abetting, facilitates its preparation or commission and that any person who, by means of a gift, promise, threat, order, or an abuse of authority or powers, provokes the commission of an offence or gives instructions to commit it, is also an accomplice.

1 Act No 2000–647 of 10 July, art 8, (2000) *Journal officiel*, 11 July.

14.367 The Penal Code, art 121–3 enables the courts to charge corporations for the offence of manslaughter. Art 221–6 specifically refers to Art 121–3 when it expressly provides[1] that manslaughter is a criminal offence. These sections act as the general legal basis. These are not specific to manslaughter in the course of work. In 2000, the New Criminal Code, art 121–3 was amended: then, the legislator introduced sub-art 4, enabling courts to charge individuals for injuries that were due to a failure to comply with an obligation of due diligence or safety. Although this subsection only governs individuals' criminal liability, it has affected the legal regime of corporate liability, extending its scope, and so from that point, directors and heads of corporations are more likely to be held responsible for such criminal offences.

1 This element is a very important, given that French law is based on the idea of *légalité des délits et des peines,* which means that 'no one may be punished for a felony or for a misdemeanour whose ingredients are not defined by law …' and 'no one may be punished with a penalty which is not provided for by the law …' (Article 111–3 of the Penal Code).

14.368 The law does not require that there be 'fault' for corporations to be held liable for such crimes; in other words the law imposes strict liability. The following conditions must be met to hold a corporation criminally liable: first, the act must be qualified as a criminal offence in law; the offence must have been committed by either a member or a representative of the corporation; and the person must have acted in the name of the corporation.

14.369 The criminal liability of legal persons does not exclude natural persons who are perpetrators or accomplices to the same act, subject to the provisions of the Penal Code, art 121–3. Public bodies may also be held liable for criminal manslaughter.

14.370 The application of corporate criminal liability is confined to a limited number of crimes.[1] Moreover, corporations can only be held liable under the French Penal Code when one of the legal representatives or organs of the corporations has positively acted. On the other hand, the violation of supervisory duties is considered to be sufficient to warrant proceedings on the basis of corporate criminal liability.[2]

> 'It is still not clear how the new legislation will work in practice or over the long term. It will be interesting to observe how the French courts will resolve essential questions – such as the level of managerial involvement necessary to trigger liability or the importance of collective knowledge within the company.'

(See R Hefendehl, 'Corporate Criminal Liability: Model Penal Code Section 2.07 and the Development in Western Legal Systems', (2001) Buffalo Criminal Review, Vol 4, pp 283–300.)

1 See G Stessens, 'Corporate Criminal Liability: A Comparative Perspective' (1994) 43 ICLQ at 496–497.

2 M Wagner, 'Corporate Criminal Liability: National and International Responses', background paper for the International Society for the Reform of Criminal Law 13th International Conference, Commercial and Financial Fraud: A Comparative Perspective, Malta, 8–12 July 1999.

14.371 Sanctions under the new provisions include fines, the closure of the company and probation. Corporate fines may be up to three times higher than those imposed on a natural person.

Italy

Introduction

14.372 The system of justice in Italy is divided into the following areas: ordinary civil and criminal, administrative, accounting, military and taxation. Criminal laws are contained within the Penal Code and other statutes and are applied throughout the whole country. Health and safety rights and obligations are contained in the Italian Constitution, which states that 'health is a fundamental individual right and a social interest' (Art 32) and that private economic enterprise 'shall not be exercised contrary to the social good or in such a way to cause harm to safety, freedom and human dignity' (Art 41). Rules on corporate liability are contained in the Civil Code. The Civil Code determines the organisation of the company and imposes more specific duties on the entrepreneur.[1]

1 See David Bergman et al, *International Comparison of Health and Safety Responsibility of Company Directors*, Initial Report of work funded by the HSE (2007).

14.373 All criminal offences (*reati*) are divided by the Penal Code (*Codice Penale*, which dates back to 1930), into two broad categories: *delitti*, which are serious criminal offences, and *contravvenzioni*, which are less serious offences. The two categories are also used to classify special law statutes. The distinction between crimes classified as *delitti* and *contravvenzioni* is based on the seriousness of the crime and on the severity of punishment. Although both categories of crime are punishable by imprisonment and/or fine, the sentences for *delitti* are more severe than those for the *contravvenzioni*. For *delitti* crimes, penalties may be imposed of up to 24 years' imprisonment, and as much as 30 years' or life imprisonment in special cases. For *contravvenzioni* crimes, the penalty imposed may be up to three years' imprisonment. Normally, sentences for *contravvenzioni* crimes are served in different types of prison facilities than those used for *delitti*.

14.374 The Penal Code generally classifies each crime under a specific heading including crimes against public safety (*delitti contro l'incolumita* *pubblica* – poisoning of food or water, drugs, arson, provoking a railway or air disaster, etc), and crimes against the person/violent crimes (*delitti contro la persona* – murder, assault, non-ransom kidnapping, defamation).[1]

1 P Marongiu, 'Italy', *Report prepared for The World Factbook of Criminal Justice Systems* under Bureau of Justice Statistics grant no 90-BJ-CX-0002 from the Bureau of Justice Statistics to the State University of New York at Albany (1993). The project director for the World Factbook of Criminal Justice was Graeme R Newman. Available at www.bjs.gov/content/pub/html/wfcj.cfm.

14.375 Health and safety laws can be found in the Constitution (see **14.372** above), the Civil Code and special statutes, which are called decrees. The Civil Code states that 'the entrepreneur shall, in carrying his business, adopt the measures which, in accordance with the particular nature of the work, experience and technology, are required for protecting the physical integrity and personality of the persons employed' (Art 2087 of the Civil Code and David Bergman et al, as above). The Decree 626/1994 transposed the European Framework Directive on health and safety and is the key legislation relating to health and safety obligations in Italy. As in other jurisdictions, the main duties are imposed on employers. However, unlike in the UK, for example, the employer under this Decree is a natural person. 'It is important to note that the term "employer" does not describe a person within the company who holds a particular title or occupies a particular post. Instead it describes a person who has effective decision-making powers and control over financial resources relevant to the safety of the workers. This person could be a board director or general manager or a senior manager.'[1]

1 David Bergman et al, *International Comparison of Health and Safety Responsibility of Company Directors*, Initial Report of work funded by the HSE (2007).

14.376 Corporate criminal liability was not recognised in Italian law until 2001. In the Italian system, criminal responsibility was limited exclusively to natural persons. Legal persons could not be subjected to any type of sanction, including even administrative ones.[1]

1 See Draft Law 689 of 1981.

14.377 Currently two decrees deal with corporate criminal liability, namely Legislative Decrees 231/2001 as amended by 61/2002. Decree 231/2001 (D Lgs 231/2001) implemented in Italy an OECD (Organisation for Economic Cooperation and Development) Convention which stated that, for some offences, the criminal courts should not only punish the individuals committing the offence, but also the legal entities (if any) benefiting from illegal acts of the individual acting on their behalf. The list of offences currently in the Decree includes not only those in the above-mentioned Convention, but also other offences to which the Italian state has decided to extend the same principle. However, these offences only cover criminal conduct associated with the company's financial dealings, such as: undue receipt of public funds; fraud against public bodies in order to obtain public funds; solicited or unsolicited corruption of public officials; fraud against the IT systems of public bodies; fraud of public bodies, etc. The basis of Decree 61/2002 (D Lgs 61/2002) is the 'legislator's attempt to rationalise criminal corporate law by restricting the number of traditional charges and by introducing new types of charges aimed at covering new "white collar crimes" which have been made possible by increasingly refined techniques, particularly as regards financial transactions'.[1]

1 See 'Directors' criminal liability under Italian law. The reform of criminal corporate law, LGS D No. 61 of 11 April 2002'.

14.378 Law 231/2001 established a generic model of criminal liability that can be applied to corporations and other organisational entities, whether or not they have legal personality. The model could potentially be invoked in respect of a wide range of crimes, even though in the first instance it was limited to crimes involving corruption, theft and fraud against the State:

> 'It had been argued that introducing such a revolutionary system to a broad number of crimes would have been difficult if not impossible for many corporations to cope with. The choice was, instead, to come up with general guidelines, limiting them to few offences at the beginning, with the view of expanding them after corporations had gained experience in setting up the types of systems of control and supervision envisaged in the legislation. True to its word, in March 2002 the Italian government extended the scope of the statute so that it applied not just to cases of property crimes committed against the State but also to similar crimes where the victim was a private party.'[1]

1 J Gobert and E Mugnai, 'Coping with Corporate Criminality: Some Lessons from Italy' [2002] Crim LR 619 at 624 and 625.

The law on corporate liability for health and safety offences and deaths in the workplace

14.379 Law 231/2001 classified offences committed by corporations as administrative ones. However, this may be misleading:

> '… Companies will be liable for criminal offences and not just administrative misdemeanours, the fact that case will be heard in the criminal courts rather than administrative tribunals, and the fact that criminal rather than administrative procedures will be in force, all suggest that the contemplated liability is more criminal than administrative in nature. Significantly, under Article 8 of the statute, a company can be held responsible even if it is not possible to identify or convict the human perpetrator of the offence.'[1]

1 J Gobert and E Mugnai, 'Coping with Corporate Criminality: Some Lessons from Italy' [2002] Crim LR 619 at 624.

14.380 According to this law, corporate liability arises in two ways. First, when the crime is committed by the head of the corporation, the director or manager, the *decreto legislativo* provides that a legal personality can be criminally liable for a lethal injury caused either by its director,[1] or a subsidiary.[2] Article 5 of Legislative Decree 231/2001 distinguishes between acts that are committed by the heads of the corporation, its directors, or managers (Art 5(1)(a)) and those that are committed by its subordinate staff (Art 5(1)(b)). Article 5(1)(b) creates a liability, which can be imposed on a corporation for the acts of those who do not constitute the directing mind of the corporation. This provision establishes a regime of vicarious liability. Article 5 requires proof that the offence either was 'an expression of corporate policy', or that it was the result of 'structural negligence within the corporation'.[3] Furthermore, Art 5(2) requires that the offence should be committed 'in the interest and for the benefit of the corporation'.

1 Article 5(1)(b), D Lgs, 231/2001.
2 Article 5(1)(a), D Lgs, 231/2001.
3 Gobert and Mugnai, note 1 to **14.379** above, at 625.

14.381 The second situation is one in which corporate liability arises 'which is based on the negligence of a corporate body in not considering that the offence which has been committed might have occurred, and in not having in place a mechanism to avert its commission'.[1] A company can be held criminally liable for manslaughter if death was caused by a management failure of the corporation. This provision allows the company to be held responsible even when it is not possible to determine which person committed the offence.

1 Gobert and Mugnai, note 1 to **14.379** above, at 625.

14.382 Articles 6 and 7 of the D Lgs 231/2001 provide for the defence of due diligence. Article 6 sets out the defence which is to be used by those who are in control of the corporation. Article 7 creates the defence which might be used by subordinates. The defence of due diligence allows the corporation to show that an appropriate control system was in place. This will discharge the onus of proof on the accused; it then falls on the prosecution to rebut the defence and to prove that the system was either 'inadequate or ineffective to prevent offences of the kind that occurred'.[1]

1 Gobert and Mugnai, note 1 to **14.379** above, at 628.

14.383 The introduction of criminal liability is a major change in Italian criminal law. The statute, however, is restricted as to its applicability. It applies solely to profit-making organisations, whether or not they are incorporated, and it does not apply to public bodies. Non-profit-making organisations are seen as 'acting for the benefit of the State and its citizens' and thus should not be prosecuted.

14.384 The most recent changes to the health and safety law were made by Decree 81, passed in April 2008. The aim of the amended law is to collect and coordinate all health and safety at work rules, contained in numerous laws though a consolidated act (*Testo Unico*) and to emphasise employer and employee obligations to reduce accidents at work.[1] The Decree establishes the fine of €8,000 on employers who fail in their duties to protect workers. In serious cases, prison sentences of 6 to 18 months could be imposed. Sanctions are even stiffer if employers are found to be responsible for causing work-related injuries or deaths. In these cases, fines of up to €1.5 million could be applied and business activities ordered to shut down.[2]

1 See Avv Francesca Lauro and Dott Davide Costa (Lauro Sovani & Associati), 'Law on Health and Safety in the Workplace Amended', American Bar Association, International Labour and Employment Law Committee Newsletter, June 2008.

2 See Michael Bradford, 'Tough Italian Law may raise safety standards', 2 June 2008, Business Insurance, at www.businessinsurance.com/cgi-bin/article.pl?article_id=25017.

14.385 Additionally, there are a number of Penal Code provisions. Article 589 relates to the offence of negligent homicide and states:

> 'Whoever by negligence causes the death of a person shall be punished by imprisonment for a period ranging from six months to five years. If the act was committed by violating the rules governing road traffic or those for the prevention of industrial accidents, the punishment shall be imprisonment from one to five years. In the event of the death of more than one person or of the death of one or more persons and personal injury to one or more persons, the punishment applied shall be that which should be inflicted for the most serious violation committed increased by up to one third, but this punishment may not exceed twelve years.'[1]

1 David Bergman et al, *International Comparison of Health and Safety Responsibility of Company Directors*, Initial Report of work funded by the HSE (2007).

14.386 A recent example where manslaughter charges were brought against company directors is connected to one of the worst aviation accidents which happened in Milan in October 2001, as a result of which 118 people died. A number of errors at the fog-bound airport led a private Cessna jet onto the runway and into the path of a Scandinavian Airlines System (SAS) passenger plane as its front wheels lifted for take-off. Causes that led to the accident included, inter alia, the following: the Cessna crew was not aided properly with correct publications, lights, markings and signs to enhance their situation awareness; official documentation failing to report the presence of unpublished markings that were unknown to air traffic controllers, thus preventing the ATC controller from interpreting the unambiguous information from the Cessna crew; instructions, training and the prevailing environmental situation prevented the ATC personnel from having full control over the aircraft movements on ground; and no functional Safety Management System was in operation.

14.387 In April 2004, four officials were found guilty of negligent manslaughter for their role in the runway crash at Linate Airport in Milan. The director of the airport received an eight-year sentence and the air traffic controller, who was on duty at the time, received a six-and-a-half-year sentence. The former Chief Executive Officer of Italy's air traffic control agency ENAV and the person who oversaw Milan's two airports were sentenced to six years and six months respectively.[1]

1 See 'World Briefing Europe: Italy: Aviation Officials Sentence in Country's Worst Crash', by Jason Horowitz, (2004) *New York Times*, 17 April.

The Netherlands

Introduction

14.388 The Kingdom of the Netherlands is a constitutional and hereditary monarchy. The Netherlands is divided into 12 provinces and 672 municipalities. Each province has an elected representative body known as the Provincial Assembly. Central legislative power is vested in the Crown and Parliament, comprising two chambers (States-General). The first or upper chamber (*Erste Kamer*) is elected by the Provincial Council; the second or lower chamber (*Tweede Kamer*) is elected under a system of direct universal suffrage and proportional representation. Both the Second Chamber and the Crown may propose legislation; the First Chamber may only approve or reject legislation without amendment.

14.389 The principal laws which guide the criminal justice system in the Netherlands are the Constitution, the Criminal or Penal Code, the Code of Criminal Procedure and Special Acts. All prohibited acts are classified either as crimes or felonies (*misdrijven*), infractions or transgressions (*overtredingen*). The legislature determines whether an offence constitutes a crime or an infraction.[1] Generally, serious offences involving physical harm are classified as crimes or felonies (murder, unintentional homicide, theft combined with violence, etc). The classification of an offence determines the level of the court that will try the case at first instance. In general, transgressions are tried in cantonal courts and crimes are tried in district courts.[2]

1 P Tak, *Criminal Justice Systems in Europe: The Netherlands*, (HEUNI (European Institute for Crime Prevention and Control, affiliated with the United Nations)) (Deventer: Kluwer Law and Taxation Publishers, 1993).

2 W Hoyng and F Schlingmann 'The Netherlands' in M Sheridan and J Cameron (eds) *EC Legal Systems: An Introductory Guide* (London: Butterworths, 1992), pp 1–42.

14.390 The Netherlands was one of the first European countries to establish criminal liability for legal persons, including liability for manslaughter. This was established in two steps. First, in 1951 law makers decided that corporate bodies could 'commit' an offence. However, at that time, legal persons could only be held liable for offences committed against public welfare. Secondly, in 1976, the Dutch Criminal Code was amended, so that it now provides that both natural persons and corporations can be held liable for a wide range of offences, including battery and involuntary manslaughter.[1]

1 Dutch Criminal Code, art 51.

14.391 General provisions for Dutch health and safety law are contained in the Working Conditions Act and the Work Conditions Decree.

The law on corporate criminal liability

14.392 The Dutch Criminal Code, art 51.2 states:

> 'Where a criminal offence is committed by a juristic person, criminal proceedings may be instituted and such penalties and measures as are prescribed by law, where applicable may be imposed: (1) against the juristic person; or (2) against those who have ordered the commission of the criminal offence, and against those in control of such unlawful behaviour; or (3) against the persons mentioned under (1) and (2) jointly.'

The explanatory memorandum to this article provides a list of offences that might be committed by a corporation, and manslaughter is included. A corporation can be thus both charged and sentenced.

14.393 Under Dutch law, criminal liability can be imposed on any legal person. For instance, in the case of *Ijzerdaad*, both the Court of Appeal[1] and the Supreme Court held that an export firm was criminally liable, although it was unincorporated and owned by a sole proprietor.

1 Hoge Raad, 23 February 1954, NJ 378.

14.394 Public bodies can also be held criminally liable under the Dutch criminal law. A good illustration of this can be seen in the case of *Rechtbank Leeuwarden*.[1] Here, the governing body of a hospital was charged 'with grossly negligently failing to ensure properly that old, redundant anaesthetic equipment was removed from the hospital or made unusable'.[2] The equipment had not been properly maintained since it was not listed in the hospital's inventory, and the hospital had failed to ensure that the replacement of old equipment was properly carried out. This equipment, therefore, was still used in an operation. In this particular instance the technicians, who acted without proper supervision, connected the tubes wrongly. This last act of negligence caused the death of a patient. On 23 December 1987, the Dutch courts convicted the hospital's governing body of negligent homicide. This was the first case in the Netherlands where manslaughter charges were successfully brought against a corporation.

1 Hospital Case, *Rechtbank Leeuwarden*, 23 December 1987, partially reported at NJ 1988, 981.
2 S Field and N Jörg, 'Corporate Liability and Manslaughter: Should We Be Going Dutch?' [1991] Crim LR 156 at 157.

14.395 According to Dutch law, corporations can be held liable irrespective of who committed the offence, whether they are mere employees or heads of corporations with the power of control. The law is not based on the principle of identification. There is a suggestion that corporate criminal liability is based on a principle of 'power and acceptance'. This is a twofold test. The court must first determine whether the company had the power to determine whether the employee carried out the reprehensible act. If the answer is positive, the court must then try to establish whether the corporation usually 'accepts' such acts.

14.396 Stewart Field and Nico Jörg stated that 'acceptance' involves making a judgement on the corporate monitoring of risks or illegal behaviour, and 'power' is a judgement on the corporate response to those risks'.[1] These principles are largely concerned with the organisation and management of the company. A company, however, is able to escape liability if it can prove that there was nothing it could have done to eliminate the risks and the unfortunate consequences were beyond its powers.[2]

1 See S Field and N Jörg, note 2 to **14.394** above, in citing the *Kabeljauw Case*, Hoge Raad, 1 July 1981, NJ 1982, 80 and the *IJzerdraad Case*, Hoge Raad, 23 February 1954, NJ 1954, 378.

2 Field and Jörg, note 2 to **14.394** above, at 157.

14.397 However, liability seems to be attributed differently depending on the size of the corporation. The smaller corporations are tested on the basis of whether they had the power to control the act in question and, if so, whether 'the act was in line with the business of the company', ie on the so-called 'power and acceptance' test.[1] A different test is applied to larger corporations: in such cases, the question focuses more on 'normal company policy' although there is no definition of what this means.[2]

1 Allens Arthur Robinson, *'Corporate Culture' as a Basis for the Criminal Liability of Corporations*, Report prepared for the United Nations Special Representative of the Secretary-General on Human Rights and Business, February 2008.

2 See Fafo AIS, 'Survey Responses, Laws of the Netherlands' (Nicola Jägers), 'Commerce Crime and Conflict: A Survey of Sixteen Jurisdictions'' (2006) 10–11.

14.398 The Dutch approach is important in that it concentrates on corporate behaviour and recognises corporate fault rather than individual fault. The arguments considered in the paper by Field and Jörg were referred to in a report published by the Accident Prevention Unit of the Health and Safety Executive in 1989, which stated:

> 'The report notes that in organisations where safety is not considered paramount individuals may be unwilling to follow good safety procedures for fear of being criticized or even disciplined. Furthermore, where priorities are confused, safety is likely to come into conflict with commercial pressures. Thus even individual acts of negligence are often identifiable as a product of collective responses.'[1]

1 Field and Jörg, note 2 to **14.394** above.

14.399 Dutch law recognises that, because the corporation's pressure can shape the behaviour of an individual who acts in accordance with corporate policies and structure, the corporation should be considered as a collective entity rather than as a group of individuals who might be identified.

Germany

Introduction

14.400 The legal system in Germany is guided by federal laws which are created by the Bundestag or Lower House of the German Parliament and approved by the Upper House (Bundesrat). Federal laws apply nationwide. Those specifically applicable to the criminal justice system are the Criminal Code (*Strafgesetzbuch*, StGB)[1] and the Code of Criminal Procedure (*Strafprozeßordnung*, StPO). Among other laws which concern the criminal justice system are the *Gesetz über Ordnungswidrigkeiten* (OWiG) or laws governing administrative or regulatory offences.

1 The Criminal Code was promulgated on 13 November 1998 (Federal Law Gazette I, pp 945, 3322).

14.401 Federal laws establish a framework for the 16 individual states or Länder.[1] The states have their own constitution and create their own special laws. The Penal Code (StGB) and the Code of Criminal Procedure (StPO), are federal Codes, which have national application. The administration of the criminal justice system (police, courts and correctional institutions) are matters left for the individual states.

1 See www.bundestag.de/htdocs_e/bundestag/function/legislation/competencies.html (last accessed 12 December 2013).

14.402 Whilst the federal government has the power to introduce legislation, the individual federal states (Länder) are responsible for ensuring that government regulations are being properly implemented.[1]

1 Marina Schroder (Head of Health and Safety, DGB, Germany), 'Occupational safety and health in Germany in pre European law reform – status and shortcomings', TUTB Newsletter, April 2004, No 22–23).

14.403 As with other European countries, criminal offences in German law can be categorised as *Verbrechen* (crimes or felonies) and *Vergehen* (misdemeanours). Less serious offences have, through a lengthy reform process, either been decriminalised, upgraded into misdemeanours, or reclassified as *Ordnungswidrigkeiten* (regulatory or administrative offences). *Verbrechen* comprise serious offences involving severe injury or extensive property damage or loss (for instance, homicide, rape, robbery, arson), which are punishable by a minimum prison sentence of up to one year, whereas *Vergehen* are offences such as simple assault, theft, vandalism, and are punishable by sentences of less than one year or a fine.[1]

1 AA Aronowitz, *Germany, Report prepared for The World Factbook of Criminal Justice Systems* under Bureau of Justice Statistics grant no 90-BJ-CX-0002 (1993), available at www.bjs.gov/content/pub/ascii/WFBCJGER.TXT.

14.404 General provisions relating to health and safety obligations and liabilities are contained in the framework legislation, which was passed by the federal Parliament in the form of the Health and Safety at Work Act (*ArbSchG*). It was enacted in mid-1996 as a result of the implementation of the European Framework Directive on health and safety of 12 June 1989. Occupational health and safety is also governed by the accident prevention regulations, under which 'employers have a duty to provide their employees with statutory accident insurance cover'.[1]

1 Marina Schroder (Head of Health and Safety, DGB, Germany), 'Occupational safety and health in Germany in pre European law reform – status and shortcomings', TUTB Newsletter, April 2004, No 22–23).

14.405 The Health and Safety at Work Act 1996 imposes duties on employers. Section 3.1 requires employers:

> 'to adopt the necessary occupational safety and health measures taking account of any circumstances affecting the safety and health of employees in the workplace. The employer must assess the effectiveness of such measures and, if need be, adjust to changing circumstances. In so doing, his goal must be to improve employees' safety and health protection.'

However, there is no offence for breach of the general duties, but only for breach of an ordinance or an enforceable order (see s 18 of the Health and Safety at Work Act). Employers are defined as either natural or legal persons. Section 13, which defines 'Responsible Persons', makes 'company directors' responsible for complying with the same duties as those imposed upon the employer.[1]

1 See David Bergman et al, *International Comparison of Health and Safety Responsibility of Company Directors*, Initial Report of work funded by the HSE (2007).

14.406 There is no principle of 'corporate liability' under German criminal law, as it does not currently provide for criminal liability for legal persons or associations of persons. 'Corporate manslaughter', for example, was not mentioned in the ruling on the ICE-disaster at Eschede, a rail accident where 101 people were killed in 1998. Although three persons were to some extent blameworthy, the court dismissed the case for formal reasons. An appeal against this decision to the Constitutional Court (*Bundesverfassungsgericht*) was unsuccessful.

14.407 According to German criminal law, criminal penalties may only be imposed on natural persons.[1] Instead of the imposition of criminal liability on legal persons, the German legal system provides for a mechanism of liability on legal persons under laws governing administrative or regulatory offences, such as health and safety offences mentioned above. As a consequence, only non-criminal fines can be imposed as penalties on corporations. Furthermore, measures of confiscation and forfeiture can be ordered against legal persons, where this is necessary for the purpose of confiscating the financial advantage gained as a result of the offence.[2]

1 G Fieberg, 'National Developments in Germany: An Overview' in Eser, Heine and Huber (eds), *Criminal Responsibility of Legal and Collective Entities* (1999), p 83.

2 B Shloer and O Filipenko, 'Criminal liability of legal persons Criminal Codes examples (within the context of the corruption offences): European states experience', Ukrainian-European Policy and Legal Advice Centre, project funded by European Union implemented by GTZ-IRZ-PLEY Consortium.

14.408 An example where a corporate failure led to prosecution of individuals was a 2002 mid-air collision between a Russian passenger plane and a DHL cargo plane over southern Germany that killed 71 people including 41 Russian children. Four senior managers from the air traffic control company Skyguide were found guilty of manslaughter. The Swiss court held that their actions had contributed to a 'climate of negligence' at Skyguide and sentenced three of the managers to a 12-month suspended prison sentence and the fourth was fined 13,500 Swiss francs (£5,500). 'An investigation into the causes of the disaster established that Peter Nielsen, 36, the air-traffic controller, was alone on duty at the time and that he had only managed to warn the doomed aircraft of their impending collision 43 seconds beforehand. Radar and phone systems at Skyguide's air-traffic control centre in Zurich were undergoing a routine service and were not operating on the night of the crash. The second controller was taking a break when the collision occurred.' Apparently, management and quality assurance of the air navigation service company tolerated for years that, during times of low traffic flow at night, only one controller worked and the other one retired to rest.

Law on corporate criminal liability

14.409 In Germany, liability on corporations may be imposed by state authorities only for administrative infractions such as health and safety offences, ie for breach of an ordinance or an enforceable order. The Health and Safety at Work Act 1996 recognises two types of offences: administrative and penal. An administrative offence is committed in two ways:

'(1) If either the employer or a natural person "intentionally or negligently" acts "contrary to an ordinance" enacted according to sections 18 and 19. The maximum fine is 5,000 Euro.

(2) If an employer or responsible person or employee "intentionally or negligently" acts contrary to an enforceable order imposed by section 22(3). The maximum fine is 25,000 Euro.'

The labour inspector has discretion of whether to impose an administrative sanction or not.[1]

1 See also David Bergman et al, *International Comparison of Health and Safety Responsibility of Company Directors*, Initial Report of work funded by the HSE (2007).

14.410 Thus a breach of health and safety ordinances, or the violation of administrative orders, can give rise to administrative sanctions, which can only be imposed on the employer as a legal entity if a 'responsible person' has committed the offence. The meaning of 'responsible person' is provided in s 13 of the Health and Safety at Work Act 1996. In addition to the employer. it includes:

'• his legal representative;
- the organ that is legally empowered to represent a legal entity;
- a shareholder of a partnership who is entitled to represent the partnership;
- persons who have been given the task of managing a company or an establishment subject to the tasks and rights assigned to them;
- other persons designated pursuant to subsection 2 hereof or an ordinance adopted under the present Law or the Accident Prevention Regulations, subject to the tasks and rights assigned to them.'

The employer can also commission reliable and expert persons to assist him with the performance of his duties under the Act (s 13(2)).

14.411 Administrative offences are governed by the Code of Criminal Procedure (StPO) and the *Ordnungswidrigkeitengesetz* (OWiG). The German law of *Ordnungswidrigkeiten* (administrative penalties) empowers administrative authorities, generally the Labour Inspectorate, as well as criminal courts to impose administrative fines (*Geldbußen*) on both natural persons and companies.[1] The key provision for the sanctioning of corporations is OWiG, s 30,[2] which calls for the imposition of fines on corporate entities. There has been much debate in legal circles about whether such penalties should be considered criminal sanctions or administrative penalties. Today, they are not perceived as criminal sanctions either by the public or by defendants, even when imposed by a criminal court.[3]

1 R Hefendehl 'Corporate Criminal Liability: Model Penal Code Section 2.07 and the Development in Western Legal Systems' (2000) 4 Buffalo Crim LR 283.

2 Several translations for this law exist, among them are Regulatory Offences Act and Law of Administrative Sanctions.
3 R Hefendehl, as note 1 above.

14.412 The class of natural persons, whose acts may make the corporation liable, is very limited.[1] In general, liability is restricted to instances in which the company's legal representatives or directors have acted improperly or failed to supervise their employees properly. So far, German policy towards corporate liability is very restrictive compared with that of other European countries. In Germany, penalties imposed on corporations may, at most, be regarded as 'quasi-criminal' sanctions.[2] There have been pressures in Germany to have 'true' corporate criminal liability because of the inadequacy of existing sanctions and of the deterrent effect of an administrative fine. However, there have also been a number of arguments against such liability, based on the notion that corporate entities are unable to act in a criminal law sense due to: the absence of a will; the belief that corporate entities are not capable of being culpable; and the inability for corporations to undergo punishment.[3]

1 See OWiG, s 30.
2 Hefendehl, note 1 to **14.411** above.
3 Wagner, 'Corporate Criminal Liability: National and International Responses', background paper for the International Society for the Reform of Criminal Law 13th International Conference, July 1999.

14.413 A penal (or criminal) offence under the Health and Safety at Work Act 1996 can be committed in two following ways: (1) by a responsible person who 'continuously' fails to comply with an enforceable order; or (2) by a responsible person or employee who 'risks life and health of an employee by intentionally failing to comply with an ordinance enacted according to sections 18 or 19'. Criminal sanctions can only be imposed on natural persons. In such cases, the inspector does not have discretion; he must report it to the police or a prosecutor.[1]

1 See David Bergman et al, above.

14.414 Additionally, employers can be liable under the law relating to statutory accident insurance, which was originally introduced by the Accident Insurance Code in 1884, reformed in 1911 under the Reich Insurance Code, and finally replaced by Part Seven of the Social Code (SGB VII). Section 209 of the Social Code VII creates a number of administrative offences against natural persons for deliberate or negligent breach of duties under accident prevention regulation,[1] or for contravening an 'enforceable order' made by the insurance inspector.[2]

1 See the BGV A1 (accident prevention regulation) which entered into force on 1 January 2004).
2 See also www.dguv.de (accessed on 19 September 2013) and **14.415** below.

14.415 Those who are found guilty of causing accidents through fault may be liable to be charged with general criminal offences under the Criminal Code,[1] such as negligent manslaughter, under s 222, or bodily injury by negligence under s 229 (Strafgesetzbuch, StGB). Criminal negligence requires proof of personal breach of a duty on the part of the employer and foreseeability of the consequences of this breach.

1 The terms Criminal Code and Penal Code are adopted from translations of existing literature. However, there is no substantial difference in meaning between the two terms.

Enforcement

14.416 The enforcement of health and safety law in Germany is the task of two bodies, the Labour Inspectorate and the Insurance Associations. Compliance with health and safety law in Germany at the state level is monitored by the Labour Inspectorates (*Gewerbeaufsicht*). The Labour Inspectorates are overseen by and are answerable to elected political representatives of the Länder. Each Land is compelled to set up a Labour Inspectorate under the Industry Act, s 139b (*Gewerbeordnung*) and is free to decide both the organisation and the functions of its Labour Inspectorate, although the fundamental principles governing the activities of the Inspectorates are similar throughout Germany. Their responsibilities include the observance of technical, medical and social regulations intended to ensure acceptable standards of health and safety at work. The Labour Inspectorates also have responsibility to monitor employers who put the health and safety of the general public at risk as a consequence of the work activities they are undertaking.

14.417 In addition to the state system, there is a further system for worker protection administered by the Employers' Liability Insurance Associations (*Berufsgenossenschaften*, BGs). The 35 German BGs, or institutions for statutory accident insurance and prevention, represent different branches of trade and industry. Every company is required by law to be a member of one of these Associations, who are the providers of statutory accident insurance and have as their main aims (1) the prevention of industrial accidents and occupational diseases, and (2) the compensation, rehabilitation and promotion of vocational retraining for persons injured at work.[1] The BGs insure employees against the consequences of occupational accidents, accidents on the way to and from work and occupational diseases. Because the employer has a statutory obligation to insure its employees against accidents with one of the BGs, it will not be held liable for the personal injury of its employees injured by accident at work unless it has caused such injury intentionally.

1 See European Commission, *Labour Inspection (Health and Safety) in the European Union, A Short Guide* (1999) accessible at: www.eurofound.europa.eu/emire/GERMANY/LABOURINSPECTORATE-DE.htm.

14.418 Both the Labour Inspectorate and the Insurance Associations have similar powers at their disposal, but the technical inspectors of the Insurance Associations are regarded as more of an advisory body, whilst the labour inspectors of the Länder are often referred to as the 'police', because they can ultimately shut down production. The Labour Inspectorates may issue admonitions, warnings, inspection reports, orders and prosecutions. Orders have a legally binding effect and, in case of non-compliance, can be enforced by an action for specific performance under the Administrative Procedure Acts of the Länder. Orders issued by the technical inspectors of the Insurance Associations are generally based on Part VII of the Social Code (SGB) and accident prevention regulation (BGV A1) and are also legally binding. Non-compliance with these orders can lead to administrative action resulting in sanctions.

14.419 The Insurance Associations are also empowered to impose administrative sanctions on the employer or employee for breaches of an accident prevention regulation intentionally or negligently[1] or committal of an administrative offence under s 209 of the Social Code (SGB) Part VII. A fine for such offence can be up to €10,000. In order for action to be taken against a company for this offence, it would be necessary to identify a legal representative to prosecute. If the legal representative is found guilty, the company can be fined up to €1 million (if there is evidence of intention) or €500,000 (in the case of negligence).[2]

1 State Insurance Act (*Reichsversicherungsordnung*, RVO) of 19 July 1911, s 710.
2 See David Bergman et al, above.

14.420 Both organisations are encouraged to advise the employers before prosecuting.[1] There has been an attempt to coordinate the activities of both enforcing bodies in Germany.[2] In practice, however, they tend to work independently of each other.

1 K Koch and N Salter, 'The Health and Safety System in the Federal Republic of Germany' (1999) 30(1) Industrial Relations Journal 61 at 67.
2 See R Wank, 'Germany' in R Baldwin and T Daintith (eds), *Harmonization and Hazard: Regulating Health and Safety in the European Workplace* (1992) pp 49–75 at p 68.

14.421 In cases of death at work, an investigation is carried out by the police, who are empowered, in this respect, by the Code of Criminal Procedure, s 163. The results of the investigation are sent to the prosecutor's office. In urgent cases, however, the case can be sent directly to the court. The sentences, which are imposed by the court, are mainly fines and imprisonment. For negligent manslaughter, a person is punished with 'imprisonment for not more than five years or a fine' (s 222 of the StGB). For negligently causing bodily harm, the sentence is 'imprisonment for not more than three years or a fine' (s 229 of the StGB). The Penal Code also refers to an offence of 'bodily injury' and states that 'whoever physically maltreats or harms the health of another person, shall be punished with imprisonment for not more than five years or a fine' (s 223), and to an offence of 'serious bodily injury' under s 226. Section 226 states that:

'(1) If the injury results in the victim:
  a. losing his sight in one eye or in both eyes, his hearing, his speech or his ability to procreate;
  b. losing or permanently losing the ability to use an important member;
  c. being permanently and seriously disfigured or contracting a lingering illness, becoming paralysed, mentally ill or disabled, then the penalty shall be imprisonment from one to ten years.

(2) If the offender intentionally or knowingly causes one of the results indicated in subsection (1), then the penalty shall be imprisonment for not less than three years.

(3) In less serious cases under subsection (1) the penalty shall be imprisonment from six months to five years, in less serious cases under subsection (2), imprisonment from one year to ten years.'

14.422 Life imprisonment is imposed for murder and may be imposed for manslaughter. Cases of negligent manslaughter are dealt with by the State Attorney. Crimes of manslaughter carry a mandatory sentence of imprisonment.[1] Only a natural person can be found guilty of manslaughter.

1 AA Aronowitz, *Germany, Report prepared for The World Factbook of Criminal Justice Systems* under Bureau of Justice Statistics grant no 90-BJ-CX-0002 (1993), available at www.bjs.gov/content/pub/ascii/WFBCJGER.TXT.

14.423 The German legislature established a working group in early 1998, which was given the task to review and improve the current situation by strengthening the role of criminal law with regard to corporate entities.[1] However, no changes have been made by the German legislature to the current state of law, which does not attribute crimes, such as manslaughter, to corporations. This traditional view is also supported by the courts.

1 Wagner, at note 2 to **14.370** above.

Finland

Introduction

14.424 Finland is a republic with a strongly centralised government. The country is divided into 12 provinces, which in turn are divided into 248 police districts, each of which generally comprises one or two municipalities. The criminal law is contained in the Criminal Code 1889 and separate statutes such as the Young Offenders Act 1939, the Narcotics Act 1972, the Traffic Act 1981 and the Conditional Sentences Act 1918. The law on criminal procedure is contained in the Code of Judicial Procedure 1734.

14.425 In Finland, there are no general distinctions or categories of crime. Rather, 'offence' distinctions are based on the expected punishment for the offence, or the 'penal latitude' defined by law.

14.426 Death at work is usually a result of the failure to comply with and follow health and safety requirements. The Act which governs protection of employees at the workplace against health and safety risks and hazards is the Occupational Safety and Health Act (738/2002). The Act sets out the employer's duties to ensure a safe working environment and emphasises the need for cooperation between the employer and employees. Provisions for enforcement of the health and safety laws are made by the Act on Occupational Safety and Health Enforcement and Cooperation on Workplace Safety and Health (44/2006). Since there is no provision in Finnish health and safety law for proceedings to be taken against a corporation, a prosecution may only be taken against an individual or individuals within a company.

14.427 According to the original Finnish criminal law, only a natural person could be held liable for the commission of a criminal offence. There were sanctions that could be imposed on corporations and the like, the most important of which were probably sanctions for damages caused by crimes committed in their operation. These sanctions, however, did not carry substantial criminal weight, rather they were considered as regulatory or administrative matters. The opportunities to develop a range of sanctions that could be imposed on crimes committed by the activities of 'artificial' persons without imposing direct criminal liability on them were, however, considered to be very limited.[1]

1 J Muhonen 'The proposed legislation for criminal liability of enterprises in Finland' (1995) 6 ICCLR 1 at 3–5.

14.428 Since then, however, the situation has changed. As a result of Act 743/1995, a new Ch 9 (amended by Act 61/2003) was introduced into the Finnish Penal Code (39/1889). A corporation may be held to be criminally liable for any criminal offence defined by Finnish criminal law.[1]

1 Chapter 9 of the Penal Code.

Law on corporate criminal liability

14.429 According to the Penal Code, Ch 9, s 2 (introduced by 61/2003), a corporation may be held liable and sentenced to a corporate fine if an offender has been an accomplice to an offence or allowed the commission of the offence, or if the care and diligence necessary for the prevention of the offence has not been observed. The offender is the person who acted on behalf or for the benefit of the corporation, and belongs to its management or is in a service or employment relationship with it or has acted on assignment by a representative of the corporation. Accordingly, the corporation may be liable on the basis of accessory conduct by the management of the corporation, where conduct may be active or passive. The corporation may also be found liable on the basis of negligence. (See Allens Arthur Robinson, as above). Section 9(2) also allows for a corporate fine to be imposed, even if the offender cannot be identified.

14.430 Additionally, s 3 (introduced by 743/1995) of Chapter 9 of the Penal Code provides for specific connection between offenders and the corporation. The offence has to be committed as part of the operation of the corporation. The offence has to be committed on its behalf or for its benefit. The offender should belong to its management or be in a service or employment relationship with it or have acted on assignment by a representative of the corporation.[1] Accordingly, the corporation must benefit from the actions of the natural person, who can be a manager or indeed any person who is employed by the corporation. It follows, therefore, that the actions of any employee of a corporation could result in a corporation being sanctioned.[2]

1 Penal Code, Ch 9, s 3 – connection between offender and corporation.
2 See Allens Arthur Robinson, as above.

14.431 According to Allens Arthur Robinson, there is an apparent tension between the concept of 'anonymous guilt' as created by s 9(2) and the requirements of s 3. It is not clear how, in the absence of an identified offender, one would satisfy the requirements of s 3. It is unclear whether s 9(2) should be taken to override s 3 to the extent of any inconsistency or whether it must still be proven that the corporation benefited in fact from the offence.[1]

1 See Allens Arthur Robinson, as above.

14.432 The above provisions of the Penal Code relating to corporate criminal liability do not exclude the liability which may be imposed on natural persons for criminal offences, such as manslaughter. The Occupational Safety and Health Act (738/2002) states that a punishment for crime against occupational safety and health is provided by Ch 47, Section 1 of the Penal Code (39/1889). Section 8 (578/1995) of the Penal Code refers to negligent homicide and states that 'a person who through negligence causes the death of another shall be sentenced for negligent homicide to a fine or to imprisonment for at most two years'. Section 9 relates to grossly negligent homicide and states that 'if in the negligent homicide the death of another is caused through gross negligence and the offence is aggravated also when assessed as a whole, the offender shall be sentenced for grossly negligent homicide to imprisonment for at least for months and at most six years'. Sections 10 and 11 refer to negligent bodily injury and grossly negligent bodily injury respectively.

Enforcement

14.433 The Occupational Safety and Health Authorities, supervised by the Ministry of Social Affairs and Health, deal with the practical enforcement of occupational safety and health. In general, in cases of wilful intent or negligence and normally following a work related accident, an inspector may draw up a report for the public prosecutor who will initiate an investigation by the police, will have the necessary powers to ensure that compliance is achieved, and will sanction the employer for breach of health and safety provisions. Any prosecution which follows may lead to an offender being fined or sent to prison.

14.434 An employer or person who intentionally or through carelessness fails to provide adequate inspections or safety and health planning, as prescribed by the Occupational Safety and Health Act (738/2002), will be issued with a fine, unless a more severe punishment is prescribed elsewhere in the law.[1] A person shall also be sentenced for violation of occupational safety and health law if they, intentionally or through carelessness, remove or ruin a device, instruction or warning intended to avoid the risk of accident, or fail to fulfil obligations relating to workers wearing adequate identification on shared sites.[2]

Punishment for a crime against occupational safety and health is provided for in Ch 47, s 1 of the Penal Code (39/1889) and includes a fine or imprisonment for a term of not more than one year.

1 Occupational Safety and Health Act (738/2002), s 63(1).
2 Occupational Safety and Health Act (738/2002), s 63(1).

14.435 Under the Finnish Penal Code the corporate fine is imposed as a lump sum. The corporate fine shall be at least €850 and at most €850,000 (Ch 9, s 5). In determining the amount of the corporate fine, the court takes into account: the nature and extent of the corporate neglect and the participation of the management in the offence; the status of the offender as a member of the organs of the corporation; the seriousness of the offence committed in the operations of the corporation and the extent of the criminal activity; the other consequences of the offence to the corporation; the measures by the corporation to prevent new offences, to prevent or remedy the effects of the offence or to further the investigation of the neglect or offence; and, where a member of the management of the corporation is sentenced to a punishment, the size of the corporation and the share of the corporation held by the offender, as well as the personal liability of the offender for the commitments of the corporation (Ch 9, s 6(1)).

14.436 When evaluating the significance of the neglect and the participation of the management, the court must take into account: the nature and seriousness of the offence; the status of the offender as a member of the organs of the corporation; whether the violation of the obligations of the corporation manifests heedlessness of the law or the orders of the authorities; as well as the bases for sentencing provided elsewhere in law. When evaluating the financial standing of the corporation, the court must take into account the size of the corporation and its solvency, as well as the earnings and the other essential indicators of the financial standing of the corporation (see Ch 9, ss 6(2)–(3)).

14.437 The public prosecutor may waive the bringing of charges against a corporation, if the corporate neglect or participation of the management are of minor significance, or if only minor damage or danger has been caused by the offence committed in the operations of the corporation and the corporation has voluntarily taken the necessary measures to prevent new offences (Ch 9, s 7).

14.438 The bringing of charges may be waived also if the member of the management of the corporation has already been sentenced to a punishment and it is to be anticipated that, for this reason, the corporation will not be sentenced to a corporate fine.[1]

1 Penal Code, Ch 9, s 7.

14.439 If a corporation is at the same time to be sentenced for two or more offences, multiple fines may be imposed, except for two offences, one of which was committed after a corporate fine had already been imposed. If charges are brought against a corporation which has been previously sentenced to a corporate fine, for an offence committed before the said sentence was passed, a joint corporate fine will not be imposed, but the prior corporate fine will be taken into account during sentencing.[1]

1 Penal Code, Ch 9, s 8.

Sweden

Introduction

14.440 Swedish law draws on Germanic, Roman, and Anglo-American law, and is neither codified to the same extent as France and other countries influenced by the Napoleonic Code, nor as dependent on judicial practice and precedents as in the UK, Australia, Canada or the US. Legislative and judicial institutions include the Swedish Parliament (Riksdag), the Supreme Court, the Supreme Administrative Court, the Labour Court, Commissions of Inquiry, the Law Council, District Courts and Courts of Appeal, the Chief Public Prosecutor, the Bar Association, and ombudsmen who oversee the application of laws with particular attention to abuses of authority.

14.441 Much of Swedish criminal law is based on legislation, while case law plays a smaller, albeit, an important role. The first Penal Code in Sweden came into force in 1734. This Penal Code was replaced in 1864.[1] The current Swedish Penal Code was adopted in 1962 and entered into force on 1 January 1965. It governs the provisions for most of the acts that constitute crimes in Sweden. It also contains general provisions on all crimes, the sanctions for crimes and the applicability of Swedish law.

1 OOH Wilkstrom and L Dolmen, *Sweden, Report prepared for The World Factbook of Criminal Justice Systems* under Bureau of Justice Statistics grant no 90-BJ-CX-0002 (1993).

14.442 The Swedish Penal Code does not differentiate between crimes and infractions. The classification of crime in the official crime statistics is based on the definitions of crimes given in the Penal Code. However, the main groups of crimes are divided into subcategories. These divisions are not systematic but are guided by general principles. The subdivisions have developed over a long period and have been determined from a pragmatic point of view.[1]

1 Swedish Penal Code 1962, as amended in 1999, English translation by Norman Bishop, Ds 1999:36.

14.443 A corporation cannot be held liable for a criminal offence, such as manslaughter, under Swedish criminal law. Liability for death at work may only be imposed on natural persons under the Swedish Penal Code.

14.444 The employer as a legal entity has duties for health and safety of employees under the Work Environment Act 1977. This Act, together with the Work Environment Ordinance and the Provisions,[1] set out the health and safety law in Sweden. However, in practice, employers' duties are placed on the most senior managers within the company. The duties can be delegated to others within the company, subject to a number of conditions established by case law.

1 These instruments are similar to the UK regulations and are made under the Work Environment Act 1977.

Law on corporate criminal liability

14.445 Under the Swedish health and safety legislation, the corporation may be liable for the infringement of health and safety provisions. These infringements are defined in work environment legislation, for example, where employers fail to take the necessary precautions to prevent their employees from being exposed to health hazards or accident risks.[1]

1 Work Environment Act 1977 no 1160 of 1977, Ch 3, s 2, as amended by Act no 585 of 2002.

14.446 The Swedish Work Environment Act 1977 and its subordinate legislation sets out those situations where the liability will arise and provides various sanctions such as penalties, fines and sometimes imprisonment. Liability arises and prosecution takes place only when a person intentionally or negligently fails to comply with an injunction or a prohibition notice issued by the Labour Inspectorate. These provisions are similar to the liability which arises under the UK's Health and Safety at Work etc Act 1974 for non-compliance with the requirement or prohibition imposed by an improvement notice or a prohibition notice issued by the Health and Safety Executive.[1] Only natural persons can be prosecuted for these infringements.

1 See Health and Safety at Work etc Act 1974, s 33.

14.447 In contrast with British legislation, the Work Environment Act 1977 does not create an offence or impose liability for contravening the general duties to take all necessary measures to safeguard employees from ill-health or injury imposed on the employer. Rather, the Swedish Act penalises the defendant for non-compliance with well-defined and prescribed provisions.[1]

1 V Howes and FB Wright, *Health and Safety Law and Environmental Law in Sweden,* report prepared for TXU plc (1999).

14.448 In cases where the infringement of health and safety provisions results in an accident from which a death results, or there are serious injuries or others are endangered, a natural person will be held criminally liable under the Penal Code (Ds 1999:36).[1] The relevant sections are contained in Part 2, Chapter 3 of the Code.

1 Ds 1999:36. The Swedish Penal Code, as amended, was adopted in 1962 and entered into force in 1965.

14.449 Section 7 of Part 2, Chapter 3 of the Penal Code provides that a person who, through carelessness, causes the death of another will be sentenced for *causing another's death* to imprisonment for up to two years or, if the crime is petty, to a fine. If the crime is gross, a term of imprisonment will be imposed of between six months and six years.

14.450 Section 8 of Part 2, Chapter 3 of the Penal Code states that a person who, through carelessness, causes another to suffer bodily injury or illness not of a petty nature will be sentenced for *causing bodily injury or illness* to a fine or imprisonment for up to six months. If the crime is gross, a term of imprisonment for up to four years will be imposed.

14.451 Section 9 of Part 2, Chapter 3 of the Penal Code provides that a person who, through gross carelessness, exposes another to mortal danger or danger of severe bodily injury or serious illness will be sentenced for *creating danger to another* to a fine or for a term of imprisonment for up to two years. And finally, s 10 states that, where a crime referred to in ss 7–9 has been committed by a person with intent or by carelessly neglecting his duty under the Work

Environment Act 1977[1] to prevent sickness or accidents, the punishment will be for an *environmental offence* and as provided for in the said provisions.

1 Law 1977:1160.

14.452 According to Swedish law, only natural persons can commit crimes. Under the Penal Code a corporate fine can be imposed on a company owner (an entrepreneur), for a crime committed in the exercise of business activities, at the instance of a public prosecutor if the crime has entailed gross disregard for the special obligations associated with the business activities or is otherwise of a serious kind, and the owner of the company has not done what could reasonably be required of him or her to ensure the prevention of the crime. These provisions, however, will not apply if the crime was directed against the owner or if it would otherwise be manifestly unreasonable to impose a corporate fine.[1]

1 Ds 1999:36, Ch 36, s 7 (Law 1986:1007).

14.453 The company owner may also be vicariously liable for the crime committed by a person working in the company, even if that person does not have a leading position. The conditions are that the crime has signified a serious disregard of the special responsibilities which are connected with the business or in any other way is of a serious kind and that the company owner has not taken reasonable measures to prevent it.

14.454 A corporate fine consists of not less than ten thousand Swedish crowns and not more than three million Swedish crowns.[1] In determining the amount of a corporate fine, special consideration is given to the nature and extent of the crime and to its relation to the business activity.[2]

1 Ds 1999:36, Ch 36, s 8 (Law 1986:118).
2 Ds 1999:36, Ch 36, s 9 (Law 1986:118).

Enforcement

14.455 Supervision of safety and health at work falls under the remit of the Work Environment Inspectorate (Arbetsmiljöinspektionen) which is part of the Swedish Work Environment Authority (Arbetsmiljöverket).[1] The Inspectorate was formed in 2001 following a merger of the Labour Inspectorate and the National Board. The Inspectorate has wide-ranging powers to order or prohibit certain measures and is empowered to visit a workplace at any time. After an inspection a notice will be issued, which will set out steps that need to be taken to improve safety and health measures. If these are not complied with, an injunction or prohibition will be issued. Anyone contravening these measures can be punished by fine or by penal sanction.

1 See www.av.se.

14.456 The Swedish Work Environment Act 1977, as amended, and its subordinate legislation provide for various sanctions including penalties, fines and, in some circumstances, imprisonment. However, it is only when there has been 'intentional' or 'negligent' non-compliance with an injunction or prohibition that a prosecution against one or more of the employer's representatives can take place.[1] Companies cannot be prosecuted under Swedish law – and, as a result, only a natural person representing the employer, including directors of the companies, may be required to pay an unlimited fine (linked to the income of the person) or to serve a term of imprisonment of up to one year for deliberate or negligent infringements of the Act or subordinate legislation.

1 See David Bergman et al, as above.

14.457 In cases of death at work, an investigation will be carried out by Sweden's Public Prosecution Office and this may result in a natural person being charged with manslaughter under the Swedish Penal Code. A prosecution may be taken against an employer where there is evidence of criminal negligence. Proceedings will be instigated by the Crown Prosecutor following an investigation by the police. The Labour Inspectorate may be asked to give technical support during this investigation. Breaches of certain legal requirements will automatically be referred to the Crown Prosecutor; for example, if pressure vessels have not been tested, or where minors have been employed in certain work or certain chemicals have been used without a permit.[1]

1 Senior Labour Inspectors Committee, European Commission, *Labour Inspection (Health and Safety) in the European Union, A Short Guide* (1999).

14.458 The sanctions, ie fines and imprisonment of up to six years, provided by the Swedish Penal Code can only be levied against physical persons, such as directors or managers of the company. It is not surprising, therefore, that in complex cases involving large corporations, when it is difficult to define a person who is responsible for the crime, the prosecution may ultimately be fruitless. A good example of this is the 2002 case where more than 50 patient deaths worldwide were caused by the dialyser filters manufactured by Baxter's plant in Ronneby, Sweden.

Norway

Introduction

14.459 Norway is a unified state in which governmental power is divided between the judiciary, executive and legislative branches, each of which is mutually independent. The executive branch is made up of the King and members of the Cabinet. Legislative power is vested in the national parliament (*Stortinget*). Although the parliament is unicameral, it is divided into two chambers (the *Lagting* and *Odelsting*) for the purpose of passing legislation. Both chambers must approve a Bill before it can be passed.

14.460 For administrative and political purposes, the country is divided into 19 counties (*fylker*) and approximately 450 municipalities (*kommuner*). The Norwegian system is similar to the legal systems of the other Nordic countries, particularly those of Denmark and Sweden. Similarly to these countries, Norwegian courts do not attach the same weight to judicial precedents as the members of the judiciary in common law countries.

14.461 Criminal law is mostly to be found in the Penal Code, which groups criminal offences into felonies (*forbrytelser*) and misdemeanours (*forseelser*). Felonies are, with some exceptions, offences which carry a maximum penalty exceeding three months' imprisonment. The majority of felonies are defined and listed in the Penal Code, Pt 2 and include murder and manslaughter. Misdemeanours are generally minor offences carrying a maximum penalty of three months' imprisonment. Examples of these types of offences are to be found in the Penal Code, Pt 3.

14.462 The first comprehensive Penal Code was enacted in 1842. This was replaced by the General Civil Penal Code of 22 May 1902, which is still in force, although it has undergone several amendments.[1] The concept of criminal liability for legal persons as opposed to natural persons was introduced into the General Civil Penal Code[2] in 1991 by including Ch 3a – Criminal liability of enterprises.

1 See L Bygrave, *Norway, Report prepared for The World Factbook of Criminal Justice Systems* under Bureau of Justice Statistics grant no 90-BJ-CX-0002 (1993).
2 Act no 10 of 22 May 1902.

14.463 Under the Norwegian Penal Code, s 48(a) and (b), a legal person can be held responsible for the contravention of any penal provision of the Code. However, it is discretionary whether the legal personality should be prosecuted or not. Where criminal charges are brought against legal persons, fines are invariably imposed. Legal persons can also be held liable for acts committed by a natural person abroad; however, this depends on whether or not Norway has jurisdiction over the criminal acts of the natural person.[1]

1 Shloer and Filipenko, at note 2 to **14.407** above.

Law on corporate criminal liability

14.464 In the context of a workplace, the main legislative provisions, which govern health and safety protection at work, are set out in Act no 62 of 17 June 2005 relating to Working Environment, Working Hours and Employment Protection, etc (the Working Environment Act).[1] By virtue of this Act a natural person can be held liable for the commission of a health and safety offence. The Act defines an 'employer' as any natural person who has engaged employees to perform work in his service.[2] Chapter 19 concerns the liability of company owners, employers and their representatives. It states that any proprietor of an establishment, employer or person managing an establishment in the employer's stead who wilfully or negligently commits a breach of the provisions or orders contained in the Act will be liable to a fine, imprisonment for up to three months, or both.

1 As amended by Act no 80 of 14 December 2012.
2 Working Environment Act, Ch 1, s1–8(2).

14.465 Under the Working Environment Act, in the event of particularly aggravating circumstances the penalty may be up to two years' imprisonment. When determining whether such circumstances exist, particular importance is attached to whether the offence involved or could have involved a serious hazard to life or health, and whether it was committed or allowed to continue notwithstanding orders or requests from public authorities, decisions adopted by the working environment committee, or notwithstanding demands or requests from safety representatives or from safety and health personnel. In the event of infringements that involved or could have involved a serious hazard to life or health, any proprietor of an establishment, employer or person managing an establishment in the employer's stead will be liable to penalty under Chapter 19, unless the person concerned has acted in a fully satisfactory manner according to his or her duties under the Act.

14.466 Employees can also be held liable under the Working Environment Act. An employee who negligently infringes the provisions or orders contained in or issued pursuant to the Act may be liable to a fine. Those found to be contributorily negligent will be subject to the same penalty. If the infringement is committed wilfully or through gross negligence, the penalty may be a fine, up to three months' imprisonment, or both. In the event of particularly aggravating circumstances, imprisonment for up to one year may be imposed. When determining whether such circumstances exist, particular importance is attached to whether the offence was contrary to special directives relating to work or safety and whether the employee understood or should have understood that the offence could have seriously endangered the life and health of others.[1]

1 Working Environment Act, s 19–2.

14.467 With regard to corporate criminal liability, it is stated in the Working Environment Act, s 19–3 that the criminal liability of corporations is regulated by the provisions of the General Civil Penal Code, ss 48a and 48b.

14.468 Section 48a of the Penal Code states that a corporation may be liable if a natural person who has acted on its behalf has contravened a provision of the Norway Penal Code. The liability on the corporation also applies even where no individual person is punishable for the contravention. The meaning of 'corporation' includes a company, society or other association, one-man enterprise, foundation, estate or public activity. The penalty imposed on the corporation is a fine. The corporation may also, by a court judgment, be deprived of the right to carry on business or may be prohibited from carrying it on in certain forms (s 48a).

14.469 The enterprise may also, by a court judgment, be deprived of the right to carry on business or may be prohibited from carrying it on in certain forms contrary to s 29 of the Penal Code. Section 29 of the Norway Penal Code states that, when it is so required in the public interest, any person who is found guilty of a criminal act that shows that the said person is unfit for (or may misuse) any position may be deprived of the position, or be deprived of the right to hold in future any position or to carry on any enterprise or activity. Loss of any right pursuant to this provision may be imposed in addition to or instead of another penalty, but may only be imposed as the only penalty if a minimum penalty of imprisonment for one year or more is not prescribed for the act.

14.470 There are also special provisions in the Norway Penal Code concerning offences against public safety.[1] In particular, any person who causes any fire, collapse, explosion, flood, maritime damage, railway accident or aircraft accident which may easily result in loss of human life or extensive destruction of another person's property, or who is an accessory thereto, will be liable to imprisonment for a term of not less than two years and not more than 21 years, but not less than five years if, as a result of the felony, any person dies or is seriously injured in body or health.[2] A person who tries to hinder the prevention or combating of any such accident will be liable to imprisonment for a term of not less than one year.[3] A person will be liable to imprisonment for a term not exceeding six years who brings about any such danger by omitting to perform any special duty incumbent on him or her, by unlawfully destroying, removing or damaging any object or guiding signal, by giving or setting a false signal, by placing any obstruction in a seaway, by interfering with the safe operation of a ship, railway, aircraft or any installations or constructions on the continental shelf, or by being accessory to any such conduct. If any such accident is caused, imprisonment for a term not exceeding 12 years will be imposed. An attempt shall be liable to the same penalty as a completed felony. If a person has committed any of the above-mentioned acts without being aware of the danger or negligently, he or she will be liable to fines or imprisonment for a term not exceeding one year. If any such fire, collapse, explosion, flood, maritime damage, railway accident or aircraft accident is caused by negligence, the offender will be liable to fines or imprisonment for a term not exceeding three years.

1 Norway Penal Code, Ch 14.

2 Norway Penal Code, s 148.
3 Norway Penal Code, s 149.

Enforcement

14.471 The Norwegian Labour Inspection Authority is a governmental agency under the Ministry of Labour and Government Administration, which has administrative, supervisory and information responsibilities in connection with the Working Environment Act. In dealing with enterprises that do not comply with the requirements of the Act, the Labour Inspection Authority may respond with orders to correct the situation within a given time limit,[1] coercive fines when the order is not complied with,[2] shutdown of operations, which may be done with immediate effect if the life and health of its employees are in imminent danger, or it may be imposed when corporations fail to comply with orders given. The Agency may report corporations to the police for serious breaches of the Act. A serious violation can result in fines, or, in the worst case, imprisonment.

1 Working Environment Act, s 18–6.
2 Working Environment Act, s 18–7.

14.472 Criminal acts committed by corporations are subject to public prosecution and are governed by the Norway Penal Code. In deciding whether a penalty should be imposed on an enterprise pursuant to s 48a, and in assessing the penalty in relation to the corporation, particular consideration is paid to: the preventive effect of the penalty; the seriousness of the offence; whether the corporation could, by guidelines, instruction, training, control or other measures, have prevented the offence; whether the offence has been committed in order to promote the interests of the corporation; whether the corporation has had or could have obtained any advantage by the offence; the corporation's economic capacity; and whether other sanctions have, as a consequence of the offence, been imposed on the corporation or on any person who has acted on its behalf, including whether a penalty has been imposed on any individual person.[1] A public prosecution will only be instituted when requested by an aggrieved person, unless the offence has resulted in someone's death, or the prosecution is required in the public interest.[2]

1 Norway Penal Code, s 48b.
2 Norway Penal Code, s 228.

Denmark

Introduction

14.473 The criminal law in Denmark naturally has very much in common with Norway, due to their geographical positions and common history. In general, the influence on the development of the criminal judicial system in Denmark has largely been Germanic compared, for example, with Sweden where the French tradition has prevailed.

14.474 Criminal offences are defined either in a special part of the Criminal Code or in separate statutes. The general conditions for imposing criminal penalties can be found in the general part of the Criminal Code, which also apply to separate statutes. The sanctions described in the general part of the Criminal Code are the same whether the criminal offence consists of a violation of the Criminal Code or of separate statutes.

14.475 The substantive Danish criminal law is monistic, meaning that violations of the law have never been divided into categories like felony/misdemeanour or crime/delicts. It does not mean, however, that major offences are treated in the same manner as petty offences in all respects.[1]

1 L Rawn, *Denmark, Report prepared for The World Factbook of Criminal Justice Systems* under Bureau of Justice Statistics grant no 90-BJ-CX-0002 (1993), available at www.bjs.gov/content/pub/ascii/WFBCJDEN.TXT.

14.476 There are three main types of punishment: ordinary imprisonment, lenient imprisonment and fines/day fines. In addition, in special cases, dangerous offenders may be sentenced to indeterminate periods of preventive detention. Terms of imprisonment may be meted out with determinate sentences from 30 days to 16 years or life, and lenient imprisonment from seven days to six months. Terms of imprisonment may be imposed in the form of suspended or non-suspended sentences. Since 1982 the rules governing probation and suspended sentences formed the basis of an experiment with community service orders, and from 1992 the scheme was made permanent.[1]

1 Rawn, at note 1 to **14.475** above.

14.477 Criminal law is to be found in the Danish Criminal Code of 1930, which has been amended on several occasions. The Danish Criminal Code is the primary, though by no means the exclusive, criminal law source. Case law plays an important role. By omitting the detailed description of offences, the Criminal Code 1930 allows the use of 'law by similarity', which provides wide scope for a judicial discretion.

14.478 Under Danish criminal law, a corporation can be held criminally liable, and legal mechanisms to impose criminal liability on corporations and other legal persons have been established in the Danish Criminal Code.[1]

1 Hefendehl, at note 1 to **14.411** above.

Law on corporate criminal liability

14.479 The law relating to health and safety at work provides for the liability of employers and employees. In 1975, the working environment rules were consolidated into a single Act, the Danish Working Environment Act, which applies to all work on the ground and, in particular, work performed for an employer. The central part of the Act is the extended safety and health concept, which means that all factors causing accidents, sickness and attrition must be taken into consideration in the prevention work. The current Working Environment Act No 1072 was adopted on 7 September 2010 (with subsequent amendments).

14.480 The Working Environment Act places the overall responsibility for safety on employers. Duties of employers are set out in Pt 4 of the Act. Section 15 states that 'it shall be the duty of the employer to ensure safe and healthy working conditions'. Special reference is made to various parts of the Act: (a) Pt 5 on the performance of the work; (b) Pt 6 on the design and fitting out of the workplace; (c) Pt 7 on technical equipment, etc; and (d) Pt 8 on substances and materials. Other sections relating to duties of employers include ss 15a, 16, 17(1) and (2), 18, 19, 20, etc. Contravention of these sections may result in a fine or imprisonment of up to one year (s 82(1)). The penalty may increase to imprisonment of up to two years if the contravention was committed intentionally or with gross negligence (s 82(2)). The Act refers to aggravating circumstances which should be taken into account when setting penalties (s 82(3)–(5)) for contravening the duties under the Act. The employer may be vicariously liable for actions of his employees and subjected to a fine, even if he has not acted intentionally of negligently.

14.481 Overall, rules made under the Working Environment Act may provide for penalties in the form of a fine or imprisonment of up to two years in respect of contravention of provisions in the rules and contravention of improvement or prohibition notices in pursuance of the rules. Furthermore, it may be laid down that an employer (but not a manager, etc) who contravenes such provisions and notices as mentioned above may be liable to a pay a fine, even if he has not acted intentionally or negligently. The liability to pay a fine is on condition that the violation can be ascribed to one or more persons associated with the enterprise or the enterprise per se (see s 84).

14.482 Criminal liability may be imposed on limited liability companies, etc (legal persons) pursuant to the rules defined in Pt V of the Danish Criminal Code introduced by Act 474 of 12 June 1996.[1] It states that a legal entity may be sanctioned with a fine if such is provided by the law or respectively adopted procedures.[2] Section 27(1) states that 'criminal liability of a legal person is conditional upon a transgression having been committed within the establishment of this person by one or more persons connected to this legal person or by the legal person himself'. The provisions concerning criminal liability of the legal entities are applicable to any entity, including joint stock companies, cooperative societies, partnerships, associations, foundations, estate compounds, municipalities and public administration authorities, unless expressly specified otherwise.[3] These provisions may also apply to privately owned enterprises.

1 Direction no 648 of 12 August 1997, as amended by Act no 403 of 26 June 1998, Act no 473 of 1 July 1998 and Act no 141 of 17 March 1999.
2 Danish Criminal Code, s 25.
3 Danish Criminal Code, s 26.

14.483 The scope of criminal liability of a legal entity depends on the criminal act perpetrated by one or more persons connected to the legal entity in question. However, unlike the identification doctrine which governed corporate manslaughter under English common law, s 27(1) is not restricted to management, senior personnel or even those who are formally employed by the company, and it encompasses agents.[1] Nielsen observes:

> 'The question of guilt is typically satisfied by an employee having negligently broken the rules. This is enough. Whether or not any blame attached to management is irrelevant under Danish law. The company will also be found guilty in cases where management has been very active in ensuring observance of the law. The fact that it is irrelevant whether management has been active or passive is probably the question attracting most debate. But, if in order to obtain judgment against the company, the prosecutor has to prove the manager is personally liable, he may as well bring the charges against him personally. If managerial negligence must be proved, company liability will lose much of its meaning.'[2]

1 Allens Arthur Robinson, *'Corporate Culture' as a Basis for the Criminal Liability of Corporations*, Report prepared for the United Nations Special Representative of the Secretary-General on Human Rights and Business, February 2008.
2 Gorm Neilsen, 'Criminal Liability of Collective Entities – the Danish Model' in Eser, Heine and Huber (eds), *Criminal Responsibility of Legal and Collective Entities* (1999) 189, n 6, cited in Allens Arthur Robinson (above).

Enforcement

14.484 The enforcement of health and safety law under the Working Environment Act is carried out under the auspices of the Ministry of Employment, but on the day-to-day level it is administered by the Danish Working Environment Authority. Its responsibilities include providing guidance in health and safety matters to the employers' associations, trade unions and public authorities and, not least, ensuring enforcement of the Act through, for example, the inspection of individual workplaces (s 72). On completion of such an inspection, its inspectors may give instructions requiring certain measures to be taken to improve safety, or issue a prohibition notice banning the continued use of highly dangerous procedures, machinery, substances and materials, etc.[1]

1 See EMIRE, Denmark, 'Arbejdsmiljo, Work Environment'.

14.485 Additionally, the Danish Working Environment Authority is empowered to penalise companies which do not comply with the working environment rules. As regards clear violations of the substantive rules of the Working Environment Act 2010, the Danish Working Environment Authority has the power to issue administrative fines. In cases of extreme danger, the Authority may also order the work to be suspended (s 77).

14.486 When death at work occurs, the case is investigated by the police and the offender is prosecuted by the Director of Public Prosecutions. The structure of prosecutions in Denmark is hierarchical. The political responsibility rests with the Minister of Justice, but in practice the Director of Public Prosecutions enjoys a high degree of independence. The Director of Public Prosecutions exercises instructive powers towards the lower prosecution instances and conducts criminal cases before the Supreme Court. Otherwise, serious criminal cases are handled by district public prosecutors and less serious cases by the chief constables.[1]

1 Rawn, at note 1 to **14.475** above.

14.487 The penalties are described in the general part of the Criminal Code. They are: fines; lenient prison (seven to 30 days); prison (one month to 16 years or imprisonment for life); and community service orders. Prison sentences and fines may be conditionally suspended. The maximum penalty is imprisonment for a lifetime, which is prescribed for murder and manslaughter. Each statute in the special part of the Criminal Code sets out the kind of penalty applicable for the crime and the upper range of its duration. This gives the court the opportunity to choose the actual penalty freely within the ranges. As in other jurisdictions, sentencing in Denmark is ultimately determined by the court.

CONCLUSION

14.488 Criminal liability of corporations has been high on the legal agenda for three decades, when corporations began to play a significant role in social and economic life.[1] In general, the law on corporate criminal liability seems to have developed along three main lines: the doctrine of identification (of the directing mind of a corporation), or direct liability; vicarious liability; and, finally, the fault of the corporation as one collective body.

1 See eg LH Leigh *The Criminal Liability of Corporations* (1977).

14.489 There has been strong opposition to the imposition of corporate liability in different legal systems. The leading legal opponent was the continental system, which had forcefully rejected this principle over many decades and was unsupportive of the imposition of such general liability, maintaining that the appropriate point of principle on this issue is that of 'societas delinquere non potest'.[1]

1 E Lederman 'Models for Imposing Corporate Criminal Liability: From Adaptation and Imitation Toward Aggregation and the Search for Self-Identity' (2000) 4(1) Buffalo Crim LR 641.

14.490 The situation has, however, changed, largely because of the new socio-, political and economic realities. Legal entities have acquired great power and control in almost all spheres of industry and commerce. Corporations have their own 'minds', and the capacity to make important decisions and to commit crimes. Policy and law-makers began to recognise that changes in law were needed to reflect the reality. As a result of this, the approach to corporate criminal liability worldwide has undergone significant transformation.

14.491 The law governing corporate criminal liability at federal level (and to some extent at State level) in the US is based on a model of vicarious liability. This law, which renders corporations liable for the criminal actions of their employees, originated in the civil law as a mechanism for holding employees liable for the torts of their employees.

14.492 It has been established in a number of cases heard by both the federal and state courts that corporations can be criminally liable for criminal acts committed by individuals on their behalf. Even at the beginning of the 20th century, some US courts began to expand the concept of corporate criminal liability to include mens rea offences,[1] a move which was confirmed by the US Supreme Court in *New York Central & Hudson River Railroad Company v United States*.[2] This case followed the passing of the Elkins Act 1903 through Congress. This Act stated that the acts and omissions of an officer acting within the scope of his employment were considered to be those of the corporation, thus promulgating the concept of vicarious liability.[3]

1 M Wagner 'Corporate Criminal Liability: National and International Responses', background paper for the International Society for the Reform of Criminal Law 13th International Conference, Commercial and Financial Fraud: A Comparative Perspective, Malta, 8–12 July 1999.
2 212 US 481 (1909).
3 G Stessens 'Corporate Criminal Liability: A Comparative Perspective' (1994) 43 ICLQ at 496–497.

14.493 In US law at federal level, we saw that a corporation can be liable for criminal offences committed by its employee if the employee commits the crime: (1) within the scope of his or her employment; (2) with the intention of benefiting the corporation; and (3) providing that the criminal acts were authorised, tolerated, or ratified by corporate management.[1] There are now only two situations in which criminal liability cannot be imposed on a corporation at federal level, namely where the offence is one for which punishment by fine is unavailable or inappropriate (remembering, of course, that fines are the principal means available for punishing a corporation), and where the crime, by its nature, cannot be committed by a corporation (for example, rape).[2]

1 See B Lowell 'Vicarious Criminal Liability of Corporations for the Acts of Their Employees and Agents' (1995) 41 Loyola LR 279.
2 See Hefendehl, at note 1 to **14.411** above, at 290.

14.494 At state level, the US position governing corporate criminal liability has gained some consistency after state legislatures began adopting positions which reflected that recommended by the American Law Institute in their Model Penal Code. In each of the five States we considered (New York, California, Kentucky, Pennsylvania and Texas), we saw that their Model Penal Codes each provided for corporate criminal liability, including liability for homicide. We also saw a number of cases in which corporations were successfully prosecuted for manslaughter.

14.495 Canada is a Commonwealth country and, unsurprisingly, its position in relation to corporate manslaughter is similar to that of the UK. In some respects, however, Canada has a federal government with a common law system, so there are also elements in common with the approach adopted by the US.

14.496 As with the UK the Canadian courts have wrestled with two main mechanisms for attributing criminal liability to a corporation, namely vicarious liability and the doctrine of identification. Canadian case law dealing with the liability of corporations for criminal actions has developed very much in line with English decisions, although the Canadian courts have adopted a slightly modified version of the doctrine of identification doctrine from that implemented by English courts. Under the Canadian approach 'The [doctrine of identification] merges the board of directors, the managing director, the superintendent, the manager or anyone else delegated by the board of directors to whom is delegated the governing executive authority of the corporation, and the conduct of any of the merged entities is thereby attributed to the corporation'.[1] The importance of this approach is that it recognises that a corporation may have more than one 'directing mind'.

1 A-M Boisvert 'Corporate Criminal Liability: A Discussion Paper' (August 1999).

14.497 This delegation theory has its limits, however.[1] Indeed, Iacobucci J held in *Rhone* (*The*) *v Peter AB Widener* (*The*) that:

> 'The key factor which distinguishes directing minds from normal employees is the capacity to exercise decision-making authority on matters of corporate policy, rather than merely to give effect such policy on an operational basis, whether at head office or across the sea.'[2]

1 See *Rhone* (*The*) *v Peter AB Widener* (*The*) [1993] 1 SCR 497.
2 [1993] 1 SCR 497.

14.498 This shows that, although a person in the company could be delegated to carry out many responsibilities, it could well be that their position was not such as to allow them to formulate corporate policies and thus they could not represent the directing mind of the company. In order to determine the directing mind of the company, the courts will seek the person who had direct control over the corporate activity in which the criminal act was committed.

14.499 Some European countries have also recently undergone significant changes in the law on corporate criminal liability. In 1988, the Council of Europe recommended to member states that they should endorse an approach allowing the imposition of criminal liability on legal bodies.[1] The Dutch courts started to adopt this direction in the mid-1970s in anticipation of the Council of Europe Recommendation.[2] The provision regarding the criminal liability of corporation can be found in the Dutch Criminal Code, art 51, which states that offences can be committed by both human beings and corporations. The Dutch approach is very far-reaching, in that it recognises that corporations act as collectives and thus they could be held liable, even if no particular individual was identifiable for a crime, but rather the corporation was at fault for failing to adopt adequate safety systems which resulted in death.

1 Committee of Ministers, Council of Europe, *Recommendation on Liability of Enterprises Having Legal Personality for Offences Committed in the Exercise of their Activities*, Recommendation no R(88) 18, October 1988.
2 See Field and Jörg, at note 2 to **14.394** above.

14.500 France, on the other hand, changed its criminal law on this issue following the recommendation, and in the early 1990s erased from its Penal Code the prohibition against rendering corporation (*personnes morales*) criminally liable.[1] As a result, strict liability offences were enforced vicariously against corporations. The reformed Penal Code states that legal entities may be liable if the offence provision specifically declares that they should be and if an employee or officer is shown to have acted on the corporation's behalf. The actions of 'rogue' employees would not be imputed to the corporate entity.

1 See Lederman, at note 1 to **14.489** above.

14.501 In Germany, the imposition of criminal liability on legal bodies was 'unthinkable' until very recently. Much attention was focused on the question of whether corporations should be held responsible under the civil or public law. There is still no legal provision in Germany which allows corporations to be held criminally liable. Corporations can only be liable for administrative offences (*Ordnungswidrigkeiten*). Roland Hefendehl has observed:

> 'Administrative remedies, which consist primarily of the imposition of fines, do not aim to establish personal responsibility or guilt but to provide preventive – as well as repressive – means for controlling and regulating risks emerging from economic processes and decisions. The focus does not lie on the personal but on the instrumental aspect of a particular wrongdoing. Thus, the difference between the law of administrative offences and the criminal law is not merely quantitative. If civil courts or economic control boards were limited to imposing future-oriented sanctions but could not sanction past acts, civil courts and economic control boards could not effectively control businesses. The purpose of administrative penalties is to fill in this gap.'[1]

1 Hefendehl, at note 1 to **14.411** above.

14.502 In Italy, a new statute was enacted in 2001 to deal with the problem created by the Italian Constitution which did not allow imposing criminal liability on legal persons. The statute classified criminal offences as administrative ones, which was misleading since, in practice and procedurally, they were considered as criminal. There are two models of corporate criminal liability created by the Italian statute: where the offence is committed by the head of the corporation (that is, direct liability), and where a corporate body is negligent in not considering and not averting the risk of the possibility of the offence (that is, failure of the corporation). The imposition of criminal liability on corporations is a key development in the Italian law, since formerly it refused to accept this type of liability, alongside countries such as Germany.

14.503 In Finland, corporations can be criminally liable under the Finnish Penal Code[1] on two grounds: for so-called 'anonymous guilt' created by s 2(2) of Ch 9 of the Code; and for the offence committed by a representative of the corporation on its behalf or for its benefit. This criminal liability is, therefore, based on two main prerequisites:[2] the liability of any person acting on behalf or for the benefit of the corporation; and the organisational failure of the corporation. The general prerequisite for liability is that the management of the corporation has itself committed the crime or served as an accomplice thereto, the management has allowed a crime to be committed, or the diligence and caution required have not been observed in the operation of the enterprise. Another prerequisite for corporate criminal liability is that a corporate fine may be imposed, even if the offender cannot be identified or otherwise is not punished, which means that the offence was committed because of the failure of the corporation to put adequate safety systems in place.

1 Finnish Penal Code 743/1995.
2 Finnish Penal Code, Ch 9, s 2(1), Prerequisites for liability.

14.504 In Sweden, a corporation cannot be held liable for a criminal offence, such as manslaughter. Liability for death at work may only be imposed on natural persons under the Swedish Penal Code. In this regard, corporations can be penalised for administrative offences and sanctioned respectively.

14.505 In Denmark, criminal liability of legal entities is governed by the Danish Criminal Code[1] and defined in Ch 5 of the Code. The criminal liability of a legal entity in Denmark is based on liability of a person who is connected with this legal entity. Similarly to France and Finland, the criminal act must be committed on behalf of a legal entity by one or more persons connected with the legal entity in question. The Code is silent on the liability for the organisational fault when no-one can be identified for the commission of the offence.

1 Direction no 648 of 12 August 1997, as amended by Act no 403 of 26 June 1998, Act no 473 of 1 July 1998 and Act no 141 of 17 March 1999.

14.506 The principles which govern corporate criminal liability in Norway are very similar to those set out in the Finnish Penal Code. Under the Norwegian Penal Code, s 48(a) and (b), a legal person can be held responsible for contravention of any penal provision of the Code. A corporation may be liable if a natural person who has acted on its behalf has contravened a provision of the Norwegian Penal Code, and it also applies where no individual person is punishable for the offence in question.

14.507 We can thus see that all the above jurisdictions have adopted different approaches when dealing with the question of corporate liability for industrial manslaughter. Many have based liability on the models of direct liability and vicarious liability. Others, such as the Netherlands and Finland, have extended the scope of liability and allowed the possibility of holding corporations liable for the managerial fault without identifying any particular individual responsible for the crime. The practice shows that this model works and perhaps deserves a closer look by legal policy-makers in other countries.

Chapter 15

Dealing with the media

Robin Stacey, Assistant News Editor, The Times

Barney Monahan, Editorial Legal Counsel, News Group Newspapers Limited

A JOURNALIST'S POINT OF VIEW

15.1 Your secretary hauls you out of meeting with a worried look on her face. She takes you to one side and slips a phone in your hand. You recognise a desperate voice on the line, a site manager you may have met but can't quite place. There's been an accident somewhere within the company. Your first aid people are doing all they can but it's not looking good. At least one person has been gravely hurt and, worse still, there could be fatalities. As if you hadn't got the message already, you hear a siren closing in in the background.

15.2 It's every managing director's worst nightmare. It could involve machinery on a production line, or it could be a nasty road or rail accident where your company is transporting people or goods. It could involve employees, members of the public, or even, heaven forbid, customers. As the news sinks in, you think to yourself, 'at least it can't get any worse than this'. Actually it can, and your priority in the coming hours must be to build a firewall around this incident to minimise the impact on your company. But how?

15.3 Your immediate concern may be for the lives and limbs of those involved, but you quickly realise there is little that you can personally do right now on that score. Dealing with the dead and injured is now in the hands of the emergency services, and your managers on the spot will naturally fall into line behind them. Your company has been badly injured too, and your priority now is to attend to that. Unheard amidst the clamour and chaos of the accident, a silent starting pistol has triggered a race to establish precisely what happened. Your workers, the emergency services and eye witnesses will all have their part to play in this, but it is the media that will publish and broadcast their accounts to a curious world. Forget cash flows and profit margins for a while: your focus now must be on finding a way to defend your company against an unruly and sceptical media. There are going to be difficult hours and days ahead. If the incident is major, you cannot hope to win the media war in the first stages, but you can most certainly lose it.

15.4 Let's be clear – a death or multiple deaths for which your company may be directly or indirectly responsible is a full-blown PR disaster which will only be neutralised by a continuous chain of correct remedial actions. All of a sudden, your entire modus operandi is facing relentless scrutiny. All of yesterday's certainties about how you do business have gone for a burton; so has any rosy outlook about the future. What's worse, no-one in your team may have the foggiest idea what to do now. Your response will have to be made up 'on the hoof', and this in a chaotic environment where nobody knows which way is 'up'. So where do you turn for inspiration?

15.5 Luckily for you, dealing with a life-and-death crisis may be rare, but for the fire service it is second nature. Every incident attended by firefighters involves events which are, by definition, dangerous. One wrong decision could result in more loss of more property or more injuries or deaths. In confronting our PR crisis, let's take a leaf out of the fire service's operating manual. How do they go about tackling a major conflagration?

15.6 For any 999 shout, the first objective is to assess the nature and scale of the incident. How large is the fire and how well established is it? Is anyone still trapped? The reason for these questions is to establish how many firefighters and fire tenders will be required and whether any specialised machinery will be needed. You'll see later on how this equates to the 'PR fire' that you may be up against one day.

15.7 The second priority is containment. A fire in one building is bad enough but, before seeking to go on the offensive, fire crews will devote their energies to establishing the boundaries of the fire and cutting off its advance. Only once that is achieved can the real fight begin.

15.8 Next comes the attack on the fire itself. To an onlooker, the drama of a fire may be its tongues of flame crazily lashing the night sky, but firefighters do not aim their hoses at these. They tackle the flames from their base deep inside the blazing building. Only once the white hot furnace of the fire has been overwhelmed with water will the flames shrink and, eventually, die away. A successful assault on the seat of the fire marks the turning point in the battle against the fire's destructive power.

15.9 Next comes what fire crews refer to as 'damping down'. It's all very well neutralising the hotspots which gave the fire its life, but so much heat is generated in a well-developed fire that any corner of the building may conceal a red-hot foundry of smouldering material that could at any moment burst into life and wreak more havoc. Damping down can carry on for many hours after the main blaze is extinguished but is essential practice if the firefighting team wants to avoid having to return to the same spot a few hours later.

15.10 Once the fire is completely out, a new danger shows itself. A building which, a day or so ago, was as safe as houses is now a dangerous shell. The next task will be to make the remaining structure safe by shoring up with scaffolding until such time as it can be re-developed and opened for business again. A new process of assessment must take place now to determine how much of the building, if any, is safe to reoccupy and how to go about rebuilding the part which was damaged. And, of course, using the adage that any setback can also be seen as an opportunity, now is the time for the owners to consider whether it is timely to review the size, shape and function of the building itself. Maybe it is time for a change.

15.11 So, there are the firefighters' priorities: assess, contain, attack, damp down, make safe, adapt. How does this work flow equate to the managing director confronted by deaths in the workplace? Let's pick up where we left off with the phone call with the incendiary message.

15.12 The first thing is to act quickly and assess exactly what has happened. Set up an incident desk and establish lines of communication with all your personnel on the ground. Make sure that every scrap of information about what happened – and is continuing to happen – is fed through this desk. We've already conceded that incoming information from the field may not be the kind of thing you would want to hear. But at least hearing the bad bits first gives you a chance to get your line of defence marshalled, just as the news goes out on air. The last thing you want is to be learning new details second-hand from a live TV report from the scene. You cannot afford to be caught on the hop now.

15.13 Choose someone reliable, articulate and genuine to head up the company's defence. This individual will become the company's public face over the coming days, and their competence at batting on a sticky wicket could determine whether the company sinks or swims. If the incident is big and on-going, send them to the scene straightaway. There may not be many answers that the company can give at this point but, in the world of 24/7 news, it is a wise move to have a presence on the ground from the word 'go'. He or she will need to find the right words to sound concerned and chastened about the incident and yet – discreetly at first – resolute about the future. Your representative will need rapidly to make themselves available to broadcast and print media and start getting the company's message across, namely 'we're not shirking our responsibility for what has gone wrong but the show goes on'. The phrase 'at this time our thoughts are with those caught up in this accident and their families' may be as empty as a bubble, but at least it conveys concern without giving anything away. If yours is a big company, it's likely someone in senior management will have had media training in readiness for this unfortunate day. If not, consider employing the services of a reputable public relations firm without delay.

15.14 Next, try to establish the boundaries of what has happened. Okay, so it's a serious mishap, but that doesn't mean your company has to completely grind to a halt. Obviously, it will be a priority to get to the bottom of what caused the accident and make changes to prevent any recurrence. It may be judicious to slow company activity down for a day or two while the dust settles, but it is equally important that any unaffected sectors resume normal work as soon as possible. The accident will cause enough loss of momentum without bringing your whole operation to a stand-still.

15.15 Now comes the crux of the PR battle itself, the hand-to-hand fighting, when your team must deal with the dangerous PR blaze at its base. In the coming hours and days, a picture will emerge about what went wrong to cause the deaths and injuries that have occurred. Snippets of apparently damning information will be plastered across the broadcast and print media. Nowadays, every Tom, Dick and Harry feels entitled to jump on any passing bandwagon and rush to throw fuel on the fire with Tweets and video clips on YouTube. Some of these posts may be relevant, but many will be misleading, incomplete or mischievous. What can you do against this screaming, shouting free-for-all of comments?

15.16 Now your spokesman must come out fighting. Thanks to the incident desk, he can always be fully apprised about what may have gone wrong. He is uniquely placed to provide accurate information about what has occurred and is therefore always in a position to put a less negative spin on any new compromising facts that emerge. If a tyre blow-out on a company lorry has caused deaths on a motorway, the spokesman will be ready with facts about how the lorry only recently passed its MOT and that the tyres had been replaced with new ones. There is no denying the essential bad news about the incident, but there are many ways to reduce its impact. Learning the full facts as soon as possible, and presenting them in a favourable light with the minimum of delay, is the way to get your PR fight off on the right foot.

15.17 On the other hand, he must never get caught out denying something which may shortly become established fact. That's the worst PR own goal. There's nothing journalists like more than catching out a company telling untruths about the true cause of a tragedy, or trying to big up its SOS response, when in fact it will soon become clear that the response was slow and inadequate. What the public cannot forgive is blunderers who won't take it on the chin. This is a sure way to PR disaster.

15.18 It's worth considering how the story will be viewed on the other side of the fence, namely in newsrooms around the country. There is a common misconception among the victims of bad news that the media unites as one with the sole objective of adding to their woes and, better still, putting them out of business altogether. The consequence is that sometimes companies clam up and abandon the media fight entirely. Nothing could be better calculated to deal a further blow to your media image. In every story you see or read, what the media is after is the truth. It may come out sooner, it may come out later, but one way or another the truth will out. In all the stories that businessmen or women complain about, it is rarely the essential facts of the story that are contested; it is the interpretation of them that complainants do not like. The job of media people is to dig out an exciting new fact or revelation and then sensationalise it to the max. Every new twist and turn generates new lines of inquiry and new questions to answer. This is why pulling down the shutters when a fatal accident has occurred is not sensible. Your lawyers may advise you to keep your own counsel and 'plead the Fifth Amendment' but, in PR terms, this is like consulting an ostrich about the art of concealment. Much better to take to the air-waves and calmly explain the true situation in the correct context, time and again if necessary.

15.19 Eventually, the media storm will burn out as drama-hungry news editors turn their attentions elsewhere. Provided your company has not been caught out trying to pull the wool over the public's eyes, the media circus will disband to await the next big story. Now is time to get your company on the front foot again. The fire-fighting may be over, but the public will not expect your firm to just carry on where it left off as if nothing had happened. A thorough assessment of what went wrong is now called for, followed by the publication of a detailed and candid report explaining how to put things right. (After the Gulf of Mexico oil spill, in which 11 people died and some 5 million barrels of crude oil were released into the ocean, BP financed a multi-million dollar fund to contribute towards the losses incurred by fishermen, shrimpers and beach tourism traders. BP provided billions of dollars to clear up the mess, and survived.) Individuals caught up in your accident may need to sue your company; and, perversely, offering to cover their legal costs may help your company's convalescence.[1] Your need now is to get this matter signed off and an end to troublesome headlines reminding staff and customers what went so badly wrong.

1 Most companies will have the benefit of insurance cover and this could be prejudiced, or the policy rendered void, if there is an admission of liability without the insurer's consent.

Accordingly, before any admission or offer is made to the victim's family, clearance should be obtained from the insurers, who may wish to take control of any potential claim from the outset.

15.20 So there you have it. Accidents will happen. Machinery breaks down, people make mistakes. Acts of God strike from nowhere. People – for which read workforce, shareholders, customers and media – will give the benefit of the doubt to any company seen to be bravely grappling with an unfortunate turn of events. But the minute a company gets caught out trying to pull a flanker on the watchful public is the minute that an honest accident begins to spin into a full-blown crisis. We all make mistakes, but most of us are smart enough not to lie about it afterwards – and certainly not if any *faux pas* will be broadcast around the nation within minutes and then spelled out in 64-point bold type on newspaper front pages the following day. Learn the facts, take responsibility, tackle the problem, show concern, publicly learn lessons, change business practices where necessary, and hopefully move on.

A LAWYER'S POINT OF VIEW: INTRODUCTION

15.21 The rolling reports and daily headlines will be unavoidable. When disaster strikes and there are fatalities involved, the public will want to know who is at fault, and the media may be quick to cast blame. While your PR strategy will undoubtedly impact on the tone of that coverage, an awareness of the laws governing what journalists can and cannot say about you will stand you in good stead.

15.22 Let us start by dispelling the myth that news reporters can report 'whatever they like'. They cannot. Journalists in print, in broadcast and online are subject to laws and regulations which, if broken, can result in costly consequences for them and their employers. A criminal prosecution or civil action is unlikely to help a journalist's career prospects and, in reality, few will risk throwing caution to the wind. Instances of media excess remain the exception. When it does happen, it is arguably down to inadequate law enforcement rather than to any inadequacy of the law. The English legal system offers you robust redress for media breaches – more than might be imagined, and more than in comparable democracies such as that of the US. If your company is caught in the media glare, you will need to have all the angles covered. This means knowing your rights and knowing a journalist's limits.

15.23 There is benefit from a sound understanding of the principal areas of law affecting the media: defamation; privacy; and contempt. Defamation laws prevent the media from unjustly attacking your reputation. The same laws can protect your company's reputation too. Contempt laws ensure that, should you or your company face prosecution, nothing is reported that may prejudice a subsequent trial.

15.24 Privacy laws can be used to deal with unlawful interferences with an individual's right to privacy. In addition, the law of confidentiality can be used by both companies and individuals over disclosures of information in breach of obligations of confidentiality.

15.25 Knowing a little of the law is just half the battle; appreciating its application is the other. You will, no doubt, want to avoid legal confrontation with journalists, if it can be helped, over what they have said about you. Court action should be a last resort. In an ideal world, the laws we examine here would work as an effective deterrent – and the media would always know where to draw the line. But one has to be realistic. Occasionally, the media will need threats of legal action to keep them in check. Occasionally, an informal reminder of your rights and the relevant laws can work wonders. This area of law is complex, so this chapter can only provide an overview, which should equip one with a basic understanding of the laws affecting the media.

DEFAMATION AND MALICIOUS FALSEHOOD

15.26 Perhaps you consider yourself to be a boss with a heart, a man or a woman of the people. Perhaps, too, your employees feel the same – they see that you have worked your way up and they know that you are in touch with their sensibilities.

But when disaster strikes and there are casualties, your reputation up to this point will count for little. Journalists will be out to find out who is to blame, and they will be looking for a name and a face to pin it to. The starting point will be the boss.

15.27 What emerges in the media may be more cliché than reality, but reporters will want you to fit the mould. Whether you are chief executive of an energy company, CEO of a collapsed bank, or chairman of a multi-national construction firm; if the story has a victim, you will find yourself playing the same part: the ruthless despot with scant regard for human life and maximum regard for profit.

There are, of course, limits on how far the media can twist the knife. If what is being said about you is patently false, you have access to legal redress. Likewise, if what is being said about you unjustly attacks your reputation in the eyes of your peers, the law can help. Defamation is a general term encompassing such reputational damage, and an understanding of how the law of defamation operates will prove invaluable.

15.28 The law of defamation seeks to protect a person's reputation, and the goodwill and reputation of a company or organisation, from unjustified attack. Every person, company or organisation has a right to take steps to protect his or her or its reputation from being disparaged by defamatory statements made about them to a third person or persons without lawful justification or excuse.

The tort of defamation can be divided into libel and slander. If a defamatory statement is made in writing or in any other permanent form (including publication on television or radio and in the performance of a dramatic piece at the theatre), the tort of libel is committed and the law presumes damage. If the defamation is spoken, or in some other transient form, it constitutes the tort of slander, which is not actionable at common law without proof of damage. There are, however, exceptions.

15.29 At the time of writing this chapter, the law of defamation was undergoing major changes. The Defamation Bill received Royal Assent on 25 April 2013 and came into force on 1 January 2014 as the Defamation Act 2013 ('2013 Act') and will sit alongside the Defamation Acts 1952 and 1996. One of the main changes to the current law is that it makes it harder for companies and individuals to sue for libel.

Who can sue?

15.30 Defamation claims can be brought by, among others:

- **Individuals**, including foreign citizens and bankrupts.[1]

- **Companies and corporations**: A company can sue for libel or slander in the same way as an individual, provided the defamatory statement injures its trading reputation or goodwill and such injury has caused or is likely to cause the company serious financial loss.[2] The imputation must reflect upon the company or corporation itself and not upon its members or officials only. A company or corporation, unlike an individual, has no feelings, so it cannot recover aggravated damages. A company can sue over a wide range of allegations, including insolvency, mismanagement of its affairs, dishonest conduct, bribery, or even manslaughter. Companies which are based abroad must be able to show that they have a trading reputation within this jurisdiction.
- **Trade unions, friendly societies and other associations**: It is not certain whether a trade union can sue in its own name for libel or slander. A registered friendly society may sue for libel and must do so in its registered name.
- **Unincorporated associations**: An unincorporated association is not a legal entity and cannot therefore sue or be sued for defamation. Officers and possibly members of the association could sue, provided that they were named or otherwise identifiable.
- **Partners**: A partnership may sue or be sued in its own name, despite the fact that it has no legal personality.
- **Local and central government**: The House of Lords ruled that a local authority could not sue for defamation in respect of governmental and administrative functions, even though it engaged in trade.[3]

1 An individual's right of action dies with the individual.
2 See **15.32** below. Individual directors could, depending on the meaning of the published words, sue in their personal capacity, which is a strategy that may be adopted in the future in order to avoid the burden of proving 'serious financial loss'.
3 *Derbyshire CC v Times Newspapers* [1993] AC 534.

What the claimant must prove

15.31 In order to bring a claim for defamation, a claimant must prove (a) that the defendant has published, or is responsible for publishing, (b) a defamatory statement, (c) which is understood to refer to the claimant.

(a) **Publication**: The claimant must prove that the defamatory statement was published.[1] This means communicating the statement to a third party, which could be to one person, as in the case of a letter, or to millions, as in the case of a national newspaper or television station. Publication can also be by electronic means, including email, web posting or Twitter.

(b) **Meaning of the words**: A defamatory statement is a statement which tends to lower the claimant in the estimation of right-thinking members of society generally or to cause him or it to be shunned or avoided or to expose him to hatred, contempt or ridicule, or to disparage him or it in his office, profession, calling, trade or business. There is a threshold of seriousness for defamation claims. Trivial or minor defamatory allegations will not be considered by the court to be sufficiently serious to warrant court action.

(c) **Identification**: The claimant must prove that he or she or it is identifiable in the material that was published. Generally, the claimant will be named and the issue will not arise, but an unnamed but identifiable claimant will be able to sue.

1 See Defamation Act 2013, s 15 for the meanings of 'publish' and 'statement'.

15.32 A statement is not defamatory unless its publication has caused or is likely to cause serious harm to the reputation of the claimant.[1] Harm to the reputation of a business (defined as a 'body that trades for profit') is not 'serious harm' unless it has caused or is likely to cause the body 'serious financial loss'.[2]

1 Defamation Act 2013, s 1(1). The Explanatory Notes to the Defamation Act 2013 state that this section 'builds on the consideration given by the courts in a series of cases to the question of what is sufficient to establish that a statement is defamatory. A recent example is *Thornton v Telegraph Media Group Ltd* [2010] EWHC 1414 in which a decision of the House of Lords in *Sim v Stretch* [1936] 2 All ER 1237 was identified as authority for the existence of a "threshold of seriousness" in what is defamatory. There is also currently potential for trivial cases to be struck out on the basis that they are an abuse of process because so little is at stake. In *Jameel v Dow Jones & Co* [2005] EWCA Civ 75 it was established that there needs to be a real and substantial tort. The section raises the bar for bringing a claim so that only cases involving serious harm to the claimant's reputation can be brought'.

2 Defamation Act 2013, s 1(2).

Defences

15.33 Once a claimant establishes a claim in defamation, it then falls on the defendant to establish one or more of the following defences:

(a) **Truth (formerly 'justification')**:[1] This means that what has been published is true or substantially true. It is a defence to an action for defamation for the defendant to show that the imputation conveyed by the statement complained of is substantially true.[2] Truth is a complete defence, so malice is irrelevant. Unlike honest opinion that protects statements of opinion, truth is a defence regarding statements of fact. The defendant must prove that the meaning of the defamatory statement is true. The standard of proof needed for truth is that used in civil cases, namely that the material must be proved true on a balance of probabilities. The jury will decide what the defamatory meaning is, and it is then for the defendant to prove the truth of that meaning. It is common for a defendant to seek to prove a less serious meaning derived from its publication than that being complained about by the claimant. A defence of truth will not be defeated simply because substantially true allegations contain some untrue words which do not seriously harm the claimant's reputation.[3] However, the truth defence has financial risks. The judge or jury may award greater damages if the defendant fails to prove the truth of the allegations at trial, as it will be perceived as a further attack on the claimant by reasserting the truth of what was published.

(b) **Honest opinion (formerly 'fair comment')**:[4] The defence of honest opinion protects opinions and not statements of fact. In order to rely on the defence, the defendant would have to show that the following conditions are met: (1) the statement complained of was a statement of opinion;[5] (2) the statement complained of indicated, whether in general or specific terms, the basis of the opinion;[6] and (3) an honest person could have held the opinion on the basis of (i) any fact which existed at the time the statement complained of was published; and (ii) anything asserted to be a fact in a privileged statement (ie publication on matter of public interest) published before the statement complained of.[7] The defence can be defeated if the claimant can show that the defendant did not hold the opinion.[8]

(c) **Privilege**: On certain occasions, the law recognises that it is better that individuals are free to speak their mind, and others to report what they say without any risk of proceedings for defamation, even if the statements turn out to be untrue. These occasions are referred to as privileged. There are, in general, two types of privilege, namely:

– *Absolute privilege*: This most commonly arises in respect of statements made either in the course of, or in fair and accurate reports of, court or parliamentary proceedings.[9] The privilege is absolute because it applies irrespective of malice (see below).

– *Qualified privilege*. This provides a lesser protection than absolute privilege, and the defence is defeated if the claimant establishes that the defendant was motivated by malice. It is available in a range of situations and arises either (a) 'where the person who makes a communication has an interest or a duty, legal, social or moral to make it to the person to whom it is made',[10] (b) for media reports on matters of public interest,[11] and (c) for fair and accurate reports of certain proceedings, documents and statements.[12] 'Malice' is a legal term which means that the defendant either knew the publication was false, was reckless in that regard, or had an otherwise improper motive for making the publication.

The new statutory defence in the 2013 Act of 'publication on a matter of public interest', which replaces the Reynolds defence, only requires the defendant to show that (a) the statement complained of was, or formed part of, a statement on a matter of public interest; and (b) the defendant reasonably believed that publishing the statement complained of was in the public interest.[13] The 2013 Act does not define 'the public interest', relying instead on the concept which is well-established in the English common law.

(d) **Other defences**: A number of other defences are available to the defendant, including: the new defence for operators of websites where a defamation action is brought against them in respect of a statement posted on the website;[14] consent (namely, the claimant consented to the publication of the defamatory statements); and accord and satisfaction (namely, if the claimant had agreed to accept an apology, he cannot then sue).

1 Defamation Act 2013, s 2. The Explanatory Notes to the Defamation Act 2013 state that this section replaces 'the common law defence of justification with a new statutory defence of truth. The section is intended broadly to reflect the current law while simplifying and clarifying certain elements'.

2 Defamation Act 2013, s 2(1).

3 Defamation Act 2013, s 2(3).

4 Defamation Act 2013, s 3. The Explanatory Notes to the Defamation Act 2013 state that this section broadly 'reflects the current law while simplifying and clarifying certain elements, but does not include the current requirement for the opinion to be on a matter of public interest'.

5 Defamation Act 2013, s 3(2).

6 Defamation Act 2013, s 3(3).

7 Defamation Act 2013, s 3(4).

8 Defamation Act 2013, s 3(5). This subsection (5) does not apply where the defendant is not the author of the statement (for example, where an action is brought against a newspaper editor in respect of a comment piece rather than against the person who wrote it) and, in these circumstances, the defence is defeated if the claimant can show that the defendant knew or ought to have known that the author did not hold the opinion (s 3(6)).

9 Defamation Act 1996, s 14, as amended by Defamation Act 2013, s 7.

10 *Adam v Ward* [1917] AC 309.

11 Defamation Act 2013, s 4(1). The Explanatory Notes to the Defamation Act 2013 state that 'It is based on the existing common law defence established in *Reynolds v Times Newspapers* [2001] 2 AC 127 and is intended to reflect the principles established in that case and in subsequent case law. Subsection (1) provides for the defence to be available in circumstances where the defendant can show that the statement complained of was, or formed part of, a statement on a matter of public interest and that he reasonably believed that publishing the statement complained of was in the public interest. The intention in this provision is to reflect

the existing common law as most recently set out in *Flood v Times Newspapers* [2012] UKSC 11. It reflects the fact that the common law test contained both a subjective element – what the defendant believed was in the public interest at the time of publication – and an objective element – whether the belief was a reasonable one for the defendant to hold in all the circumstances'.

12 Defamation Act 1996, s 15, as amended by Defamation Act 2013, s 7.

13 Defamation Act 2013, s 4(1).

14 Defamation Act 2013, s 5.

Remedies

15.34 The following remedies can be awarded by a court in a successful defamation claim:

(a) **Damages**: These are to compensate a claimant for the hurt feelings caused by publication of the defamatory words or images, to go some way towards repairing the harm caused to the claimant's reputation, and as a vindication of the claimant's reputation. These can consist of general, aggravated and/or exemplary damages.

(b) **Injunctions**: Where a claimant becomes aware of the proposed publication of defamatory material, he can in certain circumstances seek an injunction to prevent publication in the first place. If a defendant, however, indicates an intention to raise a substantive defence (eg truth, comment or privilege), an injunction cannot be granted unless it can be demonstrated that the defendant is acting in bad faith or that the proposed defence will inevitably fail.[1] Even where the allegations have been published, it is still traditional for the claimant to seek both damages and an injunction. The reason for claiming an injunction in such circumstances is that, as defamation is a claim for hurt and embarrassment or loss caused by publication of the defamatory allegations, it will be necessary to try and prevent further publication of the same allegations by that defendant again. If the claimant has been successful at the trial, the judge would normally grant a permanent injunction against repetition of the particular allegations.

1 *Bonnard v Perryman* [1891] 2 Ch 269 and *Greene v Associated Newspapers Ltd* [2004] EWCA Civ 1462, [2005] QB 972.

Malicious falsehood

15.35 A statement can be false without it being defamatory but still gives rise to a cause of action. In the event that a false statement is published, the claimant may be able to sue for malicious falsehood, provided the statement:

- is false;
- was published maliciously; and
- causes financial loss.

15.36 The claimant must not only prove that the statement is false but that the defendant published it maliciously; whereas, in defamation claims, the defendant must prove that the statement is true. As above, in order to establish malice, the claimant must demonstrate that the defendant published the statement knowing it to be false or recklessly, or with some other improper motive. Negligence, or wrongly believing the statement to be true and failing to check it, is not malice.

15.37 Malicious falsehood commonly arises where businesses make comparisons and exaggerations about the quality of the goods and services of their rivals. In such cases, it is necessary to distinguish between claims that would be taken seriously by an ordinary reader and advertising 'puffs' that should be taken with a pinch of salt.

15.38 The claimant in a malicious falsehood action is required to prove that the allegations have caused financial loss or are likely to cause such loss. The requirements differ depending on whether the words are written or spoken. It is not necessary for the claimant to prove actual damage if the words are printed and were likely to cause financial damage to the claimant, or where the words were spoken or written and were calculated to cause financial damage to the claimant in respect of any office, profession, calling, trade or business held or carried on by the claimant at the time of publication.

PRIVACY AND BREACH OF CONFIDENCE

15.39 Disaster has struck and there have been fatalities. As the public face of your company, the media will want to know all about you. Better still, they will want something unfavourable. That means delving into your background on a number of levels. Whatever term reporters give to such journalistic endeavour – 'uncovering the truth' or 'digging the dirt' – you will want to know what can be done about it. This is where the relatively new legal concept of 'misuse of private information' or breach of privacy can be a potent weapon.

15.40 Reporters may go to considerable lengths to obtain interesting details for their story. They will want personal accounts of your character and behaviour, and they will want suggestive photos. They will contact your family and friends, past and present, sometimes offering money for information. They will scour social media, internet and printed publications for any reference to you. Indeed, anything you may have said on public record will be dredged up and, where possible, used against you. While such methods are undoubtedly intrusive, are they illegal?

15.41 The answer will, of course, depend on the facts. However, if you familiarise yourself with the fundamental operation of privacy law in this country, you will be better equipped to spot potentially unlawful investigative methods or intended stories which will infringe your privacy rights in time to try to do something about it. Journalists and their employers have a right to exercise freedom of expression. But you have a right to a private life too – and the former does not trump the latter.

15.42 What follows is a basic grounding in the elements of your right to a private life under UK law. English law recognises a right to privacy – 'misuse of private information' – and will intervene to protect individuals' privacy rights. It developed from the action of breach of confidence, which is based on the principle that a person who has been given confidential information should not take unfair advantage of it.

Breach of confidence

15.43 The earlier claims for infringement of privacy were based on the more traditional cause of action, breach of confidence. The law protects confidential information from being improperly divulged. Traditionally, in order to establish a breach of confidence, information must be:

- confidential in nature;

- imparted in circumstances importing an obligation of confidence; and
- an unauthorised use of that information to the detriment of the party communicating it.[1]

1 *Coco v A N Clark (Engineers) Ltd* [1969] RPC 41, at para 47.

15.44 The law safeguards ideas and information obtained or imparted in confidential circumstances, so information that is already in the public domain or trivial is not confidential. The obligation of confidence can arise in a number of ways, including through a contractual or personal relationship or where the information has been obtained by unethical means, such as theft or listening devices. The detriment that a party suffers from a breach of confidence could be financial loss, or an adverse effect on an individual's physical or mental health, that had been caused by the unauthorised disclosure.

Privacy

15.45 The Human Rights Act 1998 (HRA 1998), which came into force on 2 October 2000, incorporated the European Convention on Human Rights (ECHR) into domestic UK law. The courts and tribunals in the UK are obliged under the HRA 1998 to act, when carrying out their functions, in a way that is compatible with Articles 8 and 10 of the ECHR,[1] as well as taking into account the decisions and opinions of the European Court of Human Rights.

1 HRA 1998, s 6(1).

15.46 Article 8 of the ECHR provides that:

> '1. Everyone has the right to respect for his private and family life, his home and his correspondence.
>
> 2. There shall be no interference by a public authority with the exercise of this right except such as is in accordance with the law and is necessary in a democratic society in the interests of national security, public safety or the economic well-being of the country, for the prevention of disorder or crime, for the protection of health or morals, or for the protection of the rights and freedoms of others.'

15.47 Article 10 of the ECHR provides that:

> '1. Everyone has the right to freedom of expression. This right shall include freedom to hold opinions and to receive and impart information and ideas without interference by a public authority and regardless of frontiers. This Article shall not prevent states from requiring the licensing of broadcasting, television or cinema enterprises.
>
> 2. The exercise of these freedoms, since it carries with it duties and responsibilities, may be subject to such formalities, conditions, restrictions or penalties as are prescribed by law and are necessary in a democratic society, in the interests of national security, territorial integrity or public safety, for the prevention of disorder or crime, for the protection of health or morals, for the protection of the reputation or rights of others, for preventing the disclosure of information received in confidence, or for the maintaining of the authority and impartiality of the judiciary.'

15.48 Courts must have particular regard to the importance of freedom of expression, guaranteed by Article 10 of the ECHR, before they consider granting an injunction banning publication of any information.[1]

1 HRA 1998, s 12.

15.49 In *Campbell v Mirror Group Newspapers Ltd*,[1] the House of Lords confirmed for the first time that the HRA 1998 had led to the establishment of 'unjustified disclosure of private information' or 'misuse of private information' as a new cause of action. It confirmed that the principal test for determining whether or not information qualifies for protection under Article 8 is to ask 'whether in respect of the disclosed facts the person in question had a reasonable expectation of privacy'. The Law Lords agreed that Ms Campbell, a supermodel, was entitled to damages after the *Daily Mirror* gave details of the therapy that she was receiving at Narcotics Anonymous and published photographs of her leaving the therapy session.

1 [2004] UKHL 22.

15.50 In 2004, the European Court of Human Rights took a wide view of the concept of 'private life' and held that publication of photographs of Princess Caroline of Monaco in her daily life, shopping and on holiday with her family, came within the scope of her private life so as to be protected under Article 8.[1] This was a significant decision for privacy law in England, due to the fact that the English courts are obliged by the HRA 1998 to take account of decisions of the European Court of Human Rights in determining any question that has arisen in connection with a right under the ECHR.

1 *Von Hannover v Germany* (Application no 59320/00), 24 June 2004.

15.51 However, in a further case involving Princess Caroline in February 2012, the European Court of Human Rights held that the publication of photographs of Princess Caroline on holiday was justified under Article 10 on the grounds that they contributed to a debate of general interest, because they accompanied an article about her father's illness.[1]

1 *Von Hannover v Germany (No 2)* (Application nos 40660/08 and 60641/08). See also *Von Hannover v Germany (No. 3)* (Application no 8772/10) in which Princess Caroline complained for a third time to the European Court of Human Rights in relation to failures by the German courts to prevent the publication of photographs that she claimed depicted her engaged in private activities. The European Court of Human Rights unanimously held that the German court's refusal to grant an injunction prohibiting any further publication of the photograph did not constitute a breach of Princess Caroline's privacy rights.

15.52 In 2005, the House of Lords made clear that neither Article 8 nor Article 10 'has precedence over the other'.[1] So, the courts have to carry out a balancing exercise between the parties' competing Article 8 and Article 10 rights. An 'intense focus' is therefore necessary upon the comparative importance of the specific rights claimed, where there is a conflict between the values under Articles 8 and 10.

1 *Re S (A Child)* [2005] 1 AC 593, per Lord Steyn.

Defences

15.53 A defendant who is sued for an alleged breach of confidence or infringement of privacy has two main defences, namely:

(a) **Information already in the public domain**: If the private or confidential information enters the public domain, then generally speaking it can no longer be protected. However, the courts treat private information concerning individuals more sensitively than, for example, confidential trade secrets. In 2011, the High Court had to consider whether an anonymity order protecting a footballer, who had sought an interim injunction to prevent disclosure of information regarding an alleged sexual relationship, should remain in place, despite the footballer's name having been disclosed on Twitter and in Parliament. The court held that, by contrast with confidential information, it was not possible to draw a line as to when private information becomes so widely available that it is no longer private, since whether there was a reasonable expectation of privacy depends on the facts. It held that despite the revelations on the internet, the injunction could still prevent intrusion and distress to the claimant and his family, particularly since 'wall-to-wall excoriation' in the national newspapers was likely to be significantly more intrusive and distressing than the availability of information on the internet.[1]

(b) **The publication was in the public interest**: This includes exposing wrongdoing, negligence or hypocrisy. In 2006, Lady Hale emphasised that a 'real public interest in communicating and receiving the information' is: 'very different from saying that it is information which interests the public – the most vapid tittle-tattle about the activities of footballers' wives and girlfriends interests large sections of the public but no-one could claim any real public interest in our being told all about it.'[2]

1 *CTB v News Group Newspapers Ltd and Imogen Thomas* [2011] EWHC 1326 (QB).
2 *Jameel v Wall Street Journal* [2006] UKHL 44.

Remedies

15.54 The same remedies are available in claims for misuse of private information and for breach of duty of confidence. These are an injunction, damages or account of profits and delivery up or destruction of the offending material. The damages awards in privacy cases have historically tended to be lower than in defamation cases, although damages awarded in the *Mosley* case,[1] at £60,000, were the highest to date. An interim injunction, now known as an 'interim non-disclosure order' (and often referred to in the press as a 'super-injunction') is the most important remedy in the vast majority of privacy cases.

1 *Mosley v News Group Newspapers* [2008] EWHC 1777 (QB).

CONTEMPT OF COURT

15.55 What if you or your company face criminal proceedings? You may feel that a trial will give you the opportunity to clear your name. Given the negative publicity surrounding your company, you will get the chance to put your point across.

But your day in court could be some way off. And what is said about you and your company between now and then could have a crucial bearing on the trial. Fortunately, the law recognises that a potential juror may be influenced by what he sees or reads. Legislation specifically sets out what can and cannot be said about pending criminal proceedings. Accordingly, news reporters are trained from the outset to obey the laws on contempt of court.

15.56 But that does not mean journalists do not push the boundaries. Indeed, sometimes they cross the line altogether. Trials have been known to collapse and newspapers face heavy fines because an article has broken the law.

In essence, nothing can be reported that may subsequently prejudice a trial. What follows is a general illustration of how the law of contempt operates. The law of contempt restricts what the media organisations can report, by protecting the administration of justice. It covers various activities which can be said to undermine the integrity of the legal process.

15.57 There are two types of contempt: statutory and common law. Both types involve interfering with legal proceedings. There is no exhaustive list of what constitutes 'legal proceedings' but it includes the main courts, eg magistrates' court, County Court, High Court as well as inquests, military courts and industrial tribunals.

A person or company guilty of contempt can be fined or imprisoned. Contempt of court is a criminal offence and carries severe penalties: an unlimited fine and/or up to two years' imprisonment of the relevant personnel responsible for the offending publication or broadcast – normally the editor. It should be noted that the law is even more strict in Scotland.

Statutory contempt

15.58 Statutory (or strict liability) contempt, which is governed by the Contempt of Court Act 1981, applies to the publication of material 'which creates a substantial risk that the course of justice in the proceedings in question will be seriously impeded or prejudiced'.[1] Statutory contempt is a strict liability offence, which means that the prosecution does not have to prove that the media organisation responsible 'intended' to create the risk, when seeking to prove that the contempt occurred. The court simply judges the actual or potential prejudicial effect of what was published. The strict liability rule applies to a publication only if the proceedings in question are 'active' at the time of the publication.[2]

1 Contempt of Court Act 1981, s 2(2).
2 Contempt of Court Act 1981, s 2(3).

15.59 Criminal proceedings are deemed active once a person is arrested, an arrest warrant has been issued, a summons is issued, or a person is charged orally.[1] Criminal proceedings cease to be active when either the arrested person is released without being charged, the case is discontinued, the defendant is acquitted or sentenced, no arrest is made within 12 months of the issue of the arrest warrant, or the defendant is found to be unfit to be tried or unfit to plead.

1 Contempt of Court Act 1981, Sch 1, para 4.

15.60 Civil proceedings are deemed to be active from the time that a date for the trial or a hearing is fixed. Civil proceedings cease to be active when the case is disposed of, abandoned, discontinued or withdrawn. However, it is very rare for civil proceedings to give rise to contempt of court issues, since the majority of cases are heard by judges rather than juries.

15.61 The 1981 Act does not define what creates a substantial risk of serious prejudice to an active case, but the cases in which media companies and their editors have been convicted of contempt show that publication of the following information is likely to be held in contempt:

- information suggesting that the person is of bad character;
- any suggestion that the accused is guilty;
- any references to the accused's previous convictions; and
- evidence linking the accused to the crimes.

Common law contempt

15.62 Common law contempt targets any other action which is intended to interfere with the administration of justice, including interfering with pending or imminent court proceedings. The prosecution would need to prove that there was 'intent' to create a substantial risk of serious prejudice. The term 'intent' could mean either deliberate intent or recklessness in publishing material which the media organisation should have foreseen would create a risk.

Common law contempt still applies to material published before proceedings become active, but it has largely been superseded by statutory contempt. Prosecutions for common law contempt are extremely rare, because the intention of the journalist or editor to create a serious risk of prejudice would need to be proved.

Defences

15.63 With statutory contempt, liability is strict, which means that the publisher cannot escape liability by arguing that it had no intention of prejudicing on-going legal proceedings. However, a limited number of defences to statutory contempt are available to a publisher, namely that the material comprised a fair and accurate contemporary report or a discussion of public affairs, and, more significantly, that publication of the potentially prejudicial material was made innocently. This last defence applies only where the publisher did not know (and had no reason to suspect) that the proceedings which featured in the material were active.

15.64 With common law contempt, intention to prejudice is key; honest mistake would be a complete defence to any prosecution.

PRESS COMPLAINTS COMMISSION AND OFCOM

15.65 The Press Complaints Commission (PCC) is an independent self-regulatory body that deals with complaints about the editorial content of newspapers and magazines, and their websites. The PCC was created to self-regulate the press and to administer the Editors' Code of Practice in order to raise standards in journalism. It will soon, however, be consigned to the history books.[1] In 2011, an official inquiry into press ethics was launched which led to the Leveson Inquiry. In his Report, Lord Justice Leveson recommended legislation that underpinned 'the independent self-organised regulatory system'.[2]

1 A number of the large publishing companies created the Independent Press Standards Organisation (IPSO) as a replacement for the PCC (see **15.72** below).
2 The Leveson Report, Chapter 7.4.

15.66 The independent regulator and competition authority for the UK communications industries ('Ofcom') adjudicates on complaints against the broadcast media.[1] All commercial broadcasters within the UK require a licence from Ofcom to transmit. The rationale for statutory regulation of broadcast media is that the emotional impact of moving images and sound is greater than that of printed text and still pictures, and the ability to air instantaneously carries greater potential to provoke disorder. The Communications Act 2003 and the Broadcasting Act 1996 require Ofcom to create standards for programme content. These standards comprise the Ofcom Broadcasting Code. Ofcom can direct that a programme should not be repeated, that the broadcaster must air a correction or a statement of its findings, and can impose a fine for serious or reckless breaches of the Code. It also has the power to shorten or suspend the broadcaster's licence.

1 Ofcom is a post-transmission regulator and can only become involved once a programme is broadcast.

15.67 Ofcom's code of rules govern impartiality when reporting on politics and social issues, accuracy, the fair treatment of people, respect for privacy and the need to avoid causing harm and offence. Ofcom permits breaches of the code where it is 'warranted' – for instance, in the public interest or by some other exceptional justification to do so.

LEVESON AND BEYOND

15.68 Let us imagine for a moment that you are not sure whether any law has been broken. This situation may not be uncommon. Imagine your company is embroiled in catastrophe and the media are all over the story. They may trot out a public interest defence in running the story. But does this defence justify their behaviour? What of the photographers 'hosing down' your staff as they arrive at the office? How about the speculative calls to your wife's mobile phone? Indeed, how on earth did they get her number? Few journalists – whether in print or broadcast – will be foolish enough to blatantly flout legislation. A fair few may, however, push the boundaries of what you might consider reasonable behaviour.

15.69 So, are you powerless in this regard? In a word, no. There are restraints imposed on the media beyond legal ones, which should, in theory at least, address conduct issues. These restraints differ for the broadcast and print industries and are currently undergoing dramatic change. You will want to be able to assess whether a journalist has crossed the line and, if he has, what you can do about it. An appreciation of how these non-legal restraints rein in the media will therefore be useful.

15.70 First, you need an idea of what is and what is not illegal. Most of the widely reported excesses of the media of late were, in fact, illegal. Phone hacking, for example, is illegal. Confidential data theft is illegal. Paying public servants for information is illegal. In other instances, however, it is harder to tell. What if a newspaper offers to limit its accusations against you in return for a full and frank interview? Isn't that a form of blackmail? Not necessarily. What if it runs a story without giving you a chance to give your side? Questionable perhaps, but not illegal. It will therefore help you to know where to turn when such situations arise.

15.71 TV and radio stations are subject to statutory regulation by Ofcom. Ofcom regulates broadcast content as well as the broadcast journalists' behaviour. If you have a complaint to make about, for example, the tone of a Sky News bulletin concerning your company, you can complain to Ofcom about it. Likewise, if you are not happy with, say, the number of calls you have been receiving from a BBC reporter, you take it up with Ofcom. Ofcom will then officiate and, if necessary, impose sanctions.

15.72 The press, however, is different. The industry is self-regulating, which means that codes of conduct and complaints procedures have traditionally been administered by the press itself. If you are unhappy with the factual accuracy of a story, you take it up with the PCC. If a journalist is bombarding your marketing team with calls on what you suspect is a fishing expedition, again, you must consider going to the PCC.

But phone hacking has caused the landscape to shift dramatically. Following closure of the *News of the World* newspaper in 2011, the Government set up the Leveson Inquiry into press methods. In reaction to the illegal activity, the spotlight now shone on all questionable activity, including that which is not technically illegal. In response to Leveson's recommendations, the Government proposed the establishment of an independent regulatory body, with powers to impose exemplary punishment and force prominent corrections to oversee the press, underpinned by legislation.[1]

It seems likely the print media will have to abide by stricter codes of behaviour or face sanctions, much like the broadcast media.

1 At the time of writing this chapter, the leaders of the three main political parties agreed to set up a new watchdog by Royal Charter, with powers to demand upfront apologies from UK publishers and administer £1 million fines on them. The newspaper industry, however, rejected the idea of 'state-sponsored regulation', and News UK, Telegraph Media Group, Associated Newspapers, Trinity Mirror and Express Newspapers drew up an alternative draft Royal Charter with a series of key differences. The publishers met in late 2013 to sign contracts establishing the new Independent Press Standards Organisation (IPSO). The new self-regulatory system is expected to be up and running by 1 May 2014.

15.73 So how does this affect you and your company? Broadly speaking, the changes should benefit you. Not only can you rely on the law to regulate what the media says about you, but you will also have powerful regulatory bodies to approach if you feel that a journalist's behaviour is unacceptable. The bottom line is this: if you or your company has genuinely misbehaved, no law can prevent the media from reporting it. But if the media has misbehaved towards you, then you will be able to take appropriate action.

Chapter 16

Sport

Graham Stoker, Cornerstone Chambers, London

Mark Gay and Matt Kyle, Burges Salmon LLP, Bristol

INTRODUCTION

16.1 In the context of sport, the main focus of health and safety activity is on ensuring the safety and comfort of spectators, rather than the protection of participants or employees and volunteers. Not that these categories of persons are unimportant or neglected. The recent report of the Hillsborough Independent Panel into the Hillsborough disaster[1] has highlighted that attempts at ensuring that sporting events are safe for spectators, particularly in football, have not until recent years been an unqualified success. On the contrary, during the 20th century there were numerous disasters (particularly at football grounds) from which the right lessons had been identified but not learnt. In the context of motor sport, the decades of the 1950s, 60s and 70s were equally undistinguished in the manner in which motor sport organisers were successful in producing safe environments for participants. As it is in these two sports in which fatal accidents have predominantly occurred, it is to these that we direct our attention.

1 'Hillsborough – The Report of the Hillsborough Independent Panel' (HC 581) 12 September 2012.

16.2 Some of the events of the past, had they occurred under the current regulatory environment, would have resulted in criminal proceedings. Officials, organisers and the police would almost certainly have found themselves facing criminal liability for corporate manslaughter or for breaches of health and safety legislation. The key point is that, since Hillsborough (in the case of football) and the death of Ayrton Senna (in the case of motor sport), massive strides have been taken to protect spectators and participants in football and motor sport. It is now possible to be reasonably confident that the health and safety environment that exists in these sports is such that the possibility of similar disasters occurring again has been massively reduced.

16.3 It is important to note that sport does not in any sense constitute an exception to corporate manslaughter or general health and safety legislation. Those who could be held responsible for causing death under the Corporate Manslaughter and Corporate Homicide Act 2007 (CMCHA 2007) or for breaches of the Health and Safety at Work etc Act 1974 (HSWA 1974) in a non-sporting context could equally be found guilty in a sporting context. In the context of health and safety, the same duties to employees[1] and the public[2] reside in sports bodies as reside in any other employer or other person owing a duty to those affected by its activity. There are, however, two interesting aspects to the interaction of sport and health and safety legislation. First, specific legislative provisions have been passed to deal with the specific context of health and safety in sport and the prevention of unnecessary deaths and injuries. These provisions are largely preventive in nature and policed in the context of general duty to ensure the health and safety of the public. Secondly, there has been a change in the application of the general legislation in sport to embrace the wider activities of sports bodies wherever they may take place.

1 See HSWA 1974, s 2.
2 See HSWA 1974, s 3.

16.4 Compliance with the specific duties set out in sports-related legislation goes a long way towards discharging the general duties imposed upon sports bodies by health and safety legislation. In understanding how corporate manslaughter and health and safety issues are dealt with in sport, it is therefore necessary to understand the specific legislation which has been put in place to prevent such instances from occurring. In doing this, it is also necessary to view the legislation in the context of disasters in sport which have given rise to such legislation.

FATAL INJURIES AT SPORTS STADIA

16.5 Disasters at sports stadia were a twentieth-century phenomenon, the earlier part of which period coincided with a massive increase in the popularity of football. On 5 April 1902, 26 people were killed and 517 injured after the collapse of a stand at Ibrox in Glasgow during a match between Scotland and England. It is interesting that this provoked no official governmental or regulatory response.

16.6 The first government report into the safety and control of spectators occurred as a result of events at the first FA Cup Final at Wembley Stadium in 1923 between West Ham United and Bolton Wanderers. Some 200,000 spectators attempted to attend the fixture, a number far in excess of the capacity even of Wembley Stadium at the time. This prompted the government to commission the Shortt Report, an inquiry chaired by former Home Secretary, the Rt Hon Edward Shortt KC.

16.7 The Shortt Commission reported in 1924. It made a number of suggestions, including improved stadium access and smaller self-contained terrace enclosures. It also highlighted the dangers posed by the threat of fire. The FA did not attend or give evidence to the Shortt Inquiry and, as noted in the report of the Hillsborough Independent Panel,[1] there is no evidence that it acknowledged or acted on the Inquiry's recommendations. This attitude of executive intransigence characterised the response of footballing authorities to the issues of crowd safety thrown up by disasters until almost the end of the 20th century.

1 See note 1 to **16.1** above, at para 1.22.

16.8 In 1946, 33 people were killed and over 500 injured at Burnden Park, the then home ground of Bolton Wanderers, when two barriers collapsed. An inquiry was commissioned, chaired by Mr Moelwyn Hughes. It reported that a crowd of 85,000 had crammed into Burnden Park, which had a maximum capacity of 60,000, an attendance which it had no way of safely accommodating. The Report suggested a system of licensing of stadia by local authorities and further:

> 'The issue of the licence would depend upon satisfying the authority as to the construction and equipment of the ground, its compliance with regulations and the proposed maximum figures on admission to the different parts.'[1]

The recommendations of the Moelwyn Hughes Report were not implemented. It took another disaster, once again at Ibrox, in 1971, for a compulsory, statutory system of licensing of sports stadia to be introduced.

1 'Enquiry into the Disaster at the Bolton Wanderers' Football Ground on 9 March 1946', Report by R Moelwyn Hughes, at p 11.

16.9 In January 1971, 66 people were killed and over 200 injured after a crush at Ibrox Park in Glasgow at an 'old firm' game between Glasgow Rangers FC and Celtic FC. Celtic were leading 1–0 in the last minutes of the game. The crush occurred when fans leaving the ground sought to return when a late equalising goal was scored by Rangers. As the crowd was leaving, the roar attending the goal drew the exiting fans back up the stairwell that they were descending from the terraces to the exit gates. The crash barriers failed to hold, and the victims were crushed at the foot of the stairwell. As we have seen, this was the second disaster at Ibrox in its history, following the one in 1902.

16.10 An inquiry into this disaster led to a report by Lord Wheatley into crowd safety at football matches.[1] The Report echoed the Moelwyn Hughes Report in recommending the licensing of stadia. Indeed, a clue as to why Moelwyn Hughes had not been implemented lay at paragraphs 66 and 67 of the Wheatley Report. Referring to the misgivings of football clubs and administrators as to the imposition of stadia licensing, it said:

> 'As I see it, their misgivings were associated with a fear that such stringent conditions may be attached to the grounds of a licence that many clubs may not be able to afford the cost and some may have to go out of business.'

Wheatley, rightly, dismissed these concerns. The Report made it clear that the primary consideration of clubs was the safety of spectators, and that the licensing of football grounds, which had been mooted over 50 years earlier, and expressly recommended by the Moelwyn Hughes Report, was unavoidable. Whether such licensing would turn out to be effective remained to be seen.

1 'Report on Crowd Safety at Sports Grounds' (Cmnd 4962, 1972).

16.11 This time, the legislature listened. In 1975 it enacted the Safety of Sports Grounds Act, which set out a statutory system for the licensing of sports stadia by local authorities.[1] Following on from the Act in 1976, the Home Office published the first 'Green Guide', being a detailed (though non-binding) manual described by the Home Office as a 'Guide to Safety at Sports Grounds'.[2] This sets out, in great detail, best and acceptable practice in relation to stadium management and safety. Prophetically, however, the first Green Guide noted that voids between stadia and the ground were a common feature in stands and made them vulnerable to fire as 'they became a resting place for paper cartons and other combustible materials which can be ignited, unnoticed by a carelessly discarded cigarette end'. The remedy proposed by the Guide was inspections before and after every event to clear such contagion.[3]

1 See **16.28–16.42** below.

2 The 5th edition (2008) is published by the Department for Culture, Media and Sport (DCMS).

3 As noted in the report of the Hillsborough Independent Panel, note 1 to **16.1** above, at para 1.31.

16.12 It is plain that the 1975 legislation and the Green Guide, while both important and necessary, did not have the effect intended. On 11 May 1985, Bradford City played Lincoln City in the final match of the season. There was a holiday atmosphere at the ground as Bradford City had been promoted. The main stand was of timber construction with a pitched roof. It ignited when a cigarette fell from the stands on to rubbish which had lain beneath the stands for what the subsequent inquiry found to be over three decades. The ensuing fireball caused panic in the stands, and spectators sought to escape through the exit gates; they were locked. Spectators successfully made their way on to the pitch which, unlike the situation that greeted spectators later at Hillsborough, was accessible to the public as there was no perimeter fencing. Nonetheless, 56 spectators were killed and 265 people were injured, many suffering severe burns.

16.13 The inquiry this time was chaired by Mr Justice Popplewell and was convened pursuant to the Safety of Sports Grounds Act 1975. The results of the inquiry and report were depressingly familiar. The report concluded:

> 'the available exits were insufficient to enable spectators safely to escape the devastating effects of the rapidly spreading fire.'

Mr Justice Popplewell also concluded that the various ways of dealing with the problems of safety at sports stadia had been frequently recommended by other reports. If there was a silver lining to the report, it was that, if there had been perimeter fences in the stands, 'casualties could have been on a substantially higher scale'.

16.14 Again, there was a legislative response to the report. In 1987, Parliament passed the Fire Safety and Safety of Places of Sport Act 1987, which was geared specifically towards measures necessary to deal with the problems of fire at stadia by updating the Fire Precautions Act 1971. In particular, this gave more flexibility to the fire authority in relation to the issue of fire certificates, led to the publication of Codes of Practice, and introduced Improvement Notices to remedy contraventions of fire certificates, as well as a number of other new offences. It also significantly amended the Safety at Sports Grounds Act 1975. Principally, it brought sporting arenas which could accommodate over 500 spectators (defined as an individual stand and termed a 'regulated stand') under the safety certificate regime. The provisions for the consideration of any application for a regulated stand safety certificate, the terms and conditions it may contain, the process of determining those terms and conditions, and the scope of any enforcement action mirrors the provisions in the Safety of Sports Grounds Act 1975 but are limited to the regulated stand, or stands, rather than the sports ground as a whole. The 1987 Act, as well as introducing a series of additional offences, significantly increased the powers of local authorities by providing them with the ability to issue a Prohibition Notice and so immediately stop the use of the ground.[1]

1 See **16.36** below.

16.15 What Mr Justice Popplewell had highlighted was the connection between policing of stadia to deal with crowd violence and the safety issues that arose therefrom. It has rightly been pointed out that:

> 'this tendency to amalgamate issues of crowd management and crowd disturbances sets a dangerous precedent.'[1]

These issues coalesced perfectly in the events at Hillsborough on 15 April 1989.

1 Gardiner, O'Leary, Welch, Boyes and Naidoo *Sports Law* (Routledge, 4th edn, 2012).

16.16 On 15 April 1989, a crowd of over 50,000 travelled to Hillsborough to watch an FA Cup Semi-Final between Liverpool and Nottingham Forest. As was common at the time, Hillsborough was a stadium comprising seating areas and standing terraces. At around the time of kick-off, it had become clear that the central area of the Leppings Lane terrace, which was already overcrowded, was causing distress to Liverpool supporters. The report of the Hillsborough Independent Panel states:

> 'in fact, the small area in which the crush occurred comprised two pens. Fans entered down a tunnel under the west stand into central pens 3 and 4. Each pen was segregated by lateral fences and a high overhanging fence between the terrace and the perimeter track around the pitch. There was a small locked gate at the front of each pen.
>
> The crush became unbearable and fans collapsed underfoot. To the front of pen 3, a safety barrier broke, creating a pile of people struggling for breath. Despite CCTV cameras transmitting images of the stress in the crowd to the ground control room and to the police control box, and the presence of officers on duty on the perimeter track, it was a while before the seriousness of what was happening was realised and rescue attempts were made.
>
> As the match was stopped and the fans were pulled from the terrace through the narrow gates onto the pitch, the enormity of the tragedy became evident. Fans tore down advertising hoardings and used them to carry the dead and dying the full length of the pitch to the stadium gymnasium.'[1]

1 Introduction, p 3.

16.17 By the end of the day, 96 people were dead and hundreds injured. The disaster was, and remains, the worse sporting disaster in the history of British sport. It was to produce, however, an outcome which was to change the safety arrangements at sporting events generally, and football venues, in particular, forever.

16.18 Two days after the disaster, on 17 April 1989, Lord Justice Taylor was appointed by the Home Secretary to conduct a judicial inquiry into the disaster. The enquiry was wide ranging and raised many issues that fall outside the scope of this chapter. On 1 August 1989, Lord Justice Taylor produced his interim report. His primary conclusion was that the immediate cause of the disaster was the failure to close access to the central pens once gate C on Leppings Lane had been opened, leading to overcrowding, injury and death. This he attributed to 'a sluggish reaction and response' by the police. Poor police leadership, including the failure to respond to the urgency of the disaster as it occurred, taken with the small number of perimeter fence gates, hindered the rescue of those dying on the terraces.

16.19 Not that South Yorkshire Police were the only party subject to criticism. Lord Justice Taylor found that Sheffield City Council had a statutory duty to issue, monitor and revise the stadium safety certificate. It had failed in its duty, as the safety certificate 'took no account of the 1981 and 1985 alterations to the ground'. Put shortly, the regulatory framework of safety certificates put in place by the 1975 Act had not worked.

16.20 This time, the Football Association, the governing body responsible for the regulation of football in England, could not escape. They were criticised for having had no procedure for checking the validity of safety certificates at venues chosen to stage FA competitions. Therefore, despite the fact that the

1975 Act had ushered in an era of safety certificates, they had been ineffectual in achieving their goal of making football stadia safe.

16.21 The most significant finding of the Taylor Report was, however, that the disaster was caused, or significantly contributed to, by the prevalence in top-flight English football of standing terraces. It recommended that all such terraces be removed and that all clubs become seating-only venues. The Football League (which at that time encompassed the top four levels of English football, including what is now known as the Premier League) and the Scottish Football League accepted these proposals, and introduced regulations that, by August 1994, compelled all clubs in the top two tiers of football to provide seating-only stadia.

16.22 In the past, issues of cost had bedevilled reform. Safety certificates had not been introduced for fear that they would drive clubs out of business. When they were introduced, it was clear that, as in the case of Hillsborough, their terms had been neither monitored nor complied with. Therefore, once again, the issue of affordability was at the fore.

16.23 This issue coincided, happily, with a revolution in the way in which football rights were sold and exploited. In 1991, the top division of football in England broke away from the Football League and formed the FA Premier League. It chose to sell and exploit its rights separately from the Football League, which now embraced the three tiers of professional football below.

16.24 Coinciding with the launch of satellite television and its desire to obtain market traction, the results were startling. In 1988, the Football League had entered into a deal for the sale of the television rights to football for the period to 1992 for £44m. This £44m (roughly £11m per year) was split between all 92 clubs in the League, albeit weighted towards what are now the Premier League clubs.

16.25 With the introduction of Sky in 1992, the rights sold on their first auction for £253.5m for the period 1992–97. On renewal in 1997, for the period 1997–2001, the rights value rose for Premier League football alone to £848m. While, undoubtedly, significant sums were spent on player transfers and on wages, it is certainly true that these sums also helped facilitate the redevelopment of old stadia, or the building of new ones.

16.26 In Lord Justice Taylor's final report, he states:[1]

> 'it is a depressing and chastening fact that mine is the 9th official report covering crowd safety of football grounds. After 8 previous reports and 3 editions of the Green Guide, it seems astounding that 96 people could die from overcrowding before the very eyes of those controlling the event.'

1 'Hillsborough Stadium Disaster Final Report' (HMSO Cm 962), at p 4.

16.27 The Taylor Report was instrumental in changing this. The structure of stadia changed as a result of the acceptance of the recommendations by the Football League and the Scottish Football League and the new resources that had come into football via the sale of media rights. The report represented a recognition that the safety certificate, as a means of prevention, would not work unless the certificates were updated, complied with and enforced. It is very significant that, since the Taylor Report, there have been no significant disasters at football stadia.

Some of the above issues will be addressed in the new inquests into the deaths at Hillsborough, scheduled to start on 31 March 2014. However, the purpose of the inquest is not to determine issues of civil or criminal liability. Whilst it is open to the coroner to make recommendations to prevent deaths occurring in similar circumstances in the future, it is likely that many of the issues will have already been addressed, as several lessons have already been learned, and more modern stadia have been constructed which reflect the new higher standard of stadium design.

THE LEGISLATIVE FRAMEWORK

16.28 The main specific obligation imposed on those operating commercial sporting events is to hold and comply with the safety certificate as set out under the Act 1975[1] and its subordinate regulations.[2] Under the Act, the Secretary of State is empowered to designate any sports ground which has accommodation of more than 10,000 spectators, or 5,000 in the Premier and Football League grounds of England and Wales, as requiring a safety certificate.

1 As amended by the Fire Safety and Safety of Places of Sport Act 1987.
2 Safety of Sports Grounds Regulations 1987, SI 1987/1941.

16.29 Most professional sports grounds will be designated (this includes those which have a retractable roof, if some open air sport takes place) or non-designated, but have a 'regulated stand' and so be covered by the Fire Safety and Safety of Places of Sport Act 1987.[1] Therefore, most professional grounds will be subject to this licensing regime.

1 Smaller sports grounds with covered stands and a capacity of 500 or more spectators ('regulated stands') require a safety certificate under the Fire Safety and Safety of Places of Sport Act 1987.

16.30 A safety certificate may be either a 'general safety certificate', which covers the use of the stadium for viewing an activity, or a number of activities, specified in the certificate for an indefinite period which starts on a specified date, or a 'special safety certificate', which covers the use of the stadium for viewing a certain specified activity or activities on a certain specified occasion or occasions.

16.31 The application is made to the local authority, although in practice the application process and some monitoring of compliance is through Safety Advisory Groups. Safety Advisory Groups include representatives from the local authority, the police, the Ambulance Service, the Fire Service and other relevant bodies (such as the organisers, venue owners and voluntary groups) convened under the authority of the local authority. These groups have no legal status or powers as such, but are important in monitoring whether event organisers are discharging their legal responsibilities.

16.32 The safety certificate will include the capacity of the ground or stand and the terms and conditions which the duty holder must comply with to maintain that capacity. The certificate requires the management of the venue to set out, in an operations manual, how reasonable safety is to be maintained. The onus is on the management to assess its own risks and tailor its response to meet its particular circumstances.

16.33 The safety certificate will also identify who in the organisation is a Qualified Person. This is a person likely to be in a position to prevent contravention of the terms and conditions of the safety certificate. This could include the Chairman, Chief Executive, Club Secretary, Ground Manager, Safety Officer or a director of the venue operator, depending on the structure of the organisation.

16.34 The Qualified Person is a likely starting point for any accident investigation, as any investigation is likely to start with an explanation from the Qualified Person as to how an incident (which may have breached the safety certificate) could have been occasioned. The venue operator needs to ensure that such persons are competent, adequately trained and sufficiently resourced. The certification process requires a planned and integrated approach to the layout of the sports ground, the safety systems and equipment, and the procedures and personnel for managing spectators. No element should be considered in isolation.

16.35 The Qualified Person can be personally criminally liable under the 1975 Act, as they are actual holder of the certificate. An offence can also be committed by a 'Responsible Person', being the person concerned in the management of the ground at the time the offence took place. The penalty on conviction in the magistrates' court is a fine of up to £5,000 and, if tried in the Crown Court, the penalty is up to two years' imprisonment and/or an unlimited fine.

16.36 Once a ground has been designated, it is an offence to admit spectators to it unless an application has been made to the certifying authority for a safety certificate. Once the certificate is granted, there are a range of measures open to the local authority to ensure compliance. The most draconian measure which can be adopted (usually in the aftermath of an incident or investigation) is to issue a Prohibition Notice. This is done in circumstances where the admission of spectators involves, or will involve, a risk so serious that the admission of spectators must be restricted or prohibited entirely until steps are taken to reduce such risks to what the local authority regards as a reasonable level. The local authority also has the power to amend or replace the safety certificate on its own initiative. It is an offence for any Responsible Person, not merely the certificate holder, to contravene the terms and conditions of a safety certificate or a Prohibition Notice.

16.37 A separate licensing system is administered and enforced by the government-sponsored Sports Ground Safety Authority ('SGSA'), formerly known as the Football Licensing Authority. Under that system, licences are issued to sports grounds which are used for designated matches, and it is an offence to admit spectators for such matches without a licence. The primary purpose of the SGSA licence is to ensure compliance with the government's policy on all-seater stadia. Conditions imposed in this licence do not apply to other (non-football) events.

16.38 The SGSA has the power to keep under review how local authorities discharge their responsibilities under the provisions of the 1975 Act. This is achieved through monitoring safety certificate compliance, records sample checks and observations. The SGSA may give notice in writing to a local authority to include in any safety certificate issued such terms and conditions as specified in the notice, and it is the duty of the local authority to comply with that notice. If venue operators fail to comply with SGSA recommendations as to the safety certificate, as transmitted through local authorities, venue operators may be liable to penalties for breach that range from words of advice (for minor and quickly remedied breaches), through formal cautions, to a prosecution for serious or persistent breaches.

16.39 Safety certification involves consultation with the relevant emergency services, in particular the Fire Authority. Compliance with the Regulatory Reform (Fire Safety) Order 2005[1] overrides any condition or requirement under a safety certificate. The Order requires that a 'Responsible Person' is employed in sports grounds to ensure that a suitable and sufficient fire risk assessment is carried out. Resulting from this, a fire safety policy is put in place which sets out evacuation procedures, how sufficient information will be provided, and staff training. Although specific to the risk of fire, this should form part of the overall safety management system.

1 SI 2005/1541.

16.40 This sports-specific legislation sits alongside the general health and safety legislative framework. In its interpretation and application of the HSWA 1974 in the field of sport, there is an increasing tendency by the HSE (and also the local authorities who retain responsibility for enforcement in certain sectors) to construe the 'workplace' in broad terms, to include venues, tracks, stadiums, racing circuits and even ad-hoc competition routes organised round the country (for example, in cycling or in car rallying). This expanded modern definition of 'workplace' brings those who have control over sporting activities within the ambit of the Act and so imposes duties upon them.[1] Fundamental to the legislation is that the 'employer' must reduce the risk to safety of those affected by its undertaking to as low as is reasonably practicable. This can be an onerous duty and is owed to employees, appointed officials, competitors in sport and also to 'members of the public and groups of people who share your workplace'.[2]

1 An example of prosecution of a sporting organisation is one brought against Aston Villa Football Club in July 2010, who were fined after a contractor fell from the roof that he was working on at their training ground.
2 HSE guidance, 'Five Steps to Risk Assessment', first published June 2011.

16.41 Linked to, but outside, the general safety legislative framework is the Corporate Manslaughter and Corporate Homicide Act 2007 (CMCHA 2007). Organisations (whether in the UK or abroad) whose activities cause harm resulting in death sustained in the UK will be subject to the offence.[1] This will cover the vast majority of professional sporting operations. Sporting organisations will be treated in the same way as any other employer. The Act has its roots in the law of negligence. For an organisation to be guilty of an offence, there must be a duty of care, a gross breach of that duty which came about due to a senior management failing, the breach must have caused the death, and the death must be foreseeable. Relevant to the determination of whether a breach was gross is whether there has been a previous failure to comply with health and safety legislation; this would include all the specific sports-related legislation, the Regulatory Reform (Fire Safety) Order 2005 and the HSWA 1974. A corporate manslaughter investigation will lead to an examination of the entire safety culture of the business and not just the systems and management processes in place which were relevant to the fatal accident.

1 'Organisations' include incorporated bodies, statutory bodies (such as local authorities and hospitals), partnerships, trade unions, certain government departments and the police.

16.42 In respect of catastrophic incidents, any prosecution would most likely be brought under the CMCHA 2007 (if death occurred) and/or the HSWA 1974. However, unaddressed breaches of a safety certificate or other relevant safety regulations may serve to provide evidence of 'senior management failings' and/or a breach of general safety duties in this context.

THE POSITION OF GOVERNING BODIES

16.43 The position of sports governing bodies, regarding health and safety considerations at venues at which no competitions are being played, is far from settled. As we have seen earlier, as far back as the Shortt Report in 1924, the FA did not even see fit to attend the inquiry or to provide any evidence, still less to listen to any of its recommendations.

16.44 In the context of the selection of Hillsborough as a venue for the semi-final of the FA Cup in 1989 between Liverpool and Nottingham Forest, Lord Justice Taylor found that the FA had no procedure put in place for checking the validity of the safety certificate of venues chosen by it.

16.45 In addition, he found that the FA:

> 'did not consider in any depth whether it [Hillsborough] was suitable for a high risk match with an attendance of 54,000 requiring to be segregated, all of whom were, in effect, amongst supporters lacking week-in week-out knowledge of the ground.'

16.46 The Football League should be exempted from such criticism, as it implemented, swiftly and without qualification, Lord Justice Taylor's recommendations for the introduction of all-seater stadia for top-level football matches.

16.47 It is clear that sports governing bodies are not immune to challenge for liability for death or personal injury caused at competitions conducted under their jurisdiction. A prime example of this derives from the facts of *Watson v British Boxing Board of Control*.[1]

1 [2001] QB 1134.

16.48 In 1991, Michael Watson fought Chris Eubank for the WBO super-middleweight title in London. The fight was stopped in the final round when Watson could not defend himself. In fact, he had sustained a serious brain haemorrhage and became unconscious in the ring. In the following chaos, it was seven minutes before Watson was examined by a doctor licensed by the British Boxing Board of Control (BBBC), and it was 30 minutes before he arrived at the local hospital by ambulance. Lack of facilities at that hospital required that he be transferred to another specialist hospital, where he was operated on, but was found to have sustained serious brain damage, which left him partially paralysed and physically and mentally disabled.

16.49 In the Court of Appeal, it was noted that the Board imposed rules on the sport of boxing, including regulating the safety standards of boxing contests. Boxers looked to the Board to stipulate the appropriate safety and medical standards. In these circumstances, it was 'difficult' for the Board to be concerned over safety but to deny a duty of care to those affected. The court found that the Board was a body with specialist knowledge which gives advice to a defined class of persons that it knew would rely upon that advice. It noted that the Board was the sole governing body of boxing in the UK, and boxers and fights were licensed by the Board. In particular, the Board produced detailed rules to secure that boxers received appropriate medical attention when injured during the course of a fight. Lord Phillips summarised the situation in these terms:

> 'The Board set out by its rules, directions and guidance to make comprehensive provision for the services to be provided to safeguard the

> health of the boxer. All involved in the boxing contest were obliged to accept and comply with the Board's requirements.'

The court noted that Mr Watson said that his understanding as a boxer was that the Board undertook responsibility for all medical aspects of boxing, including the medical supervision of boxing contests.

16.50 In the circumstances, the Court of Appeal considered there was a relationship of proximity between the Board and Mr Watson that could give rise to a duty of care. It is also very interesting to note the scope of this duty of care, as (i) it required a duty to take reasonable care to ensure personally that injuries already sustained were properly treated; and (ii) the duty was not applied directly through the Board's servants or agents, but was to make regulations imposing on others the duty to achieve this result. As the court explained:

> 'where A advises B as to action to be taken which will directly and foreseeably affect the safety or well-being of C, a situation of sufficient proximity exists to found a duty of care on the part of A towards C.'

It is interesting to note also that the court rejected a contention that the Board should not be held liable, on public policy grounds, as it was a governing body acting for the good of the whole sport, and it was a non-profit-making organisation.

16.51 The Court of Appeal therefore held that a duty of care existed, and then went on to find a breach of that duty. What happened at the venue within the first 10 minutes after a boxer suffered brain damage was seen as critical, and there should have been resuscitation equipment and proper emergency medical care provided at the venue itself.[1]

1 The outcome of this civil litigation was not without consequences for the BBBC: it has to sell its headquarters in Central London and move to Cardiff ('Boxing Board love fight with Watson' (2001) Daily Telegraph, 8 November).

16.52 The case clearly establishes civil liability on a sport's governing body for events conducted under its jurisdiction. What would be the position if *Watson* had occurred today, under current health and safety legislation and under the regime of enforcement that now exists?

16.53 In the authors' view, such a catastrophic failure to ensure proper medical care for Mr Watson could, and probably would, render the organisers of the event, and the governing body under whose rules it was staged, susceptible to prosecution under s 3 of the HSWA 1974. In the unfortunate event that Mr Watson had died, the authors consider that, in addition, and based on the principles set out earlier in this book relating to corporate liability for work-related deaths, the Corporate Manslaughter and Corporate Homicide Act 2007 would be engaged, and those organisations responsible for the regulation of Mr Watson's bout could have been subject to criminal charges.

16.54 Although the authorities in football have disclaimed any role in regulating or monitoring the safety of spectators at football grounds, the situation is different in motor sport. As we shall see later, in motor sport events, national events within Great Britain are subject to licensing by the Motor Sports Association ('MSA'), and the licence conditions attaching to the grant of such licences contain extensive obligations to ensure the health and safety of spectators and participants. At an international level, events that wish to be placed upon the international calendar are subject to the grant of a licence by the Fédération Internationale de l'Automobile ('FIA'), which itself imposes stringent detailed and specific licence conditions on any applicant seeking a licence to stage an event which is to be put on the FIA calendar.

16.55 The civil liability of the MSA and the FIA was examined in *Wattleworth v Goodwood Road Racing Co Limited, Royal Automobile Club Motor Sports Association Limited and the Fédération Internationale de l'Automobile.*[1]

1 [2004] EWHC 140 (QB).

16.56 In that case, Mr Wattleworth was killed when his car crashed during private testing at the Goodwood racing circuit in Sussex. It was alleged by Mr Wattleworth's family that his death had been caused by the design of certain structures on one of the bends at the circuit, and the sport's UK governing body, the MSA, and the world governing body, the FIA, were negligent.

16.57 The court noted that the circuit operator at Goodwood had to comply with MSA requirements if it was to obtain a licence, and that the circuit would only be licensed if, after inspection, it was considered fit to stage racing and motor sport. It was also noted that the FIA, on the other hand, was concerned with international events, and did not assume the responsibility for track safety at national events. Safety at national events was a matter for the MSA.

16.58 The case hinged on the design of a safety barrier made up of tyres located at the corner where the fatal crash occurred, and whether this was of a substandard and negligent design. The court noted that Goodwood Ltd devoted much time and attention to safety issues in the construction and design of the Goodwood circuit, but that at all times they consulted with and relied upon the MSA. The court therefore found that Goodwood Ltd had in fact discharged its duty of care by liaising closely with the MSA in the design of safety barriers at the circuit.

16.59 The authors note that this is an unusual and surprising aspect of this case, as those who own and promote sporting venues, and who commission and implement designs, clearly have a responsibility, not least under the Occupiers Liability Act 1957, to ensure that the venue is safe for lawful visitors and users of the venue, and it would be very unusual to be able to discharge this duty of care simply by relying upon the advice and licensing process of a sport's governing body.

16.60 As regards the UK governing body, the MSA, the court found that it did owe a duty of care to the racing driver who was killed, and strong reliance was placed on *Watson*.[1] The court noted that the circuit had been licensed by the MSA and, as part of that process, the safety aspects of the circuit had been scrutinised and approved by the MSA. The court therefore held that the deceased racing driver was entitled to have assumed that all due care had been exercised by the MSA in undertaking responsibility for safety matters. As the court said:

> 'the MSA assumed a responsibility going well beyond the mere authorisation of events for which an MSA Track licence and event permit was required. In my view, the MSA … was plainly giving advice to Goodwood both as to the circuit and as to the protective devices to be deployed around the track, when it contemplated and expected that such advice would be acted upon by Goodwood not only with regard to the MSA events but also with regard to other motor car uses of the circuit.'[2] [The driver had been killed at a private testing event at the circuit.]

The court continued:

> 'the MSA had, with regard to the circuit and safety barriers at Goodwood, voluntarily adopted an advisory role going well beyond the limited role of licensing or authorising MSA events. Indeed given the acknowledged expertise of the MSA and the universal desire to achieve high safety standards for a circuit that is an entirely understandable and responsible position for the MSA to have taken.'[3]

Once again, suggestions that the MSA should not be under a duty of care, because it was a non-profit governing body acting in the benefit of the sport, were rejected by the court.

1 *Watson v British Board of Boxing Control* [2001] QB 1134.
2 [2004] EWHC 140 (QB) at para 120.
3 At para 121.

16.61 It is also interesting to note that, whilst a defence of 'volenti non fit injuria' (consent) was raised, the court explained that:

> 'Of course Mr Wattleworth must be taken to have consented to the risks inherently involved in motor car racing, and I am sure that he did so. But that would have been on the basis that due steps would have been taken to see that the circuit (including the crash barriers) was reasonably safe and that those responsible for circuit safety – whoever they may be – would have taken reasonable care to provide appropriate barriers.'[1]

1 At para 174.

16.62 However, the court considered that the FIA, as the world governing body, did not owe a duty of care for this national matter. The court said of the FIA that the terms of the FIA Code, the structure of the FIA and the way in which its dealings with the relevant national sports association were conducted showed that primary responsibility with regard to safety at events for licensing purposes was left to the national sports association in question, the MSA.

16.63 As for the MSA, the court found that, whilst a duty of care existed, the duty had been discharged with reasonable skill and care. The design of the crash barrier at the bend on the circuit where the racing driver had been killed 'was a reasonable choice, designed to meet the various foreseeable types of collision that might occur at that section and appropriate for the foreseen wide range of cars, drivers and uses'.

16.64 In the authors' view, where a governing body does assume responsibility for licensing of venues and events, if death or personal injury occurs, it is at least theoretically possible that the HSWA 1974 and the CMCHA 2007 could come into play. However, the burden facing the prosecuting authorities would arguably be a very high one. What would need to be shown is that the licensing conditions, by their very nature, were such as to engage the legislation. The degree of oversight and involvement by the governing body would be very relevant as to whether responsibility had been assumed and therefore whether it should also attract liability for acts and/or omissions contributing to an unsafe situation. In practice, it is not the licensing conditions as such which give rise to liability, but their implementation or failures in their implementation. Therefore, although governing bodies such as the MSA and the FIA may impose these licensing conditions, in practice, health and safety and corporate manslaughter liability is usually more likely to reside with organisers, rather than those setting the conditions within which the organisation operates.

SAFETY AT MOTOR RACING CIRCUITS

16.65 At the start of the chapter, we mentioned the responses of the relevant authorities in football and motor sports to fatalities occurring among spectators and participants. Pre-Hillsborough and the Taylor Report, which introduced compulsory all-seater stadia, and before the death of Ayrton Senna in 1994, it is fair to say that the responses of football and motor sport organisations to fatalities was less than the public and those inquiring into these disasters (and the participants themselves) had required. However, in motor sport since the death of Ayrton Senna, real strides have been made in the elimination of fatalities among racing drivers in motor sport.

16.66 It should be noted that the mode of regulation in motor sport is radically different from that in football. Whereas football has been forced along the road of statutory regulation through a system of licensing operated by local authorities, motor sport operates through licences granted by the world governing body for motor sport, the FIA, which is headquartered in Paris and Geneva for international events, and through its affiliated members, known as ASNs, the British equivalent of which is the MSA for national events.

16.67 There is no doubt that safety has historically been an issue in motor sport. In Formula One, during the four-year period from 1963 to 1967, there were three racing drivers killed in Grand Prix races, rising to four killed in the period from 1968 to 1972, and six killed in the period from 1973 to 1977.[1] Further deaths followed in 1978, 1980 and then, in 1982, a very high-profile driver, Gilles Villeneuve, was killed. Over these periods, in addition to attempts to improve the safety of racing car designs, the protection of drivers by safety equipment, and the improvement of the safe organisation of motor races, strenuous efforts were made to improve the safety of motor racing circuits around the world.

1 Jo Schlesser (7 July 1968), Gerhard Mitter (2 August 1969), Piers Courage (7 June 1970) and Jochen Rindt (5 September 1970); Roger Williamson (29 July 1973), Francois Cervert (6 October 1973), Peter Revson (30 March 1973), Helmuth Koinigg (6 October 1974), Mark Donohue (19 August 1975) and Tom Pryce (5 March 1977).

16.68 There are a number of crucial components to striving to achieve acceptable levels of safety in motor sport: the design of racing cars (for example, 'survival cells' designed as part of the structure of the racing car, with crash testing to evaluate the suitability of a design); safety equipment for racing drivers (for example, protective racing helmets and fire-resistant clothing); the safe organisation of motor sport events (for example, fire, medical and safety vehicles available for deployment during a race); and, finally, the safety of motor racing circuits themselves.

16.69 Starting in 1963, the world governing body, the FIA, organised circuit safety inspections, with mandatory inspections from 1970, together with improvements in circuit design, such as safety catch fencing and trackside barriers, gravel speed arrestor beds from 1973, and permanent medical centres from 1980. These safety initiatives, introduced and enforced by the FIA, began to lead to tangible improvements in safety, with no drivers killed in Grand Prix from 1983 until 1994 (although Elio de Angelis was killed in a Grand Prix testing incident in 1986).

16.70 All that was to change, however, in 1994 at the San Marino Grand Prix at Imola, when Roland Ratzenberger was killed during qualifying, followed by the death of Ayrton Senna, the legendary triple World Champion during the race. The well-known commentator Murray Walker described this as 'the blackest day for Grand Prix racing I can remember'.

16.71 In an urgent and concerted response to the events of that 1994 weekend, the FIA embarked upon an even greater drive to improve safety at circuits. In this, they were helped by the dramatic advances in technology leading, for example, to video evidence of incidents, inboard data recorders located on racing cars, the testing of safety barrier designs, and the use of computers to provide circuit simulation to identify areas of danger. Responding to the events of 1994, computer analysis identified many high-risk corners at circuits worldwide that were immediately targeted by safety improvements.

16.72 The advancing technological systems now enable sophisticated computer simulation of the racing line at circuits. This enables the identification of danger areas, where racing cars can be expected to run off or potentially crash, leading to the location of asphalt run-off areas (considered much more effective than old gravel beds), and the calculation of barrier impact speeds, so that the appropriate type of energy-absorbing barrier systems are introduced, with an on-going high-speed testing programme of energy-absorbing barriers, and the development of new designs.

16.73 Regulatory approval of motor racing circuits by the FIA is now the subject of a comprehensive regulatory process. Before circuits can be licensed and recognised for inclusion on the international sporting calendar of the FIA, a detailed dossier of plans and specifications have to be submitted via the national motor sport governing body for approval by the FIA. There then follows detailed computer analysis of the design using race simulations, together with detailed inspections carried out by FIA-approved Circuit Inspectors. Once approved, any modifications to the layout or safety installations will render the circuit ineligible for FIA-approved events. The Inspection Report will specify a detailed schedule of required safety work at the circuit, and this work must be implemented under agreed works programmes, for the circuit to be granted an FIA Circuit Licence. Circuit Licences are granted by the FIA in grades from 1 to 6 on an annual basis, according to the types of racing car deemed suitable for the circuit in question, and 'the granting of a licence is a prerequisite for the submission of an application to enter any circuit event on the FIA International Sporting Calendar'. The premier championship, the FIA Formula One World Championship, requires a top-level circuit Grade 1.

16.74 By way of further illustration, in the evaluation of a particular circuit dossier by the FIA-approved Inspectors, a wide range of matters relevant to a safe motor racing circuit will be considered, including: the overall plan, width and profile of the circuit; the track edges and run-off areas; the starting straight; measures to protect the safety of spectators, and drivers, race officials and service personnel during competitions; the provision of circuit buildings (including Race Control, Marshalls' Posts, and a Medical Centre); service roads; facilities for the disabled; trackside advertising and structures; accident reporting; and circuit maintenance. This provides an up-to-date safety audit of the circuit in question, together with a regulatory regime designed to ensure the safe provision of a circuit for a particular type of racing. The licensing regime requires an on-going programme of maintenance and safety improvement, through regular inspections and the licences themselves being reviewed and reissued on an annual basis.

16.75 With the application of this comprehensive drive to improve safety in Formula One, it is worthwhile to note that there have been no further deaths in the 18 years since that terrible weekend at Imola in 1994.

THE FUTURE

16.76 Investigation of how health and safety is being managed is no longer confined to the traditional workplace; the legislation encompasses a wide range of organisations which owe health and safety duties to their employees and others affected by their undertaking. Although not in a sporting context, the conviction of the Metropolitan Police under s 3 of the HSWA 1974 for the failings which led to the death of Jean Charles de Menezes[1] is a clear departure from the origins of the legislation, and demonstrates that prosecutions will be brought for operational mistakes rather than generic policy failings, as was generally the case previously.

1 The Metropolitan Police were convicted on 1 November 2007, fined £175,000 and ordered to pay £385,000 costs.

16.77 Health and safety must be regarded as a key corporate governance issue led by the directors and senior managers. The health and safety legislation and the CMCHA 2007 impact on directors and anyone responsible for controlling safety-related activities.

16.78 Whilst there are no statutory safety duties placed on directors, joint guidance produced by the Institute of Directors and the HSE sets out what is required.[1] Good safety management requires active and visible leadership from the board, with effective two-way communication between them and the employees. For instance, health and safety should appear as a standing item on board meeting agendas, where safety issues (such as new risks identified and safety performance) are discussed. This should be supported by the provision of adequate training and resourcing, both for those implementing safety procedures and those carrying them out on a day-to-day basis. Importantly, there needs to be on-going assessment and review through comprehensive monitoring of safety performance. Scheduling in regular reviews of the policies, and ensuring that all documentation relating to these reviews is kept accurately and is up to date, will assist in showing that efforts have been made to comply with safety duties owed. The specific legislative framework and rules of the various governing bodies can prompt and assist this process, but ultimately it is for each organisation to identify its own safety risks and ensure it is operating safely.

1 'Leading Health and Safety at Work', available at www.hse.gov.uk/pubns/indg417.pdf.

16.79 Stadium safety and the roles and responsibilities of those operating at the ground, including the police and the Ambulance Service, will be brought into focus again at the new inquests into those who died at Hillsborough. These inquests are due to begin on 31 March 2014. Although the focus will be on what happened and the manner of the individuals' deaths, it will invariably spark wider public debate. The question of how the victims died will be approached in a manner broader than just an investigation into the mechanics of the individual deaths. The coroner has determined that these are inquests that engage Article 2 of the European Convention on Human Rights.[1] As such, the 'how' the deceased came by their deaths will incorporate not only 'by what means' but also 'in what circumstances'. This will require an inquiry into all factors, including the stadium construction, access, planning for the event, and the response to the overcrowding, which may have been causative of the circumstances leading to the victims' deaths.

1 Article 2 of the European Convention on Human Rights protects the right to life. The case law on Article 2 has established that there is a 'general duty' on the part of the state to protect life. This has been modified by the additional layer of an 'operational duty' where the authorities knew, or ought to have known, of a real and immediate risk to the life of an individual and where they failed to take measures which could have avoided that risk. If there is evidence of a breach of the general or operational duty, the state is obliged to investigate. This is the 'procedural duty'. This duty to investigate can be discharged either by criminal proceedings, an inquest or by other means. In *R (Middleton) v HM Coroner for Somerset* [2004] UKHL 10, the House of Lords held that, in order to be compliant with Article 2, s 11 of the Coroners Act 1988 had to be interpreted so that 'how' the deceased came by death included the wider circumstances.

16.80 Finally, as the report of the Hillsborough Independent Panel shows, it is important that the lessons of the past are not forgotten. It may be that reports written with the benefit of many years' hindsight can be of dubious usefulness in circumstances where they do nothing more than cover ground that has been covered before. However, insofar as they illuminate the mistakes of the past so as to prevent mistakes in the future, such inquiries may well do the public a very great service.

Chapter 17

Emergency services: law and liability

Ruth Barber LLB (Hons), Solicitor Advocate

INTRODUCTION

17.1 Unlike many other jurisdictions, there exists no positive duty on the citizen under English common law to rescue strangers. The absence of duty is justified as an absence of proximity (ie sufficient relationship) between the parties. If liability is imposed, there must be some additional reason why it is fair and reasonable that one person should be regarded as another's keeper. This results in the paradox that those who choose to intervene are exposed to liability if the rescue attempt goes badly, whilst those that walk by escape liability entirely. The courts are thus reluctant to find a want of care or any general duty to make things better. A defendant volunteer simply has to avoid making the situation worse. This immunity extended to the emergency services until 2001, when a positive duty of professional rescue on the ambulance service only was recognised to exist.

FRAMEWORK OF LIABILITY AND THE DUTY OF CARE

17.2 In order to establish negligence, it is necessary for a claimant to establish three requirements:

(a) the defendant owed him a duty of care, including avoidance of the damage suffered;
(b) the defendant was in breach of that duty; and
(c) the claimant suffered damage as a consequence.

17.3 Liability of the emergency services arises, just as it does for any other person, under these general principles of negligence. The duty of the emergency services is a duty to take such care as is reasonable in the circumstances. The peculiar difficulty in relation to the emergency services lies in establishing the scope of the duty, since they do not, as a rule, create loss danger but respond to it.[1]

1 See *Capital and Counties Plc v Hampshire County Council* [1997] 3 WLR 331.

17.4 In order for a duty of care to exist, there must be:[1]

(a) Foreseeability of damage
The claimant has to establish that there was a real and substantial risk or chance that something like the event which happens might occur.
In *McNern v Commissioner of Police*[2] the claimant alleged that he had suffered psychiatric injury as a result of an arrest carried out by police using a warrant. His claim was rejected. The Court of Appeal held that the police are entitled to assume that a person arrested is a person of ordinary robustness, unless told otherwise. It was not reasonably foreseeable that someone would suffer psychiatric injury from arrest.

(b) A relationship of proximity or neighbourhood between the parties
The requirement of proximity involves the notion of nearness or closeness, and embraces physical proximity between the person or property of the claimant and that of the defendant, circumstantial proximity (such as an overriding relationship of employer and employee) and what may be referred to as casual proximity (in the sense of the closeness or directness of the casual connection of the relationship between the particular act or cause of conduct and the loss or injury sustained).[3]
In *Watson v British Boxing Board of Control*,[4] the claimant suffered a severe head injury whilst engaged in an authorised fight. There was a delay in providing medical treatment, which caused a deterioration of the claimant's condition. The defendant did not have a duty to avoid causing personal injury, but to ensure that injuries were properly treated. It was not, however, involved in administering treatment, merely drawing up the rules under which treatment was given. The Court of Appeal determined that there is a distinction between cases where there is an assumption of responsibility to an individual and cases where there is an assumption of a duty to protect the general community. The Board was therefore held liable.
In contrast, the fire brigade was held to be under no duty to answer a call for help or, having done so, to exercise reasonable skill and care, since the role of the fire brigade is to protect the community.[5]
Where the police negligently released details of a police informant, the Court of Appeal concluded that there was sufficient proximity, as there had been an assumption of responsibility to a particular individual.[6]
Conversely, proximity was not established where a claim was made against the police for failure to apprehend a criminal, leading to a further death. The duty that the police owed to apprehend the criminal was owed to the public and not to the deceased.[7]

(c) That it is fair, just and reasonable that the duty should be applies in the circumstances of the case.

The first two tests relate to the facts of the incident; the 'fair, just and reasonable' test requires the courts to exercise a public policy role. In the case of *Hill v Chief Constable of West Yorkshire*,[8] it was alleged that the police had been negligent in failing to apprehend Peter Sutcliffe (the 'Yorkshire Ripper'), and this had led to the death of his last victim. The House of Lords held that public policy requires that the police should not be liable for failure to apprehend the perpetrator or crimes:

> 'In some cases, the imposition of liability may lead to the exercise of a function being carried on in a detrimentally defensive frame of mind … A great deal of police time, trouble and expense might be expected to have to be put into the preparation of the defence to the

action and the attendance of witnesses at the trial. The result would be a significant diversion of police manpower and attention from their most important function, that of the suppression of crime.'

By contrast, the Court of Appeal refused to grant immunity to the fire brigade on public policy grounds, and disagreed with arguments that a finding of liability against the fire brigade would lead to defensive firefighting and would open the floodgates of litigation.[9]

1 *Caparo Industries Plc v Dickman* [1990] 2 AC 605.
2 (unreported) 18 April 2000.
3 *Shire of Sutherland v Hayman* (1985) 60 ALR 1.
4 *Watson v British Boxing Board of Control* [2001] 1 WLR 1256.
5 *Capital & Counties plc and others v Hampshire County Council and others* [1997] 3 WLR 331.
6 *Swinney v Chief Constable of Northumbria Police* [1996] 3 WLR 968.
7 *Hill v Chief Constable of West Yorkshire* [1989] AC 53.
8 [1989] AC 53.
9 *Capital & Counties plc and others v Hampshire County Council and others* [1997] 3 WLR 331.

Causation

17.5 In order for negligence to be established, it is necessary for a claimant to satisfy the court that a defendant has been in breach of his duty of care to that claimant and that the claimant has suffered damage by reason of that breach. The test for the establishment of causation is the 'but for' test, ie that the claimant would not have suffered damage but for the defendant's negligence.

Immunity from claims

17.6 The European Court of Human Rights was unimpressed by the stance of the Court of Appeal, in relation to the question of immunity, where the court was being asked to strike out a claim based upon no reasonable prospect of success.

In *Osman v Ferguson*,[1] it was alleged that the police had been negligent in failing to apprehend someone who was harassing the Osman family. Eventually, the police decided to arrest the person; however, before they could do so, he killed a member of the Osman family and another person. Relying on the decision in *Hill*, the Court of Appeal struck out the claim. The European Court of Human Rights decided that there had been a breach of Art 6(1) of the European Convention on Human Rights which provides:

> 'In the determination of his civil rights and obligations, everyone is entitled to a fair and public hearing within a reasonable time by an independent and impartial tribunal established by law.'

The European Court considered that the conferring of what it described as a 'blanket immunity on the police for their acts and omissions during their investigation and suppression of crime' amounted to an unjustifiable restriction on the claimant's right to have a determination of his claim on the merits:

> 'It must be open to a domestic court to have regard to the presence of other public considerations which pull in the opposite direction to the application of the rule.'

1 [1993] 4 All ER 344.

17.7 This decision was revisited by the European Court of Human Rights in *Z and others v United Kingdom*,[1] a case dealing with the liability of social workers:

> 'The court considers that its reasoning in the *Osman* judgment was based on an understanding of the law of negligence … which has to be reviewed in the light of the clarifications subsequently made by the domestic courts and most notably the House of Lords.'

The court referred to Article 6 of the European Convention on Human Rights and noted that this can be relied upon by anyone who considers that he has not had the opportunity of submitting his claim to a 'fair and public hearing'. The court noted the procedure whereby a claim can be struck out without the factual matters being determined on evidence, but then said:

> 'However if as a matter of law there was no basis for the claim, the hearing of evidence would have been an expensive and time consuming process. Which would not have provided the applicant with any remedy at its conclusion. There is no need to consider the striking out procedure which rules on the existence of sustainable causes of action as per se offending the principle of access to court.'

1 [2001] 2 FLR 612.

17.8 Following this decision, the House of Lords heard the case of *Brooks v Commissioner of Police for the Metropolis*.[1] The case arose out of the death of Stephen Lawrence. The claimant, a friend of Stephen Lawrence, witnessed the incident. He claimed to have suffered post-traumatic stress disorder, which had been exacerbated by failure on the part of the police to:

(i) take reasonable steps to assess whether he was the victim of a crime and accord him reasonably appropriate protection;
(ii) give him the protection and support commonly afforded to a key witness to a serious crime of violence; and
(iii) afford reasonable weight to his account of the incident.

1 [2005] UKHL 24.

17.9 Lord Steyn commented:

> 'With hindsight not every observation in *Hill* can now be supported. Lord Keith of Kinkel observed that "from time to time [the police] make mistakes in the exercise of that function, but it is not to be doubted that they apply their best endeavours to the performance of it": Nowadays a more sceptical approach to the carrying out of all public functions is necessary.'

Lord Steyn took the view that it was desirable that police officers should treat victims and witnesses properly and with respect. However, he continued:

> 'But to convert that ethical value into general legal duties of care on the police towards victims and witnesses would be going too far. The prime function of the police is the preservation of the Queen's peace. The police must concentrate on preventing the commission of crime; protecting life and property; and apprehending criminals and preserving evidence … a retreat from the principles in *Hill* would have detrimental effects for law enforcement. Whilst focusing on investigating crime, and the arrest of suspects, police officers would in practice be required to ensure that in every contact with a potential witness or a potential victim time and resources were deployed to avoid the risk of causing harm or offence.'

17.10 The Court of Appeal decided that the observations about public policy expressed in *Hill* were of general application. As a consequence, it was not fair or reasonable that the police should be under a duty of care in the investigation of suspected crime, in the circumstances where they had been called to a shop where a burglar alarm was sounded, but formed the impression that all was secure.[1]

The case of *Hill* must now be considered with the case of *Van Colle v CC of Hertfordshire*.[2] The claimant's son, a prosecution witness, was shot dead by a former employee who was about to stand trial. The police officer in charge of the investigation was found guilty of failing to perform his duties diligently, as he had ignored increasing evidence of intimidation of the witness.

The parents of the deceased sued the police under s 7 of the Human Rights Act 1998. Mrs Justice Cox found the defendant vicariously liable for breach of the 'right to life' guaranteed by Article 2 of the European Convention on Human Rights, as well as breach of the right to family life under Article 8. Article 2 imposes a positive duty on the state to do all that could be reasonably expected to protect those whom the police know, or ought to know, are at real and immediate risk. As a witness, Van Colle was a vulnerable person and therefore entitled to greater protection than members of the public generally.

Interestingly, although this was a claim under the HRA 1998, Cox J was not persuaded that a claim in negligence would have inevitably failed. Unlike *Hill,* Van Colle, his attacker and the police were all 'proximate'.

1 *Alexandrou v Oxford (Chief Constable of Merseyside)* [1993] 4 All ER 328.
2 [2007] EWCA Civ 325.

17.11 In *Ancell v McDermott*,[1] the police observed an oil spillage on the road, but did not erect warning signs. A claim alleging negligence against the police for failing to erect warning notices was struck out. The Court of Appeal took the view that the police were under no duty of care to protect road users from, or to warn them of, hazards which they discovered while going about their duties on the highway:

> 'The extreme width and scope of such a duty would impose on a police force a potential liability of almost unlimited scope, and it would be against public policy because it would divert extensive police resources and manpower from, and hamper the performance of, ordinary police duties.'

1 [1993] 4 All ER 355.

17.12 In the case of *Elguzouli-Daf v Commissioner of Police for the Metropolis*,[1] the claimants were charged with serious offences, but the prosecutions were subsequently discontinued. The claimants argued that the Crown Prosecution Service had been negligent in failing to obtain forensic evidence. The Court of Appeal held that it was not 'fair, just and reasonable' to impose a duty on the Crown Prosecution Service.

According to Steyn LJ:

> 'Such a duty of care would tend to have an inhibiting effect on the discharge by the Crown Prosecution Service of its central function of prosecuting crime. It would in some cases lead to a defensive approach by prosecutors to their duties. It would introduce a risk that prosecutors act so as to protect themselves from claims in negligence. The CPS would have to spend valuable time and use scarce resources in order to prevent lawsuits in negligence against them. It would generate a great deal of paper to guard against the risk of lawsuits. The time and energy of CPS lawyers would be diverted from concentrating on their prime function of prosecuting offenders. That would be likely to happen not only during the prosecution

process, but also when the CPS was sued in negligence by aggrieved defendants.'

1 [1995] 2 WLR 173.

RIGHTS TO DAMAGES CLAIM

17.13 A retained firefighter, Mr Wicker, and a fire service photographer and cameraman, Mr Wembridge, were among the first at the scene of the fire at the Festival Fireworks site on 3 December 2006.

They both died in the explosion, and 20 others, mainly police and fire officers, were also injured in the massive explosion at the site, which sent fireworks and debris flying across the area. Festival Fireworks' owners – father and son, Martin and Nathan Winter – were convicted of the men's manslaughter and jailed for seven and five years respectively.

Counsel representing the families of the deceased argued that there were a number of reasons why the tragedy happened. These included lack of training for the firefighters, a lack of information about Marlie Farm, and a failure of the authorities to inspect the site regularly.

17.14 The case about compensation went to court.[1] The court found that firefighters were not prevented from recovering damages from their fire service employers for injuries sustained during the course of their work: they accepted the risks which were inherent in their work, but not the risks which the exercise of reasonable care on the part of those who owed them a duty of care could avoid. Further, there was no 'fireground immunity' at common law that would protect a fire service from suit.

The claimant firefighters and police officers sought damages from the fire service for negligence and breach of its statutory duties. They argued that the fire service was liable both in common law and for breach of four statutory workplace regulations: the Provision and Use of Work Equipment Regulations 1998,[2] the Control of Substances Hazardous to Health Regulations 2002,[3] the Dangerous Substances and Explosive Atmospheres Regulations 2002[4] and the Management of Health and Safety at Work Regulations 1999.[5] The fire service argued that the regulations did not apply to firefighters because the duties overlapped with those imposed on fire services by way of the Fire and Rescue Services Act 2004, which were only target duties and not intended to give rise to civil liability for breach; and that, in any event, there was a fireground immunity.

1 *Wembridge Claimants v Winter* [2013] EWHC 2331 (QB).
2 SI 1998/2306.
3 SI 2002/2677.
4 SI 2002/2776.
5 SI 1999/3242.

17.15 Judgment was given in favour of the claimants. Section 47(2) of the Health and Safety at Work etc. Act 1974 provides that breach of health and safety regulations is actionable, except where it specifically says that it is not. The fire and rescue services were not expressly excluded from the Act. As the police were expressly excluded from the Act, it was unlikely that any exclusion of firefighters would not also have been made explicit. The argument was therefore rejected, because the primary statute contained target duties only, and it was implied that no civil liability arose either at common law or by the imposition of regulation. Therefore, the fire service's submission, that it was not the intention of Parliament that the workplace regulations should apply to the fire service, was rejected. Neither of the 2002 Regulations applied in this case because they supposed that the employer was or should be in control of the workplace, whereas the fireground was not occupied or controlled by the employer until there was a fire. That difficulty did not arise in respect of the 1998 or 1999 Regulations: the management obligations and the obligations to provide suitable equipment were not complicated by the fireground.

17.16 Firefighters accepted the risks which were part of their work, but not the risks which the exercise of reasonable care on the part of those who owed them a duty of care could avoid. This meant that firefighters were not prevented from recovering damages from their fire service employers. Further, the police officers who were closely involved in the co-operative effort to address the fire, and who were in such close physical proximity to firefighters, would also be owed a duty of care by the fire service. It would not seem fair, just or reasonable to acknowledge liability to fire service personnel but withhold a duty of care to police officers on the fireground. There was no 'battle immunity' in relation to decisions taken in the heat of the moment by those in charge in the emergency services.

Emergency vehicles

17.17 When an emergency vehicle is being used for emergency purposes, statutory provisions imposing speed limits do not apply, but this does not relieve the driver from civil liability of any accident caused by driving at an unsafe speed, even though the speed was required to enable the vehicle to reach the emergency. The statutory provisions also do not provide immunity from prosecution for dangerous driving.

17.18 In *Marshall v Osmond*[1] a police car was pursuing a stolen vehicle. The stolen vehicle pulled into a layby and the driver ran off. The police driver intended to draw up alongside the car, but unfortunately skidded and hit a passenger alighting from the stolen vehicle. The Court of Appeal accepted that the duty of care owed to a suspected criminal was the same as the duty of care owed to anyone else, namely to exercise such care and skill as is reasonable in the circumstances.

Sir John Donaldson MR stated:

> 'There is no doubt that he made an error of judgement because in the absence of an error of judgement, there would have been no contact between the cars. But I am far from satisfied on the evidence that the police officer was negligent.'

1 [1983] 3 WLR 13.

DUTY OF CARE OWED BY THE FIRE BRIGADE

17.19 The duty of care owed by the fire brigade was considered in three cases heard together: (1) *John Munroe (Acrylics) Ltd v London Fire Brigade and Civil Defence Authority*, (2) *Church of Jesus Christ of the Latter Day Saints v West Yorkshire Fire and Civil defence authority* and (3) *Capital and Counties plc v Hampshire County Council.*[1]

1 [1997] 3 WLR 331.

17.20 In the *John Munroe* case, the fire brigade attended an emergency call but, on arrival, the fire had been extinguished already. The officers took steps to check that the fire had been extinguished and then left without inspecting the neighbouring premises, which were owned by the claimants. Later that evening, a fire broke out at the claimants' premises which were severely damaged.

In this case, it was submitted on the part of the claimants that they were entitled to rely on the doctrine of 'general reliance' as giving rise to a duty to exercise statutory powers and/or a duty of care to respond to the call for help from the public. The Court of Appeal held that there was no breach of duty on the part of the fire brigade.

17.21 The claim brought by the Church of Jesus Christ of the Latter Day Saints was concerned with the alleged inadequate provision of fire hydrants. Of the seven hydrants in the vicinity, four failed to work, and it took too long to find the remaining three to be of assistance. The Court of Appeal rejected the claim.

17.22 It is clear, however, that the fire brigade has no immunity, following the third decision by the Court of Appeal in *Capital and Counties plc v Hampshire County Council*. A fire officer ordered the sprinkler system to be turned off. When the sprinklers were disabled, the fire brigade had not yet found the seat of the fire and were not effectively fighting it. The sprinklers were therefore, at that stage, the only effective means of fighting the fire. With the sprinklers out of action, the fire rapidly grew out of control. The Court of Appeal held that there was sufficient proximity, as there had been an increase in the risk of the fire spreading, caused by the positive act of the fire brigade.

17.23 The Court of Appeal summarised its views as follows:

> 'In our judgment, the fire brigade are not under a common law duty to answer the call for help, and are not under a duty to take care to do so. If, therefore they have failed to turn up, or failed to turn up in time, because they have carelessly misunderstood the message, got lost on the way or run into a tree, they are not liable.'

17.24 The only duty owed by a fire brigade is not to negligently create an additional danger which causes injury to the individual to whose assistance it has been called.

17.25 The Fire and Rescue Services Act 2004 provides the statutory framework for the activities of the Fire Service, and established the fire and rescue authorities. Such authorities must make provision for the purpose of extinguishing fires and protecting life and property.[1]

1 Fire and Rescue Services Act 2004, s 7(1)(a), (b).

17.26 The courts have considered the question of whether the existence of the statutory power could generate a common law duty of care. In *Stovin v Wise*,[1] Lord Hoffmann observed:

> '… in one sense it is true that the fire brigade is there to protect people in situations where they could not be expected to be able to protect themselves. On the other hand they can and do protect themselves by insurance against the risk of fire. It is not obvious that there should be a right to compensation from a negligent fire authority which will ordinarily enure by right of subrogation to an insurance company. The only reason would be to provide a general deterrent against inefficiency, but there must be better ways of doing this than by compensating insurance companies out of public funds. While premiums, no doubt, take into account the existence of the fire brigade, and the likelihood it will arrive swiftly upon the scene, it is not clear they would

be very different, merely because no compensation was paid in the rare cases in which the fire authority negligently failed to perform its public duty.'

1 [1996] 3 All ER 801.

17.27 In *Gorringe v Calderdale Metropolitan Borough Council*, Lord Hoffmann stated:

'If the statute does not create a private right of action, it would be, to say the least, unusual if the mere existence of a statutory duty could generate a common law duty of care.'[1]

1 [2004] UKHL 15.

17.28 At any fire, an employee of a fire and rescue authority, who is authorised in writing by that authority, may do anything he reasonably believes to be necessary for the purpose of extinguishing the fire or protecting life or property.[1]

1 Fire and Rescue Services Act 2004, s 44(1)(a).

17.29 An employee of a fire and rescue authority, who is authorised in writing by that authority, may, if he reasonably believes a fire has started or is about to start, close a highway or stop and regulate traffic.[1]

1 Fire and Rescue Services Act 2004, s 44(2)(c), (d).

17.30 An employee of a fire and rescue authority, who is authorised in writing, can enter and, if necessary, break into premises where he reasonably believes fire to have broken out so as to extinguish the fire. The said employee can do this without the consent of the owner or occupier of the premises. He can do everything which he considers necessary to extinguish the fire, again without the consent of the owner or occupier.[1]

1 Fire and Rescue Services Act 2004, s 44(2)(a).

17.31 Three fire service incident commanders were acquitted of gross negligence manslaughter following the deaths of four firefighters at a blaze in a vegetable-packing warehouse in Warwickshire in November 2007. A jury at Stafford Crown Court acquitted the Station Manager, Timothy Woodward, and Watch Manager, Adrian Ashley, after hearing six weeks of evidence about the deaths of Ashley Stephens, Darren Yates-Badley, John Averis and Ian Reid. The prosecution had alleged that Mr Woodward and Mr Ashley, who acted as incident commanders during the blaze in Atherstone-on-Stour, Warwickshire, were criminally responsible for the ''needless'' deaths of the four-man breathing apparatus crew. The concern is that, although these men were acquitted, will this make people working in the fire and rescue service wonder why they would ever want to be incident commanders?

DUTY OF CARE OWED BY THE AMBULANCE SERVICE

17.32 There is no framework legislation setting out the duties and arrangements of the ambulance service. Many are governed as part of a NHS Trust. Following recent policy chances to deregulate and outsource former NHS services, ambulance services are increasingly being outsourced to private providers, who will be required to have suitable insurance arrangements in place to meet any claims for negligence.

17.33 The leading case on the duties of the ambulance service is *Kent v Griffiths*.[1] This case involved the late arrival of an ambulance following an emergency call. The victim was suffering an asthma attack and suffered a respiratory arrest causing brain damage before arrival at hospital. The ambulance service is distinguished from the other emergency services as providing a health service.

Woolf MR stated:

> 'The police and fire services' primary obligation is to the public at large. In protecting a particular victim of crime, the police are performing their more general role of maintaining public order and reducing crime. In the case of fire the fire service will normally be concerned not only to protect a particular property where a fire breaks out but also to prevent fire spreading. In the case of both services, there is therefore a concern to protect the public generally. The emergency services that can be summoned by a 999 call do, in the majority of situations, broadly carry out a similar function. But in reality they can be very different. The ambulance service is part of the Health Service. Its care function includes transporting patients to and from hospital when the use of an ambulance for this purpose is desirable. It is therefore appropriate to regard the LAS as providing services of the category provided by hospitals and not as providing services equivalent to those rendered by the police or the fire service. Situations could arise where there is a conflict between the interests of a particular individual and the public at large. But in the case of the ambulance service in this particular case, the only member of the public who could be adversely affected was the claimant. It was the claimant alone for whom the ambulance had been called.'

1 [2000] EWCA Civ 3017, [2000] 2 WLR 1158.

17.34 The court distinguished these circumstances from a hypothetical claim involving the allocation of resources (eg where the service has to choose between attending emergency calls), where the finding could be different.

17.35 The courts distinguish between cases where there is held to be an assumption of responsibility to an individual on the part of the defendant, on the one hand, and those cases where the defendant has accepted the role (often under statutory powers or duties) of protecting the community in general from foreseeable dangers, on the other. *Kent v Griffiths* is an example of the former, and *Capital and Counties* is an example of the latter.

17.36 In the case of *Watson v British Boxing Board of Control*,[1] Lord Phillips MR stated:

> 'The authorities support the proposition that the act of undertaking to cater for the medical needs of a victim of illness or injury will generally carry with it a duty to exercise reasonable care in addressing those needs. Whilst this may not be true of the volunteer who offers assistance at the scene of an accident, it will be true of a body whose purpose is, or includes, the provision of such assistance.'

1 [2001] 1 WLR 1256.

Breach in the performance of duties

17.37 In the case of *Barry v The National Health Service Litigation Authority*,[1] it was accepted, following *Kent v Griffiths*, that there was a duty of care; however, the question for the court was whether the defendant was in breach. The claimant was an infant who had suffered cerebral palsy as a consequence of an umbilical cord prolapse. The court was concerned with the time taken to transport the mother to hospital. The crew were trained but inexperienced in this kind of emergency. Roderick Evans J stated:

> 'I am satisfied that this was a crew of competent ambulance men. They quickly identified the nature of the emergency which faced them and there was no reason for them to delay either in the flat or on the journey to hospital. There is no evidence before me that they were engaged at any time in activities not immediately connected with the mother's welfare. They carried out appropriately the tasks which they had to perform and while it is possible that another crew might have performed those tasks in a shorter period of time, I am not persuaded that this crew were in any way negligent. There is no evidence of culpable delay either in the flat or during the journey to hospital.'

Test for breach of duty

17.38 In the case of *Bolam v Friern Hospital Management Committee*,[1] McNair J stated:

> 'A doctor is not guilty of negligence if he has acted in accordance with a practice accepted as proper by a responsible body of medical men skilled in that particular art … Putting it the other way around, a doctor is not negligent, if he is acting in accordance with such a practice, merely because there is a body of opinion that takes a contrary view.'

1 [1957] 1 WLR 582.

17.39 The test in *Bolam* was clarified in *Bolitho v City and Hackney Health Authority*,[1] where Lord Browne-Wilkinson stated that the court was required to be:

> '… satisfied that the exponents for the body of medical opinion relied upon can demonstrate that such an opinion has a logical basis. In particular, in cases involving the weighing of risks against benefits, the judge before accepting a body of opinion as being responsible, reasonable or respectable, will need to be satisfied that in forming their view, the experts have reached a defensible conclusion on the matter.'

1 *Bolitho (Deceased) v City and Hackney HA* [1998] AC 232.

DUTY OF CARE OWED BY THE POLICE

17.40 The police are governed by the arrangements set out in the Police Act 1996.

Each service will generally be held to be vicariously liable for the acts of its employees whilst at work and acting within the scope of their employment.

17.41 The chief of police is the defendant in respect of negligence claims:

> 'The chief officer of police for a police area shall be liable in respect of any unlawful conduct of constables under his direction and control in the performance or purported performance of their functions in like manner as a master is liable in respect of torts committed by his servants in the course of their employment, and accordingly shall, in the case of a tort, be treated for all purposes as a joint tortfeasor.'[1]

1 Police Act 1996, s 88.

17.42 If a negligent police officer has been seconded from one police force to another, he is deemed to be under the direction and control of the chief of police to whom he has been lent.[1]

1 Police Act 1996, s 24(3).

17.43 As previously discussed, the leading case on the duties of the police in the investigation of crime is *Hill*.[1] In that case, it was determined that there was an insufficient relationship of proximity between the police and the mother of the last victim of the Yorkshire Ripper, who alleged that the police were negligent in failing to catch the Ripper before he killed her daughter. The House of Lords determined that the duty of the police to apprehend and identify criminals was a duty owed to the public at large and not to the deceased as an individual.

Lord Templeman took the view that, even if a duty of care existed, public policy required that the police should not be liable:

> 'If this action lies, every citizen will be able to require the court to investigate the performance of every policeman. If the policeman concentrates on one crime, he may be accused of neglecting others. If the policeman does not arrest on suspicion a suspect with previous convictions, the police force may be liable for subsequent crimes. The threat of litigation against the police force would not make a policeman more efficient. The necessity for defending proceedings, successfully or unsuccessfully, would distract the policeman from his duties.'

1 *Hill v Chief Constable of West Yorkshire* [1989] AC 53.

17.44 *Hill* was confirmed by a decision of the Court of Appeal in *Osman v Ferguson and Chief Officer of the Metropolis*.[1] A teacher became infatuated with a 15-year-old boy and confessed to the police that he was at risk of doing something criminally insane. Despite attempts by the police to follow the teacher, the teacher shot both the boy and his father. The court decided that there was a relationship of proximity but, following *Hill*, concluded that it was contrary to public policy that a duty should arise.

The decision in *Osman* was criticised in the European Court of Human Rights as providing a blanket immunity for the police for their acts and omissions, and thus an unjustifiable restriction on the claimant's right to have a determination of his claim on the merits.

1 [1993] 4 All ER 344.

17.45 However, the European Court revised its view in the later case of *Z and Others v United Kingdom*:[1]

> 'In the present case, the court is led to the conclusion that the inability of the applicants to sue the local authority flowed not from an immunity but from the applicable principles governing the substantive right of action in domestic law. There was no restriction on access to a court …'

1 (Application no 29392/95) [2001] ECHR 3.

17.46 In *Alexandrou v Oxford (Chief Constable of Merseyside)*,[1] the claimant discovered that his property had been burgled, despite the fact that the police attending the scene had concluded that everything was secure.

The Court of Appeal found that there was insufficient proximity to give rise to a duty of care. Glidewell J stated:

'The communication with the police in this case was by a 999 telephone call, followed by a recorded message. If as a result of that communication the police came under a duty of care to the plaintiff, it must follow that they would be under a similar duty to any person who informs them, whether by 999 call or in some other way, that a burglary, or indeed any crime, against himself or his property, is being committed or is about to be committed. So in my view if there is a duty of care it is owed … to all members of the public who give information of a suspected crime against themselves or their property.'

1 [1990] EWCA Civ 19, [1993] 4 All ER 328.

Cases where liability was found

17.47 In *Costello v Chief Constable of Northumbria Police*,[1] a colleague had stood by whilst a police officer was assaulted by a prisoner. A relationship of proximity was found to exist between the claimant and the other police officer.

1 [1999] 1 All ER 550.

17.48 In *Wilson v Chief Constable Lothian Borders Constabulary*,[1] a drunk was released by the police into the countryside, wearing light clothing in sub-zero temperatures, and died of exposure. Liability was imposed on the basis that the police had assumed responsibility for the man's wellbeing.

1 [1989] SLT 97.

17.49 *Swinney v Chief Constable of Northumbria Police*[1] concerned a claimant who provided information which helped to identify a hit and run driver. His details were left in a car that was stolen, and the information came to the attention of the alleged criminal, resulting in threats and psychological injury. It was held that it was good public policy that informants had an expectation that their identity would be protected.

In *Smith v Chief Constable of Sussex Police*,[2] a claim was brought solely under common law negligence. It was alleged that the police had failed to protect Smith from attack by his former partner, despite being aware of numerous death threats. Sufficient proximity existed since Smith was known to be at risk. Sedley LJ concluded that it would not be right for Smith to have less protection than an informer. No blanket immunity existed for the police, but it must be shown why it would not be fair, just and reasonable for a duty of care to be imposed.

It may therefore now be possible to establish a case in negligence where a known victim's life or physical safety is at real and immediate threat from an aggressor who is also known to the police.

1 [1997] QB 464.
2 [2008] EWCA Civ 39.

EMERGENCY SERVICES – EMPLOYER'S LIABILITY

17.50 Just as all employers do, the emergency services owe a duty of care to their employees. The duty has four elements:

i) to provide a safe place of work, including safe access to that place;
ii) to provide a proper system of work, including training and supervision;
iii) to provide plant, machinery and equipment adequate for the task; and

iv) to provide competent fellow employees.

The duty is non-delegable.

17.51 A police officer holds office rather than being employed; however, for the purposes of considering liability, the relationship with the chief constable is analogous to employment.[1]

Police and ambulance services (unless they are provided by an independent provider) are exempt from the requirement to hold Employers' Liability Insurance.[2]

1 *White and Others v Chief Constable of South Yorkshire and Others* [1999] All ER 550.
2 Employers' Liability (Compulsory Insurance) Act 1969.

17.52 The duty is owed to particular employees rather than to employees as a class of persons. Therefore, an employer must take account of the peculiarities of each individual and adjust any risk assessment accordingly.[1]

The courts consider that the role of the emergency services is to perform their role in difficult and dangerous situations, where rapid decision making is required.

1 *Paris v Stepney Borough Council* [1951] AC 367.

17.53 In *Watt v Hertfordshire County Council*,[1] a member of the public was trapped under a vehicle. A mechanical jack was available at the fire station, but there was no vehicle available to carry the equipment. The senior officer told the claimant to use an ordinary lorry to carry the equipment. On the journey, the jack moved and seriously injured the claimant's foot. It was claimed that the fire brigade should have provided a suitable vehicle or phoned the nearest fire station 10 minutes away where there was a suitable vehicle.

1 [1954] 1 WLR 835.

17.54 The Court of Appeal found that the risk involved in sending the lorry out was not so great as to prohibit the attempt to save life. Denning LJ stated:

> 'It is well settled that in measuring due care one must balance the risk against the measures necessary to eliminate the risk. To that proposition there ought to be added this. One must balance the risk against the end to be achieved. If this accident had occurred in a commercial enterprise without any emergency there could be no doubt that the servant would succeed. But the commercial end to make profit is very different from the human end to save life or limb. The saving of life or limb justifies taking considerable risk … I quite agree that fire engines, ambulances and doctors' cars should not shoot past the traffic lights when they show a red light. That is because the risk is too great to warrant the incurring of the danger. It is always a question of balancing the risk against the end.'

HILLSBOROUGH

17.55 On 15 April 1989, 96 people were killed at the Hillsborough football stadium. The role that the police played in the disaster is now subject to a criminal investigation led by the former chief constable of the Durham Police Force, which is looking at the culpability of not just the police force but the Sheffield Wednesday Football Club, the city council and the Football Association. This concerns the offence (which was relevant at the time) of gross negligence manslaughter; e crime of corporate manslaughter (that entered the statute books in 2008) cannot be used retrospectively.

17.56 The investigation will look at the command and control by South Yorkshire police of the 54,000 crowd at Hillsborough, and what happened to the emergency services' response.

The investigation will examine the role of Sheffield Wednesday Football Club, which offered to host the 1989 FA Cup semi-final between Liverpool and Nottingham Forest at its ground, despite serial breaches of the Home Office guide to ground safety and a safety certificate that was 10 years out of date.

17.57 Sheffield City Council, which was statutorily responsible for licensing the stadium as safe, and the FA, which commissioned the ground for its semi-final despite Hillsborough's safety breaches and previous crushes at semi-finals there in 1981, 1987 and 1988, are also being investigated for potential manslaughter charges.

17.58 Besides unlawful killing offences, there may be other potential criminal offences under investigation, include breaches of health and safety law, and for individuals working in public bodies, including the police. The original investigation by West Midlands police, headed at the time by Geoffrey (now Lord) Dear, resulted in no prosecutions of any organisation or individual for any offence.

CPS GUIDANCE

17.59 In May 2011, the Crown Prosecution Service (CPS) issued guidance outlining the circumstances in which it will seek to enforce s 7 of the Health and Safety at Work etc Act 1974 (HSWA 1974) against individual emergency personnel.

This provision, which relates to the duty of employees to take reasonable care of their own health and safety, is perceived to conflict with the police and fire service's desire to protect the public, even if doing so might sometimes involve risks to their own safety or that of other people. This new guidance follows on from the Health and Safety Executive (HSE) guidance, *Striking the balance between operational and health and safety duties in the fire and rescue services*, and HSE guidance as to how that should be interpreted.

17.60 Section 7 of the HWSA 1974 provides that it is 'the duty of every employee to take reasonable care for the health and safety of himself and of other persons who may be affected by his acts or omissions at work'.

During the course of their duty, police officers and firefighters may be presented with situations in which they have to consider acting in a way which puts their safety at risk. Although it is likely that that any officer would attract a jury's sympathy if prosecuted for acting in such circumstances, prosecutors may be asked by police investigators to consider whether an 'heroic act' of a police officer or firefighter should be subject to prosecution under s 7 of the HSWA 1974.

17.61 The prosecution must prove beyond reasonable doubt that the acts of the individual were such that they amounted to the person not taking 'reasonable care' for the health and safety of himself or others.

However, "reasonable" may be defined by reference to all the circumstances, including why the individual acted in the way they did in the particular circumstances of the emergency as presented to him. This can mean that, in certain circumstances, an individual could take considerable disregard for their own personal safety yet still fall within the definition of 'reasonable care'.

17.62 The CPS guidance states that it is 'very difficult to envisage circumstances in which the public interest would be served by the prosecution of a police officer or firefighter who puts their own safety at risk and breaches s 7 HSWA'.

HSE GUIDANCE

17.63 The HSE views the actions of firefighters as truly heroic when it is clear:

- they have decided to act entirely of their own volition in putting themselves at risk to protect the public or colleagues; and
- there have been no orders or other directions from senior officers to do so; and
- their actions have not put other officers or members of the public at serious risk.

DEATH IN CUSTODY

17.64 Where a person is detained in an institution described in s 2(2) of the Corporate Manslaughter and Corporate Homicide Act 2007 (CMCHA 2007),[1] the organisation owes the detainee a relevant duty of care.[2] It will be for the trial judge to decide whether the organisation owed the deceased a duty of care,[3] but for the jury to determine if there was a breach, then the seriousness of it and 'how much of a risk of death it posed'.[4]

1 As amended to include UKBA customs facilities or the Ministry of Defence service custody premises.
2 Section 2(1)(d).
3 Section 2(5).
4 Section 8(2).

17.65 There is, however, an overlap between statutory duties imposed under health and safety legislation and these types of duty. For example, employers have a responsibility for the safety of their employees under the law of negligence and under health and safety law. Both statutory duties and common law duties will be owed to members of the public affected by the conduct of an organisation's activities.

17.66 There have been difficulties when prosecuting the police. Deaths in police custody are infrequent. According to the statistics for 2012/13, there were 15 deaths in police custody.[1] When there are deaths and where it is found that there has been a 'most serious neglect of duty', there is rarely a successful prosecution of individual police officers. PC Simon Harwood, who in October 2011 was charged with the manslaughter of Ian Tomlinson, was acquitted by the court in July 2012. Tomlinson had collapsed and died after he was hit by a baton and pushed to the ground by PC Harwood at the G20 protests in London. In September 2012, the Metropolitan Police disciplinary panel found PC Harwood guilty of gross misconduct and he was dismissed from the police force.

1 'Deaths during or following police contact: Statistics for England and Wales 2012/13' IPCC Research and Statistics Series: Paper 26.

17.67 Out of the 333 cases of recorded deaths in or following police custody between 1999 and 2009, prosecutions were recommended by the IPCC against 13 officers, including 14 charges for manslaughter, none of which resulted in a conviction.[1] However, following the change to the CMCHA 2007 in 2011, to permit the prosecution of police and other authorities under the Act, this could be a significant development.

1 'Deaths in or following police custody: An examination of the cases 1998/99–2008/09' IPCC Research Series Paper: 17.

17.68 Under the CMCHA 2007, a police force is deemed to be a body corporate. The police force itself can now be liable to prosecution for corporate manslaughter. This also means that companies that provide custody, both public and private, may be criminally liable for deaths of individuals being transported to and from immigration detention centres – such as the death of Jimmy Mubenga, an Angolan deportee who died after being restrained by G4S guards on a British Airways plane.

17.69 An article[1] referring to the IPCC's report 'Deaths in or following police custody' (see above) reports that, of the 333 people who died in or following police custody, 87 had been restrained, most commonly being physically held down by officers – and, in 16 of those cases, restraint was linked directly to the death, and four were classed as 'positional asphyxia'.

1 Field, S., Jones, L. 'Five years on: the impact of the Corporate Manslaughter and Corporate Homicide Act 2007: plus ca change?' *International Company and Commercial Law Review* (2013) Vol 24, Issue 6, p 329.

17.70 Any case that comes to court regarding the police will, in line with other authorities, pivot on what is 'senior management'. It is the same dilemma faced when prosecuting any organisation which has a complex layered management structure.

Chapter 18

The travel industry

Debbie Venn, Associate Solicitor, Travel Team, asb law LLP

Andrew Clinton, Managing Partner, asb law LLP

INTRODUCTION

18.1 The travel industry has been remarkably resilient in recent years in the face of economic turmoil, environmental disasters and political unrest in a number of countries. The total number of international travellers continues to rise year-on-year. Compared to the previous year, visits abroad by UK residents in the second quarter of 2013 increased by 2.7 per cent to 15.9 million, and expenditure during those visits increased by 7 per cent.[1] There are regional variations, with visits to Europe increasing, and UK residents have recently turned their focus away from some destinations such as Egypt.[2] Adventure travel is a fast-growing segment of the industry, in the form of both 'hard' adventure holidays (which can be physically challenging) and 'soft' adventure holidays. Inevitably, with so many people travelling abroad, there will be a number of accidents and, unfortunately, in some cases the accidents can result in a fatality.

1 ONS, Overseas Travel Tourism, July 2013.
2 ONS, Overseas Travel Tourism, July 2013.

18.2 This chapter takes a look at the corporate liability of businesses in the travel industry; it considers who would be liable for improper or non-performance of travel contracts, and who is liable for death, personal injury or other liabilities under the relevant regulations.

18.3 The travel industry has various different ways of contracting with customers and other third parties, which explain how corporate liability is apportioned between those parties. A business in the travel industry may contract in the following ways (this is not an exhaustive list):

- a travel organiser/tour operator, contracting with principal liability to its customers for the provision of travel products/services, which may include package holidays;
- a travel agency, putting the principal supplier of travel products/services into a contract with customers for the sale of travel products/services; or
- a supplier of travel products/services (eg airlines, hotels, ground handlers, coach companies), providing these directly to consumers or through other intermediaries or third party suppliers (eg tour operators).

The various contracting models in the travel industry, together with the regulatory framework that applies to certain transactions, raise a number of issues in terms of corporate liability.

HOW DOES THE TRAVEL INDUSTRY TRADE AND WHAT ARE THE CORPORATE RISKS?

18.4 If a travel company is contracting with a customer directly as the principal, the travel company will clearly be liable under the contract for the provision of the specified travel products/services. As with all contracts governed by English law, a party will not be able to exclude or limit its liability for death or personal injury caused by its negligence; however, other liabilities can be limited within the body of its terms and conditions (see further below).

18.5 If you are acting as an agent for a third party supplier or tour operator, your liability will not extend to the proper performance of the contract, as the principal supplier/tour operator will remain ultimately liable to the customer.

18.6 Many suppliers, such as airlines, hotels or other ground handlers (depending on their size), will generally want to contract on their own terms and conditions of supply. However, if you have bargaining power as a large corporate tour operator, you may be able to enforce your own contract terms with suppliers and seek an indemnity in favour of the tour operator in respect of any liability that the tour operator may suffer due to the supplier's non-performance or improper performance of the contract to provide the travel products/services.

18.7 Third party supplier contracts become more of an issue where those suppliers are overseas and you are dealing with different governing laws and jurisdictions and the interpretation of what is known as 'local standards' in travel contracts. This means that liability under a supplier contract would be judged according to the local standards of the country in which the services would be performed. For instance, the local standard of hotels in Greece would be the determining factor on whether or not a hotel has complied with the provision of its travel products/services under the terms of the agreement.[1]

1 *Wilson v Best Travel Ltd* [1993] 1 All ER 353. See also **18.70** onwards below.

18.8 It is therefore vital for tour operators and principal suppliers to ensure that they carry out proper due diligence on providers of accommodation and travel services, to ensure that those providers adhere to local standards and have relevant health and safety procedures in place. In addition, principal suppliers/tour operators should have their own insurance in place to cover for personal injury and public liability generally, so that, if a customer brings a claim against the tour operator/principal supplier, the provider can cover such losses. It is further advisable to obtain a copy of the provider's certificate of insurance for good measure. Certain responsibilities on principal suppliers/tour operators are imposed under UK regulations, which are explained further below.

REGULATORY FRAMEWORK EXPLAINED

18.9 Various regulations apply to the travel industry and control how consumers are protected by the companies supplying travel products/services. The various regulations set out the standard position on corporate liability in the UK, which should then be reflected in the terms and conditions with consumers. This framework sits alongside the general principles of English law. Other countries operate their own consumer protection regimes, although Members States of the European Union have implemented legislation to reflect the European Directive that governs the sale of package holidays, as explained further below.

Package Travel, Package Holidays and Package Tours Regulations 1992

18.10 Under the Package Travel, Package Holidays and Package Tours Regulations 1992 ('the PTRs'),[1] tour operators that organise packages are responsible for all aspects of providing a customer with their package holiday including, if applicable, flight arrangements. The PTRs implement the European Community Package Travel Directive ('the PTD'),[2] which aims at promoting tourism within the European Union. The PTD is currently undergoing consultation in Europe, in response to changes in the way in which people book their holidays, particularly in the digital age. As many people now book their holidays online, the revised PTD will need to find a way of bringing new ways of booking a holiday within the definition of a package holiday, so that travel companies become liable as tour operators for providing a 'holiday'. Once the PTD has been revised, this will require the UK to amend the PTRs, and the revisions are likely to increase the number of holidays protected as packages. This is not expected to happen until 2016, although consultation is due to take place at the end of 2013 (see below).

Anyone who sells or offers for sale (other than occasionally) package holidays must comply with the PTRs; and, if the package includes a flight, the Civil Aviation (Air Travel Organisers' Licensing) Regulations 2012 ('ATOL Regs')[3] apply.

The PTRs set out the responsibilities of a travel organiser to their customers and specify the remedies if there is a breach. The PTRs allow a dissatisfied holidaymaker to bring an action against a single supplier (the tour organiser) instead of against the individual suppliers, such as hoteliers or airlines separately.

1 SI 1992/3288, as amended.
2 Council Directive 90/314/EEC.
3 SI 2012/1017.

What is a package?

18.11 The definition of what constitutes a 'package' has been hotly debated in the recent past[1] and there has been controversy over how elements that create a package are sold in order to assess whether the PTRs apply.

A package is defined under the PTRs as a pre-arranged combination sold or offered for sale at an inclusive price covering a period of more than 24 hours and which includes at least two or more of the following components: transport; accommodation; and other significant tourist service (not ancillary to transport or accommodation). For example, a hotel and flight would be a package when sold together at a pre-arranged combined inclusive price, as would a combined booking for hotel and car hire sold together at a pre-arranged combined inclusive price. If a booking is for a single element (eg flight or accommodation only), this is not covered by the PTRs. A travel business needs to comply with other regulations relating to fair consumer trading and correct advertising, and regulations relating to contracting with consumers at a distance and online, if contracting through a website.

1 *R (on the application of Association of British Travel Agents Ltd (ABTA)) v Civil Aviation Authority* [2006] EWCA Civ 1356, [2007] 2 All ER (Comm) 898; *Civil Aviation Authority v Travel Republic Ltd* [2010] EWHC 1151 (Admin), [2010] CTLC 61; and *Titshall v Qwerty Travel Ltd* [2011] EWCA Civ 1569, [2012] 2 All ER 627.

Consumer protection under the PTRs

18.12 One of the main concerns of consumers is that the company they have booked their holiday with collapses after receipt of prepayments for the holiday. Under the PTRs, the seller of a package must have a guarantee (or evidence of security for the protection of prepayments) in place so that, if the seller (or any of its suppliers) ceases to trade, the consumer gets a refund of the holiday purchased, price paid or, if they are already abroad, their repatriation. In addition to the financial guarantees, the following are also required under the PTRs:

1) Advertising guidelines relating to price, airport destinations and description of what is being sold.
2) Supply of information relating to health formalities, passport and visa requirements for their journey in the country of destination.
3) Regulations on change of price for a package holiday once it has been booked, and where there is a significant change to any part of the package holiday once it has been confirmed.
4) The package organiser will be liable under the contract with the customer for the proper performance of that contract, notwithstanding that the services are to be performed by another third party, such as supplier of ground handling services, airlines or hotels. The package organiser will be liable to customers who have suffered a loss for compensation if it can be shown that the loss is due to the fault of the package organiser or the supplier.[1] Therefore, a package organiser will want to put insurance in place to cover this liability (see item 5 below). A package organiser should also put in place supplier indemnity agreements with suppliers, so that the package organiser can claim any amounts it has to pay out to a customer due to that supplier's default in supplying the travel products/services to the customer.
5) Under reg 14(2) of the PTRs, where a customer has booked a package holiday with an organiser, it is the organiser's responsibility to ensure that the package as booked by the customer is provided, or a suitable alternative arrangement is put in place (at no extra cost to the consumer) for continuation of the package, including compensation for any difference between the levels of services supplied. Regulation 14(3) states that, where it is not possible to make alternative arrangements, or those are not accepted by the customer for good reasons, the organiser

will arrange alternative transport back to the place of departure (if a customer is already abroad) and compensate the customer, if appropriate. For example, compensation would not be payable in the event that the inability to provide the services is due to a 'force majeure' reason.

6) Under reg 15 of the PTRs, the organiser who contracts with the consumer is responsible for the proper performance of the obligations under the contract, irrespective of whether such obligations are to be performed by the organiser or another third party (against whom the organiser can seek recompense). Liability for damage or losses due to non-performance or improper performance of the travel services in the package can be limited, to a reasonable extent, in accordance with various international conventions (for instance, the Montreal Convention in respect of travel by air, the Athens Convention in respect of travel by sea, the Berne Convention in respect of travel by rail, and the Paris Convention in respect of provision of accommodation).

Athens Convention: The Athens Convention relating to the Carriage of Passengers and their Luggage by Sea 1974 (Athens Convention) was brought into force by s 183 of the Merchant Shipping Act 1995 in the UK and allows a person who is injured on a ship at sea to take action in respect of that injury. In order to be able to make a claim under the Athens Convention, various conditions must be met, including that the incident that caused the death or personal injury must have occurred in the course of the carriage (ie on board the ship, or embarking or disembarking from the ship). Only a 'carrier' can be liable for death or personal injury under the Athens Convention. It is necessary to prove the fault or neglect of the carrier in order to substantiate a claim, and the burden of proving this rests with the claimant; unless the injury or death arose from (or in connection with) a collision, explosion or fire or defect in the ship or shipwreck, in which case there is a presumption of fault or neglect of the carrier.

Montreal Convention: The Convention for the Unification of Certain Rules for International Carriage by Air 1999 (Montreal Convention)[2] applies to all international carriage (art 1(1)), including where the place of departure and destination are in the same state but there is an agreed stopping place within another state, irrespective of whether the other state is a state party to the Convention (art 1(2)). It also includes embarking and disembarking from an aircraft. If a claimant can satisfy various conditions (eg the carrier is domiciled or has its place of business in a state that is party to the Montreal Convention), a carrier is liable for damage sustained for death or bodily injury of a passenger where that death or injury took place on board the aircraft or when embarking or disembarking.

7) Under reg 15(7), the organiser must also give prompt assistance to a consumer in difficulty.

1 *Griffin v My Travel UK Ltd* [2009] NIQB 98; *Titshall v Qwerty Travel Ltd* [2011] EWCA Civ 1569, [2012] 2 All ER 627; and *Moore v Hotelplan Ltd* (*t/a Inghams Travel*) [2010] EWHC 276 (QB).

2 The Montreal Convention replaced the Convention of the Unification of Certain Rules Relating to International Carriage by Air (Warsaw Convention 1929) and various protocols which supplemented the Warsaw Convention. These were brought into force in England and Wales by the Carriage by Air Acts (Implementation of the Montreal Convention 1999) Order 2002, SI 2002/263, made pursuant to s 4A of the Carriage by Air Act 1961 and s 4A of the Carriage by Air (Supplementary Provisions) Act 1962.

Consultation on the Package Travel Directive

18.13 As mentioned above, there is currently a proposal to revise the PTD at European level. This follows a review of responses submitted to the EU Commission, with proposals for policy change being issued on 9 July 2013. The proposals are generally aimed at achieving the following:

- put the onus on the organiser of a package to make alternative arrangements available;
- extend the definition of a package to bring in combined services being put together (rather than just pre-arranged combinations) and to include click-through arrangements (subject to exceptions);
- introduce a concept of 'assisted travel arrangements' which places responsibility on the organiser to provide financial protection to customers in certain circumstances;
- get rid of the use of pre-ticked boxes which automatically include additional services;
- remove business travel from the scope of package protection;
- improve consumer cancellation rights; and
- provide the consumer with new and better rights of redress.

There will be a period of scrutiny and editing of the proposal (from September 2013). The next step will be for the Commission to consider initial feedback, so a full consultation can be launched on the proposed edited PTD.

ATOL Regs

18.14 The ATOL Regs[1] came into force in the UK on 30 April 2012 and govern the sale of flights in the UK. The ATOL Regs introduce various regulatory compliance issues, including how flights are sold individually and when they are put together with another significant element of a holiday (eg accommodation), where those elements are not sold as a package (as currently defined in the PTRs). The ATOL Regs also introduce a recognisable ATOL certificate, as well as a mandatory agency agreement to be entered into between principal ATOL holders and agents acting on their behalf.

1 SI 2012/1017.

18.15 Unless a business falls within one of the exemptions, any business which makes flights available to customers must hold an ATOL licence, unless it is acting as an agent for an ATOL holder.

The ATOL Regs confirm that the following people may make flights available in the UK:

- the operator of the relevant aircraft;
- an ATOL holder who is acting in accordance with the terms of its ATOL licence;
- a person who is exempt from the need to hold an ATOL licence (as per reg 10 of the ATOL Regs); or
- a person who is exempt by the Civil Aviation Authority (CAA) under reg 11 of the ATOL Regs.

Regulation 10 confirms that a person would be exempt from the requirement to hold an ATOL licence if they act as an agent on behalf of a disclosed ATOL holder, they act as a member of an accredited body (in accordance with their membership terms and conditions), or they are an airline ticket agent. Regulation 10 provides for some other exemptions, but these are the main three.

Regulation 11 confirms that the CAA may allow any person to not hold an ATOL licence if it is satisfied that consumers will receive a level of consumer protection according to that which would otherwise be provided under the ATOL Regs. However, such an exemption must be in writing and published by the CAA, and may be subject to various conditions that it determines are appropriate.

Therefore, unless a business falls within one of the exemptions, a business which makes flights available (whether on their own, as part of a flight-plus[1] or a package) would need to hold an ATOL licence to make the flights available with the other elements (or by themselves) in accordance with the ATOL scheme.

1 A 'flight-plus' arrangement is a concept created under the ATOL Regs and is not a 'package' as such. Where a flight, plus accommodation or car hire, is booked at the request of a consumer (within a two-day window), to be taken at the same time, this will be a 'flight-plus' arrangement.

18.16 Anyone who does not hold a proper ATOL licence, or act in accordance with one of the exemptions, and makes flights available would be in breach of the ATOL Regs and, if convicted, would be subject to a fine or imprisonment not exceeding two years. The offences therefore carry criminal sanctions. It is possible to seek to rely on a 'due diligence' defence, if a business can show that it took all reasonable steps and exercised all due diligence to avoid committing the offences. Any actions brought under the ATOL Regs would also need to be brought before the end of the period of 12 months from the date of the commission of the offence.

Regulation (EC) 261/2004

18.17 Regulation (EC) 261/2004 ('the Denied Boarding Reg')[1] sets out passengers' rights against airlines when flights are delayed, cancelled or overbooked. This part of the regulatory framework for the travel industry sets out liability where there is denied boarding of a customer's flight or that flight is cancelled or delayed. Individuals are entitled to assistance (catering, communications and an overnight stay if necessary), or continuation of the trip or refund of the traveller's ticket.

1 Regulation (EC) 261/2004 establishing common rules on compensation and assistance to passengers in the event of denied boarding and of cancellation or long delay of flights.

18.18 Under the Denied Boarding Reg, customers must be provided with assistance when their flight is delayed more than two hours (depending on how long the delay is and the distance to be travelled). If a flight is cancelled, a customer must be offered a choice between a full refund and an alternative flight. If the cancellation involves the return leg of a journey, the refund should be for the unused portion, and the alternative flight would be a flight home. The passenger must always be provided with 'assistance/care', two free phone calls, emails, telexes or faxes and overnight hotel accommodation and transport to and from the hotel, as appropriate, whilst they wait for a rearranged flight.

It is therefore important to establish who will be responsible when considering the Denied Boarding Reg, and how this may affect liabilities between airlines and tour operators.

ROLE OF ABTA AND CAA

ABTA

18.19 the Association of British Travel Agents (ABTA) is probably the UK's most recognised travel industry membership organisation, providing assistance to consumers and to those businesses who are ABTA members, in order to promote accessible and responsible travel. ABTA will offer assistance to its members to help grow their business and provide assistance on mechanisms for ensuring that package travel is financially protected and that members receive all relevant business assistance through complying with ABTA's code of conduct.

ABTA seeks to raise standards in the industry, by giving guidance on issues from sustainability to health and safety, and to assist members in identifying where issues need to be addressed, eg surveys and due diligence on proposed suppliers and ensuring that appropriate terms and conditions are entered into with ABTA members. ABTA has a dedicated team working with the travel industry on raising standards, addressing health and safety issues and identifying best practice across the globe.

Civil Aviation Authority

18.20 The Civil Aviation Authority (CAA) is the UK's aviation regulator. One of its main responsibilities is to protect the consumer by ensuring the safety and management of UK airspace, as well as enforcement of consumer protection rules and management of the repatriation and refunds to holidaymakers if a travel company fails and this involves a flight. The CAA is given powers by the Secretary of State for Transport regarding the discharge of the UK's obligations under various international conventions on transportation. This is achieved in the main through a Sponsorship Statement and Directions issued by the Department of Transport. The Sponsorship Statement and Directions include how the framework of the CAA operates, as well as guidance on how it should discharge its functions. In addition to its role in protecting consumers relating to financial failure of travel companies where flights are included, the CAA also has a specific safety regulation group that helps to set safety standards and ensure these are met by the aviation industry. This includes the design, manufacture and operation/maintenance of aircraft, as well as the operation environment in airports and air traffic control systems.

INSURANCE

18.21 Many travel businesses will need to ensure that they have general liability insurance for the operation of their business, including general public liability insurance and employer's liability insurance. In addition, where a travel business sells packages, it should have in place supplier failure insurance because it remains liable to the consumer under the PTRs, notwithstanding a failure or insolvency in its supply chain, although this type of insurance can be expensive. In addition, the business should consider 'products' liability insurance, relating to the sale of the travel products/services as a package holiday.

In addition to insurance, it would also be sensible to ensure that, when a tour operator is contracting with its suppliers, as part of its due diligence process, it obtains a copy of the certificate of insurance that the supplier has in place in relation to its public liability and employer's liability.

ISSUES THAT ARISE FOLLOWING AN ACCIDENT

18.22 The logistical issues involved in responding to a fatal accident abroad can be challenging, as the legal system, language and culture may well be different. Travel companies will have health and safety procedures, as well as written incident management and communication plans, in place that set out the procedures to be followed in the event of a death of a holidaymaker. These will cover dealing with the immediate issues (such as notification of the incident to the response team, information gathering and dealing with media attention) as well as longer-term considerations. The Foreign and Commonwealth Office may become involved, as it provides support for British nationals in difficulty abroad. It is important for the people who were directly involved in the incident to record in a log, as early as possible, what happened, and for them to prepare contemporaneous statements, although ideally these will be obtained by legal representatives so that they are privileged.

When a British national dies abroad in unnatural circumstances, there will inevitably be an investigation by the appropriate authorities in that jurisdiction. There may also be an investigation by the UK police, as prosecutions for murder and manslaughter can take place here where the suspect is a British national.[1] If the body of the deceased is repatriated to this jurisdiction, the UK police may be asked to assist the coroner in the investigation and inquest.

1 Offences Against the Person Act 1861, s 9.

THE INQUEST PROCESS

18.23 Travel law claims can involve consideration of domestic and foreign regulations, conflict of law principles and the common law of negligence. There is a connection between the principles of tortious liability for negligence and criminal liability, and that connection becomes particularly relevant in the context of an inquest into a violent or unnatural death, or a sudden death of unknown cause.

18.24 Following the death of a UK citizen after an accident abroad, a coroner in England and Wales will have jurisdiction to hold an inquest, provided that the body of the deceased is now lying within that coroner's area (ie following repatriation). Inquests can also be held into the deaths of non-UK citizens if the death occurred in England and Wales. A number of far-reaching changes, which were designed to improve the consistency and efficiency of the inquest process, are set out in the Coroners and Justice Act 2009.[1] The previous system had been criticised as being a 'postcode lottery' for bereaved families, as there was no national guidance for coroners. The first Chief Coroner, which was an office created by s 35 of the Coroners and Justice Act 2009, was appointed in September 2012. The Chief Coroner has a number of responsibilities, which include setting national standards for all coroners and monitoring the system where recommendations from inquests are reported to the appropriate authorities in order to prevent further deaths (see further below).

1 A number of the provisions in the Coroners and Justice Act 2009 came into effect on 25 July 2013, along with the Coroners (Inquests) Rules 2013, SI 2013/1616.

18.25 Section 1(1) and (2) of the Coroners and Justice Act 2009 states that:

'(1) A senior coroner who is made aware that the body of a deceased person is within that coroner's area must as soon as practicable conduct an investigation into the person's death if subsection (2) applies.

(2) This subsection applies if the coroner has reason to suspect that–
  (a) the deceased died a violent or unnatural death,
  (b) the cause of death is unknown, or
  (c) the deceased died while in custody or otherwise in state detention.'

The two purposes of the investigation are: (1) to establish who the deceased was and how, when and where the deceased came by his or her death, and (2) to establish the details needed to register the death, such as the cause of death.[1]

1 Coroners and Justice Act 2009, s 5.

18.26 Article 2 of the European Convention on Human Rights (the right to life) imposes on the state both negative obligations not to take life intentionally and positive obligations to protect life. The positive duty to protect life implies a duty to investigate unnatural deaths. Article 2 has been used to expand the scope of a coroner's investigation by allowing a wider-ranging inquiry into the circumstances surrounding the death. Inquests that used to take a couple of days are now taking a lot longer, as coroners look at issues such as training, health and safety and governance.

18.27 In some respects, an inquest resembles a criminal trial: witnesses may be examined under oath, a jury may be called, and there is a thorough investigation of the facts and circumstances surrounding the death. However, the purpose of an inquest is altogether different – the determination may not be framed in such a way as to appear to determine any question of criminal liability on the part of a named person, or civil liability. Therefore, no matter how obvious it may be that there is one person who could be named as criminally responsible for the death, the coroner must not name that person.

However, it does not necessarily follow that the coroner cannot consider issues which may be central to subsequent civil or criminal proceedings. Where a verdict of unlawful killing is recorded following an inquest (and for some narrative verdicts), it is virtually impossible to ignore the implications in terms of the possibility of subsequent criminal and/or civil proceedings. An inquest is often the precursor to a criminal or civil action and represents a means of obtaining evidence that might be of assistance in proceedings that might be instituted against an individual or a corporate entity.

18.28 In view of this, it is perhaps unsurprising that the relatives of a person killed as a result of an accident, where there might appear to be at least a degree of negligence involved, often invite a coroner to consider a verdict of unlawful killing. The public and media, as well as those personally involved in a fatal accident, often want (or, at least, expect to see) a degree of accountability, and this is particularly the case where the perception is that companies put profits and commercial success above the safety of their customers. In the event of a verdict of unlawful killing, the police must open or re-open an investigation into the fatal accident and may pass a file to the Crown Prosecution Service (CPS). Those cases which require consideration of gross negligence manslaughter are dealt with by the Complex Caseworks Unit of the CPS.

18.29 The standard of proof for a verdict of unlawful killing is the criminal standard, ie beyond a reasonable doubt. The verdict of unlawful killing is of relatively recent origin (following implementation of the Coroners Rules 1984), and there is no statutory definition of 'unlawful killing'. The courts recently considered the ambit of a verdict of unlawful killing and, in particular, whether evidence of the commission of the criminal offence of causing death by careless driving (and by extension the offence of causing death by dangerous driving) was capable of justifying a verdict of unlawful killing at an inquest. There had been a divergence of views on this issue amongst coroners. In *R (on the application of Wilkinson) v HM Coroner for the Greater Manchester South District*[1] the court held that unlawful killing was restricted to the three homicide offences of murder, manslaughter (including corporate manslaughter) and infanticide. The following passage appears (at para 63 of the judgment):

> 'In our judgment, the main purpose of having a verdict of unlawful killing is to distinguish between those cases where there has been an accident of some kind (where, of course, someone may be to blame for it, even with some degree of criminality) and those cases where it would be an abuse of language to describe the events leading to death as simply an accident. Someone killed by murder, manslaughter or infanticide is killed either intentionally or by some obviously criminal state of mind on the part of, or some negligence of the grossest kind by, the author of the killing.'

The court said that bad driving cases may only be regarded as unlawful killing for inquest purposes if they satisfy the ingredients for manslaughter (gross negligence manslaughter).

1 [2012] EWHC 2755 (Admin).

18.30 Another important practical point for consideration in the context of inquests is the issue of potential conflicts of interest between a corporate entity and an individual or individuals working within that entity. This becomes particularly relevant where the coroner is tasked with considering the actions of both the individuals involved and the corporate entity, and there is a reasonable prospect that there may be a divergence of views about the corporate entity's procedures and whether an individual was following instructions. The individual and corporate entity will often require separate legal representation in order to avoid any issues if an actual conflict arises during the course of the proceedings.

18.31 Notwithstanding the reform of the inquest process, there is still no right of appeal from an inquest. The Coroners and Justice Act 2009 included a provision which set out a right of appeal for an interested party against a coroner's decision to the Chief Coroner, and a further right of appeal against the Chief Coroner to the Court of Appeal. However, that provision was repealed before it came into force. The rights of challenge therefore remain a power on the part of the High Court, on application by the Attorney General, to order a new inquest (as happened in relation to the deaths of 96 Liverpool fans in the 1989 Hillsborough disaster) or by an application for judicial review.

Prevention of Future Death Reports (formerly Rule 43 reports)

18.32 Coroners have, for some time, been able to make reports if evidence during an inquest gives rise to concern that action should be taken to prevent future deaths, which became known as 'Rule 43' reports.[1] In essence, a coroner can write to a third party setting out recommendations to prevent future deaths. For the last five years, there has been a year-on-year increase in the number of Rule 43 reports issued, with a total of 235 reports in the six-month period to March 2013. The recipients of these reports have included government departments, regulatory bodies and trade associations. It is obviously a matter of some concern to a travel company if it becomes the subject of such a report. The recommendations can be far reaching and affect how an industry operates.

1 Referring to Coroners' Rules 1984, r 43.

18.33 Rule 43 reports across all categories of death have identified communication and the lack of procedures and protocols (or the failure to follow them) as major concerns. They have also highlighted health and safety issues, including the need for first aid training and appropriate risk assessments. A common request across all categories of death is for lessons learned to be shared and implemented.[1]

It is expected that the upward trend, in terms of the volume of such reports, will continue, as the Coroners and Justice Act 2009 elevates the Rule 43 provision to primary legislation (see below) and strengthens it, by requiring coroners to report actions to prevent future deaths to relevant persons. The Chief Coroner has assumed responsibility for what are now known as Prevention of Future Death Reports.

1 'Summary of Reports and Responses under Rule 43 of the Coroners Rules' (Ministry of Justice), June 2013.

18.34 The Coroners and Justice Act 2009 provides as follows:[1]

> '(1) Where–
> (a) a senior coroner has been conducting an investigation under this Part into a person's death,
> (b) anything revealed by the investigation gives rise to a concern that circumstances creating a risk of other deaths will occur, or will continue to exist, in the future, and
> (c) in the coroner's opinion, action should be taken to prevent the occurrence or continuation of such circumstances, or to eliminate or reduce the risk of death created by such circumstances, the coroner must report the matter to a person who the coroner believes may have power to take such action.
> (2) A person to whom a senior coroner makes a report under this paragraph must give the senior coroner a written response to it.
> (3) A copy of a report under this paragraph, and of the response to it, must be sent to the Chief Coroner.'

1 Coroners and Justice Act 2009, Sch 5, para 7.

THE CORPORATE MANSLAUGHTER AND CORPORATE HOMICIDE ACT 2007

18.35 It is now easier to prosecute companies and other organisations where gross negligence leads to death. The Corporate Manslaughter and Corporate Homicide Act 2007 (CMCHA 2007) replaces the need to find a 'directing mind' with a focus on the overall management of activities. It is important to note that the offence created by the CMCHA 2007 applies only to organisations, and individual liability continues to be governed by the common law of manslaughter.

18.36 There is an interesting divergence between the extra-territorial application of manslaughter as it applies to individuals, and to corporate and other entities. It is well established that British subjects have always been susceptible to being tried in England and Wales for homicide, wherever the offence might have been committed.[1]

The question that arises in the context of travel companies is whether a company can be prosecuted for corporate manslaughter where the management failure was in England and Wales but the harm that led to the death occurred in another jurisdiction. This question has received no substantial consideration in English case law.

1 Offences Against the Person Act 1861, s 9 applies to murder and manslaughter, and provides that a subject of Her Majesty may be prosecuted for an offence committed on land out of the United Kingdom.

18.37 The CMCHA 2007 makes the offence of corporate manslaughter or homicide part of the law of each part of the United Kingdom.[1] It also limits the territorial application of corporate manslaughter by introducing a qualification that 'the harm resulting in death' must have been sustained in the United Kingdom, or within the seaward limits of the UK's territorial seas, or on a ship, aircraft or hovercraft within the definitions set out in various statutes.

1 Section 28.

18.38 In its written evidence to the Home Affairs and Work and Pensions Committee, the Centre for Corporate Accountability (CCA) suggested that the Home Office should look at the question of jurisdiction for corporate manslaughter, given that it is a sister offence to the individual crime of manslaughter. The CCA suggested that there were public policy reasons for the government to clamp down on companies who, in effect, used England and Wales as a base for causing deaths abroad. That appeal appears to have fallen on deaf ears.

18.39 One commentator has suggested that a jury could interpret 'harm' to be a decision made in this jurisdiction, as opposed to the injury that causes the death. However, it seems clear from comments made during the passage of the Bill through Parliament that there was no intention for the English courts to have jurisdiction where the physical harm was sustained outside England and Wales, even if the management failure took place within the jurisdiction.

18.40 Imagine a scenario in which a company with a corporate base in England (or, for that matter, in Wales, Scotland or Northern Ireland) operates a resort abroad and an accident occurs in the resort which results in the death of a customer. It is possible, on the current state of the law, for an individual employee who was working in the resort to be prosecuted in this jurisdiction for manslaughter, but the company could not be prosecuted for corporate manslaughter, whatever the management failures might have been.

There is, of course, a strong public interest for prosecution in the country concerned in relation to deaths abroad, just as, if a death occurs within the jurisdiction, one would expect a prosecution to take place here. However, an individual could find themselves in a situation in which two prosecuting authorities have concurrent powers to prosecute for manslaughter.

18.41 Of course, fatal accidents in other jurisdictions may become relevant to a prosecution for corporate manslaughter in this jurisdiction, when a jury comes to consider current or historical 'attitudes, policies, systems or accepted practices within the organisation that were likely to have encouraged any such failure [to comply with any health and safety legislation], or to have produced tolerance to it' under the gross breach of duty element of the offence of corporate manslaughter.[1]

1 CMCHA 2007, s 8.

18.42 The scope of 'relevant duty of care'[1] is rather broad and encompasses a number of duties owed under the law of negligence which would be likely to apply to a company involved in the organisation and provision of travel services. Further, a breach will amount to a 'gross' breach where the conduct alleged to have constituted a breach fell far below what could reasonably have been expected of the organisation in the circumstances.[2] It should be highlighted at this stage that establishing the offence is subject to a requirement that the way in which the activities of the company were managed or organised by its senior management was a 'substantial element' of the breach. 'Senior management' is defined as those people who play significant roles in the making of decisions about how the whole or a substantial part of the company's activities are to be managed or organised (or, indeed, the actual management or organisation of the same).

In this sense, it is clear that the statutory offence has made a significant departure from the common law position – the requirement to identify the 'directing mind' of the organisation has been replaced with a responsibility on the court to look at the management practices and health and safety systems in place within an organisation and to assess the adequacy and suitability of risk assessment and health and safety management in practice. It is clear that, subject to the jurisdictional issues discussed above, the activities of a tour company could become the subject of a prosecution. In November 2013 a water sports club pleaded guilty to corporate manslaughter following the death of an 11-year-old girl during a banana boat ride on a lake in Middlesex. The judge was critical of the fact that there was not an adult onlooker acting as an observer, and commented that he proposed 'to fine the company every penny that it has'. The judge said that he had no greater power than to impose a fine and he could not impose a fine greater than all of the company's assets. The fine was just under £135,000.

1 Defined by CMCHA 2007, s 2.
2 CMCHA 2007, s 1(4)(b).
3 CMCHA 2007, s 1(4)(c).

JURISDICTION: WHEN AND HOW THE ENGLISH COURTS WILL BECOME INVOLVED

18.43 It is no surprise that, in the immediate aftermath of a fatal accident involving a British national abroad, the death will be investigated by the local authorities in that country. As explained above, there may be an inquest and a police investigation in this jurisdiction, and there is also the possibility of civil or criminal proceedings. Accidents which occur overseas raise a number of different considerations regarding jurisdiction and when and how the English courts will become involved in the ensuing proceedings, if at all.

18.44 Before exploring the rules that apply to how jurisdiction in England and Wales can be established in relation to a death abroad, it is worth noting that the potential claimant in a civil action may prefer to issue proceedings in the country in which the deceased died rather than trying to establish jurisdiction in England and Wales. The rules that apply to fatal accident claims vary considerably between jurisdictions in a number of ways, including:

- the eligibility to bring a claim for damages;
- issues of liability and contributory negligence;
- the assessment of damages, particularly for non-financial losses such as bereavement or grief and sorrow; and
- limitation periods.

18.45 The legal framework in many foreign jurisdictions is often more generous, from the perspective of the potential claimant, than in England and Wales. By way of example, a claim for bereavement damages in England and Wales is restricted to a limited category of claimants (the wife, husband, civil partner or parents of the deceased) and there is a fixed amount payable, currently £12,980. The view taken under English law is that an award of bereavement damages is compensation for grief and sorrow, and it can only ever be a token payment as grief cannot be measured in monetary terms. Bereavement damages are not intended to compensate financial loss or punish the defendant. However, the damages regime in some other jurisdictions allows a wider class of people to bring a claim, and the amounts available for non-financial loss can be significantly greater.

18.46 It is not always straightforward to work out which courts have competence to hear a claim (the issue of jurisdiction) and which law applies (the issue of applicable law). It is important to note that they are separate issues and involve different considerations. It is possible, although somewhat unusual, for a court in one jurisdiction to hear a claim, even though the law that is to be applied is the law of another jurisdiction. It is not possible in this chapter to provide the full detail of what is a complex area, and reference should be made to specialist texts.

18.47 As discussed above,[1] there are a number of international conventions, such as the Athens Convention (relating to carriage by sea) and the Montreal Convention (relating to air travel), that may be relevant in a travel case. These conventions can provide the English courts with jurisdiction over accidents abroad. Outside the 'specialist' conventions, the rules on jurisdiction are mainly found in Regulation (EC) 44/2001 (the Regulation),[2] the Lugano Convention (which is similar to the Regulation, but extends to Iceland, Norway and Switzerland) and the procedural rules contained in the Civil Procedure Rules 1998 (CPR).[3] The European regime is quite different to that set out in the CPR.

1 See **18.12** onwards.
2 Regulation (EC) 44/2001 on jurisdiction and the recognition and enforcement of judgments in civil and commercial matters. The Regulation will be replaced by Regulation (EC) 1215/2012 for proceedings commenced on or after 10 January 2015.
3 SI 1998/3132.

18.48 The default position prescribed by the Regulation, which applies to civil and commercial matters in the member states of the EU, is that defendants should, regardless of their nationality, be sued in the courts of the country in which they are domiciled (Article 2). The starting point is thus that a travel company domiciled in England should be sued in England, but the default position is subject to a number of important exceptions which override the domicile 'rule', and it is useful to highlight two in particular: Article 5(3) (place where the harm occurred), and Article 24 (submission to the jurisdiction).

18.49 Article 5(3) provides that a person domiciled in a contracting state may be sued in another member state:

> 'in matters relating to tort, delict or quasi-delict, in the courts for the place where the harmful event occurred or may occur.'

The meaning of Article 5(3) has been interpreted broadly by the courts to encompass both the place where the harm occurred and the place where the events giving rise to that harm took place; although, in the context of accidents affecting holidaymakers, the likelihood is that these will be one and the same thing. The practical effect of this is that, for example, if an accident takes place in another contracting state, it is likely to be the courts of that state which have jurisdiction, irrespective of the fact that the defendant tour operator is domiciled in the UK.

18.50 Article 24 should also be considered. It essentially states that, where a defendant appears in proceedings other than to contest the jurisdiction of the court in which he appears, the courts of the member state in which those proceedings are taking place will have jurisdiction. Therefore, put simply, if a defendant takes part in proceedings before a court for any purpose other than to dispute jurisdiction, that court will have jurisdiction.

18.51 There are other Articles in the Regulation that may be relevant to a claim arising out of a fatality abroad, as the recent case of *Thomas Cook Tour Operations Ltd v Louis Hotels SA*[1] demonstrates. A tour operator brought a claim in England for breach of contract against the owners of a hotel in Greece, following the death of two children on one of its holidays as a result of carbon monoxide poisoning in a room at the hotel. The court considered the Regulation and whether the proceedings should have been in Greece. The court rejected technical arguments raised on behalf of the hotel under Article 22 of the Regulation, which provides for exclusive jurisdiction in relation to specified types of claims, including the use of immovable property. The court found that it was a claim arising from a breach of a commercial contract; and, pursuant to Article 23 of the Regulation, which relates to agreements between parties about jurisdiction, the contract conferred jurisdiction on the English courts. The court entered summary judgment on the basis that the hotel had no real prospect of successfully arguing that it was not in breach of the obligation to provide a safe hotel room, and ordered an interim payment of £1 million.

1 [2013] EWHC 2139 (QB).

18.52 Where the Regulation or Lugano Convention does not apply, which is mainly in relation to defendants outside Europe, the question of jurisdiction becomes more difficult, and there will be situations in which the courts of England and Wales do not have jurisdiction to hear the claim. Under the relevant rules, jurisdiction in England and Wales may be established in three circumstances:

- where the defendant is physically in the jurisdiction and is validly served (unless proceedings are subsequently stayed);
- where the claimant obtains permission to serve outside the jurisdiction in accordance with CPR 6.36 by demonstrating that (i) the claim falls within one of the prescribed categories set out in para 3.1 of Practice Direction 6B, (ii) that it has a reasonable prospect of success (CPR 6.37(1)(b)), and (iii) that England and Wales is the proper place to bring the claim (CPR 6.37(3)); or
- where the defendant submits to the jurisdiction of the English courts (eg by filing a defence or making an application other than to challenge jurisdiction).

18.53 Although these rules offer a valuable route to establish the jurisdiction of the English courts, difficulties may well be encountered in terms of both demonstrating that the claim falls within one of the categories within Practice Direction 6B, and that England and Wales is the proper forum. The rules are designed to enable the courts to exercise jurisdiction over foreign defendants where the subject matter of the dispute has sufficient connection to England. Paragraph 3.1(9) of Practice Direction 6B states that a claimant may serve a claim form out of the jurisdiction with the permission of the court in accordance with CPR 6.36 where:

> 'a claim is made in tort where:
>
> (a) damage was sustained within the jurisdiction; or
> (b) the damage sustained resulted from an act committed within the jurisdiction.'

Clearly, it may not be possible to show that either of these situations applies, and it may also be difficult to demonstrate that England and Wales is the correct forum if, for example, the accident occurred abroad, the initial investigation was carried out overseas, and the relevant witnesses are not present in the UK.

WHO IS LIABLE FOR THE ACCIDENT?

18.54 When an accident occurs on holiday, it is often the case that there is a contractual relationship linking the injured or deceased holidaymaker to a UK company – perhaps where they were taking part in an activity organised by a UK tour operator – in which case, it may be possible to bring a claim under the PTRs. Even where, at first glance, the PTRs appear not to apply (for example, where a service provided to a holidaymaker was not part of the original package arranged with the tour operator), it may still be possible to hold the operator liable for the injury or death.

18.55 *Parker v TUI UK Ltd*[1] involved a tobogganing accident which occurred during a skiing holiday organised by the defendant tour operator. On the particular facts, it was clear that the tobogganing was not part of the pre-arranged holiday package itself, and it was subsequently decided, at both first instance and on appeal, that the defendant had assumed no contractual obligation towards the claimant in relation to the toboggan run and that, although it owed a duty of care to the claimant, its representatives (who were present at the time of the accident) had not breached that duty. Nevertheless, had a breach occurred, the defendant tour operator would have been liable for negligence on the part of its employees because, as stated by Longmore LJ at para 14, it had 'assumed responsibility to their customers, and owed them a duty of care in tort'; and it therefore follows that, by virtue of the fact that it employed its representatives, the defendant would have been vicariously liable for their acts and omissions.

This approach follows previous case law which identifies a clear distinction between a contract of service and a contract for services, whereby a company can potentially avoid liability in negligence if the negligent party was employed under a contract *for* services rather than a contract *of* service.[2]

1 [2009] EWCA Civ 1261.
2 See *Craven v Strand Holidays (Canada) Ltd* (1982) 40 OR (2d) 186 and *Cassidy v Ministry of Health* [1951] 2 KB 343.

18.56 The approach in *Parker* can be contrasted with the more recent Court of Appeal case of *Harrison v Jagged Globe (Alpine) Ltd*,[1] in which the court examined in detail the contractual relationship existing between the consumer, the holiday company and those engaged by the company to provide particular excursions or activities. In granting permission to appeal, Smith LJ said the appeal raised issues of 'some importance to the tourist industry', particularly in relation to adventure activity trips. The defendant, Jagged Globe, was a tour operator specialising in mountaineering expeditions and arranged a mountain expedition on behalf of Sir Ranulph Fiennes, who subsequently invited the claimant to join him on the ascent. During the expedition, Sir Ranulph suggested staging a fall and rescue sequence for the purposes of the documentary film following their progress. The claimant volunteered for the role of the falling mountaineer in the stunt and subsequently sustained injuries in two separate incidents whilst performing the sequence. The claimant claimed damages from Jagged Globe under reg 15 of the PTRs, which govern the liability of tour operators for the proper performance of obligations under a holiday contract. Part of her claim was that the local guides provided by Jagged Globe had initiated the stunts.

1 [2012] EWCA Civ 835.

2 See *Craven v Strand Holidays (Canada) Ltd* (1982) 40 OR (2d) 186 and *Cassidy v Ministry of Health* [1951] 2 KB 343.

18.57 At first instance, it was found that the PTRs did not apply to the case, on the basis that the defendant had no contractual obligation which made it responsible for the stunts. It was also found that, whilst there was no evidence to suggest that the local guides had initiated the stunts, it had nevertheless been open to them to stop them from proceeding, and this assumed liability on their part (and the duty of care flowing from that) could subsequently be attributed to the defendant, under the principle of vicarious liability, who was found partially liable for the accident. However, in upholding an appeal by the defendant, Pitchford LJ found that, on the basis that the stunts were not a planned part of the expedition organised by Jagged Globe, the local guides had in fact assumed personal responsibility for the activities which led to the claimant's injuries.

The status of the guides as independent contractors engaged for the purposes of the expedition meant that, despite the guides having assumed liability for the stunts, there was no mechanism by which Jagged Globe could be held liable for the guides' failings. The duty of care owed by the guides could not be translated into a duty of care owed by Jagged Globe, and therefore a claim in negligence could not be established.

THE SOCIAL VALUE OF RECREATIONAL AND PHYSICAL ACTIVITIES

18.58 There are, of course, benefits to adventurous activity and foreign travel. Those activities inevitably carry some uncertainty, and therefore risk, in terms of their outcome, and the law recognises this. In considering the potential liability of individuals and companies for gross negligence or corporate manslaughter in the context of the travel industry, it is important to balance the need to 'regulate' activities organised by tour operators, in order to minimise risk, against the social benefits of allowing tour operators the flexibility to offer new and exciting opportunities for travellers. Clearly, undue risk should be minimised whilst still enabling activities which have a degree of social utility to be undertaken, but where is the line to be drawn? In one case, a judge mused whether an award for damages for injuries in playing a game was an example of an overprotective nanny state, robbing youth of fun simply because there was some risk involved.

18.59 Section 1 of the Compensation Act 2006 sought to ensure that potential liability in negligence did not stop or deter socially desirable activities just because they carried a degree of risk, and various cases have touched upon this issue.

18.60 In *Scout Association v Barnes*,[1] one of the issues put forward by the defendant on appeal was that the trial judge had failed, in assessing whether the defendant's agents had exercised reasonable care in organising a game in which the claimant was harmed, to take into account:

> '... (a) the social benefit of the activity in question and (b) the consequences of finding that the game was dangerous.'

The facts of the case were that a boy, then aged 13, suffered personal injuries whilst playing a game called 'Objects in the Dark' at a scout meeting. The game involved a group of players running 'full pelt' into the middle of a room when the room lights were turned off to try and grab a block. There were not enough blocks for the number of players and a player who did not get a block was eliminated from that round of the game. Rounds were played until one boy was left holding the last remaining block, and he was the winner of the game.

1 [2010] EWCA Civ 1476.

18.61 At first instance, the trial judge found that the Scout Association was liable, although he expressed a 'degree of regret' because he recognised that his decision might impinge upon the activities of others in the future. A number of authorities, which established that the foreseeable risks of certain physical recreations can be accepted because the activities have a recognised social value, were not cited to the trial judge. One of those authorities was *Bolton v Stone*,[1] in which the court had to consider whether the occupiers of a cricket ground ought to have foreseen the possibility that a cricket ball might be hit out of the ground and injure a passer-by. In that case, it was held that there was a duty of care, but there was no breach of that duty in light of the improbable nature of what happened.

1 [1951] 1 All ER 1078.

18.62 The Court of Appeal in *Scout Association v Barnes* wrestled with the issue of social value. In his dissenting judgment, Jackson LJ said that, in determining whether the Scout Association had exercised reasonable care, the trial judge had done so by reference to the traditional factors (such as the likelihood of harm, severity of harm, how the risks could be evaded and so forth) but he had failed to take into account the social value of the activity. Jackson LJ said he believed that, if the authorities had been cited to the trial judge, he would have come to the opposite conclusion. In his view, what happened was a most unfortunate accident but it did not give rise to a claim for damages.

18.63 Smith LJ said that, although the trial judge did not expressly refer to the social value of the activity in evaluating the risks and announcing his conclusion, it was clear from other passages in the judgment that he had in mind the social value of the activity, and he did not think that the increased social value of playing the game in the dark amounted to much. Smith LJ did not feel it was appropriate to interfere with the trial judge's view. She commented that:

> 'Of course, the law of tort must not interfere with activities just because they carry some risk. Of course, the law of tort must not stamp out socially desirable activities. But whether the social benefit of an activity is such that the degree of risk it entails is acceptable is a question of fact, degree and judgment, which must be decided on an individual basis and not by broad brush approach.'

18.64 Ward LJ said that he had found it 'uncommonly difficult to reach a confident judgment in this case'. He concluded that the trial judge had applied the law properly to the facts and agreed with Smith LJ that the appeal should be dismissed. He referred to a passage from the judgment of Lord Hoffmann in *Tomlinson v Congleton Borough Council*,[1] as follows:

> '... the question of what amounts to "such care as in all the circumstances of the case is reasonable" depends upon assessing, as in the case of common law negligence, not only the likelihood that someone may be injured and the seriousness of the injury which may occur, but also the social value of the activity which gives rise to the risk and the cost of preventative measures. These factors have to be balanced against each other.'

Ward LJ said that the trial judge had balanced the social value of the activity giving rise to the risk and the cost of the preventive measures, and he summarised the trial judge's thinking as:

> 'more fun playing in the dark but more risk; less fun and less risk playing with the lights on. Is the benefit of added fun worth the added risk? He decided it was not worth it. Scouting would not lose much of its value if the game was not to be played in the dark.'

1 [2003] UKHL 47, [2004] 1 AC 46, [2003] 3 All ER 1122.

18.65 It is also worth drawing attention to the comment of Lord Scott of Foscote in *Tomlinson v Congleton Borough Council* (at para 94):

> 'Of course there is some risk of accidents arising out of the joie de vivre of the young. But that is no reason for imposing a grey and dull safety regime on everyone.'

18.66 The case of *Blair Ford v CRS Adventures Ltd*[1] involved a teacher who suffered a catastrophic spinal injury when falling after throwing a wellington boot backwards through his legs during a 'welly-wanging' event at an adventure activity centre. Globe J set out a number of legal principles which included:

> '... by virtue of the common law as rehearsed in Barnes v Scout Association [2010] EWCA Civ 1476, and s.1 of the Compensation Act , the law of tort must not stamp out socially desirable activities just because an activity carries some risk.'

Globe J said he was satisfied that the defendant's activity centre was an efficient and professionally run operation for the benefit of the public that provided immense social value. He held that the risk of injury to the claimant was not such that steps should have been taken to guard against it, and he dismissed the claim.

1 [2012] EWHC 2360 (QB).

18.67 Globe J also considered the degree of judicial attention that had been given to risk assessments. The judgment states (at para 46) that:

> 'Risk assessments are meant to be an exercise by which the employer examines and evaluates all the risks entailed in his operations and takes steps

to remove or minimise those risks. Sometimes, the failure to undertake a proper risk assessment can affect or even determine the outcome of a claim and judges must be alive to that and not sweep it aside. Risk assessments remain an important feature of the health and safety landscape and can provide an opportunity for intelligent and well-informed appraisal of risk and can form a blueprint for action leading to improved safety standards.'

18.68 However, Globe J commented on the fact that there had been no formal risk assessment for the particular method used by the claimant in throwing the wellington boot, and he said (at para 62):

> 'That, though, is not decisive. As Lady Justice Smith indicated in the *Uren* case, formal written risk assessments are probably more effective in relation to static conditions or activities which are often repeated in a fairly routine way, and they may be a less effective tool where a lot of variables come into play. Here, I am satisfied that there were a lot of variables and I do not find it to be a valid criticism that there was no formal risk assessment. A dynamic risk assessment was acceptable.'

18.69 Effective risk assessment is a process which is designed to identify and then manage the risks which attach to certain activities. The courts have recognised that there is a place for formal written risk assessments, and consequently the absence of such documentation can represent a problem for a travel company. However, the courts have also stressed that the dynamic risk assessment carried out by those in charge on the ground observing what is actually happening is critically important. Travel companies need to work out an approach to risk assessment which will have various levels, from the general assessment typically recorded in writing through to the dynamic monitoring on the ground.

STANDARD OF CARE

18.70 The question of whether the required degree of care has been achieved on any particular facts will be decided by reference to all the relevant circumstances of the case and the characteristics of the persons at risk. It involves consideration of the magnitude of the risk (obviously the greater the risk, the greater the degree of care that is required), the likelihood of injury, and the consequences. The fact that there have not been any previous accidents in similar circumstances may well be relevant.

18.71 In the *Blair Ford* case, Globe J said (at para 67):

> 'Notwithstanding the submission on behalf of [the claimant] to the contrary, I am satisfied that there is good evidence of a number of people having witnessed an appreciable number of occasions of people throwing the welly backwards through the legs with two hands with no difficulty, no falling and no injury.'

18.72 So, the absence of prior accidents may support an argument that the risk was unforeseeable or so slight that it could be legitimately discounted. Conversely, if there has been a previous accident of a similar nature – either within the travel company or indeed within the wider industry – that will no doubt be relied upon by the claimant in a claim for damages. It is also a relevant factor that may be taken into account by a coroner, a judge or a regulator when considering the actions of a travel company. Travel companies need to ensure that they have systems in place to learn from accidents and 'near misses' that are relevant to their operations.

18.73 An issue that can arise in claims arising out of accidents abroad is whether the defendant is to be judged by the local standards or the standards that might apply in another country. It is an issue that has attracted a fair degree of judicial attention, and the basic approach taken by the courts is to accept that, save where there are uniform international regulations, there are bound to be different standards in different countries, and the travel company is likely to be assessed by reference to the local standards.

18.74 The point was discussed by the Court of Appeal in *Gouldbourn v Balkan Holidays Ltd*,[1] which arose following an accident on a ski slope in Bulgaria. Leveson LJ considered the judgment in the leading case on the applicable standard of care, which is that of Phillips J in *Wilson v Best Travel*.[2] Leveson LJ made the following obiter comment:

> 'It is a mistake to seek to construe the judgment of Phillips J as if it was a statute: see the observations of Richards LJ in *Evans v Kosmar Villa Holidays PLC* [2008] 1 WLR 297 at para 224 page 3068 to the effect that the case did not purport to be an exhaustive statement of the duty of care. Nevertheless it does identify a very important signpost to the correct approach to cases of this nature, which will inevitably impact on the way in which organisations from different countries provide services to UK tourists. To require such organisations to adopt a different standard of care for different tourists is quite impracticable. What might be required for American tourists may well be different to that required by a French or Western European tourist, itself different to that required by a Japanese tourist. Neither do I consider that the Regulations impose a duty on English tour operators to require a standard of care to be judged by UK criteria or necessarily western European criteria.'

1 [2010] EWCA Civ 372.
2 [1993] 1 All ER 353.

18.75 The judge at first instance in the *Gouldbourn* case was of the view that, by western European standards, the ski instructor probably 'failed to assess the claimant's ability correctly and was too quick to take her up on to this slope'. However, on the central issue of negligence, the judge said he was driven to the conclusion that the ski instructor's conduct had to be judged against the relevant local standards, and he had no evidence to satisfy himself that the ski instructor had failed to show reasonable care by reference to such standards. The judge commented that 'it may be that he fell below those standards but that is not something which I can properly infer from the evidence I have heard'. The Court of Appeal rejected the appeal, which challenged the approach taken by the judge at first instance.

18.76 The Court of Appeal also considered the issue of inadequate safety standards in the case of *Moira Japp v Virgin Holidays Limited*.[1] A holidaymaker suffered lacerations when she walked into a balcony door which shattered at a hotel in Barbados. She sued the tour operator and was awarded damages. The tour operator appealed and won on an important point of principle, which was that the duty of care should be considered by reference to the local standards at the date of construction of the hotel, rather than as at the date of the accident. The accident had occurred 14 years after the hotel had been constructed and the balcony doors installed. It was held that there was no continuing duty on the hotel to update the facilities to meet developing standards. However, as the tour operator failed in its challenge to the judge's finding of fact that the doors did not comply with local standards at the date of installation, the appeal was dismissed.

1 [2013] EWCA Civ 1371.

CONCLUSION

18.77 Many claims have been brought against travel companies for damages for personal injury and death and, on occasion, issues of criminal liability can arise, as discussed in this chapter. Domestically, there may be more unlawful killing verdicts in fatal accident cases which, in turn, increases the chances of prosecutions under the CMCHA 2007. Travel companies need clear systems, contracts and documentation in place to limit their potential exposure to corporate, and possibly personal, liabilities arising following serious and fatal accidents.

Index

References are to paragraph numbers.